Modern Methods in the Calculus of Variations: L^p Spaces

T0212360

Irene Fonseca
Giovanni Leoni

Modern Methods
in the Calculus
of Variations: L^p Spaces

Springer

Irene Fonseca

Department of Mathematical Sciences
Carnegie Mellon University
Pittsburgh, PA 15213
USA
fonseca@andrew.cmu.edu

Giovanni Leoni

Department of Mathematical Sciences
Carnegie Mellon University
Pittsburgh, PA 15213
USA
giovanni@andrew.cmu.edu

ISBN: 978-1-4419-2260-1   e-ISBN: 978-0-387-69006-3

Mathematics Subject Classification (2000): 49-00, 49-01, 49-02, 49J45, 28-01, 28-02, 28B20, 52A

springer.com

To our families

Preface

In recent years there has been renewed interest in the calculus of variations, motivated in part by ongoing research in materials science and other disciplines. Often, the study of certain material instabilities such as phase transitions, formation of defects, the onset of microstructures in ordered materials, fracture and damage, leads to the search for equilibria through a minimization problem of the type

$$\min \left\{ I\left(v\right) : v \in \mathcal{V} \right\},$$

where the class $\mathcal{V}$ of admissible functions v is a subset of some Banach space V.

This is the essence of the calculus of variations: the identification of necessary and sufficient conditions on the functional I that guarantee the existence of minimizers. These rest on certain growth, coercivity, and convexity conditions, which often fail to be satisfied in the context of interesting applications, thus requiring the relaxation of the energy. New ideas were needed, and the introduction of innovative techniques has resulted in remarkable developments in the subject over the past twenty years, somewhat scattered in articles, preprints, books, or available only through oral communication, thus making it difficult to educate young researchers in this area.

This is the first of two books in the calculus of variations and measure theory in which many results, some now classical and others at the forefront of research in the subject, are gathered in a unified, consistent way. A main concern has been to use contemporary arguments throughout the text to revisit and streamline well-known aspects of the theory, while providing novel contributions.

The core of this book is the analysis of necessary and sufficient conditions for sequential lower semicontinuity of functionals on L^p spaces, followed by relaxation techniques. What sets this book apart from existing introductory texts in the calculus of variations is twofold: Instead of laying down the theory in the one-dimensional setting for integrands $f = f(x, u, u')$, we work in N dimensions and no derivatives are present. In addition, it is self-contained in the sense that, with the exception of fundamentally basic results in measure theory that may be found in any textbook on the subject (e.g., Lebesgue dominated convergence theorem), all the statements are fully justified and proved. This renders it accessible to beginning graduate students with basic knowledge of measure theory and functional analysis. Moreover, we believe that this text is unique as a reference book for researchers, since it treats both necessary and sufficient conditions for well-posedness and lower semicontinuity of functionals, while usually only sufficient conditions are addressed.

The central part of this book is Part III, although Parts I and II contain original contributions. Part I covers background material on measure theory, integration, and L^p spaces, and it combines basic results with new approaches to the subject. In particular, in contrast to most texts in the subject, we do not restrict the context to σ-finite measures, therefore laying the basis for the treatment of Hausdorff measures, which will be ubiquitous in the setting of the second volume, in which gradients will be present. Moreover, we call attention to Section 1.1.4, on "comparison between measures", which is completely novel: The Radon–Nikodym theorem and the Lebesgue decomposition theorem are proved for positive measures without our having first to introduce signed measures, as is usual in the literature. The new arguments are based on an unpublished theorem due to De Giorgi treating the case in which the two measures in play are not σ-finite. Here, as De Giorgi's theorem states, a diffuse measure must be added to the absolutely continuous and singular parts of the decomposition. Also, we give a detailed proof of the Morse covering theorem, which does not seem to be available in other books on the subject, and we derive as a corollary the Besicovitch covering theorem instead of proving it directly.

Part II streamlines the study of convex functions, and the treatment of the direct method of the calculus of variations introduces the reader to the close connection between sequential lower semicontinuity properties and existence of minimizers. Again here we present an unpublished theorem of De Giorgi, the approximation theorem for real-valued convex functions, which provides an explicit formula for the affine functions approximating a given convex function f. A major advantage of this characterization is that additional regularity hypotheses on f are reflected immediately on the approximating affine functions.

In Part III we treat sequential lower semicontinuity of functionals defined on L^p, and we separate the cases of inhomogeneous and homogeneous functionals. The latter are studied in Chapter 5, where

$$I(u) := \int_E f(v(x))\, dx$$

with E a Lebesgue measurable subset of the Euclidean space $\mathbb{R}^N$, $f : \mathbb{R}^m \to (-\infty, \infty]$ and $v \in L^p(E; \mathbb{R}^m)$ for $1 \le p \le \infty$. This material is intended for an introductory graduate course in the calculus of variations, since it requires only basic knowledge of measure theory and functional analysis. We treat both bounded and unbounded domains E, and we address most types of strong and weak convergence. In particular, the setting in which the underlying convergence is that of $(C_b(E))'$ is new.

Chapter 6 and Chapter 7 are devoted to integrands $f = f(x,v)$ and $f = f(x,u,v)$, respectively, and are significantly more advanced, since the proofs of the necessity parts are heavily hinged on the concept of multifunctions. An important tool here is selection criteria, and the reader will benefit from a comprehensive and detailed study of this subject.

Finally, Chapter 8 describes basic properties of Young measures and how they may be used in relaxation theory.

The bibliography aims at giving the main references relevant to the contents of the book. It is by no means exhaustive, and many important contributions to the subject may have failed to be listed here.

To conclude, this text is intended as a graduate textbook as well as a reference for more-experienced researchers working in the calculus of variations, and is written with the intention that readers with varied backgrounds may access different parts of the text.

This book prepares the ground for a second volume, since it introduces and develops the basic tools in the calculus of variations and in measure theory needed to address fundamental questions in the treatment of functionals involving derivatives.

Finally, in a book of this length, typos and errors are almost inevitable. The authors will be very grateful to those readers who will write to either fonseca@andrew.cmu.edu or giovanni@andrew.cmu.edu indicating those that they have found. A list of errors and misprints will be maintained and updated at the web page http://www.math.cmu.edu/~leoni/book1.

Pittsburgh,   *Irene Fonseca*
month 2007   *Giovanni Leoni*

Contents

XIV   Contents

Part I

Measure Theory and L^p Spaces

1

Measures

> Measure what is measurable, and
> make measurable what is not so.
> (Misura ciò che è misurabile, e
> rendi misurabile ciò che non lo è)
>
> Galileo Galilei (1564–1642)

1.1 Measures and Integration

This chapter covers a wide range of properties of measures. Those that we consider basic and well known (for example the Lebesgue dominated convergence theorem) will only be stated, and the reader is referred to classical textbooks such as [DB02], [EvGa92], [Fol99], [Rao04], [Ru87], [Z67]. The reader should be warned that in some of these books outer measures are called measures.

Results that are difficult to find in the literature, that are new, or that may be presented in a more contemporary way will be proved in this text.

1.1.1 Measures and Outer Measures

Definition 1.1. *Let X be a nonempty set. A collection $\mathfrak{M} \subset \mathcal{P}(X)$ is an algebra if*

(i) $\emptyset \in \mathfrak{M}$*;*
(ii) if $E \in \mathfrak{M}$ then $X \setminus E \in \mathfrak{M}$;
(iii) if E_1, $E_2 \in \mathfrak{M}$ then $E_1 \cup E_2 \in \mathfrak{M}$.

$\mathfrak{M}$ *is said to be a σ-algebra if it satisfies (i)–(ii) and*

(iii)' if $\{E_n\} \subset \mathfrak{M}$ then $\bigcup_{n=1}^{\infty} E_n \in \mathfrak{M}$.

To highlight the dependence of the σ-algebra $\mathfrak{M}$ on X we will sometimes use the notation $\mathfrak{M}(X)$. If $\mathfrak{M}$ is a σ-algebra then the pair $(X, \mathfrak{M})$ is called a *measurable space*. For simplicity we will often apply the term measurable space only to X.

Using De Morgan's laws and (ii) and (iii)', it follows that a σ-algebra is closed under countable intersection. In particular, if E, $F \in \mathfrak{M}$ then $E \cap F \in \mathfrak{M}$, and this leads to the notion of *restriction* of $\mathfrak{M}$ to a set $E \subset X$ (not necessarily measurable), i.e., the induced σ-algebra

$$\mathfrak{M} \lfloor E := \{ E \cap F : F \in \mathfrak{M} \}.$$

Example 1.2. (i) In view of (i) and (ii), every algebra contains X. Hence the smallest algebra (respectively σ-algebra) is $\{\emptyset, X\}$ and the largest is the collection $\mathcal{P}(X)$ of all subsets of X.

(ii) If $X = [0, 1)$, the family $\mathfrak{M}$ of all finite unions of intervals of the type $[a, b) \subset [0, 1)$ is an algebra but not a σ-algebra. Indeed,

$$E := \bigcap_{n=1}^{\infty} \left[0, \frac{1}{n} \right) = \{0\} \notin \mathfrak{M}.$$

Let X be a nonempty set. Given any subset $\mathcal{F} \subset \mathcal{P}(X)$ the smallest (in the sense of inclusion) σ-algebra that contains $\mathcal{F}$ is given by the intersection of all σ-algebras on X that contain $\mathcal{F}$.

If X is a topological space, then the *Borel σ-algebra* $\mathcal{B}(X)$ is the smallest σ-algebra containing all open subsets of X. The elements of $\mathcal{B}(X)$ are called *Borel sets*. Unless indicated otherwise, in the sequel it is understood that the Euclidean space $\mathbb{R}^N$, $N \geq 1$, and the extended real line $\overline{\mathbb{R}} := [-\infty, \infty]$ are endowed with the Borel σ-algebras associated to the respective usual topologies: In $\mathbb{R}^N$ we consider the Euclidean norm

$$|x| := \sqrt{(x_1)^2 + \ldots + (x_N)^2}$$

with $x = (x_1, \ldots, x_N)$, and we take as basis of open sets in $\overline{\mathbb{R}}$ the collection of all intervals of the form (a, b), $(a, \infty]$, $[-\infty, b)$ with a, $b \in \mathbb{R}$.

Remark 1.3. If X is a topological space and $Y \subset X$ then

$$\mathcal{B}(Y) = \mathcal{B}(X) \lfloor Y. \tag{1.1}$$

Indeed, $\mathcal{B}(X) \lfloor Y$ is a σ-algebra in Y that contains

$$\{ A \cap Y : A \text{ is open in } X \} = \{ A' : A' \text{ is open in } Y \}.$$

By definition of $\mathcal{B}(Y)$ we deduce that $\mathcal{B}(Y) \subset \mathcal{B}(X) \lfloor Y$. Conversely, let

$$\mathfrak{N} := \{ F \subset X : F \cap Y \in \mathcal{B}(Y) \}.$$

Then $\mathfrak{N}$ is a σ-algebra in X that contains all open sets in X. Hence $\mathcal{B}(X) \subset \mathfrak{N}$ and so $\mathcal{B}(Y) \supset \mathcal{B}(X) \lfloor Y$.

Definition 1.4. *Let X be a nonempty set, let $\mathfrak{M} \subset \mathcal{P}(X)$ be an algebra, and let $\mu : \mathfrak{M} \to [0, \infty]$.*

(i) μ is a (positive) finitely additive measure *if*

$$\mu(\emptyset) = 0, \quad \mu(E_1 \cup E_2) = \mu(E_1) + \mu(E_2)$$

for all E_1, $E_2 \in \mathfrak{M}$ with $E_1 \cap E_2 = \emptyset$.

(ii) μ is a (positive) countably additive measure *if*

$$\mu(\emptyset) = 0, \quad \mu\left(\bigcup_{n=1}^{\infty} E_n\right) = \sum_{n=1}^{\infty} \mu(E_n)$$

for every countable collection $\{E_n\} \subset \mathfrak{M}$ of pairwise disjoint sets such that $\bigcup_{n=1}^{\infty} E_n \in \mathfrak{M}$.

Definition 1.5. *Let X be a nonempty set, let $\mathfrak{M} \subset \mathcal{P}(X)$ be a σ-algebra.*

(i) A map $\mu : \mathfrak{M} \to [0, \infty]$ is called a (positive) measure *if*

$$\mu(\emptyset) = 0, \quad \mu\left(\bigcup_{n=1}^{\infty} E_n\right) = \sum_{n=1}^{\infty} \mu(E_n)$$

for every countable collection $\{E_n\} \subset \mathfrak{M}$ of pairwise disjoint sets. The triple $(X, \mathfrak{M}, \mu)$ is said to be a measure space.

(ii) Given a measure $\mu : \mathfrak{M} \to [0, \infty]$, a set $E \in \mathfrak{M}$ has σ-finite μ measure if it can be written as a countable union of measurable sets of finite measure; μ is said to be σ-finite if X has σ-finite μ measure; μ is said to be finite *if $\mu(X) < \infty$. Analogous definitions can be given for finitely additive measures.*

(iii) A measure μ is said to be a probability measure *if $\mu(X) = 1$.*

Exercise 1.6. (i) Let X be a nonempty set. Show that the function $\mu : \mathcal{P}(X) \to [0, \infty]$ defined by

$$\mu(E) := \begin{cases} \text{card } E & \text{if } E \text{ is finite,} \\ \infty & \text{otherwise,} \end{cases}$$

is a measure. Here card is the cardinality. The measure μ is called the *counting measure*. Prove that μ is finite (respectively σ-finite) if and only if X is finite (respectively denumerable).

(ii) Given a sequence $\{x_n\}$ of nonnegative numbers we introduce $\mu : \mathcal{P}(\mathbb{N}) \to [0, \infty]$ as

$$\mu(E) := \sum_{n \in E} x_n, \quad E \subset \mathbb{N}.$$

Prove that μ is a σ-finite measure.

If $(X, \mathfrak{M}, \mu)$ is a measure space then the *restriction* of μ to a subset $E \in \mathfrak{M}$ is the measure $\mu \lfloor E : \mathfrak{M} \to [0, \infty]$ defined by

$$(\mu \lfloor E)(F) := \mu(F \cap E), \quad F \in \mathfrak{M}.$$

Often, when there is no possibility of confusion, we use the same notation $\mu \lfloor E$ to denote the restriction of the measure μ to the σ-algebra $\mathfrak{M} \lfloor E$.

Among properties of measures we single out the following monotone convergence results.

Proposition 1.7. *Let $(X, \mathfrak{M}, \mu)$ be a measure space.*

(i) If $\{E_n\}$ is an increasing sequence of subsets of $\mathfrak{M}$ then

$$\mu \left(\bigcup_{n=1}^{\infty} E_n \right) = \lim_{n \to \infty} \mu(E_n).$$

(ii) If $\{E_n\}$ is a decreasing sequence of subsets of $\mathfrak{M}$ and $\mu(E_1) < \infty$ then

$$\mu \left(\bigcap_{n=1}^{\infty} E_n \right) = \lim_{n \to \infty} \mu(E_n).$$

Proof. (i) If $\mu(E_n) = \infty$ for some $n \in \mathbb{N}$ then both sides in (i) are ∞ and there is nothing to prove. Thus assume that $\mu(E_n) < \infty$ for all $n \in \mathbb{N}$ and define

$$F_n := E_n \setminus E_{n-1},$$

where $E_0 := \emptyset$. Since $\{E_n\}$ is an increasing sequence, it follows that the sets F_n are pairwise disjoint with $\bigcup_{n=1}^{\infty} E_n = \bigcup_{n=1}^{\infty} F_n$, and so by the properties of measures we have

$$\mu \left(\bigcup_{n=1}^{\infty} E_n \right) = \mu \left(\bigcup_{n=1}^{\infty} F_n \right) = \sum_{n=1}^{\infty} \mu(F_n) = \lim_{l \to \infty} \sum_{n=1}^{l} \mu(F_n)$$

$$= \lim_{l \to \infty} \sum_{n=1}^{l} (\mu(E_n) - \mu(E_{n-1})) = \lim_{l \to \infty} \mu(E_l).$$

(ii) Apply part (i) to the increasing sequence $\{E_1 \setminus E_n\}$.

Example 1.8. Without the hypothesis $\mu(E_1) < \infty$, property (ii) may be false. Indeed, consider the counting measure defined in Exercise 1.6 with $X := \mathbb{N}$ and define $E_n := \{n, n+1, \ldots, \}$. Then $\{E_n\}$ is a decreasing sequence, $\mu(E_n) = \infty$ for all $n \in \mathbb{N}$, but

$$\mu \left(\bigcap_{n=1}^{\infty} E_n \right) = \mu(\emptyset) = 0 \neq \lim_{n \to \infty} \mu(E_n) = \infty.$$

It turns out that for a finitely additive measure, property (i) is equivalent to countable additivity. Indeed, we have the following proposition.

Proposition 1.9. *Let X be a nonempty set, let $\mathfrak{M} \subset \mathcal{P}(X)$ be an algebra, and let $\mu : \mathfrak{M} \to [0, \infty]$ be a finitely additive measure. Then μ is countably additive if and only if*

$$\mu \left(\bigcup_{n=1}^{\infty} E_n \right) = \lim_{n \to \infty} \mu(E_n) \tag{1.2}$$

for every increasing sequence $\{E_n\} \subset \mathfrak{M}$ such that $\bigcup_{n=1}^{\infty} E_n \in \mathfrak{M}$.
In addition, if μ is finite then μ is countably additive if and only if

$$\lim_{n \to \infty} \mu(E_n) = 0 \tag{1.3}$$

for every decreasing sequence $\{E_n\} \subset \mathfrak{M}$ such that $\bigcap_{n=1}^{\infty} E_n = \emptyset$.

Proof. If μ is countably additive, then (1.2) and (1.3) follow as in the proof of Proposition 1.7. Suppose now that (1.2) holds. Let $\{F_n\} \subset \mathfrak{M}$ be a sequence of mutually disjoint sets such that $\bigcup_{n=1}^{\infty} F_n \in \mathfrak{M}$, and define

$$E_n := \bigcup_{k=1}^{n} F_k.$$

Then by (1.2) we have

$$\mu \left(\bigcup_{k=1}^{\infty} F_k \right) = \mu \left(\bigcup_{n=1}^{\infty} E_n \right) = \lim_{n \to \infty} \mu(E_n)$$
$$= \lim_{n \to \infty} \sum_{k=1}^{n} \mu(F_k) = \sum_{k=1}^{\infty} \mu(F_k),$$

and with this we have shown that μ is countably additive.

Finally, assume that μ is finite and that (1.3) holds. We claim that (1.2) is satisfied. Let $\{E_n\} \subset \mathfrak{M}$ be an increasing sequence such that $\bigcup_{n=1}^{\infty} E_n \in \mathfrak{M}$. Then the sequence

$$F_n := \left(\bigcup_{k=1}^{\infty} E_k \right) \setminus E_n$$

is decreasing and $\bigcap_{n=1}^{\infty} F_n = \emptyset$. Hence from (1.3),

$$0 = \lim_{n \to \infty} \mu(F_n) = \mu \left(\bigcup_{k=1}^{\infty} E_k \right) - \lim_{n \to \infty} \mu(E_n),$$

and this proves (1.2).

Exercise 1.10. Let $X := \mathbb{N}$ and let $\mathfrak{M}$ be the algebra consisting of all finite subsets of $\mathbb{N}$ and their complements. Show that the set function $\mu : \mathfrak{M} \to \{0, \infty\}$ given by

$$\mu(E) := \begin{cases} 0 & \text{if } E \text{ is finite,} \\ \infty & \text{otherwise,} \end{cases}$$

is a finitely additive measure satisfying property (ii) of Proposition 1.7 but it is not countably additive.

Using the previous proposition one may characterize finitely additive measures that are not countably additive. This brings us to the following definition.

Definition 1.11. *Let X be a nonempty set and let $\mathfrak{M} \subset \mathcal{P}(X)$ be an algebra. A finitely additive measure $\mu : \mathfrak{M} \to [0, \infty]$ is said to be* purely finitely additive *if there exists no nontrivial countably additive measure $\nu : \mathfrak{M} \to [0, \infty]$ with $0 \le \nu \le \mu$.*

Theorem 1.12 (Hewitt–Yosida). *Let X be a nonempty set, let $\mathfrak{M} \subset \mathcal{P}(X)$ be an algebra and let $\mu : \mathfrak{M} \to [0, \infty)$ be a finitely additive measure. Then μ can be uniquely written as a sum of a countably additive measure and a purely finitely additive measure.*

Proof. For $E \in \mathfrak{M}$ define

$$\mu_p(E) := \sup \Big\{ \lim_{n \to \infty} \mu(E_n) : \{E_n\} \subset \mathfrak{M}, \; E_{n+1} \subset E_n \text{ for all } n,$$

$$E_n \subset E \text{ for all } n, \; \bigcap_{n=1}^{\infty} E_n = \emptyset \Big\},$$

$$\mu_c(E) := \mu(E) - \mu_p(E).$$

One can verify that μ_p and μ_c are finitely additive measures. We now show that μ_c is countably additive. Let $\{E_n\} \subset \mathfrak{M}$ be a decreasing sequence with $\bigcap_{n=1}^{\infty} E_n = \emptyset$. Then from the definition of μ_p for each $l \in \mathbb{N}$ we have

$$\infty > \mu(E_l) \ge \mu_p(E_l) \ge \lim_{n \to \infty} \mu(E_n),$$

and so, letting $l \to \infty$, we obtain

$$\lim_{n \to \infty} \mu_p(E_n) = \lim_{n \to \infty} \mu(E_n) < \infty,$$

which implies that $\lim_{n \to \infty} \mu_c(E_n) = 0$. The claim now follows from Proposition 1.9.

Next we show that μ_p is a purely finitely additive measure. Let $\nu : \mathfrak{M} \to [0, \infty]$ be a countably additive measure with $0 \le \nu \le \mu_p$. For any $E \in \mathfrak{M}$ set $r := \frac{1}{3}\nu(E)$. Then $\mu_p(E) \ge 3r$, and so if $r > 0$ there exists a decreasing sequence $\{E_n\} \subset \mathfrak{M}$ of subsets of E, with $\bigcap_{n=1}^{\infty} E_n = \emptyset$, such that

$$\lim_{n\to\infty} \mu\left(E_n\right) > 2r.$$

Then $\mu_p\left(E_n\right) > 2r$ for every $n \in \mathbb{N}$ (since the sequence $\{E_k\}_{k=n}^{\infty}$ is admissible in the definition of $\mu_p\left(E_n\right)$), while $\lim_{n\to\infty} \nu\left(E_n\right) = 0$. Hence

$$\lim_{n\to\infty} \nu\left(E \setminus E_n\right) = 3r,$$

and so for all n sufficiently large,

$$\mu_p\left(E\right) = \mu_p\left(E_n\right) + \mu_p\left(E \setminus E_n\right) > 2r + \nu\left(E \setminus E_n\right) > 4r.$$

Inductively we would obtain $\mu_p\left(E\right) > lr$ for every $l \in \mathbb{N}$, and this would contradict the fact that μ_p is finite. Hence $\nu \equiv 0$ and μ_p is a purely finitely additive measure.

Exercise 1.13. Prove that μ_p and μ_c are finitely additive measures and that the decomposition $\mu = \mu_p + \mu_c$ in the previous theorem is unique.

Example 1.14. The finitely additive measure μ defined in Exercise 1.10 is purely finitely additive. Indeed, if $\nu : \mathfrak{M} \to [0, \infty]$ is a countably additive measure with $0 \leq \nu \leq \mu$, then since $\mu\left(\{1, \ldots, n\}\right) = 0$ for every $n \in \mathbb{N}$, we have that $\nu\left(\{1, \ldots, n\}\right) = 0$ and so from Proposition 1.7(i) it follows that

$$\nu\left(\mathbb{N}\right) = \lim_{n\to\infty} \nu\left(\{1, \ldots, n\}\right) = 0.$$

Hence $\nu \equiv 0$ and μ is purely finitely additive.

The next proposition will be particularly useful for the analysis of derivatives of measures and in the application of the blowup method (see Theorem 5.14).

Proposition 1.15. *Let* $(X, \mathfrak{M}, \mu)$ *be a measure space with* μ *finite and let* $\{E_j\}_{j \in J} \subset \mathfrak{M}$ *be an arbitrary family of pairwise disjoint subsets of* X. *Then* $\mu\left(E_j\right) = 0$ *for all but at most countably many* $j \in J$.

Proof. Fix $k \in \mathbb{N}$ and let

$$J_k := \left\{ j \in J : \mu\left(E_j\right) > \frac{1}{k} \right\}.$$

We claim that the set J_k is finite. Indeed, if I is any finite subset of J_k, then

$$\infty > \mu\left(X\right) \geq \mu\left(\bigcup_{j \in I} E_j\right) = \sum_{j \in I} \mu\left(E_j\right) \geq \frac{1}{k}\sum_{j \in I} 1 = \frac{1}{k}\operatorname{card} I,$$

which implies that J_k cannot have more than $\lceil k\mu\left(X\right)\rceil$ elements, where $\lceil k\mu\left(X\right)\rceil$ is the integer part of $k\mu\left(X\right)$. Thus

$$\{j \in J : \mu\left(E_j\right) > 0\} = \bigcup_{k=1}^{\infty} J_k$$

is at most countable.

The definition below introduces very important properties of measures that may be perceived as some sort of Darboux continuity.

Definition 1.16. *Let $(X, \mathfrak{M}, \mu)$ be a measure space.*

(i) *The measure $\mu : \mathfrak{M} \to [0, \infty]$ is said to have the* finite subset property *or to be* semifinite *if for every $E \in \mathfrak{M}$, with $\mu(E) > 0$, there exists $F \in \mathfrak{M}$, with $F \subset E$, such that $0 < \mu(F) < \infty$.*

(ii) *A set $E \in \mathfrak{M}$ of positive measure is said to be an* atom *if for every $F \in \mathfrak{M}$, with $F \subset E$, either $\mu(F) = 0$ or $\mu(F) = \mu(E)$. The measure μ is said to be* nonatomic *if there are no atoms, that is, if for every set $E \in \mathfrak{M}$ of positive measure there exists $F \in \mathfrak{M}$, with $F \subset E$, such that $0 < \mu(F) < \mu(E)$.*

Example 1.17. (i) We will show in Remark 1.161 that the Lebesgue measure is nonatomic. An important class of non-σ-finite nonatomic measures is given by the Hausdorff measures $\mathcal{H}^s$, $s > 0$ (see [FoLe10]).

(ii) To construct an example of a measure with the finite subset property that is not σ-finite, let X be an uncountable set, and to every finite set assign a measure equal to its cardinality; to all other sets assign measure ∞.

Exercise 1.18. Let X be a nonempty set and let $f : X \to [0, \infty]$ be any function. The set function $\mu : \mathcal{P}(X) \to [0, \infty]$ defined by

$$\mu(E) := \sup \left\{ \sum_{x \in F} f(x) : F \subset E, F \text{ finite} \right\}, \quad E \subset X,$$

is a measure. Show that

(i) μ has the finite subset property if and only if $f(x) < \infty$ for all $x \in X$;

(ii) μ is σ-finite if and only if $f(x) < \infty$ for all $x \in X$ and the set $\{x \in X : f(x) > 0\}$ is countable. In the special case $f \equiv 1$ we obtain the counting measure, while if

$$f(x) := \begin{cases} 1 & \text{if } x = x_0, \\ 0 & \text{otherwise}, \end{cases}$$

for some fixed $x_0 \in X$, then for every $E \subset X$,

$$\mu(E) = \begin{cases} 1 & \text{if } x_0 \in E, \\ 0 & \text{otherwise}. \end{cases}$$

Then μ is called *Dirac delta measure* with mass concentrated at the point x_0 and is denoted by δ_{x_0}. Prove that the set $\{x_0\}$ is an atom.

Remark 1.19. (i) It follows from the definition that a nonatomic measure has the finite subset property. Another important class of measures with the finite subset property is given by σ-finite measures. Indeed, if μ is a σ-finite measure, then

$$X = \bigcup_{n=1}^{\infty} X_n$$

with $X_n \in \mathfrak{M}$ and $\mu(X_n) < \infty$. Hence if $E \in \mathfrak{M}$ and $\mu(E) > 0$, then there exists n such that $0 < \mu(E \cap X_n) \le \mu(X_n) < \infty$.

(ii) The reader should be warned that in the literature there are at least two, more restrictive, definitions of atoms. In some papers atoms are defined as above, but they are required to have finite measure, while in others a set $E \in \mathfrak{M}$ of positive measure is said to be an atom if for every $F \in \mathfrak{M}$, with $F \subset E$, either $\mu(F) = 0$ or $\mu(E \setminus F) = 0$. The main difference consists in the fact that with these two definitions there could be nonatomic measures of the form $\mu : \mathfrak{M} \to \{0, \infty\}$, while with the definition that we have adopted, any set $E \in \mathfrak{M}$ such that $\mu : \mathfrak{M} \lfloor E \to \{0, \infty\}$ is considered an atom. Note that for measures with the finite subset property (and in particular for finite or σ-finite measures) all these definitions are equivalent, since sets $E \in \mathfrak{M}$ such that $\mu : \mathfrak{M} \lfloor E \to \{0, \infty\}$ are automatically excluded. The main advantage with our approach is that nonatomic nonfinite measures preserve many of the important features of nonatomic finite measures such as the Darboux property given in the following theorem. Readers more familiar with the other definitions of atoms should simply assume in all the theorems on nonatomic measures that the finite subset property holds.

The next two results play an important role in the study of the well-posedness of energy functionals as illustrated in Theorem 5.1.

Proposition 1.20. *Let $(X, \mathfrak{M}, \mu)$ be a measure space with μ nonatomic. Then the range of μ is the closed interval $[0, \mu(X)]$.*

Proof. Fix $0 < t < \mu(X)$ and let

$$\mathcal{C} := \{E \in \mathfrak{M} : 0 < \mu(E) \le t\}.$$

We claim that $\mathcal{C}$ is nonempty. Indeed, since μ is nonatomic, there exists $X_1 \in \mathfrak{M}$ with $0 < \mu(X_1) < \mu(X)$. Using again the fact that μ is nonatomic, it is possible to partition

$$X_1 = F_1 \cup F_2,$$

where $F_i \in \mathfrak{M}$ and $0 < \mu(F_i) < \mu(X_1)$, $i = 1, 2$. Therefore one of the two sets F_1 and F_2 has measure equal to at most $\frac{1}{2}\mu(X_1)$, and we call that set E_1. By induction, assuming that $E_1, \ldots, E_n$ have been selected with

$$0 < \mu(E_n) \le \frac{1}{2^n}\mu(X_1), \tag{1.4}$$

as before we find $E_{n+1} \subset E_n$ such that $E_{n+1} \in \mathfrak{M}$ and $0 < \mu(E_{n+1}) \le \frac{1}{2}\mu(E_n)$. We have constructed a sequence $\{E_n\} \subset \mathfrak{M}$ satisfying (1.4), and so for n large enough, $E_n \in \mathcal{C}$.

Next we claim that there exists a measurable set with measure t. The proof is again by induction. Set $F_0 := \emptyset$, and suppose that F_n has been given. Define

$$t_n := \sup\{\mu(E) : E \supset F_n, \ E \in \mathfrak{M}, \ \mu(E) \le t\}$$

and let $F_{n+1} \in \mathfrak{M}$ be such that $F_{n+1} \supset F_n$ and

$$t_n - \frac{1}{n} \le \mu(F_{n+1}) \le t_n. \tag{1.5}$$

Note that $0 \le t_{n+1} \le t_n \le t$, and so there exists

$$\lim_{n \to \infty} t_n = s \le t. \tag{1.6}$$

Set

$$F_\infty := \bigcup_n F_n. \tag{1.7}$$

By Proposition 1.7(ii) and (1.5),

$$\mu(F_\infty) = \lim_{n \to \infty} \mu(F_n) = s.$$

It remains to show that $s = t$. If $s < t$ then

$$\mu(X \setminus F_\infty) = \mu(X) - \mu(F_\infty) > t - s > 0,$$

and so reasoning as in the first part of the proof with $X \setminus F_\infty$ in place of X and $t - s$ in place of t there would exist a set $E \in \mathfrak{M}$ such that $E \subset X \setminus F_\infty$ and

$$0 < \mu(E) \le t - s.$$

Thus

$$s = \mu(F_\infty) < \mu(E \cup F_\infty) \le t,$$

and so by (1.6) it would follow that $t_n < \mu(E \cup F_\infty)$ for all n sufficiently large. Since $F_n \subset E \cup F_\infty$ by (1.7), this would contradict the definition of t_n for all n sufficiently large.

A consequence of the previous theorem is the following result, which will be repeatedly used in the second part of the book (see, e.g., the proof of Theorem 5.1).

Corollary 1.21. *Let $(X, \mathfrak{M}, \mu)$ be a measure space with μ nonatomic. Let $\{r_n\}$ be a sequence of positive numbers such that*

$$\sum_{n=1}^\infty r_n \le \mu(X).$$

Then there exists a sequence of mutually disjoint measurable sets $\{E_n\}$ such that $\mu(E_n) = r_n$ for all $n \in \mathbb{N}$.

Proof. By Proposition 1.20 there exists a measurable set E_1 such that $\mu(E_1) = r_1$. Since $\mu(X \setminus E_1) \geq r_2$, again invoking the previous proposition we may find a measurable set $E_2 \subset X \setminus E_1$ such that $\mu(E_2) = r_2$. A simple induction argument yields a family of measurable sets E_n with

$$E_n \subset X \setminus \bigcup_{i=1}^{n-1} E_i, \quad \mu(E_n) = r_n.$$

This concludes the proof.

If a measure μ has atoms, then it is possible to decompose it into the sum of a nonatomic measure and a *purely atomic measure*, that is, a measure such that every set of positive measure contains an atom. Precisely, we have the following result.

Proposition 1.22. *Let* $(X, \mathfrak{M}, \mu)$ *be a measure space. Then there exist measures* $\mu_1, \mu_2 : \mathfrak{M} \to [0, \infty]$ *such that* μ_1 *is purely atomic,* μ_2 *is nonatomic, and* $\mu = \mu_1 + \mu_2$.

We begin with a preliminary result that is of interest in itself.

Lemma 1.23. *Let* $(X, \mathfrak{M}, \mu)$ *be a measure space and let* $\mathfrak{N} \subset \mathfrak{M}$ *be a family closed under finite unions and such that* $\emptyset \in \mathfrak{N}$. *Then the set functions* μ_1, $\mu_2 : \mathfrak{M} \to [0, \infty]$ *defined by*

$$\mu_1(E) := \sup\{\mu(E \cap F) : F \in \mathfrak{N}\}, \quad E \in \mathfrak{M}, \tag{1.8}$$

$$\mu_2(E) := \sup\{\mu(E \cap F) : F \in \mathfrak{M}, \mu_1(F) = 0\}, \quad E \in \mathfrak{M}, \tag{1.9}$$

are measures, $\mu = \mu_1 + \mu_2$, *and* $\mu_1(E) = \mu(E)$ *for all* $E \in \mathfrak{N}$. *Moreover, for each* $E \in \mathfrak{M}$ *the suprema in the definition of* μ_1 *and* μ_2 *are actually attained by measurable sets.*

Proof. **Step 1:** We begin by showing that μ_1 and μ_2 are measures. Observe that if $E \in \mathfrak{N}$, then the right-hand side of (1.8) coincides with $\mu(E)$, and so $\mu_1(E) = \mu(E)$. In particular, since $\emptyset \in \mathfrak{N}$ we have that $\mu_1(\emptyset) = 0$. Let $\{E_n\} \subset \mathfrak{M}$ be any sequence of pairwise disjoint sets. Then for every $F \in \mathfrak{N}$,

$$\mu\left(\left(\bigcup_{n=1}^{\infty} E_n\right) \cap F\right) = \sum_{n=1}^{\infty} \mu(E_n \cap F) \leq \sum_{n=1}^{\infty} \mu_1(E_n),$$

and taking the supremum over all $F \in \mathfrak{N}$, we obtain that

$$\mu_1\left(\bigcup_{n=1}^{\infty} E_n\right) \leq \sum_{n=1}^{\infty} \mu_1(E_n).$$

To prove the reverse inequality it suffices to assume that $\mu_1\left(\bigcup_{n=1}^{\infty} E_n\right) < \infty$. Since μ_1 is monotone with respect to inclusion, it follows that $\mu_1(E_n) < \infty$

for all $n \in \mathbb{N}$, and so for very fixed $\varepsilon > 0$ and for each $n \in \mathbb{N}$ we may find $F_n \in \mathfrak{N}$ such that

$$\mu_1(E_n) \leq \mu(E_n \cap F_n) + \frac{\varepsilon}{2^n}.$$

Hence for every $l \in \mathbb{N}$,

$$\sum_{n=1}^{l} \mu_1(E_n) \leq \sum_{n=1}^{l} \left(\mu(E_n \cap F_n) + \frac{\varepsilon}{2^n} \right) \leq \mu \left(\bigcup_{n=1}^{l} (E_n \cap F_n) \right) + \varepsilon$$

$$\leq \mu \left(\bigcup_{n=1}^{\infty} E_n \cap \left(\bigcup_{k=1}^{l} F_k \right) \right) + \varepsilon \leq \mu_1 \left(\bigcup_{n=1}^{\infty} E_n \right) + \varepsilon,$$

where we have used the fact that $\bigcup_{k=1}^{l} F_k$ belongs to $\mathfrak{N}$. By letting first $l \to \infty$ and then $\varepsilon \to 0^+$ we conclude that μ_1 is a measure.

Moreover, since $\mathfrak{N}$ is closed under finite unions, for every $E \in \mathfrak{M}$ we may find an increasing sequence of sets $\{F_n\} \subset \mathfrak{N}$ such that

$$\mu_1(E) = \mu(E \cap F_\infty), \quad \text{where } F_\infty := \bigcup_{n=1}^{\infty} F_n. \tag{1.10}$$

This shows that the supremum in the definition of μ_1 is attained.

Since μ_1 is a measure, the family $\mathfrak{M}_1 := \{F \in \mathfrak{M} : \mu_1(F) = 0\}$ is closed under finite unions and contains $\emptyset$, and so, applying what we just proved with μ_1 and $\mathfrak{N}$ replaced by μ_2 and $\mathfrak{M}_1$, respectively, we obtain that also μ_2 is a measure and that the supremum in the definition of μ_2 is attained.

Step 2: It remains to show that $\mu = \mu_1 + \mu_2$. Fix $E \in \mathfrak{M}$. Since $\mu_1(E)$, $\mu_2(E) \leq \mu(E)$, then if $\mu_1(E) = \infty$ or $\mu_2(E) = \infty$, there is nothing to prove. Thus assume that $\mu_1(E)$, $\mu_2(E) < \infty$ and let F_∞ be defined as in (1.10). We claim that $\mu_1(E \setminus F_\infty) = 0$. Indeed, if $\mu_1(E \setminus F_\infty) > 0$ then we would be able to find $F \in \mathfrak{N}$ such that

$$\mu_1(E \setminus F_\infty) \geq \mu((E \setminus F_\infty) \cap F) > 0.$$

But then

$$\mu_1(E) = \mu(E \cap F_\infty) < \mu(E \cap (F_\infty \cup F)) = \lim_{l \to \infty} \mu \left(E \cap \left(\bigcup_{n=1}^{l} F_n \cup F \right) \right),$$

and since $\bigcup_{n=1}^{l} F_n \cup F \in \mathfrak{N}$, if l is sufficiently large we would contradict the definition of $\mu_1(E)$. Hence $\mu_1(E \setminus F_\infty) = 0$, and so by (1.9) we have $\mu_2(E) \geq \mu(E \setminus F_\infty)$. Thus

$$\mu_1(E) + \mu_2(E) \geq \mu(E \cap F_\infty) + \mu(E \setminus F_\infty) = \mu(E).$$

To prove the reverse inequality find an increasing sequence of measurable sets $\{G_k\}$, with $\mu_1(G_k) = 0$, such that

$$\mu_2(E) = \mu(E \cap G_\infty), \quad \text{where } G_\infty := \bigcup_{k=1}^{\infty} G_k.$$

Note that

$$\mu_1(G_\infty) = 0. \tag{1.11}$$

We claim that

$$\mu(E \cap F_\infty \cap G_\infty) = 0.$$

Indeed, if not, then for all l sufficiently large,

$$\mu\left(E \cap \left(\bigcup_{n=1}^{l} F_n\right) \cap G_\infty\right) > 0,$$

and so

$$\mu_1(E \cap G_\infty) \geq \mu\left(E \cap \left(\bigcup_{n=1}^{l} F_n\right) \cap G_\infty\right) > 0,$$

which contradicts (1.11). Hence the claim holds, and in turn,

$$\mu_1(E) + \mu_2(E) = \mu(E \cap F_\infty) + \mu(E \cap G_\infty) = \mu(E \cap (F_\infty \cup G_\infty)) \leq \mu(E).$$

This concludes the proof.

We are now ready to prove Proposition 1.22.

Proof (Proposition 1.22). Define μ_1 and μ_2 as in (1.8) and (1.9), where $\mathfrak{N}$ is the family of all countable unions of atoms together with the empty set. By the previous lemma we have that μ_1 and μ_2 are measures, that $\mu = \mu_1 + \mu_2$, and $\mu_1(E) = \mu(E)$ for all $E \in \mathfrak{N}$. We show that μ_1 is purely atomic. Suppose that $E \in \mathfrak{M}$ and $\mu_1(E) > 0$. By (1.8) we may find a countable family $\{F_n\}$ of μ-atoms such that

$$\mu\left(E \cap \bigcup_n F_n\right) > 0.$$

Then $\mu(E \cap F_n) > 0$ for some μ-atom F_n. We claim that $E \cap F_n$ is a μ_1-atom. To see this, assume by contradiction that there exists $G \in \mathfrak{M}$ such that

$$0 < \mu_1(E \cap F_n \cap G) < \mu_1(E \cap F_n).$$

Then

$$0 < \mu_1(E \cap F_n \cap G) < \mu_1(E \cap F_n) \leq \mu(E \cap F_n) \leq \mu(F_n),$$

and by (1.8) applied to $E \cap F_n \cap G$ there exists $F \in \mathfrak{N}$ such that

$$0 < \mu(E \cap F_n \cap G \cap F) \leq \mu_1(E \cap F_n \cap G) < \mu(F_n),$$

which contradicts the fact that F_n is a μ-atom. This shows that $E \cap F_n$ is a μ_1-atom and that μ_1 is purely atomic.

Next we prove that μ_2 is nonatomic. Suppose that $E \in \mathfrak{M}$ and $\mu_2(E) > 0$. By (1.9) there is $F \in \mathfrak{M}$, with $\mu_1(F) = 0$, such that

$$\mu_2(E) \geq \mu(E \cap F) > 0. \tag{1.12}$$

Then $E \cap F$ is not a μ-atom, since otherwise, $\mu_1(E \cap F) > 0$. Since $\mu(E \cap F) > 0$ and $E \cap F$ is not a μ-atom, we may find $G \subset E \cap F$, with $G \in \mathfrak{M}$, such that

$$0 < \mu(G) < \mu(E \cap F).$$

Since $\mu_1(F) = 0$ and $G \subset E \cap F$, we have that $\mu_1(G) = 0$, and so $\mu_2(G) = \mu(G)$. Hence also by (1.12),

$$0 < \mu_2(G) < \mu(E \cap F) \leq \mu_2(E),$$

which proves that μ_2 is nonatomic.

Remark 1.24. In particular, if μ is purely atomic, then $\mu_2 \equiv 0$ and so $\mu = \mu_1$. By (1.8), for any $E \in \mathfrak{M}$ with $\mu(E) > 0$ we may find a countable family $\{F_n\}$ of μ-atoms such that

$$\mu(E) = \mu\left(E \cap \bigcup_n F_n\right).$$

Letting

$$E_n := E \cap \left(F_n \setminus \bigcup_{i=1}^{n-1} F_i\right)$$

and disregarding those E_n's, if any, having measure zero, we have that

$$\mu(E) = \mu\left(\bigcup_n E_n\right),$$

where $\{E_n\}$ is a countable collection of pairwise disjoint of atoms.

We call attention to the fact that while the finite subset property may not prevent atoms from occurring, it eliminates very degenerate measures of the form $\mu : \mathfrak{M} \to \{0, \infty\}$. Indeed, the following proposition confirms this.

Proposition 1.25. *Let $(X, \mathfrak{M}, \mu)$ be a measure space. Then μ satisfies the finite subset property if and only if for all $E \in \mathfrak{M}$ with $\mu(E) > 0$,*

$$\mu(E) = \sup\{\mu(F) : F \in \mathfrak{M}, \ F \subset E, \ 0 < \mu(F) < \infty\}. \tag{1.13}$$

Proof. If (1.13) holds, then the finite subset property follows by the definition of supremum. Conversely, let $E \in \mathfrak{M}$ with $\mu(E) > 0$ and set

$$S := \sup \{ \mu(F) : F \in \mathfrak{M}, \, F \subset E, \, 0 < \mu(F) < \infty \}.$$

If $\mu(E) < \infty$, then $\mu(E) = S$. If $\mu(E) = \infty$, then we may find a sequence of increasing sets $F_n \subset E$, $F_n \in \mathfrak{M}$, with $\mu(F_n) < \infty$, and such that

$$\lim_{n \to \infty} \mu(F_n) = S.$$

Define

$$F := \bigcup_{n=1}^{\infty} F_n.$$

Then

$$\mu(F) = \lim_{n \to \infty} \mu(F_n) = S.$$

We claim that $S = \infty$. If not, then $\mu(E \setminus F) = \infty$, and so by the finite subset property we can find $F' \subset E \setminus F$, $F' \in \mathfrak{M}$, such that $0 < \mu(F') < \infty$. Hence $F \cup F' \subset E$ with

$$S = \mu(F) < \mu(F) + \mu(F') < \infty,$$

and this contradicts the definition of S.

If a measure μ does not satisfy the finite subset property, then it is possible to construct another measure that satisfies it and that coincides with μ on sets of finite measure. Indeed, we have the following proposition.

Proposition 1.26. *Let $(X, \mathfrak{M}, \mu)$ be a measure space. Then there exist measures $\mu_1 : \mathfrak{M} \to [0, \infty]$ and $\mu_2 : \mathfrak{M} \to \{0, \infty\}$ such that μ_1 has the finite subset property and $\mu = \mu_1 + \mu_2$. In particular, $\mu_1(E) = \mu(E)$ for all $E \in \mathfrak{M}$ such that $\mu(E) < \infty$.*

Proof. Define μ_1 and μ_2 as in (1.8) and (1.9), where

$$\mathfrak{N} := \{ E \in \mathfrak{M} : \mu(E) < \infty \}.$$

Since $\mathfrak{N}$ is closed under finite unions and contains $\emptyset$, by Lemma 1.23 we have that μ_1 and μ_2 are measures, that $\mu = \mu_1 + \mu_2$, and $\mu_1(E) = \mu(E)$ for all $E \in \mathfrak{N}$. In particular, if E, $F \in \mathfrak{M}$ and $\mu(F) < \infty$, then $\mu(E \cap F) < \infty$, and so $\mu(E \cap F) = \mu_1(E \cap F)$. Hence

$$\mu_1(E) = \sup \{ \mu(E \cap F) : F \in \mathfrak{M}, \, \mu(F) < \infty \}$$
$$= \sup \{ \mu_1(E \cap F) : F \in \mathfrak{M}, \, \mu_1(F) < \infty \},$$

and so μ_1 has the finite subset property in view of Proposition 1.25.

It remains to show that $\mu_2 : \mathfrak{M} \to \{0, \infty\}$. If by contradiction

$$0 < \mu_2(E) < \infty$$

for some $E \in \mathfrak{M}$, then by (1.9) we would find $F \in \mathfrak{M}$ such that $\mu_1(F) = 0$ and

$$0 < \mu(E \cap F) < \infty;$$

but then $\mu(E \cap F) = \mu_1(E \cap F)$, which would contradict the fact that $\mu_1(F) = 0$.

In applications, the ability to exclude sets of measure zero and to ensure that subsets of sets of measure zero are still measurable is often of the utmost importance. In what follows, if μ is a measure we write that a property holds μ a.e. on a measurable set E if there exists a measurable set $F \subset E$ such that $\mu(F) = 0$ and the property holds everywhere on the set $E \setminus F$.

Definition 1.27. *Given a measure space* $(X, \mathfrak{M}, \mu)$, *the measure* μ *is said to be* complete *if for every* $E \in \mathfrak{M}$ *with* $\mu(E) = 0$ *it follows that every* $F \subset E$ *belongs to* $\mathfrak{M}$.

The next proposition shows that it always possible to render a measure complete.

Proposition 1.28. *Given a measure space* $(X, \mathfrak{M}, \mu)$ *let* $\mathfrak{M}'$ *be the family of all subsets* $E \subset X$ *for which there exist* F, $G \in \mathfrak{M}$ *such that* $\mu(G \setminus F) = 0$ *and* $F \subset E \subset G$. *Define* $\mu'(E) := \mu(G)$. *Then* $\mathfrak{M}'$ *is a* σ-algebra that contains $\mathfrak{M}$ and $\mu' : \mathfrak{M}' \to [0, \infty]$ *is a complete measure.*

Example 1.29. Possibly the most important example of completion of a measure is the Lebesgue measure defined on the Borel σ-algebra, as will be detailed at the end of the next subsection.

We now introduce the notion of outer measure.

Definition 1.30. *Let* X *be a nonempty set.*

(i) A map $\mu^* : \mathcal{P}(X) \to [0, \infty]$ *is an* outer measure *if*
    a) $\mu^*(\emptyset) = 0$;
    b) $\mu^*(E) \leq \mu^*(F)$ *for all* $E \subset F \subset X$;
    c) $\mu^*(\bigcup_{n=1}^{\infty} E_n) \leq \sum_{n=1}^{\infty} \mu^*(E_n)$ *for every countable collection* $\{E_n\} \subset \mathcal{P}(X)$;
(ii) given an outer measure $\mu^* : \mathcal{P}(X) \to [0, \infty]$, *a set* $E \subset X$ *has* σ-finite μ^* outer measure *if it can be written as a countable union of sets of finite outer measure;* μ^* *is said to be* σ-finite *if* X *has* σ-finite μ^* *outer measure;* μ^* *is said to be* finite *if* $\mu^*(X) < \infty$.

Remark 1.31. The reader should be warned that in several of books (e.g., [DuSc88], [EvGa92], [Fe69]) outer measures are called measures.

Just as in the case of measures, if $\mu^* : \mathcal{P}(X) \to [0, \infty]$ is an outer measure and if $E \subset X$, then the *restriction* of μ^* to E is the outer measure $\mu^* \lfloor E :$ $\mathcal{P}(X) \to [0, \infty]$ defined by

$$(\mu^* \lfloor E)(F) := \mu^* (F \cap E)$$

for all sets $F \subset X$.

Often, when there is no possibility of confusion, we use the same notation $\mu^* \lfloor E$ to denote the restriction of the outer measure μ^* to $\mathcal{P}(E)$.

Clearly, a measure μ on a measurable space $(X, \mathfrak{M})$ is an outer measure if $\mathfrak{M} = \mathcal{P}(X)$. Note, however, that when $\mathfrak{M} \subsetneq \mathcal{P}(X)$ it is always possible to extend μ to $\mathcal{P}(X)$ as an outer measure. Indeed, more generally we have the following proposition.

Proposition 1.32. *Let X be a nonempty set and let $\mathcal{G} \subset \mathcal{P}(X)$ be such that $\emptyset \in \mathcal{G}$ and there exists $\{E_n\} \subset \mathcal{G}$ with $X = \bigcup_{n=1}^{\infty} E_n$. Let $\rho : \mathcal{G} \to [0, \infty]$ be such that $\rho(\emptyset) = 0$. Then the map $\mu^* : \mathcal{P}(X) \to [0, \infty]$ defined by*

$$\mu^*(E) := \inf \left\{ \sum_{n=1}^{\infty} \rho(E_n) : \{E_n\} \subset \mathcal{G}, \ E \subset \bigcup_{n=1}^{\infty} E_n \right\}, \quad E \subset X, \quad (1.14)$$

is an outer measure.

Proof. Since $\emptyset \in \mathcal{G}$ we have that $\mu^*(\emptyset) = 0$. If $E \subset F \subset X$ then any sequence $\{E_n\} \subset \mathcal{G}$ admissible for F in (1.14) is also admissible for E, and so $\mu^*(E) \leq \mu^*(F)$. Finally, let $\{F_k\} \subset \mathcal{P}(X)$. If $\mu^*(F_k) = \infty$ for some $k \in \mathbb{N}$, then

$$\mu^* \left(\bigcup_{k=1}^{\infty} F_k \right) \leq \sum_{k=1}^{\infty} \mu^*(F_k) = \infty.$$

Thus assume that $\mu^*(F_k) < \infty$ for all $k \in \mathbb{N}$ and for a fixed $\varepsilon > 0$ and for each k find a sequence $\left\{ E_n^{(k)} \right\} \subset \mathcal{G}$ admissible for F_k in (1.14) and such that

$$\sum_{n=1}^{\infty} \rho \left(E_n^{(k)} \right) \leq \mu^*(F_k) + \frac{\varepsilon}{2^k}.$$

Since the sequence $\left\{ E_n^{(k)} \right\}_{k, n \in \mathbb{N}}$ is admissible for $\bigcup_{k=1}^{\infty} F_k$, we have that (see Example 1.82 below)

$$\mu^* \left(\bigcup_{k=1}^{\infty} F_k \right) \leq \sum_{k,n=1}^{\infty} \rho \left(E_n^{(k)} \right) = \sum_{k=1}^{\infty} \sum_{n=1}^{\infty} \rho \left(E_n^{(k)} \right) \leq \sum_{k=1}^{\infty} \mu^*(F_k) + \varepsilon.$$

By letting $\varepsilon \to 0^+$ we conclude the proof.

Remark 1.33. Note that if $E \in \mathcal{G}$, then taking $E_1 := E$, $E_n := \emptyset$ for all $n \geq 2$, it follows from the definition of μ^* that $\mu^*(E) \leq \rho(E)$, with the strict inequality possible. However, if ρ is countably subadditive, that is, if

$$\rho(E) \leq \sum_{n=1}^{\infty} \rho(E_n)$$

for all $E \subset \bigcup_{n=1}^{\infty} E_n$ with $E \in \mathcal{G}$, $\{E_n\} \subset \mathcal{G}$, then $\mu^* = \rho$ on $\mathcal{G}$.

The construction of the measure μ^* in the previous proposition is commonly used to build an outer measure from a family $\mathcal{G}$ of elementary sets (e.g., cubes in $\mathbb{R}^N$) for which a basic notion of measure ρ is known.

Exercise 1.34. Let $f : \mathbb{R} \to \mathbb{R}$ be increasing. Let $\mathcal{G}$ be the family of all intervals $(a, b] \subset \mathbb{R}$, and define

$$\rho((a, b]) := f(b) - f(a).$$

The outer measure μ^* given by Proposition 1.32 is called the *Lebesgue–Stieltjes outer measure* generated by f. Show that

$$\mu^*((a, b]) = f(b) - \lim_{t \to a^+} f(a) \leq f(b) - f(a) = \rho((a, b]),$$

and therefore strict inequality occurs at points where f is not continuous from the right.

Definition 1.35. *Let X be a nonempty set and let $\mu^* : \mathcal{P}(X) \to [0, \infty]$ be an outer measure. A set $E \subset X$ is said to be μ^*-measurable if*

$$\mu^*(F) = \mu^*(F \cap E) + \mu^*(F \setminus E)$$

for all sets $F \subset X$.

Remark 1.36. Note that if $\mu^*(E) = 0$, then by the monotonicity of the outer measure, $\mu^*(F \cap E) = 0$ for all sets $F \subset X$. Hence E is μ^*-measurable. Moreover, if E is an arbitrary subset of X and if $F \subset X$ is a μ^*-measurable set, then $F \cap E$ is $\mu^* \lfloor E$-measurable.

Theorem 1.37 (Carathéodory). *Let X be a nonempty set and let $\mu^* : \mathcal{P}(X) \to [0, \infty]$ be an outer measure. Then*

$$\mathfrak{M}^* := \{E \subset X : E \text{ is } \mu^*\text{-measurable}\} \tag{1.15}$$

is a σ-algebra and $\mu^* : \mathfrak{M}^* \to [0, \infty]$ is a complete measure.

From Proposition 1.32 and Remark 1.33 we have the following.

Corollary 1.38. *Let X be a nonempty set, let $\mathfrak{M} \subset \mathcal{P}(X)$ be an algebra, and let $\mu : \mathfrak{M} \to [0, \infty]$ be a finitely additive measure. Let μ^* be the outer measure defined in (1.14) (with $\mathcal{G} := \mathfrak{M}$ and $\rho := \mu$). Then every element of $\mathfrak{M}$ is μ^*-measurable. Moreover, if μ is countably additive, then $\mu^* = \mu$ on $\mathfrak{M}$.*

Proof. To show that every element of $\mathfrak{M}$ is μ^*-measurable fix $E \in \mathfrak{M}$ and let F be any subset of X. By (1.14), for any fixed $\varepsilon > 0$ find a sequence $\{E_n\} \subset \mathfrak{M}$ such that $F \subset \bigcup_{n=1}^{\infty} E_n$ and

$$\sum_{n=1}^{\infty} \mu(E_n) \leq \mu^*(F) + \varepsilon.$$

Since $\mathfrak{M}$ is an algebra we have that $\{E_n \cap E\}$, $\{E_n \setminus E\} \subset \mathfrak{M}$ are admissible sequences in (1.14) for the sets $F \cap E$ and $F \setminus E$, and so

$$\mu^*(F \cap E) + \mu^*(F \setminus E) \leq \sum_{n=1}^{\infty} \left(\mu(E_n \cap E) + \mu(E_n \setminus E) \right)$$

$$= \sum_{n=1}^{\infty} \mu(E_n) \leq \mu^*(F) + \varepsilon.$$

Given the arbitrariness of $\varepsilon > 0$, it follows that

$$\mu^*(F \cap E) + \mu^*(F \setminus E) \leq \mu^*(F).$$

Since by Carathéodory's theorem μ^* is an outer measure, the opposite inequality is immediate. Thus E is μ^*-measurable.

The last part of the statement is a consequence of Remark 1.33.

Remark 1.39. Note that the previous result implies that every countably additive measure $\mu : \mathfrak{M} \to [0, \infty]$ defined on an algebra $\mathfrak{M}$ may be extended as a measure to a σ-algebra that contains $\mathfrak{M}$. It actually turns out that when μ is finite, this extension is unique. Indeed, let $\nu : \mathfrak{M}^* \to [0, \infty]$ be any measure such that $\nu = \mu$ on $\mathfrak{M}$, where $\mathfrak{M}^*$ is the σ-algebra of all μ^*-measurable sets. Note that ν and μ^* are finite, since

$$\nu(X) = \mu^*(X) = \mu(X) < \infty.$$

Let $E \in \mathfrak{M}^*$ and consider any sequence $\{E_n\} \subset \mathfrak{M}$ such that $E \subset \bigcup_{n=1}^{\infty} E_n$. Then

$$\nu(E) \leq \nu \left(\bigcup_{n=1}^{\infty} E_n \right) \leq \sum_{n=1}^{\infty} \nu(E_n) = \sum_{n=1}^{\infty} \mu(E_n).$$

Taking the infimum over all such covers and using (1.14) yields

$$\nu(E) \leq \mu^*(E)$$

for all $E \in \mathfrak{M}^*$. Since μ^* and ν are additive on $\mathfrak{M}^*$ and coincide with μ on $\mathfrak{M}$, for any $E \in \mathfrak{M}^*$, we have

$$\mu^* (E) + \mu^* (X \setminus E) = \mu^* (X) = \mu (X) = \nu (X) = \nu (E) + \nu (X \setminus E),$$

which in view of the previous inequality (for E and $X \setminus E$) and the fact that ν and μ^* are finite implies that $\mu^* (E) = \nu (E)$.

When X is a metric space then the class of metric outer measures on X is of special interest:

Definition 1.40. *Let X be a metric space and let $\mu^* : \mathcal{P}(X) \to [0, \infty]$ be an outer measure. Then μ^* is said to be a* metric outer measure *if*

$$\mu^* (E \cup F) = \mu^* (E) + \mu^* (F)$$

for all sets E, $F \subset X$, with

$$\mathrm{dist}\,(E, F) := \inf \{d\,(x, y) : x \in E,\ y \in F\} > 0.$$

Proposition 1.41. *Let X be a metric space and let $\mu^* : \mathcal{P}(X) \to [0, \infty]$ be a metric outer measure. Then every Borel set is μ^*-measurable.*

1.1.2 Radon and Borel Measures and Outer Measures

In this subsection we study regularity properties of measures and outer measures. These will play an important role in the characterization of some dual spaces.

Definition 1.42. *An outer measure $\mu^* : \mathcal{P}(X) \to [0, \infty]$ is said to be* regular *if for every set $E \subset X$ there exists a μ^*-measurable set $F \subset X$ such that $E \subset F$ and $\mu^* (E) = \mu^* (F)$.*

An important property of regular outer measures is the fact that Proposition 1.7(i) continues to hold.

Proposition 1.43. *Let $\mu^* : \mathcal{P}(X) \to [0, \infty]$ be a regular outer measure. If $\{E_n\}$ is an increasing sequence of subsets of X then*

$$\mu^* \left(\bigcup_{n=1}^{\infty} E_n \right) = \lim_{n \to \infty} \mu^* (E_n).$$

Proof. Since μ^* is regular there exist μ^*-measurable sets $F_n \supset E_n$ such that $\mu^* (E_n) = \mu^* (F_n)$. The sets

$$G_n := \bigcap_{i=n}^{\infty} F_i$$

are μ^*-measurable, with $E_n \subset G_n \subset F_n$, and

$$\mu^*(E_n) \leq \mu^*(G_n) \leq \mu^*(F_n) = \mu^*(E_n).$$

Since $G_n \subset G_{n+1}$, we may apply Proposition 1.7 to $\mu^* : \mathfrak{M}^* \to [0, \infty]$, where $\mathfrak{M}^*$ is defined in (1.15), to obtain that

$$\mu^*\left(\bigcup_{n=1}^{\infty} E_n\right) \leq \mu^*\left(\bigcup_{n=1}^{\infty} G_n\right) = \lim_{n \to \infty} \mu^*(G_n)$$

$$= \lim_{n \to \infty} \mu^*(E_n) \leq \mu^*\left(\bigcup_{n=1}^{\infty} E_n\right),$$

from which the result follows.

Example 1.44. The previous proposition fails if μ^* is not regular. Indeed, let $X := \mathbb{N}$ and for each $E \subset \mathbb{N}$ define

$$\mu^*(E) := \begin{cases} 0 \text{ if } E = \emptyset, \\ 1 \text{ if } E \text{ is finite}, \\ 2 \text{ if } E \text{ is infinite}. \end{cases}$$

Then μ^* is an outer measure, and taking $E_n := \{1, \ldots, n\}$ we see that

$$2 = \mu^*\left(\bigcup_{n=1}^{\infty} E_n\right) \neq \lim_{n \to \infty} \mu^*(E_n) = 1.$$

The next proposition shows that any measure may be extended to a regular outer measure.

Proposition 1.45. *Let $(X, \mathfrak{M}, \mu)$ be a measure space. Let μ^* be the outer measure defined in (1.14) (with $\mathcal{G} := \mathfrak{M}$ and $\rho := \mu$). Then μ^* is a regular outer measure, and for every $E \subset X$,*

$$\mu^*(E) = \inf\{\mu(F) : F \in \mathfrak{M}, F \supset E\}. \tag{1.16}$$

Moreover, every element of $\mathfrak{M}$ is μ^*-measurable and $\mu^* = \mu$ on $\mathfrak{M}$.

Proof. In view of Corollary 1.38 we have only to show that (1.16) holds and that the outer measure μ^* is regular. For any fixed $E \subset X$ and for any $F \in \mathfrak{M}$, with $F \supset E$, we have

$$\mu^*(E) \leq \mu^*(F) = \mu(F),$$

and so

$$\mu^*(E) \leq \inf\{\mu(F) : F \in \mathfrak{M}, F \supset E\}.$$

To prove the opposite inequality consider any sequence $\{E_n\} \subset \mathfrak{M}$ such that $E \subset \bigcup_{n=1}^{\infty} E_n$. Since $\bigcup_{n=1}^{\infty} E_n \in \mathfrak{M}$ we have

$$\sum_{n=1}^{\infty} \mu\left(E_n\right) \geq \mu\left(\bigcup_{n=1}^{\infty} E_n\right) \geq \inf\left\{\mu\left(F\right) : F \in \mathfrak{M}, F \supset E\right\}.$$

Taking the infimum over all such sequences $\{E_n\} \subset \mathfrak{M}$ yields

$$\mu^*\left(E\right) \geq \inf\left\{\mu\left(F\right) : F \in \mathfrak{M}, F \supset E\right\},$$

and so (1.16) holds.

To prove regularity note that if $\mu^*\left(E\right) = \infty$ then $X \in \mathfrak{M}$, $X \supset E$ and

$$\mu\left(X\right) = \mu^*\left(X\right) \geq \mu^*\left(E\right) = \infty.$$

If $\mu^*\left(E\right) < \infty$, then by (1.16), for each $n \in \mathbb{N}$ we may find measurable sets $F_n \supset E$ such that

$$\mu\left(F_n\right) \leq \mu^*\left(E\right) + \frac{1}{n}.$$

Then the measurable set

$$F := \bigcap_{n=1}^{\infty} F_n$$

contains E, and

$$\mu^*\left(E\right) \leq \mu\left(F\right) \leq \mu\left(F_n\right) \leq \mu^*\left(E\right) + \frac{1}{n} \to \mu^*\left(E\right)$$

as $n \to \infty$. Hence $\mu^*\left(E\right) = \mu\left(F\right)$.

Remark 1.46. In view of the previous proposition and Proposition 1.28, given a measure space $(X, \mathfrak{M}, \mu)$, the two measures $\mu^* : \mathfrak{M}^* \to [0, \infty]$ and $\mu' : \mathfrak{M}' \to [0, \infty]$ are complete and extend μ. We claim that $\mathfrak{M}' \subset \mathfrak{M}^*$ and that $\mu' = \mu^*$ on $\mathfrak{M}'$. Indeed, if $E \in \mathfrak{M}'$ then there exist $F, G \in \mathfrak{M}$, with $\mu\left(G \setminus F\right) = 0$ and $F \subset E \subset G$, such that $\mu'\left(E\right) := \mu\left(G\right)$. Write $E = F \cup \left(E \setminus F\right)$. Since $E \setminus F \subset G \setminus F$, by the monotonicity of the outer measure $\mu^* : \mathcal{P}\left(X\right) \to [0, \infty]$ we have that

$$\mu^*\left(E \setminus F\right) \leq \mu^*\left(G \setminus F\right) = \mu\left(G \setminus F\right) = 0,$$

and thus, since $\mu^* : \mathfrak{M}^* \to [0, \infty]$ is complete, it follows that $E \setminus F \in \mathfrak{M}^*$ and in turn

$$E = F \cup \left(E \setminus F\right) \in \mathfrak{M}^*,$$

where we have used the fact that $F \in \mathfrak{M}$. In particular

$$\mu^*\left(E\right) = \mu^*\left(F\right) + \mu^*\left(E \setminus F\right) = \mu\left(F\right) + 0 = \mu\left(G\right) = \mu'\left(E\right).$$

Hence $\mathfrak{M}' \subset \mathfrak{M}^*$ and $\mu' = \mu^*$ on $\mathfrak{M}'$.

Exercise 1.47. Given an outer measure $\mu^* : \mathcal{P}(X) \to [0, \infty]$ it is possible to construct a regular outer measure by defining for every $E \subset X$,

$$\nu^*(E) := \inf\{\mu^*(F) : E \subset F, \ F \ \mu^*\text{-measurable}\}.$$

Prove that if E is μ^*-measurable, then E is ν^*-measurable and $\mu^*(E) = \nu^*(E)$. Conversely, show that if E is ν^*-measurable and $\nu^*(E) < \infty$, then E is μ^*-measurable.

Definition 1.48. *Let X be a topological space and let $\mu^* : \mathcal{P}(X) \to [0, \infty]$ be an outer measure.*

(i) A set $E \subset X$ is said to be inner regular *if*

$$\mu^*(E) = \sup\{\mu^*(K) : K \subset E, \ K \ compact\},$$

and it is outer regular *if*

$$\mu^*(E) = \inf\{\mu^*(A) : A \supset E, \ A \ open\}.$$

(ii) A set $E \subset X$ is said to be regular *if it is both inner and outer regular.*

The previous definition introduces concepts of regularity for subsets of X, and next we address regularity properties of outer measures.

Definition 1.49. *Let X be a topological space and let $\mu^* : \mathcal{P}(X) \to [0, \infty]$ be an outer measure.*

(i) μ^* is said to be a Borel outer measure *if every Borel set is μ^*-measurable;*
(ii) μ^* is said to be a Borel regular outer measure *if μ^* is a Borel outer measure and for every set $E \subset X$ there exists a Borel set $F \subset X$ such that $E \subset F$ and $\mu^*(E) = \mu^*(F)$.*

There is a class of Borel outer measures that plays a pivotal role in the calculus of variations. These are the Radon outer measures, as introduced next.

Definition 1.50. *Let X be a topological space, and let $\mu^* : \mathcal{P}(X) \to [0, \infty]$ be an outer measure. Then μ^* is said to be a* Radon outer measure *if*

(i) μ^* is a Borel outer measure;
(ii) $\mu^*(K) < \infty$ for every compact set $K \subset X$;
(iii) every open set $A \subset X$ is inner regular;
(iv) every set $E \subset X$ is outer regular.

We investigate the relation between Radon outer measures and Borel regular measures.

Remark 1.51. Note that a Radon outer measure is always Borel regular. Indeed, let $E \subset X$. If $\mu^*(E) = \infty$ then note that X is open, $X \supset E$, and $\mu^*(X) = \mu^*(E) = \infty$. If $\mu^*(E) < \infty$, then by outer regularity, for each $n \in \mathbb{N}$ we may find open sets $A_n \supset E$ such that

$$\mu^*(A_n) \leq \mu^*(E) + \frac{1}{n}.$$

Then the Borel set

$$B := \bigcap_{n=1}^{\infty} A_n$$

contains E, and

$$\mu^*(E) \leq \mu^*(B) \leq \mu^*(A_n) \leq \mu^*(E) + \frac{1}{n} \to \mu^*(E)$$

as $n \to \infty$.

The converse of the previous remark is false in general (see Exercise 1.59 below), but there are some partial results in that direction.

Proposition 1.52. *Let X be a locally compact Hausdorff space such that every open set $A \subset X$ is σ-compact. Let $\mu^* : \mathcal{P}(X) \to [0, \infty]$ be a Borel outer measure such that $\mu^*(K) < \infty$ for every compact set $K \subset X$. Then every Borel set is inner regular and outer regular.*

If, in addition, μ^* is a Borel regular outer measure, then it is a Radon outer measure.

Proof. Since X is σ-compact we may write

$$X = \bigcup_{n=1}^{\infty} K_n,$$

where $\{K_n\}$ is an increasing sequence of compact sets. By Theorem A.12 we may find open sets U_n such that $K_n \subset U_n \subset X$ and $\overline{U_n}$ is compact. Without loss of generality we may assume that the sequence $\{U_n\}$ is increasing.

Step 1: Fix $n \in \mathbb{N}$ and let μ_n^* denote the restriction of μ^* to $\mathcal{P}(U_n)$. We claim that every Borel subset of U_n is inner and outer regular with respect to the outer measure μ_n^*. Note that since $\overline{U_n}$ is compact, μ_n^* is a finite outer measure. Let

$$\mathfrak{M}_n := \{E \subset U_n : E \text{ is } \mu_n^* \text{ inner and outer regular}\}.$$

If $A \subset U_n$ is open, then it is outer regular. Since A is σ-compact, there exists an increasing sequence $\{C_j\}$ of compact subsets of A such that

$$A = \bigcup_{j=1}^{\infty} C_j,$$

and so by Carathéodory's theorem and Proposition 1.7,

$$\lim_{j \to \infty} \mu_n^* (C_j) = \mu_n^* (A).$$

This shows that $\mathfrak{M}_n$ contains all open subsets of U_n, and therefore the claim is proved provided we prove that $\mathfrak{M}_n$ is a σ-algebra.

The sets U_n and $\emptyset$ belong to $\mathfrak{M}_n$, and $\mathfrak{M}_n$ is closed with respect to countable unions. To prove the latter, let $\{E_i\} \subset \mathfrak{M}_n$ and let $\varepsilon > 0$. For every $i \in \mathbb{N}$ there exist

$$C_i \subset E_i \subset A_i \subset U_n$$

such that C_i is compact, A_i is open, and

$$\mu_n^* (A_i \setminus C_i) \le \frac{\varepsilon}{2^{i+1}},$$

where we have used the fact that μ_n^* restricted to the Borel sets of U_n is a finite measure by Carathéodory's theorem. Also, by Proposition 1.7,

$$\lim_{l \to \infty} \mu_n^* \left(\bigcup_{i=1}^{\infty} A_i \setminus \left(\bigcup_{j=1}^{l} C_j \right) \right) = \mu_n^* \left(\bigcup_{i=1}^{\infty} A_i \setminus \left(\bigcup_{j=1}^{\infty} C_j \right) \right)$$

$$\le \sum_{i=1}^{\infty} \mu_n^* (A_i \setminus C_i) \le \frac{\varepsilon}{2},$$

and therefore we may find l large enough that

$$\mu_n^* \left(\bigcup_{i=1}^{\infty} E_i \setminus \left(\bigcup_{j=1}^{l} C_j \right) \right) \le \mu_n^* \left(\bigcup_{i=1}^{\infty} A_i \setminus \left(\bigcup_{j=1}^{l} C_j \right) \right) \le \varepsilon.$$

Hence $\bigcup_{i=1}^{\infty} E_i$ is inner regular, and outer regularity follows from

$$\mu_n^* \left(\bigcup_{i=1}^{\infty} A_i \setminus \left(\bigcup_{j=1}^{\infty} E_j \right) \right) \le \mu_n^* \left(\bigcup_{i=1}^{\infty} A_i \setminus \left(\bigcup_{j=1}^{\infty} C_j \right) \right) \le \varepsilon.$$

It remains to show that if $E \in \mathfrak{M}_n$ then $U_n \setminus E \in \mathfrak{M}_n$. Again for a fixed $\varepsilon > 0$ there exist

$$C \subset E \subset A \subset U_n$$

such that C is compact, A is open, and

$$\mu_n^* (A \setminus C) \le \varepsilon. \tag{1.17}$$

Then $U_n \setminus C$ is open, $U_n \setminus E \subset U_n \setminus C$, and

$$\mu_n^* (U_n \setminus C) = \mu_n^* (U_n) - \mu_n^* (C) \le \mu_n^* (U_n) - \mu_n^* (A) + \varepsilon$$
$$= \mu_n^* (U_n \setminus A) + \varepsilon \le \mu_n^* (U_n \setminus E) + \varepsilon,$$

and this proves the outer regularity of $U_n \setminus E$. On the other hand, since U_n is σ-compact, it is possible to find a compact set $K \subset U_n$ such that

$$\mu_n^* (U_n) \leq \mu_n^* (K) + \varepsilon.$$

Then $K \setminus A$ is compact, $K \setminus A \subset U_n \setminus E$, and by (1.17),

$$\mu_n^* (U_n \setminus E) \leq \mu_n^* (U_n \setminus A) + \varepsilon \leq \mu_n^* (K \setminus A) + 2\varepsilon,$$

thus showing that $U_n \setminus E$ is inner regular.

Step 2: Let $B \subset X$ be any Borel subset and fix $\varepsilon > 0$. By Step 1 there exists an open set $A_1 \subset U_1$ such that $B \cap U_1 \subset A_1$ and

$$\mu^* (A_1) \leq \mu^* (B \cap U_1) + \frac{\varepsilon}{2}.$$

Inductively, assume that for $k = 1, \ldots, n$ there exist open sets $A_k \subset U_k$ such that $B \cap U_k \subset A_k$,

$$A_1 \subset A_2 \subset \ldots \subset A_n,$$

and

$$\mu^* (A_k) \leq \mu^* (B \cap U_k) + \sum_{i=1}^{k} \frac{\varepsilon}{2^i}.$$

To construct A_{n+1}, by Step 1 find an open set $A'_{n+1} \subset U_{n+1}$ such that

$$(B \setminus U_n) \cap U_{n+1} \subset A'_{n+1}$$

and

$$\mu^* (A'_{n+1}) \leq \mu^* ((B \setminus U_n) \cap U_{n+1}) + \frac{\varepsilon}{2^{n+1}}.$$

Setting $A_{n+1} := A_n \cup A'_{n+1}$, it follows that $B \cap U_{n+1} \subset A_{n+1}$ and

$$\mu^* (A_{n+1}) \leq \mu^* (A_n) + \mu^* (A'_{n+1}) \leq \mu^* (B \cap U_n)$$
$$+ \mu^* ((B \setminus U_n) \cap U_{n+1}) + \sum_{i=1}^{n} \frac{\varepsilon}{2^i} + \frac{\varepsilon}{2^{n+1}}$$
$$= \mu^* (B \cap U_{n+1}) + \sum_{i=1}^{n+1} \frac{\varepsilon}{2^i}.$$

We observe that $B \subset \bigcup_{n=1}^{\infty} A_n$, and by Carathéodory's theorem and Proposition 1.7,

$$\mu^* \left(\bigcup_{n=1}^{\infty} A_n \right) = \lim_{n \to \infty} \mu^* (A_n) \leq \lim_{n \to \infty} \mu^* (B \cap U_n) + \varepsilon = \mu^* (B) + \varepsilon.$$

We conclude that B is outer regular.

To prove inner regularity fix $t < \mu^* (B)$ and choose n large enough that

$$\mu^* (B \cap U_n) > t.$$

By Step 1 find a compact set $C \subset B \cap U_n$ such that $\mu^* (C) > t$. This shows that $\mu^* (B)$ can be approximated by the outer measures of compact subsets of B.

Step 3: If μ^* is a Borel regular outer measure, then for every set $E \subset X$ there exists a Borel set $F \subset X$ such that $E \subset F$ and $\mu^* (E) = \mu^* (F)$. Applying the first part of the proof to F, for every $\varepsilon > 0$ we may find an open set

$$A \supset F \supset E$$

such that

$$\mu^* (E) = \mu^* (F) \leq \mu^* (A) + \varepsilon.$$

Hence every set E is outer regular, and so μ^* is a Radon outer measure.

Remark 1.53. It follows from the previous proposition that if X satisfies the properties thus stated and if $\mu^* : \mathcal{P}(X) \to [0, \infty]$ is a Borel regular outer measure, then for any μ^*-measurable set $E \subset X$ with $\mu^* (E) < \infty$ the restriction $\mu^* \lfloor E$ of μ^* to E is a Radon outer measure.

Given an arbitrary Radon outer measure, in general we cannot conclude that every set is inner regular (see Example 1.58). However, the following result holds.

Proposition 1.54. *Let X be a topological space and let $\mu^* : \mathcal{P}(X) \to [0, \infty]$ be a Radon outer measure. Then every σ-finite μ^*-measurable set is inner regular.*

Proof. Let $E \subset X$ be a σ-finite μ^*-measurable set.

Step 1: Assume first that $\mu^* (E) < \infty$. By outer regularity of E, for every $\varepsilon > 0$ there exists an open set $A \supset E$ such that

$$\mu^* (A) < \mu^* (E) + \varepsilon,$$

while by inner regularity of open sets there exists a compact set $K \subset A$ such that

$$\mu^* (A) < \mu^* (K) + \varepsilon.$$

Since E is μ^*-measurable, we have that

$$\mu^* (E) + \mu^* (A \setminus E) = \mu^* (A) < \mu^* (E) + \varepsilon,$$

and so

$$\mu^* (A \setminus E) < \varepsilon.$$

Hence by outer regularity we may choose an open set $U \supset A \setminus E$ such that $\mu^{*}(U) < \varepsilon$. The compact set $C := K \setminus U$ is contained in E, and using the fact that U is μ^{*}-measurable, we get

$$\mu^{*}(C) = \mu^{*}(K \setminus U) = \mu^{*}(K) - \mu^{*}(K \cap U)$$
$$> \mu^{*}(A) - \varepsilon - \mu^{*}(U) > \mu^{*}(E) - 3\varepsilon.$$

Hence E is inner regular.

Step 2: If $\mu^{*}(E) = \infty$, then since E has σ-finite μ^{*} outer measure, we may find a sequence of sets $E_n \subset E$ such that

$$E = \bigcup_{n=1}^{\infty} E_n$$

and $\mu^{*}(E_n) < \infty$. By Remark 1.51 there exist Borel sets F_n such that $E_n \subset F_n$ and $\mu^{*}(E_n) = \mu^{*}(F_n)$. Then $F_n \cap E$ is μ^{*}-measurable,

$$\mu^{*}(F_n \cap E) \le \mu^{*}(F_n) = \mu^{*}(E_n) < \infty,$$

and

$$E = \bigcup_{n=1}^{\infty} (F_n \cap E).$$

We can now modify $\{F_n \cap E\}$ to make it increasing. In particular, it follows that

$$\infty = \mu^{*}(E) = \lim_{n \to \infty} \mu^{*}(F_n \cap E).$$

Let $M > 0$ and let n be so large that $\mu^{*}(F_n \cap E) > M$. By applying Step 1 to $F_n \cap E$ we may find a compact set $K \subset F_n \cap E$ such that $\mu^{*}(K) > M$. Hence

$$\mu^{*}(E) = \sup \{\mu^{*}(K) : K \subset E, \ K \text{ compact}\} = \infty,$$

and the proof is complete.

We now introduce analogous regularity properties for measures.

Definition 1.55. *Let $(X, \mathfrak{M}, \mu)$ be a measure space. If X is a topological space, then*

(i) μ *is a Borel measure if every Borel set is in $\mathfrak{M}$;*
(ii) μ *is a Borel regular measure if μ is a Borel measure and if for every set $E \in \mathfrak{M}$ there exists a Borel set F such that $E \subset F$ and $\mu(E) = \mu(F)$;*
(iii) *a Borel measure $\mu : \mathfrak{M} \to [0, \infty]$ is a Radon measure if*
  a) $\mu(K) < \infty$ for every compact set $K \subset X$;
  b) every open set $A \subset X$ is inner regular;
  c) every set $E \in \mathfrak{M}$ is outer regular.

Hausdorff measures $\mathcal{H}^s$, $s > 0$ (see [FoLe10]), represent an important class of regular Borel measures that are not Radon measures.

For a Borel measure it is possible to define the notion of support.

Definition 1.56. *Let X be a topological space and let $\mu : \mathfrak{M} \to [0, \infty]$ be a Borel measure. The* support *of μ is the set*

$$\operatorname{supp}\mu := \{x \in X : \ \mu(U) > 0 \text{ for all (open) neighborhoods } U \text{ of } x\}.$$

Exercise 1.57. Let X be a topological space and let $\mu : \mathfrak{M} \to [0, \infty]$ be a Borel measure. Prove that the support of μ is closed. Show also that if $E \in \mathfrak{M}$, with $E \subset X \setminus \operatorname{supp}\mu$, then $\mu(E) = 0$. Is the converse true?

Proposition 1.54, which continues to hold for Radon measures, asserts that any measurable set with σ-finite measure is inner regular. The next example shows that there exist Radon measures for which non σ-finite sets may fail to be inner regular.

Exercise 1.58. Consider $X = \mathbb{R}^2$ endowed with the following topology: A set $A \subset X$ is open if and only if for every $y \in \mathbb{R}$ the set

$$A_y := \{x \in \mathbb{R} : (x, y) \in A\}$$

is open in $\mathbb{R}$ with respect to the Euclidean topology. Show that X is a locally compact Hausdorff space. For every Borel set $B \subset X$ define

$$\mu(B) := \begin{cases} \infty & \text{if } B_y \neq \emptyset \text{ for uncountably many } y \in \mathbb{R}, \\ \sum_{y \in \mathbb{R}} \mathcal{L}^1(B_y) \text{ otherwise,} \end{cases}$$

where $\mathcal{L}^1$ is the Lebesgue measure on $\mathbb{R}$ (see the end of the subsection for its definition). Prove that μ is a Radon measure but the set $\{0\} \times \mathbb{R}$ is not inner regular, since $\mu(\{0\} \times \mathbb{R}) = \infty$, while $\mu(K) = 0$ for any compact set $K \subset \{0\} \times \mathbb{R}$.

We have proved in Remark 1.51 that Radon measures are Borel regular. The converse is not true in general, as the next example shows.

Example 1.59. Let X be an uncountable set with the discrete topology. For every $E \subset X$ define

$$\mu(E) := \begin{cases} 0 & \text{if } E \text{ is countable,} \\ \infty \text{ otherwise.} \end{cases}$$

Then $\mu : \mathcal{P}(X) \to \{0, \infty\}$ is a measure (and also an outer measure). Since every subset of X is open, μ is a Borel regular measure and every set is outer regular. Since compact sets are necessarily finite, for every compact set $K \subset X$ we have $\mu(K) = 0$, and so uncountable sets cannot be inner regular. Hence μ is not a Radon measure.

However, the following holds.

Proposition 1.60. *Let X be a locally compact Hausdorff space such that every open set $A \subseteq X$ is σ-compact. Let $\mu : \mathcal{B}(X) \to [0, \infty]$ be a (Borel) measure.*

(i) If μ is finite on compact sets, then μ is a Radon measure and every Borel set E is inner regular.

(ii) If $B \in \mathcal{B}(X)$ and $\mu(B) < \infty$, then B is inner regular.

Proof. The proof of (i) follows closely that of Proposition 1.52, where now $\mathfrak{M}_n$ is defined as

$$\mathfrak{M}_n := \{E \in \mathcal{B}(U_n) : E \text{ is } \mu_n \text{ inner and outer regular}\}$$

and where μ_n denotes the restriction of μ to $\mathcal{B}(U_n)$. We omit the details.

(ii) is a consequence of (i) applied to the finite measure $\mu : \mathcal{B}(B) \to [0, \infty)$.

Exercise 1.61. Without the assumption that open sets are σ-compact the previous proposition fails. Indeed, let Y be the set of countable ordinals, let ω_1 be the first uncountable ordinal, and define $X := Y \cup \{\omega_1\}$. In X consider the smallest topology that contains all sets of the form $\{\alpha \in X : \alpha < \beta\}$ and $\{\alpha \in X : \alpha > \beta\}$ for $\beta \in X$. Prove that X is a compact Hausdorff space, and that Y is an open set that is not σ-compact. Check that for every Borel set $B \subset X$ either $B \cup \{\omega_1\}$ or $(X \setminus B) \cup \{\omega_1\}$ (but not both) contains an uncountable closed set. Define

$$\mu(B) := \begin{cases} 1 \text{ if } B \cup \{\omega_1\} \text{ contains an uncountable closed set,} \\ 0 \text{ if } (X \setminus B) \cup \{\omega_1\} \text{ contains an uncountable closed set.} \end{cases}$$

Show that μ is a measure, but $\{\omega_1\}$ is not outer regular since $\mu(\{\omega_1\}) = 0$ but $\mu(A) = 1$ for every open set A containing ω_1. We refer to [AliBo99] for more details.

We study next the relation between regularity properties for outer measures and regularity properties of measures. If μ^* is a Borel regular (respectively Radon) outer measure, then its restriction μ to the σ-algebra of all μ^*-measurable sets is a Borel regular (respectively Radon) measure. Conversely, given a Borel regular (respectively Radon) measure μ, it is always possible to extend it to a Borel regular (respectively Radon) outer measure. This is a consequence of the following result.

Theorem 1.62 (De Giorgi–Letta). *Let (X, τ) be a topological space. Assume that $\rho : \tau \to [0, \infty]$ is an increasing set function such that*

(i) $\rho(\emptyset) = 0$;

(ii) (subadditivity) $\rho(A_1 \cup A_2) \leq \rho(A_1) + \rho(A_2)$ for all A_1, $A_2 \in \tau$;

(iii) (inner regularity) $\rho(A) = \sup\{\rho(A_1) : A_1 \subset\subset A\}$ for every $A \in \tau$.

Then the extension of ρ to every set $E \subset X$ defined by

$$\mu^*(E) := \inf\{\rho(A) : A \in \tau,\ E \subset A\} \tag{1.18}$$

is an outer measure and every set is outer regular.

Moreover, if X is a metric space and ρ satisfies the additional hypothesis

(iv) (superadditivity) $\rho(A_1 \cup A_2) \geq \rho(A_1) + \rho(A_2)$ for all A_1, $A_2 \in \tau$, with $A_1 \cap A_2 = \emptyset$;

then μ^* is a Borel outer measure and every set is outer regular.

Proof. **Step 1:** We prove that under hypotheses (i)–(iii) the set function μ^* is an outer measure. Note that, in view of the definition (1.18), $\mu^* = \rho$ on τ. Hence by (i) we have that $\mu^*(\emptyset) = 0$. If $E \subset F \subset X$, then any open set that contains F also contains E, and so $\mu^*(E) \leq \mu^*(F)$.

It remains to prove countable subadditivity for μ^*. We prove it first for ρ. Let $\{A_n\}$ be a sequence of open sets and consider any open set A' compactly contained in $\bigcup_{n=1}^{\infty} A_n$. Since the family $\{A_n\}$ is an open cover for the compact set $\overline{A'}$, there exists $l \in \mathbb{N}$ such that the finite family $\{A_n\}_{n=1}^{l}$ is still an open cover for $\overline{A'}$, and so, by the fact that ρ is increasing and nonnegative and by (ii),

$$\rho(A') \leq \rho\left(\bigcup_{n=1}^{l} A_n\right) \leq \sum_{n=1}^{l} \rho(A_n) \leq \sum_{n=1}^{\infty} \rho(A_n).$$

Taking the supremum over all open sets A' compactly contained in $\bigcup_{n=1}^{\infty} A_n$ and using (iii), we obtain

$$\rho\left(\bigcup_{n=1}^{\infty} A_n\right) \leq \sum_{n=1}^{\infty} \rho(A_n).$$

Hence ρ is countably subadditive, and we now prove that the same holds for μ^*. Let $\{E_n\} \subset \mathcal{P}(X)$. For all $n \in \mathbb{N}$ and for a fixed $\varepsilon > 0$ find A_n open, with $E_n \subset A_n$, such that

$$\rho(A_n) \leq \mu^*(E_n) + \frac{\varepsilon}{2^n}.$$

Then $\bigcup_{n=1}^{\infty} E_n \subset \bigcup_{n=1}^{\infty} A_n$, and so

$$\mu^*\left(\bigcup_{n=1}^{\infty} E_n\right) \leq \rho\left(\bigcup_{n=1}^{\infty} A_n\right) \leq \sum_{n=1}^{\infty} \rho(A_n) \leq \sum_{n=1}^{\infty} \mu^*(E_n) + \varepsilon.$$

By letting $\varepsilon \to 0^+$ we conclude that μ^* is an outer measure. Since $\mu^* = \rho$ on τ it follows from (1.18) that every set is outer regular by construction.

Step 2: We now assume that X is a metric space and (i)–(iv) hold. In view of Proposition 1.41 it suffices to prove that

$$\mu^* (E \cup F) = \mu^* (E) + \mu^* (F) \tag{1.19}$$

for all sets E, $F \subset X$, with dist $(E, F) > 0$. Fix $\varepsilon > 0$ and find an open set A, with $E \cup F \subset A$, such that

$$\rho (A) \leq \mu^* (E \cup F) + \varepsilon.$$

Since $d := \text{dist} (E, F) > 0$ we may construct open sets A_1 and A_2, with $E \subset A_1 \subset A$ and $F \subset A_2 \subset A$, and such that $A_1 \cap A_2 = \emptyset$ (e.g., take A_1 to be the union of all balls $B (x, r) \subset A$, where $x \in E$ and $0 < r < \frac{d}{2}$, and take A_2 to be the union of all balls $B (x, r) \subset A$, where $x \in F$ and $0 < r < \frac{d}{2}$). Hence by (1.18), (iv), and the fact that ρ is increasing,

$$\mu^* (E) + \mu^* (F) \leq \rho (A_1) + \rho (A_2) = \rho (A_1 \cup A_2) \leq \rho (A) \leq \mu^* (E \cup F) + \varepsilon.$$

Letting $\varepsilon \to 0^+$, and using also the fact that μ^* is an outer measure, we conclude that (1.19) holds, and this concludes the proof.

The De Giorgi–Letta theorem will often be used in [FoLe10] to study relaxation problems in Sobolev spaces, where the integral functionals under consideration are naturally defined on open sets.

As a corollary we have the following:

Proposition 1.63. *Let* (X, τ) *be a locally compact Hausdorff space and let* $\mu : \mathfrak{M} \to [0, \infty]$ *be a Radon measure. For every set* $E \subset X$ *define*

$$\mu^* (E) = \inf \{\mu (A) : A \in \tau, \, E \subset A\} .$$

Then μ^* *is a Radon outer measure, the* σ-*algebra* $\mathfrak{M}^*$ *of* μ^*-*measurable sets contains* $\mathfrak{M}$, *and* μ^* *coincides with* μ *on* $\mathfrak{M}$.

Proof. Since X is a locally compact Hausdorff space, using the inner regularity of open sets for any open set A we have

$$\mu (A) = \sup \{\mu (K) : K \subset A, \, K \text{ compact}\}$$
$$= \sup \{\mu (A_1) : A_1 \subset\subset A, \, A_1 \text{ open}\} .$$

Hence we may apply the De Giorgi–Letta theorem with $\rho := \mu$ to obtain that μ^* is an outer measure. Using the outer regularity of μ, it follows that $\mu^* = \mu$ on $\mathfrak{M}$.

It remains to show that every set $E \in \mathfrak{M}$ is μ^*-measurable. Let F be any set. Given an open set $A \supset F$ we have

$$\mu (A) = \mu (A \cap E) + \mu (A \setminus E) = \mu^* (A \cap E) + \mu^* (A \setminus E)$$
$$\geq \mu^* (F \cap E) + \mu^* (F \setminus E) .$$

Taking the infimum over all open sets containing F, we obtain

$$\mu^* (F) \geq \mu^* (F \cap E) + \mu^* (F \setminus E) ,$$

and this implies that E is μ^*-measurable.

Contemporary methods in relaxation theory exploit the structure of Radon measures, together with Radon–Nikodym-type theorems, to obtain characterizations of effective or relaxed energy densities. Therefore it is of the utmost interest to be able to determine whether a set function is (the restriction of) a Radon measure by means of a criterion easier to manipulate than Definition 1.55.

Theorem 1.64. *Let (X, τ) be a locally compact Hausdorff space. Assume that $\mu : \tau \to [0, \infty)$ is a set function and that ν is a finite Radon measure such that*

(i) $\mu(A) \leq \mu\left(A \setminus \overline{C}\right) + \mu(B)$ for all A, B, $C \in \tau$ with $\overline{C} \subset B \subset A$;
(ii) for every $\varepsilon > 0$ and for every $A \in \tau$ there exists $C \in \tau$ such that $C \subset\subset A$ and $\mu\left(A \setminus \overline{C}\right) \leq \varepsilon$;
(iii) $\nu(X) \leq \mu(X)$;
(iv) $\nu\left(\overline{A}\right) \geq \mu(A)$ for every $A \in \tau$.

Then $\mu(A) = \nu(A)$ for every $A \in \tau$.

Proof. Fix $A \in \tau$. We first prove that $\mu(A) \leq \nu(A)$. Let $\varepsilon > 0$ and choose $C \in \tau$ as in (ii). Let $B \in \tau$ be such that $\overline{C} \subset B \subset \overline{B} \subset A$. By (i), (ii), and (iv) we have

$$\mu(A) \leq \mu\left(A \setminus \overline{C}\right) + \mu(B) \leq \varepsilon + \nu\left(\overline{B}\right) \leq \varepsilon + \nu(A),$$

and it suffices to let $\varepsilon \to 0^+$.

To prove the reverse inequality, using the inner regularity of the measure ν, for every $\varepsilon > 0$ we may find a compact set $K \subset A$ such that

$$\nu(A) \leq \varepsilon + \nu(K).$$

In turn, by Theorem A.12 there exists $C \in \tau$, with $\overline{C}$ compact, such that $K \subset C \subset \overline{C} \subset A$. Since

$$\nu(A) \leq \varepsilon + \nu\left(\overline{C}\right),$$

by (iii), the first part of this proof, and (i), in this order, we have

$$\nu(A) \leq \varepsilon + \nu(X) - \nu\left(X \setminus \overline{C}\right) \leq \varepsilon + \mu(X) - \mu\left(X \setminus \overline{C}\right) \leq \varepsilon + \mu(A),$$

and the conclusion follows by letting $\varepsilon \to 0^+$.

We end this subsection by introducing in $\mathbb{R}^N$ the Lebesgue measure and the σ-algebra of Lebesgue-measurable sets. In the Euclidean space $\mathbb{R}^N$ consider the family of elementary sets

$$\mathcal{G} := \left\{Q(x, r) : x \in \mathbb{R}^N, \, 0 < r < \infty\right\} \cup \{\emptyset\}$$

and define $\rho(Q(x, r)) := r^N$ and $\rho(\emptyset) := 0$, where

$$Q\left(x,r\right):=x+\left(-\frac{r}{2},\frac{r}{2}\right)^{N}.$$

For each set $E \subset \mathbb{R}^N$ define

$$\mathcal{L}_{o}^{N}\left(E\right):=\inf\left\{\sum_{n=1}^{\infty}\left(r_{n}\right)^{N}:\ \{Q\left(x_{n},r_{n}\right)\}\subset\mathcal{G},\ E\subset\bigcup_{n=1}^{\infty}Q\left(x_{n},r_{n}\right)\right\}.$$

By Proposition 1.32, $\mathcal{L}_o^N$ is an outer measure, called the N–dimensional *Lebesgue outer measure*. Using Remark 1.33 it can be shown that

$$\mathcal{L}_{o}^{N}\left(Q\left(x,r\right)\right)=\rho\left(Q\left(x,r\right)\right)=r^{N} \tag{1.20}$$

and that $\mathcal{L}_o^N$ is translation-invariant, i.e., $\mathcal{L}_o^N\left(x+E\right)=\mathcal{L}_o^N\left(E\right)$ for all $x\in\mathbb{R}^N$ and all $E \subset \mathbb{R}^N$.

The class of all $\mathcal{L}_o^N$-measurable subsets of $\mathbb{R}^N$ is called the *σ-algebra of Lebesgue measurable sets*, and by Carathéodory's theorem, $\mathcal{L}_o^N$ restricted to this σ-algebra is a complete measure, called the N–dimensional *Lebesgue measure* and denoted by $\mathcal{L}^N$. Given a Lebesgue measurable set $E \subset \mathbb{R}^N$, we will write indifferently

$$\mathcal{L}^N\left(E\right) \ \text{ or } \ |E|$$

for the Lebesgue measure of E.

Proposition 1.65. *The Lebesgue measure $\mathcal{L}^N$ has the following properties:*

(i) $\mathcal{L}^N\left(\left[a_1,b_1\right]\times\ldots\times\left[a_N,b_N\right]\right)=\left(b_1-a_1\right)\ldots\left(b_N-a_N\right);$
(ii) $\mathcal{L}^N$ *is translation-invariant, i.e.,* $\mathcal{L}^N\left(x+E\right)=\mathcal{L}^N\left(E\right)$ *for every Lebesgue measurable set $E\subset\mathbb{R}^N$ and $x\in\mathbb{R}^N$;*
(iii) for every linear operator $L:\mathbb{R}^N\to\mathbb{R}^N$ and every Lebesgue measurable set $E\subset\mathbb{R}^N$, $\mathcal{L}^N\left(L\left(E\right)\right)=\left|\det L\right|\mathcal{L}^N\left(E\right).$

Using Proposition 1.41 we have the following proposition:

Proposition 1.66. *The outer measure $\mathcal{L}_o^N$ is a metric outer measure, so that every Borel subset of $\mathbb{R}^N$ is Lebesgue measurable.*

Moreover, if μ is any positive translation-invariant Borel measure on $\mathbb{R}^N$ finite on compact sets, then $\mu\left(B\right)=c\mathcal{L}^N\left(B\right)$ for some $c\geq0$ and for every Borel subset B of $\mathbb{R}^N$.

Remark 1.67. Using the axiom of choice it is possible to construct sets that are not Lebesgue measurable. It may also be proved that there are Lebesgue measurable sets that are not Borel sets. Hence $\mathcal{L}^N:\mathcal{B}\left(\mathbb{R}^N\right)\to\left[0,\infty\right]$ is not a complete measure.

We observe that the Lebesgue outer measure is a Radon outer measure. Indeed, outer regularity of arbitrary sets follows from (1.20) and the definition of $\mathcal{L}_o^N$, while inner regularity of open sets is an immediate consequence of the fact that each open set can be written as a countable union of closed cubes with pairwise disjoint interiors.

In view of Proposition 1.54 it turns out that every Lebesgue measurable set is the union of a Borel set with a set of Lebesgue measure zero. Hence the σ-algebra of Lebesgue measurable sets is the completion of the Borel σ-algebra for the Lebesgue measure.

Proposition 1.68. $\mathcal{L}_o^N$ *is a Radon outer measure. Moreover, every Lebesgue measurable set E is the union of a Borel set and a set of Lebesgue measure zero.*

1.1.3 Measurable Functions and Lebesgue Integration

In this subsection we introduce the notions of measurable and integrable functions.

Definition 1.69. *Let X and Y be nonempty sets, and let $\mathfrak{M}$ and $\mathfrak{N}$ be algebras on X and Y, respectively. A function $u : X \to Y$ is said to be* measurable *if $u^{-1}(E) \in \mathfrak{M}$ for every set $E \in \mathfrak{N}$.*

Remark 1.70. If $\mathfrak{M}$ is a σ-algebra on a set X and $\mathfrak{N}$ is the smallest σ-algebra that contains a given family $\mathcal{G}$ of subsets of a set Y, then $u : X \to Y$ is measurable if and only if $u^{-1}(E) \in \mathfrak{M}$ for every set $E \in \mathcal{G}$. Indeed, the family of sets

$$\left\{ E \in \mathfrak{N} : u^{-1}(E) \in \mathfrak{M} \right\}$$

is a σ-algebra that contains $\mathcal{G}$, and so it must coincide with $\mathfrak{N}$. Thus, in particular, if Y is a topological space and $\mathfrak{N} = \mathcal{B}(Y)$, then it suffices to verify that $u^{-1}(A) \in \mathfrak{M}$ for every open set $A \subset Y$.

If X and Y are topological spaces, $\mathfrak{M} := \mathcal{B}(X)$ and $\mathfrak{N} := \mathcal{B}(Y)$, then a measurable function $u : X \to Y$ will be called a *Borel function*.

If $u : X \to Y$ is measurable and if $E \in \mathfrak{M}$, then the restriction $u : X \setminus E \to Y$ is a measurable function between the measurable spaces $(X \setminus E, \mathfrak{M}\lfloor (X \setminus E))$ and $(Y, \mathfrak{N})$. Conversely, if X is a measure space with measure μ, and the function u is defined only on $X \setminus E$ with $\mu(E) = 0$, then in general, measurability of u on $X \setminus E$ does not entail the measurability of an arbitrary extension of u to X unless μ is complete.

Proposition 1.71. *Let $(X, \mathfrak{M})$ and $(Y, \mathfrak{N})$ be two measurable spaces, and let $u : X \to Y$ be a measurable function. Let $\mu : \mathfrak{M} \to [0, \infty]$ be a complete measure. If $v : X \to Y$ is a function such that $u(x) = v(x)$ for μ a.e. $x \in X$, then v is measurable.*

Going back to the setting in which $u : X \setminus E \to Y$ with $\mu(E) = 0$, since Lebesgue integration does not take into account sets of measure zero, we will see that integration of u depends mostly on its measurability on $X \setminus E$. For this reason, we extend Definition 1.69 to read as follows.

Definition 1.72. *Let $(X, \mathfrak{M})$ and $(Y, \mathfrak{N})$ be two measurable spaces, and let $\mu : \mathfrak{M} \to [0, \infty]$ be a measure. Given a function $u : X \setminus E \to Y$ where $\mu(E) = 0$, u is said to be* measurable over X *if $u^{-1}(F) \in \mathfrak{M}$ for every set $F \in \mathfrak{N}$.*

We are now in a position to introduce the notion of integral.

Definition 1.73. *Let X be a nonempty set, and let $\mathfrak{M}$ be an algebra on X. A* simple function *is a measurable function $s : X \to \mathbb{R}$ whose range consists of finitely many points.*

If $c_1, \dots, c_\ell$ are the distinct values of s, then we write

$$s = \sum_{n=1}^{\ell} c_n \chi_{E_n},$$

where χ_{E_n} is the *characteristic function* of the set $E_n := \{x \in X : s(x) = c_n\}$, i.e.,

$$\chi_{E_n}(x) := \begin{cases} 1 \text{ if } x \in E_n, \\ 0 \ \text{ otherwise.} \end{cases}$$

If μ is a finitely additive (positive) measure on X and $s \geq 0$, then for every measurable set $E \in \mathfrak{M}$ we define the *Lebesgue integral* of s over E as

$$\int_E s \, d\mu := \sum_{n=1}^{\ell} c_n \mu(E_n \cap E), \tag{1.21}$$

where if $c_n = 0$ and $\mu(E_n \cap E) = \infty$, then we use the convention

$$c_n \mu(E_n \cap E) := 0.$$

Theorem 1.74. *Let X be a nonempty set, let $\mathfrak{M}$ be an algebra on X, and let $u : X \to [0, \infty]$ be a measurable function. Then there exists a sequence $\{s_n\}$ of simple functions such that*

$$0 \leq s_1(x) \leq s_2(x) \leq \dots \leq s_n(x) \to u(x)$$

for every $x \in X$. The convergence is uniform on any set on which u is bounded from above.

Proof. For $n \in \mathbb{N}_0$ and $0 \leq k \leq 2^{2n} - 1$ define

$$E_{n,k} := \left\{ x \in X : \frac{k}{2^n} \leq u(x) < \frac{k+1}{2^{n+1}} \right\}, \quad E_n := \{x \in X : u(x) \geq 2^n\},$$

and let

$$s_n := \sum_{k=0}^{2^{2n}-1} \frac{k}{2^n} \chi_{E_{n,k}} + 2^n \chi_{E_n}.$$

If $u(x) = 0$ then $x \in E_{n,0}$, and so $s_n(x) = 0$ for all $n \in \mathbb{N}$. If $0 < u(x) < \infty$ then $x \in E_{n,\lceil 2^n u(x)\rceil}$ for all $n \in \mathbb{N}$ sufficiently large, where $\lceil \cdot \rceil$ is the integer part. Hence

$$0 \leq s_n(x) = \frac{\lceil 2^n u(x)\rceil}{2^n} \leq \frac{\lceil 2^{n+1} u(x)\rceil}{2^{n+1}} = s_{n+1}(x) \to u(x)$$

as $n \to \infty$. If $u(x) = \infty$ then $x \in E_n$ for all n, and so $s_n(x) = 2^n \nearrow \infty$.

Note that by the definition of the sets $E_{n,k}$ we have that $0 \leq u - s_n \leq \frac{1}{2^n}$ on the set where $u < 2^n$. Hence we have uniform convergence on each set on which u is bounded from above.

In the remainder of this subsection, $\mathfrak{M}$ is a σ-algebra and μ a (countably additive) measure. In view of the previous theorem, if $u : X \to [0, \infty]$ is a measurable function, then we define its *(Lebesgue) integral* over a measurable set E as

$$\int_E u \, d\mu := \sup \left\{ \int_E s \, d\mu : s \text{ simple}, \, 0 \leq s \leq u \right\}.$$

In order to extend the notion of integral to functions of arbitrary sign, consider $u : X \to [-\infty, \infty]$ and set

$$u^+ := \max\{u, 0\}, \quad u^- := \max\{-u, 0\}.$$

Note that $u = u^+ - u^-$, $|u| = u^+ + u^-$, and u is measurable if and only if u^+ and u^- are measurable. Also, if u is bounded, then so are u^+ and u^-, and in view of Theorem 1.74, u is then the uniform limit of a sequence of simple functions.

Definition 1.75. *Let $(X, \mathfrak{M}, \mu)$ be a measure space, and let $u : X \to [-\infty, \infty]$ be a measurable function. Given a measurable set $E \in \mathfrak{M}$, if at least one of the two integrals $\int_E u^+ \, d\mu$ and $\int_E u^- \, d\mu$ is finite, then we define the* (Lebesgue) *integral of u over the measurable set E by*

$$\int_E u \, d\mu := \int_E u^+ \, d\mu - \int_E u^- \, d\mu.$$

If both $\int_E u^+ \, d\mu$ and $\int_E u^- \, d\mu$ are finite, then u is said to be (Lebesgue) integrable *over the measurable set E.*

In the special case that μ is the Lebesgue measure, we denote $\int_E u \, d\mathcal{L}^N$ simply by

$$\int_E u \, dx.$$

If $(X, \mathfrak{M}, \mu)$ is a measure space, with X a topological space, and if $\mathfrak{M}$ contains $\mathcal{B}(X)$, then $u : X \to [-\infty, \infty]$ is said to be *locally integrable* if it is Lebesgue integrable over every compact set.

A measurable function $u : X \to [-\infty, \infty]$ is Lebesgue integrable over the measurable set E if and only if

$$\int_E |u| \, d\mu < \infty.$$

Finally, if $F \in \mathfrak{M}$ is such that $\mu(F) = 0$ and $u : X \setminus F \to [-\infty, \infty]$ is a measurable function in the sense of Definition 1.72, then we define the *(Lebesgue) integral of* u over the measurable set E as the Lebesgue integral of the function

$$v(x) := \begin{cases} u(x) \text{ if } x \in X \setminus F, \\ 0 \qquad \text{otherwise,} \end{cases}$$

provided $\int_E v \, d\mu$ is well-defined. Note that in this case

$$\int_E v \, d\mu = \int_E \tilde{v} \, d\mu,$$

where

$$\tilde{v}(x) := \begin{cases} u(x) \text{ if } x \in X \setminus F, \\ w(x) \text{ otherwise,} \end{cases}$$

and w is an arbitrary measurable function defined on F.

Remark 1.76. If $\mathfrak{M}$ is an algebra on X and $\mu : \mathfrak{M} \to [0, \infty]$ is a finitely additive measure on X, we can still define a notion of integral as follows: we say that a measurable function $u : X \to [-\infty, \infty]$ is *integrable* if there exists a sequence $\{s_n\}$ of measurable simple functions, each of them bounded and vanishing outside a set of finite measure (depending on n), such that

$$\lim_{n \to \infty} \mu(\{x \in X : |u(x) - s_n(x)| > \varepsilon\}) = 0$$

for each $\varepsilon > 0$, and

$$\lim_{n,l \to \infty} \int_X |s_n - s_l| \, d\mu = 0.$$

It may be shown that for every $E \in \mathfrak{M}$ the limit $\lim_{n \to \infty} \int_E s_n \, d\mu$ exists in $\mathbb{R}$ and does not depend on the particular sequence $\{s_n\}$. The *integral of* u over the measurable set E is defined by

$$\int_E u \, d\mu := \lim_{n \to \infty} \int_E s_n \, d\mu.$$

The following result follows from Theorem 1.74.

Corollary 1.77. *Let* $(X, \mathfrak{M}, \mu)$ *be a measure space, and let* $u : X \to [-\infty, \infty]$ *be a measurable function. If the set* $\{x \in X : |u(x)| > 0\}$ *has* σ*-finite measure and* u *is finite* μ *a.e., then there exists a sequence* $\{s_n\}$ *of simple functions, each of them bounded and vanishing outside a set of finite measure (depending on* n*), such that* $s_n(x) \to u(x)$ *as* $n \to \infty$ *for* μ *a.e.* $x \in X$ *and*

$$|s_n(x)| \leq |u(x)|$$

for μ *a.e.* $x \in X$ *and* $n \in \mathbb{N}$.

Proof. **Step 1:** Assume first that $u \geq 0$. Since $\{x \in X : u(x) > 0\}$ has σ-finite measure we can find an increasing sequence $\{X_n\}$ of measurable sets of finite measure such that

$$\{x \in X : u(x) > 0\} = \bigcup_n X_n.$$

For $n \in \mathbb{N}_0$ and $0 \leq k \leq 2^{2n} - 1$ define

$$E_{n,k} := \left\{ x \in X_n : \frac{k}{2^n} \leq u(x) < \frac{k+1}{2^{n+1}} \right\},$$

and let

$$s_n := \sum_{k=0}^{2^{2n}-1} \frac{k}{2^n} \chi_{E_{n,k}}.$$

Then as in the proof of Theorem 1.74, it is possible to show that at each point x where u is finite, $s_n(x) \to u(x)$ as $n \to \infty$.

Step 2: In the general case it suffices to apply Step 1 to u^+ and u^-.

Remark 1.78. Note that if u is Lebesgue integrable, then the hypotheses of the previous corollary hold, since

$$\{x \in X : |u(x)| > 0\} = \bigcup_{n=1}^{\infty} \left\{ x \in X : |u(x)| > \frac{1}{n} \right\}$$

and

$$\frac{1}{n} \mu \left(\left\{ x \in X : |u(x)| > \frac{1}{n} \right\} \right) \leq \int_X |u| \, d\mu < \infty.$$

Similarly, $\mu(\{x \in X : |u(x)| = \infty\}) = 0$.

The next two results are pivotal in the theory of integration of nonnegative functions.

Theorem 1.79 (Lebesgue monotone convergence theorem). *Let $(X, \mathfrak{M}, \mu)$ be a measure space, and let $u_n : X \to [0, \infty]$ be a sequence of measurable functions such that*

$$0 \leq u_1(x) \leq u_2(x) \leq \ldots \leq u_n(x) \to u(x)$$

for every $x \in X$. Then u is measurable and

$$\lim_{n \to \infty} \int_X u_n \, d\mu = \int_X u \, d\mu.$$

Example 1.80. The Lebesgue monotone convergence theorem does not hold in general for decreasing sequences. Indeed, consider $X = \mathbb{R}$ and let μ be the Lebesgue measure $\mathcal{L}^1$. Define

$$u_n := \frac{1}{n}\chi_{[n,\infty)}.$$

Then $u_n \geq u_{n+1}$ and

$$\lim_{n\to\infty}\int_{\mathbb{R}} u_n\,dx = \infty \neq 0 = \int_{\mathbb{R}}\lim_{n\to\infty} u_n\,dx.$$

Corollary 1.81. *Let $(X, \mathfrak{M}, \mu)$ be a measure space, and let $u_n : X \to [0,\infty]$ be a sequence of measurable functions. Then*

$$\sum_{n=1}^{\infty}\int_X u_n\,d\mu = \int_X \sum_{n=1}^{\infty} u_n\,d\mu.$$

Proof. Apply the Lebesgue monotone convergence theorem to the increasing sequence of partial sums and use linearity of the integral.

Example 1.82. Given a doubly indexed sequence $\{a_{nk}\}$, with $a_{nk} \geq 0$ for all n, $k \in \mathbb{N}$, we have

$$\sum_{n=1}^{\infty}\sum_{k=1}^{\infty} a_{nk} = \sum_{k=1}^{\infty}\sum_{n=1}^{\infty} a_{nk}.$$

To see this, it suffices to consider $X = \mathbb{N}$ with counting measure and to define $u_n : \mathbb{N} \to [0,\infty]$ by $u_n(k) := a_{nk}$. Then

$$\int_X u_n\,d\mu = \sum_{k=1}^{\infty} a_{nk},$$

and the result now follows from the previous corollary.

Lemma 1.83 (Fatou lemma). *Let $(X, \mathfrak{M}, \mu)$ be a measure space.*

(i) If $u_n : X \to [0,\infty]$ is a sequence of measurable functions, then

$$u := \liminf_{n\to\infty} u_n$$

is a measurable function and

$$\int_X u\,d\mu \leq \liminf_{n\to\infty}\int_X u_n\,d\mu;$$

(ii) if $u_n : X \to [-\infty,\infty]$ is a sequence of measurable functions such that

$$u_n \leq v$$

for some measurable function $v : X \to [0, \infty]$ with $\int_X v \, d\mu < \infty$, then

$$u := \limsup_{n \to \infty} u_n$$

is a measurable function and

$$\int_X u \, d\mu \geq \limsup_{n \to \infty} \int_X u_n \, d\mu.$$

Example 1.84. Fatou's lemma (i) fails for real-valued functions. Indeed, consider $X = \mathbb{R}$ and let μ be the Lebesgue measure $\mathcal{L}^1$. Define

$$u_n := -\frac{1}{n} \chi_{[0,n]}.$$

Then

$$\liminf_{n \to \infty} \int_{\mathbb{R}} u_n \, dx = -1 < 0 = \int_{\mathbb{R}} \lim_{n \to \infty} u_n \, dx.$$

For functions of arbitrary sign we have the following convergence result.

Theorem 1.85 (Lebesgue dominated convergence theorem). *Let $(X, \mathfrak{M}, \mu)$ be a measure space, and let $u_n : X \to [-\infty, \infty]$ be a sequence of measurable functions such that*

$$\lim_{n \to \infty} u_n (x) = u (x)$$

for μ a.e. $x \in X$. If there exists a Lebesgue integrable function v such that

$$|u_n (x)| \leq v (x)$$

for μ a.e. $x \in X$ and all $n \in \mathbb{N}$, then u is Lebesgue integrable and

$$\lim_{n \to \infty} \int_X |u_n - u| \, d\mu = 0.$$

In particular,

$$\lim_{n \to \infty} \int_X u_n \, d\mu = \int_X u \, d\mu.$$

Example 1.86. If v is not integrable then the theorem fails in general. Indeed, consider $X = [0, 1]$ and let μ be the Lebesgue measure $\mathcal{L}^1$. Define

$$u_n := n \chi_{[0, \frac{1}{n}]}.$$

Then

$$\lim_{n \to \infty} \int_0^1 u_n \, dx = 1 \neq 0 = \int_0^1 \lim_{n \to \infty} u_n \, dx.$$

The results below provide applications of the finite value property and nonatomic measures (see Proposition 1.20) in the case that the measure is given by an integral. They will be used to study well-posedness of functionals of the form $\int_E f (x, v (x)) \, dx$ in Chapter 6.

Proposition 1.87. *Let $(X, \mathfrak{M}, \mu)$ be a measure space, let μ satisfy the finite value property, and let $u, v : X \to [0, \infty]$ be measurable functions such that*

$$\int_E u \, d\mu \leq \int_E v \, d\mu$$

for all $E \in \mathfrak{M}$ with $\mu(E) < \infty$. Then $u(x) \leq v(x)$ for μ a.e. $x \in X$.

Proof. For $k \in \mathbb{N}$ define $u_k := \inf\{u, k\}$, $v_k := \inf\{v, k\}$. It is enough to show that

$$u_k(x) \leq v_k(x) \text{ for } \mu \text{ a.e. } x \in X.$$

For $n \in \mathbb{N}$ define

$$E_{n,k} := \left\{ x \in X : u_k(x) > v_k(x) + \frac{1}{n} \right\}.$$

We claim that $\mu(E_{n,k}) = 0$ for all $k, n \in \mathbb{N}$. Indeed, assume by contradiction that $\mu(E_{n,k}) > 0$ for some $k, n \in \mathbb{N}$. By the finite value property there exists a measurable set $F \subset E_{n,k}$ such that $0 < \mu(F) < \infty$. By definition of $E_{n,k}$,

$$k \geq u_k(x) > v_k(x) + \frac{1}{n}$$

for all $x \in F$, and so, since $k > v_k(x) = \inf\{v(x), k\}$, we have that

$$v(x) + \frac{1}{n} = v_k(x) + \frac{1}{n} < u_k(x) \leq u(x),$$

and upon integration over F,

$$\int_F v \, d\mu + \frac{1}{n}\mu(F) \leq \int_F u \, d\mu \leq \int_F v \, d\mu,$$

where in the last inequality we have used the hypothesis. Since $\mu(F) < \infty$ and $v \leq k$ on F, it follows that the right-hand side of the previous inequality is finite, and thus $\mu(F) = 0$, which is a contradiction.

Remark 1.88. If u is integrable, then we can dispense with the hypothesis that μ has the finite value property. Indeed, the sets $E_{n,k}$ constructed in the proof are contained in the set $\{x \in X : u(x) > 0\}$, which has σ-finite measure in view of Remark 1.78.

Proposition 1.89. *Let $(X, \mathfrak{M}, \mu)$ be a measure space, let μ be nonatomic, and let $u : X \to [0, \infty]$ be a measurable function. Then the measure*

$$\nu(E) := \int_E u \, d\mu, \quad E \in \mathfrak{M}, \tag{1.22}$$

is nonatomic if and only if

$$\mu(\{x \in X : u(x) = \infty\}) = 0.$$

Proof. Assume that ν is nonatomic. If

$$\mu\left(\{x \in X : u(x) = \infty\}\right) > 0,$$

then any subset of this set has either zero or infinite ν measure, and this violates the hypothesis.

Conversely, suppose that $\mu\left(\{x \in X : u(x) = \infty\}\right) = 0$. Let

$$E_0 := \{x \in X : 0 < u(x) < \infty\}.$$

If $\mu(E_0) = 0$, then $\nu \equiv 0$ and there is nothing to prove. Thus assume that $\mu(E_0) > 0$. Since $\nu(X \setminus E_0) = 0$, to prove that ν is nonatomic it suffices to show that this holds for $\nu : \mathfrak{M} \lfloor E_0 \to [0, \infty]$. Let $E \in \mathfrak{M}$ be a subset of E_0 with $\nu(E) > 0$. Since

$$E_0 = \bigcup_k E_k,$$

where

$$E_k := \left\{x \in X : \frac{1}{k} < u(x) < k\right\},$$

there exists k_0 such that $\mu(E \cap E_{k_0}) > 0$. Since μ is nonatomic there exists $F \in \mathfrak{M}$, $F \subset E \cap E_{k_0}$, such that

$$0 < \mu(F) < \mu(E \cap E_{k_0}).$$

If $\nu(F) = 0$ then $u = 0$ μ a.e. on F, and if $\nu((E \cap E_{k_0}) \setminus F) = 0$ then $u = 0$ μ a.e. on $(E \cap E_{k_0}) \setminus F$. By the definition of E_{k_0} these would imply that $\mu(F) = 0$ or $\mu((E \cap E_{k_0}) \setminus F) = 0$, both contradicting the choice of F. Hence

$$\nu(F), \nu((E \cap E_{k_0}) \setminus F) > 0.$$

Since $u \le k_0$ on F and $\mu(F) < \infty$, we obtain $\infty > \nu(F) > 0$. It follows that

$$0 < \nu(F) < \nu(F) + \nu((E \cap E_{k_0}) \setminus F) \le \nu(E).$$

Hence ν is nonatomic.

Corollary 1.90. *Let $(X, \mathfrak{M}, \mu)$ be a measure space with μ nonatomic.*

(i) Let $u : X \to [0, \infty]$ be a measurable function such that

$$\int_X u \, d\mu = \infty.$$

Then there exists a denumerable partition of X into disjoint sets $X_n \in \mathfrak{M}$ such that

$$\int_{X_n} u \, d\mu = \infty.$$

(ii) Let $\{u_n\}$ be a sequence of measurable functions $u_n : X \to [0, \infty]$ such that

$$\int_X u_n \, d\mu = \infty$$

for every n. Then there exists a partition of X into disjoint sets $X_n \in \mathfrak{M}$ such that

$$\int_{X_n} u_n \, d\mu = \infty.$$

Proof. **Step 1:** Let $u : X \to [0, \infty]$ be a measurable function such that

$$\int_X u \, d\mu = \infty.$$

We prove that there exists a partition of X into two disjoint sets $E_i \in \mathfrak{M}$ such that

$$\int_{E_i} u \, d\mu = \infty, \quad i = 1, 2.$$

Let

$$X_\infty := \{x \in X : u(x) = \infty\}.$$

If $\mu(X_\infty) > 0$, then since μ is nonatomic we may find $F \in \mathfrak{M}$ such that $F \subset X_\infty$ and

$$0 < \mu(F) < \mu(X_\infty).$$

It suffices to define $E_1 := F$ and $E_2 := X \setminus F \supset X_\infty \setminus F$.

If $\mu(X_\infty) = 0$, then by Proposition 1.89 the measure ν defined in (1.22) is nonatomic. Therefore by Proposition 1.20 there exists $F_1 \in \mathfrak{M}$ such that $\int_{F_1} u \, d\mu = 1$. Since $\int_{X \setminus F_1} u \, d\mu = \infty$, the same argument yields $F_2 \in \mathfrak{M}$, $F_2 \subset X \setminus F_1$, such that $\int_{F_2} u \, d\mu = 1$. Inductively, we construct a sequence of mutually disjoint sets $F_n \in \mathfrak{M}$ such that $\int_{F_n} u \, d\mu = 1$. It suffices to set

$$E_1 := \bigcup_{n=1}^\infty F_{2n}, \quad E_2 := X \setminus E_1 \supset \bigcup_{n=1}^\infty F_{2n-1}.$$

By induction the statement in (i) now follows.

Step 2: To prove (ii) for every $n \in \mathbb{N}$ and $k \in \mathbb{Z} \cup \{\infty\}$, define

$$D_{n,k} := \begin{cases} \{x \in X : \frac{1}{2^{k+1}} \leq u_n(x) < \frac{1}{2^k}\} & \text{if } k < \infty, \\ \{x \in X : u_n(x) = \infty\} & \text{if } k = \infty. \end{cases}$$

For fixed $n \in \mathbb{N}$ construct a sequence $\{C_{n,k}\}_{k \in \mathbb{Z} \cup \{\infty\}}$ such that $C_{n,k} \subset D_{n,k}$ and

$$\mu(C_{n,k}) = \frac{1}{2^n} \mu(D_{n,k}). \tag{1.23}$$

This is undertaken by induction on n. Indeed, assume that $\{C_{l,k}\}_{k \in \mathbb{Z} \cup \{\infty\}}$ has been obtained for all $l < n$. For each $l < n$ set

$$\mathcal{F}_l := \{C_{l,k} : k \in \mathbb{Z} \cup \{\infty\}\} \cup \left\{ X \setminus \bigcup_{k \in \mathbb{Z} \cup \{\infty\}} C_{l,k} \right\}.$$

Then $\mathcal{F}_l$ is a countable partition of X, because the sets $\{C_{l,k} : k \in \mathbb{Z} \cup \{\infty\}\}$ are pairwise disjoint. Hence also

$$\mathcal{G}_n := \left\{ \bigcap_{i=1}^{n-1} E_i : E_i \in \mathcal{F}_i \text{ for all } i = 1, \ldots, n-1 \right\}$$

is a countable partition of X. Write $\mathcal{G}_n = \{G_j\}$ and for $k \in \mathbb{Z} \cup \{\infty\}$ apply Proposition 1.20 to each $G_j \cap D_{n,k}$ to find a subset $H_{j,n,k} \subset G_j \cap D_{n,k}$ such that

$$\mu(H_{j,n,k}) = \frac{1}{2^n} \mu(G_j \cap D_{n,k}).$$

Set

$$C_{n,k} := \bigcup_j H_{j,n,k},$$

and observe that $C_{n,k} \subset D_{n,k}$ and for every $E \in \mathcal{G}_n$,

$$\mu(C_{n,k} \cap E) = \frac{1}{2^n} \mu(D_{n,k} \cap E). \tag{1.24}$$

In particular, (1.23) follows, and this completes the construction of the sequence $\{C_{n,k}\}_{k \in \mathbb{Z} \cup \{\infty\}}$.

Note also that if $l \in \mathbb{N}$ then each $E \in \mathcal{F}_n$, $n = 1, \ldots, l-1$, may be written as the disjoint union

$$E = \bigcup_{i \in \{1,\ldots,l-1\}\setminus\{n\}} \bigcup_{E_i \in \mathcal{F}_i} E_1 \cap \ldots \cap E_{n-1} \cap E \cap E_{n+1} \cap \ldots \cap E_{l-1}.$$

Hence for all integers $l > n$ and all $h \in \mathbb{Z} \cup \{\infty\}$, condition (1.24) (with $C_{n,k}$ replaced by $C_{l,h}$) implies

$$\mu(C_{l,h} \cap C_{n,k}) = \frac{1}{2^l} \mu(D_{l,h} \cap C_{n,k}). \tag{1.25}$$

Next we claim that for every $n \in \mathbb{N}$ and $k \in \mathbb{Z} \cup \{\infty\}$,

$$\mu\left(C_{n,k} \cap \left(\bigcup_{l=n+1}^{\infty} \bigcup_{h \in \mathbb{Z} \cup \{\infty\}} C_{l,h} \right) \right) \leq \frac{1}{2^n} \mu(C_{n,k}). \tag{1.26}$$

By (1.25) and since the collection $\{D_{l,h} : h \in \mathbb{Z} \cup \{\infty\}\}$ is a partition of X,

$$\mu\left(C_{n,k} \cap \left(\bigcup_{l=n+1}^{\infty} \bigcup_{h \in \mathbb{Z} \cup \{\infty\}} C_{l,h}\right)\right) \leq \sum_{l=n+1}^{\infty} \sum_{h \in \mathbb{Z} \cup \{\infty\}} \mu\left(C_{n,k} \cap C_{l,h}\right)$$

$$\leq \sum_{l=n+1}^{\infty} \frac{1}{2^l} \left(\sum_{h \in \mathbb{Z} \cup \{\infty\}} \mu\left(C_{n,k} \cap D_{l,h}\right)\right)$$

$$= \sum_{l=n+1}^{\infty} \frac{1}{2^l} \mu\left(C_{n,k}\right) = \frac{1}{2^n} \mu\left(C_{n,k}\right),$$

and so $C_{n,k}$ satisfies (1.26).

Step 3: For every $n \in \mathbb{N}$ and $k \in \mathbb{Z} \cup \{\infty\}$ set

$$B_{n,k} := C_{n,k} \setminus \left(\bigcup_{l=n+1}^{\infty} \bigcup_{h \in \overline{\mathbb{N}}} C_{l,h}\right).$$

Fix $n \in \mathbb{N}$. Since $B_{n,k} \subset C_{n,k}$ for all $k \in \mathbb{Z} \cup \{\infty\}$, it follows that the sets $B_{n,k}$, for $k \in \mathbb{Z} \cup \{\infty\}$, are pairwise disjoint. If $\mu\left(B_{n,\infty}\right) > 0$, then by the definition of $D_{n,\infty}$,

$$\int_{B_{n,\infty}} u_n \, d\mu = \infty.$$

If $\mu\left(B_{n,\infty}\right) = 0$, then (1.23) and (1.26) yield

$$\mu\left(B_{n,k}\right) \geq \left(1 - \frac{1}{2^n}\right) \mu\left(C_{n,k}\right) \geq \frac{1}{2^{n+1}} \mu\left(D_{n,k}\right), \qquad (1.27)$$

and by the definition of $D_{n,k}$, for all $n \in \mathbb{N}$, we have

$$\sum_{k=-\infty}^{\infty} \int_{B_{n,k}} u_n \, d\mu \geq \sum_{k=-\infty}^{\infty} \frac{1}{2^{k+1}} \mu\left(B_{n,k}\right)$$

$$\geq \frac{1}{2^{n+1}} \sum_{k=1}^{\infty} \frac{1}{2^{k+1}} \mu\left(D_{n,k}\right)$$

$$\geq \frac{1}{2^{n+1}} \frac{1}{2} \int_X u_n \, d\mu = \infty,$$

where we have used the fact that $\mu\left(B_{n,\infty}\right) = 0$ and (1.27) imply that $\mu\left(D_{n,\infty}\right) = 0$.

Hence we may take

$$X_n := \bigcup_{k \in \mathbb{Z} \cup \{\infty\}} B_{n,k}.$$

This completes the proof.

The next proposition will be used in Chapter 2 to extend the Riemann–Lebesgue lemma to finite nonatomic measures.

Proposition 1.91. *Let $(X, \mathfrak{M}, \mu)$ be a measure space with μ finite and non-atomic. Then there exists a measurable function $u : X \to [0, 1)$ such that for any Borel set $B \subset [0, 1]$,*

$$\mu\left(u^{-1}(B)\right) = \mathcal{L}^1(B)\,\mu(X). \tag{1.28}$$

Proof. **Step 1:** We claim that there exists a sequence $\{E_n\} \subset \mathfrak{M}$ such that for every $k \in \mathbb{N}$ and any k distinct positive integers $n_1, \ldots, n_k$,

$$\mu\left(E_{n_1} \cap \ldots \cap E_{n_k}\right) = \frac{1}{2^k}\mu(X). \tag{1.29}$$

The proof is by induction on n. By Proposition 1.20 we may find $E_1 \in \mathfrak{M}$ such that $\mu(E_1) = \frac{1}{2}\mu(X)$. Suppose that measurable subsets $E_1, \ldots, E_n \in \mathfrak{M}$ have been chosen so that (1.29) holds for any k distinct positive integers $n_1, \ldots, n_k$, with $1 \le n_i \le n$, $i = 1, \ldots, k$. For every set F define

$$F^{(0)} := F, \quad F^{(1)} := X \setminus F.$$

For any $\ell = 1, \ldots, n$, consider the 2^ℓ sets

$$E_{(j_1, \ldots, j_\ell)} := E_1^{(j_1)} \cap \ldots \cap E_\ell^{(j_\ell)},$$

where $j_i = 0, 1$ for all $i = 1, \ldots, \ell$. We claim that

$$\mu\left(E_{(j_1, \ldots, j_\ell)}\right) = \frac{1}{2^\ell}\mu(X). \tag{1.30}$$

The claim will be established provided we show that for every $p = 0, \ldots, n$, and for every $\ell \ge p$, equality (1.30) holds for every ℓ-tuple $(j_1, \ldots, j_\ell)$ such that

$$\sum_{i=1}^{\ell} j_i = p.$$

The proof is by induction on p. If $p = 0$ then

$$\mu\left(E_{(j_1, \ldots, j_\ell)}\right) = \mu\left(E_1 \cap \ldots \cap E_\ell\right) = \frac{1}{2^\ell}\mu(X)$$

by the induction hypothesis on n. Thus assume that (1.30) holds for p and consider $E_{(j_1, \ldots, j_\ell)}$ such that

$$\sum_{i=1}^{\ell} j_i = p + 1.$$

By relabeling the sets if necessary, we may assume that $j_i = 1$ for all $i = 1, \ldots, p+1$, and $j_i = 0$ for all $i = p+2, \ldots, \ell$. Then

$$\mu\left(E_{(j_1,\dots,j_\ell)}\right) = \mu\left((X \setminus E_1) \cap \dots \cap (X \setminus E_{p+1}) \cap E_{p+2} \cap \dots \cap E_\ell\right)$$
$$= \mu\left((X \setminus E_1) \cap \dots \cap (X \setminus E_p) \cap E_{p+2} \cap \dots \cap E_\ell\right)$$
$$- \mu\left((X \setminus E_1) \cap \dots \cap (X \setminus E_p) \cap E_{p+1} \cap \dots \cap E_\ell\right)$$
$$= \frac{1}{2^{\ell-1}}\mu(X) - \frac{1}{2^\ell}\mu(X) = \frac{1}{2^\ell}\mu(X),$$

where we have used the induction hypothesis on p. Hence (1.30) holds.

By Proposition 1.89, for each set $E_{(j_1,\dots,j_n)}$ we may find a measurable subset $F_{(j_1,\dots,j_n)}$ such that

$$\mu\left(F_{(j_1,\dots,j_n)}\right) = \frac{1}{2}\mu\left(E_{(j_1,\dots,j_n)}\right). \tag{1.31}$$

Let E_{n+1} be the union of all such sets $F_{(j_1,\dots,j_n)}$. Note that the sets $F_{(j_1,\dots,j_n)}$ are disjoint because $\{E_{(j_1,\dots,j_n)}\}$ is a family of pairwise disjoint sets.

Fix $1 \le k \le n+1$ distinct positive integers $n_1,\dots,n_k$, with $1 \le n_i \le n+1$, $i = 1,\dots,k$. We claim that (1.29) holds. It suffices to consider the case in which one of the indices n_i is $n+1$, say $n_k = n+1$. We first show that

$$\mu\left(E_{n_1}^{(j_{n_1})} \cap \dots \cap E_{n_{k-1}}^{(j_{n_{k-1}})} \cap E_{n+1}\right) = \frac{1}{2^k}\mu(X), \tag{1.32}$$

where $j_{n_i} = 0, 1$, for all $i = 1,\dots,k-1$. This is equivalent to proving that (1.32) holds for every $p = 0,\dots,n$, and for every $(k-1)$-tuple

$$\left(E_{n_1}^{(j_{n_1})}, \dots, E_{n_{k-1}}^{(j_{n_{k-1}})}\right) \tag{1.33}$$

with $k := n - p + 1$. The proof is by induction on p. If $p = 0$ then $k = n+1$, and so, by relabeling the sets if necessary,

$$\mu\left(E_{n_1}^{(j_{n_1})} \cap \dots \cap E_{n_{k-1}}^{(j_{n_{k-1}})} \cap E_{n+1}\right) = \mu\left(E_1^{(j_1)} \cap \dots \cap E_n^{(j_n)} \cap E_{n+1}\right)$$
$$= \mu\left(E_{(j_1,\dots,j_n)} \cap E_{n+1}\right) = \mu\left(F_{(j_1,\dots,j_n)}\right)$$
$$= \frac{1}{2}\mu\left(E_{(j_1,\dots,j_n)}\right) = \frac{1}{2^{n+1}}\mu(X),$$

where we have used the definition of E_{n+1}, (1.31), and (1.30), in this order. Thus assume that (1.32) holds for p and consider a $(k-1)$-tuple as in (1.33) with $k := n-p$. Let $n_k \in \{1,\dots,n\}$ be any integer distinct from $n_1,\dots,n_{k-1}$. Since $X = E_{n_k}^{(0)} \cup E_{n_k}^{(1)}$, we may decompose $E_{n_1}^{(j_{n_1})} \cap \dots \cap E_{n_{k-1}}^{(j_{n_{k-1}})} \cap E_{n+1}$ as

$$E_{n_1}^{(j_{n_1})} \cap \dots \cap E_{n_{k-1}}^{(j_{n_{k-1}})} \cap E_{n+1} = \left(E_{n_1}^{(j_{n_1})} \cap \dots \cap E_{n_{k-1}}^{(j_{n_{k-1}})} \cap E_{n_k}^{(0)} \cap E_{n+1}\right)$$
$$\cup \left(E_{n_1}^{(j_{n_1})} \cap \dots \cap E_{n_{k-1}}^{(j_{n_{k-1}})} \cap E_{n_k}^{(1)} \cap E_{n+1}\right),$$

and so

$$\mu\left(E_{n_1}^{(j_{n_1})} \cap \ldots \cap E_{n_{k-1}}^{(j_{n_{k-1}})} \cap E_{n_k}\right) = \mu\left(E_{n_1}^{(j_{n_1})} \cap \ldots \cap E_{n_{k-1}}^{(j_{n_{k-1}})} \cap E_{n_k}^{(0)} \cap E_{n+1}\right)$$

$$+ \mu\left(E_{n_1}^{(j_{n_1})} \cap \ldots \cap E_{n_{k-1}}^{(j_{n_{k-1}})} \cap E_{n_k}^{(1)} \cap E_{n+1}\right)$$

$$= \frac{1}{2^{k+1}}\mu(X) + \frac{1}{2^{k+1}}\mu(X) = \frac{1}{2^k}\mu(X),$$

where we have used the induction hypothesis and the fact that

$$\left(E_{n_1}^{(j_{n_1})}, \ldots, E_{n_{k-1}}^{(j_{n_{k-1}})}, E_{n_k}^{(0)}\right) \text{ and } \left(E_{n_1}^{(j_{n_1})}, \ldots, E_{n_{k-1}}^{(j_{n_{k-1}})}, E_{n_k}^{(1)}\right)$$

are $((k+1) - 1)$-tuples with $k + 1 = n - p + 1$.

Hence (1.32) holds, and this completes the induction argument on n.

Step 2: We claim that there exists a measurable function $u : X \to [0, 1)$ such that

$$\mu(\{x \in X : u(x) < \theta\}) = \theta\mu(X) \qquad (1.34)$$

for all $0 \le \theta \le 1$. Let $\{E_n\}$ be the sequence of measurable sets constructed in the previous step and define

$$E_* := \bigcup_{k=1}^{\infty} \bigcap_{n=k}^{\infty} E_n.$$

We claim that the function

$$u(x) := \begin{cases} 0 & \text{if } x \in E_*, \\ \displaystyle\sum_{n=1}^{\infty} \frac{1}{2^n}\chi_{E_n}(x) & \text{otherwise,} \end{cases}$$

has the desired property. To see this, for any dyadic rational $\theta = \frac{k}{2^l}$, with $l \in \mathbb{N}_0$ and $k = 1, \ldots, 2^l$, it is easy to verify that

$$\{x \in X : u(x) < \theta\} = E_* \cup \left\{x \in X : \sum_{n=1}^{l} \frac{1}{2^n}\chi_{E_n}(x) < \theta\right\}. \qquad (1.35)$$

Indeed, one inclusion is immediate, and to verify the other inclusion consider $x \in X \setminus E_*$ such that

$$\sum_{n=1}^{l} \frac{1}{2^n}\chi_{E_n}(x) < \theta = \frac{k}{2^l}.$$

Writing

$$\sum_{n=1}^{l} \frac{1}{2^n}\chi_{E_n}(x) = \frac{\ell}{2^l}$$

for some nonnegative integer ℓ, it follows that $\ell \leq k - 1$. Moreover, since $x \in X \setminus E_*$ we have that

$$\sum_{n=l+1}^{\infty} \frac{1}{2^n} \chi_{E_n}(x) < \sum_{n=l+1}^{\infty} \frac{1}{2^n} = \frac{1}{2^l},$$

and so

$$u(x) = \sum_{n=1}^{\infty} \frac{1}{2^n} \chi_{E_n}(x) < \frac{\ell}{2^l} + \frac{1}{2^l} \leq \theta,$$

and thus $u(x) < \theta$. Hence (1.35) holds.

We claim that $\left\{ x \in X : \sum_{n=1}^{l} \frac{1}{2^n} \chi_{E_n}(x) < \theta \right\}$ is the union of k pairwise disjoint sets of the form

$$E_1^{(j_1)} \cap \ldots \cap E_l^{(j_l)},$$

where $j_i = 0,\ 1$, for all $i = 1, \ldots, l$. Note that if this is true, then by (1.35) and since $\mu(E^*) = 0$, it follows that

$$\mu(\{x \in X : u(x) < \theta\}) = k\frac{\mu(X)}{2^l} = \theta\mu(X).$$

The extension of the previous formula to values of θ that are not dyadic rational follows from the countable additivity of μ. Indeed, given any $\theta \in (0, 1]$ it suffices to construct a sequence of dyadic rational numbers $\theta_n \nearrow \theta$. Then by Proposition 1.7,

$$\mu(\{x \in X : u(x) < \theta\}) = \mu\left(\bigcup_{n=1}^{\infty} \{x \in X : u(x) < \theta_n\} \right)$$

$$= \lim_{n \to \infty} \mu(\{x \in X : u(x) < \theta_n\})$$

$$= \lim_{n \to \infty} \theta_n\mu(X) = \theta\mu(X).$$

To prove the claim for $k = 1, \ldots, 2^l$, set

$$I_{k,l} := \left\{ (j_1, \ldots, j_l) : j_n = 0,\ 1, \text{ for all } n = 1, \ldots, l, \quad \sum_{n \in \{1, \ldots, l\} : j_n = 0} \frac{1}{2^n} < \frac{k}{2^l} \right\}$$

and write

$$\left\{ x \in X : \sum_{n=1}^{l} \frac{1}{2^n} \chi_{E_n}(x) < \theta = \frac{k}{2^l} \right\} = \bigcup_{(j_1, \ldots, j_l) \in I_{k,l}} E_1^{(j_1)} \cap \ldots \cap E_l^{(j_l)}.$$

Since every $\ell \in \mathbb{N}_0$, with $\ell < k$, may be written in a unique way as

$$\sum_{n \in \{1, \ldots, l\} : j_n = 0} 2^{l-n}$$

for some $(j_1, \ldots, j_l) \in I_{k,l}$, it follows that

$$\text{card } I_{k,l} = k.$$

Step 3: It remains to show that (1.28) holds. Indeed, if $[a, b) \subset [0, 1]$ then from (1.34)

$$\mu \left(\{x \in X : a \leq u(x) < b\}\right)$$
$$= \mu \left(\{x \in X : u(x) < b\}\right) \setminus \mu \left(\{x \in X : u(x) < a\}\right)$$
$$= (b - a) \mu (X).$$

In particular, if $b = a$ then $\mu \left(u^{-1} (\{a\})\right) = 0$, and so since any open set $A \subset [0, 1]$ is the disjoint union of open intervals, it follows that

$$\mu \left(u^{-1} (A)\right) = \mathcal{L}^1 (A) \mu (X)$$

for any open set $A \subset [0, 1]$. Since the sets of Borel sets $B \subset [0, 1]$ for which (1.28) holds is a σ-algebra and it contains all open sets, it follows that (1.28) holds for every Borel set $B \subset [0, 1]$.

The next proposition will be used to study well-posedness of functionals of the form $\int_E f(x, v(x)) \, dx$ in Chapter 6.

Proposition 1.92. *Let* $(X, \mathfrak{M}, \mu)$ *be a measure with* μ *satisfying the finite subset property. If* $u : X \to [0, \infty]$ *is a measurable function and if*

$$\int_X u \, d\mu > \alpha$$

for some $\alpha \geq 0$*, then there exists a nonnegative integrable simple function* s *such that* $0 \leq s \leq u$ *in* X*, with* $s < u$ *on* $\{x \in X : u(x) > 0\}$*, and*

$$\int_X s \, d\mu > \alpha.$$

Proof. By Theorem 1.74 and the Lebesgue monotone convergence theorem we can find a nonnegative simple function $\tilde{s}$ such that $0 \leq \tilde{s} \leq u$ in X and

$$\int_X \tilde{s} \, d\mu > \alpha.$$

Write

$$\tilde{s} = \sum_{n=1}^{\ell} c_n \chi_{E_n},$$

where $c_n > 0$ for $n = 1, \ldots, \ell$. Without loss of generality, we may assume that $\tilde{s}$ is integrable. Indeed, if this were not the case then there would exist $n \in \{1, \ldots, \ell\}$ such that $\mu(E_n) = \infty$. By Proposition 1.25 we may find a set

$F \subset E_n$, $F \in \mathfrak{M}$, such that $\frac{\alpha}{c_n} < \mu(F) < \infty$, and it suffices to observe that the integrable simple function $c_n \chi_F$ is below u and

$$\int_X c_n \chi_F \, d\mu > \alpha.$$

Suppose now that $\tilde{s}$ is integrable. Define

$$s := (1 - \delta) \, \tilde{s},$$

where $0 < \delta < 1$ is so small that

$$(1 - \delta) \int_X \tilde{s} \, d\mu > \alpha.$$

The function s has the desired properties.

Remark 1.93. In view of Remark 1.19, the previous proposition holds for σ-finite measures.

We conclude this subsection with regularity properties for measurable functions when the underlying measure is Radon.

Theorem 1.94 (Lusin). *Let X be a locally compact Hausdorff space, and let $\mu^* : \mathcal{P}(X) \to [0, \infty]$ be a Radon outer measure. Let $u : X \to \mathbb{R}$ be a measurable function and let $E \subset X$ be a μ^*-measurable set such that $\mu^*(E) < \infty$. Then for every $\varepsilon > 0$ there exists a compact set $K \subset X$ such that*

$$\mu^*(E \setminus K) < \varepsilon$$

and $u : K \to \mathbb{R}$ is continuous.

Remark 1.95. If X is a metric space, then Lusin's theorem continues to hold if μ^* is a Borel regular outer measure, although the set K may be only closed and not necessarily compact.

Corollary 1.96. *Let X be a locally compact Hausdorff space, and let $\mu^* : \mathcal{P}(X) \to [0, \infty]$ be a σ-finite Radon outer measure. Let $u : X \to \mathbb{R}$ be a measurable function. Then there exists a Borel function $v : X \to \mathbb{R}$ such that $u = v$ for μ^* a.e. $x \in X$.*

Next we present a regularity result for functions integrable with respect to a Radon measure.

Theorem 1.97 (Vitali–Carathéodory). *Let X be a locally compact Hausdorff space, and let $\mu : \mathcal{B}(X) \to [0, \infty]$ be a Radon measure. If $u : X \to \mathbb{R}$ is an integrable function and $\varepsilon > 0$, then there exist $v, w : X \to \mathbb{R}$ such that*

$$v(x) \le u(x) \le w(x)$$

for μ a.e. $x \in X$, v is upper semicontinuous and bounded above, w is lower semicontinuous and bounded below, and

$$\int_X (w - v) \, d\mu \le \varepsilon.$$

1.1.4 Comparison Between Measures

Given a measure space $(X, \mathfrak{M}, \mu)$ and a measurable function $u : X \to [0, \infty]$, we define

$$\nu(E) := \int_E u \, d\mu, \quad E \in \mathfrak{M}. \tag{1.36}$$

Then ν is a measure and $\nu(E) = 0$ whenever $\mu(E) = 0$. The measure ν is said to be *absolutely continuous with respect to* μ. More generally, we have the following definition.

Definition 1.98. *Let $(X, \mathfrak{M})$ be a measurable space and let $\mu, \nu : \mathfrak{M} \to [0, \infty]$ be two measures.*

(i) μ, ν are said to be mutually singular, *and we write $\nu \perp \mu$, if there exist two disjoint sets X_μ, $X_\nu \in \mathfrak{M}$ such that $X = X_\mu \cup X_\nu$ and for every $E \in \mathfrak{M}$ we have*

$$\mu(E) = \mu(E \cap X_\mu), \quad \nu(E) = \nu(E \cap X_\nu).$$

(ii) ν is said to be absolutely continuous with respect to μ, *and we write $\nu \ll \mu$, if for every $E \in \mathfrak{M}$ with $\mu(E) = 0$ we have $\nu(E) = 0$.*
(iii) ν is said to be diffuse with respect to μ *if for every set $E \in \mathfrak{M}$ with $\mu(E) < \infty$ we have $\nu(E) = 0$.*

The term absolute continuity comes from the fact that if ν is finite then $\nu \ll \mu$ if and only if

$$\lim_{\mu(E) \to 0} \nu(E) = 0.$$

More precisely, we have the following result.

Proposition 1.99. *Let $(X, \mathfrak{M})$ be a measurable space and let $\mu, \nu : \mathfrak{M} \to [0, \infty]$ be two measures with ν finite. Then ν is absolutely continuous with respect to μ if and only if for every $\varepsilon > 0$ there exists $\delta > 0$ such that*

$$\nu(E) \le \varepsilon \tag{1.37}$$

for every measurable set $E \subset X$ with $\mu(E) \le \delta$.

Proof. If (1.37) holds and if $E \in \mathfrak{M}$ is such that $\mu(E) = 0$, then necessarily $\nu(E) \le \varepsilon$ for all $\varepsilon > 0$, so that $\nu(E) = 0$. Thus $\nu \ll \mu$.

Conversely, assume by contradiction that $\nu \ll \mu$ and that (1.37) fails. Then we can find $\varepsilon > 0$ and a sequence $\{E_n\} \subset \mathfrak{M}$ such that $\mu(E_n) \le \frac{1}{2^n}$ and $\nu(E_n) \ge \varepsilon$. Define

$$F_k := \bigcup_{n=k}^{\infty} E_n, \quad F := \bigcap_{k=1}^{\infty} F_k.$$

Then

$$\mu\left(F\right) \leq \mu\left(F_k\right) \leq \sum_{n=k}^{\infty} \frac{1}{2^n} = \frac{1}{2^{k-1}} \to 0$$

as $k \to \infty$. On the other hand, since ν is finite, by Proposition 1.7(ii),

$$\nu\left(F\right) = \lim_{k \to \infty} \nu\left(F_k\right) \geq \varepsilon,$$

which contradicts the fact $\nu \ll \mu$.

Example 1.100. If ν is not finite then condition (1.37) still implies that $\nu \ll \mu$, but the converse does not hold in general. To see this, let $X = (0,1)$, $\mathfrak{M} = \mathcal{B}\left((0,1)\right)$, take μ to be the Lebesgue measure, and

$$\nu\left(E\right) := \int_E \frac{1}{x}\,dx, \quad E \in \mathfrak{M}.$$

Then $\nu \ll \mu$ but (1.37) fails.

It turns out that (1.36) gives a complete characterization of all measures ν absolutely continuous with respect to μ if μ is σ-finite (see the Radon–Nikodym theorem below).

Theorem 1.101 (Radon–Nikodym, I). *Let $(X, \mathfrak{M})$ be a measurable space and let μ, $\nu : \mathfrak{M} \to [0, \infty]$ be two measures, with μ σ-finite and ν absolutely continuous with respect to μ. Then there exists a measurable function $u : X \to [0, \infty]$ such that*

$$\nu\left(E\right) = \int_E u\,d\mu$$

for every $E \in \mathfrak{M}$. The function u is unique up to a set of μ measure zero.

The function u is called the *Radon–Nikodym derivative* of ν with respect to μ, and we write $u = \frac{d\nu}{d\mu}$.

The proof of this theorem is hinged on the following two lemmas.

Lemma 1.102. *Let $(X, \mathfrak{M})$ be a measurable space and let μ, $\nu : \mathfrak{M} \to [0, \infty]$ be two measures. For every $E \in \mathfrak{M}$ define*

$$\nu_{ac}\left(E\right) := \sup \left\{ \int_E u\,d\mu : u : X \to [0, \infty]\ \text{measurable}, \tag{1.38} \right.$$

$$\left. \int_{E'} u\,d\mu \leq \nu\left(E'\right)\ \text{for all } E' \subset E, E' \in \mathfrak{M} \right\}.$$

Then ν_{ac} is a measure, with $\nu_{ac} \ll \mu$, and for each $E \in \mathfrak{M}$ the supremum in the definition of ν_{ac} is actually attained by a function u admissible for $\nu_{ac}\left(E\right)$. Moreover, if ν_{ac} is σ-finite, then u may be chosen independently of the set E.

Proof. **Step 1:** We prove that ν_{ac} is a measure, with $\nu_{ac} \ll \mu$. It follows from (1.38) that $\nu_{ac}(\emptyset) = 0$. Let $\{E_n\} \subset \mathfrak{M}$ be a countable collection of pairwise disjoint sets and let $E := \bigcup_{n=1}^{\infty} E_n$. Let $u : X \to [0, \infty]$ be a measurable function such that $\int_{E'} u \, d\mu \leq \nu(E')$ for all $E' \subset E$, $E' \in \mathfrak{M}$. Then for every $n \in \mathbb{N}$,

$$\int_{E'} u \, d\mu \leq \nu(E')$$

for all $E' \subset E_n$, $E' \in \mathfrak{M}$, and so

$$\int_E u \, d\mu = \sum_{n=1}^{\infty} \int_{E_n} u \, d\mu \leq \sum_{n=1}^{\infty} \nu_{ac}(E_n).$$

Hence taking the supremum over all such u, we obtain

$$\nu_{ac}(E) \leq \sum_{n=1}^{\infty} \nu_{ac}(E_n). \tag{1.39}$$

Conversely, for each $\varepsilon > 0$ and $n \in \mathbb{N}$ let $u_n : X \to [0, \infty]$ be a measurable function such that $\int_{E'} u_n \, d\mu \leq \nu(E')$ for all $E' \subset E_n$, $E' \in \mathfrak{M}$, and

$$\nu_{ac}(E_n) \leq \int_{E_n} u_n \, d\mu + \frac{\varepsilon}{2^n}.$$

Define $u := \sum_{n=1}^{\infty} \chi_{E_n} u_n$. Then for all $E' \subset E$, with $E' \in \mathfrak{M}$, we have

$$\int_{E'} u \, d\mu = \sum_{n=1}^{\infty} \int_{E' \cap E_n} u_n \, d\mu \leq \sum_{n=1}^{\infty} \nu(E' \cap E_n) = \nu(E'),$$

and so for every $l \in \mathbb{N}$,

$$\sum_{n=1}^{l} \nu_{ac}(E_n) \leq \sum_{n=1}^{l} \left(\int_{E_n} u_n \, d\mu + \frac{\varepsilon}{2^n} \right) \leq \int_E u \, d\mu + \varepsilon \leq \nu_{ac}(E) + \varepsilon.$$

Letting $l \to \infty$ first and then $\varepsilon \to 0^+$, and using (1.39), we conclude that ν_{ac} is a measure.

From the definition of ν_{ac} it follows that if $E \in \mathfrak{M}$ and $\mu(E) = 0$, then $\nu_{ac}(E) = 0$, which shows that $\nu_{ac} \ll \mu$.

Step 2: We claim that for each $E \in \mathfrak{M}$ the supremum in the definition of ν_{ac} is actually reached by a measurable function u. Indeed, note first that if u, $v : X \to [0, \infty]$ are measurable functions such that

$$\int_{E'} u \, d\mu \leq \nu(E'), \qquad \int_{E'} v \, d\mu \leq \nu(E')$$

for all $E' \subset E$, $E' \in \mathfrak{M}$, then $\max\{u, v\}$ satisfies the same property, since

$$\int_{E'} \max\{u, v\} \, d\mu = \int_{E' \cap \{u \geq v\}} u \, d\mu + \int_{E' \cap \{u < v\}} v \, d\mu \qquad (1.40)$$
$$\leq \nu \left(E' \cap \{u \geq v\} \right) + \nu \left(E' \cap \{u < v\} \right)$$
$$= \nu \left(E' \right).$$

Hence we can find an increasing sequence of measurable functions $u_n : X \to [0, \infty]$ such that

$$\lim_{n \to \infty} \int_E u_n \, d\mu = \nu_{ac} \left(E \right)$$

and $\int_{E'} u_n \, d\mu \leq \nu \left(E' \right)$ for all $E' \subset E$, $E' \in \mathfrak{M}$. Define $u := \lim_{n \to \infty} u_n$. By the Lebesgue monotone convergence theorem we have

$$\int_E u \, d\mu = \nu_{ac} \left(E \right).$$

Step 3: Suppose first that ν_{ac} is finite, and let u be a function obtained in Step 2 for the set X. We claim that

$$\nu_{ac} \left(E \right) = \int_E u \, d\mu \qquad (1.41)$$

for any set $E \in \mathfrak{M}$. Since u is admissible for X, we have that $\int_{E'} u \, d\mu \leq \nu \left(E' \right)$ for all $E' \in \mathfrak{M}$. Hence

$$\nu_{ac} \left(E \right) \geq \int_E u \, d\mu.$$

If this inequality were strict for some $E \in \mathfrak{M}$, then we could find a function v admissible for $\nu_{ac} \left(E \right)$ such that

$$\int_E v \, d\mu > \int_E u \, d\mu. \qquad (1.42)$$

Define

$$\overline{u} \left(x \right) := \begin{cases} v \left(x \right) \text{ for } x \in E \cap \{v > u\}, \\ u \left(x \right) \text{ elsewhere.} \end{cases}$$

Then $\overline{u}$ is admissible for $\nu_{ac} \left(X \right)$, and so

$$\nu_{ac} \left(X \right) \geq \int_X \overline{u} \, d\mu = \int_X u \, d\mu + \int_{E \cap \{v > u\}} \left(v - u \right) \, d\mu > \int_X u \, d\mu,$$

where we have used (1.42) and the fact that $\int_X u \, d\mu < \infty$, since ν_{ac} is finite. We have reached a contradiction.

If ν_{ac} is σ-finite then we may decompose X as

$$X = \bigcup_{n=1}^{\infty} X_n,$$

where $\nu_{ac}(X_n)$ and the sets X_n are disjoint. For each $n \in \mathbb{N}$ choose a function u_n admissible for $\nu_{ac}(X_n)$ for which (1.41) holds with u_n in place of u and for every measurable subset of X_n. The function

$$u := \sum_{n=1}^{\infty} \chi_{X_n} u_n$$

has the desired property.

Lemma 1.103. *Let $(X, \mathfrak{M})$ be a measurable space and let $\mu, \nu : \mathfrak{M} \to [0, \infty)$ be two finite measures. For every $E \in \mathfrak{M}$ define*

$$(\nu - \mu)^+ (E) := \sup \{ \nu (E') - \mu (E') : \ E' \subset E, E' \in \mathfrak{M} \}. \qquad (1.43)$$

Then $(\nu - \mu)^+$ is a measure, and for every $E \in \mathfrak{M}$ we have

$$(\nu - \mu)^+ (E) = \sup \left\{ \nu (E') - \mu (E') : \ E' \subset E, E' \in \mathfrak{M}, (\mu - \nu)^+ (E') = 0 \right\}.$$

Proof. **Step 1:** We begin by showing that $(\nu - \mu)^+$ is a measure. Since

$$(\nu - \mu)^+ (\emptyset) = \nu (\emptyset) - \mu (\emptyset) = 0$$

we have $(\nu - \mu)^+ (E) \geq 0$ for all $E \in \mathfrak{M}$. Let $\{E_n\} \subset \mathfrak{M}$ be a countable collection of pairwise disjoint sets and let $E := \bigcup_{n=1}^{\infty} E_n$. Fix $\varepsilon > 0$ and let $E'_n \subset E$, $E'_n \in \mathfrak{M}$, be such that

$$\nu (E'_n) - \mu (E'_n) \geq (\nu - \mu)^+ (E_n) - \frac{\varepsilon}{2^n}.$$

Then by (1.43),

$$(\nu - \mu)^+ (E) \geq \nu \left(\bigcup_{n=1}^{\infty} E'_n \right) - \mu \left(\bigcup_{n=1}^{\infty} E'_n \right)$$

$$= \sum_{n=1}^{\infty} [\nu (E'_n) - \mu (E'_n)] \geq \sum_{n=1}^{\infty} (\nu - \mu)^+ (E_n) - \varepsilon.$$

Letting $\varepsilon \to 0^+$ we conclude that

$$(\nu - \mu)^+ (E) \geq \sum_{n=1}^{\infty} (\nu - \mu)^+ (E_n).$$

Conversely, for any $E' \subset E$, $E' \in \mathfrak{M}$, let $E'_n := E' \cap E_n$. Then

$$\nu (E') - \mu (E') = \sum_{n=1}^{\infty} [\nu (E'_n) - \mu (E'_n)] \leq \sum_{n=1}^{\infty} (\nu - \mu)^+ (E_n),$$

and taking the supremum over all such E' we obtain

$$(\nu - \mu)^+ (E) \leq \sum_{n=1}^{\infty} (\nu - \mu)^+ (E_n),$$

and this establishes that $(\nu - \mu)^+$ is a measure.

Step 2: To verify the second part of the statement it suffices to show that for $E' \in \mathfrak{M}$ with $(\mu - \nu)^+ (E') > 0$ there exists $E'' \subset E'$, $E'' \in \mathfrak{M}$, such that $(\mu - \nu)^+ (E'') = 0$ and

$$\nu (E'') - \mu (E'') \geq \nu (E') - \mu (E').$$

We proceed by induction. By (1.43) and the fact that $(\mu - \nu)^+ (E') > 0$ there exists $E_1'' \subset E'$, $E_1'' \in \mathfrak{M}$, such that

$$\mu (E_1'') - \nu (E_1'') \geq 0, \quad \mu (E_1'') - \nu (E_1'') \geq (\mu - \nu)^+ (E') - 1. \qquad (1.44)$$

Setting $E_1' := E' \setminus E_1''$, by $(1.44)_1$ we obtain

$$\nu (E_1') - \mu (E_1') = \nu (E') - \mu (E') + \mu (E_1'') - \nu (E_1'') \geq \nu (E') - \mu (E'),$$

and since by Step 1 $(\mu - \nu)^+$ is a measure, by definition of $(\mu - \nu)^+$ we have

$$(\mu - \nu)^+ (E_1') = (\mu - \nu)^+ (E') - (\mu - \nu)^+ (E_1'')$$
$$\leq (\mu - \nu)^+ (E') - (\mu (E_1'') - \nu (E_1'')) \leq 1,$$

where in the last inequality we have used $(1.44)_2$. Recursively, suppose now that $E_{n-1}' \subset E'$ has been selected such that $E_{n-1}' \in \mathfrak{M}$, and

$$\nu (E_{n-1}') - \mu (E_{n-1}') \geq \nu (E') - \mu (E'), \quad (\mu - \nu)^+ (E_{n-1}') \leq \frac{1}{n-1}.$$

If $(\mu - \nu)^+ (E_{n-1}') = 0$, then we may take $E'' := E_{n-1}'$. If $(\mu - \nu)^+ (E_{n-1}') > 0$, then by (1.43) there exists $E_n'' \subset E_{n-1}'$, $E_n'' \in \mathfrak{M}$, such that

$$\mu (E_n'') - \nu (E_n'') \geq 0, \quad \mu (E_n'') - \nu (E_n'') \geq (\mu - \nu)^+ (E_{n-1}') - \frac{1}{n}.$$

Setting $E_n' := E_{n-1}' \setminus E_n''$ one can show exactly as before that

$$\nu (E_n') - \mu (E_n') \geq \nu (E') - \mu (E'), \quad (\mu - \nu)^+ (E_n') \leq \frac{1}{n}.$$

Define $E'' := \bigcap_{n=1}^{\infty} E_n'$. Since μ, ν, and $(\mu - \nu)^+$ are finite measures, by Proposition 1.7(ii) we may let $n \to \infty$ in the previous inequalities to obtain

$$\nu (E'') - \mu (E'') \geq \nu (E') - \mu (E'), \quad (\mu - \nu)^+ (E'') = 0.$$

This concludes the proof.

We are now ready to prove the Radon–Nikodym theorem.

Proof (Radon–Nikodym theorem I). The proof is divided into several steps.

Step 1: Assume that μ and ν are finite measures. In view of Lemma 1.102, let u be a measurable function that realizes $\nu_{ac}(X)$, that is,

$$\nu_{ac}(X) = \int_X u\, d\mu.$$

Since ν is finite again by Lemma 1.102, we have that

$$\nu_{ac}(E) = \int_E u\, d\mu$$

for all $E \in \mathfrak{M}$.

By the definition of $\nu_{ac}(X)$ for all $E \in \mathfrak{M}$ we obtain that

$$\nu'(E) := \nu(E) - \int_E u\, d\mu \geq 0. \tag{1.45}$$

Then ν' is a measure and $\nu' \ll \mu$. We claim that $\nu' \equiv 0$. Indeed, if this is not the case, then there exists $E_0 \in \mathfrak{M}$ such that

$$\nu'(E_0) = \nu(E_0) - \int_{E_0} u\, d\mu > 0.$$

Since $\nu' \ll \mu$ it follows that $\mu(E_0) > 0$, and so we can find $\varepsilon > 0$ such that $\nu'(E_0) > \varepsilon\mu(E_0)$. Hence $(\nu' - \varepsilon\mu)^+(E_0) > 0$, and by Lemma 1.103 there exists $E_0' \subset E_0$, $E_0' \in \mathfrak{M}$, such that

$$\nu'(E_0') > \varepsilon\mu(E_0'), \quad (\varepsilon\mu - \nu')^+(E_0') = 0.$$

Using again the fact that $\nu' \ll \mu$, we have that $\mu(E_0') > 0$. Since

$$(\varepsilon\mu - \nu')^+(E_0') = 0,$$

it follows that for any $E'' \subset E_0'$, with $E'' \in \mathfrak{M}$, we have that

$$\varepsilon\mu(E'') \leq \nu'(E'') = \nu(E'') - \int_{E''} u\, d\mu,$$

i.e.,

$$\int_{E''} (u + \varepsilon\chi_{E_0'})\, d\mu \leq \nu(E''). \tag{1.46}$$

Therefore, by (1.45) and (1.46), for any $E \in \mathfrak{M}$,

$$\int_E (u + \varepsilon\chi_{E_0'})\, d\mu = \int_{E \setminus E_0'} u\, d\mu + \int_{E \cap E_0'} (u + \varepsilon\chi_{E_0'})\, d\mu$$
$$\leq \nu(E \setminus E_0') + \nu(E \cap E_0') = \nu(E).$$

This implies that $u + \varepsilon \chi_{E_0'}$ is an admissible function for $\nu_{ac}(X)$, and thus

$$\nu_{ac}(X) \geq \int_X \left(u + \varepsilon \chi_{E_0'}\right) d\mu = \int_X u \, d\mu + \varepsilon \mu \left(E_0'\right)$$
$$= \nu_{ac}(X) + \varepsilon \mu \left(E_0'\right).$$

Since $\nu_{ac}(X) < \infty$ and $\mu(E_0') > 0$, we have reached a contradiction, and the claim that $\nu' \equiv 0$ is proved.

To prove uniqueness, assume that there exists another measurable function $v : X \to [0, \infty]$ such that

$$\nu(E) = \int_E v \, d\mu$$

for every $E \in \mathfrak{M}$. Note that both u and v have finite integrals since ν is finite. As in (1.40), one can show that

$$\nu(E) = \int_E \max\{u, v\} \, d\mu$$

for every $E \in \mathfrak{M}$, and so

$$\int_X [\max\{u, v\} - u] \, d\mu = \int_X [\max\{u, v\} - v] \, d\mu = 0,$$

which implies that $\max\{u, v\}(x) = u(x) = v(x)$ for μ a.e. $x \in X$.

Step 2: Assume next that μ is finite and ν σ-finite. Consider a sequence of disjoint measurable sets X_n such that

$$X = \bigcup_{n=1}^{\infty} X_n$$

and $\nu(X_n) < \infty$. Applying Step 1 to the measures $\mu \lfloor X_n$ and $\nu \lfloor X_n$, we can find a unique sequence of measurable functions $u_n : X \to [0, \infty]$ such that for all $E \in \mathfrak{M}$,

$$\nu(E \cap X_n) = \int_{E \cap X_n} u_n \, d\mu.$$

Let

$$u := \sum_{n=1}^{\infty} \chi_{X_n} u_n.$$

Then for all $E \in \mathfrak{M}$,

$$\nu(E) = \sum_{n=1}^{\infty} \nu(E \cap X_n) = \sum_{n=1}^{\infty} \int_{E \cap X_n} u_n \, d\mu = \sum_{n=1}^{\infty} \int_{E \cap X_n} u \, d\mu = \int_E u \, d\mu.$$

The uniqueness of u follows from the uniqueness of u_n.

Step 3: Assume that μ is finite and ν arbitrary. Without loss of generality we may assume that $\nu(X) = \infty$. Let

$$T := \sup \{\mu(E) : E \in \mathfrak{M}, \nu(E) < \infty\}.$$

Note that $T < \infty$ since $\mu(X) < \infty$. Find a sequence of increasing sets E_n, $E_n \in \mathfrak{M}$, with $\nu(E_n) < \infty$, such that

$$\lim_{n \to \infty} \mu(E_n) = T.$$

Define

$$E_\sigma := \bigcup_{n=1}^{\infty} E_n.$$

Then

$$\mu(E_\sigma) = \lim_{n \to \infty} \mu(E_n) = T.$$

Note that $\nu : \mathfrak{M} \lfloor E_\sigma \to [0, \infty]$ is σ-finite by construction, and so by the previous step there exists a unique measurable function $u_\sigma : E_\sigma \to [0, \infty]$ such that

$$\nu(E) = \int_E u_\sigma \, d\mu \tag{1.47}$$

for every $E \in \mathfrak{M} \lfloor E_\sigma$. We claim that $\nu : \mathfrak{M} \lfloor X \setminus E_\sigma \to [0, \infty]$ takes values only in $\{0, \infty\}$ (hence it does not have the finite subset property). Indeed, if there exists $F \subset X \setminus E_\sigma$, $F \in \mathfrak{M}$, such that

$$0 < \nu(F) < \infty,$$

then $\mu(F) > 0$ (since $\nu \ll \mu$) and so

$$T = \mu(E_\sigma) < \mu(E_\sigma) + \mu(F) = \mu(E_\sigma \cup F), \tag{1.48}$$

and since $E_\sigma \cup F$ is admissible in the definition of T we arrive at a contradiction. Note also that if $\mu(F) > 0$ for some $F \subset X \setminus E_\sigma$, $F \in \mathfrak{M}$, then $\nu(F) = \infty$. Indeed, if $\nu(F) < \infty$, then again $E_\sigma \cup F$ is admissible in the definition of T, and the same argument as in (1.48) leads to a contradiction.

Define

$$u(x) := \begin{cases} u_\sigma(x) & \text{if } x \in E_\sigma, \\ \infty & \text{if } x \in X \setminus E_\sigma. \end{cases}$$

We claim that the function u has the desired properties. Let $E \in \mathfrak{M}$. If $\mu(E \cap (X \setminus E_\sigma)) > 0$, then, as we just showed, $\nu(E \cap (X \setminus E_\sigma)) = \infty$, and so since $u = \infty$ on $E \cap (X \setminus E_\sigma)$, we get

$$\infty = \nu(E) = \int_E u \, d\mu.$$

If $\mu(E \cap (X \setminus E_\sigma)) = 0$, then E is contained in E_σ up to a set of μ measure zero (and hence of ν measure zero, since $\nu \ll \mu$), and so the claim follows from (1.47).

Finally, to prove uniqueness let $v : X \to [0, \infty]$ be another measurable function such that

$$v(E) = \int_E v \, d\mu$$

for every $E \in \mathfrak{M}$. By uniqueness in the set E_σ (see Step 2) we have $v(x) = u_\sigma(x)$ for μ a.e. $x \in E_\sigma$. Thus it suffices to show that $v(x) = \infty$ for μ a.e. $x \in X \setminus E_\sigma$.

Assume that there exists a set $F \subset X \setminus E_\sigma$, $F \in \mathfrak{M}$, such that $\mu(F) > 0$ and $v < \infty$ on F. Then, as shown before, $\nu(F) = \infty$. Let

$$F_n := \{x \in F : v(x) \le n\}.$$

Since $F_n \subset F_{n+1}$ and $F = \bigcup_{n=1}^\infty F_n$, we must have

$$\lim_{n \to \infty} \mu(F_n) = \mu(F) > 0,$$

and so $\mu(F_n) > 0$ for all n sufficiently large, say $n \ge n_0$. But then $\nu(F_{n_0}) = \infty$, which is a contradiction since

$$\infty = \nu(F_{n_0}) = \int_{F_{n_0}} v \, d\mu \le n_0 \mu(F_{n_0}) < \infty.$$

This completes the proof of this step.

Step 4: In the general case in which μ is σ-finite and ν arbitrary, consider a sequence of disjoint measurable sets X_n such that

$$X = \bigcup_{n=1}^\infty X_n$$

and $\mu(X_n) < \infty$. We now proceed exactly as in Step 2, with the only difference that we apply Step 3 in place of Step 1.

Remark 1.104. Under the hypotheses of the Radon–Nikodym theorem, it can be shown that $\nu = \nu_{ac}$. Indeed, we know that

$$\nu(E) = \int_E u \, d\mu$$

for every $E \in \mathfrak{M}$ and for some measurable function $u : X \to [0, \infty]$, and so u is admissible for $\nu_{ac}(E)$. Hence

$$\nu(E) \ge \nu_{ac}(E) \ge \int_E u \, d\mu = \nu(E),$$

where the first inequality always holds in view of the definition of ν_{ac}.

Exercise 1.105. Note that the Radon–Nikodym theorem fails in general without some hypotheses on μ. This is illustrated in the next two exercises.

(i) Let X be an uncountable set and let $\mathfrak{M}$ be the family of all sets $E \subset X$ such that either E or its complement is countable. For every $E \in \mathfrak{M}$ define

$$\mu(E) := \begin{cases} \text{card } E & \text{if } E \text{ is finite,} \\ \infty & \text{otherwise,} \end{cases}$$

and

$$\nu(E) := \begin{cases} 0 & \text{if } E \text{ is countable,} \\ \infty & \text{otherwise.} \end{cases}$$

Show that $\nu \ll \mu$ but the Radon–Nikodym theorem fails.

(ii) Let $X = [0, 1]$, let $\mathfrak{M} := \mathcal{B}([0, 1])$, and let μ, ν be respectively the counting measure and the Lebesgue measure $\mathcal{L}^1$. Prove that ν is finite, $\nu \ll \mu$, but the Radon–Nikodym theorem fails.

To extend the Radon–Nikodym theorem to non-σ-finite measures we introduce the concept of supremum of a family of measurable functions:

Definition 1.106. *Let $(X, \mathfrak{M}, \mu)$ be a measure space and let $\mathcal{F} = \{u_\alpha\}_{\alpha \in J}$ be a family of measurable functions $u_\alpha : X \to [-\infty, \infty]$. A measurable function $u_0 : X \to [-\infty, \infty]$ is called the* essential supremum function *of the family $\mathcal{F}$ if*

(i) $u_0(x) \geq u_\alpha(x)$ for every $\alpha \in J$ and for μ a.e. $x \in X$;
(ii) if $u : X \to [-\infty, \infty]$ is a measurable function such that $u(x) \geq u_\alpha(x)$ for every $\alpha \in J$ and for μ a.e. $x \in X$, then $u(x) \geq u_0(x)$ for μ a.e. $x \in X$.

In an analogous way we can define the *essential infimum function* of the family $\mathcal{F}$.

Remark 1.107. (i) Condition (ii) in Definition 1.106 is actually local in the sense that if $u : X \to [-\infty, \infty]$ is a measurable function such that

$$u(x) \geq u_\alpha(x)$$

for every $\alpha \in J$ and for μ a.e. x in a measurable subset $E \subset X$, then for μ a.e. $x \in E$,

$$u(x) \geq u_0(x).$$

To see this, it suffices to apply property (ii) in Definition 1.106 to the function

$$v(x) := \begin{cases} u(x) & \text{if } x \in E, \\ u_0(x) & \text{if } x \in X \setminus E. \end{cases}$$

(ii) In the special case in which the family $\mathcal{F}$ consists of characteristic functions, that is,

$$\mathcal{F} = \{\chi_{E_\alpha}\}_{\alpha \in J}, \quad E_\alpha \in \mathfrak{M},$$

then $u_0 : X \to \{0, 1\}$, i.e., $u_0 = \chi_{E_0}$, where

$$E_0 := \{x \in X : u_0(x) = 1\}.$$

Indeed, by property (i) in Definition 1.106 we have that $u_0(x) \geq 0$ for μ a.e. $x \in X$. Now let

$$\overline{u}_0 := \chi_{E_0}.$$

Then $\overline{u}_0 \leq u_0$, and for fixed α let X_α be such that $\mu(X \setminus X_\alpha) = 0$ and for every $x \in X_\alpha$,

$$u_0(x) \geq u_\alpha(x).$$

Let $x \in X_\alpha$. If $u_\alpha(x) = 1$, then $u_0(x) \geq 1 = \overline{u}_0(x)$, and if $u_\alpha(x) = 0$, then obviously $\overline{u}_0(x) \geq u_\alpha(x)$. Hence $\overline{u}_0(x) \geq u_\alpha(x)$ for all $x \in X_\alpha$, and so by property (ii) in Definition 1.106 we deduce that $\overline{u}_0(x) \geq u_0(x)$ for μ a.e. $x \in X$. This shows that

$$u_0(x) = \chi_{E_0}(x)$$

for μ a.e. $x \in X$.

The set E_0 is called the *essential union* of the family of sets $\{E_\alpha\}_{\alpha \in J}$.

Next we prove the existence of the essential supremum function of a family $\mathcal{F}$ of measurable functions for σ-finite measures.

Theorem 1.108. *Let $(X, \mathfrak{M}, \mu)$ be a measure space, with μ a σ-finite measure, and consider a family $\mathcal{F} = \{u_\alpha\}_{\alpha \in J}$ of measurable functions $u_\alpha : X \to [-\infty, \infty]$. Then there exists a countable set $I_0 \subset J$ such that the measurable function $u_0 : X \to [-\infty, \infty]$ defined by*

$$u_0(x) := \sup_{\alpha \in I_0} u_\alpha(x), \quad x \in X, \tag{1.49}$$

is the essential supremum of $\mathcal{F}$.

Proof. If the set of indices J is countable, then it suffices to take $I_0 := J$. The function u_0 is measurable and satisfies properties (i) and (ii) of Definition 1.106.

If J is uncountable, then assume, without loss of generality, that μ is finite. Extend arctan to $[-\infty, \infty]$ by setting $\arctan(-\infty) := -\frac{\pi}{2}$ and $\arctan \infty := \frac{\pi}{2}$ and set

$$t := \sup \left\{ \int_X \arctan \left(\sup_{\alpha \in I} u_\alpha \right) d\mu : I \subset J, I \text{ countable} \right\}. \tag{1.50}$$

Note that $t \leq \frac{\pi}{2}\mu(X)$. For every $n \in \mathbb{N}$ we may find a countable set $I_n \subset J$ such that

$$t \leq \int_X \arctan \left(\sup_{\alpha \in I_n} u_\alpha \right) d\mu + \frac{1}{n}. \tag{1.51}$$

Set

$$I_0 := \bigcup_{n=1}^{\infty} I_n$$

and define u_0 as in (1.49). Property (ii) of Definition 1.106 is satisfied. To prove (i) of Definition 1.106 note that by (1.50) and (1.51), for each $n \in \mathbb{N}$,

$$t - \frac{1}{n} \leq \int_X \arctan\left(\sup_{\alpha \in I_n} u_\alpha\right) d\mu \leq \int_X \arctan\left(u_0\right) d\mu \leq t,$$

and therefore

$$\int_X \arctan\left(u_0\right) d\mu = t.$$

On the other hand, for every $\beta \in J$,

$$t = \int_X \arctan\left(u_0\right) d\mu \leq \int_X \arctan\left(\max\left\{u_0, u_\beta\right\}\right) d\mu$$

$$\leq \int_X \arctan\left(\sup_{\alpha \in I_0 \cup \{\beta\}} u_\alpha\right) d\mu \leq t,$$

where we have used again (1.50). Since all inequalities are identities, this implies that for μ a.e. $x \in X$,

$$u_0(x) \geq u_\beta(x).$$

This concludes the proof.

Definition 1.109. *Let $(X, \mathfrak{M}, \mu)$ be a measure space. The measure μ is said to be* localizable *if any family of measurable sets admits an essential union.*

Next we show that if a measure is localizable, then the existence of the essential supremum holds not only for families of characteristic functions, but for any family of measurable functions. Note that since by the previous theorem a σ-finite measure is localizable, the proposition below generalizes Theorem 1.108.

Proposition 1.110. *Let $(X, \mathfrak{M}, \mu)$ be a measure space with μ localizable. Then any family $\mathcal{F} = \{u_\alpha\}_{\alpha \in J}$ of measurable functions $u_\alpha : X \to [-\infty, \infty]$ admits an essential supremum.*

Proof. Define $E_{\alpha,r} := \{x \in X : u_\alpha(x) \geq r\}$ with $r \in \mathbb{Q}$, and let E_r be the essential union of the family $\{E_{\alpha,r}\}_{\alpha \in J}$. Set

$$u_0(x) := \sup\left\{w_r(x) : r \in \mathbb{Q}\right\},$$

where

$$w_r(x) := \begin{cases} r & \text{if } x \in \bigcup_{t > r} E_t, \\ -\infty & \text{otherwise.} \end{cases}$$

We claim that u_0 is the essential supremum of $\mathcal{F}$. Indeed, for every $r \in \mathbb{Q}$ and $\alpha \in J$ there exists $N_{\alpha,r} \in \mathfrak{M}$ such that $\mu(N_{\alpha,r}) = 0$ and $E_r \supset E_{\alpha,r} \setminus N_{\alpha,r}$. Set

$$N_\alpha := \bigcup_{r \in \mathbb{Q}} N_{\alpha,r}.$$

Then N_α has measure zero, and we show that $u_0(x) \geq u_\alpha(x)$ if $x \notin N_\alpha$. If $u_\alpha(x) = -\infty$, then there is nothing to prove. Otherwise, let $r \in \mathbb{Q}$ be such that $u_\alpha(x) > r$. Then $x \in E_{\alpha,t} \setminus N_\alpha$ for some $t > r$, $t \in \mathbb{Q}$, and so $x \in E_t$. Hence $u_0(x) \geq w_r(x) = r$. Given the arbitrariness of r we conclude that $u_0(x) \geq u_\alpha(x)$.

Next let $u : X \to [-\infty, \infty]$ be a measurable function such that

$$u(x) \geq u_\alpha(x)$$

for every $\alpha \in J$ and for all $x \notin M_\alpha$, for some set of measure zero M_α. We claim that $u(x) \geq u_0(x)$ for μ. a.e. $x \in X$.

For $r \in \mathbb{Q}$ define

$$F_r := \{x \in X : u(x) \geq r\}.$$

Since $F_r \supset E_{\alpha,r} \setminus M_\alpha$, by definition of essential union there is a set of measure zero P_r such that $F_r \supset E_r \setminus P_r$.

Let

$$P := \bigcup_{r \in \mathbb{Q}} P_r.$$

In order to show that

$$u(x) \geq u_0(x) = \sup\{w_r(x) : r \in \mathbb{Q}\}$$

for $x \notin P$, note that if $u_0(x) = -\infty$, then there is nothing to prove. Otherwise, let $r \in \mathbb{Q}$ be such that $w_r(x) = r$. Then $x \in E_t \setminus P_t$ for some $t > r$, and therefore $x \in F_t$, and so

$$u(x) \geq t > w_r(x).$$

Hence $u(x) \geq u_0(x)$.

We are now ready to prove an extension of the Radon–Nikodym theorem that will play a fundamental role in the characterization of the dual of $L^1(X)$ in the next chapter.

Theorem 1.111 (Radon-Nikodym, II). *Let $(X, \mathfrak{M})$ be a measurable space and let $\mu, \nu : \mathfrak{M} \to [0, \infty]$ be two measures, with μ localizable and ν absolutely continuous with respect to μ and such that*

$$\nu(E) = \sup\{\nu(E \cap F) : F \in \mathfrak{M}, \mu(F) < \infty\} \tag{1.52}$$

for all $E \in \mathfrak{M}$. Then there exists a measurable function $u : X \to [0, \infty]$ such that

$$\nu(E) = \int_E u \, d\mu$$

for every $E \in \mathfrak{M}$. Moreover, if μ has the finite subset property, then u is unique up to a set of μ measure zero.

Proof. Let

$$\mathfrak{M}' := \{E \in \mathfrak{M} : \mu(E) < \infty\}.$$

For every set $E \in \mathfrak{M}'$ we apply the Radon–Nikodym theorem I to obtain a unique function $u_E : E \to [0, \infty]$ such that

$$\nu(F) = \int_F u_E \, d\mu$$

for every $F \in \mathfrak{M}$, $F \subset E$. Extend u_E by zero on $X \setminus E$.

Note that by uniqueness, if $E_1, E_2 \in \mathfrak{M}'$, then $u_{E_1}(x) = u_{E_2}(x)$ for μ a.e. $x \in E_1 \cap E_2$, while if $E_1, E_2 \in \mathfrak{M}'$ with $E_1 \subset E_2$ then $u_{E_1}(x) = 0$ for μ a.e. $x \in E_2 \setminus E_1$, and so $u_{E_1}(x) \leq u_{E_2}(x)$ for μ a.e. $x \in X$.

Let u be the essential supremum of the family $\{u_E : E \in \mathfrak{M}'\}$ (the function u exists in view of the previous proposition). We claim that

$$\int_F u \, d\mu = \sup_{E \in \mathfrak{M}'} \int_F u_E \, d\mu \tag{1.53}$$

for every $F \in \mathfrak{M}$. Fix $F \in \mathfrak{M}$. By definition of essential supremum we have

$$\int_F u \, d\mu \geq \sup_{E \in \mathfrak{M}'} \int_F u_E \, d\mu.$$

To prove the reverse inequality it suffices to consider the case in which

$$\alpha := \sup_{E \in \mathfrak{M}'} \int_F u_E \, d\mu < \infty.$$

Consider a sequence $\{E_n\} \subset \mathfrak{M}'$ such that

$$\alpha = \lim_{n \to \infty} \int_F u_{E_n} \, d\mu < \infty. \tag{1.54}$$

Since

$$u_{E_n} \leq u_{F_n} \text{ for } \mu \text{ a.e. } x \in X,$$

where $F_n := \bigcup_{i=1}^n E_i \in \mathfrak{M}'$, without loss of generality we may assume that $u_{E_n} \leq u_{E_{n+1}}$ for μ a.e. $x \in X$ and for every $n \in \mathbb{N}$. Then by the Lebesgue monotone convergence theorem,

$$\alpha = \int_F u_\infty \, d\mu,$$

where $u_\infty := \lim_{n \to \infty} u_{E_n}$. We show that for all $E \in \mathfrak{M}'$,

$$u_\infty (x) \geq u_E (x) \text{ for } \mu \text{ a.e. } x \in F. \tag{1.55}$$

By Remark 1.107 (i) this will imply that $u_\infty (x) \geq u (x)$ for μ a.e. $x \in F$ and, in turn,

$$\alpha = \int_F u_\infty \, d\mu \geq \int_F u \, d\mu.$$

It remains to prove (1.55). If (1.55) fails, then there exists $F' \in \mathfrak{M}$, $F' \subset F$, with $\mu (F') > 0$ and such that $u_\infty < u_E$ in F' for some $E \in \mathfrak{M}'$. Let

$$\varepsilon := \int_{F'} u_E \, d\mu - \int_{F'} u_\infty \, d\mu > 0.$$

By (1.54) we may choose n so large that

$$\alpha \leq \int_F u_{E_n} \, d\mu + \frac{\varepsilon}{2}.$$

Then, since $u_{E_n} \leq u_\infty$,

$$\alpha \leq \int_F u_{E_n} \, d\mu + \frac{\varepsilon}{2} \leq \int_{F \backslash F'} u_{E_n} \, d\mu + \int_{F'} u_\infty \, d\mu + \frac{\varepsilon}{2}$$

$$< \int_{F \backslash F'} u_{E_n} \, d\mu + \int_{F'} u_E \, d\mu$$

$$= \int_F u_{E_n \cup E} \, d\mu \leq \alpha,$$

and we have reached a contradiction.

Thus (1.55), and in turn (1.53), holds, and so for every $F \in \mathfrak{M}$,

$$\int_F u \, d\mu = \sup_{E \in \mathfrak{M}'} \int_F u_E \, d\mu = \sup_{E \in \mathfrak{M}'} \int_{F \cap E} u_E \, d\mu$$

$$= \sup_{E \in \mathfrak{M}'} \nu (F \cap E) = \nu (F)$$

by (1.52).

Finally, to prove uniqueness assume that μ has the finite subset property and let $v : X \to [0, \infty]$ be another measurable function such that

$$\nu (E) = \int_E v \, d\mu$$

for every $E \in \mathfrak{M}$. By uniqueness in the case that μ is finite (see the Radon–Nikodym theorem I), for every $E \in \mathfrak{M}'$ we have that $v (x) = u (x)$ for μ a.e. $x \in E$, and so $v (x) \geq u_E (x)$ for μ a.e. $x \in X$ (since $u_E = 0$ on $X \backslash E$). By the properties of the essential supremum it follows that $v (x) \geq u (x)$ for μ a.e. $x \in X$. Assume that there exists a set $F \subset X$, $F \in \mathfrak{M}$, such that $\mu (F) > 0$ and $v > u$ on F. Let

$$F_n := \{x \in F : u(x) \le n\}.$$

Since $u < \infty$ on F we have that $F = \bigcup_{n=1}^{\infty} F_n$, and since $F_n \subset F_{n+1}$, then

$$\lim_{n \to \infty} \mu(F_n) = \mu(F) > 0.$$

Therefore $\mu(F_n) > 0$ for all n sufficiently large, say $n \ge n_0$. Since μ has the finite subset property there exists $F' \subset F_{n_0}$, $F' \in \mathfrak{M}$, such that $0 < \mu(F') < \infty$. But then $F' \in \mathfrak{M}'$, and so we have a contradiction since on sets of $\mathfrak{M}'$ there is uniqueness.

Remark 1.112. Note that if μ is finite or σ-finite, then in view of Proposition 1.110, μ is localizable, it has the finite subset property, and condition (1.52) holds. Thus the previous theorem is a genuine extension of Theorem 1.101.

The Radon–Nikodym theorem allows us to express ν in terms of μ when $\nu \ll \mu$. In the general case, when ν and μ are not related a priori, De Giorgi's theorem (see below) allows us to write ν as the sum of three mutually singular measures

$$\nu = \nu_{ac} + \nu_s + \nu_d,$$

where ν_{ac} is introduced in Lemma 1.102, and ν_s and ν_d are defined in the next lemma.

Lemma 1.113. *Let $(X, \mathfrak{M})$ be a measurable space and let $\mu, \nu : \mathfrak{M} \to [0, \infty]$ be two measures. For every $E \in \mathfrak{M}$ define*

$$\nu_s(E) := \sup\{\nu(E') : E' \subset E, E' \in \mathfrak{M}, \mu(E') = 0\}, \qquad (1.56)$$

$$\nu_d(E) := \sup\Big\{\nu(E') : E' \subset E, E' \in \mathfrak{M} \text{ such that} \qquad (1.57)$$

$$\text{for all } E'' \subset E', \text{ with } E'' \in \mathfrak{M} \text{ and } \nu(E'') > 0, \mu(E'') = \infty\Big\}.$$

Then ν_s and ν_d are measures, ν_d is diffuse with respect to μ, and for each $E \in \mathfrak{M}$ the suprema in the definition of ν_s and ν_d are actually attained by measurable sets.

Moreover, if ν_s is σ-finite then there exists a set $X_s \in \mathfrak{M}$ such that

$$\mu(X_s) = 0 = \nu_d(X_s) \quad \text{and} \quad \nu_s(E) = \nu(E \cap X_s)$$

for all $E \in \mathfrak{M}$. In particular, $\nu_s \perp \mu$ and $\nu_s \perp \nu_d$.

Proof. To see that ν_s is a measure and that the supremum in the definition of ν_s is attained, it suffices to observe that for every $E \in \mathfrak{M}$,

$$\nu_s(E) = \sup\{\nu(E \cap F) : F \in \mathfrak{N}\},$$

where $\mathfrak{N} := \{F \in \mathfrak{M} : \mu(F) = 0\}$. Since $\mathfrak{N}$ is closed under finite unions and contains $\emptyset$, we are in a position to apply Lemma 1.23 (with ν_s in place of μ_1).

The same reasoning applies to ν_d, since for every $E \in \mathfrak{M}$,

$$\nu_d(E) = \sup\{\nu(E \cap F) : F \in \mathfrak{N}\},$$

where now

$$\mathfrak{N} := \{F \in \mathfrak{M} : \text{for all } G \subset F \text{ with } G \in \mathfrak{M} \text{ and } \nu(G) > 0, \ \mu(G) = \infty\},$$

which is closed under finite unions and contains $\emptyset$.

It follows from (1.57) that ν_d is diffuse with respect to μ.

To address the singularity of ν_s with respect to ν_d and μ, we assume first that ν_s is finite. Choose $X_s \in \mathfrak{M}$ such that

$$\nu_s(X) = \nu(X_s) \tag{1.58}$$

and $\mu(X_s) = 0$. Given a set $E \in \mathfrak{M}$, let $E_s \in \mathfrak{M}$, $E_s \subset E$, be such that $\nu_s(E) = \nu(E_s)$ and $\mu(E_s) = 0$. We observe that $\nu(E_s \setminus X_s) = 0$, or else, since $E_s \cup X_s$ is admissible for $\nu_s(X)$ and ν_s is finite, we would have

$$\nu_s(X) \geq \nu(E_s \cup X_s) = \nu(X_s) + \nu(E_s \setminus X_s) > \nu(X_s),$$

and this contradicts (1.58). Therefore using (1.56) and the fact that $\mu(X_s) = 0$, we have

$$\nu_s(E) = \nu(E_s \cap X_s) \leq \nu(E \cap X_s) \leq \nu_s(E \cap X_s) \leq \nu_s(E).$$

The fact that $\nu_d(X_s) = 0$ follows from the fact that $\mu(X_s) = 0$.

The case that ν_s is σ-finite is straightforward.

Theorem 1.114 (De Giorgi). *Let $(X, \mathfrak{M})$ be a measurable space and let μ, $\nu : \mathfrak{M} \to [0, \infty]$ be two measures. Then ν_s, ν_{ac}, ν_d are measures and*

$$\nu = \nu_s + \nu_{ac} + \nu_d$$

with $\nu_{ac} \ll \mu$ and ν_d diffuse with respect to μ. Moreover, if ν is σ-finite, then these three measures are mutually singular and $\nu_s \perp \mu$.

A particular setting of De Giorgi's theorem addresses the case in which μ is σ-finite, and thus $\nu_d \equiv 0$.

Theorem 1.115 (Lebesgue decomposition theorem). *Let $(X, \mathfrak{M})$ be a measurable space and let μ, $\nu : \mathfrak{M} \to [0, \infty]$ be two measures, with μ σ-finite. Then*

$$\nu = \nu_{ac} + \nu_s \tag{1.59}$$

with $\nu_{ac} \ll \mu$. Moreover, if ν is σ-finite, then $\nu_s \perp \mu$ and the decomposition (1.59) is unique, that is, if

$$\nu = \bar{\nu}_{ac} + \bar{\nu}_s,$$

for some measures $\bar{\nu}_{ac}$, $\bar{\nu}_s$, with $\bar{\nu}_{ac} \ll \mu$ and $\bar{\nu}_s \perp \mu$, then

$$\nu_{ac} = \bar{\nu}_{ac} \quad \text{and} \quad \nu_s = \bar{\nu}_s.$$

When it is important to highlight the underlying measure μ, we write (1.59) as

$$\nu = \nu_{ac,\mu} + \nu_{s,\mu}. \tag{1.60}$$

Proof. In view of Lemmas 1.102 and 1.113, ν_{ac} and ν_s are measures.

Step 1: We claim that $\nu = \nu_{ac} + \nu_s$. Fix $E \in \mathfrak{M}$ and let $E_s \in \mathfrak{M}$, $E_s \subset E$, be such that $\mu(E_s) = 0$ and $\nu_s(E) = \nu(E_s)$ (see Lemma 1.113). If $\nu(E_s) = \infty$ then (1.59) is satisfied when evaluated at E.

Suppose now that $\nu(E_s) < \infty$. We claim that $\nu \lfloor E \setminus E_s$ is absolutely continuous with respect to $\mu \lfloor E \setminus E_s$. Indeed, let $F \in \mathfrak{M}$ be such that $\mu(F \cap (E \setminus E_s)) = 0$. By replacing F with $F \cap (E \setminus E_s)$, without loss of generality, we may assume that $F \subset E \setminus E_s$. If $\nu(F) > 0$, then $E_s \cup F$ is admissible in the definition of $\nu_s(E)$, and so

$$\infty > \nu_s(E) = \nu(E_s) \geq \nu(E_s \cup F) = \nu(E_s) + \nu(F) > \nu(E_s),$$

and we have reached a contradiction. Therefore $\nu(F) = 0$, and the claim is proved.

Hence the finite measure $\nu \lfloor (E \setminus E_s)$ is absolutely continuous with respect to $\mu \lfloor (E \setminus E_s)$, and thus by Remark 1.104 we have that

$$\nu(E \setminus E_s) = (\nu \lfloor (E \setminus E_s))(E \setminus E_s) = (\nu \lfloor (E \setminus E_s))_{ac}(E \setminus E_s)$$
$$= \nu_{ac}(E \setminus E_s) = \nu_{ac}(E),$$

where in the last equality we have used the fact that $\mu(E_s) = 0$ and $\nu_{ac} \ll \mu$. Hence

$$\nu(E) = \nu(E \setminus E_s) + \nu(E_s) = \nu_{ac}(E) + \nu_s(E).$$

Step 2: Suppose that ν is σ-finite. Then by Lemma 1.113 we have $\nu_s \perp \mu$.

To prove uniqueness of the decomposition, assume that

$$\nu = \nu_{ac} + \nu_s = \overline{\nu}_{ac} + \overline{\nu}_s, \tag{1.61}$$

with $\overline{\nu}_{ac} \ll \mu$ and $\overline{\nu}_s \perp \mu$. Let $X_{\overline{\nu}_s} \in \mathfrak{M}$ be such that

$$\mu(X_{\overline{\nu}_s}) = 0 \quad \text{and} \quad \overline{\nu}_s(E) = \overline{\nu}_s(E \cap X_{\overline{\nu}_s}) \tag{1.62}$$

for every $E \in \mathfrak{M}$. Then by (1.61), $\nu \lfloor X \setminus X_{\overline{\nu}_s} = \overline{\nu}_{ac} \lfloor X \setminus X_{\overline{\nu}_s}$, which is absolutely continuous with respect to μ, and so by Remark 1.104 and $(1.62)_1$, for every $E \in \mathfrak{M}$ we have

$$\overline{\nu}_{ac}(E) = \overline{\nu}_{ac}(E \setminus X_{\overline{\nu}_s}) = (\overline{\nu}_{ac} \lfloor X \setminus X_{\overline{\nu}_s})(E) = (\nu \lfloor (X \setminus X_{\overline{\nu}_s}))(E)$$
$$= (\nu \lfloor (E \setminus X_{\overline{\nu}_s}))_{ac}(E) = \nu_{ac}(E \setminus X_{\overline{\nu}_s}) = \nu_{ac}(E).$$

Hence $\overline{\nu}_{ac} = \nu_{ac}$, and so in the case that ν is finite, it follows from (1.61) that $\nu_s = \overline{\nu}_s$. If ν is σ-finite, then by restricting ν to X_n, where

$$X = \bigcup_{n=1}^{\infty} X_n, \quad \nu(X_n) < \infty,$$

we conclude that $\nu_s \lfloor X_n = \overline{\nu}_s \lfloor X_n$ for every n, which implies that $\nu_s = \overline{\nu}_s$.

Finally, we consider the case that μ is not necessarily σ-finite.

Proof (De Giorgi's Theorem). Fix $E \in \mathfrak{M}$. We claim that

$$\nu(E) = \nu_{ac}(E) + \nu_s(E) + \nu_d(E).$$

It suffices to prove this equality in the case that the right-hand side is finite. We are then in a setting in which ν_{ac} and ν_s are finite measures, and so we may apply Lemmas 1.102 and 1.113 to ensure the existence of a measurable function $u : E \to [0, \infty]$ and $E_s \subset E$, with $E_s \in \mathfrak{M}$ and $\mu(E_s) = 0$, such that

$$\nu_{ac}(E') = \int_{E'} u \, d\mu, \quad \nu_s(E') = \nu(E' \cap E_s), \tag{1.63}$$

for every $E' \subset E$, $E' \in \mathfrak{M}$. Since $\mu(E_s) = 0$ it follows from (1.57) that $\nu_d(E_s) = 0$. Define

$$E_{ac} := \bigcup_{n=1}^{\infty} E_n, \quad E_n := \left\{ x \in E \setminus E_s : u(x) > \frac{1}{n} \right\}.$$

Since $E_{ac} \subset E \setminus E_s$, by $(1.63)_2$ we have that $\nu_s(E_{ac}) = 0$. By the definition of E_n and $(1.63)_1$ it follows that

$$\frac{1}{n} \mu(E_n) \leq \int_{E_n} u \, d\mu = \nu_{ac}(E_n) \leq \nu_{ac}(E) < \infty,$$

and thus $\nu_d(E_{ac}) = 0$ by (1.57). Moreover, by the Lebesgue decomposition theorem we have that

$$\nu(E_{ac}) = \nu_{ac}(E_{ac}),$$

and in turn, $\nu_{ac}(E_{ac}) = \nu_{ac}(E)$. To conclude, we write

$$\nu(E) = \nu(E_{ac}) + \nu(E_s) + \nu(E \setminus (E_{ac} \cup E_s))$$
$$= \nu_{ac}(E) + \nu_s(E) + \nu(E \setminus (E_{ac} \cup E_s)),$$

and it suffices to prove that

$$\nu(E \setminus (E_{ac} \cup E_s)) = \nu_d(E).$$

We have

$$\nu_d(E) = \nu_d(E \setminus (E_{ac} \cup E_s)) \leq \nu(E \setminus (E_{ac} \cup E_s)).$$

If the inequality were strict, then in view of the definition of ν_d there would exist $F \subset E \setminus (E_{ac} \cup E_s)$, $F \in \mathfrak{M}$, with $\nu(F) > 0$ and $\mu(F) < \infty$. Invoking once again the Lebesgue decomposition theorem, this would imply that

$$\nu(F) = \nu_{ac}(F) + \nu_s(F) = 0,$$

and we have reached a contradiction.

Corollary 1.116. *Let $(X, \mathfrak{M})$ be a measurable space and let μ, ν, $\upsilon : \mathfrak{M} \to [0, \infty]$ be σ-finite measures, with $\mu \perp \upsilon$. Then*

$$\nu = \nu_{ac,\mu} + \nu_{ac,\upsilon} + \nu_{s,\mu+\upsilon},$$

where we are using the notation introduced in (1.60).

Proof. By the Lebesgue decomposition theorem write

$$\nu = \nu_{ac,\mu+\upsilon} + \nu_{s,\mu+\upsilon}, \tag{1.64}$$

and find $X_{\mu+\upsilon} \in \mathfrak{M}$ such that for all $E \in \mathfrak{M}$,

$$(\mu + \upsilon)(E) = (\mu + \upsilon)(E \cap X_{\mu+\upsilon}), \tag{1.65}$$
$$\nu_{s,\mu+\upsilon}(E) = \nu_{s,\mu+\upsilon}(E \cap (X \setminus X_{\mu+\upsilon})). \tag{1.66}$$

Note that by (1.64) and (1.65),

$$\nu\lfloor(X \setminus X_{\mu+\upsilon}) = \nu_{s,\mu+\upsilon}\lfloor(X \setminus X_{\mu+\upsilon}). \tag{1.67}$$

Let $X_\mu \in \mathfrak{M}$ be such that for all $E \in \mathfrak{M}$,

$$\mu(E) = \mu(E \cap X_\mu), \quad \upsilon(E) = \upsilon(E \cap (X \setminus X_\mu)). \tag{1.68}$$

Since by (1.65), $\mu\lfloor(X \setminus X_{\mu+\upsilon}) = \upsilon\lfloor(X \setminus X_{\mu+\upsilon}) = 0$, we deduce that, also from (1.68),

$$\mu(E) = \mu(E \cap X_\mu \cap X_{\mu+\upsilon}), \quad \upsilon(E) = \upsilon(E \cap (X_{\mu+\upsilon} \setminus X_\mu)) \tag{1.69}$$

for all $E \in \mathfrak{M}$. We have

$$\nu(E) = \nu(E \cap X_\mu \cap X_{\mu+\upsilon}) + \nu(E \cap (X_{\mu+\upsilon} \setminus X_\mu)) + \nu(E \cap (X \setminus X_{\mu+\upsilon})). \tag{1.70}$$

If $\mu(E) = 0$ for some $E \in \mathfrak{M}$ with $E \subset X_\mu \cap X_{\mu+\upsilon}$, then by $(1.68)_2$ we have $(\mu + \upsilon)(E) = 0$; hence $\nu_{ac,\mu+\upsilon}(E) = 0$. On the other hand, by (1.66), $\nu_{s,\mu+\upsilon}(E) = 0$, and due to (1.64) we conclude that $\nu(E) = 0$. With this we have shown that $\nu\lfloor(X_\mu \cap X_{\mu+\upsilon}) \ll \mu$, and so for all $E \in \mathfrak{M}$,

$$\nu(E \cap X_\mu \cap X_{\mu+\upsilon}) = \nu_{ac,\mu}(E \cap X_\mu \cap X_{\mu+\upsilon}). \tag{1.71}$$

In turn, by $(1.69)_1$ we get that for all $E \in \mathfrak{M}$,

$$\nu_{ac,\mu}(E \cap X_\mu \cap X_{\mu+\upsilon}) = \nu_{ac,\mu}(E),$$

and hence by (1.71), for all $E \in \mathfrak{M}$,

$$\nu(E \cap X_\mu \cap X_{\mu+\upsilon}) = \nu_{ac,\mu}(E). \tag{1.72}$$

Similarly, using $(1.69)_1$ and (1.66), we have that $\nu\lfloor(X_{\mu+\upsilon} \setminus X_\mu) \ll \upsilon$, yielding

$$\nu\left(E \cap \left(X_{\mu+\upsilon} \setminus X_\mu\right)\right) = \nu_{ac,\upsilon}\left(E\right) \tag{1.73}$$

for all $E \in \mathfrak{M}$, where we have used $(1.69)_2$. We conclude by observing that by (1.70), (1.72), (1.73), (1.66), and (1.67),

$$\nu\left(E\right) = \nu\left(E \cap X_\mu \cap X_{\mu+\upsilon}\right) + \nu\left(E \cap \left(X_{\mu+\upsilon} \setminus X_\mu\right)\right) + \nu_{s,\mu+\upsilon}\left(E \setminus X_{\mu+\upsilon}\right)$$
$$= \nu_{ac,\mu}\left(E\right) + \nu_{ac,\upsilon}\left(E\right) + \nu_{s,\mu+\upsilon}\left(E\right)$$

for all $E \in \mathfrak{M}$, which is the desired result.

We conclude this subsection by discussing the Radon–Nikodym theorem for finitely additive measures. The next example shows that the Radon–Nikodym theorem does not hold in general for finitely additive measures.

Example 1.117. Let $X = [0,1]$, let $\mathfrak{M}$ be the σ-algebra of all Lebesgue measurable sets of $[0,1]$, and let ν be a nontrivial purely finitely additive measure ν (for the existence of ν we refer to Example 2.45) with ν absolutely continuous with respect to the Lebesgue measure $\mathcal{L}^1$. Then $\mathcal{L}^1 \ll \left(\mathcal{L}^1 + \nu\right)$, and so if the Radon–Nikodym theorem were to hold, we could find a measurable function $u : [0,1] \to [0,\infty]$ such that

$$\mathcal{L}^1\left(E\right) = \int_E u\, d\left(\mathcal{L}^1 + \nu\right) = \int_E u\, d\mathcal{L}^1 + \int_E u\, d\nu \tag{1.74}$$

for all $E \in \mathfrak{M}$. Note that $0 \le u\left(x\right) \le 1$ for $\mathcal{L}^1$ a.e. $x \in [0,1]$ and, in turn, for ν a.e. $x \in [0,1]$. Hence

$$\upsilon\left(E\right) := \int_E u\, d\nu \le \nu\left(E\right)$$

for all $E \in \mathfrak{M}$, and since ν is purely finitely additive, then so must be υ. From (1.74) and the uniqueness of the decomposition in the Hewitt–Yosida theorem, it follows that $\upsilon \equiv 0$. Again by (1.74) this implies that $u\left(x\right) = 1$ for $\mathcal{L}^1$ a.e. $x \in [0,1]$ and, in turn, for ν a.e. $x \in [0,1]$. Therefore $\nu = \upsilon = 0$, which is a contradiction.

The next result gives necessary and sufficient conditions for the validity of the Radon–Nikodym theorem for finitely additive measures. Since this result will not be used in the remainder of the book we omit its proof, which may be found in [May79].

Theorem 1.118. *Let X be a nonempty set, let $\mathfrak{M} \subset \mathcal{P}\left(X\right)$ be an algebra, and let μ, $\nu : \mathfrak{M} \to [0,\infty)$ be two finite finitely additive measures. Then there exists an integrable function $u : X \to [0,\infty]$ such that*

$$\nu\left(E\right) = \int_E u\, d\mu$$

for all $E \in \mathfrak{M}$ if and only if

(i) for every $\varepsilon > 0$ there exists $\delta > 0$ such that $\nu(E) < \varepsilon$ for all $E \in \mathfrak{M}$ with $\mu(E) < \delta$;

(ii) for all $\varepsilon > 0$ there exists $E_\varepsilon \in \mathfrak{M}$, with $\mu(X \setminus E_\varepsilon) < \varepsilon$, such that

$$\sup\left\{\frac{\nu(E')}{\mu(E')} : E' \subset E_\varepsilon, E' \in \mathfrak{M}, \mu(E') > 0\right\} < \infty,$$

and for all $\eta > 0$ there is $\delta > 0$ such that for all $E \subset E_\varepsilon$, $E \in \mathfrak{M}$, there exists $F \subset E$, $F \in \mathfrak{M}$, such that $\mu(F) > \delta\mu(E)$ and

$$\sup\left\{\frac{\nu(E')}{\mu(E')} : E' \subset F, E' \in \mathfrak{M}, \mu(E') > 0\right\} < \eta.$$

1.1.5 Product Spaces

Given two measurable spaces $(X, \mathfrak{M})$ and $(Y, \mathfrak{N})$ we denote by $\mathfrak{M} \otimes \mathfrak{N} \subset \mathcal{P}(X \times Y)$ the smallest σ-algebra that contains all sets of the form $E \times F$, where $E \in \mathfrak{M}$, $F \in \mathfrak{N}$. The σ-algebra $\mathfrak{M} \otimes \mathfrak{N}$ is called the *product σ-algebra* of $\mathfrak{M}$ and $\mathfrak{N}$.

Exercise 1.119. Let X and Y be topological spaces and let $\mathcal{B}(X)$ and $\mathcal{B}(Y)$ be their respective Borel σ-algebras. Prove that

$$\mathcal{B}(X) \otimes \mathcal{B}(Y) \subset \mathcal{B}(X \times Y).$$

Show also that if X and Y are separable metric spaces, then

$$\mathcal{B}(X) \otimes \mathcal{B}(Y) = \mathcal{B}(X \times Y),$$

so that in particular, $\mathcal{B}(\mathbb{R}^N) = \mathcal{B}(\mathbb{R}) \otimes \ldots \otimes \mathcal{B}(\mathbb{R})$.

Let $(X, \mathfrak{M}, \mu)$ and $(Y, \mathfrak{N}, \nu)$ be two measure spaces. For every $E \in X \times Y$ define

$$(\mu \times \nu)^*(E) := \inf\left\{\sum_{n=1}^{\infty} \mu(F_n)\nu(G_n) : \{F_n\} \subset \mathfrak{M}, \{G_n\} \subset \mathfrak{N}, \right. \tag{1.75}$$

$$\left. E \subset \bigcup_{n=1}^{\infty}(F_n \times G_n)\right\},$$

where we define $\mu(F_n)\nu(G_n) := 0$ whenever $\mu(F_n) = 0$ or $\nu(G_n) = 0$.

By Proposition 1.32, $(\mu \times \nu)^* : \mathcal{P}(X) \to [0, \infty]$ is an outer measure, and it is called the *product outer measure* of μ and ν. By Carathéodory's theorem, the restriction of $(\mu \times \nu)^*$ to the σ-algebra $\mathfrak{M} \times \mathfrak{N}$ of $(\mu \times \nu)^*$-measurable sets is a complete measure, denoted by $\mu \times \nu$ and called the *product measure* of μ and ν.

Note that $\mathfrak{M} \times \mathfrak{N}$ is, in general, larger than the product σ-algebra $\mathfrak{M} \otimes \mathfrak{N}$.

Theorem 1.120. *Let $(X, \mathfrak{M}, \mu)$ and $(Y, \mathfrak{N}, \nu)$ be two measure spaces.*

(i) If $F \in \mathfrak{M}$ and $G \in \mathfrak{N}$, then $F \times G$ is $(\mu \times \nu)^*$-measurable and

$$(\mu \times \nu)(F \times G) = \mu(F)\nu(G);$$

(ii) if μ and ν are complete and E has σ-finite $\mu \times \nu$ measure, then for μ a.e. $x \in X$ the section

$$E_x := \{y \in Y : (x,y) \in E\}$$

belongs to the σ-algebra $\mathfrak{N}$, and for ν a.e. $y \in Y$ the section

$$E_y := \{x \in X : (x,y) \in E\}$$

belongs to the σ-algebra $\mathfrak{M}$. Moreover,

$$(\mu \times \nu)(E) = \int_Y \mu(E_y)\, d\nu(y) = \int_X \nu(E_x)\, d\mu(x).$$

The previous result is a particular case of Tonelli's theorem in the case that $u = \chi_E$.

Theorem 1.121 (Tonelli). *Let $(X, \mathfrak{M}, \mu)$ and $(Y, \mathfrak{N}, \nu)$ be two measure spaces. Assume that μ and ν are complete and σ-finite, and let $u : X \times Y \to [0, \infty]$ be an $\mathfrak{M} \times \mathfrak{N}$ measurable function. Then for μ a.e. $x \in X$ the function $u(x, \cdot)$ is measurable and the function $\int_Y u(\cdot, y)\, d\nu(y)$ is measurable. Similarly, for ν a.e. $y \in Y$ the function $u(\cdot, y)$ is measurable and the function $\int_X u(x, \cdot)\, d\mu(x)$ is measurable. Moreover,*

$$\int_{X \times Y} u(x,y)\, d(\mu \times \nu)(x,y) = \int_X \left(\int_Y u(x,y)\, d\nu(y) \right) d\mu(x)$$
$$= \int_Y \left(\int_X u(x,y)\, d\mu(x) \right) d\nu(y).$$

Remark 1.122. In the case that $u : X \times Y \to [0, \infty]$ is $\mathfrak{M} \otimes \mathfrak{N}$ measurable, then Tonelli's theorem still holds even if the measures μ and ν are not complete, and the statements are satisfied for every $x \in X$ and $y \in Y$ (as opposed to for μ a.e. $x \in X$ and for ν a.e. $y \in Y$).

A simple consequence of Tonelli's theorem is the following result.

Theorem 1.123. *Let $(X, \mathfrak{M}, \mu)$ be a measure space and let $u : X \to [0, \infty]$ be a measurable function. Then*

$$\int_X u\, d\mu = \int_0^\infty \mu(\{x \in X : u(x) > t\})\, dt. \qquad (1.76)$$

Proof. If the set $\{x \in X : u(x) > 0\}$ has non-σ-finite μ measure, then at least one of the sets

$$\left\{ x \in X : u(x) > \frac{1}{n} \right\}, \quad n \in \mathbb{N},$$

has infinite μ measure. Hence $\mu\left(\{x \in X : u\left(x\right) > t\}\right) = \infty$ for all $0 \leq t \leq \frac{1}{n}$, which implies that both sides of (1.76) are infinite.

If the set $\{x \in X : u\left(x\right) > 0\}$ has σ-finite μ measure, then by replacing μ with

$$\mu \lfloor \{x \in X : u\left(x\right) > 0\},$$

we may assume that μ is σ-finite. Applying Tonelli's theorem we obtain

$$\int_0^\infty \mu\left(\{x \in X : u\left(x\right) > t\}\right) dt = \int_0^\infty \int_X \chi_{\{y \in X : u(y) > t\}}\left(x\right) d\mu\left(x\right) dt$$

$$= \int_X \int_0^\infty \chi_{\{y \in X : u(y) > t\}}\left(x\right) dt\, d\mu\left(x\right)$$

$$= \int_X \int_0^{u(x)} dt\, d\mu\left(x\right) = \int_X u\left(x\right) d\mu\left(x\right),$$

and the proof is complete.

The version of Tonelli's theorem for integrable functions of arbitrary sign is the well–known Fubini's theorem:

Theorem 1.124 (Fubini). *Let $(X, \mathfrak{M}, \mu)$ and $(Y, \mathfrak{N}, \nu)$ be two measure spaces. Assume that μ and ν are complete, and let $u : X \times Y \to [-\infty, \infty]$ be $\mu \times \nu$-integrable. Then for μ a.e. $x \in X$ the function $u\left(x, \cdot\right)$ is ν-integrable, and the function $\int_Y u\left(\cdot, y\right) d\nu\left(y\right)$ is μ-integrable.*

Similarly, for ν a.e. $y \in Y$ the function $u\left(\cdot, y\right)$ is μ-integrable, and the function $\int_X u\left(x, \cdot\right) d\mu\left(x\right)$ is ν-integrable. Moreover,

$$\int_{X \times Y} u\left(x, y\right) d\left(\mu \times \nu\right)\left(x, y\right) = \int_X \left(\int_Y u\left(x, y\right) d\nu\left(y\right)\right) d\mu\left(x\right)$$

$$= \int_Y \left(\int_X u\left(x, y\right) d\mu\left(x\right)\right) d\nu\left(y\right).$$

Remark 1.125. In the case that $u : X \times Y \to [-\infty, \infty]$ is $\mathfrak{M} \otimes \mathfrak{N}$ measurable, then Fubini's theorem still holds even if the measures μ and ν are not complete.

We present below several examples on the validity of Tonelli's and Fubini's theorems.

Exercise 1.126. (i) This exercise shows that the σ-finiteness of μ and ν is not a necessary condition. Let $X = \mathbb{N}$ and let μ be the counting measure. Let $(Y, \mathfrak{N}, \nu)$ be any measure space with ν complete but not necessarily σ-finite. Prove that Fubini's and Tonelli's theorems continue to hold.

(ii) This exercise shows that without some condition on the measures μ and ν, Fubini's and Tonelli's theorems may fail in general. Let $X = Y = [0, 1]$, let $\mathfrak{M} = \mathfrak{N} = \mathcal{B}\left([0, 1]\right)$, let μ be the Lebesgue measure, and let ν be the counting measure. Note that ν is not σ-finite. Show that the diagonal

$$D := \{(x, x) : x \in [0, 1]\}$$

belongs to $\mathfrak{M} \otimes \mathfrak{N}$ but

$$\int_Y \mu(D_y) \, d\nu(y) \neq \int_X \nu(D_x) \, d\mu(x).$$

(iii) This exercise shows that Fubini's theorem may fail without integrability. Let $X = Y = \mathbb{N}$, let $\mathfrak{M} = \mathfrak{N} = \mathcal{P}(\mathbb{N})$, and let $\mu = \nu$ be the counting measure. Consider the function

$$u(n, l) := \begin{cases} 1 & \text{if } n = l, \\ -1 & \text{if } l = n + 1, \\ 0 & \text{otherwise.} \end{cases}$$

Prove that

$$\int_X \left(\int_Y u(x, y) \, d\nu(y) \right) d\mu(x) = \sum_{n=1}^{\infty} \sum_{l=1}^{\infty} u(n, l) = 1$$

$$\neq 0 = \sum_{l=1}^{\infty} \sum_{n=1}^{\infty} u(n, l)$$

$$= \int_Y \left(\int_X u(x, y) \, d\mu(x) \right) d\nu(y).$$

(iv) Finally, without the hypothesis that u is an $\mathfrak{M} \times \mathfrak{N}$ measurable function, Fubini's and Tonelli's theorems fail. Let $X = Y$ be the set of all ordinals less than or equal to the first uncountable ordinal ω_1, let $\mathfrak{M} = \mathfrak{N}$ be the σ-algebra consisting of all countable sets and their complements, and for every $F \in \mathfrak{M}$ define

$$\mu(F) = \nu(F) := \begin{cases} 1 \text{ if } F \text{ is countable,} \\ 0 \text{ otherwise.} \end{cases}$$

Let $E = \{(x, y) \in X \times Y : x < y\}$. Prove that the sections E_x and E_y are measurable, but

$$\int_Y \mu(E_y) \, d\nu(y) \neq \int_X \nu(E_x) \, d\mu(x).$$

Remark 1.127. Fubini's theorem fails for finitely additive measures. See Theorem 3.3 in [He-Yo52].

Next we extend Tonelli's theorem to the case that $\mu \times \nu$ has the finite subset property. We begin with a preliminary result.

Proposition 1.128. *Let $(X, \mathfrak{M}, \mu)$ and $(Y, \mathfrak{N}, \nu)$ be two measure spaces, with μ, ν nonzero. If $\mu \times \nu$ has the finite subset property then μ, ν have the finite subset property.*

Proof. Assume that $\mu \times \nu$ has the finite subset property, and let $F \in \mathfrak{M}$ be such that $\mu(F) = \infty$. Since ν is not zero, there exists $G \in \mathfrak{N}$ with $\nu(G) > 0$. Hence $(\mu \times \nu)(F \times G) = \mu(F)\nu(G) = \infty$, and by hypothesis there exists $E' \in \mathfrak{M} \times \mathfrak{N}$ with $E' \subset F \times G$ such that $0 < (\mu \times \nu)(E') < \infty$. Find two sequences $\{F_n\} \subset \mathfrak{M}$, $\{G_n\} \subset \mathfrak{N}$ admissible in the definition of $(\mu \times \nu)(E')$ (see (1.75)) such that

$$0 < \sum_{n=1}^{\infty} \mu(F_n)\nu(G_n) < \infty.$$

Since

$$E' \subset \bigcup_{n=1}^{\infty} ((F_n \cap F) \times (G_n \cap G))$$

we have

$$0 < (\mu \times \nu)(E') \le \sum_{n=1}^{\infty} \mu(F_n \cap F)\nu(G_n \cap G) \le \sum_{n=1}^{\infty} \mu(F_n)\nu(G_n) < \infty,$$

and so $0 < \mu(F_n \cap F) < \infty$ for some $n \in \mathbb{N}$. This shows that μ has the finite subset property. A similar proof yields the same conclusion for ν.

Remark 1.129. Note that if μ, ν are σ-finite measures, then $\mu \times \nu$ is also σ-finite (and so in particular it has the finite subset property). Indeed, writing

$$X = \bigcup_{n=1}^{\infty} X_n, \quad Y = \bigcup_{l=1}^{\infty} Y_l,$$

where $\{X_n\} \subset \mathfrak{M}$ and $\{Y_l\} \subset \mathfrak{N}$ are sequences of sets of finite measure, we have that

$$X \times Y = \bigcup_{n,l=1}^{\infty} X_n \times Y_l,$$

where $(\mu \times \nu)(X_n \times Y_l) = \mu(X_n)\nu(Y_l) < \infty$.

In view of the previous remark, the following result extends Tonelli's theorem.

Theorem 1.130. *Let $(X, \mathfrak{M}, \mu)$ and $(Y, \mathfrak{N}, \nu)$ be two measure spaces, with μ and ν complete. Assume that $\mu \times \nu$ has the finite subset property. Let $u : X \times Y \to [0, \infty]$ be an $\mathfrak{M} \times \mathfrak{N}$ measurable function. If one of the two iterated integrals $\int_X \left(\int_Y u(x,y)\, d\nu(y) \right) d\mu(x)$ and $\int_Y \left(\int_X u(x,y)\, d\mu(x) \right) d\nu(y)$ is well-defined and finite, then so is the other, and*

$$\int_{X \times Y} u(x,y)\, d(\mu \times \nu)(x,y) = \int_X \left(\int_Y u(x,y)\, d\nu(y) \right) d\mu(x)$$

$$= \int_Y \left(\int_X u(x,y)\, d\mu(x) \right) d\nu(y).$$

Proof. Assume that the iterated integral

$$L := \int_Y \left(\int_X u\,(x,y)\,d\mu\,(x) \right) d\nu\,(y)$$

is well-defined and finite. We claim that the set $\{(x,y) \in X \times Y : u\,(x,y) > 0\}$ has σ-finite $\mu \times \nu$ measure. Indeed, if not, then we can find $n \in \mathbb{N}$ such that the set

$$E_n := \left\{ (x,y) \in X \times Y : u\,(x,y) > \frac{1}{n} \right\}$$

has $\mu \times \nu$ infinite measure. Since $\mu \times \nu$ has the finite subset property, by Proposition 1.25 there exists $E' \subset E_n$, $E' \in \mathfrak{M} \times \mathfrak{N}$, such that

$$2Ln < (\mu \times \nu)\,(E') < \infty.$$

Then, by Fubini's theorem,

$$2L < \frac{(\mu \times \nu)\,(E')}{n} = \frac{1}{n} \int_Y \left(\int_X \chi_{E'}\,(x,y)\,d\mu\,(x) \right) d\nu\,(y)$$

$$\leq \int_Y \left(\int_X u\,(x,y)\,d\mu\,(x) \right) d\nu\,(y) = L,$$

which is a contradiction. Hence the set

$$E := \{(x,y) \in X \times Y : u\,(x,y) > 0\}$$

has σ-finite $\mu \times \nu$ measure, and so we may find a sequence $\{E'_n\}$ of pairwise disjoint sets with $E'_n \in \mathfrak{M} \times \mathfrak{N}$ such that $(\mu \times \nu)\,(E'_n) < \infty$ and

$$E = \bigcup_{n=1}^{\infty} E'_n.$$

Fix $n \in \mathbb{N}$. In view of (1.75) we may find two sequences $\left\{ F_k^{(n)} \right\} \subset \mathfrak{M}$, $\left\{ G_k^{(n)} \right\} \subset \mathfrak{N}$ such that

$$E'_n \subset \bigcup_{k=1}^{\infty} \left(F_k^{(n)} \times G_k^{(n)} \right), \quad \sum_{k=1}^{\infty} \mu \left(F_k^{(n)} \right) \nu \left(G_k^{(n)} \right) \leq (\mu \times \nu)\,(E'_n) + 1.$$

Note that if $\mu(F_k^{(n)}) = 0$ (respectively $\nu(G_k^{(n)}) = 0$) for some $k \in \mathbb{N}$, then Tonelli's theorem holds on measurable subsets of $F_k^{(n)} \times Y$ (respectively $X \times G_k^{(n)}$), since all three integrals reduce to zero. Thus, we may assume that $0 < \mu(F_k^{(n)}), \nu(G_k^{(n)}) < \infty$ for all $k \in \mathbb{N}$.

In particular, $\mu \lfloor \bigcup_{k=1}^{\infty} F_k^{(n)}$ and $\nu \lfloor \bigcup_{k=1}^{\infty} G_k^{(n)}$ are σ-finite. Hence we can apply the classical Tonelli's theorem to the function $u\chi_{E'_n}$ restricted to $\left(\bigcup_{k=1}^{\infty} F_k^{(n)} \right) \times \left(\bigcup_{k=1}^{\infty} G_k^{(n)} \right)$, and then sum over n.

Remark 1.131. Let $(X, \mathfrak{M}, \mu)$ be a measure space and let $u : X \to [0, \infty]$. If in place of the standard definition of measurability we assume that $u^{-1}\,(A) \cap F \in$

$\mathfrak{M}$ for every open set $A \subset [0, \infty]$ and for every $F \in \mathfrak{M}$ with $\mu(F) < \infty$, and define

$$\int_X u \, d\mu := \sup \left\{ \int_F u \, d\mu : F \in \mathfrak{M}, \mu(F) < \infty \right\}$$

then, interpreting iterated integrals in this new way, it is possible to prove a converse of the previous theorem, namely that if μ and ν are complete and have the finite subset property, and if Tonelli's theorem holds, then $\mu \times \nu$ has the finite subset property (see [Muk73]). Note that for σ-finite measures these new definitions of measurability and integrability coincide with the classical ones.

1.1.6 Projection of Measurable Sets

The projection of a Borel set $B \subset \mathbb{R}^2$ on the x-axis in general is not a Borel set but it is Lebesgue measurable. To prove this result we need to introduce the notion of Suslin sets. One of the major complications in what follows is the notation. Let

$$\mathcal{I} := \{\alpha = (\alpha_1, \ldots, \alpha_n) \in \mathbb{N}^n : n \in \mathbb{N}\} = \bigcup_{n=1}^{\infty} \mathbb{N}^n.$$

If $\alpha = (\alpha_1) \in \mathcal{I}$ and $\alpha_1 \in \mathbb{N}$, we write α_1 for α. Also, if $\alpha = (\alpha_1, \ldots, \alpha_n)$ and $\beta = (\beta_1, \ldots, \beta_n)$, we say that $\alpha \leq \beta$ if $\alpha_i \leq \beta_i$ for all $i = 1, \ldots, n$. If $\alpha = (\alpha_1, \ldots, \alpha_n) \in \mathbb{N}^n$ and $k \in \mathbb{N}$, we define $(\alpha, k) := (\alpha_1, \ldots, \alpha_n, k) \in \mathbb{N}^{n+1}$. Finally, if $\alpha = (\alpha_1, \ldots, \alpha_n) \in \mathbb{N}^n$ and $k \in \mathbb{N}$, with $k \leq n$, we set $\alpha|_k := (\alpha_1, \ldots, \alpha_k)$.

Let $\mathbb{N}^{\mathbb{N}} := \{f : \mathbb{N} \to \mathbb{N}\}$. To simplify the notation for every $f \in \mathbb{N}^{\mathbb{N}}$ and $n \in \mathbb{N}$ we write $f|_n$ for $(f(1), \ldots, f(n)) \in \mathcal{I}$. Hence if $f \in \mathbb{N}^{\mathbb{N}}$ and $\alpha = (\alpha_1, \ldots, \alpha_n) \in \mathbb{N}^n$, then $f|_n \leq \alpha$ means that $f(i) \leq \alpha_i$ for all $i = 1, \ldots, n$. If $f, g \in \mathbb{N}^{\mathbb{N}}$, then $f \leq g$ means that $g(i) \leq f(i)$ for all $i \in \mathbb{N}$.

Definition 1.132. *A* Suslin scheme *on a set X is a function*

$$\mathcal{E} : \mathcal{I} \to \mathcal{P}(X),$$
$$\alpha \mapsto E_\alpha.$$

Given a Suslin scheme $\mathcal{E}$, the Suslin set $A(\mathcal{E})$ *is defined as*

$$A(\mathcal{E}) := \bigcup_{f \in \mathbb{N}^{\mathbb{N}}} \bigcap_{k=1}^{\infty} E_{f|_k} = \bigcup_{f \in \mathbb{N}^{\mathbb{N}}} \bigcap_{k=1}^{\infty} E_{(f(1), \ldots, f(k))}.$$

If $\mathcal{F} \subset \mathcal{P}(X)$, a Suslin-$\mathcal{F}$ set *is any Suslin set $A(\mathcal{E})$, where $\mathcal{E} : \mathcal{I} \to \mathcal{F}$.*

Given a Suslin set $A(\mathcal{E})$, for any $\alpha = (\alpha_1, \ldots, \alpha_n) \in \mathbb{N}^n$ we define

$$S^\alpha := \bigcup_{\substack{f \in \mathbb{N}^\mathbb{N} \\ f|_n \leq \alpha}} \bigcap_{k=1}^{\infty} E_{f|_k}.$$

If $n = 1$, that is, $\alpha = (\alpha_1)$, we will write S^{α_1} for S^α. Note that the sequence $\{S^n\}_{n \in \mathbb{N}}$ is increasing and

$$A(\mathcal{E}) = \bigcup_{n=1}^{\infty} S^n. \tag{1.77}$$

Also, for any $\alpha = (\alpha_1, \ldots, \alpha_\ell) \in \mathbb{N}^\ell$ the sequence $\{S^{(\alpha,n)}\}_{n \in \mathbb{N}}$ is increasing and

$$S^\alpha = \bigcup_{n=1}^{\infty} S^{(\alpha,n)}. \tag{1.78}$$

Finally, for any $g \in \mathbb{N}^\mathbb{N}$ and $n \in \mathbb{N}$ we define

$$S_{g|_n} := \bigcup_{\substack{\alpha \in \mathbb{N}^n \\ \alpha \leq g|_n}} \bigcap_{k=1}^{n} E_{\alpha|_k} = \bigcup_{\substack{\alpha \in \mathbb{N}^n \\ \alpha \leq g|_n}} E_{(\alpha_1)} \cap E_{(\alpha_1,\alpha_2)} \cap \ldots \cap E_{(\alpha_1,\ldots,\alpha_n)}. \tag{1.79}$$

The sequence $\{S_{g|_n}\}_{n \in \mathbb{N}}$ is decreasing and

$$S_g := \bigcap_{n=1}^{\infty} S_{g|_n} = \bigcup_{\substack{f \in \mathbb{N}^\mathbb{N} \\ f \leq g}} \bigcap_{k=1}^{\infty} E_{f|_k}. \tag{1.80}$$

We prove that the class of Suslin sets is closed under countable unions and intersections.

Theorem 1.133. *Let X be a nonempty set and let $\mathcal{F} \subset \mathcal{P}(X)$ be such that $X \in \mathcal{F}$. Then the class of Suslin-$\mathcal{F}$ sets is closed under countable unions and intersections.*

Proof. Let $\{E^{(n)}\}$ be a sequence of Suslin-$\mathcal{F}$ sets, that is,

$$E^{(n)} = \bigcup_{f \in \mathbb{N}^\mathbb{N}} \bigcap_{k=1}^{\infty} E_{f|_k}^{(n)}.$$

For every $g \in \mathbb{N}^\mathbb{N}$ and $j \in \mathbb{N}$ define

$$E_{g|_j} = E_{(g(1),\ldots,g(j))} := \begin{cases} X & \text{if } j = 1, \\ E_{(g(2),\ldots,g(j))}^{(g(1))} & \text{if } j \geq 2. \end{cases}$$

Then

$$\bigcup_{g \in \mathbb{N}^{\mathbb{N}}} \bigcap_{j=1}^{\infty} E_{g|_j} = \bigcup_{g \in \mathbb{N}^{\mathbb{N}}} \bigcap_{j=2}^{\infty} E_{g|_j} = \bigcup_{g \in \mathbb{N}^{\mathbb{N}}} \bigcap_{j=2}^{\infty} E_{(g(2),...,g(j))}^{(g(1))}$$

$$= \bigcup_{n=1}^{\infty} \bigcup_{f \in \mathbb{N}^{\mathbb{N}}} \bigcap_{k=1}^{\infty} E_{(f(1),...,f(k))}^{(n)} = \bigcup_{n=1}^{\infty} E^{(n)},$$

where we have used the fact that as g varies in $\mathbb{N}^{\mathbb{N}}$ the value $g(1)$ ranges over all of $\mathbb{N}$ and we have made the "change of variables" $f(k) := g(k-1)$, $k := j - 1$. Hence $\bigcup_{n=1}^{\infty} E^{(n)}$ is a Suslin-$\mathcal{F}$ set.

Next we show that $\bigcap_{n=1}^{\infty} E^{(n)}$ is a Suslin-$\mathcal{F}$ set. For every $g \in \mathbb{N}^{\mathbb{N}}$ and $j \in \mathbb{N}$ write j in a unique way as

$$j = 2^l i, \tag{1.81}$$

where $l \in \mathbb{N}_0$ and $i \in \mathbb{N}$ is odd, and define

$$E_{g|_j} := E_{(g(2^l 1), g(2^l 3),...,g(2^l i))}^{(l+1)}. \tag{1.82}$$

We claim that

$$\bigcup_{g \in \mathbb{N}^{\mathbb{N}}} \bigcap_{j=1}^{\infty} E_{g|_j} = \bigcap_{n=1}^{\infty} E^{(n)}. \tag{1.83}$$

Indeed, if $l \in \mathbb{N}_0$ and $g \in \mathbb{N}^{\mathbb{N}}$, then

$$\bigcap_{j=1}^{\infty} E_{g|_j} \subset \bigcap_{i \text{ odd}} E_{g|_{2^l i}} = \bigcap_{i \text{ odd}} E_{(g(2^l 1), g(2^l 3),...,g(2^l i))}^{(l+1)} = \bigcap_{k=1}^{\infty} E_{(f_l(1),...,f_l(k))}^{(l+1)},$$

where $f_l \in \mathbb{N}^{\mathbb{N}}$ is defined as $f_l(p) := g(2^l(2p-1))$ for $p \in \mathbb{N}$. Hence

$$\bigcup_{g \in \mathbb{N}^{\mathbb{N}}} \bigcap_{j=1}^{\infty} E_{g|_j} \subset \bigcup_{f \in \mathbb{N}^{\mathbb{N}}} \bigcap_{k=1}^{\infty} E_{f|_k}^{(l+1)} = E^{(l+1)}$$

for all $l \in \mathbb{N}_0$, which implies that

$$\bigcup_{g \in \mathbb{N}^{\mathbb{N}}} \bigcap_{j=1}^{\infty} E_{g|_j} \subset \bigcap_{l=0}^{\infty} E^{(l+1)} = \bigcap_{n=1}^{\infty} E^{(n)}. \tag{1.84}$$

Conversely, let

$$x \in \bigcap_{n=1}^{\infty} E^{(n)} = \bigcap_{n=1}^{\infty} \bigcup_{f \in \mathbb{N}^{\mathbb{N}}} \bigcap_{k=1}^{\infty} E_{f|_k}^{(n)},$$

and for each $n \in \mathbb{N}$ choose $f^{(n)} \in \mathbb{N}^{\mathbb{N}}$ such that

$$x \in \bigcap_{n=1}^{\infty} \bigcap_{k=1}^{\infty} E_{f^{(n)}|_k}^{(n)}.$$

Using (1.81) we define $g \in \mathbb{N}^{\mathbb{N}}$ as

$$g\left(j\right) = g\left(2^{l}i\right) := f^{(l+1)}\left(\frac{i+1}{2}\right), \quad j \in \mathbb{N}.$$

Then for all $j \in \mathbb{N}$, again by (1.81), we have

$$x \in E^{(l+1)}_{\left(f^{(l+1)}(1), f^{(l+1)}(2), \ldots, f^{(l+1)}\left(\frac{i+1}{2}\right)\right)} = E^{(l+1)}_{\left(g(2^{l}1), g(2^{l}3), \ldots, g(2^{l}i)\right)} = E_{g|_{j}},$$

where we have used (1.82). Hence

$$\bigcap_{n=1}^{\infty} E^{(n)} \subset \bigcup_{g \in \mathbb{N}^{\mathbb{N}}} \bigcap_{j=1}^{\infty} E_{g|_{j}},$$

which, together with (1.84), gives (1.83).

We are now ready to study the measurability of Suslin sets. If $(X, \mathfrak{M}, \mu)$ is a measure space and $\mathcal{E} : \mathcal{I} \to \mathfrak{M}$, then in general, the Suslin-$\mathfrak{M}$ set $A\left(\mathcal{E}\right)$ does not belong to $\mathfrak{M}$. However, the next result shows that $A\left(\mathcal{E}\right)$ belongs to the σ-algebra $\mathfrak{M}_{\mu}$ of all μ^{*}-measurable subsets, where μ^{*} is the outer measure given in Corollary 1.38.

Theorem 1.134. *Let $(X, \mathfrak{M}, \mu)$ be a measure space and let $\mathfrak{M}_{\mu}$ be the σ-algebra of all μ^{*}-measurable subsets, where $\mu^{*} : \mathcal{P}\left(X\right) \to [0, \infty]$ is the outer measure*

$$\mu^{*}\left(E\right) := \inf \left\{ \sum_{n=1}^{\infty} \mu\left(E_{n}\right) : \{E_{n}\} \subset \mathfrak{M}, \ E \subset \bigcup_{n=1}^{\infty} E_{n} \right\}, \quad E \subset X.$$

Then any Suslin-$\mathfrak{M}_{\mu}$ set belongs to $\mathfrak{M}_{\mu}$.

Proof. Let $A\left(\mathcal{E}\right)$ be a Suslin-$\mathfrak{M}_{\mu}$ set, that is,

$$A\left(\mathcal{E}\right) = \bigcup_{f \in \mathbb{N}^{\mathbb{N}}} \bigcup_{k=1}^{\infty} E_{f|_{k}},$$

where $\mathcal{E} : \mathcal{I} \to \mathfrak{M}_{\mu}$. In view of Carathéodory's theorem and Proposition 1.45, μ^{*} is a regular outer measure that coincides with μ on $\mathfrak{M}$ and such that $\mathfrak{M} \subset \mathfrak{M}_{\mu}$. Moreover, for every $E \subset X$,

$$\mu^{*}\left(E\right) = \inf \left\{ \mu\left(F\right) : F \in \mathfrak{M}, F \supset E \right\}. \tag{1.85}$$

In view of the subadditivity of μ^{*}, to prove that $A\left(\mathcal{E}\right)$ belongs to $\mathfrak{M}_{\mu}$ it suffices to show that for any set $F \subset X$,

$$\mu^{*}\left(F \cap A\left(\mathcal{E}\right)\right) + \mu^{*}\left(F \setminus A\left(\mathcal{E}\right)\right) \leq \mu^{*}\left(F\right).$$

Without loss of generality we may assume that $\mu^*(F) < \infty$. Using (1.85), we may find $G \in \mathfrak{M}$ such that $G \supset F$ and $\mu^*(F) = \mu(G)$. Since the sequence $\{S^n \cap G\}_{n \in \mathbb{N}}$ is increasing, by Proposition 1.43 and (1.77),

$$\mu^*(A(\mathcal{E}) \cap G) = \lim_{n \to \infty} \mu^*(S^n \cap G),$$

and thus for any fixed $\varepsilon > 0$ there exists a positive integer n_1 such that

$$\mu^*(S^{n_1} \cap G) \geq \mu^*(A(\mathcal{E}) \cap G) - \frac{\varepsilon}{2}. \tag{1.86}$$

Inductively, if $(n_1, n_2, \ldots, n_k) \in \mathbb{N}^k$ have been defined so that

$$\mu^*\left(S^{(n_1, n_2, \ldots, n_k)} \cap G\right) \geq \mu^*\left(S^{(n_1, n_2, \ldots, n_{k-1})} \cap G\right) - \frac{\varepsilon}{2^k}, \tag{1.87}$$

using the fact that the sequence $\left\{S^{(n_1, n_2, \ldots, n_k, n)} \cap G\right\}_{n \in \mathbb{N}}$ is increasing, by Proposition 1.43 and (1.78) we have

$$\mu^*\left(S^{(n_1, n_2, \ldots, n_k)} \cap G\right) = \lim_{n \to \infty} \mu^*\left(S^{(n_1, n_2, \ldots, n_k, n)} \cap G\right),$$

and so there exists $n_{k+1} \in \mathbb{N}$ such that

$$\mu^*\left(S^{(n_1, n_2, \ldots, n_{k+1})} \cap G\right) \geq \mu^*\left(S^{(n_1, n_2, \ldots, n_k)} \cap G\right) - \frac{\varepsilon}{2^{k+1}}.$$

In this way we obtain a function $g \in \mathbb{N}^{\mathbb{N}}$ defined by $g(k) := n_k$ for $k \in \mathbb{N}$. Corresponding to g we have the sequence of sets $\{S_{g|_k}\}_{k \in \mathbb{N}}$ defined in (1.79). Note that since $\mathcal{E} : \mathcal{I} \to \mathfrak{M}_\mu$, the sets $S_{g|_k}$ belong to $\mathfrak{M}_\mu$, since they are a countable union of finite intersections of measurable sets. Using the fact that $S^{g|_k} \subset S_{g|_k}$, (1.86) and (1.87) yield

$$\mu^*\left(S_{g|_k} \cap G\right) \geq \mu^*\left(S^{g|_k} \cap G\right) = \mu^*\left(S^{(n_1, n_2, \ldots, n_k)} \cap G\right)$$
$$\geq \mu^*(A(\mathcal{E}) \cap G) - \sum_{i=1}^{k} \frac{\varepsilon}{2^i}.$$

Therefore since $S_{g|_k} \in \mathfrak{M}_\mu$ we obtain

$$\mu^*(G) = \mu^*(S_{g|_k} \cap G) + \mu^*(G \setminus S_{g|_k}) \geq \mu^*(A(\mathcal{E}) \cap G) + \mu^*(G \setminus S_{g|_k}) - \varepsilon.$$

As the sequence $\{S_{g|_k}\}_{k \in \mathbb{N}}$ decreases to $S_g \subset A(\mathcal{E})$ (see (1.80)), the sequence $G \setminus S_{g|_k}$ increases to $G \setminus S_g \supset G \setminus A(\mathcal{E})$, and so by Proposition 1.43 (or Proposition 1.7), letting $k \to \infty$ in the previous inequality gives

$$\mu^*(F) = \mu^*(G) \geq \mu^*(A(\mathcal{E}) \cap G) + \mu^*(G \setminus S_g) - \varepsilon$$
$$\geq \mu^*(A(\mathcal{E}) \cap G) + \mu^*(G \setminus A(\mathcal{E})) - \varepsilon$$
$$\geq \mu^*(A(\mathcal{E}) \cap F) + \mu^*(F \setminus A(\mathcal{E})) - \varepsilon,$$

where we have used the facts that $G \supset F$ and $\mu^*(F) = \mu(G)$. Given the arbitrariness of ε and F it follows that $A(\mathcal{E})$ is μ^*-measurable.

We now prove the first projection theorem.

Theorem 1.135. *Let $(X, \mathfrak{M}, \mu)$ be a measure space and let Y be a complete separable metric space. Let $\mathcal{R}$ be the class of all rectangles $F \times C$, where $F \in \mathfrak{M}$ and $C \subset Y$ is closed. Then the projection $\pi_X(E)$ of every Suslin-$\mathcal{R}$ set $E \subset X \times Y$ on X is a Suslin-$\mathfrak{M}$ set. In particular, $\pi_X(E)$ belongs to $\mathfrak{M}_\mu$.*

Proof. Since Y is a complete separable metric space we may find a countable family $\{y_i\} \subset Y$ dense in Y. Let $\{U_i\}$ be the countable family of closed balls centered at y_i and rational diameter less than $\frac{1}{2}$.

Since each U_i is a complete separable space we may cover it with a countable family $\{U_{ij}\}$ of closed balls of rational diameter less than $\frac{1}{2^2}$. Continuing this process for each $\alpha = (\alpha_1, \ldots, \alpha_n) \in \mathbb{N}^n$ we construct a closed ball U_α such that $\operatorname{diam} U_\alpha \in \mathbb{Q}$,

$$\operatorname{diam} U_\alpha \leq \frac{1}{2^n},$$

and

$$U_\alpha \subset \bigcup_{k=1}^{\infty} U_{(\alpha,k)}.$$

Hence we may write

$$X \times Y = \bigcup_{g \in \mathbb{N}^{\mathbb{N}}} \bigcap_{k=1}^{\infty} \left(X \times U_{g|_k} \right). \tag{1.88}$$

Let $E \subset X \times Y$ be a Suslin-$\mathcal{R}$ set. Then there exists a Suslin scheme $\mathcal{E} : \mathcal{I} \to \mathcal{R}$ such that

$$E = A(\mathcal{E}) = \bigcup_{f \in \mathbb{N}^{\mathbb{N}}} \bigcap_{k=1}^{\infty} \left(F_{f|_k} \times C_{f|_k} \right),$$

where $F_{f|_k} \in \mathfrak{M}$ and $C_{f|_k} \subset Y$ is closed. By (1.88) we have

$$E = \bigcup_{g \in \mathbb{N}^{\mathbb{N}}} \bigcup_{f \in \mathbb{N}^{\mathbb{N}}} \bigcap_{k=1}^{\infty} \left(F_{f|_k} \times \left(C_{f|_k} \cap U_{g|_k} \right) \right),$$

where we have used the fact that for each $k \in \mathbb{N}$,

$$F_{f|_k} \times C_{f|_k} = \bigcup_{g \in \mathbb{N}^{\mathbb{N}}} \left(F_{f|_k} \times \left(C_{f|_k} \cap U_{g|_k} \right) \right),$$

since by construction the family $\left\{ U_{g|_k} \right\}_{g \in \mathbb{N}^{\mathbb{N}}}$ covers Y.

With a suitable change of indexing we may write

$$E = \bigcup_{h \in \mathbb{N}^{\mathbb{N}}} \bigcap_{k=1}^{\infty} \left(F_{h|_k} \times \tilde{C}_{h|_k} \right),$$

where $\tilde{C}_{h|_k} \subset Y$ is closed and diam $\tilde{C}_{h|_k} \to 0$ as $k \to \infty$ for every $h \in \mathbb{N}^{\mathbb{N}}$. By replacing each $\tilde{C}_{h|_k}$ with the closed set

$$\hat{C}_{h|_k} := \tilde{C}_{h|_1} \cap \ldots \cap \tilde{C}_{h|_k},$$

without loss of generality we may assume that $\tilde{C}_{h|_{k+1}} \subset \tilde{C}_{h|_k}$, so that since Y is complete and diam $\tilde{C}_{h|_k} \to 0$ as $k \to \infty$, the closed set

$$\tilde{C}_h := \bigcap_{k=1}^{\infty} \tilde{C}_{h|_k}$$

is nonempty if and only if $\tilde{C}_{h|_k}$ is nonempty for each $k \in \mathbb{N}$.

Moreover, again without loss of generality, we may assume that $F_{h|_k}$ is empty whenever $\tilde{C}_{h|_k}$ is empty. Hence if $\tilde{C}_h$ is empty then so is

$$F_h := \bigcap_{k=1}^{\infty} F_{h|_k},$$

and so we may write

$$\pi_X(E) = \bigcup_{h \in \mathbb{N}^{\mathbb{N}}} \pi_X \left(F_h \times \tilde{C}_h \right) = \bigcup_{h \in \mathbb{N}^{\mathbb{N}}} \bigcap_{k=1}^{\infty} F_{h|_k},$$

where $\pi_X : X \times Y \to X$ is the projection on X. Thus $\pi_X(E)$ is a Suslin-$\mathfrak{M}$ set.

It now follows from Theorem 1.134 that $\pi_X(E) \in \mathfrak{M}_\mu$.

As a corollary of this theorem we can prove the result announced at the beginning of the subsection.

Theorem 1.136 (Projection). *Let $(X, \mathfrak{M}, \mu)$ be a measure space and let Y be a complete separable metric space. Then for every $E \in \mathfrak{M}_\mu \otimes \mathcal{B}(Y)$ the projection of E on X belongs to $\mathfrak{M}_\mu$.*

Here $\mathfrak{M}_\mu$ is the σ-algebra defined in Theorem 1.134.

Proof. Let $\mathcal{C}$ be the class of all Suslin-$\mathcal{R}$ sets whose complement is still a Suslin-$\mathcal{R}$ set. We claim that this class is a σ-algebra. Indeed, if $E \in \mathcal{C}$ then $(X \times Y) \setminus E \in \mathcal{C}$ by definition of $\mathcal{C}$. If $\{E_n\} \subset \mathcal{C}$ then E_n and $(X \times Y) \setminus E_n$ are Suslin-$\mathcal{R}$ sets, and so, by Theorem 1.133,

$$\bigcup_{n=1}^{\infty} E_n \text{ and } (X \times Y) \setminus \left(\bigcup_{n=1}^{\infty} E_n \right) = \bigcap_{n=1}^{\infty} ((X \times Y) \setminus E_n)$$

are still Suslin-$\mathcal{R}$ sets. Hence $\bigcup_{n=1}^{\infty} E_n \in \mathcal{C}$. Finally, the empty set belongs to $\mathcal{C}$ since both $X \times Y$ and $\emptyset$ are Suslin-$\mathcal{R}$ sets.

Since C contains all rectangles of the form $F \times C$, where $F \in \mathfrak{M}_\mu$ and C is closed, it must contain $\mathfrak{M}_\mu \otimes \mathcal{B}(Y)$. Hence if $E \in \mathfrak{M}_\mu \otimes \mathcal{B}(Y)$ then E is a Suslin-$\mathcal{R}$ set, and so by the previous theorem $\pi_X(E)$ is a Suslin-$\mathfrak{M}$ set. We now use Theorem 1.134 to conclude that $\pi_X(E)$ belongs to $\mathfrak{M}_\mu$.

As a direct application of the projection theorem, and of the fact that the σ-algebra of Lebesgue measurable sets in $\mathbb{R}^2$ is the completion of $\mathcal{B}(\mathbb{R}^2)$, it follows that if $E \subset \mathbb{R}^2$ is a Borel set, then the projection $\pi_1(E)$ on the x-axis is Lebesgue measurable, but not necessarily a Borel set.

1.2 Covering Theorems and Differentiation of Measures in $\mathbb{R}^N$

1.2.1 Covering Theorems in $\mathbb{R}^N$

Throughout this subsection we take as ambient space X the Euclidean space $\mathbb{R}^N$, and we present several covering theorems and their applications to the derivatives of measures.

Definition 1.137. *A set $F \subset \mathbb{R}^N$ is said to be*

  (i) *star-shaped with respect to a set $G \subset F$ if F is star-shaped with respect to each point of G, i.e., if $\theta x + (1-\theta)y \in F$ for all $x \in F$, $y \in G$, and $\theta \in (0,1)$;*
  (ii) *convex if it is star-shaped with respect to itself, i.e., if $\theta x + (1-\theta)y \in F$ for all $x,\, y \in F$ and $\theta \in (0,1)$;*
  (iii) *a γ-Morse set associated with x, with $x \in \mathbb{R}^N$ and $\gamma \geq 1$, if there exists $r > 0$ such that*

$$\overline{B(x,r)} \subset F \subset \overline{B(x,\gamma r)} \tag{1.89}$$

and F is star-shaped with respect to $\overline{B(x,r)}$.

Example 1.138. A regular pentagram is an example of a nonconvex set that is star-shaped with respect to a ball.

Given a point $x_0 \in \mathbb{R}^N$ and a function $\varphi : S^{N-1} \to [0,\infty]$, the set

$$E := \left\{ x_0 + sx : x \in S^{N-1},\, 0 \leq s < \varphi(x) \right\}$$

is star-shaped with respect to x_0.

It turns out that if φ is sufficiently regular, then E is actually star-shaped with respect to a ball centered at x_0.

Proposition 1.139. *Let $A \subset \mathbb{R}^N$ be an open, bounded set star-shaped with respect to a point $x_0 \in A$. If A is star-shaped with respect to a closed ball centered at x_0, then the function $\varphi : S^{N-1} \to [0,\infty)$, defined by*

$$\varphi(x) := \sup\{s \geq 0 : x_0 + sx \in A\} \quad x \in S^{N-1},$$

is a Lipschitz function, i.e., there exists a constant $L > 0$ such that

$$|\varphi(x) - \varphi(y)| \leq L|x - y|$$

for all $x, y \in S^{N-1}$.

The proof follows from Theorem 4.107.

Exercise 1.140. Prove the converse of the previous proposition, namely that if $A \subset \mathbb{R}^N$ is an open, bounded set star-shaped with respect to a point $x_0 \in A$ and if the function φ is Lipschitz, then the set A is star-shaped with respect to a closed ball centered at x_0.

Remark 1.141. It follows from the above proposition that an open bounded set star-shaped with respect to a closed ball $\overline{B}(x_0, r)$ may be approximated from inside by an increasing uncountable family of closed sets star-shaped with respect to the same ball. Indeed, it suffices to consider

$$A_\varepsilon = \{x_0 + sx : x \in S^{N-1}, 0 \leq s \leq \varphi(x) - \varepsilon\},$$

where $0 < \varepsilon < r$, and where we have used the fact that $\varphi \geq r$ and is finite in view of the boundedness of the set A.

Definition 1.142. *Given a set $E \subset \mathbb{R}^N$, a family $\mathcal{F}$ of nonempty subsets of $\mathbb{R}^N$ is said to be a*

(i) cover *for E if*

$$E \subset \bigcup_{F \in \mathcal{F}} F;$$

(ii) fine cover *for E if for every $x \in E$ there exists a subfamily $\mathcal{F}_x \subset \mathcal{F}$ of sets containing x such that*

$$\inf\{\operatorname{diam} F : F \in \mathcal{F}_x\} = 0; \tag{1.90}$$

(iii) Morse cover *for E if there exists $\gamma \geq 1$ such that for every $x \in E$ there exists a γ-Morse set F associated with x and such that $F \in \mathcal{F}$;*

(iv) fine Morse cover *for E if there exists $\gamma \geq 1$ such that for every $x \in E$ there exists a subfamily $\mathcal{F}_x \subset \mathcal{F}$ of γ-Morse sets associated with x for which (1.90) holds.*

Example 1.143. Given a set $E \subset \mathbb{R}^N$, the two families

$$\{B(x, r) : 0 < r < 1, x \in E\}, \quad \{Q(x, r) : 0 < r < 1, x \in E\}$$

are fine Morse covers for E. More generally, if C is any bounded, convex closed set containing the origin in its interior, then the family

$$\{x + rC : 0 < r < 1, x \in E\}$$

is a fine Morse cover for E.

We now present the main covering theorem of this subsection.

Theorem 1.144 (Morse covering theorem). *Let $E \subset \mathbb{R}^N$ and let $\mathcal{F}$ be a family of subsets of $\mathbb{R}^N$ such that*

$$\sup \{\operatorname{diam} F : F \in \mathcal{F}\} < \infty. \tag{1.91}$$

Assume that $\mathcal{F}$ is a Morse cover for E. Then there exist $\ell = \ell(\gamma, N) \in \mathbb{N}$ and $\mathcal{F}_1, \ldots, \mathcal{F}_\ell \subset \mathcal{F}$ such that each $\mathcal{F}_n$, $n = 1, \ldots, \ell$, is a countable family of disjoint sets in $\mathcal{F}$ and

$$E \subset \bigcup_{n=1}^{\ell} \bigcup_{F \in \mathcal{F}_n} F.$$

Lemma 1.145. *Let $\gamma \geq 1$, $1 < t \leq 2$, and consider $x_1, \ldots, x_n \in \mathbb{R}^N$ and $F_1, \ldots, F_n \subset \mathbb{R}^N$ such that*

(i) F_i is a γ-Morse set associated with x_i for all $1 \leq i \leq n$;
(ii) $F_i \cap F_n \neq \emptyset$ for all $1 \leq i < n$;
(iii) if $i < j$ then $x_j \notin F_i$ and $\operatorname{diam} F_j < t \operatorname{diam} F_i$.

Then there exists a constant $c = c(\gamma, N)$ such that $n < c$.

Proof. By definition of a γ-Morse set, for each $i = 1, \ldots, n$ there exists $r_i > 0$ such that

$$\overline{B(x_i, r_i)} \subset F_i \subset \overline{B(x_i, \gamma r_i)} \tag{1.92}$$

and F_i is star-shaped with respect to $\overline{B(x_i, r_i)}$.

Step 1: We start by showing that given a γ-Morse set $F \subset \mathbb{R}^N$ associated to some point x_0, points of the form $\theta y + (1 - \theta) x$ belong to the interior of F whenever $0 < \theta \leq 1$, $x \in F$ and $y \in B(x_0, r)$, with r given in the definition of a γ-Morse set.

Without loss of generality we may assume that $x = 0$, and choose $\rho > 0$ such that $B(y, \rho) \subset B(x_0, r)$. Then $\overline{B(\theta y, \theta \rho)} \subset F$ because if $|\theta y - z| \leq \theta \rho$ then $\frac{z}{\theta} \in \overline{B(x_0, r)}$, and so $z = \theta \frac{z}{\theta} + (1 - \theta) 0 \in F$ since F is star-shaped with respect to $\overline{B(x_0, r)}$.

Step 2: Without loss of generality we may assume that $x_n = 0$, and we set $r = r_n$ and $F = F_n$. We show that if i, j are such that

$$32\gamma^2 r < |x_i| \leq |x_j| \quad \text{and} \quad \left| \frac{x_i}{|x_i|} - \frac{x_j}{|x_j|} \right| \leq \frac{1}{16\gamma}, \tag{1.93}$$

then x_i belongs to the interior of F_j.

In view of (ii), find $x \in F_j \cap F$ and write

$$x_i = (1 - \theta) x + \theta y,$$

where by $(1.93)_1$,

$$y := (1 - s)\,x + s x_i, \quad s := \frac{|x_j|}{|x_i|} \geq 1, \quad \theta := \frac{1}{s}. \tag{1.94}$$

We claim that $y \in B(x_j, r_j)$. Note that by Step 1 this would entail that x_i belongs to the interior of F_j as well.

Since $F \subset \overline{B}(0, \gamma r)$, by $(1.93)_1$ we have

$$16\gamma \operatorname{diam} F \leq 32\gamma^2 r < |x_i|,$$

and thus, using the fact that $x, 0 \in F$ and (iii), we deduce that

$$|x| \leq \operatorname{diam} F \leq \min\left\{\frac{|x_i|}{16\gamma}, 2\operatorname{diam} F_j\right\}. \tag{1.95}$$

Therefore, since $|1 - s| = s - 1 \leq s$ (see (1.94)), by (1.93)–(1.95) and the fact that $x, x_j \in F_j$ we have

$$
\begin{aligned}
|y - x_j| &= \left|(1 - s)\,x + |x_j|\left(\frac{x_i}{|x_i|} - \frac{x_j}{|x_j|}\right)\right| \\
&\leq s\,|x| + \frac{|x_j|}{16\gamma} \leq s\frac{|x_i|}{16\gamma} + \frac{|x_j|}{16\gamma} \\
&= \frac{|x_j|}{8\gamma} \leq \frac{|x_j - x| + |x|}{8\gamma} < \frac{\operatorname{diam} F_j}{2\gamma} \leq r_j,
\end{aligned}
$$

where in the last inequality we have used (1.92). Hence $y \in B(x_j, r_j)$.

Step 3: Given $s > 0$, using a simple compactness argument it is possible to establish the existence of the maximum number n_s of points on the sphere S^{N-1} such that the distance between any two of those points is at least $\frac{1}{s}$. Similarly, let m_s be the maximum cardinality of any set of points in $\overline{B}(0, 1)$ including the origin and such that the distance between any two points of the set is at least $\frac{1}{s}$. We claim that

$$n \leq m_{64\gamma^3} + m_{8\gamma^2} n_{16\gamma}.$$

We decompose $\{1, \ldots, n\}$ as $J_1 \cup J_2$ where

$$J_1 := \left\{1 \leq i \leq n : |x_i| \leq 32\gamma^2 r\right\}, \quad J_2 := \left\{1 \leq i \leq n : |x_i| > 32\gamma^2 r\right\}.$$

We first show that the cardinality of J_1 is at most $m_{64\gamma^3}$. Indeed, if $i, j \in J_1$, with $i < j$, then by (iii) and (1.92),

$$|x_i - x_j| \geq r_i \geq \frac{\operatorname{diam} F_i}{2\gamma} \geq \frac{\operatorname{diam} F}{4\gamma} \geq \frac{r}{2\gamma},$$

and so

$$\left|\frac{x_i}{32\gamma^2 r} - \frac{x_j}{32\gamma^2 r}\right| \geq \frac{1}{64\gamma^3}.$$

By definition of J_1, both vectors in the previous inequality belong to $\overline{B(0,1)}$, and in view of the definition of m_s we conclude that there can be at most $m_{64\gamma^3}$ elements in J_1.

In order to prove that the cardinality of J_2 is at most $m_{8\gamma^2} n_{16\gamma}$, we will decompose J_2 as

$$J_2 = \bigcup_{i \in I} \hat{J}_i, \tag{1.96}$$

where the set I will have at most $n_{16\gamma}$ elements and the cardinality of each of the sets $\hat{J}_i$ will not exceed $m_{8\gamma^2}$.

We proceed by induction to construct I and $\hat{J}_i$. Set $\varXi_0 := J_2$ and find the smallest element i_1 in $\varXi_0$ such that $|x_{i_1}| \leq |x_j|$ for all j in $\varXi_0$. Put i_1 in I and define

$$\varXi_1 := \varXi_0 \setminus \hat{J}_{i_1},$$

where

$$\hat{J}_{i_1} := \{1 \leq i \leq n : x_{i_1} \text{ belongs to the interior of } F_i\}.$$

Suppose that $\varXi_k$ and $\hat{J}_{i_k}$ have been constructed and i_k has been selected such that

$$\varXi_k := \varXi_{k-1} \setminus \hat{J}_{i_k}.$$

If $\varXi_k$ is empty then the process is completed. Otherwise, let i_{k+1} be the smallest element in $\varXi_k$ such that $|x_{i_{k+1}}| \leq |x_j|$ for all j in $\varXi_k$. Put i_{k+1} in I and define

$$\varXi_{k+1} := \varXi_k \setminus \hat{J}_{i_{k+1}},$$

where

$$\hat{J}_{i_{k+1}} := \{1 \leq i \leq n : x_{i_{k+1}} \text{ belongs to the interior of } F_i\}.$$

Then (1.96) is satisfied. We now estimate the cardinality of I. Note that by construction, if $i, j \in I$ and $i < j$ then $|x_i| \leq |x_j|$ and x_i does not belong to the interior of F_j, and so by Step 2,

$$\left| \frac{x_i}{|x_i|} - \frac{x_j}{|x_j|} \right| > \frac{1}{16\gamma}.$$

Hence, by definition of n_s, we deduce that I has at most $n_{16\gamma}$ elements.

Finally, it remains to show that the cardinality of each $\hat{J}_i$ is less than or equal to $m_{8\gamma^2}$. Fix $i \in I$ and consider $k, j \in \hat{J}_i$ with $k < j$. Then x_i belongs to the interior of F_k; hence

$$|x_k - x_i| \leq \operatorname{diam} F_k \leq 2 \operatorname{diam} F_i \leq 4\gamma r_i, \tag{1.97}$$

where we have used (1.92) and (iii), and similarly,

$$|x_j - x_i| \leq 4\gamma r_i. \tag{1.98}$$

Also, since $k < j$ by (iii) we have that $x_j \notin F_k$, and so

$$|x_k - x_j| \geq r_k \geq \frac{\operatorname{diam} F_k}{2\gamma} \geq \frac{|x_k - x_i|}{2\gamma},$$

where we have used (1.92) and (1.97). In turn, by (iii), since $x_i \in F_k$ we must have $i < k$, and therefore again by (iii), $x_k \notin F_i$, and in particular, $|x_k - x_i| \geq r_i$. We conclude that

$$|x_k - x_j| \geq \frac{r_i}{2\gamma}. \tag{1.99}$$

Using (1.97) and (1.98), we have

$$\frac{x_k - x_i}{4\gamma r_i}, \frac{x_j - x_i}{4\gamma r_i} \in \overline{B}(0,1),$$

while by (1.99),

$$\left| \frac{x_k - x_i}{4\gamma r_i} - \frac{x_j - x_i}{4\gamma r_i} \right| = \frac{|x_k - x_j|}{4\gamma r_i} \geq \frac{1}{8\gamma^2}.$$

This and the definition of m_s entail that the cardinality of $\hat{J}_i$ is at most $m_{8\gamma^2}$.

We are now in a position to prove the Morse covering theorem

Proof (Theorem 1.144). For every $x \in E$ select $F(x) \in \mathcal{F}$ such that $F(x)$ is a γ-Morse set associated with x. We divide the proof into three steps.

Step 1: Assume first that E is bounded. We construct a countable subfamily of $\mathcal{F}$ that still covers E. By (1.91) we may choose $x_1 \in E$ such that

$$\operatorname{diam} F(x_1) > \frac{3}{4} \sup_{x \in E} \operatorname{diam} F(x)$$

and set $E_2 := E \setminus F(x_1)$. By induction, assuming that $x_1, \ldots, x_n$ have been chosen, define

$$E_{n+1} := E \setminus \bigcup_{i=1}^{n} F(x_i). \tag{1.100}$$

If E_{n+1} is empty then set $J := \{1, \ldots, n\}$ and observe that

$$E \subset \bigcup_{i=1}^{n} F(x_i). \tag{1.101}$$

Otherwise, again by (1.91), select $x_{n+1} \in E_{n+1}$ such that

$$\operatorname{diam} F(x_{n+1}) > \frac{3}{4} \sup_{x \in E_{n+1}} \operatorname{diam} F(x). \tag{1.102}$$

If $E_n \neq \emptyset$ for every n then we set $J := \mathbb{N}$, and we claim that

$$E \subset \bigcup_{n=1}^{\infty} F(x_n). \tag{1.103}$$

This follows easily from the fact that

$$\operatorname{diam} F(x_n) \to 0. \tag{1.104}$$

Indeed, if (1.104) holds and if $x \in E$, then find n large enough such that

$$\operatorname{diam} F(x_{n+1}) < \frac{3}{4} \operatorname{diam} F(x).$$

By (1.102) it follows that $x \notin E_{n+1}$, and so, in view of (1.100), we conclude that

$$x \in \bigcup_{i=1}^{n} F(x_i).$$

In order to prove (1.104), for every $n \in J$ let $F_n := F(x_n)$ and let $r_n > 0$ be such that

$$\overline{B(x_n, r_n)} \subset F_n \subset \overline{B(x_n, \gamma r_n)},$$

and

$$\sup_{n} r_n \le \frac{1}{2} \sup_{x \in E} \operatorname{diam} F(x) =: R < \infty, \tag{1.105}$$

where we have used (1.91). Note that by construction if $i, j \in J$ and $i < j$ then

$$x_j \notin F_i \text{ and } \operatorname{diam} F_i > \frac{3}{4} \operatorname{diam} F_j, \tag{1.106}$$

and so $r_i > \frac{3r_j}{4\gamma}$. Moreover, the balls $B\left(x_i, \frac{r_i}{\beta}\right)$ and $B\left(x_j, \frac{r_j}{\beta}\right)$ are disjoint, where $\beta := \max\{3, 2\gamma\}$. Indeed, since $x_j \notin F_i$, then

$$|x_i - x_j| > r_i = \frac{r_i}{3} + \frac{2r_i}{3} \ge \frac{r_i}{3} + \frac{r_j}{2\gamma} \ge \frac{r_i}{\beta} + \frac{r_j}{\beta}.$$

Since

$$\bigcup_{n=1}^{\infty} B\left(x_n, \frac{r_n}{\beta}\right) \subset \left\{x \in \mathbb{R}^N : \operatorname{dist}(x, E) < \frac{R}{\beta}\right\},$$

we conclude that

$$\sum_{n=1}^{\infty} \left(\frac{r_n}{\beta}\right)^N < \infty,$$

where we have used the fact that E is bounded. Hence (1.104) holds.

Step 2: Let $\{F_i\}_{i \in J}$ be the countable subfamily of $\mathcal{F}$ constructed in Step 1 and that still covers the bounded set E. If $J = \{1\}$ then there is nothing left to prove. Otherwise, for any fixed integer $k > 1$ let

$$I_k := \{i \in J : 1 \le i < k, F_i \cap F_k \ne \emptyset\}.$$

By Lemma 1.145 and in view of (1.106) there exists an integer $c(\gamma, N)$ such that the cardinality of I_k is less than $c(\gamma, N)$.

Construct a function $\sigma : J \to \{1, \ldots, c(\gamma, N)\}$ in the following way. If $j \le c(\gamma, N)$ set $\sigma(j) := j$. For $j > c(\gamma, N)$ we define $\sigma(j)$ recursively: assuming that $\sigma(1), \ldots, \sigma(j-1)$ have been assigned, set $\sigma(j) := l$, where $l \in \{1, \ldots, c(\gamma, N)\}$ satisfies the property

$$F_j \cap F_i = \emptyset \tag{1.107}$$

whenever $i \in \{1, \ldots, j-1\}$ is such that $\sigma(i) = l$. Note that there is at least one such number l. Indeed, if not, then for every $l \in \{1, \ldots, c(\gamma, N)\}$ there would exist $i_l \in \{1, \ldots, j-1\}$ such that $\sigma(i_l) = l$ and $F_j \cap F_{i_l} \ne \emptyset$. In particular, $i_l \in I_{j-1}$. Since $l \ne l'$ implies that $i_l \ne i_{l'}$, this would entail that the cardinality of I_{j-1} would be at least $c(\gamma, N)$, and this is a contradiction.

For $n \in \{1, \ldots, c(\gamma, N)\}$ let

$$\mathcal{F}_n := \{F_i : \sigma(i) = n\}.$$

Since if $i \in J$ then $F_i \in \mathcal{F}_{\sigma(i)}$, by (1.101) and (1.103) we have

$$E \subset \bigcup_{n=1}^{c(\gamma,N)} \bigcup_{F \in \mathcal{F}_n} F.$$

It remains to show that distinct elements of $\mathcal{F}_n$ are mutually disjoint. Indeed, if $i < j$ and $\sigma(i) = \sigma(j)$, then j is strictly bigger than $c(\gamma, N)$, and so $F_j \cap F_i = \emptyset$ by (1.107).

Step 3: If the set E is unbounded, then for each $k \in \mathbb{N}$ apply the previous steps to the set

$$E_k := E \cap (B(0, k\gamma R) \setminus B(0, (k-1)\gamma R)),$$

where R is defined in (1.105), to find

$$E_k \subset \bigcup_{n=1}^{c(\gamma,N)} \bigcup_{F \in \mathcal{F}_n^{(k)}} F,$$

where the elements of $\mathcal{F}_n^{(k)}$ are mutually disjoint. Without loss of generality, we may remove from each family $\mathcal{F}_n^{(k)}$ those elements that do not intersect E_k. Hence by (1.105), if $F \in \mathcal{F}_n^{(k)}$ and $F' \in \mathcal{F}_n^{(k+2)}$ then $F \cap F' = \emptyset$. For $n \in \{1, \ldots, c(\gamma, N)\}$ define

$$\mathcal{F}_n := \left\{ F \in \mathcal{F}_n^{(2k)} : k \in \mathbb{N} \right\},$$

and if $n \in \{c(\gamma, N) + 1, \ldots, 2c(\gamma, N)\}$ set

$$\mathcal{F}_n := \left\{ F \in \mathcal{F}_n^{(2k-1)} : k \in \mathbb{N} \right\}.$$

Then

$$E \subset \bigcup_{n=1}^{2c(\gamma,N)} \bigcup_{F \in \mathcal{F}_n} F,$$

and this cover satisfies the desired properties.

As a corollary of the Morse covering theorem we have the following theorem of Besicovitch:

Theorem 1.146 (Besicovitch covering theorem). *There exists a constant* ℓ, *depending only on the dimension* N *of* $\mathbb{R}^N$, *such that for any collection* $\mathcal{F}$ *of (nondegenerate) closed balls with*

$$\sup \{\operatorname{diam} B : B \in \mathcal{F}\} < \infty$$

there exist $\mathcal{F}_1, \ldots, \mathcal{F}_\ell \subset \mathcal{F}$ *such that each* $\mathcal{F}_n$, $n = 1, \ldots, \ell$, *is a countable family of disjoint balls in* $\mathcal{F}$ *and*

$$E \subset \bigcup_{n=1}^{\ell} \bigcup_{B \in \mathcal{F}_n} B,$$

where E *is the set of centers of balls in* $\mathcal{F}$.

Proof. It suffices to take $\gamma = 1$ and for each $x \in E$ define $\mathcal{F}_x$ to be the family of all balls in $\mathcal{F}$ that are centered at x.

An important consequence of the Morse covering theorem is the following result, which will be crucial in the characterization of Radon–Nikodym derivatives in the next subsections.

Theorem 1.147 (Morse measure covering theorem). *Let* $E \subset \mathbb{R}^N$ *and let* $\mathcal{F}$ *be a fine Morse cover of closed subsets for* E. *Let* $\mu^* : \mathcal{P}(\mathbb{R}^N) \to [0, \infty]$ *be a Radon outer measure. Then for every open set* $A \subset \mathbb{R}^N$ *there exists a countable family* $\mathcal{F}_0 \subset \mathcal{F}$ *of pairwise disjoint subsets such that*

$$\bigcup_{F \in \mathcal{F}_0} F \subset A, \quad \mu^* \left((E \cap A) \setminus \bigcup_{F \in \mathcal{F}_0} F \right) = 0.$$

Proof. Let $\ell = \ell(\gamma, N)$ be the number given in the Morse covering theorem and choose $1 - \frac{1}{\ell} < \theta < 1$.

Step 1: Assume that E is bounded. We show that there exists a finite subfamily $\{F_1, \ldots, F_{m_1}\} \subset \mathcal{F}$ of mutually disjoint subsets of A such that

$$\mu^* \left((E \cap A) \setminus \bigcup_{i=1}^{m_1} F_i \right) \leq \theta \mu^* (E \cap A). \tag{1.108}$$

Indeed, let

$$\mathcal{F}^{(1)} := \{ F \in \mathcal{F} : \operatorname{diam} F \leq 1 \text{ and } F \subset A \}.$$

Since $\mathcal{F}$ is a fine cover of $E \cap A$, it follows that $\mathcal{F}^{(1)}$ satisfies the hypotheses of the Morse covering theorem, and so there exist $\mathcal{F}_1^{(1)}, \ldots, \mathcal{F}_\ell^{(1)} \subset \mathcal{F}^{(1)}$ such that each $\mathcal{F}_n^{(1)}$ is a countable family of disjoint sets in $\mathcal{F}^{(1)}$ and

$$E \cap A \subset \bigcup_{n=1}^{\ell} \bigcup_{F \in \mathcal{F}_n^{(1)}} F.$$

Hence

$$\mu^* (E \cap A) \leq \sum_{n=1}^{\ell} \mu^* \left(E \cap A \cap \left(\bigcup_{F \in \mathcal{F}_n^{(1)}} F \right) \right),$$

and thus there exists $j \in \{1, \ldots, \ell\}$ such that

$$\mu^* \left(E \cap A \cap \bigcup_{F \in \mathcal{F}_j^{(1)}} F \right) \geq \frac{1}{\ell} \mu^* (E \cap A).$$

Writing

$$\mathcal{F}_j^{(1)} = \left\{ F_{i,j}^{(1)} \right\},$$

in view of Proposition 1.43, and since $1 - \theta < \frac{1}{\ell}$, we find m_1 so large that

$$\mu^* \left(E \cap A \cap \bigcup_{i=1}^{m_1} F_{i,j}^{(1)} \right) \geq (1 - \theta) \mu^* (E \cap A). \tag{1.109}$$

Since $\bigcup_{i=1}^{m_1} F_{i,j}^{(1)}$ is closed, therefore μ^*-measurable, we have

$$\mu^* (E \cap A) = \mu^* \left(E \cap A \cap \bigcup_{i=1}^{m_1} F_{i,j}^{(1)} \right) + \mu^* \left(E \cap A \setminus \bigcup_{i=1}^{m_1} F_{i,j}^{(1)} \right),$$

which, together with (1.109), establishes (1.108) with $F_i := F_{i,j}^{(1)}$, $i = 1, \ldots, m_1$. Note that here we have used the fact that E is bounded and μ^* is a Radon outer measure, so that $\mu^* (E) < \infty$.

Set

$$A_2 := A \setminus \bigcup_{i=1}^{m_1} F_i$$

and

$$\mathcal{F}^{(2)} := \{F \in \mathcal{F} : \operatorname{diam} F \leq 1 \text{ and } F \subset A_2\}.$$

Just as before, we may find a finite subfamily $\{F_{m_1+1}, \ldots, F_{m_2}\} \subset \mathcal{F}^{(2)}$ of mutually disjoint subsets of A such that

$$\mu^* \left((E \cap A) \setminus \bigcup_{i=1}^{m_2} F_i \right) = \mu^* \left((E \cap A_2) \setminus \bigcup_{i=m_1+1}^{m_2} F_i \right)$$
$$\leq \theta \mu^* (E \cap A_2) \leq \theta^2 \mu^* (E \cap A).$$

By induction, we construct a countable family $\mathcal{F}_0 = \{F_i\} \subset \mathcal{F}$ of pairwise disjoint subsets contained in A such that for every $k \in \mathbb{N}$,

$$\mu^* \left((E \cap A) \setminus \bigcup_{F \in \mathcal{F}_0} F \right) \leq \mu^* \left((E \cap A) \setminus \bigcup_{i=1}^{m_k} F_i \right) \leq \theta^k \mu^* (E \cap A).$$

Since $\mu^* (E) < \infty$ we conclude the proof in this case by letting $k \to \infty$.

Step 2: If E is unbounded, then applying Proposition 1.15 to the Radon measure $\mu^* : \mathcal{B}(\mathbb{R}^N) \to [0, \infty]$, choose an increasing sequence of radii $\{r_n\}$ such that $r_n \to \infty$ and

$$\mu^* (\partial B(0, r_n)) = 0. \tag{1.110}$$

Applying Step 1 to

$$E_n := E \cap \left(B(0, r_{n+1}) \setminus \overline{B(0, r_n)} \right)$$

and

$$A_n := A \cap \left(B(0, r_{n+1}) \setminus \overline{B(0, r_n)} \right),$$

we find a countable family of mutually disjoint sets $\left\{ F_i^{(n)} \right\}_{i \in \mathbb{N}}$ contained in A_n and that cover E_n up to a set of μ^* outer measure zero. Therefore the collection of

$$\left\{ F_i^{(n)} \right\}_{i, \, n \in \mathbb{N}}$$

is a family of mutually disjoint sets contained in A, and in view of (1.110) it covers E up to a set of μ^* outer measure zero.

Remark 1.148. (i) In the special case that E is open and μ^* is the Lebesgue measure, the previous theorem still holds for a fine cover $\mathcal{F}$ of closed, not necessarily star-shaped, sets satisfying the inclusion property (1.89). The proof follows closely that of the previous theorem. We note first that since E is open, without loss of generality we may assume that $\overline{B(x, \gamma r)} \subset E$ whenever $x \in E$ and there exists $F \in \mathcal{F}$ such that (1.89) holds. Applying the previous theorem to E and to this cover $\left\{ \overline{B(x, \gamma r)} \right\}$ we may find a disjoint family $\left\{ \overline{B(x_i, \gamma r_i)} \right\}$ such that

$$\left| E \setminus \bigcup_i \overline{B\left(x_i, \gamma r_i\right)} \right| = 0.$$

Choose m_1 large enough that

$$\left| E \setminus \bigcup_{i=1}^{m_1} \overline{B\left(x_i, \gamma r_i\right)} \right| \le \frac{1}{2} |E|.$$

Then

$$\left| \bigcup_{i=1}^{m_1} \overline{B\left(x_i, \gamma r_i\right)} \right| \ge \frac{1}{2} |E|.$$

It follows that

$$\left| E \setminus \bigcup_{i=1}^{m_1} \overline{B\left(x_i, r_i\right)} \right| = |E| - \frac{1}{\gamma^N} \left| \bigcup_{i=1}^{m_1} \overline{B\left(x_i, \gamma r_i\right)} \right| \le \left(1 - \frac{1}{2\gamma^N} \right) |E|,$$

and so

$$\left| E \setminus \bigcup_{i=1}^{m_1} F_i \right| \le \left(1 - \frac{1}{2\gamma^N} \right) |E|.$$

Reasoning as in the previous theorem, for every $k \in \mathbb{N}$ we may find a finite collection of disjoint sets contained in E such that

$$\left| E \setminus \bigcup_{i=1}^{m_k} F_i \right| \le \left(1 - \frac{1}{2\gamma^N} \right)^k |E|.$$

It now suffices to let $k \to \infty$. Note that a similar argument holds, more generally, for Radon measures $\mu : \mathcal{B}\left(\mathbb{R}^N\right) \to [0, \infty]$ with the *doubling property*, that is, such that

$$\mu\left(B\left(x, 2r\right)\right) \le c\mu\left(B\left(x, r\right)\right)$$

for all $x \in \mathbb{R}^N$ and $r > 0$, and for some $c > 0$.

(ii) As a simple application of (i), given any open set $\Omega \subset \mathbb{R}^N$, a bounded open set A containing the origin with $|\partial A| = 0$, and $\delta > 0$, it is possible to find a countable family $\{x_i + r_i A\}$ of mutually disjoint subsets of Ω, with $0 < r_i < \delta$, such that

$$\left| \Omega \setminus \bigcup_i \left(x_i + r_i A\right) \right| = 0. \tag{1.111}$$

Indeed, it suffices to set

$$\mathcal{F} := \left\{ \overline{\left(x + rA\right)} : x \in \Omega, \, 0 < r < \min\{r_x, \delta\} \right\},$$

where $r_x > 0$ is such that $B(x, r_x \gamma R) \subset \Omega$ and

$$B(0, R) \subset A \subset B(0, \gamma R),$$

for some $R > 0$ and $\gamma \geq 1$. This property will be used in the study of quasiconvexity in [FoLe10]. Note that the fact that $|\partial A| = 0$ has been used to replace closed sets with open ones in (1.111).

By taking as Morse covers the families of cubes or balls as in Example 1.143, we obtain the following immediate corollaries.

Corollary 1.149. *Let* $E \subset \mathbb{R}^N$ *and let* $\mathcal{F}$ *be a family of closed cubes satisfying the property that for all* $x \in E$ *and for all* $\delta > 0$ *there exists* $\overline{Q(z, r)} \in \mathcal{F}$ *such that* $0 < r \leq \delta$ *and* $x \in Q\left(z, \frac{r}{2}\right)$. *Let* $\mu^* : \mathcal{P}\left(\mathbb{R}^N\right) \rightarrow [0, \infty]$ *be a Radon outer measure. Then there exists a countable family* $\mathcal{F}_0 \subset \mathcal{F}$ *of closed cubes with pairwise disjoint interior such that*

$$\mu^* \left(E \setminus \bigcup_{F \in \mathcal{F}_0} F \right) = 0.$$

Proof. We observe that $\mathcal{F}$ is a fine Morse cover, and therefore we may apply the Morse covering theorem.

Theorem 1.150 (Vitali–Besicovitch covering theorem). *Let* $E \subset \mathbb{R}^N$ *and let* $\mathcal{F}$ *be a family of closed balls such that each point of* E *is the center of arbitrarily small balls, that is,*

$$\inf \left\{ r : \overline{B(x, r)} \in \mathcal{F} \right\} = 0 \text{ for every } x \in E.$$

Let $\mu^* : \mathcal{P}\left(\mathbb{R}^N\right) \rightarrow [0, \infty]$ *be a Radon outer measure. Then there exists a countable family* $\mathcal{F}_0 \subset \mathcal{F}$ *of closed balls with pairwise disjoint interior such that*

$$\mu^* \left(E \setminus \bigcup_{F \in \mathcal{F}_0} F \right) = 0.$$

Remark 1.151. In general, one cannot replace closed balls with open ones. This is possible, however, if we assume that for each $x \in E$ and for each $\delta > 0$ the number of closed balls in $\mathcal{F}$ centered at x and with radius less than δ is uncountable. This is a consequence of Proposition 1.15.

Remark 1.152. Given $x \in \mathbb{R}^N$ and $\gamma \geq 1$, let $F \subset \mathbb{R}^N$ be an open γ-Morse set associated with x, and let $r > 0$ be such that

$$\overline{B(x, r)} \subset F \subset \overline{B(x, \gamma r)}$$

and F is star-shaped with respect to $\overline{B(x, r)}$. If $\mu^* : \mathcal{P}\left(\mathbb{R}^N\right) \rightarrow [0, \infty]$ is a Radon outer measure and $\mu^*(F) < \infty$, then reasoning as in Remark 1.141, it is possible to construct an increasing uncountable family $\{F_t\}_{t>0}$ of closed γ-Morse sets associated with x such that $\mu^* (\partial F_t) = 0$,

$$\overline{B(x,r)} \subset F_t \subset \overline{B(x,\gamma r)},$$

and

$$\bigcup_{t>0} F_t = F.$$

1.2.2 Differentiation Between Radon Measures in $\mathbb{R}^N$

As it will be explained in Chapter 5, when we study lower semicontinuity and relaxation properties of integral functionals, the blowup method allows us to reduce the problem to the characterization of the Radon–Nikodym derivative $\frac{d\nu}{d\mu}$, where μ is the underlying measure of integration (usually the Lebesgue measure), and ν is a measure generated by a sequence of admissible fields. This is the subject of this subsection.

Theorem 1.153 (Besicovitch derivation theorem). *Let*

$$\mu, \nu : \mathfrak{M}\left(\mathbb{R}^N\right) \to [0, \infty]$$

be Radon measures. Then there exists a Borel set $M \subset \mathbb{R}^N$, with $\mu(M) = 0$, such that for any $x \in \mathbb{R}^N \setminus M$,

$$\frac{d\nu_{ac}}{d\mu}(x) = \lim_{r \to 0^+} \frac{\nu(x + rC)}{\mu(x + rC)} \in \mathbb{R} \tag{1.112}$$

and

$$\lim_{r \to 0^+} \frac{\nu_s(x + rC)}{\mu(x + rC)} = 0,$$

where

$$\nu = \nu_{ac} + \nu_s, \quad \nu_{ac} \ll \mu, \quad \nu_s \perp \mu, \tag{1.113}$$

and C is any bounded, convex closed set containing the origin in its interior.

Note that (1.113) holds in view of the Lebesgue decomposition theorem, since every Radon measure in $\mathbb{R}^N$ is σ-finite.

Proof. Let

$$M_1 := \left\{ x \in \mathbb{R}^N : \mu(B(x,r)) = 0 \text{ for some } r > 0 \right\}.$$

Note that the set M_1 is open. Using the density of the rationals in the reals it is possible to cover M_1 with countably many balls of measure zero. Therefore $\mu(M_1) = 0$. Moreover, since $\nu_s \perp \mu$ there exist two disjoint sets E_{ν_s}, $E_\mu \in \mathfrak{M}$ such that $\mathbb{R}^N = E_{\nu_s} \cup E_\mu$ and $\mu(E_{\nu_s}) = \nu_s(E_\mu) = 0$. It suffices to prove (1.112) for μ a.e. $x \in \mathbb{R}^N \setminus (M_1 \cup E_{\nu_s})$.

Fix $\gamma \geq 1$ and for any $x \in \mathbb{R}^N \setminus (M_1 \cup E_{\nu_s})$ and $r > 0$ let $\mathcal{F}_{x,r}$ be the family of all closed sets $F \subset \mathbb{R}^N$ such that

$$\overline{B(x,r)} \subset F \subset \overline{B(x,\gamma r)} \tag{1.114}$$

and F is star-shaped with respect to $\overline{B(x,r)}$ and define

$$u_\gamma^-(x) := \liminf_{r \to 0^+} \inf_{F \in \mathcal{F}_{x,r}} \frac{\nu(F)}{\mu(F)}, \quad u_\gamma^+(x) := \limsup_{r \to 0^+} \sup_{F \in \mathcal{F}_{x,r}} \frac{\nu(F)}{\mu(F)}. \tag{1.115}$$

Then $u_\gamma^- \leq u_\gamma^+$. We claim that

$$u_\gamma^- = u_\gamma^+ = \frac{d\nu_{ac}}{d\mu} \tag{1.116}$$

μ a.e. in $\mathbb{R}^N \setminus (M_1 \cup E_{\nu_s})$. Given $t > 0$, let E_t be a bounded subset of

$$\left\{ x \in \mathbb{R}^N \setminus (M_1 \cup E_{\nu_s}) : \frac{d\nu_{ac}}{d\mu}(x) < t < u_\gamma^+(x) \right\}.$$

Extend μ and ν as Radon outer measures μ^* and ν^*, respectively (see Proposition 1.63), and let A be any open set that contains E_t. Set

$$\mathcal{F} := \left\{ F \in \mathcal{F}_{x,r} : x \in E_t, \ \overline{B(x,\gamma r)} \subset A, \ t\mu(F) < \nu(F) \right\}.$$

We claim that $\mathcal{F}$ is a fine Morse cover for E_t. Indeed, if $x \in E_t$, then, since $u_\gamma^+(x) > t$, we may find a sequence $r_n \to 0^+$ such that

$$\sup_{F \in \mathcal{F}_{x,r_n}} \frac{\nu(F)}{\mu(F)} > t,$$

and in turn, there exists a sequence of sets $F_n \in \mathcal{F}_{x,r_n}$ such that $t\mu(F_n) < \nu(F_n)$. It follows from (1.114) that $\operatorname{diam} F_n \to 0$ as $r_n \to 0^+$, and so we can assume that $\overline{B(x,\gamma r_n)} \subset A$ for all n. Thus (1.90) holds, and so $\mathcal{F}$ is a fine Morse cover for E_t.

By the Morse measure covering theorem there exists a countable family $\mathcal{F}_0 \subset \mathcal{F}$ of pairwise disjoint subsets such that

$$\mu^* \left(E_t \setminus \bigcup_{F \in \mathcal{F}_0} F \right) = 0.$$

Therefore, using the definition of $\mathcal{F}$,

$$t\mu^*(E_t) \leq \sum_{F \in \mathcal{F}_0} t\mu(F) \leq \sum_{F \in \mathcal{F}_0} \nu(F) \leq \nu(A).$$

By the outer regularity of ν^* it follows that $t\mu^*(E_t) \leq \nu^*(E_t)$. We claim that this implies that $\mu^*(E_t) = 0$. Indeed, if this is not the case then let $G_t \supset E_t$ be a Borel set such that (see Proposition 1.63 and Remark 1.51)

$$0 < \mu^* (E_t) = \mu (G_t) < \infty.$$

Since $\frac{d\nu_{ac}}{d\mu}$ is a measurable function, without loss of generality we may assume that

$$E_t \subset G_t \subset \left\{ x \in \mathbb{R}^N \setminus (M_1 \cup E_{\nu_s}) : \frac{d\nu_{ac}}{d\mu} (x) < t \right\}.$$

Hence, using (1.113) and the fact that $\nu_s (\mathbb{R}^N \setminus E_{\nu_s}) = 0$, we have

$$\nu^* (E_t) \le \nu (G_t) = \int_{G_t} \frac{d\nu_{ac}}{d\mu} \, d\mu < t\mu (G_t) = t\mu^* (E_t) \le \nu^* (E_t),$$

which is a contradiction. Therefore $\mu^* (E_t) = \mu (G_t) = 0$. Since

$$\left\{ x \in \mathbb{R}^N \setminus (M_1 \cup E_{\nu_s}) : \frac{d\nu_{ac}}{d\mu} (x) < u_\gamma^+ (x) \right\}$$

$$= \bigcup_{t \in \mathbb{Q}} \bigcup_{n \in \mathbb{N}} \left\{ x \in B (0, n) \setminus (M_1 \cup E_{\nu_s}) : \frac{d\nu_{ac}}{d\mu} (x) < t < u_\gamma^+ (x) \right\},$$

using the subadditivity of μ^*, we conclude that

$$\mu^* \left(\left\{ x \in \mathbb{R}^N \setminus (M_1 \cup E_{\nu_s}) : \frac{d\nu_{ac}}{d\mu} (x) < u_\gamma^+ (x) \right\} \right) = 0.$$

Similarly, one can show that

$$\mu^* \left(\left\{ x \in \mathbb{R}^N \setminus (M_1 \cup E_{\nu_s}) : \frac{d\nu_{ac}}{d\mu} (x) > u_\gamma^- (x) \right\} \right) = 0.$$

In view of Proposition 1.63 there exists a Borel set B_γ such that

$$B_\gamma \supset \left\{ x \in \mathbb{R}^N : \frac{d\nu_{ac}}{d\mu} < u_\gamma^+ \text{ or } \frac{d\nu_{ac}}{d\mu} > u_\gamma^- \right\} \cup M_1 \cup E_{\nu_s}$$

and $\mu (B_\gamma) = 0$. Then for all $x \in \mathbb{R}^N \setminus B_\gamma$ we have

$$\frac{d\nu_{ac}}{d\mu} (x) \le u_\gamma^- (x) \le u_\gamma^+ (x) \le \frac{d\nu_{ac}}{d\mu} (x),$$

which proves (1.116).

Next we observe that for any compact set K,

$$\infty > \nu (K) \ge \nu_{ac} (K) = \int_K \frac{d\nu_{ac}}{d\mu} \, d\mu,$$

and so $\frac{d\nu_{ac}}{d\mu} \in L^1_{loc} (\mathbb{R}^N, \mu)$. In particular, $\frac{d\nu_{ac}}{d\mu} (x) \in \mathbb{R}$ for μ a.e. $x \in \mathbb{R}^N$. Hence, without loss of generality, we may assume that the set B_γ also contains the set

$$\left\{ x \in \mathbb{R}^N : \left| \frac{d\nu_{ac}}{d\mu}(x) \right| = \infty \right\}.$$

Choose a sequence $\gamma_k \nearrow \infty$ and let

$$M := \bigcup_{k=1}^{\infty} B_{\gamma_k}.$$

If C is any bounded, convex closed set containing the origin in its interior, choose k so large that

$$\overline{B(0, 1/\gamma_k)} \subset C \subset \overline{B(x, \gamma_k)}.$$

By (1.115) and (1.116), for all $x \in \mathbb{R}^N \setminus M$ we have

$$\frac{d\nu_{ac}}{d\mu}(x) = u^-_{\gamma_k}(x) \le \liminf_{r \to 0^+} \frac{\nu(x + rC)}{\mu(x + rC)}$$

$$\le \limsup_{r \to 0^+} \frac{\nu(x + rC)}{\mu(x + rC)} \le u^+_{\gamma_k}(x) = \frac{d\nu_{ac}}{d\mu}(x),$$

and so (1.112) is satisfied.

Applying (1.112) also to ν_{ac}, we obtain that for μ a.e. $x \in \mathbb{R}^N$,

$$\frac{d\nu_{ac}}{d\mu}(x) = \lim_{r \to 0^+} \frac{\nu_{ac}(x + rC)}{\mu(x + rC)} = \lim_{r \to 0^+} \frac{\nu(x + rC)}{\mu(x + rC)} \in \mathbb{R},$$

which yields

$$\lim_{r \to 0^+} \frac{\nu_s(x + rC)}{\mu(x + rC)} = 0.$$

This concludes the proof.

Remark 1.154. (i) The set M is independent of the choice of C. Moreover, it follows from the proof that a much stronger statement holds, namely if $x \in \mathbb{R}^N \setminus M$, then for any sequence $r_n \to 0^+$ and for every family of closed sets $F_n \subset \mathbb{R}^N$ such that

$$\overline{B(x, r_n)} \subset F_n \subset \overline{B(x, \gamma r_n)}$$

and F_n is star-shaped with respect to $\overline{B(x, r_n)}$ we have that

$$\frac{d\nu_{ac}}{d\mu}(x) = \lim_{n \to \infty} \frac{\nu(F_n)}{\mu(F_n)}.$$

(ii) The Besicovitch derivation theorem holds also for bounded, convex open sets C containing the origin. Indeed, it suffices to replace $\mathcal{F}_{x,r}$ with the family $\mathcal{G}_{x,r}$ of all open sets $D \subset \mathbb{R}^N$ such that

$$B(x, r) \subset D \subset B(x, \gamma r)$$

and D is star-shaped with respect to $B(x,r)$. By Remark 1.152, for any open set $D \in \mathcal{G}_{x,r}$ such that $t\mu(D) < \nu(D)$ one can find a closed set $F \in \mathcal{F}_{x,r}$ such that $F \subset D$ and $t\mu(F) < \nu(F)$. Hence $\mathcal{F}$ is still a fine Morse cover for E_t.

It is possible to obtain the following local version of the Besicovitch derivation theorem.

Theorem 1.155. *Let $E \subset \mathbb{R}^N$ be a Borel set and let $\mu, \nu : \mathcal{B}(E) \to [0,\infty]$ be two Radon measures. Then*

$$\nu = \nu_{ac} + \nu_s, \quad \nu_{ac} \ll \mu, \quad \nu_s \perp \mu,$$

and there exists $M \in \mathcal{B}(E)$, with $\mu(M) = 0$, such that for any $x \in E \setminus M$,

$$\frac{d\nu_{ac}}{d\mu}(x) = \lim_{r \to 0^+} \frac{\nu((x+rC) \cap E)}{\mu((x+rC) \cap E)} \in \mathbb{R} \tag{1.117}$$

and

$$\lim_{r \to 0^+} \frac{\nu_s((x+rC) \cap E)}{\mu((x+rC) \cap E)} = 0, \tag{1.118}$$

where C is any bounded, convex closed set containing the origin in its interior.

Proof. Consider the extensions $\bar{\mu}, \bar{\nu} : \mathcal{B}(\mathbb{R}^N) \to [0,\infty]$ defined by $\bar{\mu}(F) := \mu(F \cap E)$ and $\bar{\nu}(F) := \nu(F \cap E)$ for every $F \in \mathcal{B}(\mathbb{R}^N)$. Note that $\bar{\mu}, \bar{\nu}$ are well-defined, since by Remark 1.3,

$$\{F \cap E : F \in \mathcal{B}(\mathbb{R}^N)\} = \mathcal{B}(E).$$

Moreover, if $K \subset \mathbb{R}^N$ is compact then $K \cap E$ is a compact set of E for the induced topology, and so $\bar{\mu}(K)$ and $\bar{\nu}(K)$ are finite. By Proposition 1.60 we deduce that $\bar{\mu}$ and $\bar{\nu}$ are Radon measures.

Applying the Besicovitch derivation theorem to the extensions, we have

$$\frac{d\bar{\nu}_{ac}}{d\bar{\mu}}(x) = \lim_{r \to 0^+} \frac{\bar{\nu}(x+rC)}{\bar{\mu}(x+rC)} = \lim_{r \to 0^+} \frac{\nu((x+rC) \cap E)}{\mu((x+rC) \cap E)}$$

for all $x \in \mathbb{R}^N \setminus M_0$ and for some Borel set M_0 with $\bar{\mu}(M_0) = 0$. Setting $M := M_0 \cap E$, by (1.1) we have $M \in \mathcal{B}(E)$.

In view of the Lebesgue and Radon–Nikodym theorems we may write

$$\bar{\nu} = \bar{\nu}_{ac} + \bar{\nu}_s,$$

where $\bar{\nu}_{ac}, \bar{\nu}_s : \mathcal{B}(\mathbb{R}^N) \to [0,\infty]$, with $\bar{\nu}_{ac} \ll \bar{\mu}$, $\bar{\nu}_s \perp \bar{\mu}$, and similarly

$$\nu = \nu_{ac} + \nu_s,$$

where $\nu_{ac}, \nu_s : \mathcal{B}(E) \to [0,\infty]$, with $\nu_{ac} \ll \mu$, $\nu_s \perp \mu$.

Since $\overline{\nu}_s \lfloor E : \mathcal{B}(E) \to [0, \infty]$ and μ are mutually singular, and $\overline{\nu}_{ac} \lfloor E : \mathcal{B}(E) \to [0, \infty]$ is absolutely continuous with respect to μ, due to the uniqueness of the decomposition we conclude that $\overline{\nu}_s \lfloor E = \nu_s$ and $\overline{\nu}_{ac} \lfloor E = \nu_{ac}$ on $\mathcal{B}(E)$. In particular,

$$\nu_{ac}(F) = \int_F \frac{d\,(\overline{\nu}_{ac})\,\lfloor E}{d\overline{\mu}}\,d\mu = \int_F \frac{d\overline{\nu}_{ac}}{d\overline{\mu}}\,d\mu$$

for every $F \in \mathcal{B}(E)$, and so $\frac{d\nu_{ac}}{d\mu}(x) = \frac{d\overline{\nu}_{ac}}{d\overline{\mu}}(x)$ for μ a.e. $x \in E$.

Equality (1.112) motivates the following definition:

Definition 1.156. *Let $E \subset \mathbb{R}^N$ and let μ, $\nu : \mathcal{B}(E) \to [0, \infty]$ be two Radon measures. We define the* Radon–Nikodym derivative *of μ with respect to ν as*

$$\frac{d\nu}{d\mu}(x) = \lim_{r \to 0^+} \frac{\nu\,((x + r\,C) \cap E)}{\mu\,((x + r\,C) \cap E)}$$

whenever the limit exists, is finite, and is independent of the choice of the bounded, convex closed set containing the origin in its interior C.

Remark 1.157. In view of Theorem 1.155, if $\nu \ll \mu$, then the notion introduced in the previous definition coincides μ a.e. with the Radon–Nikodym derivative of μ with respect to ν as in the Radon–Nikodym theorem.

Also, by (1.117) and (1.118), $\frac{d\nu}{d\mu}(x)$, $\frac{d\nu_{ac}}{d\mu}(x)$, $\frac{d\nu_s}{d\mu}(x)$ exist and

$$\frac{d\nu}{d\mu}(x) = \frac{d\nu_{ac}}{d\mu}(x), \quad \frac{d\nu_s}{d\mu}(x) = 0$$

for μ a.e. $x \in E$. In particular,

$$\nu = \frac{d\nu}{d\mu}\mu + \nu_s.$$

The next theorem leads to the definition of Lebesgue points, which are a central concept for the study of L^p spaces in the next chapter.

Theorem 1.158 (Lebesgue differentiation theorem). *Let*

$$\mu : \mathfrak{M}\left(\mathbb{R}^N\right) \to [0, \infty]$$

be a Radon measure, and let $u : \mathbb{R}^N \to [-\infty, \infty]$ be a locally integrable function. Then there exists a Borel set $M \subset \mathbb{R}^N$, with $\mu(M) = 0$, such that

$$\mathbb{R}^N \setminus M \subset \{x \in \mathbb{R}^N : u(x) \in \mathbb{R}\}$$

and for any $x \in \mathbb{R}^N \setminus M$,

$$\lim_{r \to 0^+} \frac{1}{\mu\,(B\,(x, r))} \int_{B(x,r)} u\,(y)\,d\mu\,(y) = u\,(x).$$

Proof. Define

$$\nu(E) := \int_E u^+ \, d\mu, \quad E \in \mathfrak{M}\left(\mathbb{R}^N\right).$$

Since u is locally integrable it follows that ν is a Radon measure absolutely continuous with respect to μ. Since every Radon measure in $\mathbb{R}^N$ is σ-finite, we can apply the Radon–Nikodym theorem to conclude that

$$\frac{d\nu}{d\mu}(x) = u^+(x) \in \mathbb{R}$$

for μ a.e. $x \in \mathbb{R}^N$. In turn, by the Besicovitch differentiation theorem,

$$\lim_{r \to 0^+} \frac{1}{\mu(B(x,r))} \int_{B(x,r)} u^+(y) \, d\mu(y) = u^+(x) \in \mathbb{R}$$

for μ a.e. $x \in \mathbb{R}^N$, and similarly

$$\lim_{r \to 0^+} \frac{1}{\mu(B(x,r))} \int_{B(x,r)} u^-(y) \, d\mu(y) = u^-(x) \in \mathbb{R}$$

for μ a.e. $x \in \mathbb{R}^N$. The conclusion now follows.

By enlarging the "bad" set M we can strengthen the conclusion of the previous theorem.

Corollary 1.159. *Let* $\mu : \mathfrak{M}\left(\mathbb{R}^N\right) \to [0, \infty]$ *be a Radon measure and let* $u : \mathbb{R}^N \to [-\infty, \infty]$ *be a locally integrable function. Then there exists a Borel set* $M \subset \mathbb{R}^N$, *with* $\mu(M) = 0$, *such that*

$$\mathbb{R}^N \setminus M \subset \left\{ x \in \mathbb{R}^N : u(x) \in \mathbb{R} \right\},$$

and for any $x \in \mathbb{R}^N \setminus M$,

$$\lim_{r \to 0^+} \frac{1}{\mu(B(x,r))} \int_{B(x,r)} |u(y) - u(x)| \, d\mu(y) = 0. \tag{1.119}$$

Proof. Write $\mathbb{Q} = \{r_n\}$. Applying the previous theorem to the function $v_n(x) := |u(x) - r_n|$, we may find a Borel set $M_n \subset \mathbb{R}^N$, with $\mu(M_n) = 0$, such that

$$\lim_{r \to 0^+} \frac{1}{\mu(B(x,r))} \int_{B(x,r)} |u(y) - r_n| \, d\mu(y) = |u(x) - r_n| \tag{1.120}$$

for all $x \in \mathbb{R}^N \setminus M_n$. Since u is locally integrable and μ is σ-finite, the measurable set

$$M_0 := \left\{ x \in \mathbb{R}^N : |u(x)| = \infty \right\}$$

has zero μ measure. Set

$$M := \bigcup_{n=0}^{\infty} M_n.$$

If $x \in \mathbb{R}^N \setminus M$ then for every $n \in \mathbb{N}$,

$$\limsup_{r \to 0^+} \frac{1}{\mu(B(x,r))} \int_{B(x,r)} |u(y) - u(x)| \, d\mu(y)$$

$$= \limsup_{r \to 0^+} \frac{1}{\mu(B(x,r))} \int_{B(x,r)} |u(y) + r_n - r_n - u(x)| \, d\mu(y)$$

$$\leq \limsup_{r \to 0^+} \frac{1}{\mu(B(x,r))} \int_{B(x,r)} |u(y) - r_n| \, d\mu(y) + |u(x) - r_n|$$

$$= 2 |u(x) - r_n|,$$

where we have used (1.120). Since $u(x) \in \mathbb{R}$, it now suffices to select a subsequence of $\{r_n\}$ converging to $u(x)$.

Remark 1.160. It actually follows from Remark 1.154 that if $x \in \mathbb{R}^N \setminus M$, then for any sequence $r_n \to 0^+$ and for every family of closed sets $F_n \subset \mathbb{R}^N$ such that

$$\overline{B(x, r_n)} \subset F_n \subset \overline{B(x, \gamma r_n)}$$

and F_n is star-shaped with respect to $\overline{B(x, r_n)}$, we have that

$$\lim_{n \to \infty} \frac{1}{\mu(F_n)} \int_{F_n} |u(y) - u(x)| \, d\mu(y) = 0.$$

Moreover, if $x \in \mathbb{R}^N \setminus M$, then for any family of Borel subsets $\{E_r\}_{r>0}$ such that $E_r \subset B(x, r)$ and

$$\mu(E_r) > \alpha \mu(B(x,r))$$

for some constant $\alpha > 0$ independent of $r > 0$, we have that

$$\limsup_{r \to 0^+} \frac{1}{\mu(E_r)} \int_{E_r} |u(y) - u(x)| \, dy$$

$$\leq \limsup_{r \to 0^+} \frac{1}{\mu(E_r)} \int_{B(x,r)} |u(y) - u(x)| \, dy$$

$$\leq \frac{1}{\alpha} \lim_{r \to 0^+} \frac{1}{\mu(B(x,r))} \int_{B(x,r)} |u(y) - u(x)| \, d\mu(y) = 0.$$

Note that the sets E_r need not contain x.

Any point $x \in \mathbb{R}^N$ for which (1.119) holds is called a *Lebesgue point* of u. Given a Radon measure $\mu : \mathfrak{M}(\mathbb{R}^N) \to [0, \infty]$ and a measurable set $E \subset \mathbb{R}^N$, by applying the Lebesgue differentiation theorem to χ_E we obtain that

$$\lim_{r \to 0^+} \frac{\mu(B(x,r) \cap E)}{\mu(B(x,r))} = \chi_E(x)$$

for μ a.e. $x \in \mathbb{R}^N$. A point $x \in E$ for which the previous limit is one is called a *point of density one for E*. More generally, for any $t \in [0,1]$ a point $x \in \mathbb{R}^N$ such that

$$\lim_{r \to 0^+} \frac{\mu(B(x,r) \cap E)}{\mu(B(x,r))} = t$$

is called a *point of density t for E*.

Remark 1.161. If $\mu : \mathfrak{M}(\mathbb{R}^N) \to [0,\infty]$ is a Radon measure such that for μ a.e. $x \in \mathbb{R}^N$,

$$\mu(\{x\}) = 0, \tag{1.121}$$

then μ is nonatomic. In particular, the Lebesgue measure $\mathcal{L}^N$ is nonatomic.

Indeed, if $E \in \mathfrak{M}(\mathbb{R}^N)$ has positive measure, then let x_0 be a point of density one of E for which (1.121) holds. By Proposition 1.7(i),

$$\mu(B(x_0,r)) \to \mu(\{x_0\}) = 0$$

as $r \to 0^+$, and so $\mu(B(x_0,r)) < \mu(E)$ for all $r > 0$ sufficiently small. On the other hand, by (1.121), taking $r > 0$ even smaller if necessary, we have

$$0 < \frac{1}{2}\mu(B(x_0,r)) \leq \mu(B(x_0,r) \cap E) \leq \mu(B(x_0,r)),$$

where we have used the fact that x_0 is a point of density one of E. Hence for all $r > 0$ sufficiently small the set $F := B(x_0,r) \cap E$ satisfies the condition

$$0 < \mu(F) < \mu(E).$$

For measurable functions that are not necessarily integrable, the analogous concept of Lebesgue points is given by points of approximate continuity.

Definition 1.162. *Let $\mu : \mathfrak{M}(\mathbb{R}^N) \to [0,\infty]$ be a Radon measure, and let $u : \mathbb{R}^N \to [-\infty,\infty]$ be a measurable function. The function u is said to be approximately continuous at $x_0 \in \mathbb{R}^N$ if*

$$\lim_{r \to 0^+} \frac{\mu(B(x_0,r) \cap u^{-1}(A))}{\mu(B(x_0,r))} = 1 \tag{1.122}$$

for every open set A in the extended real line $[-\infty,\infty]$ that contains $u(x_0)$. The point x_0 is called a point of approximate continuity *for u.*

Note that if $u(x_0) \in \mathbb{R}$ then (1.122) is equivalent to requiring that for every $\varepsilon > 0$,

$$\lim_{r \to 0^+} \frac{\mu(B(x_0,r) \cap \{x \in \mathbb{R}^N : |u(x) - u(x_0)| \geq \varepsilon\})}{\mu(B(x_0,r))} = 0.$$

We now study the relation between Lebesgue points and points of approximate continuity.

Proposition 1.163. *Let* $\mu : \mathfrak{M}\left(\mathbb{R}^N\right) \to [0,\infty]$ *be a Radon measure, and let* $u : \mathbb{R}^N \to [-\infty,\infty]$ *be a locally integrable function. If* $x_0 \in \mathbb{R}^N$ *is a Lebesgue point for* u, *then* u *is approximately continuous at* x_0. *Conversely, if* $u : \mathbb{R}^N \to \mathbb{R}$ *is bounded on compact sets, measurable, and is approximately continuous at* x_0, *then* x_0 *is a Lebesgue point for* u.

Proof. Assume that $x_0 \in \mathbb{R}^N$ is a Lebesgue point for u. Then

$$\frac{\mu\left(\{x \in B\left(x_0,r\right) : |u\left(x\right) - u\left(x_0\right)| \geq \varepsilon\}\right)}{\mu\left(B\left(x_0,r\right)\right)}$$

$$\leq \frac{1}{\varepsilon} \frac{1}{\mu\left(B\left(x_0,r\right)\right)} \int_{B(x_0,r)} |u\left(x\right) - u\left(x_0\right)| \, d\mu \to 0$$

as $r \to 0^+$. Hence u is approximately continuous at x_0.

Conversely assume that $u : \mathbb{R}^N \to \mathbb{R}$ is locally bounded and is approximately continuous at x_0. Then for every $\varepsilon > 0$,

$$\frac{1}{\mu\left(B\left(x_0,r\right)\right)} \int_{B(x_0,r)} |u\left(x\right) - u\left(x_0\right)| \, d\mu$$

$$\leq \frac{1}{\mu\left(B\left(x_0,r\right)\right)} \int_{B(x_0,r) \cap \{ |u(y)-u(x_0)|\geq\varepsilon\}} |u\left(x\right) - u\left(x_0\right)| \, d\mu$$

$$+ \frac{1}{\mu\left(B\left(x_0,r\right)\right)} \int_{B(x_0,r) \cap \{ |u(y)-u(x_0)|\leq\varepsilon\}} |u\left(x\right) - u\left(x_0\right)| \, d\mu$$

$$\leq 2 \left(\sup_{B(x_0,1)} |u| \right) \frac{\mu\left(\{x \in B\left(x_0,r\right) : |u\left(x\right) - u\left(x_0\right)| \geq \varepsilon\}\right)}{\mu\left(B\left(x_0,r\right)\right)} + \varepsilon.$$

Letting $r \to 0^+$ we have

$$\limsup_{r\to 0^+} \frac{1}{\mu\left(B\left(x_0,r\right)\right)} \int_{B(x_0,r)} |u\left(x\right) - u\left(x_0\right)| \, d\mu \leq \varepsilon,$$

which, in view of the arbitrariness of $\varepsilon > 0$, shows that x_0 is a Lebesgue point for u.

Exercise 1.164. The converse of the previous proposition does not hold in general without the assumption of local boundedness of u. Indeed, let μ be the Lebesgue measure $\mathcal{L}^1$ and consider the function

$$u\left(x\right) := \sum_{n=1}^{\infty} 2^n \chi_{\left(\frac{1}{2^n}, \frac{1}{2^n}\left(1+\frac{1}{n^2}\right)\right)}\left(x\right), \quad x \in \mathbb{R}.$$

Prove that u is integrable, approximately continuous at $x = 0$, but 0 is not a Lebesgue point for u.

Using Lusin's theorem it is possible to give the following characterization of approximate continuity.

Theorem 1.165. *Let* $\mu : \mathfrak{M}\left(\mathbb{R}^N\right) \to [0,\infty]$ *be a Radon measure. A function* $u : \mathbb{R}^N \to [-\infty,\infty]$ *is measurable in the sense of Definition 1.72 if and only if* u *is approximately continuous at* μ *a.e.* $x \in \mathbb{R}^N$.

1.3 Spaces of Measures

In this section we introduce the notion of signed measures and we study several spaces of signed measures that arise as duals of various spaces of bounded and continuous functions.

1.3.1 Signed Measures

Definition 1.166. *Let $(X, \mathfrak{M})$ be a measurable space. A* signed measure *is a function $\lambda : \mathfrak{M} \to [-\infty, \infty]$ such that*

(i) $\lambda(\emptyset) = 0$;
(ii) λ takes at most one of the two values ∞ and $-\infty$, that is, either $\lambda : \mathfrak{M} \to (-\infty, \infty]$ or $\lambda : \mathfrak{M} \to [-\infty, \infty)$;
(iii) for every countable collection $\{E_i\} \subset \mathfrak{M}$ of pairwise disjoint sets we have

$$\lambda \left(\bigcup_{n=1}^{\infty} E_n \right) = \sum_{n=1}^{\infty} \lambda(E_n).$$

Remark 1.167. We call the attention to the fact that in this text signed measures are not necessarily finite, as is often considered in the literature. Indeed, the way to read (iii) is that whenever $\{E_i\} \subset \mathfrak{M}$ is a countable collection of pairwise disjoint sets, then the sequence of partial sums $\left\{ \sum_{n=1}^{l} \lambda(E_n) \right\}_l$ is convergent in $[-\infty, \infty]$ to $\lambda(\bigcup_{n=1}^{\infty} E_n)$.

The following result is the analogue of Proposition 1.9 for signed measures.

Proposition 1.168. *Let $(X, \mathfrak{M})$ be a measurable space. A set function $\lambda : \mathfrak{M} \to [-\infty, \infty]$ is a signed measure if and only it satisfies (ii) of Definition 1.166, is finitely additive, and for every increasing sequence $\{E_n\} \subset \mathfrak{M}$,*

$$\lambda \left(\bigcup_{n=1}^{\infty} E_n \right) = \lim_{n \to \infty} \lambda(E_n). \tag{1.123}$$

Proof. One implication follows from the previous remark and (iii). Suppose now that λ is finitely additive and that (1.123) holds. Let $\{F_n\} \subset \mathfrak{M}$ be a sequence of mutually disjoint measurable sets, and define

$$E_n := \bigcup_{k=1}^{n} F_k.$$

Then by (1.123) we have

$$\lambda \left(\bigcup_{k=1}^{\infty} F_k \right) = \lim_{n \to \infty} \lambda(E_n) = \lim_{n \to \infty} \sum_{k=1}^{n} \lambda(F_k) = \sum_{k=1}^{\infty} \lambda(F_k),$$

and with this we have shown that λ is a signed measure.

As we will see below, in order to study signed measures it is convenient to write them as differences of (positive) measures.

Definition 1.169. *Let $(X, \mathfrak{M})$ be a measurable space and let $\lambda : \mathfrak{M} \to [-\infty, \infty]$ be a signed measure. A set $E \in \mathfrak{M}$ is said to be* positive *(respectively* negative*) if $\lambda(F) \geq 0$ (respectively $\lambda(F) \leq 0$) for all $F \subset E$ with $F \in \mathfrak{M}$.*

Note that we may have sets with positive measure that are not positive.

Example 1.170. Let $X = \mathbb{R}$, and let $u : \mathbb{R} \to \mathbb{R}$ be an odd function, integrable with respect to the Lebesgue measure, and such that $u(x) > 0$ for $x > 0$. Then

$$\lambda(E) := \int_E u(x) \, dx, \quad E \in \mathcal{B}(\mathbb{R}),$$

is a signed measure and any set of the form $[-a, b]$, $0 < a < b$, has positive λ measure without being a positive set.

However, it is possible to prove the following result.

Proposition 1.171. *Let $(X, \mathfrak{M})$ be a measurable space and let $\lambda : \mathfrak{M} \to [-\infty, \infty]$ be a signed measure. Let $E \in \mathfrak{M}$ be such that $0 < \lambda(E) < \infty$. Then there exists a positive measurable subset $F \subset E$ with $\lambda(F) > 0$.*

Proof. **Step 1:** We begin by showing that if $E \in \mathfrak{M}$ is such that $|\lambda(E)| < \infty$, then for any $F \subset E$, with $F \in \mathfrak{M}$, we have $|\lambda(F)| < \infty$. Without loss of generality we may assume that $\lambda : \mathfrak{M} \to [-\infty, \infty)$ (the case $\lambda : \mathfrak{M} \to (-\infty, \infty]$ being analogous). Let $F \subset E$, with $F \in \mathfrak{M}$. If $\lambda(F) \geq 0$ then there is nothing to prove. Thus assume that $\lambda(F) < 0$. Then

$$0 < -\lambda(F) = \lambda(E \setminus F) - \lambda(E) < \infty,$$

where we have used that facts that $|\lambda(E)| < \infty$ and that $\lambda : \mathfrak{M} \to [-\infty, \infty)$.

Step 2: Let $E \in \mathfrak{M}$ be such that $0 < \lambda(E) < \infty$. By the previous step, $\lambda : \mathfrak{M} \lfloor E \to \mathbb{R}$ is finite. For every $F \in \mathfrak{M} \lfloor E$ define

$$\lambda_+(F) := \sup \{\lambda(F') : F' \subset F, F' \in \mathfrak{M}\}, \quad (1.124)$$
$$\lambda_-(F) := -\inf \{\lambda(F') : F' \subset F, F' \in \mathfrak{M}\}.$$

Then, exactly as in the proof of Lemma 1.103 (with $\nu - \mu$ replaced by λ), we have that $\lambda_+ : \mathfrak{M} \lfloor E \to [0, \infty)$ and $\lambda_- : \mathfrak{M} \lfloor E \to [0, \infty)$ are finite measures, and for every $F \in \mathfrak{M} \lfloor E$ we have

$$\lambda_+(F) = \sup \{\lambda(F') : F' \subset F, F' \in \mathfrak{M}, \lambda_-(F') = 0\} \quad (1.125)$$
$$= \sup \{\lambda(F') : F' \subset F, F' \in \mathfrak{M}, F' \text{ positive}\}.$$

Since $0 < \lambda(E) < \infty$ it follows that $\lambda_+(E) > 0$, and so we may find a positive measurable subset $F \subset E$ with $\lambda(F) > 0$.

As a consequence of the previous proposition we have the following important theorem.

Theorem 1.172 (Hahn decomposition theorem). *Let* $(X, \mathfrak{M})$ *be a measurable space and let* $\lambda : \mathfrak{M} \to [-\infty, \infty]$ *be a signed measure. Then* X *can be decomposed as* $X = X^+ \cup X^-$, *where* X^+ *is positive and* X^- *is negative.*

Proof. Without loss of generality we may assume that $\lambda : \mathfrak{M} \to [-\infty, \infty)$ (the case $\lambda : \mathfrak{M} \to (-\infty, \infty]$ being analogous). Let

$$\lambda_+(X) = \sup \{\lambda(E) : E \in \mathfrak{M}, E \text{ positive}\}.$$

Find a sequence of increasing positive sets E_n, $E_n \in \mathfrak{M}$, such that

$$\lim_{n \to \infty} \lambda(E_n) = \lambda_+(X).$$

Define

$$X^+ := \bigcup_{n=1}^{\infty} E_n.$$

Then X^+ is positive and

$$\lambda(X^+) = \lim_{n \to \infty} \lambda(E_n) = \lambda_+(X).$$

We claim that $X^- := X \setminus X^+$ is negative. Indeed, if not, then there exists $F \subset X^-$, $F \in \mathfrak{M}$, such that

$$0 < \lambda(F) < \infty.$$

By the previous proposition there exists a positive measurable subset $F' \subset F$ with $\lambda(F') > 0$. But then

$$\lambda_+(X) = \lambda(X^+) < \lambda(X^+) + \lambda(X^+) = \lambda(X^+ \cup F') < \infty,$$

and since $X^- \cup F'$ is admissible in the definition of $\lambda_+(X)$ we arrive at a contradiction.

Example 1.173. Note that the Hahn decomposition may not be unique. As an example let $X = [-1, 1]$, let $\mathfrak{M} = \mathcal{B}([-1, 1])$, and let

$$\lambda(E) := \int_E x \, dx, \quad E \in \mathcal{B}([-1, 1]).$$

Then the pairs $([0, 1], [-1, 0))$ and $((0, 1], [-1, 0])$ are both Hahn decompositions of λ.

Remark 1.174. In general, if (X^+, X^-) and (X_1^+, X_1^-) are Hahn decompositions of a signed measure $\lambda : \mathfrak{M} \to [-\infty, \infty]$, then $\lambda(E) = 0$ for every $E \in \mathfrak{M}$ with

$$E \subset (X^+ \Delta X_1^+) \cup (X^- \Delta X_1^-).$$

Here Δ stands for the *symmetric difference* between two sets, i.e.,

$$F \Delta G := (F \setminus G) \cup (G \setminus F).$$

Let $(X, \mathfrak{M})$ be a measurable space and let $\lambda : \mathfrak{M} \to [-\infty, \infty]$ be a signed measure. By the Hahn decomposition theorem we may uniquely (in the sense of Remark 1.174) decompose X as $X = X^+ \cup X^-$, where X^+ is positive and X^- is negative. For $E \in \mathfrak{M}$ define

$$\lambda^+ (E) := \lambda \left(E \cap X^+ \right), \quad \lambda^- (E) := -\lambda \left(E \cap X^- \right).$$

Then λ^+, $\lambda^- : \mathfrak{M} \to [0, \infty]$ are measures, with at least one of them finite, and by construction, they are mutually singular. Moreover, for any $E \in \mathfrak{M}$ we have

$$\lambda (E) = \lambda^+ (E) - \lambda^- (E). \tag{1.126}$$

The measures λ^+, λ^- are called, respectively, the *upper* and *lower variation* of λ, while the measure

$$\|\lambda\| := \lambda^+ + \lambda^-$$

is called the *total variation* of λ . Note that both λ^+, λ^- are absolutely continuous with respect to $\|\lambda\|$.

Remark 1.175. Note that for any $E \in \mathfrak{M}$,

$$\begin{aligned}
\lambda^+ (E) &= \sup \left\{ \lambda (F) : F \subset E, F \in \mathfrak{M}, F \text{ positive} \right\} \\
&= \sup \left\{ \lambda (F) : F \subset E, F \in \mathfrak{M} \right\}
\end{aligned}$$

and

$$\begin{aligned}
\lambda^- (E) &= -\inf \left\{ \lambda (F) : F \subset E, F \in \mathfrak{M}, F \text{ negative} \right\} \\
&= -\inf \left\{ \lambda (F) : F \subset E, F \in \mathfrak{M} \right\}.
\end{aligned}$$

To see this, assume, without loss of generality, that $\lambda : \mathfrak{M} \to [-\infty, \infty)$. Then, since X^+ is positive, it follows that

$$\begin{aligned}
\lambda^+ (E) &= \lambda \left(E \cap X^+ \right) \\
&= \sup \left\{ \lambda (F) : F \subset E, F \in \mathfrak{M}, F \text{ positive} \right\} \\
&= \lambda_+ (E),
\end{aligned}$$

where λ_+ has been defined in (1.124) and we have used (1.125) (which holds, since λ does not take the value ∞). Similarly, since X^- is negative, it follows that

$$\begin{aligned}
\lambda^- (E) &= \lambda \left(E \cap X^- \right) \\
&= -\inf \left\{ \lambda (F) : F \subset E, F \in \mathfrak{M}, F \text{ negative} \right\}.
\end{aligned}$$

If $|\lambda (E)| < \infty$ then

$$\begin{aligned}
\lambda^- (E) &= \lambda^+ (E) - \lambda (E) = \lambda_- (E) - \lambda (E) \\
&= \sup \left\{ \lambda (F) - \lambda (E) : F \subset E, F \in \mathfrak{M} \right\} \\
&= \sup \left\{ -\lambda (E \setminus F) : F \subset E, F \in \mathfrak{M} \right\} \\
&= -\inf \left\{ \lambda (G) : G \subset E, F \in \mathfrak{M} \right\}.
\end{aligned}$$

If $\lambda(E) = -\infty$ then, since $\lambda^+(E) < \infty$, it follows that

$$\lambda^-(E) = \lambda^+(E) - \lambda(E) = \infty,$$

and so

$$\lambda^-(E) = -\inf\{\lambda(F) : F \subset E, F \in \mathfrak{M}\} = -\lambda(F) = \infty.$$

This completes the proof.

Definition 1.176. *Let $(X, \mathfrak{M})$ be a measurable space and let $\lambda : \mathfrak{M} \to [-\infty, \infty]$ be a signed measure. A set $E \in \mathfrak{M}$ has σ-finite measure λ if it has σ-finite measure $\|\lambda\|$; λ is said to be σ-finite if X has σ-finite measure λ.*

Proposition 1.177. *Let $(X, \mathfrak{M})$ be a measurable space and let $\lambda : \mathfrak{M} \to [-\infty, \infty]$ be a signed measure. Then for every $E \in \mathfrak{M}$ we have that*

$$\|\lambda\|(E) = \sup\left\{\sum_{n=1}^{\infty} |\lambda(E_n)| : \{E_n\} \subset \mathfrak{M} \text{ partition of } E\right\}.$$

It turns out that the decomposition (1.126) is unique.

Theorem 1.178 (Jordan decomposition theorem). *Let $(X, \mathfrak{M})$ be a measurable space and let $\lambda : \mathfrak{M} \to [-\infty, \infty]$ be a signed measure. Then there exists a unique pair (λ^+, λ^-) of mutually singular (nonnegative) measures, one of which is finite, such that $\lambda = \lambda^+ - \lambda^-$.*

Proof. It remains only to prove the uniqueness of the decomposition. We leave this as an exercise.

In view of the previous theorem, given a measurable space $(X, \mathfrak{M})$, a signed measure $\lambda : \mathfrak{M} \to [-\infty, \infty]$, a measurable function $u : X \to [-\infty, \infty]$, and a measurable set $E \in \mathfrak{M}$, if at least one of the two integrals $\int_E u \, d\lambda^+$ and $\int_E u \, d\lambda^-$ is finite then we define the *Lebesgue integral of u* over the measurable set E as

$$\int_E u \, d\lambda := \int_E u \, d\lambda^+ - \int_E u \, d\lambda^-.$$

If both $\int_E u \, d\lambda^+$ and $\int_E u \, d\lambda^-$ are finite, then u is said to be *Lebesgue integrable* over the measurable set E.

We now extend The Lebesgue decomposition theorem to signed measures.

Definition 1.179. *Let $(X, \mathfrak{M})$ be a measurable space and let $\lambda, \varsigma : \mathfrak{M} \to [-\infty, \infty]$ be two signed measures.*

(i) λ, ς are said to be mutually singular, *and we write $\lambda \perp \varsigma$, if $\|\lambda\|$ and $\|\varsigma\|$ are mutually singular.*

(ii) λ is said to be absolutely continuous with respect to ς, *and we write $\lambda \ll \varsigma$, if $\|\lambda\|$ is absolutely continuous with respect to $\|\varsigma\|$.*

If $(X, \mathfrak{M})$ is a measurable space, $\lambda : \mathfrak{M} \to [-\infty, \infty]$ is a signed measure, and $\mu : \mathfrak{M} \to [0, \infty]$ is a σ-finite (positive) measure, then by applying the Lebesgue decomposition theorem to λ^+ and μ (respectively to λ^- and μ) we can write

$$\lambda^+ = \left(\lambda^+\right)_{ac} + \left(\lambda^+\right)_s, \quad \lambda^- = \left(\lambda^-\right)_{ac} + \left(\lambda^-\right)_s,$$

where the measures $\left(\lambda^+\right)_{ac}$ and $\left(\lambda^+\right)_s$ are defined in (1.38) and (1.56), and $\left(\lambda^+\right)_{ac}$, $\left(\lambda^-\right)_{ac} \ll \mu$. Hence we can apply the Radon–Nikodym theorem to find two measurable functions u^+, $u^- : X \to [0, \infty]$ such that

$$\left(\lambda^+\right)_{ac}(E) = \int_E u^+ \, d\mu, \quad \left(\lambda^-\right)_{ac}(E) = \int_E u^- \, d\mu$$

for every $E \in \mathfrak{M}$. The functions u^+ and u^- are unique up to a set of μ measure zero

Since either λ^+ or λ^- is finite we may define

$$\lambda_{ac} := \left(\lambda^+\right)_{ac} - \left(\lambda^-\right)_{ac}, \quad \lambda_s := \left(\lambda^+\right)_s - \left(\lambda^-\right)_s, \quad u := u^+ - u^-.$$

Then λ_{ac} is a signed measure with $\lambda_{ac} \ll \mu$. Note that if λ is positive then so are λ_{ac} and λ_s.

By the Radon–Nikodym theorem, the Lebesgue decomposition theorem, and the Jordan decomposition theorem we have proved the following result.

Theorem 1.180 (Lebesgue decomposition theorem). *Let $(X, \mathfrak{M})$ be a measurable space, let $\lambda : \mathfrak{M} \to [-\infty, \infty]$ be a signed measure, and let $\mu : \mathfrak{M} \to [0, \infty]$ be a σ-finite (positive) measure. Then*

$$\lambda = \lambda_{ac} + \lambda_s$$

with $\lambda_{ac} \ll \mu$, and

$$\lambda_{ac}(E) = \int_E u \, d\mu$$

for all $E \in \mathfrak{M}$. Moreover, if λ is σ-finite then $\lambda_s \perp \mu$ and the decomposition is unique, that is, if

$$\lambda = \overline{\lambda}_{ac} + \overline{\lambda}_s,$$

for some signed measures $\overline{\lambda}_{ac}$, $\overline{\lambda}_s$, with $\overline{\lambda}_{ac} \ll \mu$ and $\overline{\lambda}_s \perp \mu$, then

$$\lambda_{ac} = \overline{\lambda}_{ac} \quad and \quad \lambda_s = \overline{\lambda}_s.$$

We call λ_{ac} and λ_s, respectively, the *absolutely continuous part* and the *singular part of λ* with respect to μ, and often we write

$$u = \frac{d\lambda_{ac}}{d\mu}.$$

A particular case of the Lebesgue decomposition theorem is that in which $\mu = \|\lambda\|$, and here $\lambda = \lambda_{ac}$, $\lambda_s = 0$.

Corollary 1.181. *Let $(X, \mathfrak{M})$ be a measurable space and let $\lambda : \mathfrak{M} \to [-\infty, \infty]$ be a σ-finite signed measure. Then there exists a measurable function $u : X \to [-\infty, \infty]$ such that $|u(x)| = 1$ for $\|\lambda\|$ a.e. $x \in X$ and*

$$u = \frac{d\lambda}{d\|\lambda\|}.$$

The previous equation is called the *polar decomposition* of λ.

Definition 1.182. *Let $(X, \mathfrak{M})$ be a measurable space and let $\lambda : \mathfrak{M} \to [-\infty, \infty]$ be a signed measure. If X is a topological space then λ is a signed Radon measure if $\|\lambda\| : \mathfrak{M} \to [0, \infty]$ is a Radon measure.*

If X is a topological space then $\mathcal{M}(X; \mathbb{R})$ is the space of all signed finite Radon measures $\lambda : \mathcal{B}(X) \to \mathbb{R}$ endowed with the total variation norm. It can be verified that $\mathcal{M}(X; \mathbb{R})$ is a Banach space.

In applications we will also consider vector-valued measures.

Definition 1.183. *Let $(X, \mathfrak{M})$ be a measurable space. A set function $\lambda = (\lambda_1, \ldots, \lambda_m) : \mathfrak{M} \to \mathbb{R}^m$ is a vectorial measure if each component $\lambda_i : \mathfrak{M} \to \mathbb{R}$ is a signed measure, $i = 1, \ldots, m$. The total variation of λ is defined by*

$$\|\lambda\| := \|\lambda_1\| + \ldots + \|\lambda_m\|.$$

If X is a topological space, then $\lambda : \mathfrak{M} \to \mathbb{R}^m$ is a vectorial Radon measure if each component $\lambda_i : \mathfrak{M} \to \mathbb{R}$ is a signed Radon measure, $i = 1, \ldots, m$, and $\mathcal{M}(X; \mathbb{R}^m)$ is the space of all vectorial Radon measures $\lambda : \mathcal{B}(X) \to \mathbb{R}^m$ endowed with the total variation norm.

Exercise 1.184. In some applications it is more convenient to define the total variation of $\lambda : \mathfrak{M} \to \mathbb{R}^m$ as

$$\|\lambda\|_1(E) := \sup \left\{ \sum_{n=1}^{\infty} |\lambda(E_n)| : \{E_n\} \subset \mathfrak{M} \text{ partition of } E \right\}, \quad E \in \mathfrak{M}.$$
$$(1.127)$$

Prove that $\|\lambda\|_1 : \mathfrak{M} \to [0, \infty)$ is a (positive) measure. Note that in view of Proposition 1.177, the definition of $\|\lambda\|_1$ is consistent with the case $m = 1$.

1.3.2 Signed Finitely Additive Measures

In this subsection we introduce some important spaces of finitely additive signed measures. Let X be a nonempty set and let $\mathfrak{M} \subset \mathcal{P}(X)$ be an algebra. Let $B(X, \mathfrak{M})$ denote the space of all bounded measurable functions $u : X \to \mathbb{R}$.

Exercise 1.185. Prove that $B(X, \mathfrak{M})$ is a Banach space endowed with the norm

$$\|u\|_\infty := \sup_{x \in X} |u(x)|.$$

By Theorem 1.74 it turns out that $B(X, \mathfrak{M})$ consists of all uniform limits of finite linear combinations of characteristic functions of sets in $\mathfrak{M}$.

Definition 1.186. *Let X be a nonempty set and let $\mathfrak{M} \subset \mathcal{P}(X)$ be an algebra. The space* ba$(X, \mathfrak{M})$ *of bounded finitely additive signed measures is composed of all set functions $\lambda : \mathfrak{M} \to \mathbb{R}$ such that*

(i) $\lambda(\emptyset) = 0$;
(ii) λ *is finitely additive, that is,*

$$\lambda(E_1 \cup E_2) = \lambda(E_1) + \lambda(E_2)$$

for all E_1, $E_2 \in \mathfrak{M}$ with $E_1 \cap E_2 = \emptyset$;
(iii) λ *is bounded, that is, its total variation norm*

$$\|\lambda\|(X) := \sup \left\{ \sum_{n=1}^{l} |\lambda(E_n)| : \{E_n\} \subset \mathfrak{M} \text{ finite partition of } X \right\}$$

is finite.

Exercise 1.187. Prove that ba$(X, \mathfrak{M})$ is a Banach space endowed with the total variation norm $\|\cdot\|(X)$.

Analogously to the Hahn decomposition theorem for signed measures, it is possible to prove that if $\lambda \in$ ba$(X, \mathfrak{M})$, then it may be decomposed as

$$\lambda = \lambda^+ - \lambda^-,$$

where λ^+, λ^- are (positive) finitely additive measures. Indeed, we have the following result.

Theorem 1.188. *Let X be a nonempty set, let $\mathfrak{M} \subset \mathcal{P}(X)$ be an algebra, and let $\lambda \in$ ba$(X, \mathfrak{M})$. Then*

$$\lambda^+(E) := \sup\{\lambda(F) : F \subset E, F \in \mathfrak{M}\}, \quad E \in \mathfrak{M},$$
$$\lambda^-(E) := -\inf\{\lambda(F) : F \subset E, F \in \mathfrak{M}\}, \quad E \in \mathfrak{M},$$

are (positive) finitely additive measures and $\lambda = \lambda^+ - \lambda^-$.

Proof. Let $E_1, E_2 \in \mathfrak{M}$, with $E_1 \cap E_2 = \emptyset$. Since λ is bounded, for $\varepsilon > 0$ there exist $F_i \subset E_i$, $F_i \in \mathfrak{M}$, such that

$$\lambda^+(E_i) \leq \lambda(F_i) + \varepsilon,$$

$i = 1, 2$. Hence

$$\lambda^+(E_1 \cup E_2) \geq \lambda(F_1 \cup F_2) = \lambda(F_1) + \lambda(F_2) \geq \lambda^+(E_1) + \lambda^+(E_2) - 2\varepsilon,$$

and by the arbitrariness of ε we deduce that

$$\lambda^+ (E_1 \cup E_2) \geq \lambda^+ (E_1) + \lambda^+ (E_2).$$

To prove the reverse inequality, for $\varepsilon > 0$ we may find a set $F \subset E_1 \cup E_2$, with $F \in \mathfrak{M}$, such that

$$\lambda^+ (E_1 \cup E_2) \leq \lambda (F) + \varepsilon.$$

Since $E_1 \cap E_2 = \emptyset$ it follows that

$$\lambda^+ (E_1 \cup E_2) \leq \lambda (F) + \varepsilon = \lambda (F \cap E_1) + \lambda (F \cap E_2) + \varepsilon$$
$$\leq \lambda^+ (E_1) + \lambda^+ (E_2) + \varepsilon,$$

and by the arbitrariness of ε we conclude that λ^+ is finitely additive. Since $\lambda^- = (-\lambda)^+$, we obtain that λ^- is finitely additive.

Since $\lambda^+ \geq \lambda$ and $-\lambda^- \leq \lambda$ for all $E \in \mathfrak{M}$ we have

$$0 \leq \lambda^+ (E) - \lambda (E) = \sup \{\lambda (F) - \lambda (E) : F \subset E, F \in \mathfrak{M}\}$$
$$= \sup \{-\lambda (E \setminus F) : F \subset E, F \in \mathfrak{M}\}$$
$$= - \inf \{\lambda (G) : G \subset E, F \in \mathfrak{M}\} = \lambda^- (E).$$

This concludes the proof.

The previous theorem, together with (1.21), Theorem 1.74, and Remark 1.76, leads to the definition of the Lebesgue integral of $u \in B (X, \mathfrak{M})$, namely

$$\int_E u \, d\lambda := \lim_{n \to \infty} \int_E s_n \, d\lambda \tag{1.128}$$

for any sequence $\{s_n\} \subset B (X, \mathfrak{M})$ of simple functions that converges uniformly to u. It may be verified that the integral does not depend on the particular approximating sequence $\{s_n\}$.

We consider next the case that X is a normal space.

Definition 1.189. *Let X be a normal space and let $\mathfrak{M}$ be the smallest algebra that contains all open sets of X. The space* rba $(X, \mathfrak{M})$ *of all regular bounded finitely additive signed measures consists of all set functions $\lambda : \mathfrak{M} \to \mathbb{R}$ such that*

(i) $\lambda \in$ ba $(X, \mathfrak{M})$;
(ii) *for every $E \in \mathfrak{M}$,*

$$\lambda^+ (E) = \inf \{\lambda^+ (A) : A \text{ open}, A \supset E\} \tag{1.129}$$
$$= \sup \{\lambda^+ (C) : C \text{ closed}, C \subset E\} \tag{1.130}$$

and

$$\lambda^- (E) = \inf \{\lambda^- (A) : A \text{ open } A \supset E\}$$
$$= \sup \{\lambda^- (C) : C \text{ closed } C \subset E\},$$

where λ^+ and λ^- are defined in Theorem 1.188.

Exercise 1.190. Show that the space $\mathrm{rba}\,(X,\mathfrak{M})$ is a closed subspace of $\mathrm{ba}\,(X,\mathfrak{M})$ and hence a Banach space with the total variation norm $\|\cdot\|\,(X)$.

In the special case that X is compact, then regular finitely additive measures are measures. Indeed, we have the following:

Theorem 1.191 (Alexandroff). *Let X be a compact topological space and let $\mathfrak{M}$ be an algebra that contains all open sets. If $\lambda \in \mathrm{rba}\,(X,\mathfrak{M})$ then λ is countably additive.*

Proof. In view of Theorem 1.188 it suffices to show that the finitely additive measures λ^+ and λ^- defined are countably additive. Let $\{E_n\} \subset \mathfrak{M}$ be a sequence of mutually disjoint sets such that $\bigcup_{n=1}^{\infty} E_n \in \mathfrak{M}$. Then

$$\sum_{n=1}^{l} \lambda^+ (E_n) = \lambda^+ \left(\bigcup_{n=1}^{l} E_n \right) \le \lambda^+ \left(\bigcup_{n=1}^{\infty} E_n \right)$$

and letting $l \to \infty$, we get

$$\sum_{n=1}^{\infty} \lambda^+ (E_n) \le \lambda^+ \left(\bigcup_{n=1}^{\infty} E_n \right).$$

To prove the reverse inequality, let C be any closed set contained in $\bigcup_{n=1}^{\infty} E_n$. For $n \in \mathbb{N}$ and for every $\varepsilon > 0$, by (1.129) there exists an open set $A_n \supset E_n$ such that

$$\lambda^+ (A_n) \le \lambda^+ (E_n) + \frac{\varepsilon}{2^n}.$$

Since C is a closed subset of the compact space X we have that C is also compact, and since $\{A_n\}$ is an open cover for C, there exist $A_1, \dots, A_l$ whose union still covers C. Hence

$$\lambda^+ (C) \le \lambda^+ \left(\bigcup_{n=1}^{l} A_n \right) \le \sum_{n=1}^{l} \lambda^+ (A_n) \le \sum_{n=1}^{l} \lambda^+ (E_n) + \varepsilon \le \sum_{n=1}^{\infty} \lambda^+ (E_n) + \varepsilon.$$

Taking the supremum over all closed sets $C \subset \bigcup_{n=1}^{\infty} E_n$ and using (1.130), we obtain

$$\lambda^+ \left(\bigcup_{n=1}^{\infty} E_n \right) \le \sum_{n=1}^{\infty} \lambda^+ (E_n) + \varepsilon,$$

and letting $\varepsilon \to 0^+$ gives the desired inequality.

With exactly the same proof it may be shown that λ^- is countably additive.

Corollary 1.192. *If in addition to the hypotheses of the previous theorem, we assume also that X is a Hausdorff space, then λ can be extended in a unique way as a signed Radon measure to the smallest σ-algebra that contains $\mathfrak{M}$.*

Proof. Since $\lambda^+ : \mathfrak{M} \to [0, \infty)$ is a finite countably additive measure, in view of Remark 1.39, the finite measure $(\lambda^+)^* : \mathfrak{M}^* \to [0, \infty)$ defined in (1.14) (with $\mathcal{G} := \mathfrak{M}$ and $\rho := \lambda^+$) is the unique extension of λ^+ to the σ-algebra $\mathfrak{M}^*$ of all $(\lambda^+)^*$-measurable sets. Note that since $\mathfrak{M}$ contains all open sets, then necessarily $\mathcal{B}(X) \subset \mathfrak{M}^*$. Thus it remains to show that the Borel measure $(\lambda^+)^* : \mathfrak{M}^* \to [0, \infty)$ is a Radon measure. In view of Remark A.10 in the appendix, the class of closed sets of X coincides with the class of compact sets. Hence for any open set $A \subset X$, by (1.130),

$$
\begin{aligned}
(\lambda^+)^*(A) = (\lambda^+)(A) &= \sup \left\{ \lambda^+(C) : C \text{ closed } C \subset A \right\} \\
&= \sup \left\{ \lambda^+(K) : K \text{ compact } K \subset A \right\} \\
&= \sup \left\{ (\lambda^+)^*(K) : K \text{ compact } K \subset A \right\},
\end{aligned}
$$

and so every open set is inner regular.

We now show that any $E \in \mathfrak{M}^*$ is outer regular. By (1.14), for any fixed $\varepsilon > 0$ we may find a sequence $\{E_n\} \subset \mathfrak{M}$ such that $E \subset \bigcup_{n=1}^{\infty} E_n$ and

$$
\sum_{n=1}^{\infty} \lambda^+(E_n) \leq (\lambda^+)^*(E) + \varepsilon. \tag{1.131}
$$

By (1.129), for every $n \in \mathbb{N}$ we may find an open set $A_n \supset E_n$ such that

$$
\lambda^+(A_n) \leq \lambda^+(E_n) + \frac{\varepsilon}{2^n}. \tag{1.132}
$$

Since $E \subset \bigcup_{n=1}^{\infty} A_n$ by (1.131) and (1.132), we have that

$$
\lambda^+\left(\bigcup_{n=1}^{\infty} A_n \right) \leq \sum_{n=1}^{\infty} \lambda^+(A_n) \leq \sum_{n=1}^{\infty} \lambda^+(E_n) + \varepsilon \leq (\lambda^+)^*(E) + 2\varepsilon.
$$

In turn,

$$
(\lambda^+)^*(E) \leq \inf \left\{ \lambda^+(A) : A \text{ open } A \supset E \right\} \leq (\lambda^+)^*(E) + 2\varepsilon,
$$

and given the arbitrariness of $\varepsilon > 0$ we deduce that E is outer regular. This concludes the proof.

1.3.3 Spaces of Measures as Dual Spaces

Let X be a nonempty set and let $\mathfrak{M}$ be an algebra. In this subsection we identify the dual of certain subspaces of $B(X, \mathfrak{M})$ with $\mathrm{ba}(X, \mathfrak{M})$, $\mathrm{rba}(X, \mathfrak{M})$ and $\mathcal{M}(X; \mathbb{R})$.

It turns out that the dual of $B(X, \mathfrak{M})$ may be identified with $\mathrm{ba}(X, \mathfrak{M})$:

Theorem 1.193 (Riesz representation theorem in B). *Let X be a nonempty set and let $\mathfrak{M} \subset \mathcal{P}(X)$ be an algebra. Then every bounded linear functional $L : B(X, \mathfrak{M}) \to \mathbb{R}$ is represented by a unique $\lambda \in \mathrm{ba}(X, \mathfrak{M})$, in the sense that*

$$L(u) = \int_X u \, d\lambda \quad \text{for every } u \in B(X, \mathfrak{M}). \tag{1.133}$$

Moreover, the norm of L coincides with $\|\lambda\|(X)$. Conversely, every functional of the form (1.133), where $\lambda \in \mathrm{ba}(X, \mathfrak{M})$, is a bounded linear functional on $B(X, \mathfrak{M})$.

Proof. **Step 1:** Let $L \in (B(X, \mathfrak{M}))'$ and for every $E \in \mathfrak{M}$ define

$$\lambda(E) := L(\chi_E).$$

Since L is linear and $L(0) = 0$ it follows that $\lambda : \mathfrak{M} \to \mathbb{R}$ is a finitely additive signed measure. To prove that λ is bounded, let $\{E_n\}_{n=1}^{l} \subset \mathfrak{M}$ be a finite partition of X and define

$$s := \sum_{n=1}^{l} c_n \chi_{E_n},$$

where $c_n := \mathrm{sgn}(L(\chi_{E_n}))$. Then $s \in B(X, \mathfrak{M})$ and

$$\sum_{n=1}^{l} |\lambda(E_n)| = \sum_{n=1}^{l} \mathrm{sgn}(L(\chi_{E_n})) L(\chi_{E_n})$$

$$= \left| \sum_{n=1}^{l} c_n L(\chi_{E_n}) \right| = |L(s)|$$

$$\leq \|L\|_{(B(X,\mathfrak{M}))'} \|s\|_\infty \leq \|L\|_{(B(X,\mathfrak{M}))'},$$

where we have used the fact that $\|s\|_\infty \leq 1$. Hence

$$\|\lambda\|(X) \leq \|L\|_{(B(X,\mathfrak{M}))'}, \tag{1.134}$$

which shows that $\lambda \in \mathrm{ba}(X, \mathfrak{M})$.

We now show that (1.133) holds. Fix $u \in B(X, \mathfrak{M})$ and $\varepsilon > 0$, and partition the interval $[-\|u\|_\infty, \|u\|_\infty]$ into l intervals I_n, $n = 1, \ldots, l$, of length less than or equal to ε. For each $n = 1, \ldots, l$, define

$$E_n := u^{-1}(I_n) \in \mathfrak{M}$$

and let

$$s := \sum_{n=1}^{l} c_n \chi_{E_n},$$

where c_n is any fixed number in I_n. Note that for $x \in E_n$ we have that $u(x) \in I_n$, and so

$$|u(x) - s(x)| \le \varepsilon.$$

Hence $\|u - s\|_\infty \le \varepsilon$, while by the linearity of L,

$$
\begin{aligned}
\left| L(u) - \int_X s \, d\lambda \right| &= \left| L(u) - \sum_{n=1}^{l} c_n \lambda(E_n) \right| \\
&= \left| L(u) - \sum_{n=1}^{l} c_n L(\chi_{E_n}) \right| \\
&= |L(u-s)| \le \|L\|_{(B(X,\mathfrak{M}))'} \|u - s\|_\infty \\
&\le \|L\|_{(B(X,\mathfrak{M}))'} \varepsilon.
\end{aligned}
\tag{1.135}
$$

By taking $\varepsilon := \frac{1}{k}$ we can construct a sequence $\{s_k\}$ of simple functions converging uniformly to u such that

$$\lim_{k \to \infty} \int_X s_k \, d\lambda = L(u).$$

By (1.128) this implies that

$$L(u) = \int_X u \, d\lambda.$$

Moreover, from (1.135) and the fact that $\|u - s\|_\infty \le \varepsilon$ we have that

$$
\begin{aligned}
|L(u)| &\le \left| \int_X s \, d\lambda \right| + \|L\|_{(B(X,\mathfrak{M}))'} \varepsilon \\
&\le \sum_{n=1}^{l} |c_n| \, |\lambda(E_n)| + \|L\|_{(B(X,\mathfrak{M}))'} \varepsilon \\
&\le (\|u\|_\infty + \varepsilon) \sum_{n=1}^{l} |\lambda(E_n)| + \|L\|_{(B(X,\mathfrak{M}))'} \varepsilon \\
&\le (\|u\|_\infty + \varepsilon) \|\lambda\|(X) + \|L\|_{(B(X,\mathfrak{M}))'} \varepsilon,
\end{aligned}
$$

and so, by the arbitrariness of ε, we get

$$|L(u)| \le \|u\|_\infty \|\lambda\|(X),$$

which, together with (1.134), implies that $\|\lambda\|(X) = \|L\|_{(B(X,\mathfrak{M}))'}$.

Step 2: Conversely, given $\lambda \in \text{ba}(X, \mathfrak{M})$, for $u \in B(X, \mathfrak{M})$ define

$$L(u) := \int_X u \, d\lambda.$$

Then $L \in (B(X, \mathfrak{M}))'$ and $\|\lambda\|(X) = \|L\|_{(B(X,\mathfrak{M}))'}$. We leave the details as an exercise.

We consider next the case that X is a normal Hausdorff space. Let $C_b(X)$ denote the space of all continuous bounded functions $u : X \to \mathbb{R}$. It is well known that $C_b(X)$ is a Banach space with the norm $\|\cdot\|_\infty$.

Theorem 1.194. *Let X be a normal space. Then $C_b(X)$ is separable if and only if X is a compact metrizable space.*

It turns out that the dual of $C_b(X)$ may be identified with rba$(X, \mathfrak{M})$:

Theorem 1.195 (Riesz representation theorem in C_b). *Let X be a normal Hausdorff space and let $\mathfrak{M}$ be the smallest algebra that contains all open sets of X. Then every bounded linear functional $L : C_b(X) \to \mathbb{R}$ is represented by a unique $\lambda \in$ rba$(X, \mathfrak{M})$ in the sense that*

$$L(u) = \int_X u \, d\lambda \quad \text{for every } u \in C_b(X). \tag{1.136}$$

Moreover, the norm of L coincides with the total variation norm $\|\lambda\|(X)$. Conversely, every functional of the form (1.136), where $\lambda \in$ rba$(X, \mathfrak{M})$, is a bounded linear functional on $C_b(X)$.

The proof of this theorem is rather lengthy, and so we omit it. We refer the reader to [AliBur99].

Let X be a compact Hausdorff space and let $C(X) = C_b(X)$ be the space of all continuous functions. In view of Corollary 1.192 and the Riesz representation theorem in C_b, we have that the dual of $C(X)$ may be identified with $\mathcal{M}(X; \mathbb{R})$:

Theorem 1.196 (Riesz representation theorem in C). *Let X be a compact Hausdorff space. Then every bounded linear functional $L : C(X) \to \mathbb{R}$ is represented by a unique finite signed Radon measure $\lambda \in \mathcal{M}(X; \mathbb{R})$ in the sense that*

$$L(u) = \int_X u \, d\lambda \quad \text{for every } u \in C(X). \tag{1.137}$$

Moreover, the norm of L coincides with the total variation norm $\|\lambda\|$. Conversely, every functional of the form (1.137), where $\lambda \in \mathcal{M}(X; \mathbb{R})$, is a bounded linear functional on $C(X)$.

Finally, we consider the dual space of the space of continuous functions "that vanish at infinity". Let X be a topological space, let $C_c(X)$ be the collection of all continuous functions $u : X \to \mathbb{R}$ whose support is compact, and let $C_0(X)$ be the completion of $C_c(X)$ relative by the supremum norm $\|\cdot\|_\infty$.

Theorem 1.197. *Let X be a normal space. Then $C_0(X)$ is separable if and only if X is a σ-compact metrizable space.*

Remark 1.198. It can be shown that a normal, separable, locally compact metrizable space is σ-compact.

If X is compact then

$$C_0(X) = C_c(X) = C(X).$$

If X is a locally compact Hausdorff space, then it can be shown that $u \in C_0(X)$ if and only if $u \in C(X)$ and for every $\varepsilon > 0$ there exists a compact set $K \subset X$ such that

$$|u(x)| < \varepsilon \text{ for all } x \in X \setminus K.$$

Note that the duals of $C_c(X)$ and $C_0(X)$ "coincide" in the sense that if $L \in (C_0(X))'$ then $L \lfloor C_c(X) \in (C_c(X))'$, and conversely, if $L \in (C_c(X))'$ then there exists a unique $\overline{L} \in (C_0(X))'$ such that

$$\overline{L} \lfloor C_c(X) = L, \quad \|\overline{L}\| = \|L\|.$$

Exercise 1.199 (Alexandroff compactification). Let X be a locally compact Hausdorff space. Let ∞ denote a point not in X and consider $X^\infty := X \cup \{\infty\}$. Let $\tau \subset \mathcal{P}(X^\infty)$ be the collection of all subsets $A \subset X^\infty$ such that either A is an open set of X or $\infty \in A$ and $X^\infty \setminus A$ is a compact set of X. Prove that (X^∞, τ) is a compact Hausdorff space, and that $v \in C(X^\infty)$ if and only if

$$v = u\chi_X + c\chi_{\{\infty\}}$$

for some $u \in C_0(X)$ and $c \in \mathbb{R}$. Show also that the decomposition of v is unique, and

$$\|v\|_{C(X^\infty)} = \max\left\{\|u\|_{C_0(X)}, |c|\right\}. \tag{1.138}$$

We now show that the dual of $C_0(X)$ may be identified with $\mathcal{M}(X;\mathbb{R})$.

Theorem 1.200 (Riesz representation theorem in C_0). *Let X be a locally compact Hausdorff space. Then every bounded linear functional $L : C_0(X) \to \mathbb{R}$ is represented by a unique finite signed Radon measure $\lambda \in \mathcal{M}(X;\mathbb{R})$ in the sense that*

$$L(u) = \int_X u \, d\lambda \quad \text{for every } u \in C_0(X). \tag{1.139}$$

Moreover, the norm of L coincides with the total variation norm $\|\lambda\|(X)$. Conversely, every functional of the form (1.139), where $\lambda \in \mathcal{M}(X;\mathbb{R})$, is a bounded linear functional on $C_0(X)$.

Proof. It is immediate to verify that given $\lambda \in \mathcal{M}(X;\mathbb{R})$, then (1.139) determines L as an element of $(C_0(X))'$.

Conversely, given $L \in (C_0(X))'$, we use the *one-point compactification* or *Alexandroff compactification* defined in the previous exercise to find $\lambda \in \mathcal{M}(X; \mathbb{R})$ satisfying (1.139). In $C(X^\infty)$ consider the norm

$$p(v) := \|L\|_{(C_0(X))'} \|v\|_{C(X^\infty)}.$$

In view of (1.138) we have

$$|L(u)| \leq p(u)$$

for all $u \in C_0(X)$, and thus by the Hahn–Banach theorem we may extend L to a bounded linear functional L_1 on $C(X^\infty)$ in such a way that $\|L\|_{(C_0(X))'} = \|L_1\|_{(C(X^\infty))'}$. By the Riesz representation theorem in $C(X^\infty)$ there is a unique finite signed Radon measure $\lambda \in \mathcal{M}(X^\infty; \mathbb{R})$ with

$$\|\lambda\|(X^\infty) = \|L_1\|_{(C(X^\infty))'},$$

and such that for every $v \in C(X^\infty)$,

$$L_1(v) = \int_{X^\infty} v \, d\lambda = \int_X v \, d\lambda + v(\infty)\lambda(\{\infty\}).$$

In particular, it follows that for every $u \in C_0(X)$ we have that

$$L_1(u) = L(u) = \int_X u \, d\lambda. \qquad (1.140)$$

Moreover, $\lambda : \mathcal{B}(X) \to \mathbb{R}$ is in $\mathcal{M}(X; \mathbb{R})$, and we have

$$\|\lambda\|(X) \leq \|\lambda\|(X^\infty) = \|L_1\|_{(C(X^\infty))'} = \|L\|_{(C_0(X))'} \leq \|\lambda\|(X),$$

where the last inequality follows from (1.140). This proves that $\|L\|_{(C_0(X))'} = \|\lambda\|(X)$. Note that this implies that $\lambda(\{\infty\}) = 0$.

Finally, to prove the uniqueness of the representation (1.140) assume that

$$L(u) = \int_X u \, d\lambda = \int_X u \, d\varsigma$$

for all $u \in C_0(X)$ and for some $\varsigma \in \mathcal{M}(X; \mathbb{R})$. Extend ς to $\varsigma_1 \in \mathcal{M}(X^\infty; \mathbb{R})$ by setting $\varsigma_1(\{\infty\}) := 0$. Then

$$L_1(v) = \int_X v \, d\lambda = \int_X v \, d\varsigma_1$$

for all $v \in C(X^\infty)$, and so by the Riesz representation theorem in C it follows that $\varsigma_1 = \lambda$, in particular $\varsigma_1 \lfloor X = \varsigma = \lambda \lfloor X$.

In the study of distributions in [FoLe10] we will also need the following variants of the previous theorem.

Theorem 1.201 (Riesz representation theorem in C_c). *Let X be a locally compact Hausdorff space and let $L : C_c(X) \to \mathbb{R}$ be a linear functional. Then*

(i) if L is positive, *that is, $L(v) \geq 0$ for all $v \in C_c(X)$ with $v \geq 0$, then there exists a unique (positive) Radon measure $\mu : \mathcal{B}(X) \to [0, \infty]$ such that*

$$L(u) = \int_X u \, d\mu \quad \text{for every } u \in C_c(X);$$

(ii) if L is locally bounded, *that is, for every compact set $K \subset X$ there exists a constant $C_K > 0$ such that*

$$|L(u)| \leq C_K \|v\|_{C_c(X)}$$

for all $v \in C_c(X)$ with $\operatorname{supp} v \subset K$, then there exist two (positive) Radon measures μ_1, $\mu_2 : \mathcal{B}(X) \to [0, \infty]$ such that

$$L(u) = \int_X u \, d\mu_1 - \int_X u \, d\mu_2 \quad \text{for every } u \in C_c(X).$$

Note that since both μ_1 and μ_2 could have infinite measure, their difference is not defined in general, although on every compact set it is a well-defined finite signed measure. The proof of part (i) of the previous theorem is classical (see, e.g., [Ru87]), while (ii) follows from (i) by writing L as the difference of two positive functionals (see Step 3 of the proof of Theorem 2.37 below for an entirely similar argument).

1.3.4 Weak Star Convergence of Measures

Let X be a normal space and let $\mathfrak{M}$ be an algebra that contains all open sets of X. If $C_b(X)$ is separable, i.e., if X is a compact metrizable space (see Theorem 1.194), then in view of Theorem A.55 and the Riesz representation theorem, any bounded sequence of measures $\{\lambda_n\} \subset \operatorname{rba}(X, \mathfrak{M})$ is sequentially weakly star precompact. In particular, if $u \in C_b(X)$ then

$$\lim_{k \to \infty} \int_X u \, d\lambda_{n_k} = \int_X u \, d\lambda \tag{1.141}$$

for a subsequence $\{\lambda_{n_k}\}$ and for some $\lambda \in \operatorname{rba}(X, \mathfrak{M})$.

Similarly, invoking now Theorem 1.197 and the Riesz representation theorem in $C_0(X)$, we have the following result.

Proposition 1.202. *Let X be a σ-compact metric space, and let $\{\lambda_n\} \subset \mathcal{M}(X; \mathbb{R})$ be a sequence of signed Radon measures such that*

$$\sup_n \|\lambda_n\| < \infty.$$

Then there exist a subsequence $\{\lambda_{n_k}\}$ and a signed Radon measure λ such that

$$\lambda_{n_k} \xrightarrow{*} \lambda \quad in \; \mathcal{M}\left(X;\mathbb{R}\right),$$

i.e.,

$$\lim_{n\to\infty} \int_X u\, d\lambda_n = \int_X u\, d\lambda$$

for every $u \in C_0\left(X\right)$.

The following properties will be frequently used in the sequel.

Proposition 1.203. *Let X be a locally compact Hausdorff space and let $\{\mu_n\} \subset \mathcal{M}\left(X;\mathbb{R}\right)$ be a sequence of (positive) Radon measures such that $\mu_n \xrightarrow{*} \mu$ in $\mathcal{M}\left(X;\mathbb{R}\right)$. Then*

(i) if $A \subset X$ is open then

$$\mu\left(A\right) \leq \liminf_{n\to\infty} \mu_n\left(A\right);$$

(ii) if $K \subset X$ is compact then

$$\mu\left(K\right) \geq \limsup_{n\to\infty} \mu_n\left(K\right);$$

(iii) if $A \subset X$ is open, $\overline{A}$ is compact, and $\mu\left(\partial A\right) = 0$, then

$$\mu\left(A\right) = \lim_{n\to\infty} \mu_n\left(A\right).$$

Proof. (i) Let $K \subset A$ be a compact subset. By Corollary A.13 there exists a cutoff function $\varphi \in C_c\left(X\right)$ such that $0 \leq \varphi \leq 1$, $\varphi \equiv 1$ on K, and $\varphi \equiv 0$ on $X \setminus A$. Then

$$\mu\left(K\right) \leq \int_A \varphi\, d\mu = \lim_{n\to\infty} \int_A \varphi\, d\mu_n \leq \liminf_{n\to\infty} \mu_n\left(A\right).$$

Since μ is a Radon measure it is inner regular, and so taking the supremum over all compact subsets of A, we have

$$\mu\left(A\right) = \sup\left\{\mu\left(K\right) : K \subset A\right\} \leq \liminf_{n\to\infty} \mu_n\left(A\right).$$

(ii) Let $A \supset K$ be an open set and consider a cutoff function $\varphi \in C_c\left(X\right)$ such that $0 \leq \varphi \leq 1$, $\varphi \equiv 1$ on K, and $\varphi \equiv 0$ on $X \setminus A$. Then

$$\limsup_{n\to\infty} \mu_n\left(K\right) \leq \lim_{n\to\infty} \int_A \varphi\, d\mu_n = \int_A \varphi\, d\mu \leq \mu\left(A\right).$$

Since μ is a Radon measure, taking the supremum over all open supersets of K and using outer regularity, we have

$$\limsup_{n \to \infty} \mu_n (K) \leq \mu(K) = \inf \{\mu(A) : K \subset A\}.$$

(iii) If $\mu(\partial A) = 0$ then from (i) and (ii),

$$\limsup_{n \to \infty} \mu_n (A) \leq \limsup_{n \to \infty} \mu_n (\overline{A}) \leq \mu(\overline{A}) = \mu(A) \leq \liminf_{n \to \infty} \mu_n (A).$$

This concludes the proof.

As a corollary of the previous proposition we obtain the following result.

Corollary 1.204. *Let X be a locally compact Hausdorff space, and let $\{\lambda_n\} \subset \mathcal{M}(X; \mathbb{R})$ be a sequence of signed Radon measures such that $\lambda_n \overset{*}{\rightharpoonup} \lambda$ in $\mathcal{M}(X; \mathbb{R})$ and $\|\lambda_n\| \overset{*}{\rightharpoonup} \nu$ in $\mathcal{M}(X; \mathbb{R})$. If $A \subset X$ is open, $\overline{A}$ is compact, and $\nu(\partial A) = 0$, then*

$$\int_A u \, d\lambda_n \to \int_A u \, d\lambda$$

for all $u \in C_b(X)$. In particular,

$$\lambda_n (A) \to \lambda(A).$$

Proof. Fix $u \in C_b(X)$. Since ν is outer regular and $\nu(\partial A) = 0$ for any fixed $\varepsilon > 0$, there exists an open set $U \supset \partial A$ such that

$$\nu(U) \leq \frac{\varepsilon}{2\left(1 + \|u\|_{C_b(X)}\right)}. \tag{1.142}$$

Since $U \cup A$ is an open set containing the compact set $\overline{A}$, by Corollary A.13 there exists a cutoff function $\varphi \in C_c(X)$ such that $0 \leq \varphi \leq 1$, $\varphi \equiv 1$ on $\overline{A}$, and $\varphi \equiv 0$ on $X \setminus (U \cup A)$. For any $u \in C_b(X)$ we have that $\varphi u \in C_c(X)$, and so

$$\int_A u \, d\lambda_n - \int_A u \, d\lambda = \int_X \varphi u \, d\lambda_n - \int_X \varphi u \, d\lambda + \int_{U \setminus A} \varphi u \, d\lambda_n - \int_{U \setminus A} \varphi u \, d\lambda.$$

In turn,

$$\left| \int_A u \, d\lambda_n - \int_A u \, d\lambda \right| \leq \left| \int_X \varphi u \, d\lambda_n - \int_X \varphi u \, d\lambda \right|$$
$$+ \|u\|_{C_b(X)} \left(\|\lambda_n\| (\operatorname{supp} \varphi \setminus A) + \|\lambda\| (\operatorname{supp} \varphi \setminus A) \right).$$

Letting $n \to \infty$ and using part (ii) of the previous proposition, the facts that $\lambda_n \overset{*}{\rightharpoonup} \lambda$ in $\mathcal{M}(X; \mathbb{R})$ and $\|\lambda\| \leq \nu$, and (1.142) yield

$$\limsup_{n \to \infty} \left| \int_A u \, d\lambda_n - \int_A u \, d\lambda \right| \leq 2 \|u\|_{C_b(X)} \nu(\operatorname{supp} \varphi \setminus A)$$
$$\leq 2 \|u\|_{C_b(X)} \nu(U) \leq \varepsilon.$$

The last part of the theorem follows by taking $u \equiv 1$.

Remark 1.205. The previous corollary continues to hold with the obvious changes for a sequence $\{\lambda_n\} \subset \mathcal{M}(X; \mathbb{R}^m)$ of vector-valued Radon measures.

If X is a topological space, then $C_0(X) \subset C_b(X)$, with the strict inclusion possible. Hence, if a sequence of (positive) Radon measures $\{\mu_n\}$ converges weakly star to some Radon measure μ in $\mathcal{M}(X; \mathbb{R}) = (C_0(X))'$, then in general we cannot conclude that $\mu_n \overset{*}{\rightharpoonup} \mu$ in $(C_b(X))'$. The next results provide two sufficient conditions for this to happen.

Proposition 1.206. *Let X be a topological space and let μ, $\mu_n : \mathcal{B}(X) \to [0, \infty)$, $n \in \mathbb{N}$, be finite Borel measures such that*

$$\lim_{n \to \infty} \mu_n(X) = \mu(X) \tag{1.143}$$

and

$$\liminf_{n \to \infty} \mu_n(A) \geq \mu(A) \tag{1.144}$$

for every open set $A \subset X$. Then $\mu_n \overset{*}{\rightharpoonup} \mu$ in $(C_b(X))'$.

In particular, if X is a locally compact Hausdorff space, μ, μ_n, $n \in \mathbb{N}$, are (positive) finite Radon measures, and $\mu_n \overset{*}{\rightharpoonup} \mu$ in $\mathcal{M}(X; \mathbb{R})$, then $\mu_n \overset{*}{\rightharpoonup} \mu$ in $(C_b(X))'$ if and only if (1.143) holds.

The proof of the proposition makes use of the following exercise.

Exercise 1.207. Let $\{x_n\}$ and $\{y_n\}$ be two sequences of real numbers such that

$$\liminf_{n \to \infty} x_n \geq x, \quad \liminf_{n \to \infty} y_n \geq y,$$

$$\limsup_{n \to \infty} (x_n + y_n) \leq x + y,$$

for some x, $y \in \mathbb{R}$. Prove that $\lim_{n \to \infty} x_n = x$ and $\lim_{n \to \infty} y_n = y$.

Proof (Proposition 1.206). Let $u \in C_b(X)$. Assume first that $0 \leq u \leq 1$. Then by Theorem 1.123 and Fatou's lemma

$$
\begin{aligned}
\liminf_{n \to \infty} \int_X u \, d\mu_n &= \liminf_{n \to \infty} \int_0^1 \mu_n(\{x \in X : u(x) > t\}) \, dt \\
&\geq \int_0^1 \liminf_{n \to \infty} \mu_n(\{x \in X : u(x) > t\}) \, dt \\
&\geq \int_0^1 \mu(\{x \in X : u(x) > t\}) \, dt \\
&= \int_X u \, d\mu,
\end{aligned}
$$

where we have used (1.144) and that fact that the sets $\{x \in X : u(x) > t\}$ are open, since u is continuous.

Similarly,

$$\liminf_{n\to\infty} \int_X (1-u)\, d\mu_n \geq \int_X (1-u)\, d\mu.$$

Let

$$x_n := \int_X u\, d\mu_n, \quad y_n := \int_X (1-u)\, u\, d\mu_n,$$

$$x := \int_X u\, d\mu, \quad y := \int_X (1-u)\, d\mu.$$

By (1.143) and the previous exercise we conclude that

$$\lim_{n\to\infty} \int_X u\, d\mu_n = \int_X u\, d\mu$$

for all $u \in C_b(X)$, with $0 \leq u \leq 1$.

For a general function $u \in C_b(X) \setminus \{0\}$ it suffices to apply the first part of the proof to the function

$$v := \frac{u - \inf_X u}{\sup_X u - \inf_X u}$$

and use (1.143).

Finally, if X is a locally compact Hausdorff space and $\mu_n \overset{*}{\rightharpoonup} \mu$ in $\mathcal{M}(X; \mathbb{R})$, then property (1.144) holds in view of Proposition 1.203(i). Since $1 \in C_b(X)$, by the first part of the proposition we have that $\mu_n \overset{*}{\rightharpoonup} \mu$ in $(C_b(X))'$ if and only if (1.143) holds.

Using the previous proposition we may prove an important theorem, usually stated for probability measures.

Theorem 1.208 (Prohorov). *Let X be a metric space and let $\{\mu_n\}$ be a sequence of Borel measures, $\mu_n : \mathcal{B}(X) \to [0, \infty)$, such that*

$$\sup_n \mu_n(X) < \infty. \tag{1.145}$$

Assume that for all $\varepsilon > 0$ there exists a compact set $K \subset X$ such that

$$\sup_n \mu_n(X \setminus K) \leq \varepsilon. \tag{1.146}$$

Then there exist a subsequence $\{\mu_{n_k}\}$ of $\{\mu_n\}$ and a Borel measure $\mu :$ $\mathcal{B}(X) \to [0, \infty)$ such that $\mu_{n_k} \overset{*}{\rightharpoonup} \mu$ in $(C_b(X))'$.

In particular, if X is a locally compact metric space, μ, μ_n, $n \in \mathbb{N}$, are (positive) finite Radon measures, and $\mu_n \overset{*}{\rightharpoonup} \mu$ in $\mathcal{M}(X; \mathbb{R})$, then $\mu_n \overset{*}{\rightharpoonup} \mu$ in $(C_b(X))'$ if and only if (1.146) holds.

The proof makes use of the following exercise.

Exercise 1.209. Let X be a metric space and let $\left\{K^{(j)}\right\} \subset X$ be an increasing sequence of compact sets. Prove that the set $\bigcup_{j=1}^{\infty} K^{(j)}$ is separable and construct a countable family $\mathcal{A}$ of open sets with the property that for every open set $U \subset X$ and for every $x \in U \cap \left(\bigcup_{j=1}^{\infty} K^{(j)}\right)$ there exists $A \in \mathcal{A}$ such that $x \in A$ and $\overline{A} \subset U$.

Proof (Prohorov's theorem). By extracting a subsequence (not relabeled) we may assume that

$$\lim_{n \to \infty} \mu_n (X) = M < \infty.$$

If $M = 0$ then $\mu_n \overset{*}{\rightharpoonup} 0$ in $(C_b (X))'$ and there is nothing to prove. If $M > 0$, then without loss of generality we may assume that $\mu_n (X) = 1$ for all $n \in \mathbb{N}$. Indeed, if we can prove that a subsequence $\left\{\frac{\mu_{n_k}}{\mu_{n_k}(X)}\right\}$ of $\left\{\frac{\mu_n}{\mu_n(X)}\right\}$ converges weakly star to some Borel measure ν in $(C_b (X))'$, then $\{\mu_{n_k}\}$ converges weakly star to the Borel measure $M\nu$ in $(C_b (X))'$.

By (1.146) we may construct an increasing sequence of compact sets $\left\{K^{(j)}\right\} \subset X$ such that

$$\mu_n \left(K^{(j)}\right) > 1 - \frac{1}{j} \tag{1.147}$$

for all j, $n \in \mathbb{N}$. Let $\mathcal{A}$ be the countable family of open sets given in the previous exercise and let $\mathcal{K}$ be the countable family of compact sets formed by $\emptyset$ and finite unions of sets of the form $\overline{A} \cap K^{(j)}$, $A \in \mathcal{A}$, $j \in \mathbb{N}$.

Using a diagonal procedure, we may extract a subsequence of $\{\mu_n\}$ (not relabeled) such that the limit $\lim_{n \to \infty} \mu_n (K)$ exists for all $K \in \mathcal{K}$. Set

$$\mu^* (K) := \lim_{n \to \infty} \mu_n (K), \quad K \in \mathcal{K}, \tag{1.148}$$

and observe that $0 \leq \mu^* \leq 1$, $\mu^* (\emptyset) = 0$, and for all $K_1, K_2 \in \mathcal{K}$,

$$\mu^* (K_1) \leq \mu^* (K_2) \text{ if } K_1 \subset K_2; \tag{1.149}$$
$$\mu^* (K_1 \cup K_2) = \mu^* (K_1) + \mu^* (K_2) \text{ if } K_1 \cap K_2 = \emptyset; \tag{1.150}$$
$$\mu^* (K_1 \cup K_2) \leq \mu^* (K_1) + \mu^* (K_2). \tag{1.151}$$

For every open set $U \subset X$ define

$$\mu^* (U) := \sup \{\mu^* (K) : K \subset U, K \in \mathcal{K}\}, \tag{1.152}$$

while for an arbitrary set $E \subset X$ define

$$\mu^* (E) := \inf \{\mu^* (U) : E \subset U, U \text{ open}\}. \tag{1.153}$$

Step 1: We prove that μ^* is an outer measure.

Substep 1a: We claim that if $C \subset U \subset X$, where C is closed and U is open, and $C \subset K$ for some $K \in \mathcal{K}$, then there exists $K_0 \in \mathcal{K}$ such that $C \subset K_0 \subset U$. To see this, for every $x \in C$ choose $A_x \in \mathcal{A}$ such that $x \in A_x$ and $\overline{A_x} \subset U$. Since C is a closed subset of the compact set K and the family $\{A_x\}_{x \in C}$ is an open cover of C, we may find a finite subcover $A_{x_1} \ldots, A_{x_\ell}$ of C. Using the definition of $\mathcal{K}$ and fact that $\{K^{(j)}\}$ is an increasing sequence of compact sets, there exists $j \in \mathbb{N}$ so large that $C \subset K^{(j)}$. Then the set

$$K_0 := \bigcup_{n=1}^{\ell} \left(\overline{A_{x_n}} \cap K^{(j)} \right)$$

belongs to $\mathcal{K}$ and $C \subset K_0 \subset U$.

Substep 1b: We prove that if $U_1, U_2 \subset X$ are open, then

$$\mu^* (U_1 \cup U_2) \leq \mu^* (U_1) + \mu^* (U_2) .$$

Fix any $K \in \mathcal{K}$, with $K \subset U_1 \cup U_2$, and define the closed sets

$$C_1 := \{x \in K : \text{dist} (x, X \setminus U_1) \geq \text{dist} (x, X \setminus U_2)\} ,$$
$$C_2 := \{x \in K : \text{dist} (x, X \setminus U_2) \geq \text{dist} (x, X \setminus U_1)\} .$$

Note that $C_1 \subset U_1$. Indeed, if by contradiction there existed $x_1 \in C_1$ such that $x \notin U_1$, then $x \in U_2$ and, since $X \setminus U_2$ is closed,

$$0 = \text{dist} (x, X \setminus U_1) < \text{dist} (x, X \setminus U_2) ,$$

which contradicts the definition of C_1. Hence $C_1 \subset U_1$, and similarly $C_2 \subset U_2$.

Since $C_i \subset U_i$ and $C_i \subset K \in \mathcal{K}$, $i = 1, 2$, we may apply Substep 1a to find $K_i \in \mathcal{K}$ such that $C_i \subset K_i \subset U_i$, $i = 1, 2$. Thus by (1.149), (1.151), and (1.152)

$$\mu^* (K) \leq \mu^* (K_1 \cup K_2) \leq \mu^* (K_1) + \mu^* (K_2) \leq \mu^* (U_1) + \mu^* (U_2) .$$

Taking the supremum over all admissible $K \subset U_1 \cup U_2$ and using (1.152) yields the desired inequality.

Substep 1c: We claim that if $\{U_n\} \subset X$ is a sequence of open sets, then

$$\mu^* \left(\bigcup_{n=1}^{\infty} U_n \right) \leq \sum_{n=1}^{\infty} \mu^* (U_n) .$$

To see this, fix any $K \in \mathcal{K}$, with $K \subset \bigcup_{n=1}^{\infty} U_n$. Since K is compact, we may find $\ell \in \mathbb{N}$ such that $K \subset \bigcup_{n=1}^{\ell} U_n$. By (1.152) and the previous substep we have that

$$\mu^* (K) \leq \mu^* \left(\bigcup_{n=1}^{\ell} U_n \right) \leq \sum_{n=1}^{\ell} \mu^* (U_n) \leq \sum_{n=1}^{\infty} \mu^* (U_n) .$$

Taking the supremum over all admissible $K \subset \bigcup_{n=1}^{\infty} U_n$ and using (1.152) proves the claim.

Substep 1d: Finally, we prove that μ^* is an outer measure. By (1.149), (1.152), and (1.153), μ^* is a monotone set function. Since $\mu^*(\emptyset) = 0$, it remains to prove that μ^* is countably subadditive. Let $\{E_n\} \subset X$ be a sequence of arbitrary sets. Fix $\varepsilon > 0$, and by (1.153) for each $n \in \mathbb{N}$ find an open set $U_n \subset X$ such that $E_n \subset U_n$ and

$$\mu^*(U_n) \leq \mu^*(E_n) + \frac{\varepsilon}{2^n}. \tag{1.154}$$

Then $\bigcup_{n=1}^{\infty} E_n \subset \bigcup_{n=1}^{\infty} U_n$, and so by (1.153) and the previous substep

$$\mu^*\left(\bigcup_{n=1}^{\infty} E_n\right) \leq \mu^*\left(\bigcup_{n=1}^{\infty} U_n\right) \leq \sum_{n=1}^{\infty} \mu^*(U_n) \leq \sum_{n=1}^{\infty} \mu^*(E_n) + \varepsilon,$$

where in the last inequality we have used (1.154). Letting $\varepsilon \to 0^+$ gives the desired result.

Step 2: We prove that every closed set $C \subset X$ is μ^*-measurable. Fix a closed set $C \subset X$. We begin by showing that for any open set $U \subset X$,

$$\mu^*(U) \geq \mu^*(U \cap C) + \mu^*(U \cap (X \setminus C)). \tag{1.155}$$

Fix $\varepsilon > 0$. Since $U \cap (X \setminus C)$ is open, by (1.152) we may find $K_1 \in \mathcal{K}$, with $K_1 \subset U \cap (X \setminus C)$, such that

$$\mu^*(U \cap (X \setminus C)) \leq \mu^*(K_1) + \varepsilon. \tag{1.156}$$

Using (1.152) once more, find $K_2 \in \mathcal{K}$, with $K_2 \subset U \cap (X \setminus K_1)$, such that

$$\mu^*(U \cap (X \setminus K_1)) \leq \mu^*(K_2) + \varepsilon. \tag{1.157}$$

Since the sets K_1 and K_2 are disjoint and are contained in U, by (1.152), (1.150), (1.156), and (1.157),

$$\begin{aligned}
\mu^*(U) &\geq \mu^*(K_1 \cup K_2) = \mu^*(K_1) + \mu^*(K_2) \\
&\geq \mu^*(U \cap (X \setminus K_1)) + \mu^*(U \cap (X \setminus C)) - 2\varepsilon \\
&\geq \mu^*(U \cap C) + \mu^*(U \cap (X \setminus C)) - 2\varepsilon,
\end{aligned}$$

where in the last inequality we have used the fact that $C \subset X \setminus K_1$ and the monotonicity of μ^*. Letting $\varepsilon \to 0^+$ yields (1.155).

Now consider an arbitrary set $E \subset X$ and let $U \subset X$ be any open set that contains it. By (1.155) and the monotonicity of μ^* we have that

$$\begin{aligned}
\mu^*(U) &\geq \mu^*(U \cap C) + \mu^*(U \cap (X \setminus C)) \\
&\geq \mu^*(E \cap C) + \mu^*(E \cap (X \setminus C)).
\end{aligned}$$

Taking the infimum over all open sets that contain E and using (1.153) gives

$$\mu^*(E) \geq \mu^*(E \cap C) + \mu^*(E \cap (X \setminus C)),$$

and so C is μ^*-measurable.

Step 3: By the previous step the σ-algebra of all μ^*-measurable sets contains $\mathcal{B}(X)$. Hence, by the Carathéodory theorem $\mu^* : \mathcal{B}(X) \to [0,1]$ is a measure. In view of the previous proposition and that fact that $\mu_n(X) = 1$ for all $n \in \mathbb{N}$, it remains to prove that $\mu^*(X) = 1$ and that

$$\liminf_{n \to \infty} \mu_n(U) \geq \mu^*(U) \tag{1.158}$$

for every open set $U \subset X$. Fix $j \in \mathbb{N}$. Reasoning as in Substep 1a, with $C := K^{(j)}$, we have that $K^{(j)} \in \mathcal{K}$. Moreover, by (1.147) and (1.148)

$$\mu^*\left(K^{(j)}\right) = \lim_{n \to \infty} \mu_n\left(K^{(j)}\right) \geq 1 - \frac{1}{j},$$

and so, by (1.152)

$$1 \geq \mu^*(X) \geq \mu^*\left(K^{(j)}\right) \geq 1 - \frac{1}{j} \to 1$$

as $j \to \infty$.

Finally, to prove (1.158), fix an open set $U \subset X$. For any $K \in \mathcal{K}$, with $K \subset U$, by (1.148) and the monotonicity of each μ_n, we have

$$\mu^*(K) = \lim_{n \to \infty} \mu_n(K) \leq \liminf_{n \to \infty} \mu_n(U).$$

Taking the supremum over all admissible $K \subset U$ and using (1.152) yields (1.158) and completes the proof.

If the metric space X has additional properties, then the converse of the previous theorem holds.

Exercise 1.210. Let X be a complete, separable metric space and let μ, $\mu_n : \mathcal{B}(X) \to [0, \infty)$, $n \in \mathbb{N}$, be finite Borel measures such that $\mu_n \overset{*}{\rightharpoonup} \mu$ in $(C_b(X))'$.

(i) Assume that $\mu_n(X) = 1$ for all $n \in \mathbb{N}$ and prove that if $\{U_j\}$ is an increasing sequence of open sets such that $\bigcup_{j=1}^{\infty} U_j = X$, then for every $\varepsilon > 0$ there exists $j \in \mathbb{N}$ such that

$$\mu_n(U_j) \geq 1 - \varepsilon$$

for all $n \in \mathbb{N}$. Construct a sequence of open balls $\left\{ B\left(x_i^{(k)}, \frac{1}{k}\right) \right\}_{i \in \mathbb{N}}$ that covers X and prove that there exists $i_k \in \mathbb{N}$ such that

$$\mu_n\left(\bigcup_{i=1}^{i_k} B\left(x_i^{(k)}, \frac{1}{k}\right) \right) \geq 1 - \frac{\varepsilon}{2^k}$$

for all $n \in \mathbb{N}$. Prove that (1.146) holds.

(ii) Prove that (1.146) holds without the additional hypothesis that $\mu_n(X) = 1$ for all $n \in \mathbb{N}$.

The next exercise provides a simpler proof of Prohorov's theorem in the case $X = \mathbb{R}$. For every finite Borel measure $\mu : \mathcal{B}(\mathbb{R}) \to [0, \infty)$ define the function $f_\mu : \mathbb{R} \to [0, \infty)$

$$f_\mu(x) := \mu((-\infty, x]), \quad x \in \mathbb{R}.$$

Exercise 1.211. Let $\{\mu_n\}$ be a sequence of finite Borel measures, $\mu_n : \mathcal{B}(\mathbb{R}) \to [0, \infty)$.

(i) Prove that if $\mu : \mathcal{B}(\mathbb{R}) \to [0, \infty)$ is a finite Borel measure, then $\mu_n \overset{*}{\rightharpoonup} \mu$ in $(C_b(\mathbb{R}))'$ if and only if $f_{\mu_n}(x) \to f_\mu(x)$ for every $x \in \mathbb{R}$ at which f_μ is continuous.

(ii) Assume that $\{\mu_n\}$ satisfies conditions (1.145) and (1.146), and construct a subsequence $\{\mu_{n_k}\}$ of $\{\mu_n\}$ such that the limit

$$f(q) := \lim_{k \to \infty} f_{\mu_{n_k}}(q)$$

exists in $\mathbb{R}$ for all rational numbers $q \in \mathbb{Q}$. Extend f to $\mathbb{R}$ by setting

$$f(x) := \lim_{q \in \mathbb{Q}, \, q \searrow x} f_{\mu_{n_k}}(q).$$

Prove that $f : \mathbb{R} \to [0, \infty)$ is increasing, right-continuous, $\lim_{x \to \infty} f(x) < \infty$, and that $f_{\mu_n}(x) \to f(x)$ for every $x \in \mathbb{R}$ at which f_μ is continuous. Let $\mu : \mathcal{B}(\mathbb{R}) \to [0, \infty)$ be the Lebesgue–Stieltjes measure associated to f (see Exercise 1.34) and prove that $\mu_{n_k} \overset{*}{\rightharpoonup} \mu$ in $(C_b(\mathbb{R}))'$.

L^p Spaces

> This new integral of Lebesgue is proving itself a wonderful tool. I might compare it with a modern Krupp gun, so easily does it penetrate barriers which were impregnable.
>
> E. B. van Vleck, 1916

2.1 Abstract Setting

In this section we introduce the L^p spaces and study their properties. Proofs of standard results, such as Hölder's and Minkowski's inequalities, will be omitted, and we refer the reader to [Bar95], [DB02], [DuSc88], [EvGa92], [Fol99], [Rao04], [Ru87].

2.1.1 Definition and Main Properties

Let $(X, \mathfrak{M}, \mu)$ be a measure space. Given two measurable functions $u, v : X \to [-\infty, \infty]$, we say that u is *equivalent* to v, and we write

$$u \sim v \text{ if } u(x) = v(x) \text{ for } \mu \text{ a.e. } x \in X. \tag{2.1}$$

Note that $\sim$ is an equivalence relation in the class of measurable functions. With an abuse of notation, from now on we identify a measurable function $u : X \to [-\infty, \infty]$ with its equivalence class $[u]$.

Definition 2.1. *Let $(X, \mathfrak{M}, \mu)$ be a measure space and let $1 \leq p < \infty$. Then*

$$L^p (X, \mathfrak{M}, \mu) := \left\{ u : X \to [-\infty, \infty] : u \text{ measurable and } \|u\|_{L^p(X, \mathfrak{M}, \mu)} < \infty \right\},$$

where

$$\|u\|_{L^p(X,\mathfrak{M},\mu)} := \left(\int_X |u|^p \, d\mu\right)^{1/p}.$$

If $p = \infty$ then

$$L^\infty(X,\mathfrak{M},\mu) := \left\{u : X \to [-\infty,\infty] : u \text{ measurable and } \|u\|_{L^\infty(X,\mathfrak{M},\mu)} < \infty\right\},$$

where $\|u\|_{L^\infty(X,\mathfrak{M},\mu)}$ is the essential supremum esssup $|u|$ of the function $|u|$, that is,

$$\|u\|_{L^\infty(X,\mathfrak{M},\mu)} = \text{esssup } |u| := \inf\left\{\alpha \in \mathbb{R} : |u(x)| < \alpha \text{ for } \mu \text{ a.e. } x \in X\right\}.$$

For simplicity, and when there is no possibility of confusion, we denote the spaces $L^p(X,\mathfrak{M},\mu)$ simply by $L^p(X,\mu)$ or $L^p(X)$ and the norms $\|u\|_{L^p(X,\mathfrak{M},\mu)}$ by $\|u\|_{L^p(X)}$, $\|u\|_{L^p}$ or $\|u\|_p$.

We denote by $L^p(X,\mathfrak{M},\mu;\mathbb{R}^m)$ (or more simply by $L^p(X;\mathbb{R}^m)$) the space of all functions $u : X \to \mathbb{R}^m$ whose components are in $L^p(X,\mathfrak{M},\mu)$. We will endow $L^p(X,\mathfrak{M},\mu;\mathbb{R}^m)$ with the norm

$$\|u\|_{L^p(X,\mathfrak{M},\mu;\mathbb{R}^m)} := \sum_{i=1}^m \|u_i\|_{L^p(X,\mathfrak{M},\mu)}.$$

For $1 \le p < \infty$ sometimes it will be more convenient to use the equivalent norm

$$\|u\|_{L^p(X,\mathfrak{M},\mu;\mathbb{R}^m)} := \left(\int_X |u|^p \, d\mu\right)^{1/p}.$$

Remark 2.2. In the special case that $X = \mathbb{N}$ and μ is the counting measure, the spaces $L^p(\mathbb{N})$ are also denoted by ℓ^p, and we have

$$\ell^p = \left\{\{x_n\}_{n\in\mathbb{N}} : \sum_{n=1}^\infty |x_n|^p < \infty\right\}$$

for $1 \le p < \infty$, while

$$\ell^\infty = \left\{\{x_n\}_{n\in\mathbb{N}} : \sup_{n\in\mathbb{N}} |x_n| < \infty\right\}.$$

Let q be the Hölder conjugate exponent of p, i.e.,

$$q := \begin{cases} \frac{p}{p-1} & \text{if } 1 < p < \infty, \\ \infty & \text{if } p = 1, \\ 1 & \text{if } p = \infty. \end{cases}$$

Note that, with an abuse of notation, we have

$$\frac{1}{p} + \frac{1}{q} = 1.$$

In the sequel, the Hölder conjugate exponent of p will often be denoted by p'.

Theorem 2.3 (Hölder's inequality). *Let $(X, \mathfrak{M}, \mu)$ be a measure space, let $1 \leq p \leq \infty$, and let q be its Hölder conjugate exponent. If $u, v : X \to [-\infty, \infty]$ are measurable functions then*

$$\|uv\|_{L^1} \leq \|u\|_{L^p} \|v\|_{L^q}. \tag{2.2}$$

In particular, if $u \in L^p(X)$ and $v \in L^q(X)$ then $uv \in L^1(X)$.

Remark 2.4. (i) If $1 < p_1, \ldots, p_n < \infty$, with $\frac{1}{p_1} + \ldots + \frac{1}{p_n} = 1$, and $u_i \in L^{p_i}(X)$, $i = 1, \ldots, n$, then

$$\left\| \prod_{i=1}^{n} u_i \right\|_{L^1} \leq \prod_{i=1}^{n} \|u_i\|_{L^{p_i}}.$$

This inequality follows by applying Hölder's inequality with $u := u_1$, $v := \prod_{i=2}^{n} u_i$, and then using an induction argument.

(ii) Another consequence of Hölder's inequality is

$$\|u\|_{L^q} \leq \left(\|u\|_{L^p} \right)^\theta \left(\|u\|_{L^r} \right)^{1-\theta},$$

which holds for $u \in L^p(X)$, and where $1 < p < q < r < \infty$, with $\frac{1}{q} = \frac{\theta}{p} + \frac{(1-\theta)}{r}$. To prove it, apply Hölder's inequality, with $\frac{p}{\theta q}$, $|u|^{\theta q}$, and $|u|^{(1-\theta)q}$ in place of p, u, and v, respectively, to obtain

$$\int_X |u|^q \, d\mu = \int_X |u|^{\theta q} |u|^{(1-\theta)q} \, d\mu$$

$$\leq \left(\int_X |u|^{\theta q \left(\frac{p}{\theta q} \right)} \, d\mu \right)^{1/\left(\frac{p}{\theta q} \right)} \left(\int_X |u|^{(1-\theta)q \left(\frac{p}{\theta q} \right)'} \, d\mu \right)^{1/\left(\frac{p}{\theta q} \right)'}$$

$$= \left(\int_X |u|^p \, d\mu \right)^{\theta q/p} \left(\int_X |u|^r \, d\mu \right)^{(1-\theta)q/r}.$$

(iii) If $u \neq 0$ and the right-hand side of (2.2) is finite, then the equality in (2.2) holds if and only if there exists $c \geq 0$ such that

  a) $|v| = c|u|^{p-1}$ if $1 < p < \infty$;
  b) $|v| \leq c$ and $|v(x)| = c$ whenever $u(x) \neq 0$ if $p = 1$;
  c) $|u| \leq c$ and $|u(x)| = c$ whenever $v(x) \neq 0$ if $p = \infty$.

We now turn to the relation between different L^p spaces.

Theorem 2.5. *Let $(X, \mathfrak{M}, \mu)$ be a measure space. Suppose that $1 \leq p < q < \infty$. Then*

(i) $L^p(X)$ is not contained in $L^q(X)$ if and only if X contains measurable sets of arbitrarily small positive measure;

(ii) $L^q(X)$ is not contained in $L^p(X)$ if and only if X contains measurable sets of arbitrarily large finite measure.

Proof. (i) Assume that $L^p(X)$ is not contained in $L^q(X)$. Then there exists $u \in L^p(X)$ such that

$$\int_X |u|^q \, d\mu = \infty. \tag{2.3}$$

For each $n \in \mathbb{N}$ let

$$E_n := \{x \in X : |u(x)| > n\}.$$

Then

$$\mu(E_n) \leq \frac{1}{n^p} \int_X |u|^p \, d\mu \to 0$$

as $n \to \infty$. Thus, it suffices to show that $\mu(E_n) > 0$ for all n sufficiently large. If to the contrary, $\mu(E_n) = 0$ for infinitely many n, we have that

$$\int_X |u|^q \, d\mu = \int_{\{|u| \leq n\}} |u|^q \, d\mu \leq n^{q-p} \int_{\{|u| \leq n\}} |u|^p \, d\mu < \infty,$$

which is a contradiction with (2.3). Hence, X contains measurable sets of arbitrarily small positive measure.

Conversely, assume that X contains measurable sets of arbitrarily small positive measure. Then it is possible to construct a sequence of pairwise disjoint sets $\{E_n\} \subset \mathfrak{M}$ such that $\mu(E_n) > 0$ for all $n \in \mathbb{N}$ and

$$\mu(E_n) \searrow 0.$$

Let

$$u := \sum_{n=1}^{\infty} c_n \chi_{E_n},$$

where $c_n \nearrow \infty$ are chosen such that

$$\sum_{n=1}^{\infty} c_n^q \mu(E_n) = \infty, \quad \sum_{n=1}^{\infty} c_n^p \mu(E_n) < \infty. \tag{2.4}$$

Then $u \in L^p(X) \setminus L^q(X)$.

(ii) Assume that $L^q(X)$ is not contained in $L^p(X)$. Then there exists $u \in L^q(X)$ such that

$$\int_X |u|^p \, d\mu = \infty. \tag{2.5}$$

For each $n \in \mathbb{N}$ let

$$F_n := \left\{ x \in X : \frac{1}{n+1} < |u(x)| \leq \frac{1}{n} \right\}$$

and let

$$F_\infty := \{x \in X : 0 < |u(x)| \leq 1\} = \bigcup_{n=1}^{\infty} F_n.$$

If $\mu(F_\infty) < \infty$, then

$$\int_X |u|^p \, d\mu = \int_{\{|u|\leq 1\}} |u|^p \, d\mu + \int_{\{|u|>1\}} |u|^p \, d\mu$$

$$\leq \mu(F_\infty) + \int_{\{|u|>1\}} |u|^q \, d\mu < \infty,$$

which contradicts (2.5). Hence, $\mu(F_\infty) = \infty$. On the other hand, since for every $n \in \mathbb{N}$,

$$\infty > \int_X |u|^q \, d\mu \geq \int_{\{\frac{1}{n+1}<|u|\leq\frac{1}{n}\}} |u|^q \, d\mu \geq \frac{1}{(n+1)^q}\mu(F_n),$$

it follows that X contains measurable sets of arbitrarily large finite measure. Indeed, setting

$$G_n := \bigcup_{k=1}^n F_k,$$

we have that $\mu(G_n) < \infty$, while by Proposition 1.7(i),

$$\mu(G_n) \to \mu(F_\infty) = \infty.$$

Conversely, assume that X contains measurable sets of arbitrarily large finite measure. Then it is possible to construct a sequence of pairwise disjoint sets $\{E_n\} \subset \mathfrak{M}$ of finite measure such that

$$\mu(E_n) \nearrow \infty.$$

Let

$$u := \sum_{n=1}^\infty c_n \chi_{E_n},$$

where $c_n \searrow 0$ are chosen such that

$$\sum_{n=1}^\infty c_n^q \mu(E_n) < \infty, \quad \sum_{n=1}^\infty c_n^p \mu(E_n) = \infty. \tag{2.6}$$

Then $u \in L^q(X) \setminus L^p(X)$.

Exercise 2.6. (i) Let $X = [0,1]$ and let μ be the Lebesgue measure. Show that for every $1 \leq p < \infty$ the function

$$u(x) = \frac{1}{x^{1/p} \log^{2/p}\left(\frac{2}{x}\right)}$$

is in $L^p([0,1])$ but not in $L^q([0,1])$ for all $q > p$.
(ii) Construct sequences $c_n \nearrow \infty$ and $c_n \searrow 0$ for which conditions (2.4) and (2.6) hold, respectively.

Theorem 2.7 (Minkowski's inequality). *Let $(X, \mathfrak{M}, \mu)$ be a measure space, let $1 \leq p \leq \infty$, and let $u, v \in L^p(X)$. Then $u + v \in L^p(X)$ and*

$$\|u + v\|_{L^p} \leq \|u\|_{L^p} + \|v\|_{L^p}.$$

By identifying functions with their equivalence classes $[u]$, it follows from Minkowski's inequality that $\|\cdot\|_{L^p}$ is a norm on $L^p(X)$.

Theorem 2.8. *Let $(X, \mathfrak{M}, \mu)$ be a measure space. Then*

(i) $L^p(X)$ is a Banach space for $1 \leq p \leq \infty$;
(ii) $L^2(X)$ is a Hilbert space.

Exercise 2.9. When $0 < p < 1$ we can still define the space

$$L^p(X, \mathfrak{M}, \mu) := \left\{ u : X \to [-\infty, \infty] : u \text{ measurable and } \int_X |u|^p \, d\mu < \infty \right\}.$$

This is no longer a normed space. Using the elementary inequality

$$(a + b)^p \leq a^p + b^p,$$

where $a, b \geq 0$, show that

$$d_p(u, v) := \int_X |u - v|^p \, d\mu$$

is a metric (provided we identify measurable functions that coincide μ a.e.) and that $(L^p(X, \mathfrak{M}, \mu), d_p)$ is a complete metric space. Prove also that the family of balls is a base for a topology that renders $L^p(X, \mathfrak{M}, \mu)$ a topological vector space.

Next we study some density results for $L^p(X)$ spaces.

Theorem 2.10. *Let $(X, \mathfrak{M}, \mu)$ be a measure space. Then the family of all simple functions in $L^p(X)$ is dense in $L^p(X)$ for $1 \leq p \leq \infty$.*

Proof. Assume first that $1 \leq p < \infty$. Since $(u^+)^p$, $(u^-)^p \in L^1(X)$, by Theorem 1.74 there exist increasing sequences $\{s_n\}$ and $\{t_n\}$ of simple functions, each of which is bounded in X and vanishes except on a set of finite measure μ, such that $\{(s_n(x))^p\}$ converges monotonically to $(u^+)^p(x)$ for μ a.e. $x \in X$ and $\{(t_n(x))^p\}$ converges monotonically to $(u^-)^p(x)$ for μ a.e. $x \in X$. Then for each $n \in \mathbb{N}$ the function $S_n := s_n - t_n$ is still simple, belongs to $L^p(X)$, and

$$
\begin{aligned}
|u(x) - S_n(x)|^p &= \left| u^+(x) - s_n(x) - (u^-(x) - t_n(x)) \right|^p \\
&\leq 2^{p-1} \left(u^+(x) - s_n(x) \right)^p + 2^{p-1} \left(u^-(x) - t_n(x) \right)^p \\
&\leq 2^{p-1} \left(u^+(x) \right)^p + 2^{p-1} \left(u^-(x) \right)^p
\end{aligned}
$$

for μ a.e. $x \in X$. Since $u(x) - S_n(x) \to 0$ as $n \to \infty$ for μ a.e. $x \in X$, we may apply the Lebesgue dominated convergence theorem to conclude that $S_n \to u$ in $L^p(X)$.

For $p = \infty$ the result follows by the second part of Theorem 1.74 applied to the bounded functions u^+, u^-.

The next result gives conditions on X and μ that ensure the density of continuous functions in $L^p(X)$.

Theorem 2.11. *Let $(X, \mathfrak{M}, \mu)$ be a measure space, with X a normal space and μ a Borel measure such that*

$$\mu(E) = \sup\{\mu(C): C \text{ closed}, C \subset E\} = \inf\{\mu(A): A \text{ open}, A \supset E\}$$

for every set $E \in \mathfrak{M}$ with finite measure. Then $L^p(X) \cap C_b(X)$ is dense in $L^p(X)$ for $1 \le p < \infty$.

Proof. Since by Theorem 2.10 simple functions in $L^p(X)$ are dense in $L^p(X)$, it suffices to approximate in $L^p(X)$ functions χ_E, with $E \in \mathfrak{M}$ and $\mu(E) < \infty$, by functions in $L^p(X) \cap C_b(X)$. Thus, fix $E \in \mathfrak{M}$ with $\mu(E) < \infty$, and for any $\varepsilon > 0$ find an open set $A \supset E$ and a closed set $C \subset E$ such that

$$\mu(A \setminus C) \le \varepsilon^p.$$

By Urysohn's lemma there exists a continuous function $u : X \to [0,1]$ such that $u \equiv 1$ in C and $u \equiv 0$ in $X \setminus A$. Since $\operatorname{supp} u \subset A$ and $\mu(A) < \infty$, it follows that $u \in L^p(X) \cap C_b(X)$. Moreover,

$$\int_X |\chi_E - u|^p \, d\mu = \int_{A \setminus C} |\chi_E - u|^p \, d\mu \le \mu(A \setminus C) \le \varepsilon^p,$$

and the result follows.

Definition 2.12. *Let $(X, \mathfrak{M}, \mu)$ be a measure space, with X a topological space, $\mu : \mathfrak{M} \to [0, \infty]$ a Borel measure, and $1 \le p \le \infty$. A measurable function $u : X \to [-\infty, \infty]$ is said to belong to $L^p_{\mathrm{loc}}(X)$ if $u \in L^p(K)$ for every compact set $K \subset X$. A sequence $\{u_n\} \subset L^p_{\mathrm{loc}}(X)$ is said to converge to u in $L^p_{\mathrm{loc}}(X)$ if $u_n \to u$ in $L^p(K)$ for every compact set $K \subset X$.*

Definition 2.13. *A measurable space $(X, \mathfrak{M})$ is called* separable *if there exists a sequence $\{E_n\} \subset \mathfrak{M}$ such that the smallest σ-algebra that contains all the sets E_n is $\mathfrak{M}$. In this case $\mathfrak{M}$ is said to be* generated *by the sequence $\{E_n\}$.*

Example 2.14. The σ-algebra of all Lebesgue measurable sets in $\mathbb{R}^N$ is generated by the countable family of cubes with centers in $\mathbb{Q}^N$ and rational side length.

Exercise 2.15. Prove that if X is a separable metric space and $\mathfrak{M}$ is the Borel σ-algebra, then X is a separable measurable space.

Theorem 2.16. *Let $(X, \mathfrak{M})$ be a separable measurable space with $\mathfrak{M}$ generated by a sequence $\{E_n\} \subset \mathfrak{M}$, and assume that μ is σ-finite. Let $\mathfrak{N}$ be the smallest algebra containing $\{E_n\}$. Then simple functions of the form*

$$\sum_{i=1}^{n} c_i \chi_{F_i},$$

where $n \in \mathbb{N}$, $c_i \in \mathbb{Q}$, and $F_i \in \mathfrak{N}$, $\mu(F_i) < \infty$, $i = 1, \ldots, n$, form a countable dense subset of $L^p(X)$ for $1 \le p < \infty$. In particular, $L^p(X)$ is separable for $1 \le p < \infty$.

Proof. Since by Theorem 2.10 simple functions in $L^p(X)$ are dense in $L^p(X)$, it suffices to approximate in $L^p(X)$ functions χ_E, with $E \in \mathfrak{M}$ and $\mu(E) < \infty$, by χ_{F_n} for some $F_n \in \mathfrak{N}$.

Step 1: Assume that $\mu(X) < \infty$. Let $\mathfrak{M}'$ be the family of sets $G \in \mathfrak{M}$ for which there exists a sequence $\{F_j\} \subset \mathfrak{N}$ such that $\chi_{F_j} \to \chi_G$ in $L^p(X)$ as $j \to \infty$. We claim that $\mathfrak{M}'$ is a σ-algebra.

Note that $\mathfrak{N} \subset \mathfrak{M}'$, and so $\emptyset$, $X \in \mathfrak{M}'$. Moreover, if $G \in \mathfrak{M}'$ then $X \setminus G \in \mathfrak{M}'$. Indeed, let $\{F_j\} \subset \mathfrak{N}$ be such that $\chi_{F_j} \to \chi_G$ in $L^p(X)$ as $j \to \infty$. Then $\{X \setminus F_j\} \subset \mathfrak{N}$ and

$$\chi_{X \setminus F_j} = 1 - \chi_{F_j} \to 1 - \chi_G = \chi_{X \setminus G}$$

in $L^p(X)$ as $j \to \infty$.

Next, assume that $G_1, G_2 \in \mathfrak{M}'$ and let $\left\{F_j^{(i)}\right\} \subset \mathfrak{N}$, $i = 1, 2$, be such that $\chi_{F_j^{(i)}} \to \chi_{G_i}$ in $L^p(X)$ as $j \to \infty$. By selecting a subsequence if necessary, we may assume (see Theorem 2.20 below) that $\chi_{F_j^{(i)}}(x) \to \chi_{G_i}(x)$ as $j \to \infty$ for μ a.e. $x \in X$, $i = 1, 2$, and so $\chi_{F_j^{(1)} \cup F_j^{(2)}}(x) \to \chi_{G_1 \cup G_2}(x)$ for μ a.e. $x \in X$. By the Lebesgue dominated convergence theorem it follows that $\chi_{F_j^{(1)} \cup F_j^{(2)}} \to \chi_{G_1 \cup G_2}$ in $L^p(X)$, and so $G_1 \cup G_2 \in \mathfrak{M}'$. Hence $\mathfrak{M}'$ is an algebra.

Finally, to show that $\mathfrak{M}'$ is a σ-algebra, consider $\{G_i\} \subset \mathfrak{M}'$ and define

$$G_\infty := \bigcup_{i=1}^{\infty} G_i, \quad G_i' := G_i \setminus \bigcup_{l=1}^{i-i} G_l.$$

Since $\mathfrak{M}'$ is an algebra, it follows that $G_i' \in \mathfrak{M}'$, and since the sets G_i' are pairwise disjoint, we have that

$$\infty > \mu(G_\infty) = \mu\left(\bigcup_{i=1}^{\infty} G_i'\right) = \sum_{i=1}^{\infty} \mu(G_i').$$

Thus, for any fixed $\varepsilon > 0$ we can find a positive integer i_ε so large that

$$\sum_{i=i_\varepsilon+1}^{\infty} \mu(G_i') \le \left(\frac{\varepsilon}{2}\right)^p.$$

Since $G_\varepsilon := \bigcup_{i=1}^{i_\varepsilon} G_i \in \mathfrak{M}'$, there exists a sequence $\left\{ F_j^{(\varepsilon)} \right\} \subset \mathfrak{N}$ such that $\chi_{F_j^{(\varepsilon)}} \to \chi_{G_\varepsilon}$ in $L^p(X)$ as $j \to \infty$, and so for all j sufficiently large,

$$\left\| \chi_{F_j^{(\varepsilon)}} - \chi_{G_\varepsilon} \right\|_{L^p} \le \frac{\varepsilon}{2}.$$

In turn, by Minkowski's inequality,

$$\left\| \chi_{F_j^{(\varepsilon)}} - \chi_{G_\infty} \right\|_{L^p} = \left\| \chi_{F_j^{(\varepsilon)}} - \chi_{G_\varepsilon} - \chi_{\bigcup_{i>i_\varepsilon} G_i'} \right\|_{L^p}$$

$$\le \left\| \chi_{F_j^{(\varepsilon)}} - \chi_{G_\varepsilon} \right\|_{L^p} + \left(\sum_{i=i_\varepsilon+1}^\infty \mu(G_i') \right)^{1/p} \le \varepsilon.$$

By taking $\varepsilon = \frac{1}{n}$ and using a diagonalization argument, we can construct a sequence $\left\{ F_{j_n}^{(1/n)} \right\} \subset \mathfrak{N}$ such that $\chi_{F_{j_n}^{(1/n)}} \to \chi_{G_\infty}$ in $L^p(X)$ as $n \to \infty$. This shows that $\bigcup_{i=1}^\infty G_i \in \mathfrak{M}'$. Hence $\mathfrak{M}' \subset \mathfrak{M}$ is a σ-algebra containing $\mathfrak{N}$, and so by the hypothesis that $\mathfrak{N}$ is generated by $\{E_n\}$, it must coincide with $\mathfrak{M}$.

Step 2: Here we remove the additional hypothesis that $\mu(X) < \infty$. Since μ is σ-finite, let $\{G_j\} \subset \mathfrak{M}$ be a sequence of pairwise disjoint sets of finite measure such that

$$X = \bigcup_{j=1}^\infty G_j.$$

Applying the results of Step 1 to $(G_j, \mathfrak{M} \lfloor G_j)$, we may find $F_n^{(j)} \in \mathfrak{N}$ such that $\chi_{F_n^{(j)} \cap G_j} \to \chi_{E \cap G_j}$ in $L^p(G_j)$ as $n \to \infty$. Fix $\varepsilon > 0$ and find j_0 such that

$$\sum_{j=j_0+1}^\infty \mu(E \cap G_j) \le \frac{\varepsilon}{2}.$$

Let n_0 be such that for every $n \ge n_0$ and for all $j = 1, \dots, j_0$,

$$\int_X \left| \chi_{E \cap G_j} - \chi_{F_n^{(j)} \cap G_j} \right|^p d\mu = \int_{G_j} \left| \chi_{E \cap G_j} - \chi_{F_n^{(j)} \cap G_j} \right|^p d\mu \le \frac{\varepsilon}{2j_0}.$$

Since the sets G_j are pairwise disjoint, we have that

$$\int_X \left| \chi_E - \chi_{\bigcup_{k \le j_0} F_n^{(k)} \cap G_k} \right|^p d\mu = \sum_{j=1}^\infty \int_{G_j} \left| \chi_E - \chi_{\bigcup_{k \le j_0} F_n^{(k)} \cap G_k} \right|^p d\mu$$

$$= \sum_{j=j_0+1}^\infty \int_{G_k} |\chi_E|^p d\mu + \sum_{j=1}^{j_0} \int_{G_j} \left| \chi_E - \chi_{F_n^{(j)} \cap G_j} \right|^p d\mu \le \varepsilon,$$

and this concludes the proof.

Step 3: We show that $\mathfrak{N}$ is countable. Recursively, we define the families of sets $\mathcal{C}_l \subset \mathcal{P}(X)$ as follows:

$$\mathcal{C}_1 := \{E_n\} \cup \{X, \emptyset\},$$
$$\mathcal{C}_{l+1} := \mathcal{C}_l \cup \{X \setminus E : E \in \mathcal{C}_l\} \cup \{E \cup F : E, F \in \mathcal{C}_l\}.$$

By induction, each $\mathcal{C}_l$ is countable, and so is

$$\mathcal{C}_\infty = \bigcup_{l=1}^\infty \mathcal{C}_l.$$

We claim that $\mathcal{C}_\infty = \mathfrak{N}$. Indeed, $\mathcal{C}_\infty$ contains all E_n, X, and $\emptyset$, and for any E_1, $E_2 \in \mathcal{C}_\infty$ there are l_1, l_2 with $E_i \in \mathcal{C}_{l_i}$ and hence $E_1 \cup E_2 \in \mathcal{C}_{\max\{l_1, l_2\}+1}$ and $X \setminus E_1 \in \mathcal{C}_{l_1+1}$. Hence, $\mathcal{C}_\infty$ is an algebra that contains $\{E_n\}$, and so $\mathfrak{N} \subset \mathcal{C}_\infty$. We conclude that $\mathcal{C}_\infty \subset \mathfrak{N}$.

Remark 2.17. From the proof of the previous theorem it follows that given any $E \in \mathfrak{M}$, its characteristic function may be approximated in L^p by characteristic functions of sets in $\mathfrak{N}$.

2.1.2 Strong Convergence in L^p

In this subsection we study different modes of convergence and their relation to one another.

Definition 2.18. *Let* $(X, \mathfrak{M}, \mu)$ *be a measure space and let* u_n, $u : X \to \mathbb{R}$ *be measurable functions.*

(i) $\{u_n\}$ *is said to* converge to u almost uniformly *if for every* $\varepsilon > 0$ *there exists a set* $E \in \mathfrak{M}$ *such that* $\mu(E) < \varepsilon$ *and* $\{u_n\}$ *converges to* u *uniformly in* $X \setminus E$;

(ii) $\{u_n\}$ *is said to* converge to u in measure *if for every* $\varepsilon > 0$,

$$\lim_{n \to \infty} \mu(\{x \in X : |u_n(x) - u(x)| > \varepsilon\}) = 0.$$

Definition 2.19. *Let* $(X, \mathfrak{M}, \mu)$ *be a measure space with* μ σ-finite, *and let* u_n, $u \in L^\infty(X)$, $n \in \mathbb{N}$. *The sequence* $\{u_n\}$ *is said to* converge to u with respect to the Mackey topology *if* $\{u_n\}$ *converges to* u *in measure and*

$$\sup_{n \in \mathbb{N}} \|u_n\|_{L^\infty} < \infty.$$

The next theorem relates the types of convergence introduced in Definition 2.18 with convergence in L^p.

Theorem 2.20. *Let* $(X, \mathfrak{M}, \mu)$ *be a measure space and let* u_n, $u : X \to \mathbb{R}$ *be measurable functions.*

(i) If $\{u_n\}$ converges to u almost uniformly, then it converges to u in measure and pointwise μ almost everywhere;

(ii) if $\{u_n\}$ converges to u in measure, then there exists a subsequence $\{u_{n_k}\}$ such that $\{u_{n_k}\}$ converges to u almost uniformly (and hence pointwise μ almost everywhere);

(iii) if $\{u_n\}$ converges to u in L^p, $1 \le p \le \infty$, then it converges to u in measure and there exist a subsequence $\{u_{n_k}\}$ and a function $v \in L^p$ such that $\{u_{n_k}\}$ converges to u almost uniformly (and hence pointwise μ almost everywhere) and $|u_{n_k}(x)| \le v(x)$ for μ a.e. $x \in X$ and for all $k \in \mathbb{N}$.

All the other implications fail in general.

Exercise 2.21. (i) Pointwise convergence μ almost everywhere does not imply convergence in measure or in L^p, $1 \le p < \infty$. To see this, let $X = \mathbb{N}$, $\mathfrak{M} = \mathcal{P}(\mathbb{N})$, and let μ be the counting measure. Define $u_n := \chi_{\{n\}}$. Show that $\{u_n\}$ converges to zero pointwise but neither in measure nor in L^p, $1 \le p < \infty$.

(ii) Convergence in measure or in L^p, $1 \le p < \infty$, does not imply pointwise convergence μ almost everywhere. Let $X = [0, 1)$, $\mathfrak{M} = \mathcal{B}([0,1))$, and let μ be the Lebesgue measure. Consider the sequence of intervals $[0, \frac{1}{2})$, $[\frac{1}{2}, 1)$, $[0, \frac{1}{3})$, $[\frac{1}{3}, \frac{2}{3})$, $[\frac{2}{3}, 1)$, $[0, \frac{1}{4})$, $[\frac{1}{4}, \frac{2}{4})$, $[\frac{2}{4}, \frac{3}{4})$, $[\frac{3}{4}, 1)$, ... Let u_n be the characteristic function of the nth interval of this sequence. Show that $\{u_n\}$ converges to zero in measure and in L^p, $1 \le p < \infty$, but the limit

$$\lim_{n \to \infty} u_n(x)$$

does not exist for every $x \in [0, 1)$.

(iii) Pointwise convergence μ almost everywhere does not imply almost uniform convergence. Let $X = \mathbb{N}$, $\mathfrak{M} = \mathcal{P}(\mathbb{N})$, and let μ be the counting measure. Define $u_n := \chi_{\{1,...,n\}}$. Show that $\{u_n\}$ converges to one pointwise but not almost uniformly.

Concerning (iii) in the example above, we remark that when the measure is finite, then convergence μ almost everywhere implies almost uniform convergence. This follows from the next theorem.

Theorem 2.22 (Egoroff). Let $(X, \mathfrak{M}, \mu)$ be a measure space with μ finite and let $u_n : X \to \mathbb{R}$ be measurable functions converging pointwise μ almost everywhere. Then $\{u_n\}$ converges almost uniformly (and hence in measure).

In order to characterize convergence in $L^p(X)$ for $1 \le p < \infty$ we need to introduce the notion of equi-integrability.

Definition 2.23. Let $(X, \mathfrak{M}, \mu)$ be a measure space. A family $\mathcal{F}$ of measurable functions $u : X \to [-\infty, \infty]$ is said to be

(i) equi-integrable *if for every $\varepsilon > 0$ there exists $\delta > 0$ such that*

$$\int_E |u|\, d\mu \le \varepsilon$$

for all $u \in \mathcal{F}$ and for every measurable set $E \subset X$ with $\mu(E) \le \delta$.
(ii) *p-equi-integrable, $p > 0$, if the family of functions $\{|u|^p : u \in \mathcal{F}\}$ is equi-integrable.*

Theorem 2.24 (Vitali convergence theorem). *Let $(X, \mathfrak{M}, \mu)$ be a measure space, let $1 \le p < \infty$, and let $u_n, u : X \to \mathbb{R}$ be measurable functions. Then $\{u_n\}$ converges to u in $L^p(X)$ if and only if*

   (i) *$\{u_n\}$ converges to u in measure;*
   (ii) *$\{u_n\}$ is p-equi-integrable;*
   (iii) *for every $\varepsilon > 0$ there exists $E \subset X$ with $E \in \mathfrak{M}$ such that $\mu(E) < \infty$ and*

$$\int_{X \setminus E} |u_n|^p\, d\mu \le \varepsilon$$

for all n.

Remark 2.25. Note that condition (iii) is automatically satisfied when X has finite measure.

Example 2.26. Let $X = \mathbb{R}$, $\mathfrak{M} = \mathcal{B}(\mathbb{R})$, and let μ be the Lebesgue measure. The sequence $u_n = n\chi_{[0, \frac{1}{n}]}$ converges in measure to zero, satisfies (iii), but is not equi-integrable, while the sequence $u_n = \frac{1}{n}\chi_{[n, 2n]}$ converges in measure to zero, is equi-integrable, but does not satisfy (iii).

    Conditions (ii) and (iii) hold if the sequence is dominated by a function $v \in L^p(X)$. Precisely, we have the following result, which will be used to prove the decomposition lemma below.

Proposition 2.27. *Let $(X, \mathfrak{M}, \mu)$ be a measure space and let $u_n : X \to [-\infty, \infty]$ be measurable functions. If there exists $v \in L^p(X)$ such that*

$$|u_n(x)| \le v(x) \tag{2.7}$$

for all n and for μ a.e. $x \in X$, then $\{u_n\}$ satisfies conditions (ii) and (iii) of Vitali's theorem.

Proof. Define

$$\nu(E) := \int_E |v|^p\, d\mu, \quad E \in \mathfrak{M}.$$

Then ν is a finite measure absolutely continuous with respect to μ, and so by Proposition 1.99, for every $\varepsilon > 0$ there exists $\delta > 0$ such that

$$\int_E |v|^p \, d\mu \le \varepsilon \tag{2.8}$$

for every measurable set $E \subset X$ with $\mu(E) \le \delta$, while by Proposition 1.7(i),

$$\lim_{t \to \infty} \int_{\{x \in X: \frac{1}{t} < |v| < t\}} |v|^p \, d\mu = \int_{\{x \in X: 0 < |v| < \infty\}} |v|^p \, d\mu \tag{2.9}$$

$$= \int_X |v|^p \, d\mu,$$

where we have used the fact that the set $\{x \in X : |v(x)| = \infty\}$ has μ measure zero (see Remark 1.78).

Fix $\varepsilon > 0$ and let $\delta > 0$ be such that (2.8) is satisfied.

Then for every measurable set $E \subset X$ with $\mu(E) \le \delta$, by (2.7) and (2.8) for each $n \in \mathbb{N}$ we have

$$\int_E |u_n|^p \, d\mu \le \int_E |v|^p \, d\mu \le \varepsilon.$$

Hence $\{u_n\}$ is p-equi-integrable.

Next, by (2.9) there exists $t > 0$ so large that

$$\int_X |v|^p \, d\mu - \int_{\{x \in X: \frac{1}{t} < |v| < t\}} |v|^p \, d\mu = \int_{X \setminus \{x \in X: \frac{1}{t} < |v| < t\}} |v|^p \, d\mu \le \varepsilon.$$

Set

$$E_\varepsilon := \left\{ x \in X : \frac{1}{t} < |v(x)| < t \right\}.$$

Then $\mu(E_\varepsilon) < \infty$ by Remark 1.78 and by (2.7),

$$\int_{X \setminus E_\varepsilon} |u_n|^p \, d\mu \le \int_{X \setminus E_\varepsilon} |v|^p \, d\mu \le \varepsilon$$

for all n.

Remark 2.28. Note that any finite family $\{u_1, \ldots, u_l\}$ of functions in $L^p(X)$ satisfies the hypothesis of the previous proposition. Indeed, it suffices to define $v := \max\{|u_1|, \ldots, |u_l|\}$.

In view of Vitali's theorem it becomes important to understand equi-integrability.

Theorem 2.29. *Let $(X, \mathfrak{M}, \mu)$ be a measure space, let $1 \le p < \infty$, and let $\mathcal{F} \subset L^p(X)$ be a bounded set. Then the following conditions are equivalent:*

(i) $\mathcal{F}$ is p-equi-integrable;
(ii)

$$\lim_{t \to \infty} \sup_{u \in \mathcal{F}} \int_{\{x \in X: |u| > t\}} |u|^p \, d\mu = 0; \tag{2.10}$$

(iii) **(De la Vallée Poussin)** there exists an increasing function $\gamma : [0, \infty) \rightarrow$
[0, ∞], with

$$\lim_{t \to \infty} \frac{\gamma(t)}{t} = \infty \tag{2.11}$$

such that

$$\sup_{u \in \mathcal{F}} \int_X \gamma(|u|^p) \, d\mu < \infty. \tag{2.12}$$

Proof. By replacing the family $\mathcal{F}$ with $\{|u|^p : u \in \mathcal{F}\}$, without loss of generality, we may assume that $p = 1$. Let

$$C := \sup_{u \in \mathcal{F}} \|u\|_{L^1(X)} < \infty. \tag{2.13}$$

Step 1: Assume that $\mathcal{F}$ is equi-integrable and, given $\varepsilon > 0$ let $\delta > 0$ be such that

$$\sup_{u \in \mathcal{F}} \int_F |u| \, d\mu \leq \varepsilon$$

for every measurable set $F \subset X$ with $\mu(F) \leq \delta$. Choose $t_\varepsilon > 0$ such that $\frac{C}{t_\varepsilon} \leq \delta$. Then, also by (2.13), for every $u \in \mathcal{F}$ and for all $t \geq t_\varepsilon$ we have

$$\mu(\{x \in X : |u| > t\}) \leq \frac{1}{t} \|u\|_{L^1(X)} \leq \frac{C}{t} \leq \delta,$$

and so

$$\sup_{u \in \mathcal{F}} \int_{\{x \in X : |u| > t\}} |u| \, d\mu \leq \varepsilon,$$

and this validates (ii).

Conversely, suppose that (ii) holds, fix $\varepsilon > 0$, and choose $t_\varepsilon > 0$ such that

$$\sup_{u \in \mathcal{F}} \int_{\{x \in X : |u| > t_\varepsilon\}} |u| \, d\mu \leq \frac{\varepsilon}{2}.$$

Then for every measurable set $F \subset X$ with $\mu(F) \leq \frac{\varepsilon}{2t_\varepsilon}$ and for all $u \in \mathcal{F}$ we have

$$\int_F |u| \, d\mu = \int_{\{x \in F : |u| > t_\varepsilon\}} |u| \, d\mu + \int_{\{x \in F : |u| \leq t_\varepsilon\}} |u| \, d\mu$$

$$\leq \frac{\varepsilon}{2} + t_\varepsilon \mu(F) \leq \varepsilon.$$

Step 2: Assume that (ii) holds and construct an increasing sequence of nonnegative integers $\{k_i\}$ such that

$$\sup_{u \in \mathcal{F}} \int_{\{x \in X : |u| > k_i\}} |u| \, d\mu \leq \frac{1}{2^{i+1}}.$$

For each $l \in \mathbb{N}_0$ let b_l be the number of nonnegative integers i such that $k_i < l$. Note that $b_l \nearrow \infty$ as $l \to \infty$. Define

$$\gamma(t) := tb_l \quad \text{if } t \in [l, l+1).$$

Then

$$\frac{\gamma(t)}{t} \geq b_{\lceil t \rceil} \to \infty$$

as $t \to \infty$, where $\lceil t \rceil$ is the integer part of t. Moreover, for all $u \in \mathcal{F}$, by Example 1.82,

$$\int_X \gamma(|u|) \, d\mu = \sum_{l=0}^{\infty} \int_{\{l \leq |u| < l+1\}} \gamma(|u|) \, d\mu$$

$$= \sum_{l=0}^{\infty} b_l \int_{\{l \leq |u| < l+1\}} |u| \, d\mu = \sum_{l=0}^{\infty} \sum_{i: k_i < l} \int_{\{l \leq |u| < l+1\}} |u| \, d\mu$$

$$= \sum_{i=0}^{\infty} \sum_{l > k_i} \int_{\{l \leq |u| < l+1\}} |u| \, d\mu \leq \sum_{i=0}^{\infty} \int_{\{|u| > k_i\}} |u| \, d\mu \leq \sum_{i=0}^{\infty} \frac{1}{2^{i+1}}.$$

Conversely, assume that (iii) holds and let

$$M := \sup_{u \in \mathcal{F}} \int_X \gamma(|u|) \, d\mu < \infty.$$

By (2.11), for every $\varepsilon > 0$ there exists $t_\varepsilon > 0$ such that

$$\gamma(t) \geq \frac{2tM}{\varepsilon} \quad \text{for all } t \geq t_\varepsilon.$$

Then for every measurable set $F \subset X$ with $\mu(F) \leq \frac{\varepsilon}{2t_\varepsilon}$ and for all $u \in \mathcal{F}$ we have

$$\int_F |u| \, d\mu = \int_{\{x \in F: |u| > t_\varepsilon\}} |u| \, d\mu + \int_{\{x \in F: |u| \leq t_\varepsilon\}} |u| \, d\mu$$

$$\leq \frac{\varepsilon}{2M} \int_X \gamma(|u|) \, d\mu + t_\varepsilon \mu(F) \leq \varepsilon.$$

Hence $\mathcal{F}$ is equi-integrable.

Remark 2.30. Often the sufficient condition (2.12) for equi-integrability is stated for a continuous, finite, increasing function γ. Note that if γ takes the value ∞ at some $t_0 \geq 0$, then $\gamma \equiv \infty$ on $[t_0, \infty)$. Therefore condition (2.12) implies that the set $\mathcal{F}$ is bounded in L^∞. Moreover, if γ is finite, then it is possible to construct a finite, nonnegative, increasing, continuous, convex function $\tilde{\gamma}$ below γ that still satisfies (2.11) (see Remark 4.99 below), and therefore it suffices to apply the previous theorem with $\tilde{\gamma}$ in place of γ.

An important consequence of Theorem 2.29 is the following result.

Lemma 2.31 (Decomposition lemma, I). *Let $(X, \mathfrak{M}, \mu)$ be a measure space, and let $u_n : X \to [-\infty, \infty]$ be measurable functions such that*

$$\sup_n \|u_n\|_{L^1} < \infty.$$

For $r > 0$ consider the truncation $\tau_r : \mathbb{R} \to \mathbb{R}$ defined by

$$\tau_r(z) := \begin{cases} z & \text{if } |z| \leq r, \\ \dfrac{z}{|z|} r & \text{if } |z| > r. \end{cases}$$

Then there exist a subsequence $\{u_{n_k}\}$ of $\{u_n\}$ and an increasing sequence of positive integers $j_k \to \infty$ such that the truncated sequence $\{\tau_{j_k} \circ u_{n_k}\}$ is equi-integrable and

$$\mu\left(\{x \in X : u_{n_k}(x) \neq (\tau_{j_k} \circ u_{n_k})(x)\}\right) \to 0$$

as $k \to \infty$.

We first state and prove a simple auxiliary criterion for equi-integrability.

Lemma 2.32. *Let $(X, \mathfrak{M}, \mu)$ be a measure space, and let $u_n : X \to [-\infty, \infty]$ be measurable functions such that $\sup_n \|u_n\|_{L^1} < \infty$. Then $\{u_n\}$ is equi-integrable if and only if*

$$\lim_{\substack{j \to \infty \\ j \in \mathbb{N}}} \sup_n \int_{\{x \in X : |u_n| > j\}} (|u_n| - j) \, d\mu = 0.$$

Proof. If $\{u_n\}$ is equi-integrable, then

$$\sup_n \int_{\{x \in X : |u_n| > j\}} (|u_n| - j) \, d\mu \leq \sup_n \int_{\{x \in X : |u_n| > j\}} |u_n| \, d\mu \to 0$$

as $j \to \infty$, where we have used (2.10).

Conversely, fix $\varepsilon > 0$, let

$$c := \sup_n \|u_n\|_{L^1},$$

and choose $j \in \mathbb{N}$ so large that

$$\sup_n \int_{\{x \in X : |u_n| > j\}} (|u_n| - j) \, d\mu \leq \frac{\varepsilon}{2}.$$

If $t \geq \frac{2cj}{\varepsilon}$, then

$$\int_{\{x \in X : |u_n| > t\}} |u_n| \, d\mu = \int_{\{x \in X : |u_n| > t\}} (|u_n| - j) \, d\mu + j\mu\left(\{x \in X : |u_n| > t\}\right)$$

$$\leq \frac{\varepsilon}{2} + \frac{j}{t} \int_X |u_n| \, d\mu \leq \varepsilon.$$

In view of Theorem 2.29 we conclude that the sequence is equi-integrable.

Proof (Lemma 2.31). Without loss of generality we may assume that

$$\sup_n \int_X |u_n|\, d\mu \leq 1.$$

For $j \in \mathbb{N}$ define

$$\varphi(j) := \sup_n \int_{\{|u_n| \geq j\}} (|u_n| - j)\, d\mu, \quad \varphi_\infty := \limsup_{j \to \infty} \varphi(j). \tag{2.14}$$

If $\varphi_\infty = 0$, then by the previous lemma we deduce that $\{u_n\}$ is equi-integrable and the statement of the decomposition lemma is verified with $n_k := k$ and $j_k := k$. If $\varphi_\infty > 0$, then find $j_1, n_1 \in \mathbb{N}$ so large that

$$\int_{\{|u_{n_1}| \geq j_1\}} (|u_{n_1}| - j_1)\, d\mu \geq \frac{\varphi_\infty}{2}.$$

By induction, assume that positive integers $n_1 < \ldots < n_{k-1}$ and $j_1 < \ldots < j_{k-1}$ have been chosen, and we claim that there exist $j_k > j_{k-1}$ and $n_k > n_{k-1}$, $j_k, n_k \in \mathbb{N}$, such that

$$\int_{\{|u_{n_k}| \geq j_k\}} (|u_{n_k}| - j_k)\, d\mu \geq \left(1 - \frac{1}{2^k}\right)\varphi_\infty. \tag{2.15}$$

Indeed, if this were false, then

$$\limsup_{j \to \infty} \sup_{n > n_{k-1}} \int_{\{|u_n| \geq j\}} (|u_n| - j)\, d\mu \leq \left(1 - \frac{1}{2^k}\right)\varphi_\infty,$$

and in turn,

$$\limsup_{j \to \infty} \sup_{n \leq n_{k-1}} \int_{\{|u_n| \geq j\}} (|u_n| - j)\, d\mu = \varphi_\infty > 0,$$

and this, in view of the previous lemma, implies that the finite family $\{u_1, \ldots, u_{n_{k-1}}\}$ is not equi-integrable. In view of Remark 2.28 we have reached a contradiction, and hence the claim is established.

Next we show that the sequence $\{\tau_{j_k} \circ u_{n_k}\}$ satisfies the properties requested. Note first that

$$\mu(\{x \in X : u_{n_k}(x) \neq (\tau_{j_k} \circ u_{n_k})(x)\})$$

$$\leq \mu(\{x \in X : |u_{n_k}(x)| \geq j_k\}) \leq \frac{1}{j_k} \int_X |u_n|\, d\mu \leq \frac{1}{j_k} \to 0$$

as $k \to \infty$.

In order to prove equi-integrability of the sequence $\{\tau_{j_k} \circ u_{n_k}\}$, fix $l \in \mathbb{N}$ and let

$$k_l := \min\{k \in \mathbb{N} : j_k \geq l\}.$$

Then $k_l \to \infty$ as $l \to \infty$. Note that if $k < k_l$, then $\{x \in X : |\tau_{j_k} \circ u_{n_k}| \geq l\}$ is empty, while if $k \geq k_l$, then $j_k \geq l$, and so

$$\min \{|\tau_{j_k} \circ u_{n_k}|, l\} = \min \{|u_{n_k}|, l\}.$$

Therefore

$$\int_{\{|\tau_{j_k} \circ u_{n_k}| \geq l\}} (|\tau_{j_k} \circ u_{n_k}| - l) \, d\mu$$

$$= \int_X (|\tau_{j_k} \circ u_{n_k}| - \min \{|\tau_{j_k} \circ u_{n_k}|, l\}) \, d\mu$$

$$= \int_X (|u_{n_k}| - \min \{|u_{n_k}|, l\}) \, d\mu - \int_X (|u_{n_k}| - |\tau_{j_k} \circ u_{n_k}|) \, d\mu$$

$$= \int_{\{|u_{n_k}| \geq l\}} (|u_{n_k}| - l) \, d\mu - \int_{\{|u_{n_k}| \geq j_k\}} (|u_{n_k}| - j_k) \, d\mu$$

$$\leq \varphi(l) - \left(1 - \frac{1}{2^k}\right) \varphi_\infty \leq \varphi(l) - \left(1 - \frac{1}{2^{k_l}}\right) \varphi_\infty,$$

where we have used (2.15) and the fact that $k \geq k_l$. We conclude that

$$\sup_{k \in \mathbb{N}} \int_{\{|\tau_{j_k} \circ u_{n_k}| \geq l\}} (|\tau_{j_k} \circ u_{n_k}| - l) \, d\mu \leq \varphi(l) - \left(1 - \frac{1}{2^{k_l}}\right) \varphi_\infty. \tag{2.16}$$

Note that the right-hand side of the previous inequality converges to zero as $l \to \infty$, and we invoke the previous lemma to conclude the equi-integrability of the sequence $\{\tau_{j_k} \circ u_{n_k}\}$.

Remark 2.33. (i) In Chapter 8, using Young measures techniques we give an alternative proof of this result in the particular case that X is a bounded Lebesgue measurable subset of $\mathbb{R}^N$.

(ii) Note that taking the limit inferior in the inequality (2.16) and using the fact that the left-hand side is nonnegative yields

$$0 \leq \liminf_{l \to \infty} \varphi(l) - \varphi_\infty,$$

which, in view of the definition of φ_∞ in (2.14), implies that φ_∞ is actually a limit.

2.1.3 Dual Spaces

In view of Hölder's inequality, if $1 \leq p < \infty$ and q is its Hölder conjugate exponent, then for every $v \in L^q(X)$ the functional

$$u \in L^p(X) \mapsto \int_X uv \, d\mu$$

is linear and continuous and thus belongs to $(L^p(X))'$.

Under appropriate hypotheses on the measure μ it is possible to identify the dual of $L^p(X)$ with $L^q(X)$ for $1 \leq p < \infty$. We begin with some preliminary results, which give the converse of Hölder's inequality.

Theorem 2.34. *Let $(X, \mathfrak{M}, \mu)$ be a measure space, let $1 \leq p \leq \infty$, and let q be its Hölder conjugate exponent. Let $u : X \to [-\infty, \infty]$ be measurable and assume that uv is integrable for every $v \in L^q(X)$. If the set $\{x \in X : u(x) \neq 0\}$ has σ-finite measure, then $u \in L^p(X)$.*

Proof. We begin by showing that

$$\mu(\{x \in X : |u(x)| = \infty\}) = 0.$$

Indeed, if not, then since $\{x \in X : u(x) \neq 0\}$ has σ-finite measure we would be able to find a measurable set

$$E \subset \{x \in X : |u(x)| = \infty\}$$

with $0 < \mu(E) < \infty$. Since $v := \chi_E \in L^q(X)$, we would have that

$$\infty > \int_X |uv| \, d\mu = \int_E |u| \, d\mu = \infty,$$

which is a contradiction. Hence we are in a position to apply Corollary 1.77 to construct a sequence of simple functions $\{s_n\}$, each of which is bounded and vanishes except on a set of finite measure, such that $\{|s_n(x)|\}$ converges monotonically to $|u(x)|$ for μ a.e. $x \in X$. The functionals

$$L_n(v) := \int_X s_n v \, d\mu, \quad v \in L^q(X),$$

are linear and continuous with (see the proof of (2.19) below)

$$\|L_n\|_{(L^q(X))'} = \|s_n\|_{L^p(X)}.$$

Moreover, since $|s_n(x)| \leq |u(x)|$ for μ a.e. $x \in X$, for any $v \in L^q(X)$ we have that

$$\sup_n |L_n(v)| \leq \sup_n \int_X |s_n v| \, d\mu \leq \int_X |uv| \, d\mu < \infty$$

by the hypothesis on u. It now follows from the Banach–Steinhaus theorem that there exists $M > 0$ such that

$$\sup_n \|L_n\|_{(L^q(X))'} = \sup_n \|s_n\|_{L^p(X)} \leq M.$$

If $1 \leq p < \infty$, then by the Lebesgue monotone convergence theorem we now have

$$M^p \geq \lim_{n \to \infty} \int_X |s_n|^p \, d\mu = \int_X |u|^p \, d\mu,$$

so that $u \in L^p(X)$, while if $p = \infty$ for μ a.e. $x \in X$, we have

$$M \geq \lim_{n \to \infty} |s_n(x)| = |u(x)|,$$

which implies that $\|u\|_{L^\infty(X)} \leq M$.

As a consequence of the previous result we may now prove the converse of Hölder's inequality.

Theorem 2.35 (Converse Hölder's inequality). *Let $(X, \mathfrak{M}, \mu)$ be a measure space, let $1 < p < \infty$, and let q be its Hölder conjugate exponent. Then the following are equivalent:*

(i) for every measurable function $u : X \to [-\infty, \infty]$ such that $uv \in L^1(X)$ for all $v \in L^q(X)$, $u \in L^p(X)$;
(ii) μ has the finite subset property.

Proof. **Step 1:** Assume that (i) holds. If (ii) fails, then we can find a set $E \in \mathfrak{M}$, with $\mu(E) = \infty$, such that $\mu(F) \in \{0, \infty\}$ for all $F \in \mathfrak{M}$, with $F \subset E$. Then the function $u = \chi_E$ is not in $L^p(X)$. On the other hand, if $v \in L^q(X)$, then

$$\{x \in X : v(x) \neq 0\} = \bigcup_{n=1}^{\infty} \left\{ x \in X : |v(x)| \geq \frac{1}{n} \right\}$$

and

$$\frac{1}{n^q} \mu \left(\left\{ x \in X : |v(x)| \geq \frac{1}{n} \right\} \right) \leq \int_X |v|^q \, d\mu < \infty,$$

and so $\mu\left(\left\{x \in E : |v(x)| \geq \frac{1}{n}\right\}\right) = 0$, which implies that $uv = 0$ for μ a.e. in X. Hence $uv \in L^1(X)$ for all $v \in L^q(X)$ and $u \notin L^p(X)$, which contradicts (i).

Step 2: Assume that μ has the finite subset property and let $u : X \to [-\infty, \infty]$ be such that $uv \in L^1(X)$ for all $v \in L^q(X)$. We claim that $u \in L^p(X)$. In view of Theorem 2.34 it suffices to show that the set $\{x \in X : u(x) \neq 0\}$ has σ-finite measure. Let

$$E_n := \left\{ x \in X : |u(x)| \geq \frac{1}{n} \right\}.$$

Since

$$\{x \in X : u(x) \neq 0\} = \bigcup_{n=1}^{\infty} E_n,$$

it remains to show that $\mu(E_n) < \infty$ for all $n \in \mathbb{N}$. Assume by contradiction that $\mu(E_l) = \infty$ for some $l \in \mathbb{N}$. Then, as in the proof of Proposition 1.25, it is possible to show that the set E_l contains a set $F \in \mathfrak{M}$ with σ-finite measure such that $\mu(F) = \infty$. The function $w := u\chi_F$ is such that $wv \in L^1(X)$ for all $v \in L^q(X)$ and the set $\{x \in X : w(x) \neq 0\}$ has σ-finite measure. Hence by Theorem 2.34, $w \in L^q(X)$, which is in contradiction with the fact that $|w(x)| \geq \frac{1}{n}$ on the set F of infinite measure.

Remark 2.36. Note that the nonzero function $f = \chi_E$ constructed in Step 1 belongs to $L^\infty(X)$, while $uv = 0$ for μ a.e. in X and for all $v \in L^1(X)$.

Using the previous results one can show that the dual of $L^p(X)$ may be identified with $L^q(X)$ for $1 < p < \infty$.

Theorem 2.37 (Riesz representation theorem in L^p). *Let $(X, \mathfrak{M}, \mu)$ be a measure space, let $1 < p < \infty$, and let q be its Hölder conjugate exponent. Then every bounded linear functional $L : L^p(X) \to \mathbb{R}$ is represented by a unique $v \in L^q(X)$ in the sense that*

$$L(u) = \int_X uv\,d\mu \quad \text{for every } u \in L^p(X). \tag{2.17}$$

Moreover, the norm of L coincides with $\|v\|_{L^q}$. Conversely, every functional of the form (2.17), where $v \in L^q(X)$, is a bounded linear functional on $L^p(X)$. In particular, $L^p(X)$ is reflexive.

Proof. **Step 1:** Assume first that μ is finite and that $L : L^p(X) \to \mathbb{R}$ is linear, bounded, and $L \geq 0$, that is, $L(u) \geq 0$ whenever $u \geq 0$. Then for every $E \in \mathfrak{M}$ the function χ_E is in $L^q(X)$, and so the set function

$$\nu(E) := L(\chi_E), \quad E \in \mathfrak{M},$$

is well-defined. We claim that ν is countably additive. By the linearity of L it follows that μ is finitely additive. Let $\{E_n\} \subset \mathfrak{M}$ be a sequence of pairwise disjoint sets and denote by E their union. Then

$$\chi_{\bigcup_{n=1}^l E_n} \to \chi_E \quad \text{in } L^p(X),$$

and hence, by the continuity of L,

$$\sum_{n=1}^l \nu(E_n) = \nu\left(\bigcup_{n=1}^l E_n\right) = L\left(\chi_{\bigcup_{n=1}^l E_n}\right) \to L(\chi_E) = \nu\left(\bigcup_{n=1}^\infty E_n\right)$$

as $l \to \infty$. Since ν is absolutely continuous with respect to μ we may apply the Radon–Nikodym theorem to find a unique nonnegative function $v \in L^1(X)$ such that

$$L(\chi_E) = \nu(E) = \int_X \chi_E v\,d\mu \quad \text{for every } E \in \mathfrak{M}.$$

By the linearity of L we conclude that

$$L(s) = \int_X sv\,d\mu$$

for every nonnegative simple function s.

If $u \in L^p(X)$ is a nonnegative function, then by Theorem 1.74 we may find an increasing sequence $\{s_n\}$ of nonnegative simple functions converging to u μ a.e.

Since $0 \le s_n^p \le u^p$, by the Lebesgue dominated convergence theorem $s_n \to u$ in $L^p(X)$, and so $L(s_n) \to L(u)$. On the other hand, by the Lebesgue monotone convergence theorem

$$L(s_n) = \int_X s_n v\,d\mu \to \int_X uv\,d\mu,$$

and so

$$L(u) = \int_X uv\,d\mu \tag{2.18}$$

for all nonnegative $u \in L^p(X)$.

For a general $u \in L^p(X)$ it suffices to write $u = u^+ - u^-$ and apply (2.18) to conclude that

$$L(u) = \int_X uv\,d\mu$$

for all $u \in L^p(X)$. Moreover, since $uv \in L^1(X)$ for all $u \in L^p(X)$, it follows from the converse Hölder's inequality that $v \in L^q(X)$. Next we show that the norm of L coincides with $\|v\|_{L^q(X)}$. By Hölder's inequality,

$$|L(u)| \le \|u\|_{L^p(X)} \|v\|_{L^q(X)}$$

for all $u \in L^p(X)$, and so

$$\|L\|_{(L^p(X))'} \le \|v\|_{L^q(X)}.$$

To prove the reverse inequality, take $u := v^{q-1}$. Then

$$\int_X u^p\,d\mu = \int_X v^{p(q-1)}\,d\mu = \int_X v^q\,d\mu < \infty,$$

while

$$|L(u)| = \int_X uv\,d\mu = \int_X v^q\,d\mu,$$

and so

$$\|L\|_{(L^p(X))'} = \|v\|_{L^q(X)}. \tag{2.19}$$

Step 2: We now remove the extra assumption that μ is finite. Let

$$\mathfrak{M}' := \{E \in \mathfrak{M} : \mu(E) < \infty\}.$$

Note that $\mathfrak{M}'$ is not an algebra. For $E \in \mathfrak{M}'$ define

$$M_E := \{u \in L^p(X) : u = 0 \text{ outside } E\}.$$

By identifying M_E with $L^p(E)$ we can apply Step 1 to the positive linear continuous functional $L : M_E \to \mathbb{R}$ to find a unique function $v_E \in L^q(E)$ such that

$$L(u) = \int_X u v_E \, d\mu \quad \text{for every } u \in M_E \tag{2.20}$$

and

$$\|v_E\|_{L^q(E)} = \sup_{u \in M_E \setminus \{0\}} \frac{|L(u)|}{\|u\|_{L^p(E)}} = \sup_{u \in M_E \setminus \{0\}} \frac{|L(u)|}{\|u\|_{L^p(X)}} \le \|L\|_{(L^p(X))'}. \tag{2.21}$$

Extend v_E to be zero outside E.

Note that in view of the uniqueness of v_E, if $E, F \in \mathfrak{M}'$ then for μ a.e. $x \in E \cap F$,

$$v_E(x) = v_{E \cap F}(x) = v_F(x).$$

In particular,

$$v_{E \cup F} = \max\{v_E, v_F\}, \tag{2.22}$$

while if $E \subset F$, then

$$0 \le v_E \le v_F.$$

By (2.21),

$$\sup_{E \in \mathfrak{M}'} \|v_E\|_{L^q(X)} \le \|L\|_{(L^p(X))'} < \infty,$$

and so we may find a sequence of sets $\{E_n\} \subset \mathfrak{M}'$ such that

$$\lim_{n \to \infty} \|v_{E_n}\|_{L^q(X)} = \sup_{E \in \mathfrak{M}'} \|v_E\|_{L^q(X)}.$$

Replacing v_{E_n} with $v_{\bigcup_{i=1}^n E_i}$, by (2.22) we may assume without loss of generality that the sequence $\{E_n\}$ is increasing (and the same holds for $\{v_{E_n}\}$). Hence for μ a.e. $x \in X$ there exists

$$\lim_{n \to \infty} v_{E_n}(x) =: v(x),$$

and by the Lebesgue monotone convergence theorem,

$$\|v\|_{L^q(X)} = \lim_{n \to \infty} \|v_{E_n}\|_{L^q(X)} = \sup_{E \in \mathfrak{M}'} \|v_E\|_{L^q(X)} \le \|L\|_{(L^p(X))'} < \infty.$$

Next we claim that if $E \in \mathfrak{M}'$ then

$$v_E(x) = v(x) \tag{2.23}$$

for μ a.e. $x \in X$. Indeed, by (2.22),

$$\sup_{E \in \mathfrak{M}'} \|v_E\|_{L^q(X)} = \|v\|_{L^q(X)} = \lim_{n \to \infty} \|v_{E_n}\|_{L^q(X)}$$

$$\leq \lim_{n \to \infty} \|v_{E_n \cup E}\|_{L^q(X)} = \lim_{n \to \infty} \|\max\{v_{E_n}, v_E\}\|_{L^q(X)}$$

$$\leq \sup_{E \in \mathfrak{M}'} \|v_E\|_{L^q(X)}.$$

Hence

$$\|v\|_{L^q(X)} = \lim_{n \to \infty} \|\max\{v_{E_n}, v_E\}\|_{L^q(X)} = \|\max\{v, v_E\}\|_{L^q(X)},$$

and so the claim holds.

By (2.20) and (2.23), for every simple function s vanishing outside a set of finite measure we have

$$L(s) = \int_X sv_{\{s \neq 0\}} \, d\mu = \int_X sv \, d\mu,$$

and by Corollary 1.77, the continuity of L, and the Lebesgue monotone convergence theorem, we conclude that

$$L(u) = \int_X uv \, d\mu$$

for all $u \in L^p(X)$. The fact that $\|L\|_{(L^p(X))'} = \|v\|_{L^q(X)}$ follows exactly as in Step 1.

Step 3: Finally, to remove the assumption that L is positive, for any linear, bounded functional $L : L^p(X) \to \mathbb{R}$, for $u \in L^p(X)$, $u \geq 0$, define

$$L^+(u) := \sup\{L(v) : 0 \leq v \leq u\}, \quad L^-(u) := -\inf\{L(v) : 0 \leq v \leq u\}.$$

We claim that L^+ is additive. Indeed, let $u_1, u_2 \in L^p(X)$ be nonnegative. For any $v_i \in L^p(X)$, with $0 \leq v_i \leq u_i$, $i = 1, 2$, we have

$$L(v_1) + L(v_2) = L(v_1 + v_2) \leq L^+(u_1 + u_2),$$

and by the arbitrariness of v_1 and v_2 we get

$$L^+(u_1) + L^+(u_2) \leq L^+(u_1 + u_2).$$

To prove the opposite inequality, let $v \in L^p(X)$, with $0 \leq v \leq u_1 + u_2$, and define

$$v_1 := \min\{v, u_1\}, \quad v_2 := v - v_1.$$

Then $0 \leq v_i \leq u_i$, $i = 1, 2$, and so

$$L(v) = L(v_1) + L(v_2) \leq L^+(u_1) + L^+(u_2).$$

By the arbitrariness of v we obtain

$$L^+ (u_1 + u_2) \leq L^+ (u_1) + L^+ (u_2) ,$$

and so the claim follows.

Next observe that by the linearity of L, for $t \geq 0$ and $u \in L^p (X)$, $u \geq 0$,

$$L^+ (tu) := \sup \{ L (v) : 0 \leq v \leq tu \} = \sup \{ L (tw) : 0 \leq w \leq u \}$$
$$= t \sup \{ L (w) : 0 \leq w \leq u \} = tL^+ (u) ,$$

while for any $v \in L^p (X)$, with $0 \leq v \leq u$, we have

$$L (v) \leq \| L \|_{(L^p (X))'} \| v \|_{L^p (X)} \leq \| L \|_{(L^p (X))'} \| u \|_{L^p (X)} ,$$

and so

$$L^+ (u) \leq \| L \|_{(L^p (X))'} \| u \|_{L^p (X)} .$$

Similar properties hold for L^-.

Finally, for all $u \in L^p (X)$, $u \geq 0$,

$$L^+ (u) - L (u) = \sup \{ L (v) - L (u) : 0 \leq v \leq u \}$$
$$= \sup \{ -L (u - v) : 0 \leq v \leq u \}$$
$$= - \inf \{ L (u - v) : 0 \leq u - v \leq u \} = L^- (u) .$$

Hence the functionals

$$L^+ (u) := L^+ (u^+) - L^+ (u^-) , \quad L^- (u) := L^- (u^+) - L^- (u^-) , \quad u \in L^p (X) ,$$

are linear, continuous, positive, and

$$L = L^+ - L^- .$$

Applying Step 2 to L^+, L^-, we may find unique functions v_1, $v_2 \in L^q (X)$ such that

$$L^+ (u) = \int_X uv_1 \, d\mu \quad L^- (u) = \int_X uv_2 \, d\mu \quad \text{for every } u \in L^p (X) .$$

The function $v := v_1 - v_2$ satisfies

$$L (u) = \int_X uv \, d\mu \quad \text{for every } u \in L^p (X) ,$$

and once again the fact that $\| L \|_{(L^p (X))'} = \| v \|_{L^q (X)}$ follows as in Step 1 with the only change that one should take now $u := |v|^{q-2} v$.

The previous theorem fails if $p = 1$, as the following exercise shows:

Exercise 2.38. Let $X = (0,1)$, let $\mathfrak{M}$ be the σ-algebra consisting of all countable subsets of $(0,1)$ and their complements, and for every $E \in \mathfrak{M}$ define $\mu(E)$ as the cardinality of E. Prove that $L^1(X)$ consists of all functions that vanish outside a denumerable set, and whose remaining values form an absolutely convergent series. Show that if $u \in L^1(X)$, then the function $v(x) := xu(x)$ is also in $L^1(X)$ and the functional

$$L(u) := \int_X xu(x)\, d\mu(x) \quad u \in L^1(X)$$

is linear and continuous. Since this functional takes on an uncountable set of distinct values on the family of characteristic functions of singletons, prove that it cannot be of the form

$$L(u) = \int_X uv\, d\mu \quad \text{for every } u \in L^1(X)$$

for some measurable function v. Indeed, any measurable function must be constant except for a countable set.

Theorem 2.39 (Riesz representation theorem in L^1). *Let $(X, \mathfrak{M}, \mu)$ be a measure space and let*

$$\mathfrak{M}' := \{E \in \mathfrak{M} : \mu(E) < \infty\}.$$

Then for every bounded linear functional $L : L^1(X) \to \mathbb{R}$ there exists a unique signed measure $\lambda : \mathfrak{M}' \to \mathbb{R}$ absolutely continuous with respect to μ such that

$$\|\lambda\| := \sup\left\{\frac{|\lambda(E)|}{\mu(E)} : E \in \mathfrak{M}, \, 0 < \mu(E) < \infty\right\} < \infty \qquad (2.24)$$

and

$$L(u) = \int_X u\, d\lambda \quad \text{for every } u \in L^1(X). \qquad (2.25)$$

Moreover, the norm of L coincides with $\|\lambda\|$. Conversely, every functional of the form (2.25), where λ is as above, is a bounded linear functional on $L^1(X)$.[1]

---

[1] Since $\mathfrak{M}'$ is not an algebra (unless μ is finite), here by a measure $\nu : \mathfrak{M}' \to [0,\infty]$ we mean that

$$\nu(\emptyset) = 0, \quad \nu\left(\bigcup_{n=1}^{\infty} E_n\right) = \sum_{n=1}^{\infty} \nu(E_n)$$

for every countable collection $\{E_n\} \subset \mathfrak{M}'$ of pairwise disjoint sets such that $\bigcup_{n=1}^{\infty} E_n \in \mathfrak{M}'$, while by a signed measure $\lambda : \mathfrak{M}' \to \mathbb{R}$ we mean that λ is the difference of two measures $\nu_1, \nu_2 : \mathfrak{M}' \to [0,\infty)$. In the proof of the theorem we will show that ν_1 and ν_2 may be extended as measures on the σ-algebra $\mathfrak{M}$. Hence, the integral $\int_X u\, d\lambda$ should be understood as

$$\int_X u\, d\lambda := \int_X u\, d\tilde{\nu}_1 - \int_X u\, d\tilde{\nu}_2,$$

provided the right-hand side is well-defined, and where $\tilde{\nu}_1$ and $\tilde{\nu}_2$ are the extensions of ν_1 and ν_2, respectively.

Proof. **Step 1:** Assume that $L : L^1(X) \to \mathbb{R}$ is linear, bounded and $L \geq 0$, that is, $L(u) \geq 0$ whenever $u \geq 0$. As in the previous theorem, the set function

$$\nu(E) := L(\chi_E), \quad E \in \mathfrak{M}',$$

is countably additive, absolutely continuous with respect to μ, with

$$\|L\|_{(L^1(X))'} \geq \frac{L(\chi_E)}{\|\chi_E\|_{L^1(X)}} = \frac{\nu(E)}{\mu(E)} \quad \text{for all } E \in \mathfrak{M}, \, 0 < \mu(E) < \infty.$$

Hence

$$\|L\|_{(L^1(X))'} \geq \|\nu\|. \tag{2.26}$$

To extend ν to the σ-algebra $\mathfrak{M}$ define

$$\tilde{\nu}(E) := \sup\{\nu(E \cap F) : F \in \mathfrak{M}'\}, \quad E \in \mathfrak{M}. \tag{2.27}$$

As in the proof of Lemma 1.23, one can verify that $\tilde{\nu}$ is a measure, absolutely continuous with respect to μ. Moreover, $\tilde{\nu} = \nu$ on $\mathfrak{M}'$.

For any simple function s of the form

$$s = \sum_{i=1}^{\ell} c_i \chi_{E_i}, \tag{2.28}$$

where $c_i \in \mathbb{R}$ and $E_i \in \mathfrak{M}'$, we have

$$L(s) = \int_X s \, d\nu.$$

Given a nonnegative function $u \in L^1(X) = L^1(X, \mathfrak{M}, \mu)$, in view of Theorem 1.74 one can construct a sequence of nonnegative simple functions $\{s_n\}$, each of which is bounded and vanishes except on a set of finite measure μ, such that $\{s_n(x)\}$ converges monotonically to $u(x)$ for μ a.e. $x \in X$. Write

$$s_n = \sum_{i=1}^{\ell_n} c_i^{(n)} \chi_{E_i^{(n)}},$$

where $c_i^{(n)} \geq 0$ and $E_i^{(n)} \in \mathfrak{M}'$. Then

$$\int_X s_n \, d\nu = \sum_{i=1}^{\ell_n} c_i^{(n)} \nu\left(E_i^{(n)}\right) \leq \|\nu\| \sum_{i=1}^{\ell_n} c_i^{(n)} \mu\left(E_i^{(n)}\right) = \|\nu\| \int_X s_n \, d\mu,$$

and so by the Lebesgue monotone convergence theorem,

$$\int_X u \, d\tilde{\nu} \leq \|\nu\| \int_X u \, d\mu, \tag{2.29}$$

which implies that $u \in L^1(X, \mathfrak{M}, \tilde{\nu})$. Using Corollary 1.77 and the Lebesgue dominated convergence theorem, it follows that

$$L(u) = \int_X u \, d\tilde{\nu}$$

for every nonnegative $u \in L^1(X)$.

If now $u \in L^1(X) = L^1(X, \mathfrak{M}, \mu)$, we can apply the previous part to u^+ and u^- to conclude that

$$L(u) = \int_X u \, d\tilde{\nu} \quad \text{for every } u \in L^1(X), \tag{2.30}$$

with (see (2.29))

$$\int_X |u| \, d\tilde{\nu} \le \|\nu\| \int_X |u| \, d\mu, \tag{2.31}$$

which implies that $u \in L^1(X, \mathfrak{M}, \tilde{\nu})$. By (2.30) and (2.31) we obtain

$$\|L\|_{(L^1(X))'} \le \|\nu\|,$$

which, together with (2.26), yields $\|L\|_{(L^1(X))'} = \|\nu\|$.

Step 2: If $L : L^1(X) \to \mathbb{R}$ is linear, bounded as in the previous theorem, we write

$$L = L^+ - L^-,$$

where for $u \in L^1(X)$, $u \ge 0$,

$$L^+(u) := \sup\{L(v) : 0 \le v \le u\}, \quad L^-(u) := -\inf\{L(v) : 0 \le v \le u\},$$

while for $u \in L^1(X)$,

$$L^+(u) := L^+(u^+) - L^+(u^-), \quad L^-(u) := L^-(u^+) - L^-(u^-).$$

One can verify that L^+ and L^- are linear, continuous, and nonnegative with

$$\|L\|_{(L^1(X))'} = \|L^+\|_{(L^1(X))'} + \|L^-\|_{(L^1(X))'}.$$

Hence by the previous step we can find two measures $\lambda^+, \lambda^- : \mathfrak{M}' \to [0, \infty)$ such that

$$L^{\pm}(u) = \int_X u \, d\lambda^{\pm} \quad \text{for every } u \in L^1(X)$$

and

$$\|\lambda^{\pm}\| = \sup\left\{\frac{|\lambda^{\pm}(E)|}{\mu(E)} : E \in \mathfrak{M}, \, 0 < \mu(E) < \infty\right\} = \|L^{\pm}\|_{(L^1(X))'}.$$

It suffices to define $\lambda := \lambda^+ - \lambda^-$.

Exercise 2.40. Prove that the integral $\int_X u\, d\tilde{\nu}$ constructed in Step 1 of the proof of the previous theorem does not depend on the particular extension $\tilde{\nu}$ of ν. Using (2.24) show that

$$\|\lambda\| = \|\lambda^+\| + \|\lambda^-\|,$$

where λ^+, λ^-, and λ are defined in Step 2 of the proof of the previous theorem.

Using the Radon–Nikodym theorem II we can deduce from the previous theorem the following result.

Corollary 2.41. *Let $(X, \mathfrak{M}, \mu)$ be a measure space. Then the dual of $L^1(X)$ may be identified with $L^\infty(X)$ if and only if the measure μ is localizable and has the finite subset property.*

Proof. In the remainder of the book we will use only the fact if the measure μ is localizable and has the finite subset property, then the dual of $L^1(X)$ may be identified with $L^\infty(X)$. Thus we prove here only this implication, and we refer to [Rao04] and [Z67] for the converse one.

Hence assume that μ is localizable and has the finite subset property and let $L : L^1(X) \to \mathbb{R}$ be linear, bounded, and positive. Let ν and $\tilde{\nu}$ be defined as in the proof of step 1 of the previous theorem. Since $\tilde{\nu}$ is absolutely continuous with respect to μ and satisfies (1.52) (see (2.27)), we are in a position to apply the Radon–Nikodym theorem II to obtain a unique measurable function $v : X \to [0, \infty]$ such that

$$\tilde{\nu}(E) = \int_E v\, d\mu$$

for all $E \in \mathfrak{M}$. We claim that $v \in L^\infty(X)$ with

$$\|\nu\| = \|v\|_{L^\infty(X)},$$

where $\|\nu\|$ is defined in (2.24). Indeed, let

$$E_0 := \{x \in X : v(x) > \|\nu\|\}.$$

Assume by contradiction that $\mu(E_0) > 0$. Since μ has the finite subset property, if $\mu(E_0) = \infty$ we may find a measurable subset $F \subset E_0$ with $0 < \mu(F) < \infty$ (if $\mu(E_0) < \infty$ take $F := E_0$). Then F is admissible in the definition of (2.24), and so

$$\|\nu\| = \|\nu\| \frac{\mu(F)}{\mu(F)} < \frac{1}{\mu(F)} \int_F v\, d\mu = \frac{\nu(F)}{\mu(F)} \le \|\nu\|,$$

which is a contradiction. Hence $\mu(E_0) = 0$, and so $v \in L^\infty(X)$ and $\|v\|_{L^\infty(X)} \le \|\nu\|$. In turn, for every $E \in \mathfrak{M}$, with $0 < \mu(E) < \infty$, we have

$$\frac{\nu(E)}{\mu(E)} = \frac{1}{\mu(E)} \int_E v\, d\mu \le \|v\|_{L^\infty(X)},$$

and so, taking the supremum over all such E we get $\|\nu\| \le \|v\|_{L^\infty(X)}$. This proves the claim.

Since

$$L(u) = \int_X u\, d\tilde{\nu} \quad \text{for every } u \in L^1(X),$$

by taking u to be a simple function s as in (2.28) we get

$$L(s) = \int_X s\, d\nu = \sum_{i=1}^{\ell} c_i \nu(E_i) = \sum_{i=1}^{\ell} c_i \int_{E_i} v\, d\mu = \int_X sv\, d\mu.$$

The continuity of L, Corollary 1.77, and the Lebesgue dominated convergence theorem yield

$$L(u) = \int_X uv\, d\mu \quad \text{for every } u \in L^1(X).$$

This concludes the proof in the case that L is positive. The general case may be treated by the usual decomposition $L = L^+ - L^-$. We omit the details.

Remark 2.42. Note that if μ is σ-finite, then, in view of Proposition 1.110, μ is localizable, and it has the finite subset property. Hence for σ-finite measures the identification $\left(L^1(X)\right)' = L^\infty(X)$ is valid.

The next exercise shows that the dual of $L^p(X)$, $0 < p < 1$, is nontrivial if and only if μ has no atoms of finite measure.

Exercise 2.43 (The dual of $L^p(X)$, $0 < p < 1$). Let $(X, \mathfrak{M}, \mu)$ be a measure space and let $0 < p < 1$.

(i) Assume that μ has no atoms of finite measure, that is, if for every set $E \in \mathfrak{M}$ of positive *finite* measure there exists $F \in \mathfrak{M}$, with $F \subset E$, such that $0 < \mu(F) < \mu(E)$, and let $u \in L^p(X) \setminus \{0\}$. Prove that the measure

$$\nu(E) := \int_E |u|^p\, d\mu, \quad E \in \mathfrak{M},$$

has no atoms, so that the range of ν is $\left[0, \int_X |u|^p\, d\mu\right]$.

(ii) Under the same hypothesis on μ prove that if $U \subset L^p(X)$ is open, convex, and $0 \in U$, then $U = L^p(X)$. Conclude that $\left(L^p(X)\right)' = \{0\}$.

(iii) Conversely, let $E \in \mathfrak{M}$ be an atom with positive finite measure. Prove that the set

$$U := \{u \in L^p(X) : |u(x)| < 1 \text{ for } \mu \text{ a.e. } x \in E\}$$

is open, convex, and $0 \in U$, but $U \ne L^p(X)$. Hence $\left(L^p(X)\right)'$ is nontrivial.

Finally, we study the dual of $L^\infty(X)$. Note that in general,

$$(L^\infty(X))' \supsetneqq L^1(X).$$

Indeed, the dual of $L^\infty(X)$ may be identified with $\mathrm{ba}(X, \mathfrak{M}, \mu)$, which is the space of all *bounded finitely additive signed measures absolutely continuous with respect to μ*, that is, all maps $\lambda : \mathfrak{M} \to \mathbb{R}$ such that

(i) $\lambda \in \mathrm{ba}(X, \mathfrak{M})$;
(ii) $\lambda(E) = 0$ whenever $E \in \mathfrak{M}$ and $\mu(E) = 0$.

Theorem 2.44 (Riesz representation theorem in L^∞). *Let $(X, \mathfrak{M}, \mu)$ be a measure space. Then every bounded linear functional $L : L^\infty(X) \to \mathbb{R}$ is represented by a unique $\lambda \in \mathrm{ba}(X, \mathfrak{M}, \mu)$ in the sense that*

$$L(u) = \int_X u \, d\lambda \quad \text{for every } u \in L^\infty(X). \tag{2.32}$$

Moreover, the norm of L coincides with $\|\lambda\|$. Conversely, every functional of the form (2.32), where $\lambda \in \mathrm{ba}(X, \mathfrak{M}, \mu)$, is a bounded linear functional on $L^\infty(X)$.

Proof. Let $L \in (L^\infty(X))'$. The elements of $L^\infty(X)$ are equivalence classes $[u]$ (see (2.1)) of measurable functions bounded μ almost everywhere. Given a function $u \in B(X, \mathfrak{M})$ we have that $[u] \in L^\infty(X)$ and

$$\sup_{x \in X} |u| \geq \|[u]\|_{L^\infty}.$$

Define the linear functional

$$L_1(u) := L([u]), \quad u \in B(X, \mathfrak{M}).$$

Then

$$|L_1(u)| = |L([u])| \leq \|L\|_{(L^\infty(X))'} \|[u]\|_{L^\infty} \leq \|L\|_{(L^\infty(X))'} \sup_{x \in X} |u|,$$

and so

$$\|L_1\|_{(B(X, \mathfrak{M}))'} \leq \|L\|_{(L^\infty(X))'}.$$

On the other hand, by the definition of essential supremum, from each equivalence class $[u] \in L^\infty(X)$ we can select a measurable representative $\tilde{u} : X \to \mathbb{R}$ such that

$$\sup_{x \in X} |\tilde{u}| = \|[u]\|_{L^\infty(X)}.$$

Hence

$$\frac{|L([u])|}{\|[u]\|_{L^\infty(X)}} = \frac{|L_1(\tilde{u})|}{\sup_{x \in X} |\tilde{u}|} \leq \|L_1\|_{(B(X, \mathfrak{M}))'},$$

which shows that $\|L_1\|_{(B(X,\mathfrak{M}))'} = \|L\|_{(L^\infty(X))'}$. It follows, in particular, that $L_1 \in (B(X,\mathfrak{M}))'$, and so by the Riesz representation theorem in $B(X,\mathfrak{M})$ there exists a unique $\lambda \in \mathrm{ba}(X,\mathfrak{M})$ such that

$$L_1(u) = \int_X u \, d\lambda \quad \text{for every } u \in B(X,\mathfrak{M})$$

and

$$\|L\|_{(L^\infty(X))'} = \|L_1\|_{(B(X,\mathfrak{M}))'} = \|\lambda\|.$$

Note that if $u, v \in B(X,\mathfrak{M})$ and $u = v$ μ a.e. everywhere, then

$$\int_X u \, d\lambda = L_1(u) = L([u]) = L([v]) = L_1(v) = \int_X v \, d\lambda.$$

In particular, if $E \in \mathfrak{M}$ is such that $\mu(E) = 0$, then $\chi_E \in [0]$, and so

$$\lambda(E) = \int_X \chi_E \, d\lambda = L([0]) = 0.$$

This shows that $\lambda \in \mathrm{ba}(X,\mathfrak{M},\mu)$.

Conversely given any $\lambda \in \mathrm{ba}(X,\mathfrak{M},\mu)$, since λ is absolutely continuous with respect to μ, the functional

$$L([u]) := \int_X u \, d\lambda \quad \text{for every } [u] \in L^\infty(X)$$

is well-defined, linear, and bounded.

The next example shows that $\mathrm{ba}(X,\mathfrak{M},\mu)$ strictly contains the subspace of finite signed Radon measures absolutely continuous with respect to μ.

Exercise 2.45. Let $X = [0,1]$, let $\mathfrak{M} = \mathcal{B}([0,1])$, and let μ be the Lebesgue measure. Fix $0 < a < 1$ and for each $u \in L^\infty([0,1])$ define

$$p(u) := \lim_{\varepsilon \to 0^+} \underset{a<x<a+\varepsilon}{\mathrm{esssup}} \ u - \lim_{\varepsilon \to 0^+} \underset{a-\varepsilon<x<a}{\mathrm{essinf}} \ u,$$

where, for $0 \le c < d \le 1$,

$$\underset{c<x<d}{\mathrm{esssup}} \ u := \inf \left\{ \alpha \in \mathbb{R} : u(x) < \alpha \text{ for } \mathcal{L}^1 \text{ a.e. } x \in (c,d) \right\},$$

$$\underset{c<x<d}{\mathrm{essinf}} \ u := \sup \left\{ \alpha \in \mathbb{R} : u(x) > \alpha \text{ for } \mathcal{L}^1 \text{ a.e. } x \in (c,d) \right\}.$$

(i) Show that $p(u+v) \le p(u) + p(v)$ for all $u, v \in L^\infty([0,1])$, that $p(tu) = tp(u)$ for all $t > 0$ and $u \in L^\infty([0,1])$, and that $p(u) = 0$ for all $u \in C([0,1])$.

(ii) Define
$$L(u) := 0, \quad u \in C([0,1]).$$

Prove that L may be extended to a linear functional defined on $L^\infty([0,1])$ with the properties that

$$-p(u) \le L(u) \le p(u)$$

for all $u \in L^\infty([0,1])$, and $L\left(\chi_{[a,1]}\right) = 1$.

(iii) Let $\lambda \in \mathrm{ba}\left([0,1], \mathcal{B}([0,1]), \mathcal{L}^1\right)$ be the finitely additive measure such that

$$L(u) = \int_{[0,1]} u\, d\lambda \quad \text{for every } u \in L^\infty([0,1]),$$

representing L via (2.32). Show that λ cannot be countably additive.

2.1.4 Weak Convergence in L^p

Let $1 \le p \le \infty$ and let q be the Hölder conjugate exponent of p. If $p = 1$ or $p = \infty$ then assume in addition that μ is σ-finite. By Proposition A.49 and the Riesz representation theorem in $L^p(X)$, a sequence $\{u_n\} \subset L^p(X)$ converges weakly (weakly star if $p = \infty$) to some function u in $L^p(X)$ if and only if

$$\int_X u_n v\, d\mu \to \int_X uv\, d\mu$$

for every $v \in L^q(X)$.

Propositions A.68, A.57, Theorem 2.37, Remark 2.42, and Corollaries A.55 and A.66 yield the following results.

Proposition 2.46. *Let $(X, \mathfrak{M}, \mu)$ be a measure space, let $1 \le p \le \infty$, and assume that the measure μ is σ-finite when $p = \infty$. Let $u_n \in L^p(X)$, $n \in \mathbb{N}$.*

(i) If $1 < p < \infty$, $\sup_n \|u_n\|_{L^p} < \infty$ and if $\{u_n\}$ converges to u pointwise μ a.e. or in measure, then $u \in L^p(X)$ and $u_n \rightharpoonup u$ in $L^p(X)$.

(ii) If $u_n \rightharpoonup u$ in $L^p(X)$ ($\overset{*}{\rightharpoonup}$ if $p = \infty$), then

$$\|u\|_{L^p} \le \liminf_{n \to \infty} \|u_n\|_{L^p} \le \sup_n \|u_n\|_{L^p} < \infty.$$

(iii) If $1 < p < \infty$, $u_n \rightharpoonup u$ in $L^p(X)$, and if $\|u\|_{L^p} = \lim_{n \to \infty} \|u_n\|_{L^p}$, then $u_n \to u$ in $L^p(X)$.

(iv) If $1 < p < \infty$ and if $\sup_n \|u_n\|_{L^p} < \infty$, then there exists a subsequence $\{u_{n_k}\}$ such that $u_{n_k} \rightharpoonup u$ in $L^p(X)$ for some $u \in L^p(X)$. This property still holds in $L^\infty(X)$ with respect to the weak star convergence, if, in addition, $L^1(X)$ is separable.

Remark 2.47. Concerning property (iv) above, we recall that $L^1(X)$ is separable if in addition, X is a separable measurable space (see Theorem 2.16).

Exercise 2.48. Using part (iv) prove part (i) of the previous proposition for $1 < p < \infty$.

Corollary 2.49. *Let $(X, \mathfrak{M}, \mu)$ be a measure space, let $1 < p \le \infty$, and assume that the measure μ is σ-finite when $p = \infty$. Let $u_n,\ u \in L^p(X)$. Then $u_n \rightharpoonup u$ in $L^p(X)$ ($\stackrel{*}{\rightharpoonup}$ if $p = \infty$) if and only if*

(i) $\sup_n \|u_n\|_{L^p} < \infty$;

(ii) $\int_E u_n \, d\mu \to \int_E u \, d\mu$ *for every $E \in \mathfrak{M}$ with $\mu(E) < \infty$.*

Proof. The result follows from the fact that simple functions are dense in $L^{p'}(X)$.

Exercise 2.50. Let $X = [0,1]$ with the Lebesgue measure and let $u_n,\ u \in L^p([0,1])$, $1 < p < \infty$. Prove that $u_n \rightharpoonup u$ in $L^p([0,1])$ if and only if

$$\sup_n \|u_n\|_{L^p([0,1])} < \infty$$

and

$$\lim_{n \to \infty} \int_0^x u_n(t) \, dt = \int_0^x u(t) \, dt$$

for all $x \in [0,1]$.

Remark 2.51. Proposition 2.46(iv) is false for $p = 1$ (see the following exercise); precisely, if

$$\sup_n \|u_n\|_{L^1} < \infty$$

then there is no guarantee that some subsequence will converge weakly in L^1. However, if in addition X is a σ-compact metric space and μ is a Radon measure, considering the measures defined by

$$\lambda_n(E) := \int_E u_n \, d\mu$$

for every Borel set $E \subset X$ (or, simply, $\lambda_n := u_n\mu$), then by Proposition 1.202 there exist a subsequence not relabeled and a signed Radon measure λ such that $\lambda_n \stackrel{*}{\rightharpoonup} \lambda$, i.e.,

$$\int_X \varphi u_n \, d\mu \to \int_X \varphi \, d\lambda$$

for every $\varphi \in C_0(X)$.

Exercise 2.52. Let $X = \mathbb{R}$, $\mathfrak{M} = \mathcal{B}(\mathbb{R})$, and let μ be the Lebesgue measure. Show that the sequence $u_n = n\chi_{[0,\frac{1}{n}]}$ converges in measure to zero, satisfies

$$\|u_n\|_{L^1(\mathbb{R})} = 1,$$

but does not converge weakly in L^1. However, by considering

$$\lambda_n (E) := n \int_{E \cap [0, \frac{1}{n}]} dx$$

for every Borel set $E \subset \mathbb{R}$, prove that $\lambda_n \overset{*}{\rightharpoonup} \delta_0$, the Dirac delta measure with mass concentrated at the origin.

The next result will play a crucial role in the characterization of weakly compact sets in L^1. It gives a property of uniform absolute continuity of a particular sequence of signed measures (see Proposition 1.99).

Theorem 2.53 (Vitali–Hahn–Saks). *Let* $(X, \mathfrak{M}, \mu)$ *be a measure space and let* $\{\lambda_n\}$ *be a sequence of finite signed measures such that* $\lim_{n \to \infty} \lambda_n (E)$ *exists in* $\mathbb{R}$ *for all* $E \in \mathfrak{M}$ *with* $\mu (E) < \infty$. *If each* λ_n *is absolutely continuous with respect to* μ, *then for every* $\varepsilon > 0$ *there exists* $\delta > 0$ *such that*

$$|\lambda_n (E)| \leq \varepsilon$$

for all $n \in \mathbb{N}$ *and for every measurable set* $E \subset X$ *with* $\mu (E) \leq \delta$.

Proof. Let

$$\mathfrak{M}' := \{ E \in \mathfrak{M} : \mu (E) < \infty \}.$$

By identifying sets that differ by a set of μ measure zero we can regard $\mathfrak{M}'$ as a closed subset $\mathcal{F}$ of $L^1 (X, \mathfrak{M}, \mu)$ through the mapping

$$E \in \mathfrak{M}' \mapsto \chi_E.$$

Fix $\varepsilon > 0$ and for $k \in \mathbb{N}$ define the sets

$$\mathcal{F}_k := \left\{ \chi_E : E \in \mathfrak{M}', \ \sup_{n, l \geq k} |\lambda_n (E) - \lambda_l (E)| \leq \varepsilon \right\}.$$

We claim that the sets $\mathcal{F}_k$ are closed in $L^1 (X, \mathfrak{M}, \mu)$. Indeed, if $\{\chi_{E_j}\} \subset \mathcal{F}_k$ converges in $L^1 (X, \mathfrak{M}, \mu)$, then, since $\mathcal{F}$ is closed, $\chi_{E_j} \to \chi_{E_\infty}$ as $j \to \infty$ for some $E_\infty \in \mathfrak{M}'$. By extracting a subsequence (not relabeled), if necessary, we may assume that $\chi_{E_j} (x) \to \chi_{E_\infty} (x)$ for μ a.e. $x \in X$ (see Theorem 2.20). Let

$$Y := E_\infty \cup \bigcup_{j=1}^{\infty} E_j.$$

Then $\mu : \mathfrak{M} \lfloor Y \to [0, \infty]$ is σ-finite, and so for any $n \in \mathbb{N}$ and by the Radon–Nikodym theorem applied to $(\lambda_n)^+$ and $(\lambda_n)^-$ restricted to $\mathfrak{M} \lfloor Y$, there exists $v_n \in L^1 (Y, \mathfrak{M} \lfloor Y, \mu)$ such that

$$\lambda_n (E) = \int_E v_n \, d\mu, \quad E \in \mathfrak{M} \lfloor Y.$$

By the Lebesgue dominated convergence theorem, for each $n \in \mathbb{N}$,

$$\lim_{j \to \infty} \lambda_n\left(E_j\right) = \lambda_n\left(E_\infty\right).$$

Since $\{\chi_{E_j}\} \subset \mathcal{F}_k$ for any fixed $n, l \geq k$ and for all $j \in \mathbb{N}$, we have

$$\left|\lambda_n\left(E_j\right) - \lambda_l\left(E_j\right)\right| \leq \varepsilon,$$

and so letting $j \to \infty$ we obtain that

$$\left|\lambda_n\left(E_\infty\right) - \lambda_l\left(E_\infty\right)\right| \leq \varepsilon,$$

which shows that $\mathcal{F}_k$ is closed in $L^1\left(X, \mathfrak{M}, \mu\right)$.

By hypothesis, $\lim_{n \to \infty} \lambda_n\left(E\right)$ exists in $\mathbb{R}$ for all $E \in \mathfrak{M}'$, and so

$$\mathcal{F} = \bigcup_{k=1}^{\infty} \mathcal{F}_k.$$

Applying the Baire category theorem to the complete metric space $\mathcal{F}$, at least one of the sets $\mathcal{F}_k$ has nonempty interior. Hence there exist $\delta_1 > 0$, $k \in \mathbb{N}$, and $\chi_{E_0} \in \mathcal{F}_k$ such that if $\chi_E \in \mathcal{F}$ and if

$$\int_X \left|\chi_E - \chi_{E_0}\right| d\mu < \delta_1, \tag{2.33}$$

then $\chi_E \in \mathcal{F}_k$, that is,

$$\sup_{n,\, l \geq k} \left|\lambda_n\left(E\right) - \lambda_l\left(E\right)\right| \leq \varepsilon. \tag{2.34}$$

By Proposition 1.99 applied to $(\lambda_n)^+$ and $(\lambda_n)^-$, we may find $0 < \delta < \delta_1$ such that

$$\left|\lambda_n\left(E\right)\right| \leq \varepsilon \tag{2.35}$$

for all $1 \leq n \leq k$ and for every measurable set $E \subset X$ with $\mu\left(E\right) \leq \delta$.

Fix $E \in \mathfrak{M}'$ with $\mu\left(E\right) \leq \delta$. Then

$$E = \left(E \cup E_0\right) \setminus \left(E_0 \setminus E\right),$$

with

$$\int_X \left|\chi_{E \cup E_0} - \chi_{E_0}\right| d\mu, \int_X \left|\chi_{E_0 \setminus E} - \chi_{E_0}\right| d\mu < \delta_1,$$

and so by (2.33) and (2.34),

$$\sup_{n,\, l \geq k} \left|\lambda_n\left(E \cup E_0\right) - \lambda_l\left(E \cup E_0\right)\right| \leq \varepsilon,$$

$$\sup_{n,\, l \geq k} \left|\lambda_n\left(E_0 \setminus E\right) - \lambda_l\left(E_0 \setminus E\right)\right| \leq \varepsilon.$$

It follows that for any $n \geq k$,

$$
\begin{aligned}
|\lambda_n(E)| &\leq |\lambda_k(E)| + |\lambda_k(E) - \lambda_n(E)| \\
&\leq \varepsilon + |\lambda_k(E \cup E_0) - \lambda_n(E \cup E_0)| \\
&\quad + |\lambda_k(E_0 \setminus E) - \lambda_n(E_0 \setminus E)| \\
&\leq 3\varepsilon,
\end{aligned}
$$

where we have also used (2.35). This completes the proof.

In view of Remark 2.51, L^1 weak compactness of L^1 bounded sequences requires additional conditions, and this brings us to the following result.

Theorem 2.54 (Dunford–Pettis). *Let $(X, \mathfrak{M}, \mu)$ be a measure space and let $\mathcal{F} \subset L^1(X)$. Then $\mathcal{F}$ is weakly sequentially precompact if and only if*

(i) $\mathcal{F}$ is bounded in $L^1(X)$;
(ii) $\mathcal{F}$ is equi-integrable and for every $\varepsilon > 0$ there exists $E \subset X$ with $E \in \mathfrak{M}$ such that $\mu(E) < \infty$ and

$$
\sup_{u \in \mathcal{F}} \int_{X \setminus E} |u| \, d\mu \leq \varepsilon. \tag{2.36}
$$

Proof. **Step 1:** Assume that $\mathcal{F}$ is weakly sequentially precompact.

Substep 1a: If $\mathcal{F}$ is not bounded in $L^1(X)$, then it contains a sequence $\{u_n\} \subset \mathcal{F}$ such that $\|u_n\|_{L^1(X)} \to \infty$. But since $\mathcal{F}$ is weakly sequentially precompact, a subsequence is weakly convergent in $L^1(X)$, and this contradicts Proposition A.68.

Substep 1b: Assume by contradiction that $\mathcal{F}$ is not equi-integrable. Then we may find a sequence $\{u_n\} \subset \mathcal{F}$ and sets $\{E_n\} \subset \mathfrak{M}$, with $\mu(E_n) \to 0$, such that for all $n \in \mathbb{N}$,

$$
\int_{E_n} |u_n| \, d\mu \geq \varepsilon_0 \tag{2.37}
$$

for some $\varepsilon_0 > 0$. Since $\mathcal{F}$ is weakly sequentially precompact, there exists a subsequence of $\{u_n\}$ (not relabeled) such that $u_n \rightharpoonup u$ in $L^1(X)$. In particular, for every $E \in \mathfrak{M}$, with $\mu(E) < \infty$, we have that

$$
\lim_{n \to \infty} \int_E u_n \, d\mu = \int_E u \, d\mu,
$$

and so the signed measures

$$
\lambda_n(E) := \int_E u_n \, d\mu, \quad E \in \mathfrak{M},
$$

satisfy the hypotheses of the Vitali-Hahn-Saks theorem. Hence there is $\delta > 0$ such that

$$\left| \int_E u_n \, d\mu \right| \leq \frac{\varepsilon_0}{3}$$

for all $n \in \mathbb{N}$ and for every measurable set $E \subset X$ with $\mu(E) \leq \delta$. By taking $E_n \cap \{u_n \geq 0\}$ and $E_n \cap \{u_n < 0\}$ for all n so large that $\mu(E_n) \leq \delta$, we conclude that

$$\int_{E_n} |u_n| \, d\mu \leq \frac{2\varepsilon_0}{3},$$

which contradicts (2.37).

Substep 1c: Assume by contradiction that (2.36) is violated. Then we may find $\varepsilon_0 > 0$ with the property that for any measurable set $E \subset X$, with $\mu(E) < \infty$, there exists $u \in \mathcal{F}$ such that

$$\int_{X \setminus E} |u| \, d\mu \geq \varepsilon_0.$$

Fix any $E_1 \in \mathfrak{M}$ with $\mu(E_1) < \infty$ and let $u_1 \in \mathcal{F}$ be such that

$$\int_{X \setminus E_1} |u_1| \, d\mu \geq \varepsilon_0.$$

By Remark 2.28 there exists a measurable set $E_2 \supset E_1$, with $\mu(E_2) < \infty$, such that

$$\int_{X \setminus E_2} |u_1| \, d\mu < \frac{\varepsilon_0}{2}.$$

By an induction argument we may find an increasing sequence $\{E_n\} \subset \mathfrak{M}$, with $\mu(E_n) < \infty$ for all $n \in \mathbb{N}$, and $\{u_n\} \subset \mathcal{F}$ such that for all $n \in \mathbb{N}$,

$$\int_{E_{n+1} \setminus E_n} |u_n| \, d\mu \geq \frac{\varepsilon_0}{2}. \tag{2.38}$$

Since $\mathcal{F}$ is weakly sequentially precompact there exists a subsequence of $\{u_n\}$ (not relabeled) such that $u_n \rightharpoonup u$ in $L^1(X)$. Let

$$Y := \bigcup_{n=1}^{\infty} E_n.$$

Then $\mu : \mathfrak{M} \lfloor Y \to [0, \infty]$ is σ-finite. Hence we may find a sequence $\{F_j\} \subset \mathfrak{M} \lfloor Y$ of pairwise disjoint sets of positive finite measure such that $Y = \bigcup_{j=1}^{\infty} F_j$. Construct a sequence of positive numbers $\{a_j\}$ such that

$$\sum_{j=1}^{\infty} a_j \mu(F_j) < \infty$$

and define $v : Y \to (0, \infty)$ as

$$v := \sum_{j=1}^{\infty} a_j \chi_{F_j}.$$

Then $v \in L^1(Y, \mathfrak{M} \lfloor Y, \mu)$. Set

$$\nu(E) := \int_E v \, d\mu, \quad E \in \mathfrak{M} \lfloor Y.$$

Since $v > 0$ in Y it follows that the measures $\lambda_n : \mathfrak{M} \lfloor Y \to \mathbb{R}$,

$$\lambda_n(E) := \int_E u_n \, d\mu, \quad E \in \mathfrak{M} \lfloor Y,$$

are absolutely continuous with respect to ν. Moreover,

$$\lambda_n(E) \to \int_E u \, d\mu$$

as $n \to \infty$ for all $E \in \mathfrak{M} \lfloor Y$. By the Vitali–Hahn–Saks theorem there exists $\delta > 0$ such that

$$\left| \int_E u_n \, d\mu \right| \leq \frac{\varepsilon_0}{8}$$

for all $n \in \mathbb{N}$ and for every measurable set $E \in \mathfrak{M} \lfloor Y$ with $\nu(E) \leq \delta$. Since ν is finite, we may apply Proposition 1.7 to conclude that

$$\lim_{n \to \infty} \nu(Y \setminus E_n) = 0,$$

and so for n sufficiently large,

$$\nu(Y \setminus E_n) \leq \delta,$$

and in turn,

$$\left| \int_E u_n \, d\mu \right| \leq \frac{\varepsilon_0}{8}$$

for all measurable subsets $E \subset Y \setminus E_n$. As in the previous step we conclude that

$$\int_{Y \setminus E_n} |u_n| \, d\mu \leq \frac{\varepsilon_0}{4},$$

which contradicts (2.38).

Step 2: Assume that $\{u_n\} \subset \mathcal{F}$ satisfies (i) and (ii). Since $u_n^+ + u_n^- = |u_n|$ it follows that also $\{u_n^+\}$ and $\{u_n^-\}$ satisfy (i) and (ii), and so without loss of generality it suffices to assume that $u_n \geq 0$.

Substep 2a: Assume that the space $L^q(X)$ is separable for some $1 < q < \infty$ and let p be its Hölder conjugate exponent. Let

$$C := \sup_n \|u_n\|_{L^1} < \infty, \qquad (2.39)$$

and for each $k \in \mathbb{N}$ consider the truncation $T_k(s) := \min\{s, k\}$, $s \geq 0$. Then

$$\int_X (T_k(u_n))^p \, d\mu \leq k^{p-1} \int_X u_n \, d\mu \leq k^{p-1} C.$$

Hence for every fixed $k \in \mathbb{N}$ the sequence $\{T_k(u_n)\}_{n\in\mathbb{N}}$ is bounded in $L^p(X)$.

Let $\{h_j\}$ be a dense sequence in $L^q(X)$. Since the sequence $\{T_1(u_n)\}_{n\in\mathbb{N}}$ is bounded in $L^p(X)$ by Corollary A.66, we can extract a subsequence $\{u_{n^{(1)}}\} \subset \{u_n\}$ such that the truncated sequence $\{T_1(u_{n^{(1)}})\}$ is weakly convergent in $L^p(X)$ to some function g_1. In particular,

$$\lim_{n^{(1)} \to \infty} \int_X T_1(u_{n^{(1)}}) h_j \, d\mu = \int_X g_1 h_j \, d\mu$$

for all $j \in \mathbb{N}$. Similarly, since the sequence $\{T_2(u_{n^{(1)}})\}$ is bounded in $L^p(X)$ we can extract a subsequence $\{u_{n^{(2)}}\} \subset \{u_{n^{(1)}}\}$ such that the truncated sequence $\{T_2(u_{n^{(2)}})\}$ is weakly convergent in $L^p(X)$ to some function g_2. In particular,

$$\lim_{n^{(2)} \to \infty} \int_X T_k(u_{n^{(2)}}) h_j \, d\mu = \int_X g_k h_j \, d\mu$$

for all $j \in \mathbb{N}$ and $k = 1, 2$.

Recursively, for each $i \in \mathbb{N}$ we can find a subsequence $\{u_{n^{(i)}}\} \subset \{u_{n^{(i-1)}}\}$ such that the truncated sequence $\{T_i(u_{n^{(i)}})\}$ is weakly convergent in $L^p(X)$ to some function g_i. In particular,

$$\lim_{n^{(i)} \to \infty} \int_X T_k(u_{n^{(i)}}) h_j \, d\mu = \int_X g_k h_j \, d\mu$$

for all $j \in \mathbb{N}$ and $k = 1, \ldots, i$.

Hence for each $i \in \mathbb{N}$ there exists an integer m_i such that

$$\left| \int_X T_k(u_{n^{(i)}}) h_j \, d\mu - \int_X g_k h_j \, d\mu \right| \leq \frac{1}{i} \qquad (2.40)$$

for all $j, k = 1, \ldots, i$ and for all $n^{(i)} \geq m_i$.

Let n_i be the first natural number $n^{(i)}$ greater than or equal to m_i. We claim that the subsequence $\{u_{n_i}\}$ has the property that

$$T_k(u_{n_i}) \rightharpoonup g_k \text{ in } L^p(X) \qquad (2.41)$$

as $i \to \infty$ for all $k \in \mathbb{N}$. Fix $k \in \mathbb{N}$. Since $L^q(X)$ is separable and the sequence $\{T_k(u_{n_i})\}_{i\in\mathbb{N}}$ is bounded in $L^p(X)$, it suffices to show that

$$\lim_{i \to \infty} \int_X T_k(u_{n_i}) h_j \, d\mu = \int_X g_k h_j \, d\mu$$

for all $j \in \mathbb{N}$. Fix $\varepsilon > 0$, $j \in \mathbb{N}$, and let $i_0 \in \mathbb{N}$ be so large that

$$i_0 \geq \max \left\{ \frac{1}{\varepsilon}, k, j \right\}.$$

Then by (2.40), for all $i \geq i_0$,

$$\left| \int_X T_k (u_{n_i}) h_j \, d\mu - \int_X g_k h_j \, d\mu \right| \leq \frac{1}{i} \leq \varepsilon$$

for all $i \geq i_0$. This proves the claim.

For every $E \in \mathfrak{M}$ with $\mu(E) < \infty$ we have that $\chi_E \in L^q(X)$, and so

$$0 \leq \int_E g_k \, d\mu = \lim_{i \to \infty} \int_E T_k (u_{n_i}) \, d\mu \leq \lim_{i \to \infty} \int_E T_{k+1} (u_{n_i}) \, d\mu$$

$$= \int_E g_{k+1} \, d\mu \leq C,$$

where in the last inequality we have used (2.39) and the fact that $T_{k+1}(u_{n_i}) \leq u_{n_i}$. Since $g_k \in L^p(X)$, the set $\{x \in X : g_k(x) \neq 0\}$ is σ-finite. Hence the inequalities

$$0 \leq \int_E g_k \, d\mu \leq \int_E g_{k+1} \, d\mu \leq C$$

hold for all $E \in \mathfrak{M}$. We conclude that

$$\int_X g_k \, d\mu \leq C \tag{2.42}$$

and that $0 \leq g_k(x) \leq g_{k+1}(x)$ for μ a.e. $x \in X$. Thus we may find a measurable function g such that $g_k \nearrow g$. By the Lebesgue monotone convergence theorem it follows from (2.42) that $g \in L^1(X)$.

We claim that $u_{n_i} \rightharpoonup g$ in $L^1(X)$. Indeed, fix $\varepsilon > 0$, $\lambda \in \left(L^1(X) \right)'$, and by hypothesis, find $E_\varepsilon \in \mathfrak{M}$ and $\delta > 0$ such that $\mu(E_\varepsilon) < \infty$,

$$\int_{X \setminus E_\varepsilon} (g + u_{n_i}) \, d\mu \leq \frac{\varepsilon}{1 + \|\lambda\|} \tag{2.43}$$

for all $i \in \mathbb{N}$, and

$$\int_E u_{n_i} \, d\mu \leq \frac{\varepsilon}{1 + \|\lambda\|} \tag{2.44}$$

for all $i \in \mathbb{N}$ and for every Lebesgue measurable set $E \subset X$ with $\mu(E) \leq \delta$. Then

$$\int_X (u_{n_i} - g) \, d\lambda = \int_{E_\varepsilon} (T_k(u_{n_i}) - g_k) \, d\lambda + \int_{E_\varepsilon} (g_k - g) \, d\lambda$$

$$+ \int_{E_\varepsilon \cap \{u_{n_i} > k\}} (u_{n_i} - k) \, d\lambda + \int_{X \setminus E_\varepsilon} (u_{n_i} - g) \, d\lambda, \tag{2.45}$$

where $k \in \mathbb{N}$ is chosen so large that

$$\int_{E_\varepsilon} |g - g_k| \, d\mu \le \frac{\varepsilon}{1 + \|\lambda\|}, \quad \mu\left(\{u_{n_i} > k\}\right) \le \frac{1}{k} \int_X u_{n_i} \, d\mu \le \delta. \qquad (2.46)$$

Hence by (2.44),

$$\left| \int_{E_\varepsilon \cap \{u_{n_i} > k\}} (u_{n_i} - k) \, d\lambda \right| \le \|\lambda\| \int_{E_\varepsilon \cap \{u_{n_i} > k\}} (u_{n_i} - k) \, d\mu$$

$$\le \|\lambda\| \int_{E_\varepsilon \cap \{u_{n_i} > k\}} u_{n_i} \, d\mu \le \varepsilon,$$

and so, also from (2.45), $(2.46)_1$, (2.43), we get that

$$\left| \int_X (u_{n_i} - g) \, d\lambda \right| \le \left| \int_{E_\varepsilon} (T_k(u_{n_i}) - g_k) \, d\lambda \right| + 3\varepsilon.$$

Since $\mu(E_\varepsilon) < \infty$ by (2.41), it follows that

$$\lim_{i \to \infty} \int_{E_\varepsilon} (T_k(u_{n_i}) - g_k) \, d\lambda = 0,$$

and so we conclude that

$$\limsup_{i \to \infty} \left| \int_X (u_{n_i} - g) \, d\lambda \right| \le 3\varepsilon.$$

Given the arbitrariness of ε the proof is concluded in this case.

Substep 2b: We now remove the additional hypothesis on the separability of $L^q(X)$. Let $\mathfrak{M}_1$ be the smallest σ-algebra containing all the sets $\{x \in X : u_n(x) > t\}$ for $t \in \mathbb{Q}$ and $n \in \mathbb{N}$ and let

$$Y := \bigcup_{n=1}^\infty \{x \in X : u_n(x) \ne 0\}.$$

Then $(Y, \mathfrak{M}_1 \lfloor Y)$ is a separable measurable space and $\mu : \mathfrak{M}_1 \lfloor Y \to [0, \infty]$ is σ-finite. Hence $L^q(Y, \mathfrak{M}_1 \lfloor Y, \mu)$ is separable for all $1 \le q < \infty$ by Theorem 2.16. Moreover, the restriction of u_n to Y belongs to $L^1(Y, \mathfrak{M}_1 \lfloor Y, \mu)$ and still satisfies the equi-integrability condition as well as hypothesis (ii) in Vitali's theorem. Hence by Substep 2a there exists a subsequence $\{u_{n_i}\}$ of $\{u_n\}$ and a function $g \in L^1(Y, \mathfrak{M}_1 \lfloor Y, \mu)$ such that u_{n_i} restricted to Y converges to g in $L^1(Y, \mathfrak{M}_1 \lfloor Y, \mu)$. Extend g by zero outside Y. Since $\mathfrak{M}_1 \subset \mathfrak{M}$ we have that $g \in L^1(X)$. We claim that $u_{n_i} \rightharpoonup g$ in $L^1(X)$. Indeed, $\lambda \in (L^1(X))'$, and using again the fact that $\mathfrak{M}_1 \subset \mathfrak{M}$, we have that

$$\sup\left\{\frac{|\lambda\,(E)|}{\mu\,(E)} : E \in \mathfrak{M}_1, 0 < \mu\,(E) < \infty\right\}$$

$$\leq \sup\left\{\frac{|\lambda\,(E)|}{\mu\,(E)} : E \in \mathfrak{M}, 0 < \mu\,(E) < \infty\right\} < \infty,$$

and so $\lambda \in \left(L^1\,(Y, \mathfrak{M}_1 \lfloor Y, \mu)\right)'$. Hence, since $u_{n_i} = g = 0$ outside Y, we have

$$\lim_{i\to\infty} \int_X u_{n_i}\, d\lambda = \lim_{i\to\infty} \int_Y u_{n_i}\, d\lambda = \int_Y g\, d\lambda = \int_X g\, d\lambda.$$

This concludes the proof.

Exercise 2.55. Prove that if the measure μ is finite and nonatomic in the Dunford–Pettis theorem, then condition (i) is implied by (ii).

Remark 2.56. By comparing the previous theorem with Vitali's convergence theorem, it follows that if a sequence $\{u_n\}$ weakly converges to some function u in $L^1\,(X)$, then it converges strongly to u if and only if it converges in measure to u. Thus typical examples of sequences converging weakly but not strongly in $L^1\,(X)$ are given by oscillating sequences.

Exercise 2.57. Let $X = [0, 2\pi]$ with the Lebesgue measure. Prove that the sequence $u_n\,(x) := \sin nx$ converges weakly to zero in $L^1\,([0, 2\pi])$ but for all x irrational the limit

$$\lim_{n\to\infty} \sin nx$$

does not exist and the sequence $\{u_n\}$ does not converge in measure.

Corollary 2.58. *Let $(X, \mathfrak{M}, \mu)$ be a measure space, and let u_n, $u \in L^1\,(X)$, $n \in \mathbb{N}$. Then $u_n \rightharpoonup u$ in $L^1\,(X)$ if and only if*

(i) $\sup_n \|u_n\|_{L^1} < \infty$;
(ii) $\int_E u_n\, d\mu \to \int_E u\, d\mu$ *for every $E \in \mathfrak{M}$.*

Proof. Assume that (i) and (ii) hold. Let $\{u_{n_k}\}$ be a subsequence of $\{u_n\}$. Reasoning as in the proof of Substeps 1b and 1c of the Dunford–Pettis theorem, we can conclude that $\{u_{n_k}\}$ is equi-integrable and satisfies (2.36). Hence $\{u_{n_k}\}$ is weakly sequentially precompact, and so up to a further subsequence (not relabeled) we may assume that $u_{n_k} \rightharpoonup v$ in $L^1\,(X)$. In view of (ii) we have that

$$\int_E (u - v)\, d\mu = 0$$

for every $E \in \mathfrak{M}$, which implies that $u\,(x) = v\,(x)$ for μ a.e. $x \in X$. Thus we have shown that $u_{n_k} \rightharpoonup u$ in $L^1\,(X)$. Hence the whole sequence converges weakly to u in $L^1\,(X)$.

The converse implication follows from Proposition 2.46 and the fact that $\chi_E \in \left(L^1\,(X)\right)'$ for every $E \in \mathfrak{M}$.

The following result allows us to identify the weak limit.

Theorem 2.59. *Let $(X, \mathfrak{M}, \mu)$ be a measure space with μ finite and let $\{u_n\}$ be a sequentially weakly compact sequence in $L^1(X)$. Then $u_n \rightharpoonup u$ in $L^1(X)$ if and only if*

$$\int_X |u + v| \, d\mu \leq \liminf_{n \to \infty} \int_X |u_n + v| \, d\mu$$

for every $v \in L^1(X)$.

Proof. If $u_n \rightharpoonup u$ in $L^1(X)$ then $u_n + v \rightharpoonup u + v$ in $L^1(X)$, and so, by Proposition 2.46 (i),

$$\int_X |u + v| \, d\mu \leq \liminf_{n \to \infty} \int_X |u_n + v| \, d\mu.$$

Conversely, assume that a subsequence of $\{u_n\}$ (not relabeled) converges weakly in $L^1(X)$ to some function w, and let

$$X^+ := \{x \in X : w(x) > u(x)\}, \quad X^- := \{x \in X : w(x) \leq u(x)\}.$$

Since $\{u_n - u\}$ is equi-integrable, by Theorem 2.29, given $\varepsilon > 0$ there exists $t > 0$ such that

$$\int_{\{x \in X : |u_n - u| > t\}} |u_n - u| \, d\mu \leq \varepsilon$$

for all $n \in \mathbb{N}$. Set

$$v := -u - t\chi_{X^+} + t\chi_{X^-}.$$

Since μ is finite, it follows that $v \in L^1(X)$, and

$$|u_n + v| \leq t - (u_n - u)\chi_{X^+} + (u_n - u)\chi_{X^-} + 2|u_n - u|\chi_{\{|u_n - u| > t\}}.$$

Then

$$t\mu(X) = \int_X |u + v| \, d\mu \leq \liminf_{n \to \infty} \int_X |u_n + v| \, d\mu$$

$$\leq t\mu(X) - \int_{X^+} (w - u) \, d\mu + \int_{X^-} (w - u) \, d\mu + 2\varepsilon$$

$$= t\mu(X) - \int_X |w - u| \, d\mu + 2\varepsilon,$$

and so

$$\int_X |w - u| \, d\mu \leq 2\varepsilon.$$

It suffices to let $\varepsilon \to 0^+$.

Exercise 2.60. Can you extend the previous theorem to the case in which μ is not finite?

If $u_n \rightharpoonup u$ in $L^p(X)$ ($\overset{*}{\rightharpoonup}$if $p = \infty$) and $v_n \to v$ strongly in $L^q(X)$, where p and q are conjugate exponents, then

$$u_n v_n \rightharpoonup uv \text{ in } L^1(X). \tag{2.47}$$

In the case $p = 1$, using the equi-integrability we may improve this result in the sense that we may replace strong convergence in $L^\infty(X)$ with convergence with respect to the Mackey topology.

Proposition 2.61. *Let* $(X, \mathfrak{M}, \mu)$ *be a measure space with* μ *finite. If* $u_n \rightharpoonup u$ *in* $L^1(X)$, $v_n \to v$ *pointwise for* μ *a.e.* $x \in X$, *and* $\sup_n \|v_n\|_{L^\infty} < \infty$, *then* $u_n v_n \rightharpoonup uv$ *in* $L^1(X)$.

Proof. In view of Corollary 2.58, it suffices to show that

$$\int_E u_n v_n \, d\mu \to \int_E uv \, d\mu$$

for every $E \in \mathfrak{M}$. Fix $\varepsilon > 0$. Since $\{u_n\}$ is equi-integrable, there exists $\delta > 0$ such that

$$\int_F |u_n| \, d\mu \leq \frac{\varepsilon}{6\left(1 + \sup_k \|v_k\|_{L^\infty}\right)} \tag{2.48}$$

for all n and for every measurable set $F \subset X$ with $\mu(F) \leq \delta$. On the other hand, by Egoroff's theorem there exists a measurable set $X_\varepsilon \subset X$ with $\mu(X_\varepsilon) \leq \delta$ such that

$$v_n \to v \text{ strongly in } L^\infty(X \setminus X_\varepsilon).$$

Let n_0 be large enough so that for all $n \geq n_0$,

$$\|v_n - v\|_{L^\infty(X \setminus X_\varepsilon)} \leq \frac{\varepsilon}{3\left(1 + \sup_k \|u_k\|_{L^1}\right)} \tag{2.49}$$

and

$$\left| \int_{E \setminus X_\varepsilon} (u_n - u) v \, d\mu \right| \leq \frac{\varepsilon}{3}, \tag{2.50}$$

where in the last inequality we have used the fact that $u_n \rightharpoonup u$ in $L^1(X)$.

Writing

$$\int_E u_n v_n \, d\mu - \int_E uv \, d\mu = \int_{E \setminus X_\varepsilon} (u_n - u) v \, d\mu + \int_{E \setminus X_\varepsilon} u_n (v_n - v) \, d\mu$$

$$+ \int_{E \cap X_\varepsilon} u_n v_n \, d\mu - \int_{E \cap X_\varepsilon} uv \, d\mu,$$

we obtain

$$\left| \int_E u_n v_n \, d\mu - \int_E uv \, d\mu \right|$$

$$\leq \left| \int_{E \backslash X_\varepsilon} (u_n - u) \, v \, d\mu \right| + \| v_n - v \|_{L^\infty (E \backslash X_\varepsilon)} \sup_k \int_X |u_k| \, d\mu$$

$$+ \sup_k \| v_k \|_{L^\infty (X)} \int_{X_\varepsilon} |u_n| \, d\mu + \| v \|_{L^\infty (X)} \int_{X_\varepsilon} |u| \, d\mu \leq \varepsilon,$$

where we have used (2.50), (2.49), and (2.48) twice, in this order. Note that to estimate the last term we also used the facts that $\| v \|_{L^\infty (X)} \leq \liminf_{n \to \infty} \| v_n \|_{L^\infty (X)}$ and, by Proposition 2.46(ii),

$$\int_{X_\varepsilon} |u| \, d\mu \leq \liminf_{n \to \infty} \int_{X_\varepsilon} |u_n| \, d\mu.$$

The proof is complete.

Exercise 2.62. Can you extend the previous theorem to the case in which μ is not finite?

2.1.5 Biting Convergence

As we observed already in Remark 2.51, if $\{u_n\}$ is a bounded sequence in $L^1 (X)$ there is no guarantee that it admits a weakly convergent subsequence. However, this is possible, provided we exclude a decreasing sequence of measurable sets $\{E_j\}$ with $\mu (E_j) \to 0$.

Lemma 2.63 (Biting lemma). *Let $(X, \mathfrak{M}, \mu)$ be a measure space with μ finite and let $\{u_n\}$ be a sequence of functions bounded in $L^1 (X)$. Then there exist a function $u \in L^1 (X)$, a subsequence $\{u_{n_k}\}$ of $\{u_n\}$, and a decreasing sequence of measurable sets $\{E_j\} \subset X$, with $\mu (E_j) \to 0$, such that*

$$u_{n_k} \rightharpoonup u \text{ in } L^1 (X \backslash E_j) \text{ for every } j \in \mathbb{N}.$$

Proof. By the decomposition lemma there exist a subsequence of $\{u_n\}$ (not relabeled) and an increasing sequence of numbers $r_n \to \infty$ such that the truncated sequence $\{\tau_{r_n} \circ u_n\}$ is equi-integrable and

$$\mu (\{x \in X : u_n (x) \neq (\tau_{r_n} \circ u_n) (x)\}) \to 0$$

as $n \to \infty$. By selecting a further subsequence if necessary, we may assume that

$$\mu (\{x \in X : u_n (x) \neq (\tau_{r_n} \circ u_n) (x)\}) \leq \frac{1}{2^n}$$

for all $n \in \mathbb{N}$. For $j \in \mathbb{N}$ set

$$E_j := \bigcup_{n=j}^{\infty} \{x \in X : u_n (x) \neq (\tau_{r_n} \circ u_n) (x)\}.$$

Since $u_n = \tau_{r_n} \circ u_n$ on $X \setminus E_j$ for all $n \geq j$, it follows that $\{u_n\}_{n \in \mathbb{N}}$ is equi-integrable on $X \setminus E_j$.

Since μ is finite, by the Dunford–Pettis theorem, the sequence $\{u_n\}$ admits a subsequence $\{u_{n,1}\}$ converging weakly in $L^1(X \setminus E_1)$ to a function v_1. By induction, assuming that $\{u_{n,k}\}$ has been selected, we extract a subsequence $\{u_{n,k+1}\} \subset \{u_{n,k}\}$ converging weakly in $L^1(X \setminus E_{k+1})$ to a function v_{k+1}. Using a diagonal argument we may select a subsequence $\{u_{n_k}\}$ of $\{u_{n,k}\}$ such that u_{n_k} converges weakly to v_j in $L^1(X \setminus E_j)$ for each fixed j. By the uniqueness of the weak limit, there exists a measurable function u such that

$$u(x) = v_j(x) \quad \text{if } x \in X \setminus E_j \text{ for some } j \in \mathbb{N}.$$

Moreover, $u \in L^1(X)$, since by Proposition 2.46(ii),

$$\int_{X \setminus E_j} |u| \, d\mu \leq \sup_n \int_X |u_n| \, d\mu < \infty$$

for every $j \in \mathbb{N}$.

Note that in general we cannot conclude that the entire sequence $\{u_n\}$ converges in the sense of the biting lemma.

Exercise 2.64. Let $X := (0,1)$ with the Lebesgue measure and for n, $k \in \mathbb{N}$, $n \neq k$, define

$$u_{n,k}(x) := \begin{cases} \dfrac{1}{r_n} & \text{if } x \in (r_k - r_n, r_k + r_n), \\ 0 & \text{otherwise}, \end{cases}$$

where $\{r_n\}$ is an enumeration of the rational numbers in $(0, \infty)$. Note that

$$\int_0^1 |u_{n,k}| \, dx \leq 2.$$

Show that for any measurable set $E \subset (0,1)$, with $\mathcal{L}^1(E) > 0$, and for any $\ell \in \mathbb{N}$ there exists an infinite number of pairs of n, $k \in \mathbb{N}$ such that

$$\int_{\{|u_{n,k}| \geq \ell\} \cap E} |u_{n,k}| \, dx \geq 1.$$

Hence the entire sequence does not converge in the sense of the biting lemma.

The biting lemma suggests the following definition.

Definition 2.65. *Let $(X, \mathfrak{M}, \mu)$ be a measure space. Given $\{u_n\}$ a bounded sequence in $L^1(X)$, $u \in L^1(X)$, we say that $\{u_n\}$ converges weakly to u in the biting sense, and we write $u_n \overset{b}{\rightharpoonup} u$, if there exists a decreasing sequence of Lebesgue measurable sets $\{E_j\} \subset X$, with $\mu(E_j) \to 0$, such that*

$$u_n \rightharpoonup u \text{ in } L^1(X \setminus E_j) \text{ for every } j \in \mathbb{N}.$$

It can be checked that if the biting limit exists then it is unique. Indeed, assume that

$$u_n \overset{b}{\rightharpoonup} u, \quad u_n \overset{b}{\rightharpoonup} v.$$

Then $u_n \rightharpoonup u$ in $L^1(X \setminus E_j)$ for every $j \in \mathbb{N}$ and $u_n \rightharpoonup v$ in $L^1(X \setminus F_j)$ for every $j \in \mathbb{N}$, where $\mu(E_j)$, $\mu(F_j) \to 0$. By the uniqueness of the weak limit, it follows that $u = v$ for μ a.e. $x \in X \setminus (E_j \cup F_i)$ for every $i, j \in \mathbb{N}$. Hence $u = v$ for μ a.e. $x \in X \setminus F$, where

$$F := \left(\bigcap_{j=1}^{\infty} E_j \right) \cup \left(\bigcap_{j=1}^{\infty} F_j \right).$$

Since $\mu(F) \leq \mu(E_j) + \mu(F_j) \to 0$, we conclude that $u = v$ for μ a.e. $x \in X$.

Remark 2.66. The previous result continues to hold if $\{u_n\}$ is a sequence of functions uniformly bounded in $L^1(X; Y)$, where X is a measurable space with a finite positive measure μ, and Y is a reflexive Banach space (see Section 2.3 for the definition of $L^1(X; Y)$). We refer to [BaMu89] for more details.

Proposition 2.67. *Let $(X, \mathfrak{M}, \mu)$ be a measure space and let $\{u_n\}$ be a sequence of nonnegative functions in $L^1(X)$ such that $u_n \overset{b}{\rightharpoonup} u$. Then a subsequence converges weakly to u in $L^1(X)$ if and only if*

$$\liminf_{n \to \infty} \int_X u_n \, d\mu \leq \int_X u \, d\mu.$$

In addition, $u_n \rightharpoonup u$ in $L^1(X)$ if and only if

$$\lim_{n \to \infty} \int_X u_n \, d\mu = \int_X u \, d\mu.$$

Proof. Assume that a subsequence $\{u_{n_k}\}$ of $\{u_n\}$ converges weakly to u in $L^1(X)$. Then

$$\liminf_{n \to \infty} \int_X u_n \, d\mu \leq \lim_{k \to \infty} \int_X u_{n_k} \, d\mu = \int_X u \, d\mu.$$

Conversely, assume that

$$\liminf_{n \to \infty} \int_X u_n \, d\mu \leq \int_X u \, d\mu$$

and extract a subsequence $\{u_{n_k}\}$ of $\{u_n\}$ such that

$$\liminf_{n \to \infty} \int_X u_n \, d\mu = \lim_{k \to \infty} \int_X u_{n_k} \, d\mu.$$

By definition of biting convergence there exists a decreasing sequence $\{E_j\} \subset X$ of measurable sets with $\mu(E_j) \to 0$ such that

$$u_{n_k} \rightharpoonup u \text{ in } L^1(X \setminus E_j) \text{ for every } j \in \mathbb{N}.$$

In view of Corollary 2.49, it suffices to prove that for every measurable set $F \subset X$ we have

$$\lim_{k \to \infty} \int_F u_{n_k} \, d\mu = \int_F u \, d\mu. \tag{2.51}$$

Indeed, since $u_{n_k} \geq 0$ we have

$$\int_{F \setminus E_j} u \, d\mu = \lim_{k \to \infty} \int_{F \setminus E_j} u_{n_k} \, d\mu \leq \liminf_{k \to \infty} \int_F u_{n_k} \, d\mu,$$

and so, letting $j \to \infty$, we conclude that

$$\int_F u \, d\mu \leq \liminf_{k \to \infty} \int_F u_{n_k} \, d\mu.$$

Similarly,

$$\int_X u \, d\mu - \int_F u \, d\mu = \int_{X \setminus F} u \, d\mu \leq \liminf_{k \to \infty} \int_{X \setminus F} u_{n_k} \, d\mu$$

$$= \lim_{k \to \infty} \int_X u_{n_k} \, d\mu - \limsup_{k \to \infty} \int_F u_{n_k} \, d\mu$$

$$\leq \int_X u \, d\mu - \limsup_{k \to \infty} \int_F u_{n_k} \, d\mu,$$

i.e.,

$$\limsup_{k \to \infty} \int_F u_{n_k} \, d\mu \leq \int_F u \, d\mu,$$

and (2.51) is proved.

To prove the second statement of the proposition, assume that $u_n \rightharpoonup u$ in $L^1(X)$. Since $1 \in (L^1(X))'$ we have that

$$\lim_{n \to \infty} \int_X u_n \, d\mu = \int_X u \, d\mu.$$

Conversely, assume that

$$\lim_{n \to \infty} \int_X u_n \, d\mu = \int_X u \, d\mu.$$

This implies that

$$\liminf_{k \to \infty} \int_X u_{n_k} \, d\mu = \int_X u \, d\mu$$

for every subsequence $\{u_{n_k}\}$ of $\{u_n\}$. From the first part we obtain that from every subsequence $\{u_{n_k}\}$ of $\{u_n\}$ we may extract a further subsequence $\{u_{n_{k_j}}\}$ converging weakly to u in $L^1(X)$. Therefore the whole sequence $\{u_n\}$ converges weakly to u in $L^1(X)$.

Remark 2.68. Let X be a σ-compact metric space, let μ be a Radon measure, let $\{u_n\}$ be a sequence of functions uniformly bounded in $L^1(X)$. By Remark 2.51, Proposition 1.202, and the biting lemma we may extract a subsequence $\{u_{n_k}\}$ such that $u_{n_k}\mu \overset{*}{\rightharpoonup} \lambda$ in $\mathcal{M}(X;\mathbb{R})$ and $u_{n_k} \overset{b}{\rightharpoonup} u$, for some function $u \in L^1(X)$ and a signed Radon measure λ. As the next exercise shows, in general there is no obvious relation between λ and u, and even when $\lambda \ll \mu$ we cannot conclude that

$$\frac{d\lambda}{d\mu} = u. \tag{2.52}$$

Exercise 2.69. Let $X := (0,1)$, let $\mu := \mathcal{L}^1$, and for $n \in \mathbb{N}$, $n \geq 2$, define

$$u_n(x) := \begin{cases} \frac{n^2}{2}, & \frac{k}{n+1} - \frac{1}{n^3} < x < \frac{k}{n+1} + \frac{1}{n^3}, \quad k = 1,\ldots,n, \\ 0, & \text{otherwise.} \end{cases}$$

Prove that $u_n \mathcal{L}^1 \overset{*}{\rightharpoonup} \mathcal{L}^1$ in $\mathcal{M}(X;\mathbb{R})$, while $u_n \overset{b}{\rightharpoonup} 0$. Note that from the uniqueness of the biting limit, this implies that the sets E_j cannot be chosen to be closed (see Proposition 2.70 below).

In the next proposition we show that (2.52) holds if the sets E_j are closed.

Proposition 2.70. *Let X be a locally compact Hausdorff space, let $\mu : \mathcal{B}(X) \to [0,\infty]$ be a Borel measure, and let $\{u_n\} \subset L^1(X)$ be a sequence of functions such that*

$$u_n\mu \overset{*}{\rightharpoonup} \lambda \quad \text{in } \mathcal{M}(X;\mathbb{R}) \quad \text{and} \quad u_n \overset{b}{\rightharpoonup} u.$$

If the sets E_j are closed, then

$$\frac{d\lambda}{d\mu} = u \quad \mu \text{ a.e. in } X.$$

Proof. Since $X \setminus E_j$ is open, for any $\varphi \in C_c(X \setminus E_j)$ we have

$$\int_{X\setminus E_j} u\varphi \, d\mu = \lim_{n\to\infty} \int_{X\setminus E_j} u_n\varphi \, d\mu = \lim_{n\to\infty} \int_X u_n\varphi \, d\mu$$

$$= \int_X \varphi \, d\lambda = \int_{X\setminus E_j} \varphi \, d\lambda.$$

Thus $u\mu \lfloor (X \setminus E_j) = \lambda \lfloor (X \setminus E_j)$ for every $j \in \mathbb{N}$, and so

$$\frac{d\lambda}{d\mu}(x) = u(x)$$

for μ a.e. $x \in X \setminus F$, where

$$F := \bigcap_{j=1}^{\infty} E_j,$$

$\mu(F) = 0$, and also $\text{supp}(\lambda_s) \cap (X \setminus E_j) = \emptyset$, i.e., $\text{supp}(\lambda_s) \subset F$.

The last result of this section shows that if (2.52) holds and $u_n \geq 0$ for all $n \in \mathbb{N}$, then we actually have weak convergence on compact sets.

Proposition 2.71. *Let* X *be a locally compact Hausdorff space, let* $\mu :$ $\mathcal{B}(X) \rightarrow [0, \infty]$ *be a Borel measure, and let* $\{u_n\} \subset L^1(X)$ *be a sequence of nonnegative functions such that*

$$u_n \mu \overset{*}{\rightharpoonup} u\mu \quad in \; \mathcal{M}(X; \mathbb{R}) \quad and \quad u_n \overset{b}{\rightharpoonup} u. \tag{2.53}$$

Then $u_n \rightharpoonup u$ *in* $L^1(K)$ *for each compact set* $K \subset X$.

Proof. By definition of biting convergence there exists a decreasing sequence $\{E_j\} \subset X$ of measurable sets with $\mu(E_j) \rightarrow 0$ such that

$$u_n \rightharpoonup u \; in \; L^1(X \setminus E_j) \; for \; every \; j \in \mathbb{N}.$$

Let $K \subset X$ be a compact set and let $\varphi \in C_0(X)$ be such that $\varphi \equiv 1$ on K, and $0 \leq \varphi \leq 1$. Then

$$\begin{aligned}
\limsup_{n \to \infty} \int_{K \cap E_j} u_n \, d\mu &\leq \limsup_{n \to \infty} \int_{E_j} \varphi u_n \, d\mu \\
&= \limsup_{n \to \infty} \left(\int_X \varphi u_n \, d\mu - \int_{X \setminus E_j} \varphi u_n \, d\mu \right) \\
&= \int_X \varphi u \, d\mu - \int_{X \setminus E_j} \varphi u \, d\mu \\
&= \int_{E_j} \varphi u \, d\mu \leq \int_{E_j} u \, d\mu,
\end{aligned}$$

where we have used (2.53).

Let $v \in L^\infty(K)$. By the previous inequality and (2.53) once more we have

$$\begin{aligned}
\limsup_{n \to \infty} &\left| \int_K v(u_n - u) \, d\mu \right| \\
&\leq \lim_{n \to \infty} \left| \int_{K \setminus E_j} v(u_n - u) \, d\mu \right| + \limsup_{n \to \infty} \left| \int_{K \cap E_j} v(u_n - u) \, d\mu \right| \\
&\leq \|v\|_{L^\infty(K)} \limsup_{n \to \infty} \int_{K \cap E_j} (u_n + u) \, d\mu \\
&\leq 2 \|v\|_{L^\infty(K)} \int_{E_j} u \, d\mu.
\end{aligned}$$

Given the arbitrariness of j, and since $\mu(E_j) \rightarrow 0$, we conclude that

$$\lim_{n \to \infty} \int_K v u_n \, d\mu = \int_K v u \, d\mu$$

for all $v \in L^\infty(K)$. Hence $u_n \rightharpoonup u$ in $L^1(K)$.

2.2 Euclidean Setting

Throughout this section the ambient space is the euclidean space $\mathbb{R}^N$, and we consider the Lebesgue measure $\mathcal{L}^N$. Also, $E \subset \mathbb{R}^N$ will denote a Lebesgue measurable set, not necessarily bounded. For a *multi-index* $\alpha = (\alpha_1, \ldots, \alpha_N) \in (\mathbb{N}_0)^N$ we set

$$\frac{\partial^\alpha}{\partial x^\alpha} := \frac{\partial^{|\alpha|}}{\partial x_1^{\alpha_1} \ldots \partial x_N^{\alpha_N}}, \quad |\alpha| := \alpha_1 + \ldots + \alpha_N,$$

where $x = (x_1, \ldots, x_N)$. If $\Omega \subset \mathbb{R}^N$ is any open set (not necessarily bounded), for any nonnegative integer $l \in \mathbb{N}_0$ we denote by $C^l(\Omega)$ the vector space of all functions that are continuous together with their partial derivatives up to order l. We set $C^\infty(\Omega) := \bigcap_{l=0}^{\infty} C^l(\Omega)$ and we define $C_c^l(\Omega)$ and $C_c^\infty(\Omega)$ as the subspaces of $C^l(\Omega)$ and $C^\infty(\Omega)$, respectively, consisting of all functions that have compact support.

2.2.1 Approximation by Regular Functions

Given a nonnegative bounded function $\varphi : \mathbb{R}^N \to [0, \infty)$ with

$$\operatorname{supp} \varphi \subset \overline{B(0,1)}, \quad \int_{\mathbb{R}^N} \varphi(x)\, dx = 1, \tag{2.54}$$

for $u \in L^1_{\text{loc}}(\mathbb{R}^N)$ and $0 < \varepsilon < 1$, define the *mollification*

$$u_\varepsilon(x) := (u * \varphi_\varepsilon)(x) = \int_{\mathbb{R}^N} \varphi_\varepsilon(x - y)\, u(y)\, dy, \quad x \in \mathbb{R}^N,$$

where

$$\varphi_\varepsilon(x) := \frac{1}{\varepsilon^N} \varphi\left(\frac{x}{\varepsilon}\right), \quad x \in \mathbb{R}^N.$$

The functions φ_ε are called *mollifiers*. Note that $\operatorname{supp} \varphi_\varepsilon \subset \overline{B(0, \varepsilon)}$.

Remark 2.72. In the applications we will consider two special cases:

(i) φ is the (renormalized) characteristic function of the unit ball, that is,

$$\varphi(x) := \frac{1}{\alpha_N} \chi_{B(0,1)}(x), \quad x \in \mathbb{R}^N;$$

(ii) φ is the C_c^∞ function

$$\varphi(x) := \begin{cases} c \exp\left(\frac{1}{|x|^2 - 1}\right) & \text{if } |x| < 1, \\ 0 & \text{if } |x| \geq 1, \end{cases} \tag{2.55}$$

where we choose $c > 0$ such that (2.54) is satisfied. In this case, the functions φ_ε are called *standard mollifiers*.

The first main result of this subsection is the following theorem.

Theorem 2.73. *Let* $\varphi : \mathbb{R}^N \to [0, \infty)$ *be a nonnegative bounded function satisfying* (2.54), *and let* $u \in L^1_{\text{loc}}(\mathbb{R}^N)$.

(i) If $u \in C(\mathbb{R}^N)$ *then* $u_\varepsilon \to u$ *as* $\varepsilon \to 0^+$ *uniformly on compact sets.*
(ii) $u_\varepsilon(x) \to u(x)$ *as* $\varepsilon \to 0^+$ *for every Lebesgue point* $x \in \mathbb{R}^N$ *of* u *(and so for* $\mathcal{L}^N$ *a.e.* $x \in \mathbb{R}^N$*).*
(iii) If $u \in L^p(\mathbb{R}^N)$, $1 \le p < \infty$, *then*

$$\|u_\varepsilon\|_{L^p(\mathbb{R}^N)} \le \|u\|_{L^p(\mathbb{R}^N)} \tag{2.56}$$

and

$$u_\varepsilon \to u \quad \text{in } L^p(\mathbb{R}^N) \text{ as } \varepsilon \to 0^+.$$

Proof. (i) Let $K \subset \mathbb{R}^N$ be a compact set. For any fixed $\eta > 0$ let

$$K_\eta := \{x \in \mathbb{R}^N : \text{dist}(x, K) \le \eta\}.$$

Then K_η is compact and since u is uniformly continuous on K_η, for any $\rho > 0$ there exists $\delta = \delta(\eta, K, \rho) > 0$ such that

$$|u(x) - u(y)| \le \frac{\rho}{1 + \|\varphi\|_\infty} \tag{2.57}$$

for all $x, y \in K_\eta$, with $|x - y| \le \delta$. Let $0 < \varepsilon < \min\{\delta, \eta\}$. Then for all $x \in K$,

$$|u_\varepsilon(x) - u(x)| = \left| \int_{\mathbb{R}^N} \varphi_\varepsilon(x - y) u(y) \, dy - u(x) \right| \tag{2.58}$$

$$= \frac{1}{\varepsilon^N} \left| \int_{B(x,\varepsilon)} \varphi\left(\frac{x-y}{\varepsilon}\right) [u(y) - u(x)] \, dy \right|$$

$$\le \|\varphi\|_\infty \frac{1}{\varepsilon^N} \int_{B(x,\varepsilon)} |u(y) - u(x)| \, dy,$$

where we have used (2.54) and the fact that $\text{supp}\,\varphi_\varepsilon \subset \overline{B(0,\varepsilon)}$. It follows by (2.57) that

$$|u_\varepsilon(x) - u(x)| \le \rho$$

for all $x \in K$, and so $\|u_\varepsilon - u\|_{C(K)} \le \rho$.

(ii) Let $x \in \mathbb{R}^N$ be a Lebesgue point of u, that is,

$$\lim_{\varepsilon \to 0^+} \frac{1}{\varepsilon^N} \int_{B(x,\varepsilon)} |u(y) - u(x)| \, dy = 0.$$

Then from (2.58) it follows that $u_\varepsilon(x) \to u(x)$ as $\varepsilon \to 0^+$.

(iii) Let $u \in L^p(\mathbb{R}^N)$, $1 \le p < \infty$. We prove (2.56). By Hölder's inequality and (2.54), for all $x \in \mathbb{R}^N$,

$$|u_\varepsilon(x)| = \left| \int_{\mathbb{R}^N} (\varphi_\varepsilon(x-y))^{1/p'} (\varphi_\varepsilon(x-y))^{1/p} u(y) \, dy \right|$$

$$\leq \left(\int_{\mathbb{R}^N} \varphi_\varepsilon(x-y) \, dy \right)^{1/p'} \left(\int_{\mathbb{R}^N} \varphi_\varepsilon(x-y) |u(y)|^p \, dy \right)^{1/p}$$

$$= \left(\int_{\mathbb{R}^N} \varphi_\varepsilon(x-y) |u(y)|^p \, dy \right)^{1/p},$$

and so by Fubini's theorem and (2.54) once more,

$$\int_{\mathbb{R}^N} |u_\varepsilon(x)|^p \, dx \leq \int_{\mathbb{R}^N} \int_{\mathbb{R}^N} \varphi_\varepsilon(x-y) |u(y)|^p \, dy \, dx$$

$$= \int_{\mathbb{R}^N} |u(y)|^p \left(\int_{\mathbb{R}^N} \varphi_\varepsilon(x-y) \, dx \right) dy$$

$$= \int_{\mathbb{R}^N} |u(y)|^p \, dy.$$

It remains to show that $u_\varepsilon \to u$ in $L^p(\mathbb{R}^N)$. Fix $\rho > 0$ and find a function $v \in C_c(\mathbb{R}^N)$ such that

$$\|u - v\|_{L^p(\mathbb{R}^N)} \leq \rho.$$

Since $K := \operatorname{supp} v$ is compact, it follows from part (i) that for every $\eta > 0$, the mollification v_ε of v converges to v uniformly in the compact set

$$K_\eta := \{x \in \mathbb{R}^N : \operatorname{dist}(x, K) \leq \eta\}.$$

Since $v_\varepsilon = v = 0$ in $\mathbb{R}^N \setminus K_\eta$ for $0 < \varepsilon < \eta$, we have that

$$\int_{\mathbb{R}^N} |v_\varepsilon - v|^p \, dx = \int_{K_\eta} |v_\varepsilon - v|^p \, dx \leq \left(\|v_\varepsilon - v\|_{C(K_\eta)} \right)^p |K_\eta| \leq \rho,$$

provided $\varepsilon > 0$ is sufficiently small. By Minkowski's inequality,

$$\|u_\varepsilon - u\|_{L^p(\mathbb{R}^N)} \leq \|u_\varepsilon - v_\varepsilon\|_{L^p(\mathbb{R}^N)} + \|v_\varepsilon - v\|_{L^p(\mathbb{R}^N)} + \|v - u\|_{L^p(\mathbb{R}^N)}$$

$$\leq 2\|u - v\|_{L^p(\mathbb{R}^N)} + \|v_\varepsilon - v\|_{L^p(\mathbb{R}^N)} \leq 3\rho,$$

where we have used (2.56) for the function $u - v$.

More can be said about the regularity of u_ε if we restrict our attention to standard mollifiers.

Theorem 2.74. *Let φ be defined as in (2.55), and let $u \in L^1_{\mathrm{loc}}(\mathbb{R}^N)$. Then $u_\varepsilon \in C^\infty(\mathbb{R}^N)$ for all $0 < \varepsilon < 1$, and for any multi-index α,*

$$\frac{\partial^\alpha u_\varepsilon}{\partial x^\alpha}(x) = \left(u * \frac{\partial^\alpha \varphi_\varepsilon}{\partial x^\alpha} \right)(x) = \int_{\mathbb{R}^N} \frac{\partial^\alpha \varphi_\varepsilon}{\partial x^\alpha}(x-y) u(y) \, dy \qquad (2.59)$$

for all $x \in \mathbb{R}^N$.

Proof. Fix $x \in \mathbb{R}^N$, $\eta > 0$, and $0 < \varepsilon < 1$. Let e_i, $i = 1, \ldots, N$, be an element of the canonical basis of $\mathbb{R}^N$, and for any $h \in \mathbb{R}$, with $0 < |h| \leq \eta$, consider

$$
\frac{u_\varepsilon (x + he_i) - u_\varepsilon (x)}{h} - \int_{\mathbb{R}^N} \frac{\partial \varphi_\varepsilon}{\partial x_i} (x - y) u(y) \, dy
$$

$$
= \int_{\mathbb{R}^N} \left[\frac{\varphi_\varepsilon (x + he_i - y) - \varphi_\varepsilon (x - y)}{h} - \frac{\partial \varphi_\varepsilon}{\partial x_i} (x - y) \right] u(y) \, dy
$$

$$
= \int_{\mathbb{R}^N} \left(\frac{1}{h} \int_0^h \frac{\partial \varphi_\varepsilon}{\partial x_i} (x - y + te_i) \, dt - \frac{\partial \varphi_\varepsilon}{\partial x_i} (x - y) \right) u(y) \, dy
$$

$$
= \frac{1}{h} \int_0^h \left(\int_{B(x, \varepsilon + \eta)} \left[\frac{\partial \varphi_\varepsilon}{\partial x_i} (x - y + te_i) - \frac{\partial \varphi_\varepsilon}{\partial x_i} (x - y) \right] u(y) \, dy \right) dt,
$$

where we have used Fubini's theorem and the fact that $\operatorname{supp} \varphi_\varepsilon \subset \overline{B(0, \varepsilon)}$.

Since $\varphi_\varepsilon \in C_c^\infty (\mathbb{R}^N)$, its partial derivatives are uniformly continuous. Hence for any for any $\rho > 0$ there exists $\delta = \delta(\eta, x, \rho, \varepsilon) > 0$ such that

$$
\left| \frac{\partial \varphi_\varepsilon}{\partial x_i} (z) - \frac{\partial \varphi_\varepsilon}{\partial x_i} (w) \right| \leq \frac{\rho}{1 + \|u\|_{L^1 (B(x, \varepsilon + \eta))}}
$$

for all z, $w \in B(x, \varepsilon + \eta)$, with $|z - w| \leq \delta$. Then for $0 < |h| \leq \min\{\eta, \delta\}$ we have

$$
\left| \frac{u_\varepsilon (x + he_i) - u_\varepsilon (x)}{h} - \int_{\mathbb{R}^N} \frac{\partial \varphi_\varepsilon}{\partial x_i} (x - y) u(y) \, dy \right| \leq \rho,
$$

which shows that

$$
\frac{\partial u_\varepsilon}{\partial x_i} (x) = \int_{\mathbb{R}^N} \frac{\partial \varphi_\varepsilon}{\partial x_i} (x - y) u(y) \, dy.
$$

Note that the only properties that we have used about the function φ_ε are that $\varphi_\varepsilon \in C_c^\infty (\mathbb{R}^N)$ with $\operatorname{supp} \varphi_\varepsilon \subset \overline{B(0, \varepsilon)}$. Hence the same proof carries over if we replace φ_ε with $\psi_\varepsilon := \frac{\partial \varphi_\varepsilon}{\partial x_i}$. Thus by induction we may prove that for any multi-index α,

$$
\frac{\partial^\alpha u_\varepsilon}{\partial x^\alpha} (x) = \left(u * \frac{\partial^\alpha \varphi_\varepsilon}{\partial x^\alpha} \right) (x) = \int_{\mathbb{R}^N} \frac{\partial^\alpha \varphi_\varepsilon}{\partial x^\alpha} (x - y) u(y) \, dy.
$$

This completes the proof. ∎

If the function u is defined only in an open set $\Omega \subset \mathbb{R}^N$, with $u \in L_{\text{loc}}^1 (\Omega)$, then since $\operatorname{supp} \varphi_\varepsilon \subset \overline{B(0, \varepsilon)}$, we may still define

$$
u_\varepsilon (x) := (u * \varphi_\varepsilon)(x) = \int_\Omega \varphi_\varepsilon (x - y) u(y) \, dy
$$

for $x \in \Omega_\varepsilon$, where the open set Ω_ε is given by

$$
\Omega_\varepsilon := \{x \in \Omega : \operatorname{dist}(x, \partial \Omega) > \varepsilon\}.
$$

In this case an analogous version of Theorem 2.73 holds.

Theorem 2.75. *Let $\Omega \subset \mathbb{R}^N$ be an open set, let $\varphi : \mathbb{R}^N \to [0, \infty)$ be a nonnegative bounded function satisfying (2.54), and let $u \in L^1_{\text{loc}}(\Omega)$.*

(i) *If $u \in C(\Omega)$, then $u_\varepsilon \to u$ as $\varepsilon \to 0^+$ uniformly on compact subsets of Ω.*
(ii) *$u_\varepsilon(x) \to u(x)$ as $\varepsilon \to 0^+$ for every Lebesgue point $x \in \Omega$ of u.*
(iii) *If $u \in L^p_{\text{loc}}(\Omega)$, $1 \le p < \infty$, then*

$$\|u_\varepsilon\|_{L^p(\Omega_\varepsilon)} \le \|u\|_{L^p(\Omega)}$$

and for every compact set $K \subset \Omega$,

$$u_\varepsilon \to u \quad in \ L^p(K) \ as \ \varepsilon \to 0^+.$$

Proof. The proof is very similar to that of Theorem 2.73. In the proof of (i), for any compact $K \subset \Omega$ one should take

$$0 < \eta < \text{dist}(K, \partial\Omega),$$

so that $K_\eta \subset \Omega$. Moreover, for $\varepsilon > 0$ sufficiently small we have that $K_\eta \subset \Omega_\varepsilon$. A similar change can be made in the proof of (iii). We omit the details.

The analogue of Theorem 2.74 is the following.

Theorem 2.76. *Let $\Omega \subset \mathbb{R}^N$ be an open set, let φ be defined as in (2.55), and let $u \in L^1_{\text{loc}}(\Omega)$. Then $u_\varepsilon \in C^\infty(\Omega_\varepsilon)$ for all $0 < \varepsilon < 1$, and for any multi-index α the formula (2.59) holds for all $x \in \Omega_\varepsilon$.*

Proof. The proof is very similar to that of Theorem 2.74. Since now $x \in \Omega_\varepsilon$, we have that $\text{dist}(x, \partial\Omega) > \varepsilon$, and so one should take

$$0 < \eta < \text{dist}(x, \partial\Omega) - \varepsilon.$$

We omit the details.

An important application of the theory of mollifiers is the existence of smooth partitions of unity.

Theorem 2.77 (Smooth partition of unity). *Let $\Omega \subset \mathbb{R}^N$ be an open set and let $\{U_\alpha\}_{\alpha \in I}$ be an open cover of Ω. Then there exists a sequence $\{\psi_n\} \subset C^\infty_c(\Omega)$ of nonnegative functions such that*

(i) *each ψ_n has support in some U_α;*
(ii) *$\sum_{n=1}^\infty \psi_n(x) = 1$ for all $x \in \Omega$;*
(iii) *for every compact set $K \subset \Omega$ there exist an integer $\ell \in \mathbb{N}$ and an open set U, with $K \subset U \subset \Omega$, such that*

$$\sum_{n=1}^\ell \psi_n(x) = 1$$

for all $x \in U$.

Proof. Let S be a countable dense set in Ω, e.g., $S := \{x \in \mathbb{Q}^N \cap \Omega\}$, and consider the countable family $\mathcal{F}$ of closed balls

$$\mathcal{F} := \left\{ \overline{B(x,r)} : r \in \mathbb{Q}, 0 < r < 1, x \in S, \right.$$
$$\left. \overline{B(x,r)} \subset U_\alpha \cap \Omega \text{ for some } \alpha \in I \right\}.$$

Since $\mathcal{F}$ is countable we may write $\mathcal{F} = \left\{ \overline{B(x_n, r_n)} \right\}$. Since $\{U_\alpha\}_{\alpha \in I}$ is an open cover of Ω, by the density of S and of the rational numbers we have that

$$\Omega = \bigcup_{n=1}^{\infty} B\left(x_n, \frac{r_n}{2}\right). \tag{2.60}$$

For each $n \in \mathbb{N}$ consider

$$\phi_n := \varphi_{\frac{r_n}{4}} * \chi_{B\left(x_n, \frac{3}{4}r_n\right)},$$

where $\varphi_{\frac{r_n}{4}}$ are standard mollifiers (with $\varepsilon := \frac{r_n}{4}$). By Theorem 2.74, $\phi_n \in C^\infty\left(\mathbb{R}^N\right)$. Moreover, if $x \in B\left(x_n, \frac{r_n}{2}\right)$ then

$$\phi_n(x) = \int_{\mathbb{R}^N} \varphi_{\frac{r_n}{4}}(x - y) \chi_{B\left(x_n, \frac{3}{4}r_n\right)}(y) \, dy$$
$$= \int_{B\left(x, \frac{r_n}{4}\right)} \varphi_{\frac{r_n}{4}}(x - y) \chi_{B\left(x_n, \frac{3}{4}r_n\right)}(y) \, dy$$
$$= \int_{B\left(x, \frac{r_n}{4}\right)} \varphi_{\frac{r_n}{4}}(x - y) \, dy = 1,$$

where we have used (2.54) and the fact that if $x \in B\left(x_n, \frac{r_n}{2}\right)$ then

$$B\left(x, \frac{r_n}{4}\right) \subset B\left(x_n, \frac{3}{4}r_n\right).$$

Since $0 \le \chi_{B\left(x_n, \frac{3}{4}r_n\right)} \le 1$, a similar calculation shows that $0 \le \phi_n \le 1$. On the other hand, if $x \notin \overline{B(x_n, r_n)}$ then

$$\phi_n(x) = \int_{\mathbb{R}^N} \varphi_{\frac{r_n}{4}}(x - y) \chi_{B\left(x_n, \frac{3}{4}r_n\right)}(y) \, dy$$
$$= \int_{B\left(x, \frac{r_n}{4}\right)} \varphi_{\frac{r_n}{4}}(x - y) \chi_{B\left(x_n, \frac{3}{4}r_n\right)}(y) \, dy = 0,$$

where we have used the fact that if $x \notin \overline{B(x_n, r_n)}$ then

$$B\left(x, \frac{r_n}{4}\right) \cap B\left(x_n, \frac{3}{4}r_n\right) = \emptyset.$$

In particular, $\phi_n \in C_c^\infty\left(\mathbb{R}^N\right)$ and $\operatorname{supp}\phi_n \subset \overline{B\left(x_n, r_n\right)}$. Note that in view of the definition of $\mathcal{F}$, $\operatorname{supp}\phi_n \subset U_\alpha \cap \Omega$ for some $\alpha \in I$.

Define $\psi_1 := \phi_1$ and

$$\psi_n := (1 - \phi_1)\ldots(1 - \phi_{n-1})\,\phi_n \tag{2.61}$$

for $n \geq 2$, $n \in \mathbb{N}$. Since $0 \leq \phi_k \leq 1$ and $\operatorname{supp}\phi_k \subset \overline{B\left(x_k, r_k\right)}$ for all $k \in \mathbb{N}$, we have that $0 \leq \psi_n \leq 1$ and $\operatorname{supp}\psi_n \subset \overline{B\left(x_n, r_n\right)}$. This gives (i). To prove (ii) we prove by induction that

$$\psi_1 + \ldots + \psi_n = 1 - (1 - \phi_1)\ldots(1 - \phi_n) \tag{2.62}$$

for all $n \in \mathbb{N}$. The relation (2.62) is true for $n = 1$, since $\psi_1 := \phi_1$. Assume that (2.62) holds for n. Then by (2.61),

$$\begin{aligned}
\psi_1 + \ldots + \psi_n + \psi_{n+1} &= 1 - (1 - \phi_1)\ldots(1 - \phi_n) + \psi_{n+1} \\
&= 1 - (1 - \phi_1)\ldots(1 - \phi_n) + (1 - \phi_1)\ldots(1 - \phi_n)\,\phi_{n+1} \\
&= 1 - (1 - \phi_1)\ldots(1 - \phi_{n+1}).
\end{aligned}$$

Hence (2.62) holds for all $n \in \mathbb{N}$.

Since $\phi_k = 1$ in $B\left(x_k, \frac{r_k}{2}\right)$ for all $k \in \mathbb{N}$, by (2.62), for all n, $m \in \mathbb{N}$ with $m \geq n$ we have that

$$\psi_1(x) + \ldots + \psi_m(x) = 1 \quad \text{for all } x \in \bigcup_{k=1}^{n} B\left(x_k, \frac{r_k}{2}\right). \tag{2.63}$$

Thus, in view of (2.60), property (ii) holds.

Finally, if $K \subset \Omega$ is compact, again by (2.60), we may find $\ell \in \mathbb{N}$ so large that

$$\bigcup_{k=1}^{\ell} B\left(x_k, \frac{r_k}{2}\right) \supset K,$$

and so (iii) follows by (2.63).

Using mollifiers it is possible to improve the density result of Theorem 2.99:

Theorem 2.78. *Let $\Omega \subset \mathbb{R}^N$ be an open set. Then the space $C_c^\infty\left(\Omega\right)$ is dense in $L^p\left(\Omega\right)$ for $1 \leq p < \infty$.*

Proof. Let $u \in L^p\left(\Omega\right)$ and extend it to be zero outside Ω. Define

$$K_n := \left\{x \in \mathbb{R}^N : |x| \leq n,\ \operatorname{dist}\left(x, \mathbb{R}^N \setminus \Omega\right) \geq \frac{2}{n}\right\}.$$

Then K_n is compact, $K_n \subset K_{n+1}$, and

$$\bigcup_{n=1}^{\infty} K_n = \Omega.$$

Define $v_n := u\chi_{K_n}$ and $u_n := \varphi_{\frac{1}{n}} * v_n$, where $\varphi_{\frac{1}{n}}$ are standard mollifiers (with $\varepsilon := \frac{1}{n}$). Since $\operatorname{supp}\varphi_{\frac{1}{n}} \subset \overline{B\left(0, \frac{1}{n}\right)}$ it follows that

$$\operatorname{supp} u_n \subset \left\{ x \in \mathbb{R}^N : \operatorname{dist}(x, K_n) \leq \frac{1}{n} \right\} \subset \Omega.$$

Hence by Theorem 2.76, $u_n \in C_c^{\infty}(\Omega)$. Moreover, by Minkowski's inequality,

$$\|u_n - u\|_{L^p(\Omega)} = \|u_n - u\|_{L^p(\mathbb{R}^N)}$$
$$\leq \left\|\varphi_{\frac{1}{n}} * v_n - \varphi_{\frac{1}{n}} * u\right\|_{L^p(\mathbb{R}^N)} + \left\|\varphi_{\frac{1}{n}} * u - u\right\|_{L^p(\mathbb{R}^N)}$$
$$\leq \|v_n - u\|_{L^p(\mathbb{R}^N)} + \left\|\varphi_{\frac{1}{n}} * u - u\right\|_{L^p(\mathbb{R}^N)}$$
$$= \|u\chi_{K_n} - u\|_{L^p(\mathbb{R}^N)} + \left\|\varphi_{\frac{1}{n}} * u - u\right\|_{L^p(\mathbb{R}^N)},$$

where we have used (2.56). It now follows from the Lebesgue dominated convergence theorem that

$$\|u\chi_{K_n} - u\|_{L^p(\mathbb{R}^N)} \to 0$$

as $n \to \infty$, while

$$\left\|\varphi_{\frac{1}{n}} * u - u\right\|_{L^p(\mathbb{R}^N)} \to 0$$

by Theorem 2.73. This completes the proof.

Given a (positive) Radon measure $\mu : \mathcal{B}(\mathbb{R}^N) \to [0, \infty]$ and a standard mollifier φ_ε, $\varepsilon > 0$, we define

$$(\varphi_\varepsilon * \mu)(x) := \int_{\mathbb{R}^N} \varphi_\varepsilon(x - y) \, d\mu(y), \quad x \in \mathbb{R}^N.$$

Theorem 2.79. *Let $\mu : \mathcal{B}(\mathbb{R}^N) \to [0, \infty]$ be a (positive) Radon measure. Then for all $u \in C_c(\mathbb{R}^N)$,*

$$\int_{\mathbb{R}^N} u(x)(\varphi_\varepsilon * \mu)(x) \, dx \to \int_{\mathbb{R}^N} u(x) \, d\mu(x) \tag{2.64}$$

as $\varepsilon \to 0^+$. If, in addition, $\mu(\mathbb{R}^N) < \infty$, then (2.64) holds for all uniformly continuous bounded functions u.

Proof. Let $u \in C_c(\mathbb{R}^N)$. Using Fubini's theorem and (2.54) we have

$$\int_{\mathbb{R}^N} u(x) (\varphi_\varepsilon * \mu)(x) \, dx - \int_{\mathbb{R}^N} u(y) \, d\mu(y)$$

$$= \int_{\mathbb{R}^N} u(x) \int_{\mathbb{R}^N} \varphi_\varepsilon(x - y) \, d\mu(y) \, dx - \int_{\mathbb{R}^N} u(y) \, d\mu(y)$$

$$= \int_{\mathbb{R}^N} \left(\int_{\mathbb{R}^N} u(x) \varphi_\varepsilon(x - y) \, dx - \int_{\mathbb{R}^N} u(y) \varphi_\varepsilon(x) \, dx \right) d\mu(y)$$

$$= \int_{\mathbb{R}^N} \left(\int_{\mathbb{R}^N} u(x + y) \varphi_\varepsilon(x) \, dx - \int_{\mathbb{R}^N} u(y) \varphi_\varepsilon(x) \, dx \right) d\mu(y)$$

$$= \int_{\mathbb{R}^N} \left(\int_{B(0,\varepsilon)} (u(x + y) - u(y)) \varphi_\varepsilon(x) \, dx \right) d\mu(y).$$

Since $\int_{\mathbb{R}^N} \varphi_\varepsilon(x) \, dx = 1$, and since u is uniformly continuous and bounded, if either u has compact support or $\mu(\mathbb{R}^N) < \infty$, by the Lebesgue dominated convergence theorem we conclude (2.64).

Remark 2.80. By considering separately the positive and negative parts of each component, it follows that the previous theorem still applies to vector-valued Radon measures.

2.2.2 Weak Convergence in L^p

In this subsection we exploit properties of $\mathbb{R}^N$ and the Lebesgue measure in order to give deeper insight into the notion of weak convergence. The next two results relate to Corollary 2.49 and the Dunford–Pettis theorem, where now the arbitrariness of the measurable set of integration is reduced to testing on cubes.

Proposition 2.81. *Let $E \subset \mathbb{R}^N$ be a Lebesgue measurable set. If $1 < p \leq \infty$, then $u_n \rightharpoonup u$ in $L^p(E)$ ($\overset{*}{\rightharpoonup}$ if $p = \infty$) if and only if*

(i) $\int_{Q \cap E} u_n \, dx \to \int_{Q \cap E} u \, dx$ *for every cube $Q \subset \mathbb{R}^N$;*
(ii) $\sup_n \|u_n\|_{L^p} < \infty$.

When $p = 1$ conditions (i) and (ii) of the previous proposition are still necessary but no longer sufficient.

Theorem 2.82 (Dunford–Pettis). *Let $E \subset \mathbb{R}^N$ be a Lebesgue measurable set. A sequence u_n converges weakly to u in $L^1(E)$ if and only if*

(i) $\int_{Q \cap E} u_n \, dx \to \int_{Q \cap E} u \, dx$ *for every cube $Q \subset \mathbb{R}^N$;*
(ii) $\{u_n\}$ *is equi-integrable;*
(iii) *for every $\varepsilon > 0$ there exists $F \subset E$ with $\mathcal{L}^N(F) < \infty$ such that*

$$\int_{E \setminus F} |u_n| \, dx \leq \varepsilon$$

for all n.

Remark 2.83. Conditions (ii) and (iii) imply (but they are not implied by) $\sup_n \|u_n\|_{L^1} < \infty$ (see Exercise 2.55).

Definition 2.84. *A function $u : \mathbb{R}^N \to \mathbb{R}^d$ is said to be Q-periodic if $u(x + e_i) = u(x)$ for all $x \in \mathbb{R}^N$ and every $i = 1, \ldots, N$, where $(e_1, \ldots, e_N)$ is the canonical basis of $\mathbb{R}^N$. More generally, u is said to be kQ-periodic, $k \in \mathbb{N}$, if $u(k \cdot)$ is Q-periodic.*

Lemma 2.85 (Riemann–Lebesgue lemma). *Let $u \in L^p_{\text{loc}}(\mathbb{R}^N)$, $1 \le p \le \infty$, be kQ-periodic. For every $\varepsilon > 0$ and $x \in \mathbb{R}^N$ set*

$$u_\varepsilon(x) := u\left(\frac{x}{\varepsilon}\right).$$

Then $u_\varepsilon \rightharpoonup \bar{u}$ in $L^p(E)$ ($\overset{*}{\rightharpoonup}$ if $p = \infty$) for every bounded measurable set $E \subset \mathbb{R}^N$, where

$$\bar{u}(x) \equiv const := \frac{1}{k^N} \int_{Q(0,k)} u(y)\, dy.$$

Proof. Without loss of generality we may assume that $k = 1$.

Step 1: Suppose first that $u \in L^\infty(Q(0,1))$. Then $\{u_\varepsilon\}$ is a bounded sequence in $L^\infty(\mathbb{R}^N)$, and so there exists a subsequence (not relabeled) such that $u_\varepsilon \overset{*}{\rightharpoonup} U$ in $L^\infty(\mathbb{R}^N)$. We claim that

$$U(x) = \bar{u}(x) = \int_{Q(0,1)} u(y)\, dy$$

for $\mathcal{L}^N$ a.e. $x \in Q(0,1)$. Indeed, let x_0 be a Lebesgue point for U. Fix $\delta > 0$ and let k_ε be the integer part of δ/ε ($0 < \varepsilon << \delta$), so that $\delta/\varepsilon - 1 < k_\varepsilon \le \delta/\varepsilon$. Decompose

$$\frac{1}{\varepsilon} Q(x_0, \delta) = \bigcup_{i=1}^{k_\varepsilon^N} Q\left(a_i^{(\varepsilon)}, 1\right) \cup E_\varepsilon,$$

where

$$\mathcal{L}^N(E_\varepsilon) = \left(\frac{\delta}{\varepsilon}\right)^N - k_\varepsilon^N \le \left(\frac{\delta}{\varepsilon}\right)^N - \left(\frac{\delta}{\varepsilon} - 1\right)^N = O_\delta\left(\frac{1}{\varepsilon^{N-1}}\right).$$

Then

$$\left| \int_{Q(x_0,\delta)} (U(x) - \overline{u})\, dx \right|$$

$$= \lim_{\varepsilon \to 0^+} \left| \int_{Q(x_0,\delta)} \left(u\left(\frac{x}{\varepsilon}\right) - \overline{u} \right) dx \right|$$

$$= \lim_{\varepsilon \to 0^+} \left| \varepsilon^N \int_{\frac{Q(x_0,\delta)}{\varepsilon}} (u(y) - \overline{u})\, dy \right|$$

$$= \lim_{\varepsilon \to 0^+} \left| \varepsilon^N \sum_{i=1}^{k_\varepsilon^N} \int_{Q\left(a_i^{(\varepsilon)},1\right)} (u(y) - \overline{u})\, dy + \varepsilon^N \int_{E_\varepsilon} (u(y) - \overline{u})\, dy \right|$$

$$\leq 2 \|u\|_{L^\infty(Q(0,1))} \limsup_{\varepsilon \to 0^+} \varepsilon^N \mathcal{L}^N(E_\varepsilon) = 0,$$

where we have used the fact that $\int_{Q\left(a_i^{(\varepsilon)},1\right)} (u(y) - \overline{u})\, dy = 0$, due to the Q-periodicity of u. Hence

$$0 = \lim_{\delta \to 0^+} \frac{1}{\delta^N} \int_{Q(x_0,\delta)} (U(x) - \overline{u})\, dx = U(x_0) - \overline{u}.$$

Step 2: In the general case let $u \in L^p_{\text{loc}}(\mathbb{R}^N)$, $1 \leq p \leq \infty$, and consider standard mollifiers of the form

$$\varphi_n(x) := n^N \varphi(nx), \quad x \in \mathbb{R}^N, \quad n \in \mathbb{N},$$

where φ is defined in (2.55). For $x \in \mathbb{R}^N$ set

$$u_n(x) := \int_{\mathbb{R}^N} \varphi_n(x - y) u(y)\, dy \quad \text{and} \quad u_{n,\varepsilon}(x) := u_n\left(\frac{x}{\varepsilon}\right).$$

Note that u_n is Q-periodic because for all $x \in \mathbb{R}^N$ and $i = 1, \ldots, N$,

$$u_n(x + e_i) = \int_{\mathbb{R}^N} \varphi_n(x + e_i - y) u(y)\, dy = \int_{\mathbb{R}^N} \varphi_n(y) u(x + e_i - y)\, dy$$

$$= \int_{\mathbb{R}^N} \varphi_n(y) u(x - y)\, dy = u_n(x).$$

Hence, since u_n is continuous and Q-periodic, we have that $u_n \in L^\infty(\mathbb{R}^N)$ with $\|u_n\|_{L^\infty(\mathbb{R}^N)} = \|u_n\|_{L^\infty(Q(0,1))}$ and $u_n \to u$ in $L^p_{\text{loc}}(\mathbb{R}^N)$.

Now if Q' is a cube in $\mathbb{R}^N$ and $v \in L^{p'}(Q')$, where p' is the conjugate exponent of p, then by Hölder's inequality,

$$\left| \int_{Q'} (u_\varepsilon v - \overline{u}v) \, dx \right| \leq \int_{Q'} |u_\varepsilon - u_{n,\varepsilon}| \, |v| \, dx + \left| \int_{Q'} (u_{n,\varepsilon} - \overline{u}_n) v \, dx \right|$$

$$+ \int_{Q'} |\overline{u}_n - \overline{u}| \, |v| \, dx \qquad (2.65)$$

$$\leq \left| \int_{Q'} (u_{n,\varepsilon} - \overline{u}_n) v \, dx \right|$$

$$+ \|v\|_{L^{p'}(Q')} \left\{ \|u_\varepsilon - u_{n,\varepsilon}\|_{L^p(Q')} + \|\overline{u}_n - \overline{u}\|_{L^p(Q')} \right\}.$$

We may decompose

$$\frac{1}{\varepsilon} Q' \subset \bigcup_{i=1}^{L} \overline{Q\left(a_i^{(\varepsilon)}, 1\right)},$$

where the cubes $Q\left(a_i^{(\varepsilon)}, 1\right)$ are disjoint and $L = \mathrm{O}\left(\frac{1}{\varepsilon^N}\right)$, and so by periodicity,

$$\|u_\varepsilon - u_{n,\varepsilon}\|_{L^p(Q')}^p = \varepsilon^N \int_{\frac{1}{\varepsilon}Q'} |u(x) - u_n(x)|^p \, dx$$

$$\leq \varepsilon^N \sum_{i=1}^{L} \int_{Q\left(a_i^{(\varepsilon)}, 1\right)} |u(x) - u_n(x)|^p \, dx \qquad (2.66)$$

$$= \varepsilon^N L \|u - u_n\|_{L^p(Q(0,1))}^p \leq C \|u - u_n\|_{L^p(Q(0,1))}^p,$$

for some $C > 1$.

Fix $\eta > 0$ and let n_0 be such that

$$\|u_{n_0} - u\|_{L^p(Q(0,1))} \leq \frac{\eta}{3\left(1 + \|v\|_{L^{p'}(Q')}\right)\left(C^{1/p} + |Q'|^{1/p}\right)}. \qquad (2.67)$$

Then

$$\|\overline{u}_{n_0} - \overline{u}\|_{L^p(Q')} = \left| \int_{Q(0,1)} (u_{n_0} - u) \, dx \right| |Q'|^{1/p} \qquad (2.68)$$

$$\leq \|u_{n_0} - u\|_{L^p(Q(0,1))} |Q'|^{1/p} \leq \frac{\eta}{3\left[1 + \|v\|_{L^{p'}(Q')}\right]}.$$

Finally, by Step 1 we may choose $\varepsilon_0 = \varepsilon_0(n_0) = \varepsilon_0(\eta)$ such that for all $0 < \varepsilon < \varepsilon_0$,

$$\left| \int_{Q'} (u_{n_0,\varepsilon} - \overline{u}_{n_0}) v \, dx \right| \leq \frac{\eta}{3}.$$

This together with (2.65)–(2.68) yields

$$\left| \int_{Q'} (u_\varepsilon v - \overline{u}v) \, dx \right| \leq \eta$$

for all $0 < \varepsilon < \varepsilon_0$. Hence we have shown that $u_\varepsilon \rightharpoonup \overline{u}$ in $L^p(Q')$ ($\overset{*}{\rightharpoonup}$ if $p = \infty$) and in turn (given the arbitrariness of Q') in $L^p(E)$ for every bounded measurable set $E \subset \mathbb{R}^N$.

Example 2.86. A typical application of this lemma is to u the characteristic function of the interval $(0, \theta)$ extended periodically to all of $\mathbb{R}$ with period 1. Then $u_\varepsilon \overset{*}{\rightharpoonup} \overline{u}$ in $L^\infty_{\text{loc}}(\mathbb{R})$, where

$$\overline{u}(x) \equiv \int_0^1 \chi_{(0,\theta)}(y) \, dy = \theta.$$

As a consequence, if $z \in S^{N-1} \subset \mathbb{R}^N$ then

$$\int_E \chi_{(0,\theta)}\left(\frac{x \cdot z}{\varepsilon}\right) v(x) \, dx \to \theta \int_E v(x) \, dx \tag{2.69}$$

for every $v \in L^1(E)$ and for every bounded measurable set $E \subset \mathbb{R}^N$. Said differently, (2.69) may be written as

$$\chi_{E_\varepsilon} \overset{*}{\rightharpoonup} \theta \text{ in } L^\infty(E), \tag{2.70}$$

where

$$E_\varepsilon := \left\{ x \in \mathbb{R}^N : \frac{x \cdot z}{\varepsilon} \in \bigcup_{k \in \mathbb{Z}} (k, k + \theta) \right\}.$$

Property (2.70) may be extended to measures other than the Lebesgue measure.

Proposition 2.87. *Let $(X, \mathfrak{M}, \mu)$ be a measure space with μ finite. Then μ is nonatomic if and only if for every $E \in \mathfrak{M}$ and $\theta \in [0, 1]$ there exists a sequence $\{E_n\} \subset \mathfrak{M}$ with $E_n \subset E$ such that $\chi_{E_n} \overset{*}{\rightharpoonup} \theta \chi_E$ in $L^\infty(X)$.*

Proof. To prove that μ is nonatomic, let $E \in \mathfrak{M}$ be such that $\mu(E) > 0$. With $\theta = \frac{1}{2}$ find $\{E_n\} \subset \mathfrak{M}$, with $E_n \subset E$, such that $\chi_{E_n} \overset{*}{\rightharpoonup} \frac{1}{2}\chi_E$ in $L^\infty(X)$. In particular,

$$\mu(E_n) \to \frac{1}{2}\mu(E),$$

and so for n sufficiently large we have

$$0 < \mu(E_n) < \mu(E).$$

Conversely, assume that μ is nonatomic and let $E \in \mathfrak{M}$ and $\theta \in [0, 1]$. Without loss of generality we may assume that $E = X$.

Step 1: Assume first that $\mathfrak{M}$ is generated by a countable family of sets $\{F_j\}$. Let $\mathfrak{N}$ be the algebra generated by $\{F_j\}$. By Theorem 2.16, $\mathfrak{N}$ is countable, and so we may write $\mathfrak{N} = \{G_j\}$. Fix $l \in \mathbb{N}$ and let $\mathcal{I}^{(l)}$ be the family of sets I of the form

$$I = \{i_1, \ldots, i_j\} \subset \mathbb{N}^j,$$

where $1 \le j \le l$ and $1 \le i_1 < \ldots < i_j \le l$. For every $I \in \mathcal{I}^{(l)}$ let

$$G_I^{(l)} := \left(\bigcap_{j \in I} G_j \right) \setminus \left(\bigcup_{j \notin I} G_j \right).$$

The sets $G_I^{(l)}$, $I \in \mathcal{I}^{(l)}$, belong to $\mathfrak{N}$, are pairwise disjoint, and for every $1 \le j \le l$,

$$G_j = \bigcup_{I \in \mathcal{I}^{(l)}, j \in I} G_I^{(l)}. \tag{2.71}$$

For simplicity in the notation we write

$$\left\{ G_I^{(l)} : I \in \mathcal{I}^{(l)} \right\} = \left\{ H_j^{(l)} : j = 1, \ldots, \ell_l \right\}.$$

By Proposition 1.91 there exist measurable functions $u_j^{(l)} : H_j^{(l)} \to [0, 1)$ such that for any Borel set $B \subset [0, 1]$,

$$\mu \left(\left(u_j^{(l)} \right)^{-1} (B) \right) = \mathcal{L}^1 (B) \mu \left(H_j^{(l)} \right). \tag{2.72}$$

Define

$$u_l := \sum_{j=1}^{\ell_l} \chi_{H_j^{(l)}} u_j^{(l)}.$$

As in the previous example we may construct a sequence of Borel sets $\{B_n\} \subset [0, 1]$ such that $\chi_{B_n} \overset{*}{\rightharpoonup} \theta$ in $L^\infty_{\text{loc}} (\mathbb{R}, \mathcal{L}^1)$. Set

$$E_{n,l} := u_l^{-1} (B_n).$$

For every $l \in \mathbb{N}$ and $1 \le j \le l$,

$$\int_{H_j^{(l)}} \chi_{E_{n,l}} \, d\mu = \mu \left(E_{n,l} \cap H_j^{(l)} \right) = \mu \left(u_l^{-1} (B_n) \cap H_j^{(l)} \right)$$

$$= \mu \left(\left(u_j^{(l)} \right)^{-1} (B_n) \cap H_j^{(l)} \right) = \mu \left(\left(u_j^{(l)} \right)^{-1} (B_n) \right)$$

$$= \mathcal{L}^1 (B_n) \mu \left(H_j^{(l)} \right) \to \theta \mu \left(H_j^{(l)} \right)$$

as $n \to \infty$, where we have used (2.72).

In turn, by (2.71) for every $l \in \mathbb{N}$ and $1 \le j \le l$,

$$\int_{G_j} \chi_{E_{n,l}} \, d\mu \to \theta \mu (G_j)$$

as $n \to \infty$.

Hence for every $l \in \mathbb{N}$ we may find n_l so large that

$$\left| \int_{G_j} \chi_{E_{n_l,l}} \, d\mu - \theta \mu (G_j) \right| \leq \frac{1}{l}$$

for all $1 \leq j \leq l$.

We claim that the sets $E_l := E_{n_l,l}$ have the desired property. Indeed, for any $\varepsilon > 0$ and for every $j \in \mathbb{N}$ choose $l_0 \geq j$ so large that $\frac{1}{l_0} \leq \varepsilon$. Then for all $l \geq l_0$,

$$\left| \int_{G_j} \chi_{E_l} \, d\mu - \theta \mu (G_j) \right| \leq \varepsilon.$$

The result now follows from Remark 2.17.

Step 2: We now remove the additional separability assumption. For every $l \in \mathbb{N}$, by Corollary 1.21 we partition X into l disjoint subsets $H_n^{(l)} \in \mathfrak{M}$, $n = 1, \ldots, l$, such that

$$\mu \left(H_n^{(l)} \right) = \frac{1}{l} \mu (X)$$

for all $n = 1, \ldots, l$. Let $\mathfrak{M}' \subset \mathfrak{M}$ be the σ-algebra generated by the collection

$$\left\{ H_n^{(l)} : l \in \mathbb{N}, 1 \leq n \leq l \right\}.$$

Then $\mathfrak{M}'$ is generated by a countable family of sets, and we claim that $\mu : \mathfrak{M}' \to [0, \infty)$ is nonatomic. Indeed, let $F \in \mathfrak{M}'$ with $\mu (F) > 0$. We show that there exists $G \in \mathfrak{M}'$, $G \subset F$, with $0 < \mu (G) < \mu (F)$. Choose l so large that

$$\frac{1}{l} \mu (X) < \mu (F).$$

Since

$$F = \bigcup_{n=1}^{l} F \cap H_n^{(l)},$$

there exists $n \in \{1, \ldots, l\}$ such that

$$\mu \left(F \cap H_n^{(l)} \right) > 0.$$

Set $G := F \cap H_n^{(l)} \subset F$. Then

$$0 < \mu (G) \leq \mu \left(H_n^{(l)} \right) \leq \frac{1}{l} \mu (X) < \mu (F),$$

and so the claim holds.

In view of Step 1 there exist $E_n \in \mathfrak{M}'$ such that $\chi_{E_n} \overset{*}{\rightharpoonup} \theta$ in $L^\infty (X, \mathfrak{M}', \mu)$. We show that $\chi_{E_n} \overset{*}{\rightharpoonup} \theta$ in $L^\infty (X, \mathfrak{M}, \mu)$. Let $u \in L^1 (X, \mathfrak{M}, \mu)$ and define the signed measure

$$\lambda_u\left(F\right) := \int_F u \, d\mu, \quad F \in \mathfrak{M}'.$$

Then λ_u is a finite signed measure absolutely continuous with respect to $\mu : \mathfrak{M}' \rightarrow [0, \infty)$, and so by the Radon–Nikodym theorem there exists $v \in L^1\left(X, \mathfrak{M}', \mu\right)$ such that

$$\lambda_u\left(F\right) = \int_F v \, d\mu, \quad F \in \mathfrak{M}'.$$

Therefore

$$\int_{E_n} u \, d\mu = \lambda_u\left(E_n\right) = \int_{E_n} v \, d\mu \rightarrow \theta \int_X v \, d\mu = \theta \lambda_u\left(X\right) = \theta \int_X u \, d\mu.$$

This concludes the proof.

We conclude this section with a well-known compactness condition that will be used to prove the Rellich-Kondrachov theorem in [FoLe10].

Theorem 2.88. *Let $1 \le p < \infty$ and let $\{u_n\}$ be a sequence of functions converging weakly in $L^p\left(\mathbb{R}^N\right)$. Then $\{u_n\}$ converges strongly in $L^p\left(\mathbb{R}^N\right)$ if and only if*

$$\lim_{h \to 0} \sup_{n \in \mathbb{N}} \int_{\mathbb{R}^N} |u_n\left(x + h\right) - u_n\left(x\right)|^p \, dx = 0, \tag{2.73}$$

$$\lim_{R \to \infty} \sup_{n \in \mathbb{N}} \int_{\mathbb{R}^N \setminus B(0, R)} |u_n\left(x\right)|^p \, dx = 0. \tag{2.74}$$

Proof. **Step 1:** Assume that (2.73) and (2.74) hold and consider standard mollifiers of the form

$$\varphi_k\left(x\right) := k^N \varphi\left(k \, x\right), \quad x \in \mathbb{R}^N, \quad k \in \mathbb{N},$$

where φ is defined in (2.55). For $u \in L^p\left(\mathbb{R}^N\right)$ and $x \in \mathbb{R}^N$ set

$$u^{(k)}\left(x\right) := \int_{\mathbb{R}^N} \varphi_k\left(x - y\right) u\left(y\right) \, dy.$$

By Hölder's inequality and (2.54) we have

$$\left|u^{(k)}\left(x\right) - u\left(x\right)\right|^p \le \int_{\mathbb{R}^N} \varphi_k\left(x - y\right) |u\left(y\right) - u\left(x\right)|^p \, dy \tag{2.75}$$

$$\le ck^N \int_{B\left(0, \frac{1}{k}\right)} |u\left(x + h\right) - u\left(x\right)|^p \, dh.$$

Hence by Fubini's theorem,

$$\int_{\mathbb{R}^N} \left|u^{(k)}\left(x\right) - u\left(x\right)\right|^p \, dx \le ck^N \int_{B\left(0, \frac{1}{k}\right)} \int_{\mathbb{R}^N} |u\left(x + h\right) - u\left(x\right)|^p \, dx \, dh. \tag{2.76}$$

Fix $\varepsilon > 0$. Taking into account (2.73) we have that

$$\sup_{n\in\mathbb{N}} \int_{\mathbb{R}^N} |u_n(x+h) - u_n(x)|^p \, dx \leq \varepsilon$$

for all h sufficiently small. Hence, by applying (2.76) to u_n we get

$$\sup_{n\in\mathbb{N}} \int_{\mathbb{R}^N} \left|(u_n)^{(k)}(x) - u_n(x)\right|^p \, dx$$

$$\leq ck^N \int_{B\left(0,\frac{1}{k}\right)} \sup_{n\in\mathbb{N}} \int_{\mathbb{R}^N} |u_n(x+h) - u_n(x)|^p \, dx \, dh \leq c\varepsilon$$

for all k sufficiently large, and so

$$\lim_{k\to\infty} \sup_{n\in\mathbb{N}} \int_{\mathbb{R}^N} \left|(u_n)^{(k)}(x) - u_n(x)\right|^p \, dx = 0. \tag{2.77}$$

Let u_0 be the weak limit of $\{u_n\}$. Then we have

$$\|u_n - u_0\|_{L^p} \leq \left\|(u_n)^{(k)} - u_n\right\|_{L^p} + \left\|(u_n)^{(k)} - (u_0)^{(k)}\right\|_{L^p} + \left\|(u_0)^{(k)} - u_0\right\|_{L^p}.$$

Fix $\epsilon > 0$. By (2.77) and Theorem 2.73(iii) there exists $\bar{k}$ depending only on ϵ such that for all $k \geq \bar{k}$ and all $n \in \mathbb{N}$ the first and last terms in the previous inequality are both bounded by ϵ, and so

$$\|u_n - u_0\|_{L^p} \leq \left\|(u_n)^{(k)} - (u_0)^{(k)}\right\|_{L^p} + 2\epsilon$$

for all $k \geq \bar{k}$ and all $n \in \mathbb{N}$. Hence to complete the proof it suffices to show that

$$\lim_{n\to\infty} \left\|(u_n)^{(\bar{k})} - (u_0)^{(\bar{k})}\right\|_{L^p} = 0. \tag{2.78}$$

By (2.74) and the fact that $u_0 \in L^p(\mathbb{R}^N)$, there exists $R > 1$ such that

$$\sup_{n\in\mathbb{N}} \int_{\mathbb{R}^N \setminus B(0,R-1)} |u_n(x)|^p \, dx + \int_{\mathbb{R}^N \setminus B(0,R-1)} |u_0(x)|^p \, dx \leq \epsilon,$$

and so reasoning as in (2.75) and (2.76),

$$\int_{\mathbb{R}^N \setminus B(0,R)} \left|(u_n)^{(\bar{k})} - (u_0)^{(\bar{k})}\right|^p \, dx \tag{2.79}$$

$$\leq c\bar{k}^N \int_{B\left(0,\frac{1}{k}\right)} \int_{\mathbb{R}^N \setminus B(0,R)} |u_n(x+h) - u_0(x+h)|^p \, dx \, dh$$

$$\leq c\bar{k}^N \int_{B\left(0,\frac{1}{k}\right)} \int_{\mathbb{R}^N \setminus B(0,R-1)} |u_n(y) - u_0(y)|^p \, dy \, dh \leq c\epsilon,$$

where we have used the change of variables $y = x + h$. Since $u_n \rightharpoonup u_0$ in $L^p(\mathbb{R}^N)$ it follows that for all $x \in \mathbb{R}^N$,

$$(u_n)^{(\bar{k})}(x) = \int_{\mathbb{R}^N} \varphi_{\bar{k}}(x - y) u_n(y) \, dy$$

$$\rightarrow \int_{\mathbb{R}^N} \varphi_{\bar{k}}(x - y) u_0(y) \, dy = (u_0)^{(\bar{k})}(x)$$

as $n \to \infty$. Moreover, reasoning as in (2.75) and since $\{u_n\}$ is bounded in $L^p(\mathbb{R}^N)$, we get

$$\left| (u_n)^{(\bar{k})}(x) - (u_0)^{(\bar{k})}(x) \right|^p \le c\bar{k}^N \int_{B\left(0, \frac{1}{k}\right)} |u_n(x + h) - u_0(x + h)|^p \, dh$$

$$\le c\bar{k}^N$$

for all $x \in \mathbb{R}^N$ and all $n \in \mathbb{N}$. Therefore, by the Lebesgue dominated convergence theorem we have

$$\lim_{n \to \infty} \int_{B(0,R)} \left| (u_n)^{(\bar{k})} - (u_0)^{(\bar{k})} \right|^p \, dx = 0,$$

which, combined with (2.79), yields (2.78).

Step 2: Conversely, assume that $\{u_n\}$ converges strongly in $L^p(\mathbb{R}^N)$. We begin by showing that (2.73) holds for a single function $u \in L^p(\mathbb{R}^N)$, that is,

$$\lim_{h \to 0} \int_{\mathbb{R}^N} |u(x + h) - u(x)|^p \, dx = 0. \tag{2.80}$$

Fix $\varepsilon > 0$. By Theorem 2.78 there exists a function $v \in C_c^\infty(\mathbb{R}^N)$ such that

$$\|v - u\|_{L^p} \le \varepsilon.$$

Hence, by Minkowski's inequality, and the change of variable $y = x + h$,

$$\left(\int_{\mathbb{R}^N} |u(x + h) - u(x)|^p \, dx \right)^{1/p}$$

$$\le \left(\int_{\mathbb{R}^N} |u(x + h) - v(x + h)|^p \, dx \right)^{1/p}$$

$$+ \left(\int_{\mathbb{R}^N} |v(x + h) - v(x)|^p \, dx \right)^{1/p}$$

$$+ \left(\int_{\mathbb{R}^N} |u(x) - v(x)|^p \, dx \right)^{1/p}$$

$$= 2 \|u - v\|_{L^p} + \left(\int_{\mathbb{R}^N} |v(x + h) - v(x)|^p \, dx \right)^{1/p}$$

$$\le 2\varepsilon + \left(\int_{\mathbb{R}^N} |v(x + h) - v(x)|^p \, dx \right)^{1/p}.$$

Since v is uniformly continuous and has compact support, we may find $\delta = \delta(\varepsilon) > 0$ such that

$$\left(\int_{\mathbb{R}^N} |v(x+h) - v(x)|^p \, dx \right)^{1/p} \le \varepsilon$$

for all $|h| \le \delta$. Hence (2.80) holds.

Since $\{u_n\}$ converges to u_0 in $L^p(\mathbb{R}^N)$, there exists an integer $\bar{n}$ such that

$$\|u_n - u_0\|_{L^p} \le \varepsilon$$

for all $n \ge \bar{n}$. Hence by Minkowski's inequality and the change of variable $y = x + h$,

$$\left(\int_{\mathbb{R}^N} |u_n(x+h) - u_n(x)|^p \, dx \right)^{1/p}$$

$$\le \left(\int_{\mathbb{R}^N} |u_n(x+h) - u_0(x+h)|^p \, dx \right)^{1/p}$$

$$+ \left(\int_{\mathbb{R}^N} |u_0(x+h) - u_0(x)|^p \, dx \right)^{1/p}$$

$$+ \left(\int_{\mathbb{R}^N} |u_n(x) - u_0(x)|^p \, dx \right)^{1/p}$$

$$= 2\|u_n - u_0\|_{L^p} + \left(\int_{\mathbb{R}^N} |u_0(x+h) - u_0(x)|^p \, dx \right)^{1/p}$$

$$\le 4\varepsilon + \left(\int_{\mathbb{R}^N} |u_0(x+h) - u_0(x)|^p \, dx \right)^{1/p}$$

for all $n \ge \bar{n}$. Using (2.80) with $u_0, \ldots, u_{\bar{n}-1}$ in place of u we conclude that (2.73) holds.

The proof of (2.74) follows a similar argument (see also Vitali's convergence theorem) and therefore we omit it.

2.2.3 Maximal Functions

In this subsection we introduce the notion of maximal function and study its properties. Maximal functions will play an important role in the Sobolev space setting in [FoLe10] but they will not be used in the remainder of this volume.

Throughout this subsection we consider L^p spaces in the case that the underlying measure is the Lebesgue measure.

Definition 2.89. Let $u \in L^1_{\text{loc}}(\mathbb{R}^N)$. The (Hardy–Littlewood) maximal function of u is defined by

$$M\left(u\right)\left(x\right) := \sup_{r>0} \frac{1}{\left|B\left(x,r\right)\right|} \int_{B(x,r)} \left|u\left(y\right)\right| dy$$

for all $x \in \mathbb{R}^N$. For $0 < R \leq \infty$ we also define

$$M_R\left(u\right)\left(x\right) := \sup_{0<r<R} \frac{1}{\left|B\left(x,r\right)\right|} \int_{B(x,r)} \left|u\left(y\right)\right| dy.$$

Note that $M = M_\infty$.

Exercise 2.90. Prove that the sets

$$\left\{x \in \mathbb{R}^N : M\left(u\right)\left(x\right) > t\right\}, \quad \left\{x \in \mathbb{R}^N : M_R\left(u\right)\left(x\right) > t\right\}$$

are open (and so Lebesgue measurable).

Theorem 2.91. Let $u \in L^p\left(\mathbb{R}^N\right)$, $1 \leq p \leq \infty$. Then

(i) $u\left(x\right) \leq M\left(u\right)\left(x\right) < \infty$ for $\mathcal{L}^N$ a.e. $x \in \mathbb{R}^N$;
(ii) if $p = 1$ then for any $t > 0$,

$$\left|\left\{x \in \mathbb{R}^N : M\left(u\right)\left(x\right) > t\right\}\right| \leq \frac{3^N}{t} \int_{\mathbb{R}^N} \left|u\left(x\right)\right| dx; \qquad (2.81)$$

(iii) if $1 < p \leq \infty$ then $M\left(u\right) \in L^p\left(\mathbb{R}^N\right)$ and

$$\left\|M\left(u\right)\right\|_{L^p} \leq C\left(N,p\right) \left\|u\right\|_{L^p}.$$

The proof of the theorem is hinged on the following covering result.

Lemma 2.92 (Vitali). Let $E \subset \mathbb{R}^N$ be the union of a finite number of balls $B\left(x_i, r_i\right)$, $i = 1, \ldots, \ell$. Then there exists a subset $I \subset \{1, \ldots, \ell\}$ such that the balls $B\left(x_i, r_i\right)$ with $i \in I$ are pairwise disjoint and

$$E \subset \bigcup_{i \in I} B\left(x_i, 3r_i\right).$$

Proof. Without loss of generality we may assume that

$$r_1 \geq r_2 \geq \ldots \geq r_\ell.$$

Put $i_1 := 1$ and discard all the balls that intersect $B\left(x_1, r_1\right)$. Let i_2 be the first integer, if it exists, such that $B\left(x_{i_2}, r_{i_2}\right)$ does not intersect $B\left(x_1, r_1\right)$. If i_2 does not exist then set $I := \{i_1\}$, while if i_2 exists discard all the balls $B\left(x_i, r_i\right)$, with $i > i_2$, that intersect $B\left(x_{i_2}, r_{i_2}\right)$. Let $i_3 > i_2$ be the first integer, if it exists, such that $B\left(x_{i_3}, r_{i_3}\right)$ does not intersect $B\left(x_{i_2}, r_{i_2}\right)$. If i_3 does not exist then set $I := \{i_1, i_2\}$; if i_3 exists continue the process. Since there is only a finite number of balls the process stops after a finite number of steps and we obtain the desired set $I \subset \{1, \ldots, \ell\}$.

By construction, the balls $B(x_i, r_i)$ with $i \in I$ are pairwise disjoint. To prove the last statement let $x \in E$. Then there exists $i = 1, \ldots, \ell$ such that $x \in B(x_i, r_i)$. If $i \in I$ then there is nothing to prove. If $i \notin I$ then $B(x_i, r_i)$ is one of the balls that has been discarded and thus there exists $j \in I$ such that $r_j \geq r_i$ and $B(x_j, r_j) \cap B(x_i, r_i) \neq \emptyset$. Let $z \in B(x_j, r_j) \cap B(x_i, r_i)$. Then

$$|x - x_j| \leq |x - x_i| + |x_i - z| + |z - x_j| < r_i + r_i + r_j \leq 3r_j,$$

and so $x \in B(x_j, 3r_j)$ and the proof is complete.

We now turn to the proof of Theorem 2.91:

Proof (Theorem 2.91). We begin by proving (ii). Let

$$A_t := \left\{ x \in \mathbb{R}^N : \mathrm{M}(u)(x) > t \right\}.$$

By the definition of $\mathrm{M}(u)$, for every $x \in A_t$ we can find a ball $B(x, r_x)$, with $r_x > 0$, such that

$$\frac{1}{|B(x, r_x)|} \int_{B(x, r_x)} |u(y)| \, dy > t. \tag{2.82}$$

Let $K \subset A_t$ be a compact set. Since $\{B(x, r_x)\}_{x \in A_t}$ is an open cover for K, we may find a finite subcover. By the previous lemma there exists a disjoint finite subfamily $\{B(x_i, r_i)\}_{i=1}^n$, such that

$$K \subset \bigcup_{i=1}^{n} B(x_i, 3r_i).$$

Hence, by (2.82),

$$|K| \leq 3^N \sum_{i=1}^{n} |B(x_i, r_i)| \leq \frac{3^N}{t} \sum_{i=1}^{n} \int_{B(x_i, r_i)} |u| \, dy \leq \frac{3^N}{t} \int_{\mathbb{R}^N} |u| \, dy.$$

Using the inner regularity of the Lebesgue measure, together with Exercise 2.90, we obtain that

$$\left| \left\{ x \in \mathbb{R}^N : \mathrm{M}(u)(x) > t \right\} \right| \leq \frac{3^N}{t} \int_{\mathbb{R}^N} |u| \, dy.$$

Note that this implies, in particular, that $\mathrm{M}(u)(x) < \infty$ for $\mathcal{L}^N$ a.e. $x \in \mathbb{R}^N$. Thus (i) is proved for $p = 1$.

In order to prove (iii) it suffices to consider $1 < p < \infty$, since the case $p = \infty$ is immediate from the definition of $\mathrm{M}(u)$. For $t > 0$ define

$$u_t(x) := \begin{cases} u(x) & \text{if } |u(x)| > \frac{t}{2}, \\ 0 & \text{otherwise.} \end{cases}$$

We claim that $u_t \in L^1(\mathbb{R}^N)$. Indeed,

$$\int_{\mathbb{R}^N} |u_t| \, dy = \int_{\left\{x \in \mathbb{R}^N : |u(x)| > \frac{t}{2}\right\}} |u| \, dy$$

$$\leq \left(\frac{2}{t}\right)^{p-1} \int_{\left\{x \in \mathbb{R}^N : |u(x)| > \frac{t}{2}\right\}} |u|^p \, dy < \infty.$$

Moreover, since $|u| \leq |u_t| + \frac{t}{2}$ we have that $\mathrm{M}(u) \leq \mathrm{M}(u_t) + \frac{t}{2}$, and so

$$\left\{x \in \mathbb{R}^N : \mathrm{M}(u)(x) > t\right\} \subset \left\{x \in \mathbb{R}^N : \mathrm{M}(u_t)(x) > \frac{t}{2}\right\}.$$

Part (ii) applied to $u_t \in L^1(\mathbb{R}^N)$ now yields

$$\left|\left\{x \in \mathbb{R}^N : \mathrm{M}(u)(x) > t\right\}\right| \leq \frac{3^N 2}{t} \int_{\mathbb{R}^N} |u_t| \, dy \qquad (2.83)$$

$$= \frac{3^N 2}{t} \int_{\left\{x \in \mathbb{R}^N : |u(x)| > \frac{t}{2}\right\}} |u| \, dy.$$

Hence, using Theorem 1.123 and Fubini's theorem we obtain

$$\int_{\mathbb{R}^N} (\mathrm{M}(u)(x))^p \, dx = \int_0^\infty \left|\left\{x \in \mathbb{R}^N : (\mathrm{M}(u)(x))^p > s\right\}\right| \, ds$$

$$= p \int_0^\infty t^{p-1} \left|\left\{x \in \mathbb{R}^N : \mathrm{M}(u)(x) > t\right\}\right| \, dt$$

$$\leq p 3^N 2 \int_0^\infty t^{p-2} \int_{\left\{x \in \mathbb{R}^N : |u(x)| > \frac{t}{2}\right\}} |u(y)| \, dy \, dt$$

$$= p 3^N 2 \int_{\mathbb{R}^N} |u(y)| \left(\int_0^{2|u(y)|} t^{p-2} \, dt\right) dy$$

$$= \frac{p 3^N 2^p}{p-1} \int_{\mathbb{R}^N} |u(x)|^p \, dx.$$

This proves (iii) and in turn (i) for $p > 1$.

Property (ii) in the previous theorem is usually called the *weak L^1 inequality* because, as opposed to the case $p > 1$ (see (iii)), for $p = 1$ the operator

$$u \in L^1(\mathbb{R}^N) \mapsto \mathrm{M}(u)$$

is not bounded in L^1. Actually, if u is not identically zero then $\mathrm{M}(u) \notin L^1(\mathbb{R}^N)$. To see this, it suffices to assume, without loss of generality, that

$$\int_{B(0,1)} |u| \, dx > 0.$$

Then if $|x| \geq 1$ it follows that $B(0,1) \subset B(x, 2|x|)$, and thus for some $c > 0$,

$$\mathrm{M}\left(u\right)\left(x\right) \geq \frac{1}{\left|B\left(x, 2\left|x\right|\right)\right|} \int_{B\left(x, 2\left|x\right|\right)} \left|u\left(y\right)\right| dy \geq \frac{c}{\left|x\right|^N} \int_{B\left(0,1\right)} \left|u\right| dx,$$

with $\left|x\right|^{-N} \notin L^1\left(\mathbb{R}^N \setminus B\left(0,1\right)\right)$. Moreover, even local integrability of $\mathrm{M}\left(u\right)$ does not hold in general. However, it can be shown that $\mathrm{M}\left(u\right) \in L^1_{\mathrm{loc}}\left(\mathbb{R}^N\right)$, provided u belongs to the space $L \log L\left(\mathbb{R}^N\right)$:

Definition 2.93. *Let $E \subset \mathbb{R}^N$ be a Lebesgue measurable set. We say that a measurable function $u : E \to \mathbb{R}$ belongs to the space $L \log L\left(E\right)$ if*

$$\int_E \left|u\left(x\right)\right| \log \left(2 + \left|u\left(x\right)\right|\right) dx < \infty.$$

Note that

$$L \log L\left(\mathbb{R}^N\right) \subsetneqq L^1\left(\mathbb{R}^N\right).$$

Proposition 2.94. *Assume that $u \in L \log L\left(\mathbb{R}^N\right)$. Then $\mathrm{M}\left(u\right)$ is integrable over measurable sets of finite measure.*

Proof. Let $E \subset \mathbb{R}^N$ be any measurable set of finite measure. By Theorem 1.123,

$$\int_E \mathrm{M}\left(u\right)\left(x\right) dx = \int_0^\infty \left|\{x \in E : \mathrm{M}\left(u\right) > t\}\right| dt$$

$$\leq \left|E\right| + \int_1^\infty \left|\{x \in E : \mathrm{M}\left(u\right) > t\}\right| dt.$$

Now (2.83) and Fubini's theorem yield

$$\int_E \mathrm{M}\left(u\right)\left(x\right) dx \leq \left|E\right| + C \int_1^\infty \frac{1}{t} \left(\int_{\{x \in \mathbb{R}^N : \left|u(x)\right| > t/2\}} \left|u\left(y\right)\right| dy\right) dt$$

$$= \left|E\right| + C \int_{\{x \in \mathbb{R}^N : \left|u(x)\right| > \frac{1}{2}\}} \left|u\left(y\right)\right| \left(\int_1^{2\left|u(y)\right|} \frac{1}{t} dt\right) dy$$

$$= \left|E\right| + C \int_{\{x \in \mathbb{R}^N : \left|u(x)\right| > \frac{1}{2}\}} \left|u\left(y\right)\right| \log \left(2\left|u\left(y\right)\right|\right) dy,$$

and so the proof is complete.

The previous proposition admits a partial converse, which relies on the following decomposition theorem:

Theorem 2.95 (Calderón–Zygmund). *Let $u \in L^1\left(\mathbb{R}^N\right)$ be a nonnegative function, and let $t > 0$. Then there exists a countable family $\{Q_n\}$ of open mutually disjoint cubes such that*

$$u\left(x\right) \leq t \quad \text{for } \mathcal{L}^N \text{ a.e. } x \in \mathbb{R}^N \setminus \bigcup_{n=1}^\infty \overline{Q_n}, \tag{2.84}$$

and for every $n \in \mathbb{N}$,

$$t < \frac{1}{|Q_n|} \int_{Q_n} u(x) \, dx \le 2^N t. \tag{2.85}$$

Proof. Fix $t > 0$ and choose $L > 0$ large enough that

$$\int_{\mathbb{R}^N} u(x) \, dx \le t L^N.$$

Decompose $\mathbb{R}^N$ into a rectangular grid such that each cube Q of the partition has side length L, and thus

$$\frac{1}{|Q|} \int_Q u(x) \, dx \le t. \tag{2.86}$$

Fix one such cube Q and subdivide it into 2^N congruent subcubes. Let Q' be one of these subcubes. If

$$\frac{1}{|Q'|} \int_{Q'} u(x) \, dx > t,$$

and in view of the fact that

$$\frac{1}{|Q'|} \int_{Q'} u(x) \, dx \le \frac{2^N}{|Q|} \int_Q u(x) \, dx \le 2^N t,$$

then (2.85) is satisfied and therefore Q' will be selected as one of the Q_n. On the other hand, if

$$\frac{1}{|Q'|} \int_{Q'} u(x) \, dx \le t$$

(note that by (2.86) there is at least one), then we subdivide Q' into 2^N congruent subcubes and we repeat the process.

In this way we construct a family of cubes $\{Q_n\}$ for which (2.85) is satisfied, and it remains to prove (2.84). It can be seen from the construction that the cubes that were not selected to belong to the family $\{Q_n\}$ form a fine covering $\mathcal{F}$ of $\mathbb{R}^N \setminus \bigcup_{n=1}^\infty \overline{Q_n}$. Therefore if $x \in \mathbb{R}^N \setminus \bigcup_{n=1}^\infty \overline{Q_n}$ is a Lebesgue point for u, then by Remark 1.160(ii) we have

$$u(x) = \limsup_{\mathrm{diam}\, F \to 0,\, F \in \mathcal{F},\, x \in F} \frac{1}{|F|} \int_F u(y) \, dy \le t$$

and the proof is completed.

Remark 2.96. A local version of the previous theorem holds. Precisely, if Q is a cube, $u \in L^1(Q)$ is nonnegative, and if

$$t \ge \frac{1}{|Q|} \int_Q u(x) \, dx,$$

then it follows from the above proof that there exists a countable family $\{Q_n\} \subset Q$ of open mutually disjoint cubes such that

$$u(x) \le t\, \mathcal{L}^N \text{ a.e. in } Q \setminus \bigcup_{n=1}^{\infty} \overline{Q_n},$$

and for every $n \in \mathbb{N}$,

$$t < \frac{1}{|Q_n|} \int_{Q_n} u(x)\, dx \le 2^N t.$$

As a consequence of the Calderón–Zygmund theorem we obtain a partial reverse of the weak L^1 inequality (2.81).

Corollary 2.97. *Let $u \in L^1_{\mathrm{loc}}\left(\mathbb{R}^N\right)$. Then there exists a constant C depending on N such that for every cube Q in $\mathbb{R}^N$ and for every $t > 0$ such that*

$$t > \frac{1}{|Q|} \int_Q |u(x)|\, dx, \tag{2.87}$$

we have

$$|\{x \in Q : \mathrm{M}(u)(x) > Ct\}| \ge \frac{C}{t} \int_{\{y \in Q :\, |u(y)| > t\}} |u(x)|\, dx, \tag{2.88}$$

where

$$C := \min\left\{ \frac{1}{\alpha_N N^{\frac{N}{2}}},\ \frac{1}{2^N} \right\}.$$

Proof. Fix t as in (2.87) and apply the Calderón–Zygmund theorem together with Remark 2.96 to $|u|$ and t to find a family $\{\overline{Q_n}\} \subset Q$ of closed cubes with mutually disjoint interiors such that

$$|u(x)| \le t \text{ for } \mathcal{L}^N \text{ a.e. } x \in Q \setminus \bigcup_{n=1}^{\infty} \overline{Q_n}, \tag{2.89}$$

and for every $n \in \mathbb{N}$,

$$t < \frac{1}{|Q_n|} \int_{Q_n} |u(y)|\, dy \le 2^N t. \tag{2.90}$$

If $x \in Q_n$ then the ball centered at x with radius $R_n := \sqrt{N}\,|Q_n|^{\frac{1}{N}}$ contains Q_n, and so it follows from (2.90) that

$$\mathrm{M}(u)(x) \ge \frac{1}{|B(x, R_n)|} \int_{B(x, R_n)} |u(y)|\, dy$$

$$\ge \frac{1}{|B(x, R_n)|} \int_{Q_n} |u(y)|\, dy \ge \frac{1}{\alpha_N N^{\frac{N}{2}}} t.$$

Therefore, in view of (2.89) and (2.90),

$$\left| \left\{ x \in Q : \, \mathrm{M}\left(u\right)\left(x\right) > \frac{1}{\alpha_N N^{\frac{N}{2}}} t \right\} \right| \geq \sum_{n=1}^{\infty} |Q_n|$$

$$\geq \frac{1}{2^N t} \sum_{n=1}^{\infty} \int_{Q_n} |u\left(x\right)| \, dx = \frac{1}{2^N t} \int_{\{y \in Q: \, |u(y)| > t\}} |u\left(x\right)| \, dx.$$

It suffices to set

$$C := \min \left\{ \frac{1}{\alpha_N N^{\frac{N}{2}}}, \frac{1}{2^N} \right\}.$$

This concludes the proof. ∎

We are now ready to prove a partial converse of Proposition 2.94.

Proposition 2.98. *If Q is a cube in $\mathbb{R}^N$ and $u \in L^1_{\mathrm{loc}}\left(\mathbb{R}^N\right)$ is such that $\mathrm{M}\left(u\right) \in L^1\left(Q\right)$, then $u \in L \log L\left(Q\right)$ and*

$$\int_Q |u\left(x\right)| \log\left(2 + |u\left(x\right)|\right) \, dx$$

$$\leq C\left(N, |Q|\right) \|\mathrm{M}\left(u\right)\|_{L^1(Q)} \log\left(2 + \|\mathrm{M}\left(u\right)\|_{L^1(Q)}\right).$$

Proof. Fix

$$t := 1 + \frac{1}{|Q|} \int_Q |u\left(y\right)| \, dy.$$

We have

$$\int_Q |u| \log\left(2 + |u|\right) \, dx$$

$$\leq \log\left(2 + t\right) \int_{\{x \in Q: |u| \leq t\}} |u| \, dx + \int_{\{x \in Q: |u| > t\}} |u| \log\left(2 + |u|\right) \, dx$$

$$\leq \log\left(2 + t\right) \int_{\{x \in Q: |u| \leq t\}} \mathrm{M}\left(u\right) \, dx + \int_{\{x \in Q: |u| > t\}} |u| \log\left(3 |u|\right) \, dx$$

$$\tag{2.91}$$

$$\leq \left(\log\left(2 + t\right) + \log 3\right) \int_Q \mathrm{M}\left(u\right) \, dx + \int_{\{x \in Q: |u| > t\}} |u| \log |u| \, dx,$$

where we have used the fact that $|u| \leq \mathrm{M}\left(u\right) \, \mathcal{L}^N$ a.e. in Q.

Using Fubini's theorem and Theorem 1.123, together with (2.88), we obtain

$$\int_{\{x\in Q:|u|>t\}} |u|\log|u|\,dx = \int_{\{x\in Q:|u|>t\}} |u|\left(\int_1^{|u(y)|}\frac{1}{s}\,ds\right)dx$$

$$= \int_t^\infty \frac{1}{s}\left(\int_{\{x\in Q:|u|>s\}}|u(x)|\,dx\right)ds$$

$$\leq \frac{1}{C}\int_t^\infty |\{x\in Q: M(u)(x)>Cs\}|\,ds$$

$$= \frac{1}{C^2}\int_{\{x\in Q: M(u)>Ct\}} M(u)\,dx \leq \frac{1}{C^2}\int_Q M(u)\,dx.$$

Hence by (2.91),

$$\int_Q |u|\log(2+|u|)\,dx \leq \left(\log(2+t)+\log 3+\frac{1}{C^2}\right)\int_Q M(u)\,dx$$

$$\leq C(N,|Q|)\|M(u)\|_{L^1(Q)}\log\left(2+\|M(u)\|_{L^1(Q)}\right),$$

where we have used the fact that

$$t \leq 1+\frac{1}{|Q|}\int_Q M(u)\,dx.$$

The proof is now complete.

Remark 2.99. A simple covering argument yields an analogue to Proposition 2.98 for arbitrary compact sets. Precisely, if $\Omega \subset \mathbb{R}^N$ is an open set and $\Omega' \subset\subset \Omega$ then there exists a constant $C = C(N,\Omega',\Omega)>0$ such that

$$\int_{\Omega'} |u|\log(2+|u|)\,dx \leq C\|M(u)\|_{L^1(\Omega)}\log\left(2+\|M(u)\|_{L^1(\Omega)}\right).$$

We conclude this subsection with an alternative proof of the Lebesgue differentiation theorem for the Lebesgue measure using maximal functions.

Theorem 2.100 (Lebesgue differentiation theorem). *Let $u \in L^1_{loc}(\mathbb{R}^N)$. Then there exists a Borel set $E \subset \mathbb{R}^N$, with $|E|=0$, such that*

$$\mathbb{R}^N \setminus E \subset \{x\in\mathbb{R}^N: u(x)\in\mathbb{R}\}$$

and for any $x \in \mathbb{R}^N \setminus E$,

$$\lim_{r\to 0^+}\frac{1}{|B(x,r)|}\int_{B(x,r)} u(y)\,dy = u(x). \tag{2.92}$$

Proof. Since the result is local, we may assume, without loss of generality, that $u \in L^1(\mathbb{R}^N)$. Since for every $x \in \mathbb{R}^N$

$$\left| \liminf_{r \to 0^+} \frac{1}{|B(x,r)|} \int_{B(x,r)} u(y) \, dy \right| \leq \mathrm{M}(u)(x), \qquad (2.93)$$

$$\left| \limsup_{r \to 0^+} \frac{1}{|B(x,r)|} \int_{B(x,r)} u(y) \, dy \right| \leq \mathrm{M}(u)(x),$$

it follows from Theorem 2.91(i)–(ii) that the function

$$\widetilde{u}(x) := \limsup_{r \to 0^+} \frac{1}{|B(x,r)|} \int_{B(x,r)} u(y) \, dy$$

$$- \liminf_{r \to 0^+} \frac{1}{|B(x,r)|} \int_{B(x,r)} u(y) \, dy$$

is well-defined (and in turn finite) for $\mathcal{L}^N$ a.e. $x \in \mathbb{R}^N$, and for every $t > 0$,

$$\left| \{ x \in \mathbb{R}^N : \widetilde{u}(x) > 2t \} \right| \leq \left| \{ x \in \mathbb{R}^N : \mathrm{M}(u)(x) > t \} \right|$$

$$\leq \frac{3^N}{t} \int_{\mathbb{R}^N} |u(x)| \, dx.$$

If we now replace in the previous inequality u with $u - v$, where $v \in C_c(\mathbb{R}^N)$, and observe that $\widetilde{u}(x) = \widetilde{(u - v)}(x)$ for $\mathcal{L}^N$ a.e. $x \in \mathbb{R}^N$, then we obtain that

$$\left| \{ x \in \mathbb{R}^N : \widetilde{u}(x) > 2t \} \right| \leq \left| \{ x \in \mathbb{R}^N : \mathrm{M}(u - v)(x) > t \} \right|$$

$$\leq \frac{3^N}{t} \int_{\mathbb{R}^N} |u(x) - v(x)| \, dx.$$

Since $C_c(\mathbb{R}^N)$ is dense in $L^1(\mathbb{R}^N)$, the right-hand side of the previous inequality can be made arbitrarily small, and so

$$\left| \{ x \in \mathbb{R}^N : \widetilde{u}(x) > 2t \} \right| = 0$$

for all $t > 0$. Hence $\widetilde{u}(x) = 0$ for $\mathcal{L}^N$ a.e. $x \in \mathbb{R}^N$, and in turn, by (2.93) we have that there exists

$$\lim_{r \to 0^+} \frac{1}{|B(x,r)|} \int_{B(x,r)} u(y) \, dy \in \mathbb{R} \qquad (2.94)$$

for all $x \in \mathbb{R}^N \setminus E_1$, where E_1 is a set of Lebesgue measure zero.

Take $r = \frac{1}{n}$, and apply Theorem 2.73 (see also Remark 2.72) to conclude that the sequence of functions

$$v_n(x) := \frac{1}{|B(x,\frac{1}{n})|} \int_{B(x,\frac{1}{n})} u(y) \, dy$$

converges in L^1 to u. In view of Theorem 2.20(iii), we may extract a subsequence $\{v_{n_k}\}$ converging pointwise to u except on a set E_2 of Lebesgue measure zero. Therefore if $x \notin E := E_1 \cup E_2$ then (2.92) follows from (2.94).

2.3 L^p Spaces on Banach Spaces

In this section we briefly introduce L^p spaces on Banach spaces. These will play an important role in the study of Young measures in the last chapter. We will omit most of the proofs, since the only results that will be used in the sequel are Theorem 2.108 and part (i) of the Riesz representation theorem in $L^p(X;Y)$, which we will prove. We refer to [DuSc88], [DieU77], [Ed95], and [SY05] for more information on the subject and for the proofs omitted here.

Definition 2.101. *Let $(X, \mathfrak{M}, \mu)$ be a measure space and let Y be a Banach space. A (measurable) simple function s is a function $s : X \to Y$ of the form*

$$s = \sum_{i=1}^{\ell} c_i \chi_{E_i},$$

where $\ell \in \mathbb{N}$, $c_1, \ldots, c_\ell \in Y$ are distinct and the sets $E_i \subset X$ are measurable and mutually disjoint.

(i) A function $u : X \to Y$ is said to be strongly measurable *if there exists a sequence $\{s_n\}$ of (measurable) simple functions $s_n : X \to Y$ such that*

$$\lim_{n \to \infty} \|s_n(x) - u(x)\|_Y = 0 \quad \text{for } \mu \text{ a.e. } x \in X.$$

(ii) A function $u : X \to Y$ is said to be weakly measurable *if for any $L \in Y'$ the function*

$$x \in X \mapsto L(u(x)) \text{ is measurable.}$$

(iii) A function $u : X \to Y'$ is said to be weakly star measurable *if for any $y \in Y$ the map*

$$x \in X \mapsto \langle u(x), y \rangle_{Y',Y} = u(x)(y) \text{ is measurable,}$$

where $\langle \cdot, \cdot \rangle_{Y',Y} : Y' \times Y \to \mathbb{R}$ is the duality pairing.

Exercise 2.102. Let $(X, \mathfrak{M}, \mu)$ be a measure space and let Y be a Banach space. Prove that if a function $u : X \to Y$ is strongly measurable then the function $x \in X \mapsto \|u(x)\|_Y$ is measurable.

Exercise 2.103 (Egoroff). Let $(X, \mathfrak{M}, \mu)$ be a measure space with μ finite, let Y be a Banach space, and let $u, u_n : X \to Y$, $n \in \mathbb{N}$, be strongly measurable functions such that
$$\lim_{n \to \infty} \|u_n(x) - u(x)\|_Y = 0$$

for μ a.e. $x \in X$. Prove that for every $\varepsilon > 0$ there exists a measurable set $E \in \mathfrak{M}$, with $\mu(X \setminus E) \le \varepsilon$, such that

$$\lim_{n \to \infty} \|u_n(x) - u(x)\|_Y = 0$$

uniformly on E: (Hint: For every n, $k \in \mathbb{N}$ define

$$E_{n,k} := \bigcap_{i,j>n} \left\{ x \in X : \|u_i(x) - u_j(x)\|_Y < \frac{1}{k} \right\}$$

and prove that there exists an increasing sequence $\{n_k\}_{k \in \mathbb{N}}$ such that the set

$$E := \bigcap_{k=1}^{\infty} E_{n_k,k}$$

has the desired properties.)

The relation between weak and strong measurability is given by the following theorem.

Theorem 2.104 (Pettis). *Let $(X, \mathfrak{M}, \mu)$ be a measure space with μ finite and let Y be a Banach space. A function $u : X \to Y$ is strongly measurable if and only if it is weakly measurable and there exists $E \in \mathfrak{M}$, with $\mu(E) = 0$, such that the set $u(X \setminus E)$ is a separable set of Y (in the norm sense).*

Proof. We prove only the easy direction of the theorem, because it will be needed in Theorem 2.108. Thus assume that $u : X \to Y$ is strongly measurable. Then there exists a sequence $\{s_n\}$ of simple functions $s_n : X \to Y$ such that

$$\lim_{n \to \infty} \|s_n(x) - u(x)\|_Y = 0 \quad \text{for } \mu \text{ a.e. } x \in X.$$

Hence for any $L \in Y'$ we have that

$$\lim_{n \to \infty} L(s_n(x)) = L(u(x)) \quad \text{for } \mu \text{ a.e. } x \in X.$$

Since the function $L \circ s_n : X \to \mathbb{R}$ is a simple function it follows that the real-valued function

$$x \in X \mapsto L(u(x))$$

is measurable as the pointwise limit of measurable functions.

By Egoroff's theorem, for every $k \in \mathbb{N}$ there exists a measurable set $E_k \in \mathfrak{M}$, with $\mu(E_k) \le \frac{1}{k}$, such that

$$\lim_{n \to \infty} \|s_n(x) - u(x)\|_Y = 0$$

uniformly on $X \setminus E_k$. Since the range of each s_n is finite, the set $\bigcup_{n=1}^{\infty} s_n(X)$ is countable. Hence for every $k \in \mathbb{N}$ the set $u(X \setminus E_k)$ is separable, and so is the set

$$u\left(X \setminus \left(\bigcap_{k=1}^{\infty} E_k \right) \right) = u\left(\bigcup_{k=1}^{\infty} (X \setminus E_k) \right) = \bigcup_{k=1}^{\infty} u((X \setminus E_k)).$$

To conclude this part of the proof we observe that for every $j \in \mathbb{N}$,

$$0 \le \mu\left(\bigcap_{k=1}^{\infty} E_k\right) \le \mu(E_j) \le \frac{1}{j} \to 0$$

as $j \to \infty$.

Example 2.105. Let μ be the Lebesgue measure on $[0,1]$ and define the function $u : [0,1] \to \ell^{\infty}$ by

$$u(x) := \left\{\frac{\operatorname{sgn}(\sin(2^n \pi x)) + 1}{2}\right\}_{n \in \mathbb{N}}, \quad x \in [0,1].$$

By Remark 2.42 we may identify ℓ^{∞} with the dual $(\ell^1)'$, so for every $y = \{y_n\}_{n \in \mathbb{N}} \in \ell^1$ we have that

$$x \in [0,1] \mapsto (u(x))(y) = \sum_{n=1}^{\infty} y_n \frac{\operatorname{sgn}(\sin(2^n \pi x)) + 1}{2}$$

is well-defined. This is a measurable function; hence u is weakly star measurable. It may be verified that there is no set $E \in \mathfrak{M}$, with $\mu(E) = 0$, such that the set $u(X \setminus E)$ is a separable set of ℓ^{∞} and, even more, that there exists $L \in (\ell^{\infty})'$ such that the function

$$x \in [0,1] \mapsto L(u(x)) \text{ is not measurable.}$$

Hence u is not weakly measurable (and so, by the Pettis theorem, it is not strongly measurable). We refer to [DieU77] for more details.

Definition 2.106. *Let $(X, \mathfrak{M}, \mu)$ be a measure space and let Y be a Banach space. A (measurable) simple function $s : X \to Y$ is (Bochner) integrable if it has the form*

$$s = \sum_{i=1}^{\ell} c_i \chi_{E_i},$$

where $c_1, \ldots, c_{\ell} \in Y$, $\ell \in \mathbb{N}$, are distinct, the sets $E_i \subset X$ are mutually disjoint, and $c_i = 0$ whenever $\mu(E_i) = \infty$. For any measurable set $E \in \mathfrak{M}$ the Bochner integral of s over E is defined by

$$\int_E s \, d\mu := \sum_{i=1}^{\ell} c_i \mu(E_i \cap E),$$

where $c_i \mu(E_i \cap E)$ is set to be zero whenever $c_i = 0$ and $\mu(E_i \cap E) = \infty$.

Definition 2.107. *Let $(X, \mathfrak{M}, \mu)$ be a measure space and let Y be a Banach space. A strongly measurable function $u : X \to Y$ is (Bochner) integrable if there exists a sequence $\{s_n\}$ of simple integrable functions such that*

$$\lim_{n \to \infty} \|s_n(x) - u(x)\|_Y = 0 \quad for \ \mu \ a.e. \ x \in X$$

and

$$\lim_{n \to \infty} \int_X \|s_n - u\|_Y \, d\mu = 0.$$

For any measurable set $E \in \mathfrak{M}$ the Bochner integral *of u over E is defined by*

$$\int_E u \, d\mu := \lim_{n \to \infty} \int_E s_n \, d\mu.$$

It may be verified that this limit exists and that is independent of the particular sequence $\{s_n\}$.

Theorem 2.108. *Let $(X, \mathfrak{M}, \mu)$ be a measure space with μ finite and let Y be a Banach space. A strongly measurable function $u : X \to Y$ is (Bochner) integrable if and only if $\|u\|_Y$ is Lebesgue integrable over X.*

Proof. Assume that the real-valued function $x \in X \mapsto \|u(x)\|_Y$ is integrable. Since u is strongly measurable, by the Pettis theorem there exists $E \in \mathfrak{M}$, with $\mu(E) = 0$, such that the set $u(X \setminus E)$ is a separable subset of Y. Let $\{y_n\} \subset Y$ be a dense set in $u(X \setminus E)$. By Exercise 2.102, for each $n \in \mathbb{N}$ the function $x \in X \mapsto \|u(x) - y_n\|_Y$ is measurable, and so for each $k \in \mathbb{N}$ the set

$$E_{n,k} := \left\{ x \in X \setminus E : \|u(x) - y_n\|_Y < \frac{1}{k} \right\}$$

is measurable. By the density of $\{y_n\}$ in $u(X \setminus E)$ we have that

$$\bigcup_{n=1}^{\infty} E_{n,k} = X \setminus E$$

for every $k \in \mathbb{N}$. For every $n, k \in \mathbb{N}$ define

$$F_{n,k} := E_{n,k} \setminus \bigcup_{j=1}^{n-1} E_{j,k}.$$

Then for each fixed $k \in \mathbb{N}$, the measurable sets $\{F_{n,k}\}_{n \in \mathbb{N}}$ are pairwise disjoint with

$$\bigcup_{n=1}^{\infty} F_{n,k} = X \setminus E.$$

Since $x \in X \mapsto \|u(x)\|_Y$ is integrable and

$$\mu(X) = \mu(X \setminus E) = \sum_{n=1}^{\infty} \mu(F_{n,k}),$$

we may find an integer $n_k \in \mathbb{N}$ so large that

$$\int_{\underset{n=n_k+1}{\overset{\infty}{\bigcup}} F_{n,k}} \|u\|_Y \, d\mu \le \frac{1}{k}, \qquad \sum_{n=n_k+1}^{\infty} \mu\left(F_{n,k}\right) \le \frac{1}{k}. \qquad (2.95)$$

Define the simple function

$$s_k\left(x\right) := \begin{cases} y_n & \text{if } x \in F_{n,k}, \ n = 1, \dots, n_k, \\ 0 & \text{otherwise.} \end{cases}$$

Then

$$\|s_k\left(x\right) - u\left(x\right)\|_Y \le \frac{1}{k} \text{ for all } x \in \bigcup_{n=1}^{n_k} F_{n,k},$$

and so by $(2.95)_2$,

$$\lim_{k \to \infty} \|s_k\left(x\right) - u\left(x\right)\|_Y = 0 \quad \text{for } \mu \text{ a.e. } x \in X,$$

while by $(2.95)_1$,

$$\int_X \|s_k - u\|_Y \, d\mu = \sum_{n=1}^{n_k} \int_{F_{n,k}} \|s_k - u\|_Y \, d\mu + \int_{\underset{n=n_k+1}{\overset{\infty}{\bigcup}} F_{n,k}} \|u\|_Y \, d\mu$$

$$\le \frac{1}{k} \sum_{n=1}^{n_k} \mu\left(F_{n,k}\right) + \frac{1}{k} \le \frac{1}{k}\left(\mu\left(X\right) + 1\right) \to 0$$

as $k \to \infty$. Thus $u : X \to Y$ is Bochner integrable.

Conversely, assume that $u : X \to Y$ is Bochner integrable and consider a sequence $\{s_n\}$ of simple integrable functions such that

$$\lim_{n \to \infty} \|s_n\left(x\right) - u\left(x\right)\|_Y = 0 \quad \text{for } \mu \text{ a.e. } x \in X$$

and

$$\lim_{n \to \infty} \int_X \|s_n - u\|_Y \, d\mu = 0.$$

Then for $x \in X$ and for $\ell, n \in \mathbb{N}$

$$0 \le |\|s_n\left(x\right)\|_Y - \|s_\ell\left(x\right)\|_Y| \le \|s_n\left(x\right) - s_\ell\left(x\right)\|_Y$$
$$\le \|s_n\left(x\right) - u\left(x\right)\|_Y + \|s_\ell\left(x\right) - u\left(x\right)\|_Y,$$

and so

$$\int_X |\|s_n\|_Y - \|s_\ell\|_Y| \, d\mu \le \int_X \|s_n - u\|_Y \, d\mu + \int_X \|s_\ell - u\|_Y \, d\mu \to 0$$

as $\ell, n \to \infty$. Thus $\{\|s_n\|_Y\}$ is a Cauchy sequence in $L^1\left(X, \mathfrak{M}, \mu\right)$, and so it converges in $L^1\left(X, \mathfrak{M}, \mu\right)$ (and up to a subsequence also pointwise μ a.e. in X) to a function $v : X \to \mathbb{R}$. On the other hand, for μ a.e. $x \in X$,

$$0 \le |\|s_n\left(x\right)\|_Y - \|u\left(x\right)\|_Y| \le \|s_n\left(x\right) - u\left(x\right)\|_Y \to 0$$

as $n \to \infty$. Hence $v\left(x\right) = \|u\left(x\right)\|_Y$ for μ a.e. $x \in X$, which shows that $\|u\|_Y$ is integrable.

Definition 2.109. *Let* $(X, \mathfrak{M}, \mu)$ *be a measure space, let* Y *be a Banach space, and let* $1 \leq p < \infty$. *Then*

$$L^p \left((X, \mathfrak{M}, \mu) \,; Y \right) := \left\{ u : X \to Y : u \ strongly \ measurable, \right.$$

$$\left. \|u\|_{L^p((X,\mathfrak{M},\mu);Y)} < \infty \right\},$$

where

$$\|u\|_{L^p((X,\mathfrak{M},\mu);Y)} := \left(\int_X \|u\|_Y^p \, d\mu \right)^{1/p}.$$

If $p = \infty$ *then*

$$L^\infty \left((X, \mathfrak{M}, \mu) \,; Y \right) := \left\{ u : X \to Y : u \ strongly \ measurable, \right.$$

$$\left. \|u\|_{L^\infty((X,\mathfrak{M},\mu);Y)} < \infty \right\},$$

where

$$\|u\|_{L^\infty((X,\mathfrak{M},\mu);Y)} = \operatorname*{esssup}_{x \in X} \|u(x)\|_Y$$

$$:= \inf \left\{ \alpha \in \mathbb{R} : \|u(x)\|_Y < \alpha \ \mu \ a.e. \ x \in X \right\}.$$

When there is no possibility of confusion we abbreviate $L^p \left((X, \mathfrak{M}, \mu) \,; Y \right)$ as $L^p (X; Y)$. As in the case $Y = \mathbb{R}$ we identify functions with their equivalence classes.

Theorem 2.110. *Let* $(X, \mathfrak{M}, \mu)$ *be a measure space and let* Y *be a Banach space.*

(i) $L^p (X; Y)$ *is a Banach space for* $1 \leq p \leq \infty$;
(ii) the family of all integrable simple functions is dense in $L^p (X; Y)$ *for* $1 \leq p < \infty$;
(iii) if X *is a separable metric space,* μ *is a* σ*-finite Radon measure, and if* Y *is separable, then* $L^p (X; Y)$ *is separable for* $1 \leq p < \infty$.

A major problem in the theory of L^p spaces on Banach spaces is the identification of the dual of $L^p (X; Y)$. Here we will consider only the two important special cases in which Y is either separable or reflexive.

Definition 2.111. *Let* $(X, \mathfrak{M}, \mu)$ *be a measure space, let* Y *be a Banach space, and let* $1 \leq p \leq \infty$. *Then the space* $L^p_w (X; Y')$ *is the space of all (equivalence classes of) weakly star measurable functions* $u : X \to Y'$ *such that* $\|u\|_{Y'} \in L^p (X; \mathbb{R})$. *The space* $L^p_w (X; Y')$ *is endowed with the norm*

$$\|u\|_{L^p_w(X;Y')} := \left(\int_X \|u\|_{Y'}^p \, d\mu \right)^{1/p}$$

for $1 \leq p < \infty$, *and*

$$\|u\|_{L^\infty_w(X;Y')} := \underset{x \in X}{\mathrm{esssup}} \, \|u(x)\|_{Y'}$$

for $p = \infty$.

Theorem 2.112 (Riesz representation theorem in L^p). *Let* $(X, \mathfrak{M}, \mu)$ *be a measure space with* μ σ*-finite, let* Y *be a Banach space, let* $1 \leq p < \infty$, *and let* q *be its Hölder conjugate exponent.*

(i) Assume that Y *is separable. If* $L \in (L^p(X;Y))'$ *then there exists a unique* $v \in L^q_w(X;Y')$ *such that*

$$L(u) = \int_X \langle v, u \rangle_{Y',Y} \, d\mu \tag{2.96}$$

for every $u \in L^p(X;Y)$. *Moreover, the norm of* L *coincides with* $\|v\|_{L^q_w(X;Y')}$. *Conversely, every functional of the form (2.96), where* $v \in L^q_w(X;Y')$, *is a bounded linear functional on* $L^p(X;Y)$.

(ii) Assume that Y *is reflexive. Then for* $L \in (L^p(X;Y))'$ *there exists a unique* $v \in L^q(X;Y')$ *such that*

$$L(u) = \int_X \langle v, u \rangle_{Y',Y} \, d\mu \tag{2.97}$$

for every $u \in L^p(X;Y)$. *Moreover, the norm of* L *coincides with* $\|v\|_{L^q(X;Y')}$. *Conversely, every functional of the form (2.97), where* $v \in L^q(X;Y')$, *is a bounded linear functional on* $L^p(X;Y)$.

Proof. We prove only (i). Let $L \in (L^p(X;Y))'$. Since Y is separable there exists $\{y_n\} \subset Y$ such that $\overline{\{y_n\}} = Y$. For each $n \in \mathbb{N}$ the functional

$$L_n(u) := L(uy_n) \quad u \in L^p(X) = L^p(X;\mathbb{R})$$

is linear and

$$\|L_n\|_{(L^p(X))'} = \sup_{u \in L^p(X) \setminus \{0\}} \frac{|L(uy_n)|}{\|u\|_{L^p(X)}} \leq \|L\|_{(L^p(X;Y))'} \|y_n\|_Y < \infty, \tag{2.98}$$

and so by the Riesz representation theorem in $L^p(X)$ there exists $v_{y_n} \in L^{p'}(X)$ such that

$$L(uy_n) = \int_X u v_{y_n} \, d\mu \quad \text{for all } u \in L^p(X) \tag{2.99}$$

and (see (2.98))

$$\|L_n\|_{(L^p(X))'} = \|v_{y_n}\|_{L^{p'}(X)} \leq \|L\|_{(L^p(X;Y))'} \|y_n\|_Y. \tag{2.100}$$

We claim that

$$\left\| \sup_{n \in \mathbb{N}} \frac{|v_{y_n}|}{\|y_n\|_Y} \right\|_{L^{p'}(X)} \le \|L\|_{(L^p(X;Y))'} . \tag{2.101}$$

If $p = 1$ then from (2.100),

$$|v_{y_n}(x)| \le \|L\|_{(L^p(X;Y))'} \|y_n\|_Y \quad \text{for } \mathcal{L}^N \text{ a.e. } x \in X \text{ and for all } n \in \mathbb{N},$$

and so the claim follows.

If $p > 1$ fix $l \in \mathbb{N}$ and write

$$\sup_{1 \le n \le l} \frac{|v_{y_n}(x)|}{\|y_n\|_Y} = \sum_{n=1}^{l} \chi_{E_n}(x) \frac{|v_{y_n}(x)|}{\|y_n\|_Y}, \quad x \in X,$$

where the sets $E_1, \ldots, E_l$ are measurable and pairwise disjoint. Note that the function

$$u := \sum_{n=1}^{l} \chi_{E_n} \frac{|v_{y_n}|^{p'-2}}{\|y_n\|_Y^{p'}} v_{y_n} y_n$$

belongs to $L^p(X;Y)$. Indeed,

$$\|u\|_{L^p(X;Y)}^p = \int_X \left\| \sum_{n=1}^{l} \chi_{E_n} \frac{|v_{y_n}|^{p'-2}}{\|y_n\|_Y^{p'}} v_{y_n} y_n \right\|_Y^p d\mu$$

$$= \sum_{n=1}^{l} \int_{E_n} \frac{|v_{y_n}|^{p'}}{\|y_n\|_Y^{p'}} d\mu = \sum_{n=1}^{l} \int_{E_n} \left(\sup_{1 \le k \le l} \frac{|v_{y_k}|}{\|y_k\|_Y} \right)^{p'} d\mu \tag{2.102}$$

$$= \int_X \left(\sup_{1 \le k \le l} \frac{|v_{y_k}|}{\|y_k\|_Y} \right)^{p'} d\mu.$$

Then, by the linearity of L and (2.99),

$$|L(u)| = \left| \sum_{n=1}^{l} L\left(\chi_{E_n} \frac{|v_{y_n}|^{p'-2}}{\|y_n\|_Y^{p'}} v_{y_n} y_n \right) \right| = \sum_{n=1}^{l} \int_{E_n} \frac{|v_{y_n}|^{p'}}{\|y_n\|_Y^{p'}} d\mu$$

$$= \int_X \left(\sup_{1 \le k \le l} \frac{|v_{y_k}|}{\|y_k\|_Y} \right)^{p'} d\mu \le \|L\|_{(L^p(X;Y))'} \|u\|_{L^p(X;Y)}$$

$$= \|L\|_{(L^p(X;Y))'} \left(\int_X \left(\sup_{1 \le k \le l} \frac{|v_{y_k}|}{\|y_k\|_Y} \right)^{p'} d\mu \right)^{1/p},$$

where we used (2.102), and so

$$\left(\int_X \left(\sup_{1 \le k \le l} \frac{|v_{y_k}|}{\|y_k\|_Y} \right)^{p'} d\mu \right)^{1/p'} \le \|L\|_{(L^p(X;Y))'},$$

and by letting $l \to \infty$ the claim follows from the Lebesgue monotone convergence theorem.

We are now ready to construct the function $v \in L_w^q(X; Y')$ in the statement. Since $\{y_n\}$ is dense in Y, for every $y \in Y$ we may find a subsequence $\{y_{n_j}\}$ converging to y. For all $u \in L^p(X)$ and $k > j$ we have

$$\left| \int_X u \left(v_{y_{n_k}} - v_{y_{n_j}} \right) d\mu \right| = \left| L \left(u y_{n_k} - u y_{n_j} \right) \right|$$
$$\leq \|L\|_{(L^p(X;Y))'} \left\| y_{n_k} - y_{n_j} \right\|_Y \|u\|_{L^p(X)},$$

and so by Corollary A.42,

$$\left\| v_{y_{n_k}} - v_{y_{n_j}} \right\|_{L^{p'}(X)} \leq \|L\|_{(L^p(X;Y))'} \left\| y_{n_k} - y_{n_j} \right\|_Y \to 0$$

as $j \to \infty$. Hence $\left\{ v_{y_{n_j}} \right\}$ is a Cauchy sequence in $L^{p'}(X)$; thus there exists $v_y \in L^{p'}(X)$ such that $v_{y_{n_j}} \to v_y$ in $L^{p'}(X)$. Similar reasoning also shows that v_y does not depend on the particular approximating subsequence. From (2.99) and (2.100), taking n_j in place of n and letting $j \to \infty$, we obtain

$$L(uy) = \int_X u v_y \, d\mu \quad \text{for all } u \in L^p(X) \tag{2.103}$$

and

$$\|v_y\|_{L^{p'}(X)} \leq \|L\|_{(L^p(X;Y))'} \|y\|_Y. \tag{2.104}$$

Define $v : X \to Y'$ as

$$v(x) : Y \to \mathbb{R},$$
$$y \mapsto v_y(x).$$

Note that $v(x)$ is linear in view of the linearity of L and (2.103). Using once more the fact that $\{y_n\}$ is dense in Y, we have that

$$\|v(x)\|_{Y'} = \sup_n \frac{|v(x)(y_n)|}{\|y_n\|_Y} = \sup_n \frac{|v_{y_n}(x)|}{\|y_n\|_Y},$$

which is finite for μ a.e. $x \in X$ by (2.101).

Moreover, for any $y \in Y$ the map $x \in X \mapsto v(x)(y) = v_y(x)$ is measurable, and so v is weakly star measurable, and by (2.101),

$$\|v\|_{L_w^p(X;Y')} = \left\| \sup_{n \in \mathbb{N}} \left(\frac{|v_{y_n}|}{\|y_n\|_Y} \right) \right\|_{L^{p'}(X)} \leq \|L\|_{(L^p(X;Y))'}.$$

To prove the reverse inequality we observe that the class $\mathcal{S}$ of simple functions of the form

$$s = \sum_{i=1}^{n} \chi_{F_i} c_i y_i, \tag{2.105}$$

where $c_i \in \mathbb{R}$ and $F_i \in \mathfrak{M}$ for all $i = 1, \ldots, n$, is dense in $L^p(X; Y)$, and so

$$\|L\|_{(L^p(X;Y))'} = \sup_{s \in \mathcal{S} \setminus \{0\}} \frac{|L(s)|}{\|s\|_{L^p(X;Y)}}.$$

For any $s \in \mathcal{S}$ of the form (2.105), by (2.99) and Hölder's inequality we have

$$|L(s)| = \left| \sum_{i=1}^{n} \int_{F_i} c_i v_{y_i} \, d\mu \right| \leq \sum_{i=1}^{n} \int_{F_i} |c_i| \|y_i\|_Y \frac{|v_{y_i}|}{\|y_i\|_Y} \, d\mu$$

$$\leq \int_X \left(\sum_{i=1}^{n} \chi_{F_i} |c_i| \|y_i\|_Y \right) \sup_k \frac{|v_{y_k}|}{\|y_k\|_Y} \, d\mu$$

$$\leq \left(\sum_{i=1}^{n} |c_i|^p \|y_i\|_Y^p \, \mu(F_i) \right)^{1/p} \left\| \sup_{k \in \mathbb{N}} \left(\frac{|v_{y_k}|}{\|y_k\|_Y} \right) \right\|_{L^{p'}(X)}$$

$$= \|s\|_{L^p(X;Y)} \left\| \sup_{k \in \mathbb{N}} \left(\frac{|v_{y_k}|}{\|y_k\|_Y} \right) \right\|_{L^{p'}(X)}.$$

Hence

$$\|L\|_{(L^p(X;Y))'} = \sup_{s \in \mathcal{S} \setminus \{0\}} \frac{|L(s)|}{\|s\|_{L^p(X;Y)}} \leq \left\| \sup_{k \in \mathbb{N}} \left(\frac{|v_{y_k}|}{\|y_k\|_Y} \right) \right\|_{L^{p'}(X)} = \|v\|_{L_w^p(X;Y')},$$

and so

$$\|L\|_{(L^p(X;Y))'} = \|v\|_{L_w^p(X;Y')}.$$

Finally, for $s \in \mathcal{S}$ of the form (2.105), by the linearity of $y \mapsto v_y(x)$ for every $x \in X$, we have

$$L(s) = \sum_{i=1}^{n} \int_{F_i} c_i v_{y_i}(x) \, d\mu(x) = \sum_{i=1}^{n} \int_{F_i} v_{c_i y_i}(x) \, d\mu(x) = \sum_{i=1}^{n} \int_{F_i} v_{s(x)} \, d\mu(x)$$

$$= \int_X v_{s(x)} \, d\mu(x) = \int_X v(x)(s(x)) \, d\mu(x) = \int_X \langle v(x), s(x) \rangle_{Y',Y} \, d\mu(x).$$

For any $u \in L^p(X; Y)$ find a sequence $\{s_j\} \subset \mathcal{S}$ of simple functions such that

$$\lim_{j \to \infty} \|s_j(x) - u(x)\|_Y = 0 \quad \text{for } \mu \text{ a.e. } x \in X.$$

Then

$$L(u) = \lim_{j \to \infty} L(s_j) = \lim_{j \to \infty} \int_X v_{s_j(x)} \, d\mu(x).$$

From (2.104), and again using the linearity of $y \mapsto v_y(x)$ for μ a.e. $x \in X$, we deduce that

$$\left\|v_{s_j(x)} - v_{u(x)}\right\|_{L^{p'}(X)} \leq \|L\|_{(L^p(X;Y))'} \|s_j(x) - u(x)\|_Y \to 0,$$

and so

$$L(u) = \int_X v_{u(x)} \, d\mu(x) = \int_X \langle v(x), u(x) \rangle_{Y',Y} \, d\mu(x),$$

and this concludes the proof.

The Direct Method and Lower Semicontinuity

3

The Direct Method and Lower Semicontinuity

> Although this may seem a paradox, all exact science is dominated by the idea of approximation.

> Bertrand Russell (1872–1970)

The direct method in the calculus of variations is an important tool in relaxation theory and it relies heavily on the notion of lower semicontinuity, which is the subject of the next section.

3.1 Lower Semicontinuity

Definition 3.1. *Let V be a topological space and let $I : V \to [-\infty, \infty]$.*

(i) I is lower semicontinuous if the set $\{v \in V : I(v) \leq t\}$ is closed for every $t \in \mathbb{R}$.

(ii) I is sequentially lower semicontinuous *if the set $\{v \in V : I(v) \leq t\}$ is sequentially closed for every $t \in \mathbb{R}$.*

(iii) I is upper semicontinuous *(respectively sequentially upper semicontinuous) if $-I$ is lower semicontinuous (respectively sequentially lower semicontinuous).*

Remark 3.2. Note that (i) implies (ii). To see this, assume that I is lower semicontinuous, fix $t \in \mathbb{R}$, and let $\{v_n\} \subset V$ be any sequence converging to v_0 and such that $I(v_n) \leq t$. We claim that $I(v_0) \leq t$. Indeed, if $t < I(v_0)$ then the set $I^{-1}((t, \infty])$ is an open set containing v_0, and thus $v_n \in I^{-1}((t, \infty])$ for all n sufficiently large, which is a contradiction. Hence $I(v_0) \leq t$, and so the set $\{v \in V : I(v) \leq t\}$ is sequentially closed.

In general, the converse is not true (see Exercise A.34 below). However, it can be shown that (ii) implies (i) if V satisfies the first axiom of countability (see Proposition 3.12 below).

The following characterizations of lower semicontinuity will be of use in the sequel. We recall the definition of epigraph and of limit inferior.

Definition 3.3. *Let (V, τ) be a topological space and let $I : V \to [-\infty, \infty]$.*

(i) The epigraph *of I is the set*

$$\operatorname{epi} I := \{(v, t) \in V \times \mathbb{R} : I(v) \le t\} \,.$$

(ii) For any $v_0 \in V$ the limit inferior *of I as v tends to v_0 is defined as*

$$\liminf_{v \to v_0} I(v) := \sup_{A \in \tau(v_0)} \inf_{v \in A \setminus \{v_0\}} I(v) \,,$$

where $\tau(v_0)$ stands for the collection of all open sets in τ that contain v_0.[1]

Proposition 3.4. *Let V be a topological space and let $I : V \to [-\infty, \infty]$. Then the following are equivalent:*

(i) I is lower semicontinuous;
(ii) $\operatorname{epi} I$ is closed;
(iii) for every $v_0 \in V$

$$I(v_0) \le \liminf_{v \to v_0} I(v) \,.$$

Similarly, the following are equivalent:

(i)' I is sequentially lower semicontinuous;
(ii)' $\operatorname{epi} I$ is sequentially closed;
(iii)' for every $v_0 \in V$

$$I(v_0) \le \liminf_{n \to \infty} I(v_n)$$

for every sequence $\{v_n\}$ converging to v_0.

Proof. **Step 1:** We prove that (i) is equivalent to (ii). Assume that I is lower semicontinuous and let

$$D := (V \times \mathbb{R}) \setminus \operatorname{epi} I = \{(v, t) \in V \times \mathbb{R} : I(v) > t\} \,. \tag{3.1}$$

Fix $(v_0, t_0) \in D$ and let $0 < \varepsilon < I(v_0) - t_0$. Then the set

---

[1] In several books, $\sup_{A \in \tau(v_0)} \inf_{v \in A} I(v)$ is used as a definition for the limit inferior $\liminf_{v \to v_0} I(v)$. The definition we use here is in accordance with the definition of limit, in which the value of the function at the point v_0 plays no role. In particular, with our definition we recover the fact that the limit exists at v_0 if and only if the limit inferior and superior coincide, while with the other definition one would get that the limit inferior and superior at v_0 coincide if and only if the function I is continuous at v_0.

$$U := I^{-1}\left((t_0 + \varepsilon, \infty]\right)$$

is open and contains v_0, and so $U \times (t_0 - \varepsilon, t_0 + \varepsilon)$ is a neighborhood of (v_0, t_0). If $(v, t) \in U \times (t_0 - \varepsilon, t_0 + \varepsilon)$ then $I(v) > t_0 + \varepsilon > t$, which implies that $(v, t) \in D$. Hence D is open and its complement, epi I, is closed.

Conversely, assume that epi I is closed (and so the set D defined in (3.1) is open), and for $t_0 \in \mathbb{R}$ consider the set

$$U := \{v \in V : I(v) > t_0\}.$$

If $v_0 \in U$ then for any fixed $0 < \varepsilon_0 < I(v_0) - t_0$ the pair $(v_0, t_0 + \varepsilon_0)$ belongs to the open set D, and so we may find a neighborhood $U_0 \subset V$ of v_0 and $0 < \varepsilon \le \varepsilon_0$ such that

$$U_0 \times (t_0 + \varepsilon_0 - \varepsilon, t_0 + \varepsilon_0 + \varepsilon) \subset D.$$

In particular, if $v \in U_0$ then $I(v) > t_0 + \varepsilon_0 - \varepsilon > t_0$, and so $U_0 \subset U$, which shows that U is open and in turn proves the lower semicontinuity of I.

Step 2: We show that (i) is equivalent to (iii). Let I be lower semicontinuous and assume by contradiction that there exists $v_0 \in V$ such that

$$I(v_0) > \liminf_{v \to v_0} I(v).$$

Fix $I(v_0) > t > \liminf_{v \to v_0} I(v)$. Then the open set

$$A_t := \{v \in V : I(v) > t\} \tag{3.2}$$

belongs to $\tau(v_0)$, and so

$$\liminf_{v \to v_0} I(v) = \sup_{A \in \tau(v_0)} \inf_{v \in A \setminus \{v_0\}} I(v) \ge \inf_{v \in A_t \setminus \{v_0\}} I(v) \ge t,$$

which is a contradiction.

Conversely, assume that (iii) holds and fix $t \in \mathbb{R}$. If the set A_t defined in (3.2) is nonempty, fix any v_0 that belongs to it. Since

$$\liminf_{v \to v_0} I(v) \ge I(v_0) > t,$$

by the definition of limit inferior there exists $A \in \tau(v_0)$ such that

$$\inf_{v \in A \setminus \{v_0\}} I(v) > t,$$

which, together with the fact that $I(v_0) > t$, implies that $A \subset A_t$. Hence (i) holds.

Step 3: We prove that (i)$'$ and (iii)$'$ are equivalent. Let I be sequentially lower semicontinuous and assume by contradiction that there exist $v_0 \in V$ and a sequence $\{v_n\}$ converging to v_0 such that

$$I\left(v_0\right) > \liminf_{n \to \infty} I\left(v_n\right).$$

Fix $I\left(v_0\right) > t > \liminf_{n \to \infty} I\left(v_n\right)$ and select a subsequence (not relabeled) such that $I\left(v_n\right) \le t$ for all $n \in \mathbb{N}$. Since the set

$$C_t := \{v \in V : I\left(v\right) \le t\} \tag{3.3}$$

is sequentially closed, it follows that $v_0 \in C_t$, which is a contradiction.

Conversely, assume that (iii)$'$ holds and fix $t \in \mathbb{R}$. If the set C_t defined in (3.3) is nonempty, consider any sequence $\{v_n\} \subset C_t$ converging to some v_0. Then

$$I\left(v_0\right) \le \liminf_{n \to \infty} I\left(v_n\right) \le t,$$

which implies that $v_0 \in C_t$. Thus C_t is sequentially closed.

Step 4: Finally, we show that (ii)$'$ and (iii)$'$ are equivalent. Suppose that epi I is sequentially closed, fix $v_0 \in V$, and let $\{v_n\} \subset V$ be any sequence converging to v_0. If

$$\liminf_{n \to \infty} I\left(v_n\right) = \infty,$$

then there is nothing to show. Hence assume that the limit inferior is finite and find a subsequence $\{v_{n_k}\}$ of $\{v_n\}$ such that

$$s := \liminf_{n \to \infty} I\left(v_n\right) = \lim_{k \to \infty} I\left(v_{n_k}\right).$$

Then $\{(v_{n_k}, I\left(v_{n_k}\right))\} \subset \text{epi } I$ and $(v_{n_k}, I\left(v_{n_k}\right)) \to (v_0, s)$, and so $(v_0, s) \in \text{epi } I$ by hypothesis. This implies that

$$I\left(v_0\right) \le s = \liminf_{n \to \infty} I\left(v_n\right) = \lim_{k \to \infty} I\left(v_{n_k}\right),$$

and so (iii)$'$ is satisfied.

Finally, assume that (iii)$'$ holds and let $\{(v_n, t_n)\} \subset \text{epi } I$ be such that $v_n \to v$ and $t_n \to t$ for some $v \in V$ and $t \in \mathbb{R}$. Since $I\left(v_n\right) \le t_n$, letting $n \to \infty$ and using the sequential lower semicontinuity of I yields

$$I\left(v\right) \le \liminf_{n \to \infty} I\left(v_n\right) \le \lim_{n \to \infty} t_n = t.$$

Hence $(v, t) \in \text{epi } I$, and so epi I is sequentially closed.

An important property of lower semicontinuous functions is presented in the next proposition.

Proposition 3.5. *Let V be a topological space and let $\{I_\alpha\}$ be a (possibly uncountable) family of lower semicontinuous (respectively sequentially lower semicontinuous) functions, $I_\alpha : V \to [-\infty, \infty]$. Then the function*

$$I_+ := \sup_\alpha I_\alpha$$

is still lower semicontinuous (respectively sequentially lower semicontinuous). In addition, if the family $\{I_\alpha\}$ is finite, then the function

$$I_- := \min_\alpha I_\alpha$$

is still lower semicontinuous (respectively sequentially lower semicontinuous).

Proof. **Step 1:** Assume that all the functions I_α are lower semicontinuous. Using the definition of supremum, it follows that for every $t \in \mathbb{R}$,

$$\{v \in V : I_+(v) > t\} = \bigcup_\alpha \{v \in V : I_\alpha(v) > t\},$$

and thus the set $\{v \in V : I_+(v) > t\}$ is open, since it is a union of open sets.

Suppose next that all the functions I_α are sequentially lower semicontinuous, fix $v_0 \in V$, and let $\{v_n\} \subset V$ be any sequence converging to v_0. Then for every α,

$$I_\alpha(v_0) \le \liminf_{n \to \infty} I_\alpha(v_n) \le \liminf_{n \to \infty} I_+(v_n),$$

and it now suffices to take the supremum over all α.

Step 2: Assume that the family $\{I_\alpha\}$ is finite, say $\{I_\alpha\} = \{I_1, \dots, I_l\}$, and that each I_j, $j = 1, \dots, l$, is lower semicontinuous. For any $t \in \mathbb{R}$,

$$\{v \in V : I_-(v) > t\} = \bigcap_{j=1}^{l} \{v \in V : I_j(v) > t\},$$

and so the set $\{v \in V : I_-(v) > t\}$ is open, since it is the finite intersection of open sets.

If instead, the functions $I_1, \dots, I_l$ are sequentially lower semicontinuous, fix $v_0 \in V$, let $\{v_n\} \subset V$ be any sequence converging to v_0, and find a subsequence $\{v_{n_k}\}$ of $\{v_n\}$ such that

$$s := \liminf_{n \to \infty} I_-(v_n) = \lim_{k \to \infty} I_-(v_{n_k}).$$

Then there exists an integer $j_0 \in \{1, \dots, l\}$ such that

$$\min_{1 \le j \le l} I_j(v_{n_k}) = I_{j_0}(v_{n_k})$$

for infinitely many k. Hence, selecting a further subsequence (not relabeled), we may assume that

$$s = \liminf_{n \to \infty} I_-(v_n) = \lim_{k \to \infty} I_{j_0}(v_{n_k}) \ge I_{j_0}(v_0) \ge I_-(v_0),$$

and this concludes the proof.

We now show that Weierstrass's theorem continues to hold for lower semicontinuous functions. This fact alone explains the importance of this class of functions in minimization problems.

Theorem 3.6 (Weierstrass). *Let V be a topological space, let $K \subset V$ be compact (respectively sequentially compact), and let $I : V \to [-\infty, \infty]$ be a lower semicontinuous (respectively sequentially lower semicontinuous) function. Then there exists $v_0 \in K$ such that*

$$I(v_0) = \min_{v \in K} I(v).$$

Proof. Assume first that K is compact and I lower semicontinuous and let

$$t := \inf_{v \in K} I(v).$$

If the infimum is not attained, then for every $v \in K$ we may find $t < t_v < I(v)$. Then the family of open sets

$$U_v := \{w \in V : I(w) > t_v\}, \quad v \in K,$$

is an open cover for the compact set K, and so we may find a finite cover $U_{v_1}, \ldots, U_{v_l}$ of the set K. But then for all $v \in K$,

$$I(v) \geq \min_{i=1,\ldots,l} t_{v_i} > t = \inf_{w \in K} I(w),$$

which contradicts the definition of t.

Suppose instead that $K \subset V$ is sequentially compact and $I : V \to [-\infty, \infty]$ is a sequentially lower semicontinuous function. Let $\{v_n\} \subset K$ be such that

$$\inf_{w \in K} I(w) \leq I(v_n) \leq \inf_{w \in K} I(w) + \frac{1}{n}$$

for all n. By the sequential compactness of K there exists a subsequence $\{v_{n_k}\}$ of $\{v_n\}$ that converges to some $v_0 \in K$, and so

$$\inf_{w \in K} I(w) \leq I(v_0) \leq \liminf_{k \to \infty} I(v_{n_k}) \leq \inf_{w \in K} I(w),$$

which implies that v_0 is the desired minimizer.

The next result will be used in the proof of Theorem 6.49. Note that for metric spaces, lower semicontinuity is equivalent to sequential lower semicontinuity; see also Proposition 3.12 below.

Proposition 3.7. *Let V be a separable metric space and let $I : V \to [-\infty, \infty]$ be a lower semicontinuous function. Then there exists a countable set $Y \subset V$ such that for each $v \in V$,*

$$I(v) = \inf \left\{ \liminf_{n \to \infty} I(v_n) : \{v_n\} \subset Y, \ v_n \to v \right\}.$$

Proof. In view of the lower semicontinuity of I, it is enough to find a countable set $Y \subset V$ such that for each $v \in V$,

$$I(v) \geq \inf \left\{ \liminf_{n \to \infty} I(v_n) : \{v_n\} \subset Y, \, v_n \to v \right\}.$$

For each rational number q let $V_q := \{v \in V : I(v) \leq q\}$. Since V is separable, then so is V_q, and thus if $V_q \neq \emptyset$, then there exists a countable set $Y_q \subset V_q$ that is dense in V_q. Set

$$Y := \left(\bigcup_{q \in \mathbb{Q}} Y_q \right) \cup Y_\infty,$$

where Y_∞ is a countable dense subset of V. Fix $v \in V$. If $I(v) = \infty$, then in view of the lower semicontinuity of I it follows that

$$I(v) = \lim_{n \to \infty} I(v_n)$$

for any sequence $\{v_n\} \subset Y_\infty$ converging to v. If $I(v) < \infty$, then for any rational number $q > I(v)$ we have that $v \in V_q$, and so, since Y_q is dense in V_q, for any $\varepsilon > 0$ there exists $w \in Y_q$ such that $I(w) \leq q$ and $d(v, w) \leq \varepsilon$. Hence

$$\liminf_{w \in Y, w \to v} I(w) \leq q,$$

and it now suffices to let $q \searrow I(v)$.

When I is not lower semicontinuous often in applications it is of interest to study the greatest lower semicontinuous function below I.

Definition 3.8. *Let V be a topological space and let $I : V \to [-\infty, \infty]$.*

(i) The lower semicontinuous envelope $\mathrm{lsc}\, I : V \to [-\infty, \infty]$ *of I is defined by*

$$\mathrm{lsc}\, I(v) := \sup \{ H(v) : H : V \to [-\infty, \infty]$$
$$\text{is lower semicontinuous, } H \leq I \}.$$

(ii) The sequentially lower semicontinuous envelope $\mathrm{slsc}\, I : V \to [-\infty, \infty]$ *of I is defined by*

$$\mathrm{slsc}\, I(v) := \sup \{ H(v) : H : V \to [-\infty, \infty] \text{ is sequentially}$$
$$\text{lower semicontinuous, } H \leq I \}.$$

In view of Proposition 3.5, $\mathrm{lsc}\, I$ is lower semicontinuous and $\mathrm{slsc}\, I$ is sequentially lower semicontinuous. When it is important to highlight the underlying topology τ, we write $\mathrm{lsc}_\tau I$ and $\mathrm{slsc}_\tau I$.

Exercise 3.9. Let (V, τ) be a topological space and let $I : V \to [-\infty, \infty]$. Prove that for all $v \in V$,

$$\operatorname{lsc} I(v) = \sup_{A \in \tau(v)} \inf_{w \in A} I(w) = \min \left\{ I(v_0), \liminf_{v \to v_0} I(v) \right\},$$

where, we recall, $\tau(v_0)$ stands for the collection of all open sets in τ that contain v_0.

In view of Proposition 3.4 the following proposition is not surprising.

Proposition 3.10. *Let V be a topological space and let $I : V \to [-\infty, \infty]$. Then*

$$\operatorname{epi}(\operatorname{lsc} I) = \overline{\operatorname{epi} I}.$$

Proof. Since $\operatorname{lsc} I \leq I$ we have

$$\operatorname{epi} I \subset \operatorname{epi}(\operatorname{lsc} I),$$

and therefore $\overline{\operatorname{epi} I} \subset \overline{\operatorname{epi}(\operatorname{lsc} I)} = \operatorname{epi}(\operatorname{lsc} I)$, where in the last equality we used Proposition 3.4.

Conversely, define

$$g(v) := \inf \left\{ t : (v, t) \in \overline{\operatorname{epi} I} \right\}, \quad v \in V.$$

We show that $g \leq I$. Let $v \in V$ be such that $I(v) < \infty$. Then for any $t > I(v)$ we have $(v, t) \in \operatorname{epi} I$, and therefore $g(v) \leq t$. Consequently, $g(v) \leq I(v)$.

Note also that by the definition of g it follows that $\overline{\operatorname{epi} I} \subset \operatorname{epi} g$. We prove equality, i.e., that $\overline{\operatorname{epi} I} = \operatorname{epi} g$. Let $(v, t) \in \operatorname{epi} g$ and let $\varepsilon > 0$. By the definition of g we may find $t' < t + \varepsilon$ such that $(v, t') \in \overline{\operatorname{epi} I}$. The arbitrariness of ε implies that $(v, t) \in \overline{\operatorname{epi} I}$.

Since $\operatorname{epi} g$ is a closed set, we conclude that g is lower semicontinuous by Proposition 3.4, and thus $\operatorname{lsc} I \geq g$.

Finally, we obtain

$$\operatorname{epi}(\operatorname{lsc} I) \subset \operatorname{epi} g = \overline{\operatorname{epi} I},$$

and this completes the proof.

Remark 3.11. Note that $\operatorname{lsc} I$ is lower semicontinuous and $\operatorname{slsc} I$ is sequentially lower semicontinuous. By Remark 3.2 the function $\operatorname{lsc} I$ is also sequentially lower semicontinuous, and so

$$\operatorname{lsc} I \leq \operatorname{slsc} I \leq \inf_{\{v_n\}} \left\{ \liminf_{n \to \infty} I(v_n) : \{v_n\} \subset V, \ v_n \to v \right\}. \tag{3.4}$$

Equality holds if V satisfies the first axiom of countability, as shown in the next proposition.

Proposition 3.12. *Let V be a topological space satisfying the first axiom of countability and let $I : V \to [-\infty, \infty]$. Then*

$$\mathrm{lsc}\, I\,(v) = \mathrm{slsc}\, I\,(v) = \inf_{\{v_n\}} \left\{ \liminf_{n\to\infty} I\,(v_n) : \{v_n\} \subset V,\ v_n \to v \right\}$$

for every $v \in V$. Moreover, the infimum is attained, that is, for every $v \in V$ there exists a sequence $\{v_n\}$ converging to v such that

$$\mathrm{lsc}\, I\,(v) = \lim_{n\to\infty} I\,(v_n)\,.$$

Proof. **Step 1:** Fix $v \in V$. Define

$$\Phi\,(v) := \inf \left\{ \liminf_{n\to\infty} I\,(v_n) : \{v_n\} \subset V,\ v_n \to v \right\}.$$

We prove that

$$\Phi\,(v) = \lim_{n\to\infty} I\,(w_n)$$

for some $\{w_n\} \subset V$, $w_n \to v$. Let $\{U_i\}_{i\in\mathbb{N}}$ be a countable basis of open neighborhoods of v. Find a sequence $\left\{ v_n^{(1)} \right\}$ such that $v_n^{(1)} \to v$ and

$$\liminf_{n\to\infty} I\left(v_n^{(1)} \right) \le \Phi\,(v) + \frac{1}{2}.$$

Choose $w_1 := v_{n_1}^{(1)} \in U_1$ such that

$$I\,(w_1) \le \Phi\,(v) + 1.$$

Recursively, with $\left\{ v_n^{(k)} \right\}$ converging to v and

$$\liminf_{n\to\infty} I\left(v_n^{(k)} \right) \le \Phi\,(v) + \frac{1}{k+1},$$

choose $w_k := v_{n_k}^{(k)} \in \bigcap_{i=1}^{k} U_i$ such that

$$I\,(w_k) \le \Phi\,(v) + \frac{1}{k}. \tag{3.5}$$

Then $w_k \to v$, and so

$$\Phi\,(v) \le \liminf_{k\to\infty} I\,(w_k)\,.$$

This, together with (3.5), entails

$$\Phi\,(v) = \lim_{k\to\infty} I\,(w_k)\,.$$

Step 2: In view of (3.4), all that remains to be proved is that $\Phi \le \mathrm{lsc}\, I$, or, equivalently, that $\mathrm{epi}\,(\mathrm{lsc}\, I) \subset \mathrm{epi}\,\Phi$. By Proposition 3.10 we have that epi (lsc I) $= \overline{\text{epi} \, I}$, and so it suffices to prove that $\overline{\text{epi} \, I} \subset \text{epi} \, \Phi$. Choose $(v, t) \in \overline{\text{epi} \, I}$ and, due to the first axiom of countability, find $(v_n, t_n) \in \text{epi} \, I$ such that $(v_n, t_n) \to (v, t)$. Then $t_n \geq I (v_n)$, and so

$$t = \lim_{n \to \infty} t_n \geq \liminf_{n \to \infty} I (v_n) \geq \Phi (v).$$

We conclude that $(v, t) \in \text{epi} \, \Phi$.

In applications V is often a normed space endowed with a weaker topology τ. In this book we will focus on the following three cases:

τ is the weak topology on V and V' is separable; $\qquad\qquad$ (3.6)

τ is the weak topology on V and V is reflexive; $\qquad\qquad$ (3.7)

τ is the weak star topology on $V \subset Y'$ with Y a separable normed space. $\qquad\qquad$ (3.8)

Remark 3.13. (i) An important application of this theory concerns the situation in which $V = L^p (X, \mathfrak{M}, \mu)$, $1 \leq p \leq \infty$. When $1 < p < \infty$ and V is endowed with the weak topology, then condition (3.7) is always satisfied, and if in addition, μ is σ-finite and X a separable measurable space (see the Riesz representation theorem and Theorem 2.16), then we are also under condition (3.6).

(ii) If X is a σ-compact metric space and μ is a Radon measure, then condition (3.8) is satisfied by $L^1 (X, \mathfrak{M}, \mu)$ (identified with a subspace of $(C_0 (X))'$) with the weak star topology on finite signed Radon measures, provided $C_0 (X)$ is separable (see Theorem 1.197). If $(X, \mathfrak{M}, \mu)$ is a measure space and μ is σ-finite with X a separable measurable space (see the Riesz representation theorem and Theorem 2.16), then (3.8) applies to $L^\infty (X, \mathfrak{M}, \mu) = \left(L^1 (X, \mathfrak{M}, \mu)\right)'$, endowed with the weak star topology.

We recall that under (3.6) or (3.8) the unit ball of V is metrizable with respect to τ (see Theorems A.59 and A.54).

Let $I : V \to [-\infty, \infty]$ and for all $v \in V$ define

$$\text{lsc}_\tau \, I (v) := \sup \{ \Phi (v) : \ \Phi \leq I, \ \Phi \text{ lower}$$
$$\text{semicontinuous with respect to } \tau\}$$

and

$$\text{slsc}_\tau \, I (v) := \sup \{ \Phi (v) : \ \Phi \leq I, \ \Phi \text{ sequentially lower}$$
$$\text{semicontinuous with respect to } \tau\}.$$

In applications it is often most useful to ensure that $\text{slsc}_\tau \, I$ coincides with the (somewhat friendlier) functional $\mathcal{I} : V \to [-\infty, \infty]$, defined for all $v \in V$ as

$$\mathcal{I} (v) := \inf_{\{v_n\}} \left\{ \liminf_{n \to \infty} I (v_n) : \ \{v_n\} \subset V, \ v_n \overset{\tau}{\to} v \right\}.$$

This will be established in Proposition 3.16 below under the assumption of *coercivity.*

Note that by Remark 3.11 we have

$$\mathrm{lsc}_\tau\, I \le \mathrm{slsc}_\tau\, I \le \mathcal{I}. \tag{3.9}$$

Definition 3.14. *Let V be a normed space and let $I : V \to [-\infty, \infty]$. The map I is said to be coercive if*

$$I\left(v\right) \to \infty \ as \ \|v\| \to \infty.$$

Proposition 3.15. *Let V be a normed space and let $I : V \to [-\infty, \infty]$. Then $\mathrm{slsc}_\tau\, I$ is sequentially lower semicontinuous with respect to τ, and if I is coercive then $\mathrm{slsc}_\tau\, I$ is also coercive.*

Proof. The sequential lower semicontinuity of $\mathrm{slsc}_\tau\, I$ follows from the fact that $\mathrm{slsc}_\tau\, I$ is the supremum of sequentially lower semicontinuous functions. Assume that I is coercive. We divide the proof of the coercivity of $\mathrm{slsc}_\tau\, I$ into three steps.

Step 1: We claim that for every $s \in \mathbb{R}$,

$$\{v \in V : \mathrm{slsc}_\tau\, I\left(v\right) \le s\} \subset \overline{\{v \in V : I\left(v\right) \le t\}}^{\,\tau}$$

for all $t > s$. If this were not true, then there would exist $s, t \in \mathbb{R}$, with $t > s$, $v \in V$, and $A \in \tau\left(v\right)$, such that $\mathrm{slsc}_\tau\, I\left(v\right) \le s$ and $I\left(w\right) > t$ for all $w \in A$. Using (3.9) and Exercise 3.9 we get

$$s \ge \mathrm{slsc}_\tau\, I\left(v\right) \ge \mathrm{lsc}_\tau\, I\left(v\right) \ge \inf_{w \in A} I\left(w\right) \ge t,$$

and this contradicts the inequality $t > s$.

Step 2: We claim that for every $s \in \mathbb{R}$,

$$C_s := \{v \in V : \mathrm{slsc}_\tau\, I\left(v\right) \le s\} \subset \overline{B\left(0, R\right)}$$

for some $R > 0$. Let $t > s$. By the coercivity of I there exists $R > 0$ such that

$$\{v \in V : I\left(v\right) \le t\} \subset B\left(0, R\right),$$

and so

$$\overline{\{v \in V : I\left(v\right) \le t\}}^{\,\tau} \subset \overline{B\left(0, R\right)}^{\,\tau} = \overline{B\left(0, R\right)},$$

where we have used the fact that the τ-closure of a convex set coincides with the strong closure (see Theorem A.47).

By Step 1 we conclude that $C_s \subset \overline{B\left(0, R\right)}$.

Step 3: We claim that $\mathrm{slsc}_\tau\, I$ is coercive. Indeed, if

$$\liminf_{\|v\| \to \infty} \mathrm{slsc}_\tau\, I\left(v\right) =: M < \infty,$$

then for any $s > M$ we may find a sequence $\{v_n\} \subset C_s$ such that $\|v_n\| \to \infty$. This contradicts the previous step.

In the next proposition we give conditions under which $\mathrm{lsc}_\tau\, I, \mathcal{I}$, and $\mathrm{slsc}_\tau\, I$ coincide.

Proposition 3.16. *Let V be a normed space satisfying either condition (3.6) or (3.8). If $I : V \to [-\infty, \infty]$ is coercive then $\mathrm{lsc}_\tau\, I = \mathrm{slsc}_\tau\, I = \mathcal{I}$.*

Proof. In view of (3.9) it suffices to show that $\mathrm{lsc}_\tau\, I \geq \mathcal{I}$. Fix $v_0 \in V$. If $\mathrm{lsc}_\tau\, I\,(v_0) = \infty$ there is nothing to prove. Thus assume that $\mathrm{lsc}_\tau\, I\,(v_0) < \infty$ and let $\mathrm{lsc}_\tau\, I\,(v_0) < t < \infty$. By the coercivity of I there exists $R > 0$ such that
$$\{v \in V : I\,(v) < t\} \subset B\,(0, R),$$
and so as in Step 2 of the previous proof we have that
$$K := \overline{\{v \in V : I\,(v) < t\}}^{\,\tau} \subset \overline{B\,(0, R)}^{\,\tau} = \overline{B\,(0, R)}.$$

By (3.6) or (3.8) the set K is metrizable (see Theorems A.59 and A.54) with a metric d compatible with τ restricted to K. Let $I|_K : K \to [-\infty, \infty]$ be the restriction of I to the set K. We claim that $v_0 \in K$. Indeed, if not, then $v_0 \in V \setminus K$, and since $V \setminus K$ is open with respect to τ it would follow that $V \setminus K \in \tau\,(v_0)$ with
$$\mathrm{lsc}_\tau\, I\,(v_0) \geq \inf_{w \in V \setminus K} I\,(w) \geq t,$$
which is a contradiction by Exercise 3.9.

Let $\mathrm{lsc}_{\tau_K}\, I|_K$ be the lower semicontinuous envelope of I_K with respect to d and let τ_K be the induced topology on K. Then again by Exercise 3.9,
$$\mathrm{lsc}_{\tau_K}\, I|_K\,(v_0) = \sup_{A \in \tau_K(v_0)} \inf_{w \in A} I|_K\,(w) = \sup_{A \in \tau(v_0)} \inf_{w \in A \cap K} I\,(w) = \mathrm{lsc}_\tau\, I\,(v_0),$$
where we have used the facts that $v_0 \in K$ and $\mathrm{lsc}_\tau\, I\,(v_0) < t$, so that for every $A \in \tau\,(v_0)$ we have
$$\inf_{w \in A \cap K} I\,(w) = \inf_{w \in A} I\,(w).$$
On the other hand, since in K we have a metric, by Proposition 3.12 there exists a sequence $\{v_n\} \subset K$ converging to v_0 with respect to d (and so with respect to τ) such that
$$\mathrm{lsc}_\tau\, I\,(v_0) = \mathrm{lsc}_{\tau_K}\, I|_K\,(v_0) = \lim_{n \to \infty} I|_K\,(v_n) = \lim_{n \to \infty} I\,(v_n) \geq \mathcal{I}\,(v_0).$$

This completes the proof.

When I is not coercive then $\mathcal{I}$ may fail to be sequentially lower semicontinuous with respect to τ.

Using Remark A.85, for every countable ordinal α we define a functional $\mathcal{E}_\alpha$ as follows: For every $v \in V$,

$$\mathcal{E}_0(v) := I(v),$$

$$\mathcal{E}_{\alpha+1}(v) := \inf\left\{\liminf_{n\to\infty}\mathcal{E}_\alpha(v_n) : \{v_n\} \subset V, \ v_n \xrightarrow{\tau} v\right\},$$

$$\mathcal{E}_\alpha(v) := \inf\{\mathcal{E}_\beta(v) : \beta < \alpha, \ \beta \text{ not a limit ordinal}\} \text{ if } \alpha \text{ is a limit ordinal.}$$

Proposition 3.17. *Let V be a normed space and let $I : V \to [-\infty, \infty]$. Then*

$$\mathrm{slsc}_\tau\, I(v) = \inf\{\mathcal{E}_\alpha(v) : \alpha \text{ countable ordinal}\}$$

for each $v \in V$.

Proof. For each $v \in V$ define

$$\mathcal{H}(v) := \inf\{\mathcal{E}_\alpha(v) : \alpha \in \Lambda\}.$$

If Φ is any functional sequentially lower semicontinuous with respect to τ, and $\Phi \leq I$, then by Remark A.83, $\Phi \leq \mathcal{E}_\alpha$ for any countable ordinal α. Hence $\Phi \leq \mathcal{H}$ and in turn $\mathrm{slsc}_\tau\, I \leq \mathcal{H}$.

To prove the reverse inequality, note first that if $\alpha \geq \beta$ then $\mathcal{E}_\alpha \leq \mathcal{E}_\beta$, and so $\mathcal{H} \leq I$. By the definition of $\mathrm{slsc}_\tau\, I$ it is now enough to prove that $\mathcal{H}$ is sequentially lower semicontinuous with respect to τ. To see this, let $\{v_n\} \subset V$ be such that $v_n \xrightarrow{\tau} v$ as $n \to \infty$. By the definition of $\mathcal{H}(v_n)$ we may find countable ordinals α_n such that

$$\mathcal{E}_{\alpha_n}(v_n) \leq \mathcal{H}(v_n) + \frac{1}{n}.$$

Consider the countable ordinal $\alpha_\infty := \sup\{\alpha_n : n \in \mathbb{N}\}$. By the previous inequality we have

$$\mathcal{H}(v) \leq \mathcal{E}_{\alpha_\infty+1}(v) = \inf\left\{\lim_{k\to\infty}\mathcal{E}_{\alpha_\infty}(w_k) : \{w_k\} \subset V, \ w_k \xrightarrow{\tau} v\right\}$$

$$\leq \lim_{n\to\infty}\mathcal{E}_{\alpha_\infty}(v_n) \leq \lim_{n\to\infty}\mathcal{E}_{\alpha_n}(v_n) \leq \lim_{n\to\infty}\mathcal{H}(v_n).$$

This concludes the proof.

Finally, we observe that when $p = 1$ and we consider the weak topology in $L^1(X, \mathfrak{M}, \mu)$, we are unable to use Proposition 3.16 because the dual of $L^1(X, \mathfrak{M}, \mu)$ is $L^\infty(X, \mathfrak{M}, \mu)$ (under suitable hypotheses on $(X, \mathfrak{M}, \mu)$), which is not separable in general. However, as it turns out, we can still assert that $\mathrm{slsc}_\tau\, I = \mathcal{I}$, provided we have a condition stronger than coercivity:

Proposition 3.18. *Let X be a separable, locally compact metric space, and let μ be a finite Radon measure defined on $\mathcal{B}(X)$. Let*

$$I : L^1(X, \mathcal{B}(X), \mu) \to [-\infty, \infty]$$

satisfy

$$I(v) \geq \int_X \gamma(|v(x)|) \, d\mu$$

for every $v \in L^1(X, \mathcal{B}(X), \mu)$ and for some increasing function $\gamma : [0, \infty) \to [0, \infty]$ with

$$\lim_{s \to \infty} \frac{\gamma(s)}{s} = \infty. \tag{3.10}$$

Then $\mathrm{slsc}_\tau \, I = \mathcal{I}$, where τ is the $L^1(X, \mathcal{B}(X), \mu)$ weak topology.

Proof. Since $\mathrm{slsc}_\tau \, I$ is sequentially lower semicontinuous with respect to the weak topology in $L^1(X, \mathcal{B}(X), \mu)$ and $\mathrm{slsc}_\tau \, I \leq I$, we have $\mathrm{slsc}_\tau \, I \leq \mathcal{I}$. To prove the opposite inequality, since $\mathcal{I} \leq I$ (take $v_n \equiv v$), it suffices to show that $\mathcal{I}$ is sequentially lower semicontinuous with respect to weak convergence in $L^1(X, \mathcal{B}(X), \mu)$. Let $\{v_n\} \subset L^1(X, \mathcal{B}(X), \mu)$ be such that $v_n \rightharpoonup v$ in $L^1(X, \mathcal{B}(X), \mu)$. Without loss of generality we may assume that

$$\liminf_{n \to \infty} \mathcal{I}(v_n) = \lim_{n \to \infty} \mathcal{I}(v_n) < \infty.$$

Let

$$M := \sup_n \mathcal{I}(v_n) < \infty,$$

fix $0 < \varepsilon < 1$, and for every n find a sequence $\{v_{n,k}\}_k \subset L^1(X, \mathcal{B}(X), \mu)$ such that $v_{n,k} \rightharpoonup v_n$ in $L^1(X, \mathcal{B}(X), \mu)$ as $k \to \infty$ and

$$\mathcal{I}(v_n) \geq \lim_{k \to \infty} I(v_{n,k}) - \varepsilon.$$

Without loss of generality we may assume that

$$v_{n,k} \in \left\{ w \in L^1(X, \mathcal{B}(X), \mu) : I(w) \leq M + 1 \right\}$$

for every n, k. Hence

$$\sup_{n,k} \int_X \gamma(|v_{n,k}(x)|) \, d\mu \leq M + 1. \tag{3.11}$$

By (3.10) we have

$$\sup_{n,k} \int_X |v_{n,k}(x)| \, d\mu < \infty,$$

and so, identifying L^1 functions with measures in $(C_0(X))'$, we are back to the setting (3.8). By Theorem A.54 we may find a metric d compatible with weak star convergence in a large ball of $(C_0(X))'$.

In particular, we have

$$\lim_{n \to \infty} \lim_{k \to \infty} d(v_{n,k}, v) = 0.$$

Let

$$\ell := \liminf_{n \to \infty} \lim_{k \to \infty} I\left(v_{n,k}\right).$$

We now use a *diagonalization argument* as follows. Choose n_1 such that

$$\left| \ell - \lim_{k \to \infty} I\left(v_{n_1,k}\right) \right| + \lim_{k \to \infty} d\left(v_{n_1,k}, v\right) \leq 1,$$

and let k_1 be such that

$$\left| I\left(v_{n_1,k_1}\right) - \lim_{k \to \infty} I\left(v_{n_1,k}\right) \right| + \left| d\left(v_{n_1,k_1}, v\right) - \lim_{k \to \infty} d\left(v_{n_1,k}, v\right) \right| \leq 1.$$

Then

$$\left| \ell - I\left(v_{n_1,k_1}\right) \right| + d\left(v_{n_1,k_1}, v\right) \leq 2.$$

Choose $n_2 > n_1$ such that

$$\left| \ell - \lim_{k \to \infty} I\left(v_{n_2,k}\right) \right| + \lim_{k \to \infty} d\left(v_{n_2,k}, v\right) \leq \frac{1}{2},$$

and let $k_2 > k_1$ be such that

$$\left| I\left(v_{n_2,k_2}\right) - \lim_{k \to \infty} I\left(v_{n_2,k}\right) \right| + \left| d\left(v_{n_2,k_2}, v\right) - \lim_{k \to \infty} d\left(v_{n_2,k}, v\right) \right| \leq \frac{1}{2}.$$

Note that

$$\left| \ell - I\left(v_{n_2,k_2}\right) \right| + d\left(v_{n_2,k_2}, v\right) \leq 1.$$

Continuing in this way we construct a diagonalized sequence $\left\{ v_{n_j,k_j} \right\}$ such that

$$\lim_{j \to \infty} I\left(v_{n_j,k_j}\right) = \lim_{n \to \infty} \lim_{k \to \infty} I\left(v_{n,k}\right), \quad \lim_{j \to \infty} d\left(v_{n_j,k_j}, v\right) = 0.$$

In view of (3.11) and by Theorem 2.29 and the Dunford–Pettis theorem, we deduce $v_{n_j,k_j} \rightharpoonup v$ in L^1, and so by the definition of $\mathcal{I}$ we conclude that

$$\lim_{n \to \infty} \mathcal{I}\left(v_n\right) + \varepsilon \geq \liminf_{n \to \infty} \lim_{k \to \infty} I\left(v_{n,k}\right) = \lim_{j \to \infty} I\left(v_{n_j,k_j}\right) \geq \mathcal{I}\left(v\right).$$

It now suffices to let $\varepsilon \to 0^+$ to conclude the proof.

3.2 The Direct Method

Let V be a normed space and let $I : V \to [-\infty, \infty]$ be not identically equal to ∞. Tonelli's direct method provides conditions on V and I that ensure the existence of a minimum point for I. The procedure may be reduced to four steps:

Step 1 Consider a *minimizing sequence* $\{v_n\} \subset V$, that is, a sequence such that
$$\lim_{n\to\infty} I(v_n) = \inf_{v\in V} I.$$

Step 2 Prove that $\{v_n\}$ admits a subsequence $\{v_{n_k}\}$ that converges with respect to some (possibly weaker) topology τ to some point $v_0 \in V$.

Step 3 Establish the sequential lower semicontinuity of I with respect to τ.

Step 4 In view of Steps 1–3, conclude that v_0 is a minimum of I because

$$\inf_{v\in V} I = \lim_{n\to\infty} I(v_n) = \lim_{k\to\infty} I(v_{n_k}) \geq I(v_0) \geq \inf_{v\in V} I.$$

In the above program the challenge usually resides in guaranteeing that Step 3 is satisfied, while, as we will see below, Step 2 usually can be easily proved through natural growth and coercivity conditions of I. Indeed, if I is coercive and if $\{v_n\}$ is a minimizing sequence, then $\{v_n\}$ must be bounded, and under condition (3.7) or (3.8) of the previous subsection, this suffices to ensure sequential compactness with respect to τ.

In most applications Step 3 fails for I, and this leads us to an important topic at the core of the calculus of variations, namely the introduction of a relaxed function $\mathcal{E}'$ that is related to I as follows:

(a) $\mathcal{E}'$ is sequentially lower semicontinuous with respect to τ;
(b) $\mathcal{E}'$ inherits coercivity from I;
(c) $\min_{v\in V} \mathcal{E}' = \inf_{v\in V} I$.

Note that in (c) the existence of a minimum $v_0 \in V$ of $\mathcal{E}'$ is guaranteed by (a), (b), and the direct method. In view of Proposition 3.15 (see also Proposition 3.16), a natural candidate for $\mathcal{E}'$ is the function $\mathrm{slsc}_\tau I$ introduced in the previous subsection. Indeed, under the hypotheses of Proposition 3.16

$$\inf_{v\in V} I \geq \min_{v\in V} \mathrm{slsc}_\tau I = \mathrm{slsc}_\tau I(v_0)$$
$$= \inf_{\{v_n\}} \left\{ \liminf_{n\to\infty} I(v_n) : \{v_n\} \subset V, \, v_n \xrightarrow{\tau} v_0 \right\} \geq \inf_{v\in V} I,$$

where in the second equality we have used Proposition 3.16.

The next task is then to study under what conditions

$$I(v_0) = \mathrm{slsc}_\tau I(v_0),$$

i.e.,

$$I(v_0) = \min_{v\in V} I.$$

We will not address this topic in this text (see, however, Section 5.5).

4

Convex Analysis

Where there is matter, there is geometry. (Ubi materia, ibi geometria)

Johannes Kepler (1571–1630)

4.1 Convex Sets

We recall that a subset E of a vector space V is *convex* if for all $v_1, v_2 \in E$ and $\theta \in (0, 1)$ we have

$$\theta v_1 + (1 - \theta) v_2 \in E.$$

An induction argument shows that if $v_i \in E$, $i = 1, \ldots, n$, then $z := \sum_{i=1}^{n} \theta_i v_i \in E$, whenever $\sum_{i=1}^{n} \theta_i = 1$, $\theta_i \geq 0$. The vector z is called a *convex combination* of $v_1, \ldots, v_n$.

The entire space V and the empty set are two examples of convex sets. The arbitrary intersection of convex sets is still convex, but in general the union is not (the simplest example is the union of two disjoint closed segments on the real line).

Similarly, E is said to be *affine* if for all $v_1, v_2 \in E$ and $\theta \in \mathbb{R}$ we have

$$\theta v_1 + (1 - \theta) v_2 \in E.$$

Remark 4.1. Every affine set $E \subset V$ can be written as $v_0 + W$, where $v_0 \in E$ and W is a vector subspace of V.

Given any set $E \subset V$, the *convex hull* $\mathrm{co}\,(E)$ is the intersection of all convex sets that contain E.

Proposition 4.2. *Let V be a vector space and let $E \subset V$. Then*

$$\mathrm{co}\,(E) = \left\{ \sum_{i=1}^{n} \theta_i v_i \,:\, n \in \mathbb{N},\ \sum_{i=1}^{n} \theta_i = 1,\ \theta_i \geq 0,\ v_i \in E,\ i = 1, \ldots, n \right\}.$$

$$(4.1)$$

Proof. If F is any convex set that contains E, then it must contain all convex combinations of elements of E, and so

$$F \supset \left\{ \sum_{i=1}^{n} \theta_i v_i \,:\, n \in \mathbb{N},\ \sum_{i=1}^{n} \theta_i = 1,\ \theta_i \geq 0,\ v_i \in E,\ i = 1, \ldots, n \right\} =: G.$$

Since this holds for all convex sets containing E, it follows that

$$\mathrm{co}\,(E) \supset G.$$

To prove the opposite inclusion it suffices to show that G is convex and contains E. The latter assertion follows from the fact that if $u \in E$, then we can take $n = 1$ and $\theta_1 = 1$. To show that G is convex, let $0 \leq \theta \leq 1$ and let u, $v \in V$ be of the form

$$u = \sum_{i=1}^{n} \theta_i v_i, \quad v = \sum_{j=1}^{l} s_j w_j,$$

where $\sum_{i=1}^{n} \theta_i = \sum_{j=1}^{l} s_j = 1$, $\theta_i, s_j \geq 0$, $v_i, w_j \in E$, $i = 1, \ldots, n$, $j = 1, \ldots, l$. Note that without loss of generality, we may always assume that $n = l$. Indeed, if $n \neq l$, say $n > l$, then it suffices to set $s_{l+1}, \ldots, s_n := 0$ and $w_{l+1}, \ldots, w_n := w_1$. Then

$$\theta u + (1 - \theta) v = \sum_{i=1}^{n} \theta \theta_i v_i + \sum_{i=1}^{n} (1 - \theta)\, s_i w_i,$$

which is still a convex combination of elements of E, and so it belongs to G.

Note that without loss of generality, in (4.1) one may consider only positive coefficients θ_i.

Remark 4.3. If the space V is a topological vector space, then the convex hull of an open set $U \subset V$ is open. To see this, for any $n \in \mathbb{N}$ fix $\Lambda := \{\theta_1, \ldots, \theta_n\}$ with $\sum_{i=1}^{n} \theta_i = 1$ and $\theta_i > 0$ for all $i = 1, \ldots, n$ and consider the set

$$U_\Lambda := \left\{ \sum_{i=1}^{n} \theta_i v_i \,:\, v_i \in U,\ i = 1, \ldots, n \right\}.$$

Since addition and positive scalar multiplication are both open maps, it follows that U_Λ is an open set. By the previous proposition,

$$\mathrm{co}\,(U) = \bigcup_\Lambda U_\Lambda,$$

and so $\mathrm{co}(U)$ is open.

Analogously, the *affine hull* aff (E) is the intersection of all affine sets that contain E. Reasoning as in the proof of Proposition 4.2, it can be shown that

$$\text{aff}\,(E) = \left\{ \sum_{i=1}^{n} \theta_i v_i : n \in \mathbb{N},\ \sum_{i=1}^{n} \theta_i = 1,\ \theta_i \in \mathbb{R},\ v_i \in E,\ i = 1,\dots,n \right\}.$$

Exercise 4.4. Prove that if V is a topological vector space, then the interior and the closure of a convex set are also convex sets.

Given any set E of a topological vector space V, the intersection of all closed convex sets that contain E coincides with the closure of the convex hull co (E), and is denoted by $\overline{\text{co}\,(E)}$. The *relative interior of E with respect to* aff (E), denoted by $\text{ri}_{\text{aff}}\,(E)$, is the set of points $v \in E$ such that $A \cap \text{aff}\,(E) \subset E$ for some open set $A \subset V$ with $v \in A$. The *relative boundary of E with respect to* aff (E), denoted by $\text{rb}_{\text{aff}}\,(E)$, is the set $\overline{E} \setminus \text{ri}_{\text{aff}}\,(E)$.

Exercise 4.5. Let $C \subset \mathbb{R}^N$ be a nonempty convex set and let $z \in C$ and $z_0 \in \text{ri}_{\text{aff}}\,(C)$.

(i) Prove that $z_0 + t\,(z_0 - z) \in \text{aff}\,(C)$ for all $t \in \mathbb{R}$.
(ii) Prove that the function $g : \mathbb{R} \to \text{aff}\,(C)$, defined by $g\,(t) := z_0 + t\,(z_0 - z)$, $t \in \mathbb{R}$, is continuous.
(iii) Prove that $z_0 + t\,(z_0 - z) \in \text{ri}_{\text{aff}}\,(C)$ for all t sufficiently small.

Exercise 4.6. Let $C \subset \mathbb{R}^N$ be a nonempty convex set and let $z \in C$. Prove that

$$\text{ri}_{\text{aff}}\,(C) = z + \text{ri}_{\text{aff}}\,(-z + C).$$

Proposition 4.7. *Let $C \subset \mathbb{R}^N$ be a nonempty convex set. Then $\text{ri}_{\text{aff}}\,(C)$ is convex and nonempty. Moreover, if $z \in \overline{C}$ and $z_0 \in \text{ri}_{\text{aff}}\,(C)$, then*

$$\theta z + (1 - \theta)\,z_0 \in \text{ri}_{\text{aff}}\,(C)$$

for all $0 \leq \theta < 1$.

Proof. **Step 1:** If C is a singleton, then $\text{ri}_{\text{aff}}\,(C) = C$, and so there is nothing to prove. Thus assume that C has at least two elements. By Exercise 4.6, without loss of generality, we may assume that $0 \in C$, so that aff (C) is a subspace of $\mathbb{R}^m$. In particular, $\text{ri}_{\text{aff}}\,(C)$ is relatively open in aff (C). Since $\mathbb{R}^N$ is finite-dimensional, there is a finite maximal (with respect to inclusion) set of linearly independent vectors $z_1,\dots,z_k \in \text{ri}_{\text{aff}}\,(C)$. The vectors $z_1,\dots,z_k$ are a basis in aff (C), and thus we can define on aff (C) the norm

$$\|w\|_{\text{aff}(C)} := \max_{1 \leq i \leq k} |s_i|,$$

where

$$w = \sum_{i=1}^{k} s_i z_i \in \text{aff}\,(C).$$

Since $0 \in C$, any point of the form

$$w = \sum_{i=1}^{k} \theta_i z_i = 0 \left(1 - \sum_{i=1}^{k} \theta_i\right) + \sum_{i=1}^{k} \theta_i z_i,$$

where $\sum_{i=1}^{k} \theta_i \leq 1$ and $\theta_i \geq 0$ for all $i = 1, \ldots, k$, belongs to C. In particular, the point

$$w_0 := \sum_{i=1}^{k} \frac{1}{2k} z_i$$

belongs to C. We claim that $w_0 \in \mathrm{ri}_{\mathrm{aff}}(C)$. To see this, consider the open ball

$$B_{\mathrm{aff}(C)}\left(w_0, \frac{1}{2k}\right) := \left\{w \in \mathrm{aff}(C) : \|w - w_0\|_{\mathrm{aff}(C)} < \frac{1}{2k}\right\}.$$

If $w \in B_{\mathrm{aff}(C)}\left(w_0, \frac{1}{2k}\right)$, then we may write $w = \sum_{i=1}^{k} s_i z_i$, where

$$\max_{1 \leq i \leq k} \left|s_i - \frac{1}{2k}\right| < \frac{1}{2k}.$$

Then $s_i \geq 0$ for all $i = 1, \ldots, k$ and

$$\sum_{i=1}^{k} s_i < \sum_{i=1}^{k} \frac{1}{k} = 1,$$

which shows that $w \in C$. Thus $B_{\mathrm{aff}(C)}\left(w_0, \frac{1}{2k}\right) \subset C$, and so w_0 belongs to the interior of C relative to $\mathrm{aff}(C)$, or equivalently, $w_0 \in \mathrm{ri}_{\mathrm{aff}}(C)$.

Step 2: To prove the second part, let $z \in \overline{C}$ and $z_0 \in \mathrm{ri}_{\mathrm{aff}}(C)$. Fix $0 < \theta < 1$. If C is a singleton, then $\mathrm{ri}_{\mathrm{aff}}(C) = C = \overline{C}$, and so there is nothing to prove. Thus assume that C has at least two elements. Hence, as in the previous step there is no loss of generality in assuming that $0 \in C$. Let $z_1, \ldots, z_k \in \mathrm{ri}_{\mathrm{aff}}(C)$ and $\|\cdot\|_{\mathrm{aff}(C)}$ be defined as in the previous step. Since $z_0 \in \mathrm{ri}_{\mathrm{aff}}(C)$, there exists an open ball $B_{\mathrm{aff}(C)}(z_0, r) \subset C$. Note that since the affine hull of C is a closed set, it must contain $\overline{C}$, and consequently it coincides with the affine hull of $\overline{C}$. Thus $z \in \mathrm{aff}(C)$, and we may find a sequence $\{w_n\} \subset C$ such that $\|z - w_n\|_{\mathrm{aff}(C)} \to 0$. We claim that

$$B_{\mathrm{aff}(C)}(\theta z_0 + (1 - \theta) z, \theta r) \subset C.$$

To see this, note that if $w \in B_{\mathrm{aff}(C)}(\theta z_0 + (1 - \theta) z, \theta r)$, then

$$\|w - (\theta z_0 + (1 - \theta) z)\|_{\mathrm{aff}(C)} < \theta r,$$

or equivalently,

$$\left\|\frac{w}{\theta} - \frac{(1 - \theta)}{\theta} z - z_0\right\|_{\mathrm{aff}(C)} < r.$$

Since $\|z - w_n\|_{\text{aff}(C)} \to 0$ we may find $n \in \mathbb{N}$ so large that

$$\left\| \frac{w}{\theta} - \frac{(1 - \theta)}{\theta} w_n - z_0 \right\|_{\text{aff}(C)} < r.$$

Thus $\frac{w}{\theta} - \frac{(1-\theta)}{\theta} w_n \in B_{\text{aff}(C)} (z_0, r) \subset C$, which implies that

$$\frac{w}{\theta} - \frac{(1 - \theta)}{\theta} w_n = \xi \in C.$$

In turn, $w = (1 - \theta) w_n + \theta \xi \in C$, and the proof is complete.

Example 4.8. Note that if $C_1 \subset C_2$ are two nonempty convex sets, then in general one cannot conclude that

$$\text{ri}_{\text{aff}} (C_1) \subset \text{ri}_{\text{aff}} (C_2).$$

Indeed, let C_2 be the closed unit cube in $\mathbb{R}^3$ and let C_1 be one of its faces. Then $\text{ri}_{\text{aff}} (C_2)$ is the open unit cube, while $\text{ri}_{\text{aff}} (C_1)$ is the face without its four edges. Hence $\text{ri}_{\text{aff}} (C_1)$ and $\text{ri}_{\text{aff}} (C_2)$ are disjoint and nonempty.

Proposition 4.9. *Let C_1, C_2 be two nonempty convex sets of $\mathbb{R}^m$. Then the following three conditions are equivalent:*

(i) $\overline{C_1} = \overline{C_2}$;
(ii) $\text{ri}_{\text{aff}} (C_1) = \text{ri}_{\text{aff}} (C_2)$;
(iii) $\text{ri}_{\text{aff}} (C_1) \subset C_2 \subset \overline{C_1}$.

Proof. **Step 1:** We show that for any nonempty convex set $C \subset \mathbb{R}^N$,

$$\overline{C} = \overline{\text{ri}_{\text{aff}} (C)}, \quad \text{ri}_{\text{aff}} \left(\overline{C} \right) = \text{ri}_{\text{aff}} (C).$$

Since $\text{ri}_{\text{aff}} (C) \subset C$, we have that $\overline{\text{ri}_{\text{aff}} (C)} \subset \overline{C}$. To prove the converse inclusion, let $z \in \overline{C}$ and let $z_0 \in \text{ri}_{\text{aff}} (C)$. Note that in view of Proposition 4.7 such z_0 exists and

$$\theta z + (1 - \theta) z_0 \in \text{ri}_{\text{aff}} (C)$$

for all $0 \le \theta < 1$. This implies that $z \in \overline{\text{ri}_{\text{aff}} (C)}$. Hence the first identity is established.

Since the affine hull of C is a closed set, it must contain $\overline{C}$, and consequently, it coincides with the affine hull of $\overline{C}$. Therefore

$$\text{ri}_{\text{aff}} (C) \subset \text{ri}_{\text{aff}} \left(\overline{C} \right).$$

To prove the converse inclusion, let $z \in \text{ri}_{\text{aff}} \left(\overline{C} \right)$ and let $z_0 \in \text{ri}_{\text{aff}} (C)$. If $z = z_0$ then $z \in \text{ri}_{\text{aff}} (C)$, and there is nothing to prove. Thus assume that $z \ne z_0$ and let

$$w_t := tz + (1 - t) z_0, \quad t > 1.$$

Since $z \in \text{ri}_{\text{aff}}(\overline{C})$, if $t - 1$ is sufficiently small, then $w_t \in \text{ri}_{\text{aff}} \overline{C} \subset \overline{C}$. Hence

$$z = \frac{1}{t} w_t + \left(1 - \frac{1}{t}\right) z_0,$$

and so $z \in \text{ri}_{\text{aff}}(C)$, again by Proposition 4.7.

Step 2: If (i) holds, then by Step 1,

$$\text{ri}_{\text{aff}}(C_1) = \text{ri}_{\text{aff}}(\overline{C_1}) = \text{ri}_{\text{aff}}(\overline{C_2}) = \text{ri}_{\text{aff}}(C_2),$$

that is, (ii) is true.

Similarly, if (ii) holds, then again by Step 1,

$$\overline{C_1} = \overline{\text{ri}_{\text{aff}}(C_1)} = \overline{\text{ri}_{\text{aff}}(C_2)} = \overline{C_2},$$

which is (i).

If (iii) holds, then by Step 1,

$$\overline{\text{ri}_{\text{aff}}(C_1)} \subset \overline{C_2} \subset \overline{C_1} = \overline{\text{ri}_{\text{aff}}(C_1)},$$

and (i) is satisfied.

Finally, if (i) holds, then so does (ii), and we deduce that

$$\text{ri}_{\text{aff}}(C_1) = \text{ri}_{\text{aff}}(C_2) \subset C_2 \subset \overline{C_2} = \overline{C_1},$$

which is (iii).

Exercise 4.10. Let C_1, C_2 be two nonempty convex sets of $\mathbb{R}^m$ and let $t \in \mathbb{R}$. Prove that

(i) $\text{ri}_{\text{aff}}(tC_1) = t \, \text{ri}_{\text{aff}}(C_1)$;
(ii) $\text{ri}_{\text{aff}}(C_1 + C_2) = \text{ri}_{\text{aff}}(C_1) + \text{ri}_{\text{aff}}(C_2)$.

In the remainder of this section we discuss some properties of convex cones that will be used to prove Lemma 7.7 in Chapter 7.

Definition 4.11. *A subset K of $\mathbb{R}^m$ is a* convex cone *if it is closed under addition and positive scalar multiplication. The* polar K^* *of a convex cone is the set*

$$K^* := \{z \in \mathbb{R}^m : z \cdot \xi \leq 0 \text{ for all } \xi \in K\}.$$

Note that in general there is no guarantee that a convex cone contains the origin.

Remark 4.12. Given a set $E \subset \mathbb{R}^m$, the smallest convex cone K containing E is given by

$$K' := \left\{ \sum_{i=1}^{n} t_i z_i : n \in \mathbb{N}, \, t_i > 0, \, z_i \in E, \, i = 1, \ldots, n \right\}.$$

Indeed, since K is closed under addition and positive scalar multiplication, it must contain K'. On the other hand, it follows from its definition that K' is a convex cone containing E.

Proposition 4.13. *If $K \subset \mathbb{R}^m$ is a convex cone, then*

$$(K^*)^* = \overline{K}.$$

Proof. If $\xi \in K^*$ then

$$K \subset \{z \in \mathbb{R}^m : z \cdot \xi \le 0\},$$

and thus

$$K \subset \bigcap_{\xi \in K^*} \{z \in \mathbb{R}^m : z \cdot \xi \le 0\} = (K^*)^*.$$

Since the polar set is always closed, we conclude that

$$\overline{K} \subset (K^*)^*.$$

Conversely, suppose that $z_0 \notin \overline{K}$. By the second geometric form of the Hahn–Banach theorem there exist $z' \in \mathbb{R}^m$, $\alpha \in \mathbb{R}$, and $\varepsilon > 0$ such that

$$z' \cdot z \le \alpha - \varepsilon \quad \text{for all } z \in \overline{K} \text{ and } z' \cdot z_0 \ge \alpha + \varepsilon.$$

Since K is a cone, if $z \in K$ and $t > 0$, then $tz \in K$, and so

$$z' \cdot tz \le \alpha - \varepsilon. \tag{4.2}$$

Hence

$$z' \cdot z \le \frac{\alpha - \varepsilon}{t},$$

and letting $t \to \infty$ we conclude that $z' \cdot z \le 0$ for all $z \in K$, and so $z' \in K^*$. Also, letting $t \to 0^+$ in (4.2), it follows that $\alpha \ge \varepsilon$, and therefore $z' \cdot z_0 \ge 2\varepsilon > 0$. We conclude that $z_0 \notin (K^*)^*$.

Definition 4.14. *Given a set $E \subset \mathbb{R}^m$, the* convex cone generated by E *is the convex cone obtained by adjoining the origin to the smallest convex cone containing E.*

Theorem 4.15. *Let $C \subset \mathbb{R}^m$ be a nonempty closed convex set not containing the origin, and let K be the convex cone generated by C. Then*

$$\overline{K} = \{tz : t > 0,\ z \in C\} \cup \{z : tz + C \subset C \text{ for all } t \ge 0\}.$$

Proof. **Step 1:** We claim that

$$K = \{tz : t > 0,\ z \in C\} \cup \{0\}.$$

By Remark 4.12,

$$K = \left\{ \sum_{i=1}^{n} t_i z_i : n \in \mathbb{N},\ t_i \ge 0,\ z_i \in C,\ i = 1, \ldots, n \right\}.$$

If $z \in K \setminus \{0\}$ then

$$z = \sum_{i=1}^{n} t_i z_i$$

with $t_i > 0$, $z_i \in C$, $i = 1, \ldots, n$, for some $n \in \mathbb{N}$. Setting $t := \sum_{k=1}^{n} t_k$, then $t > 0$, and with

$$\theta_i := \frac{t_i}{t} \quad i = 1, \ldots, n, \quad z_0 := \sum_{i=1}^{n} \theta_i z_i,$$

it follows that $z_0 \in C$ and $z = t z_0$. This proves the claim.

Step 2: We show first that

$$\overline{K} \supset \{z : tz + C \subset C \text{ for all } t \geq 0\}. \tag{4.3}$$

Indeed, fix $z \in \mathbb{R}^m$ for which $tz + C \subset C$ for all $t \geq 0$ and let $w \in C$. Then $nz + w \in C$ for all $n \in \mathbb{N}$, and so

$$z + \frac{1}{n} w \in K$$

for all n. By letting $n \to \infty$ we obtain that z must belong to $\overline{K}$. Thus (4.3) holds, and so, also by Step 1, we have

$$\overline{K} \supset \{tz : t > 0, z \in C\} \cup \{z : tz + C \subset C \text{ for all } t \geq 0\} =: D. \tag{4.4}$$

To prove the converse inclusion, consider $z \in \overline{K}$. If $z = 0$ then $tz + C \subset C$ for all $t \geq 0$.

If $z \neq 0$, then by Step 1 there exist sequences $\{t_n\} \subset (0, \infty)$ and $\{z_n\} \subset C$ such that $t_n z_n \to z$. If $\{t_n\}$ were unbounded, then (up to the extraction of a subsequence) $z_n \to 0$, and since C is closed we would get $0 \in C$, which is in contradiction with the hypotheses. Therefore, up to a subsequence, $t_n \to t_0$ for some $t_0 \geq 0$. Consider first the case $t_0 > 0$. Since $z_n \to \frac{z}{t_0}$ we deduce that $\frac{z}{t_0} \in C$, and so $z = t_0 \frac{z}{t_0}$ belongs to the set D defined in (4.4). If $t_0 = 0$, then we claim that given $t > 0$ and $w \in C$ we have $tz + w \in C$. Indeed,

$$tz + w = \lim_{n \to \infty} t t_n z_n + w = \lim_{n \to \infty} (t t_n z_n + (1 - t t_n) w),$$

which, together with the fact that $t t_n z_n + (1 - t t_n) w \in C$ for all n so large that $t t_n \in [0, 1)$, and because C is closed, entails $tz + w \in C$.

4.2 Separating Theorems

In this section we prove some separation theorems for convex sets in $\mathbb{R}^N$. Their counterparts in the infinite-dimensional setting are the Hahn–Banach theorems, which are stated in the appendix.

Theorem 4.16. *Let C, $K \subset \mathbb{R}^m$ be nonempty disjoint convex sets, with C closed and K compact. Then there exist a vector $b \in \mathbb{R}^m \setminus \{0\}$ and two numbers $\alpha \in \mathbb{R}$ and $\varepsilon > 0$ such that*

$$b \cdot z \leq \alpha - \varepsilon \quad \text{for all } z \in C \text{ and } b \cdot z \geq \alpha + \varepsilon \quad \text{for all } z \in K.$$

Proof. The distance function

$$z \in \mathbb{R}^m \mapsto \text{dist}\,(z, C)$$

is a continuous function (it is actually Lipschitz), and so by the Weierstrass theorem it admits a minimum on K at some point $z_0 \in K$. Moreover, since C is closed, there exists a point $w_0 \in C$ such that

$$|z_0 - w_0| = \text{dist}\,(z_0, C).$$

Define $b = z_0 - w_0$ and observe that $b \neq 0$, since C and K are disjoint sets. Then

$$0 < |b|^2 = b \cdot (z_0 - w_0),$$

so that $b \cdot w_0 < b \cdot z_0$. We claim that

$$b \cdot w \leq b \cdot w_0 \quad \text{for all } w \in C \text{ and } b \cdot z_0 \leq b \cdot z \quad \text{for all } z \in K.$$

To see this, fix $w_1 \in C$. Since w_0 minimizes the distance (and so also the square of the distance) to z_0 over C, we have that

$$(z_0 - w) \cdot (z_0 - w) \geq (z_0 - w_0) \cdot (z_0 - w_0)$$

for all $w \in C$. In particular, taking $w := w_0 + \theta\,(w_1 - w_0) \in C$, $0 < \theta \leq 1$, in the previous inequality yields

$$0 \geq (z_0 - w_0) \cdot (z_0 - w_0) - (z_0 - w) \cdot (z_0 - w)$$
$$= 2\theta\,(z_0 - w_0) \cdot (w_1 - w_0) - \theta^2\,(w_1 - w_0) \cdot (w_1 - w_0).$$

Dividing by $\theta > 0$ and letting $\theta \to 0^+$ gives

$$0 \geq 2\,(z_0 - w_0) \cdot (w_1 - w_0) = 2b \cdot (w_1 - w_0),$$

which shows that $b \cdot w_1 \leq b \cdot w_0$.

A similar argument, which is left as an exercise, gives the analogous inequality for K and completes the proof.

Remark 4.17. The previous proof continues to hold for Hilbert spaces.

Exercise 4.18. Let $z_1, \dots, z_k \in \mathbb{R}^m$ be linearly independent vectors and let s_i, $s_i^{(n)} \in \mathbb{R}$, $i = 1, \dots, k$, $n \in \mathbb{N}$. Prove that if

$$\lim_{n \to \infty} \sum_{i=1}^{k} s_i^{(n)} z_i = \sum_{i=1}^{k} s_i z_i,$$

then $\lim_{n \to \infty} s_i^{(n)} = s_i$ for every $i = 1, \dots, k$.

Theorem 4.19. *Let $C_1, C_2 \subset \mathbb{R}^m$ be nonempty convex sets. Then there exist a vector $b \in \mathbb{R}^m \setminus \{0\}$ and $\alpha \in \mathbb{R}$ such that*

$$b \cdot z \leq \alpha \quad \text{for all } z \in C_1 \text{ and } b \cdot z \geq \alpha \quad \text{for all } z \in C_2,$$

and $C_1 \cup C_2$ is not contained in the hyperplane $\{z \in \mathbb{R}^m : b \cdot z = \alpha\}$ if and only if $\mathrm{ri}_{\mathrm{aff}}(C_1) \cap \mathrm{ri}_{\mathrm{aff}}(C_2) = \emptyset$.

Proof. **Step 1:** Assume that $C_1, C_2 \subset \mathbb{R}^N$ are nonempty convex sets with $\mathrm{ri}_{\mathrm{aff}}(C_1) \cap \mathrm{ri}_{\mathrm{aff}}(C_2) = \emptyset$. Define

$$C := C_1 - C_2.$$

By Exercise 4.10,

$$\mathrm{ri}_{\mathrm{aff}}(C) = \mathrm{ri}_{\mathrm{aff}}(C_1) - \mathrm{ri}_{\mathrm{aff}}(C_2),$$

and so by hypothesis $0 \notin \mathrm{ri}_{\mathrm{aff}}(C)$. To complete the proof in this case it suffices to prove that there exists $b \in \mathbb{R}^m \setminus \{0\}$ such that $b \cdot z \geq 0$ for all $z \in C$ with strict inequality for at least one element of C. There are two cases. If $0 \notin \overline{C}$ then it suffices to apply Theorem 4.16 to the closed set $\overline{C}$ and the compact set $\{0\}$. Thus assume that $0 \in \overline{C}$. Define the set

$$E = \bigcup_{t>0} t \, \mathrm{ri}_{\mathrm{aff}}(C).$$

Then E is convex, $\mathrm{ri}_{\mathrm{aff}}(C) \subset E \subset \mathrm{aff}(E)$, and $0 \notin E$. Since $\mathrm{ri}_{\mathrm{aff}}(C)$ is nonempty by Proposition 4.7 and $\mathbb{R}^N$ is finite-dimensional, there is a finite maximal (with respect to inclusion) set of linearly independent vectors $z_1, \ldots, z_k \in \mathrm{ri}_{\mathrm{aff}}(C)$. Again by Proposition 4.7 we have that $\mathrm{ri}_{\mathrm{aff}}(C)$ is convex, and so the vector

$$z_0 := \sum_{i=1}^{k} \frac{1}{k} z_i$$

belongs to $\mathrm{ri}_{\mathrm{aff}}(C)$. We claim that $-z_0 \notin \overline{E}$. Indeed, assume by contradiction that $-z \in \overline{E}$ and find a sequence $\{w_n\} \subset E$ converging to $-z_0$. Then by definition of E we may write each w_n as $w_n = t_n \xi_n$, where $t_n > 0$ and $\xi_n \in \mathrm{ri}_{\mathrm{aff}}(C)$. Since $z_1, \ldots, z_k$ form a maximal set of linearly independent vectors in $\mathrm{ri}_{\mathrm{aff}}(C)$, each ξ_n can be written as their linear combination (and so can $w_n = t_n \xi_n$). Thus we may write

$$w_n = \sum_{i=1}^{k} s_i^{(n)} z_i$$

for some $s_i^{(n)} \in \mathbb{R}$, $i = 1, \ldots, k$. Hence

$$\lim_{n \to \infty} \sum_{i=1}^{k} s_i^{(n)} z_i = \lim_{n \to \infty} w_n = -z_0 = \sum_{i=1}^{k} \left(-\frac{1}{k}\right) z_i.$$

By Exercise 4.18 this implies that $\lim_{n\to\infty} s_i^{(n)} = -\frac{1}{k}$ for every $i = 1,\ldots,k$. Fix $n \in \mathbb{N}$ so large that $s_i^{(n)} < 0$ for every $i = 1,\ldots,k$ and set

$$s := \sum_{i=1}^{k} s_i^{(n)} < 0.$$

By the convexity of the set E we have that

$$0 = \frac{1}{1-s} w_n + \sum_{i=1}^{k} \left(-\frac{s_i^{(n)}}{1-s} \right) z_i \in E,$$

which is a contradiction.

This shows that $-z_0 \notin \overline{E}$. We are now in a position to apply Theorem 4.16 to the closed set $\overline{E}$ and the compact set $\{-z_0\}$ to find a vector $b \in \mathbb{R}^m \setminus \{0\}$ and two numbers $\alpha \in \mathbb{R}$ and $\varepsilon > 0$ such that

$$b \cdot z \le \alpha - \varepsilon \quad \text{for all } z \in \overline{E} \text{ and } b \cdot (-z_0) \ge \alpha + \varepsilon.$$

By the definition of E, for any $z \in \mathrm{ri}_{\mathrm{aff}}(C)$ and $t > 0$ we have that $tz \in E$, and so from the previous inequality we get

$$b \cdot z \le \frac{\alpha - \varepsilon}{t} \quad \text{for all } t > 0.$$

Letting $t \to 0^+$ and $t \to \infty$ yield $\alpha - \varepsilon \ge 0$ and $b \cdot z \le 0$, respectively. Moreover, $b \cdot z_0 \le -(\alpha + \varepsilon) < 0$. Hence we have proved that $b \cdot z \le 0$ for all $z \in \mathrm{ri}_{\mathrm{aff}}(C)$ with the strict inequality at $z_0 \in \mathrm{ri}_{\mathrm{aff}}(C)$.

Step 2: To prove the converse implication assume that there exist a vector $b \in \mathbb{R}^m \setminus \{0\}$ and $\alpha \in \mathbb{R}$ such that

$$b \cdot z \le \alpha \quad \text{for all } z \in C_1 \text{ and } b \cdot z \ge \alpha \quad \text{for all } z \in C_2 \qquad (4.5)$$

and $C_1 \cup C_2$ is not contained in the hyperplane $\{z \in \mathbb{R}^m : b \cdot z = \alpha\}$. As in the previous step define

$$C := C_1 - C_2.$$

By Exercise 4.10,

$$\mathrm{ri}_{\mathrm{aff}}(C) = \mathrm{ri}_{\mathrm{aff}}(C_1) - \mathrm{ri}_{\mathrm{aff}}(C_2).$$

Thus it suffices to show that $0 \notin \mathrm{ri}_{\mathrm{aff}}(C)$. By (4.5),

$$b \cdot z \le 0 \quad \text{for all } z \in C$$

with the strict inequality for at least one element $z_0 \in C$. Assume by contradiction that $0 \in \mathrm{ri}_{\mathrm{aff}}(C)$. Then by Exercise 4.5,

$$0 + \varepsilon (0 - z_0) \in C$$

for all $\varepsilon > 0$ sufficiently small. This implies that $b \cdot (-\varepsilon z_0) \le 0$, which is a contradiction since $b \cdot z_0 < 0$.

As a consequence of the previous separation theorem we obtain the following result.

Corollary 4.20. *Let $C_1 \subset C_2 \subset \mathbb{R}^N$ be nonempty convex sets. Then there exist $b \in \mathbb{R}^N \setminus \{0\}$ and $\alpha \in \mathbb{R}$ such that*

$$b \cdot z = \alpha \quad \text{for all } z \in C_1 \text{ and } b \cdot z \geq \alpha \quad \text{for all } z \in C_2$$

if and only $C_1 \cap \mathrm{ri}_{\mathrm{aff}}(C_2) = \emptyset$.

4.3 Convex Functions

We now turn to the study of convex functions.

Definition 4.21. *A function $f : V \to [-\infty, \infty]$ is said to be*

(i) convex *if*
$$f(\theta v_1 + (1 - \theta) v_2) \leq \theta f(v_1) + (1 - \theta) f(v_2) \tag{4.6}$$

for all v_1, $v_2 \in V$ and $\theta \in (0, 1)$ for which the right-hand side is well-defined;

(ii) strictly convex *if*

$$f(\theta v_1 + (1 - \theta) v_2) < \theta f(v_1) + (1 - \theta) f(v_2)$$

for all v_1, $v_2 \in V$, $v_1 \neq v_2$, and $\theta \in (0, 1)$ for which the right-hand side is well-defined;

(iii) proper *if it is convex, does not take the value $-\infty$, and is not identically ∞;*

(iv) concave *(respectively strictly concave) if $-f$ is convex (respectively strictly convex).*

In (i) and (ii) the right-hand side is not defined only when $f(v_1) = \pm\infty$ and $f(v_2) = \mp\infty$.

If E is a subset of the vector space V, then a function $f : E \to [-\infty, \infty]$ is said to be *convex* if the extension

$$\bar{f}(v) := \begin{cases} f(v) & \text{if } v \in E, \\ \infty & \text{if } v \notin E, \end{cases}$$

is a convex function in V. Analogous definitions apply to the concept of strict convexity, concavity, and strict concavity.

Proposition 4.22. *Let V be a vector space. A function $f : V \to [-\infty, \infty]$ is convex if and only if epi f is a convex set.*

Proof. Assume that f is convex and let (v_1, s), $(v_2, t) \in \text{epi} f$ and $\theta \in (0, 1)$. We claim that

$$\theta(v_1, s) + (1 - \theta)(v_2, t) \in \text{epi} f.$$

Indeed, since $s \geq f(v_1)$, $t \geq f(z)$, it follows that

$$\theta s + (1 - \theta) t \geq \theta f(v_1) + (1 - \theta) f(v_2) \geq f(\theta v_1 + (1 - \theta) v_2),$$

where we have used the convexity of f. Hence the claim is proved.

Conversely, let $\text{epi} f$ be a convex set, let $v_1, v_2 \in V$, and let $\theta \in (0, 1)$. If $f(v_1) = \pm\infty$ and $f(v_2) = \mp\infty$, then the right-hand side of (4.6) is not well-defined.

If $f(v_1) = \infty$ and $f(v_2) > -\infty$ or $f(v_1) > -\infty$ and $f(v_2) = \infty$, then (4.6) holds. Thus assume that $f(v_1)$, $f(v_2) < \infty$ and let $s \geq f(v_1)$, $t \geq f(v_2)$. Since (v_1, s), $(v_2, t) \in \text{epi} f$ it follows from the convexity of $\text{epi} f$ that $\theta(v_1, s) + (1 - \theta)(v_2, t) \in \text{epi} f$, that is,

$$f(\theta v_1 + (1 - \theta) v_2) \leq \theta s + (1 - \theta) t.$$

The convexity of f follows by letting $s \searrow f(v_1)$ and $t \searrow f(v_2)$. $\quad\blacksquare$

Let V be a vector space. The *effective domain* of a function $f : V \to [-\infty, \infty]$ is the set

$$\text{dom}_e f := \{v \in V : f(v) < \infty\}.$$

We observe that if f is convex, then the effective domain of f is a convex set.

Remark 4.23. Note that if V is a topological vector space and if $f : V \to [-\infty, \infty]$ is a convex function, finite at some point in the interior of $\text{dom}_e f$, then f is proper. Indeed, suppose that $f(v_0) \in \mathbb{R}$ for some v_0 in the interior of $\text{dom}_e f$, and that $f(v) = -\infty$ for some $v \in V$. Let U be an open neighborhood of v_0 with $U \subset \text{dom}_e f$ and set $W := -v_0 + U$. Since W is a neighborhood of zero, it is absorbing, and so there exists $r > 0$ such that $r(v_0 - v) \in W$, i.e., $v_0 + r(v_0 - v) \in U$. Set $\theta := \frac{r}{1+r}$. Note that

$$v_0 = \theta v + (1 - \theta)(v_0 + r(v_0 - v)),$$

and due to the convexity of f we conclude that $f(v_0) = -\infty$, which is a contradiction.

Exercise 4.24. Prove that:

(i) The function $f : \mathbb{R}^m \to [0, \infty)$ defined by $f(z) := |z|^p$, $p > 0$, is convex if and only if $p \geq 1$, and is strictly convex if and only if $p > 1$. In particular, if $m = 1$, a, $b \geq 0$, and $p \geq 1$, then by the convexity of f,

$$\left(\frac{1}{2}a + \frac{1}{2}b\right)^p \leq \frac{1}{2}a^p + \frac{1}{2}b^p,$$

or equivalently,

$$(a + b)^p \leq 2^{p-1}(a^p + b^q).$$

(ii) The function $f : \mathbb{R}^m \to [0, \infty)$ defined by $f(z) := \sqrt{|z|^2 + 1}$ is strictly convex.

(iii) The function $f : \mathbb{R}^m \to \mathbb{R}$ defined by

$$f(z) := Az \cdot z,$$

where A is a symmetric matrix in $\mathbb{R}^{m \times m}$, is convex if and only if A is positive semidefinite.

(iv) The function

$$f(z) := \log z \quad \text{if } z > 0$$

is strictly concave. In particular, if a, $b > 0$, $1 < p < \infty$, and q is its conjugate exponent, then by the concavity of f,

$$\log(ab) = \frac{1}{p} \log a^p + \frac{1}{q} \log b^q \le \log\left(\frac{1}{p} a^p + \frac{1}{q} b^q\right),$$

or equivalently,

$$ab \le \frac{1}{p} a^p + \frac{1}{q} b^q.$$

This is known as *Young's inequality*. Note that, in view of the strict concavity of f, equality holds if and only if $a^p = b^q$.

(v) If V is a vector space, the *indicator function* of a set $E \subset V$ defined by

$$f(v) = I_E(v) := \begin{cases} 0 & \text{if } v \in E, \\ \infty & \text{if } v \notin E, \end{cases}$$

is a convex function if and only if the set E is convex.

Note that if f is convex and $c > 0$, then cf is still convex, the sum of two convex functions f and g is convex (we set $(f + g)(v) := +\infty$ whenever $f(v) = \pm\infty$ and $g(v) = \mp\infty$), and the pointwise supremum of an arbitrary family of convex functions is again a convex function. If f is convex and if $g : [-\infty, \infty] \to [-\infty, \infty]$ is convex and nondecreasing, then $g \circ f$ is convex.

Proposition 4.25. *Let V be a topological vector space and let $f : V \to [-\infty, \infty]$ be convex. Then lsc f is convex, and*

$$\operatorname{epi}(\operatorname{lsc} f) = \overline{\operatorname{epi} f}.$$

Proof. Since f is convex, then epi f is convex by Proposition 4.22, and by Proposition 3.10 we have

$$\operatorname{epi}(\operatorname{lsc} f) = \overline{\operatorname{epi} f};$$

hence epi $(\operatorname{lsc} f)$ is convex because it is the closure of a convex set, i.e., lsc f is convex.

Proposition 4.26. *Let V be a locally convex topological vector space and let $f : V \to [-\infty, \infty]$ be a lower semicontinuous convex function. Then*

(i) f is weakly lower semicontinuous;
(ii) if f assumes the value $-\infty$ at some point, then $f : V \to \{-\infty, \infty\}$.

Proof. (i) By Propositions 3.4 and 4.22 the set epi f is closed and convex, and so by Theorem A.47 it is weakly closed. Using once more Proposition 3.4 but now with the weak topology, we conclude that f is weakly lower semicontinuous.

(ii) Assume by contradiction that there exists $v_0 \in V$ such that $f(v_0) \in \mathbb{R}$. Since epi f is closed and convex by Propositions 3.10 and 4.22, we may apply the second geometric form of the Hahn–Banach theorem to the sets $K := \{(v_0, t_0)\}$, where $t_0 < f(v_0)$, and $C := $ epi f to find two numbers $\alpha \in \mathbb{R}$ and $\varepsilon > 0$ and a continuous linear functional $L : V \times \mathbb{R} \to \mathbb{R}$ of the form $L(v, t) = \langle v', v \rangle_{V', V} + \alpha_0 t$ for some $v' \in V'$ and $\alpha_0 \in \mathbb{R}$ such that

$$\langle v', v \rangle_{V', V} + \alpha_0 t \geq \alpha + \varepsilon \text{ for all } (v, t) \in \text{epi} f \text{ and } \langle v', v_0 \rangle_{V', V} + \alpha_0 t_0 \leq \alpha - \varepsilon.$$

In particular, since $(v_0, f(v_0)) \in \text{epi} f$, we have that

$$\langle v', v_0 \rangle_{V', V} + \alpha_0 f(v_0) \geq \alpha + \varepsilon > \alpha - \varepsilon \geq \langle v', v_0 \rangle_{V', V} + \alpha_0 t_0,$$

which gives

$$\alpha_0 (f(v_0) - t_0) > 0.$$

Since $t_0 < f(v_0)$, we conclude that $\alpha_0 > 0$, and so

$$t \geq \frac{\alpha + \varepsilon}{\alpha_0} - \frac{1}{\alpha_0} \langle v', v \rangle_{V', V} \text{ for all } (v, t) \in \text{epi} f.$$

In turn,

$$f(v) \geq \frac{\alpha + \varepsilon}{\alpha_0} - \frac{1}{\alpha_0} \langle v', v \rangle_{V', V} \text{ for all } v \in \text{dom}_e f,$$

which contradicts the fact that f assumes the value $-\infty$ at some point. $\quad\blacksquare$

Definition 4.27. *Let $V_1, \ldots, V_m$ be vector spaces. A function*

$$f : V_1 \times \ldots \times V_m \to [-\infty, \infty]$$

is separately convex if it is convex in each variable, i.e., if for every $i = 1, \ldots, m$, and for all $v_j \in V_j$, $j = 1, \ldots, m$, $j \neq i$, the function

$$f(v_1, \ldots, v_{i-1}, \cdot, v_{i+1}, \ldots, v_m) : V_i \to [-\infty, \infty]$$

is convex.

If $E \subset V_1 \times \ldots \times V_m$, we say that a function $f : E \to [-\infty, \infty]$ is separately convex if the function obtained by extending f as ∞ outside E is separately convex.

Note that a convex function defined on a Cartesian product of vector spaces is separately convex but the converse is not true in general.

Example 4.28. (i) Let $f : \mathbb{R}^2 \to \mathbb{R}$ be defined as

$$f(z) = f(z_1, z_2) := z_1 z_2.$$

Then f is convex (actually linear) in each variable but not convex.
(ii) Let $E := \{(z_1, z_2) \in \mathbb{R}^2 : z_1 z_2 \geq 0\}$ and let $f : \mathbb{R}^2 \to \mathbb{R}$ be defined as

$$f(z_1, z_2) := \begin{cases} 0 & \text{if } (z_1, z_2) \in E, \\ \infty & \text{otherwise.} \end{cases}$$

Then the restriction of f to each line through the origin is convex, but f is not convex.

4.4 Lipschitz Continuity in Normed Spaces

In order to study the regularity of convex functions we need to introduce the notion of Lipschitz and Hölder continuity.

Definition 4.29. *Let V be a normed space and let $E \subset V$. A function $f : E \to \mathbb{R}$ is said to be*

(i) Lipschitz continuous *if*

$$\text{Lip}(f; E) := \sup \left\{ \frac{|f(v) - f(w)|}{\|v - w\|} : v, w \in E, \; v \neq w \right\} < \infty;$$

(ii) locally Lipschitz continuous *if for every compact set $K \subset E$,*

$$\sup \left\{ \frac{|f(v) - f(w)|}{\|v - w\|} : v, w \in K, \; v \neq w \right\} < \infty;$$

(iii) Hölder continuous *with exponent $0 < \alpha < 1$ if*

$$|f|_{C^{0,\alpha}(E)} := \sup \left\{ \frac{|f(v) - f(w)|}{\|v - w\|^\alpha} : v, w \in E, \; v \neq w \right\} < \infty.$$

Exercise 4.30. The Weierstrass function

$$f(s) := \sum_{n=1}^{\infty} \frac{1}{2^n} \sin 2^n s, \quad s \in \mathbb{R},$$

satisfies the estimate

$$|f(s) - f(t)| \leq C |s - t| \log \frac{1}{|s - t|}$$

for all $s, t \in \mathbb{R}$, with $0 < |s - t| < 1$, and hence provides an example of a function that is Hölder continuous of any order $\alpha < 1$. Prove that f is not Lipschitz continuous, and actually it is nowhere differentiable (see [Ha16]).

Given a subset E of a normed space V and a function $f : E \to \mathbb{R}$, we define its *modulus of continuity* by

$$\omega(s) := \sup \{|f(v) - f(w)| : v, w \in E, \ \|v - w\| < s\}, \quad s > 0.$$

Note that f is uniformly continuous if and only if $\omega(s) \to 0$ as $s \to 0^+$. The *concave modulus of continuity* $\varpi : [0, \infty] \to [0, \infty]$ is defined as the smallest concave function above ω.

Remark 4.31. If $E = V$ and f is uniformly continuous, then it is always possible to replace ω with ϖ. Indeed, we claim that

$$\omega(s) \le \varpi(s) \le 2\omega(s) \tag{4.7}$$

for all $s \ge 0$. To see this, note that if $s_0 > 0$, then for all $n \in \mathbb{N}_0$ and $ns_0 \le s \le (n+1)s_0$ we have

$$\omega(s) \le \omega((n+1)s_0) \le (n+1)\omega(s_0) \le \omega(s_0) + s\frac{\omega(s_0)}{s_0},$$

where we have used the fact that ω is subadditive (this is a consequence of the definition of ω and the fact that $E = V$). Since the function

$$s \mapsto \omega(s_0) + s\frac{\omega(s_0)}{s_0}$$

is concave and above ω, it follows from the definition of ϖ that

$$\varpi(s) \le \omega(s_0) + s\frac{\omega(s_0)}{s_0}$$

for all $s \ge 0$, and taking $s = s_0$ we deduce (4.7) for all $s > 0$. For $s = 0$ it suffices to observe that ω is continuous and $\omega(0) = 0$, and so letting $s \to 0^+$ in (4.7) we obtain that $\varpi(0) = 0$.

The next result shows that any uniformly continuous function may be extended to all of V with the same concave modulus of continuity.

Theorem 4.32. *Let E be a subset of a normed space V and let $f : E \to \mathbb{R}$ be a uniformly continuous function such that*

$$\omega(s) \le a + bs \quad \text{for all } s \ge 0, \tag{4.8}$$

for some $a, b \ge 0$. Then there exists a uniformly continuous function $g : V \to \mathbb{R}$ with the same concave modulus of continuity ϖ and such that

$$g(v) = f(v) \quad \text{for all } v \in E, \tag{4.9}$$

and

$$\inf_V g = \inf_E f, \quad \sup_V g = \sup_E f. \tag{4.10}$$

Proof. **Step 1:** We claim that ϖ is nondecreasing, $\varpi(0) = 0$, and ϖ is subadditive, that is,

$$\varpi(s_1 + s_2) \leq \varpi(s_1) + \varpi(s_2) \tag{4.11}$$

for all s_1, $s_2 \geq 0$. Since ϖ is concave, if ϖ were decreasing in some interval, then it would be unbounded from below, and this would contradict the fact that it is nonnegative.

To prove that $\varpi(0) = 0$, note first that if $a = 0$, then this follows from the fact that, by (4.8), $\varpi(s) \leq bs$ for all $s \geq 0$. If $a > 0$, then fix $0 < \varepsilon < a$, and since $w(s) \to 0$ as $s \to 0^+$, there exists $\delta_\varepsilon > 0$ such that

$$w(s) \leq \varepsilon \quad \text{for all } 0 \leq s \leq \delta_\varepsilon.$$

The function $\phi : [0, \infty] \to \mathbb{R}$, defined by

$$\phi(s) := \begin{cases} a + bs & \text{if } s > \delta_\varepsilon, \\ \varepsilon + \dfrac{a + b\delta_\varepsilon - \varepsilon}{\delta_\varepsilon} s & \text{if } 0 \leq s \leq \delta_\varepsilon, \end{cases}$$

is concave and above w. Therefore $\varpi(0) \leq \phi(0) = \varepsilon$, and given the arbitrariness of ε we conclude that $\varpi(0) = 0$.

It remains to show that ϖ is subadditive. Let s_1, $s_2 > 0$. Then by concavity,

$$\varpi(s_1) \geq \frac{s_1}{s_1 + s_2} \varpi(s_1 + s_2) + \frac{s_2}{s_1 + s_2} \varpi(0) = \frac{s_1}{s_1 + s_2} \varpi(s_1 + s_2),$$

$$\varpi(s_2) \geq \frac{s_2}{s_1 + s_2} \varpi(s_1 + s_2) + \frac{s_1}{s_1 + s_2} \varpi(0) = \frac{s_2}{s_1 + s_2} \varpi(s_1 + s_2),$$

and summing the two inequalities, we obtain (4.11).

Step 2: Since ϖ is nondecreasing and subadditive, for all $v_1, v_2, w \in V$,

$$\varpi(|v_2 - w|) \leq \varpi(|v_1 - w| + |v_1 - v_2|) \leq \varpi(|v_1 - w|) + \varpi(|v_1 - v_2|). \tag{4.12}$$

Moreover, from the definition of w and ϖ it follows that

$$\varpi(|w_1 - w_2|) \geq |f(w_1) - f(w_2)| \quad \text{for all } w_1, w_2 \in E. \tag{4.13}$$

Define

$$h(v) := \inf \{f(w) + \varpi(|v - w|) : w \in E\}, \quad v \in V. \tag{4.14}$$

We first claim that $h(v)$ is finite for every $v \in V$. Fix $w_0 \in E$. By (4.12) and (4.13), if $w \in E$ then

$$f(w) + \varpi(|v - w|) - f(w_0) \geq \varpi(|v - w|) - \varpi(|w - w_0|)$$
$$\geq -\varpi(|v - w_0|),$$

and so

$$h(v) = \inf \left\{ f(w) + \varpi \left(|v - w| \right) : w \in E \right\}$$
$$\geq f(w_0) - \varpi \left(|v - w_0| \right) > -\infty.$$

Note that if $v \in E$, then we can choose $w_0 := v$ in the previous inequality to obtain $h(v) \geq f(v)$, since $\varpi(0) = 0$. On the other hand, taking $w = v$ in (4.14) yields $h(v) \leq f(v)$, and so (4.9) holds for h (in place of g). In particular,

$$\inf_V h \leq \inf_E f,$$

and since $\varpi \geq 0$, we have that

$$h \geq \inf_E f,$$

and thus we conclude that

$$\inf_V h = \inf_E f.$$

Next we claim that

$$|h(v_1) - h(v_2)| \leq \varpi \left(|v_1 - v_2| \right) \tag{4.15}$$

for all $v_1, v_2 \in V$. Indeed, let $v_1, v_2 \in V$, $\varepsilon > 0$. By (4.14) there exists $w_1 \in E$ such that

$$h(v_1) \geq f(w_1) + \varpi \left(|v_1 - w_1| \right) - \varepsilon.$$

Since $h(v_2) \leq f(w_1) + \varpi \left(|v_2 - w_1| \right)$, we get

$$h(v_1) - h(v_2) \geq \varpi \left(|v_1 - w_1| \right) - \varpi \left(|v_2 - w_1| \right) - \varepsilon$$
$$\geq -\varpi \left(|v_1 - v_2| \right) - \varepsilon$$

by (4.12). Given the arbitrariness of $\varepsilon > 0$ and interchanging the roles of v_1 and v_2, we have proved the claim.

Finally, define

$$g(v) := \inf \left\{ h(v), \sup_E f \right\}.$$

Then (4.9) and (4.10) hold for g. We now show that (4.15) continues to hold for g. Indeed, this is immediate if either $g(v_i) = h(v_i)$ for $i = 1, 2$, or if $g(v_i) = \sup_E f$ for $i = 1, 2$. Thus the only remaining case is $g(v_1) = h(v_1) < \sup_E f$ and $g(v_2) = \sup_E f \leq h(v_2)$. From (4.15) we obtain

$$0 \leq g(v_2) - g(v_1) = \sup_E f - h(v_1) \leq h(v_2) - h(v_1) \leq \varpi \left(|v_1 - v_2| \right),$$

which establishes that the concave modulus of continuity of g, denoted by ϖ_g, is below ϖ. Conversely, since g extends f, it follows that the modulus of continuity of f is less than or equal to that of g, and so $\varpi_g = \varpi$.

Remark 4.33. (i) The previous result may be extended to metric spaces with the obvious adaptations.

(ii) In the special case that $\omega(s) \leq Ls^{\alpha}$, $0 < \alpha \leq 1$, $L > 0$, we have that $\varpi(s) \leq Ls^{\alpha}$. Hence if $0 < \alpha < 1$, then we can extend f to a Hölder continuous function g on $\mathbb{R}^N$ such that

$$\left.|g|\right._{C^{0,\alpha}(\mathbb{R}^N)} = \left.|f|\right._{C^{0,\alpha}(E)}, \quad \inf_{\mathbb{R}^N} g = \inf_E f, \quad \text{and} \quad \sup_{\mathbb{R}^N} g = \sup_E f,$$

while if $\alpha = 1$ we can extend f to a Lipschitz continuous function g on V such that

$$\text{Lip}\, g = \text{Lip}\, f, \quad \inf_{\mathbb{R}^N} g = \inf_E f, \quad \text{and} \quad \sup_{\mathbb{R}^N} g = \sup_E f.$$

4.5 Regularity of Convex Functions

In this section we address continuity and differentiability properties of convex functions.

In the first result we prove that real-valued separately convex functions on finite-dimensional spaces are locally Lipschitz. This is hinged on the following characterization of convex functions on the real line.

Proposition 4.34. *Let $J \subset \mathbb{R}$ be an interval and let $g : J \to [-\infty, \infty]$. If g is convex, then for every $t_0 \in J$ such that $g(t_0) \in \mathbb{R}$, the difference quotient*

$$t \mapsto \frac{g(t) - g(t_0)}{t - t_0}$$

is nondecreasing in $J \setminus \{t_0\}$. The converse is true if $g : J \to (-\infty, \infty]$.

Exercise 4.35. Prove the previous proposition.

If E is a subset of $\mathbb{R}^m$ and $f : E \to \mathbb{R}$, then the *oscillation* of f on E is defined by

$$\text{osc}\,(f; E) := \sup \left\{ |f(z_1) - f(z_2)| : z_1, z_2 \in E \right\}.$$

The proof of the next theorem is particularly simple for convex functions (see Remark 4.37 below), but in view of applications to rank one convex functions in [FoLe10], we need a version for separately convex functions.

Theorem 4.36. *If $f : B(z_0, 2r) \subset \mathbb{R}^m \to \mathbb{R}$ is separately convex, then*

$$\text{Lip}\,(f; B(z_0, r)) \leq \sqrt{m}\,\frac{\text{osc}\,(f; B(z_0, 2r))}{r}.$$

In particular, any separately convex function $f : \mathbb{R}^m \to (-\infty, \infty]$ is locally Lipschitz in the interior of its effective domain $\text{dom}_e\, f$.

Proof. **Step 1:** Without loss of generality we may assume that $z_0 = 0$.[1] Let $L := \operatorname{osc}(f; B(0, 2r))/r$. Assume first that $w, z \in B(0, r)$ are such that $(z - w)/|z - w| = \pm e^{(i)}$ for some element $e^{(i)}$ of the canonical basis of $\mathbb{R}^m$. For simplicity we consider the case $(z - w)/|z - w| = e^{(i)}$. Fix $0 < \varepsilon < r$. Let ξ be a point of intersection of $\partial B(0, 2r - \varepsilon)$ with the ray from w through z, so that z belongs to the segment of endpoints w and ξ, and $|w - \xi| \geq r - \varepsilon$. Define $g(t) := f(w + te_i)$. Since f is separately convex, then g is convex, and so by the previous proposition and the fact that $|\xi - w| \geq |z - w|$ we have

$$\frac{g(|z - w|) - g(0)}{|z - w|} \leq \frac{g(|\xi - w|) - g(0)}{|\xi - w|},$$

or equivalently

$$\frac{f(z) - f(w)}{|z - w|} \leq \frac{f(\xi) - f(w)}{|\xi - w|} \leq \frac{\operatorname{osc}(f; B(0, 2r))}{r - \varepsilon},$$

where we have used the fact that $|w - \xi| \geq r - \varepsilon$. Letting $\varepsilon \to 0^+$ yields

$$\frac{f(z) - f(w)}{|z - w|} \leq \frac{\operatorname{osc}(f; B(0, 2r))}{r} = L. \tag{4.16}$$

Next we show that any $w = (w_1, \ldots, w_m)$, $z = (z_1, \ldots, z_m) \in B(0, r)$ can be joined by a chain $\xi^{(0)} := w$, $\xi^{(1)}, \ldots, \xi^{(m)} := z$ of points in $B(0, r)$ such that $\xi^{(i)} - \xi^{(i-1)} = (w_{\sigma(i)} - z_{\sigma(i)}) e^{(\sigma(i))}$, $i = 1, \ldots, m$, for a suitable bijection $\sigma : \{1, \ldots, m\} \to \{1, \ldots, m\}$.

Let $J_+ := \{i \in \{1, \ldots, m\} : |w_i| \geq |z_i|\}$. Construct a bijection σ of $\{1, \ldots, m\}$ onto itself in such a way that $\sigma(i) \in J_+$ if and only if $i \leq \operatorname{card}(J_+)$. Define

$$\xi^{(i)} := \xi^{(0)} - \sum_{j=1}^{i} (w_{\sigma(j)} - z_{\sigma(j)}) e^{(\sigma(j))}, \quad i = 1, \ldots, m - 1.$$

Then $\xi^{(i)} - \xi^{(i-1)} = (w_{\sigma(i)} - z_{\sigma(i)}) e^{(\sigma(i))}$ for all $i = 1, \ldots, m$ and

$$\left| \xi^{(i)} \right| \leq \begin{cases} |w| < r & \text{if } 1 \leq i \leq \operatorname{card}(J_+), \\ |z| < r & \text{if } \operatorname{card}(J_+) < i \leq m. \end{cases}$$

Since $\xi^{(i)} \in B(0, r)$ and $\xi^{(i)} - \xi^{(i-1)} = (w_{\sigma(i)} - z_{\sigma(i)}) e^{(\sigma(i))}$ we can apply (4.16) to obtain

---

[1] In the proof of this theorem, $z^{(i)}$ denotes a vector of $\mathbb{R}^m$, while z_i is the ith component of a vector $z \in \mathbb{R}^m$.

$$|f(w) - f(z)| \le \sum_{i=1}^{m} \left| f\left(\xi^{(i)}\right) - f\left(\xi^{(i-1)}\right) \right| \le L \sum_{i=1}^{m} \left| \xi^{(i)} - \xi^{(i-1)} \right|$$

$$= L \sum_{i=1}^{m} \left| w_{\sigma(i)} - z_{\sigma(i)} \right|$$

$$\le L \left(\sum_{i=1}^{m} 1^2 \right)^{1/2} \left(\sum_{i=1}^{m} \left(w_{\sigma(i)} - z_{\sigma(i)} \right)^2 \right)^{1/2}$$

$$= L\sqrt{m} \, |w - z| \, .$$

Step 2: Here we prove that if z_0 belongs to the interior of $\mathrm{dom}_e \, f$, then there exists a neighborhood of z_0 on which f is bounded and thus its oscillation is finite.

Without loss of generality we may assume that $z_0 = 0$, and consider in $\mathbb{R}^m$ the equivalent norm

$$\|z\|_\infty := \max\left\{ |z_i| : i = 1, \ldots, m \right\}.$$

Let $\varepsilon > 0$ be such that $B_\infty(0, 2\varepsilon) \subset \mathrm{dom}_e \, f$ and set

$$a := \max\left\{ f(z) : z_i \in \{-\varepsilon, 0, \varepsilon\}, \, i = 1, \ldots, m \right\}.$$

We claim that

$$f(z) \le a \quad \text{for all } z \in \overline{B_\infty(0, \varepsilon)}. \tag{4.17}$$

Indeed, let $z, w \in \overline{B_\infty(0, \varepsilon)}$ with $w_i \in \{-\varepsilon, 0, \varepsilon\}$, $i = 1, \ldots, m$. If $z_m \ne 0$ write

$$z_m = \frac{|z_m|}{\varepsilon} (\mathrm{sgn} \, z_m) \varepsilon + \left(1 - \frac{|z_m|}{\varepsilon} \right) 0.$$

By the separate convexity of f we have

$$f(w_1, \ldots, w_{m-1}, z_m) \le \frac{|z_m|}{\varepsilon} f(w_1, \ldots, w_{m-1}, (\mathrm{sgn} \, z_m) \varepsilon)$$

$$+ \left(1 - \frac{|z_m|}{\varepsilon} \right) f(w_1, \ldots, w_{m-1}, 0) \le a.$$

The same inequality holds if $z_m = 0$. Similarly, if $z_{m-1} \ne 0$, then we have

$$f(w_1, \ldots, w_{m-2}, z_{m-1}, z_m)$$

$$\le \frac{|z_{m-1}|}{\varepsilon} f(w_1, \ldots, w_{m-2}, (\mathrm{sgn} \, z_{m-1}) \varepsilon, z_m)$$

$$+ \left(1 - \frac{|z_{m-1}|}{\varepsilon} \right) f(w_1, \ldots, w_{m-2}, 0, z_m) \le a,$$

where we have used the previous inequality. Recursively we obtain (4.17).

Next we show that

$$f(z) \geq 2^m f(0) - (2^m - 1) a \quad \text{for all } z \in \overline{B_\infty(0,\varepsilon)}. \tag{4.18}$$

Writing

$$0 = \frac{1}{2} z_m + \frac{1}{2}(-z_m),$$

we have

$$f(0) \leq \frac{1}{2} f(0, \ldots, 0, z_m) + \frac{1}{2} f(0, \ldots, 0, -z_m) \leq \frac{1}{2} f(0, \ldots, 0, z_m) + \frac{1}{2} a,$$

and so

$$f(0, \ldots, 0, z_m) \geq 2f(0) - a,$$

where we have used (4.17). Similarly,

$$2f(0) - a \leq f(0, \ldots, 0, z_m)$$
$$\leq \frac{1}{2} f(0, \ldots, 0, z_{m-1}, z_m) + \frac{1}{2} f(0, \ldots, 0, -z_{m-1}, z_m)$$
$$\leq \frac{1}{2} f(0, \ldots, 0, z_{m-1}, z_m) + \frac{1}{2} a,$$

and thus

$$f(0, \ldots, 0, z_{m-1}, z_m) \geq 2^2 f(0) - (2^2 - 1) a.$$

Proceeding by induction we deduce (4.18).

Remark 4.37. (i) It follows from the previous proof (see (4.18)) that if $f : B_\infty(z_0, 2r) \subset \mathbb{R}^m \to \mathbb{R}$ is separately convex with $f(z_0) = 0$, then

$$\inf_{B_\infty(z_0,r)} f + (2^m - 1) \sup_{B_\infty(z_0,r)} f \geq 0.$$

(ii) This proof provides a sharper estimate for the Lipschitz constant of f as compared with simpler arguments already available in the literature (see [Dac89, Theorem 2.3]).

(iii) If $f : B(z_0, 2r) \subset \mathbb{R}^m \to \mathbb{R}$ is convex, then the sharper estimate

$$\mathrm{Lip}(f; B(z_0, r)) \leq \frac{\mathrm{osc}(f; B(z_0, 2r))}{r}$$

holds. To see this, assume that the right-hand side is finite and let w, $z \in B(z_0, r)$. Fix $0 < \varepsilon < r$. Suppose that $f(z) \geq f(w)$ and choose ξ to be a point of intersection of $\partial B(z_0, 2r - \varepsilon)$ with the ray from w through z such that $|w - \xi| \geq r - \varepsilon$. Then as in the proof of (4.16) we have

$$\frac{f(z) - f(w)}{|z - w|} \leq \frac{f(\xi) - f(w)}{|\xi - w|} \leq \frac{\mathrm{osc}(f; B(0, 2r))}{r - \varepsilon}.$$

It suffices to let $\varepsilon \to 0^+$.

Corollary 4.38. *Let* $f : \mathbb{R}^m \to (-\infty, \infty]$ *be a proper convex function. Then the restriction of* f *to* $\mathrm{ri}_{\mathrm{aff}} (\mathrm{dom}_e f)$ *is locally Lipschitz. In particular, if* $\mathrm{dom}_e f$ *is affine, then the restriction of* f *to* $\mathrm{dom}_e f$ *is locally Lipschitz.*

Proof. If $\mathrm{dom}_e f$ consists of a point, then there is nothing to prove. If $\mathrm{dom}_e f$ has more than one point, by Remark 4.1,

$$\mathrm{aff} (\mathrm{dom}_e f) = z_0 + W,$$

where $z_0 \in \mathrm{dom}_e f$ and W is a subspace of $\mathbb{R}^m$ of dimension $1 \leq \ell \leq m$. By a change of coordinates, without loss of generality we may assume that $\mathrm{aff} (\mathrm{dom}_e f) = \mathbb{R}^\ell \times \{0\}$. Define

$$g (w) := f ((w,0)), \quad w \in \mathbb{R}^\ell.$$

Then $\mathrm{ri}_{\mathrm{aff}} (\mathrm{dom}_e g)$ reduces simply to the interior of $\mathrm{dom}_e g$. The result now follows from Theorem 4.36 applied to g.

Exercise 4.39. The previous corollary cannot be improved in general. Indeed, let $m = 2$ and consider the function

$$f (z) = f (z_1, z_2) = \begin{cases} \dfrac{(z_2)^2}{2z_1} & \text{if } z_1 > 0, \\ 0 & \text{if } z_1 = z_2 = 0, \\ \infty & \text{otherwise.} \end{cases}$$

Prove that f is convex in $\mathbb{R}^2$ and continuous everywhere except at the origin.

Corollary 4.38 implies in particular that the lower semicontinuous envelope of a proper convex function coincides with the function except at most on the relative boundary of its effective domain. Indeed, we have the following result.

Theorem 4.40. *Let* $f : \mathbb{R}^m \to (-\infty, \infty]$ *be a proper convex function. Then* $\mathrm{lsc}\, f$ *agrees with* f *everywhere except possibly on* $\mathrm{rb}_{\mathrm{aff}} (\mathrm{dom}_e f)$. *Moreover, for any fixed* $z_0 \in \mathrm{ri}_{\mathrm{aff}} (\mathrm{dom}_e f)$ *and for any* $z \in \mathbb{R}^m$,

$$f (z) \geq \mathrm{lsc}\, f (z) = \lim_{\theta \to 1^-} f ((1 - \theta) z_0 + \theta z). \tag{4.19}$$

Proof. **Step 1:** Fix any

$$z \notin \mathrm{rb}_{\mathrm{aff}} (\mathrm{dom}_e f) = \overline{\mathrm{dom}_e f} \setminus \mathrm{ri}_{\mathrm{aff}} (\mathrm{dom}_e f).$$

By Proposition 3.12, for every $z \in \mathbb{R}^m$,

$$\mathrm{lsc}\, f (z) = \inf_{\{z_n\}} \left\{ \liminf_{n \to \infty} f (z_n) : \{z_n\} \subset \mathbb{R}^m,\ z_n \to z \right\},$$

and so we may find $\{z_n\} \subset \mathbb{R}^m$ such that $z_n \to z$ and

$$\operatorname{lsc} f(z) = \lim_{n \to \infty} f(z_n).$$

We now distinguish two cases.

If $z \in \operatorname{ri_{aff}}(\operatorname{dom}_e f)$, then since

$$\infty > f(z) \geq \operatorname{lsc} f(z) = \lim_{n \to \infty} f(z_n),$$

it follows that $z_n \in \operatorname{ri_{aff}}(\operatorname{dom}_e f)$ for all n sufficiently large, and thus, using the continuity of f in $\operatorname{ri_{aff}}(\operatorname{dom}_e f)$ (see Corollary 4.38), we obtain that

$$\operatorname{lsc} f(z) = \lim_{n \to \infty} f(z_n) = f(z).$$

If $z \notin \overline{\operatorname{dom}_e f}$, then $z_n \notin \operatorname{dom}_e f$ for all n sufficiently large, and so

$$f(z) \geq \operatorname{lsc} f(z) = \lim_{n \to \infty} f(z_n) = \infty,$$

which concludes the first part of the theorem.

Step 2: To prove (4.19) fix any $z_0 \in \operatorname{ri_{aff}}(\operatorname{dom}_e f)$ (note that since f is proper, by Proposition 4.7 we may always find at least one). We again distinguish two cases.

If $z \in \overline{\operatorname{dom}_e f}$, then in view of the convexity of $\operatorname{dom}_e f$, by Proposition 4.7 we have

$$\theta z + (1 - \theta) z_0 \in \operatorname{ri_{aff}}(\operatorname{dom}_e f)$$

for all $0 \leq \theta < 1$. Hence by Step 1,

$$f(\theta z + (1 - \theta) z_0) = \operatorname{lsc} f(\theta z + (1 - \theta) z_0)$$

for all $0 \leq \theta < 1$. The lower semicontinuous function

$$g(\theta) := \operatorname{lsc} f(\theta z + (1 - \theta) z_0), \quad \theta \in [0,1],$$

is convex and real-valued (except possibly at $\theta = 1$), and so continuous in $[0,1]$. Therefore

$$f(z) \geq \operatorname{lsc} f(z) = g(1) = \lim_{\theta \to 1^-} g(\theta)$$
$$= \lim_{\theta \to 1^-} \operatorname{lsc} f(\theta z + (1 - \theta) z_0) = \lim_{\theta \to 1^-} f(\theta z + (1 - \theta) z_0).$$

Finally, if $z \notin \overline{\operatorname{dom}_e f}$, then $\theta z + (1 - \theta) z_0 \notin \operatorname{dom}_e f$ for all θ sufficiently close to one, and so again by Step 1,

$$\infty = f(\theta z + (1 - \theta) z_0) = \operatorname{lsc} f(\theta z + (1 - \theta) z_0)$$

for all θ sufficiently close to one, which yields (4.19).

Corollary 4.41. *Let $f : \mathbb{R}^m \to (-\infty, \infty]$ be a proper convex function and let $C \subset \mathrm{dom}_e\, f$ be such that $\overline{C}$ is convex. Then*

$$\inf_C f = \inf_{\overline{C}} f.$$

Proof. It is enough to show that if $t \in \mathbb{R}$ is such that $f(z_1) < t$ for some $z_1 \in \overline{C}$, then $f(z_2) < t$ for some $z_2 \in C$. Define

$$g(z) := \begin{cases} f(z) & \text{if } z \in \overline{C}, \\ \infty & \text{otherwise.} \end{cases}$$

Since

$$\mathrm{ri}_{\mathrm{aff}}(C) \subset C \subset \mathrm{dom}_e\, g \subset \overline{C},$$

it follows from Proposition 4.9 that

$$\mathrm{ri}_{\mathrm{aff}}(\mathrm{dom}_e\, g) = \mathrm{ri}_{\mathrm{aff}}(C).$$

Fix any $z_0 \in \mathrm{ri}_{\mathrm{aff}}(C)$. By (4.19),

$$t > f(z_1) = g(z_1) \geq \lim_{\theta \to 1^-} g((1-\theta)z_0 + \theta z_1),$$

and thus the result follows by Proposition 4.7.

Next we extend some of the previous continuity results to the infinite-dimensional setting.

Proposition 4.42. *Let $V_1, \ldots, V_m$ be locally convex topological vector spaces and let $f : V_1 \times \ldots \times V_m \to [-\infty, \infty]$ be separately convex. If f is bounded from above in a neighborhood of a point v_0 such that $f(v_0) \in \mathbb{R}$, then f is continuous at v_0.*

Proof. Without loss of generality we may assume that $v_0 = 0$ and that $f(0) = 0$. For every $i = 1, \ldots, m$, let U_i be an open convex neighborhood of the origin in V_i such that

$$f(v) \leq a \quad \text{for all } v \in U_1 \times \ldots \times U_m$$

for some $a \in \mathbb{R}$. We symmetrize the neighborhoods as $W_i := U_i \cap (-U_i)$, $i = 1, \ldots, m$. Given $0 < \varepsilon < 1$, we prove that

$$|f(v)| \leq m\varepsilon a \quad \text{for all } v \in \varepsilon(W_1 \times \ldots \times W_m). \tag{4.20}$$

The argument is similar to the one used in Step 2 of the proof of Theorem 4.36. Fix $v \in \varepsilon(W_1 \times \ldots \times W_m)$ and write

$$v = (v_1, \ldots, v_m),$$

where $v_i \in V_i$, $i = 1, \ldots, m$. By the separate convexity of f we have

$$f(0,\ldots,0,v_m) \le (1-\varepsilon) f(0) + \varepsilon f\left(0,\ldots,0,\frac{v_m}{\varepsilon}\right) \le \varepsilon a,$$

where we have used the fact that $\left(0,\ldots,0,\frac{v_m}{\varepsilon}\right) \in U_1 \times \ldots \times U_m$. Again by separate convexity we obtain

$$f(0,\ldots,0,v_m) \ge (1+\varepsilon) f(0) - \varepsilon f\left(0,\ldots,0,-\frac{v_m}{\varepsilon}\right) \ge -\varepsilon a,$$

and thus

$$|f(0,\ldots,0,v_m)| \le \varepsilon a. \tag{4.21}$$

Similarly,

$$
\begin{aligned}
f(0,&\ldots,0,v_{m-1},v_m) \\
&\le (1-\varepsilon) f(0,\ldots,0,v_m) + \varepsilon f\left(0,\ldots,0,\frac{v_{m-1}}{\varepsilon},v_m\right) \\
&\le 2\varepsilon a
\end{aligned}
$$

by (4.21) and the fact that $\left(0,\ldots,0,\frac{v_{m-1}}{\varepsilon},v_m\right) \in U_1 \times \ldots \times U_m$. Here we have used the convexity of U_m to deduce that $v_m \in \varepsilon U_m \subset U_m$. Recursively we obtain (4.20).

More can be said about regularity when f is convex.

Theorem 4.43. *Let V be a locally convex topological vector space and let $f : V \to [-\infty, \infty]$ be convex. Then the following are equivalent:*

(i) f is a proper function and is continuous in the interior of $\mathrm{dom}_e\, f$, which is nonempty;

(ii) there exists a nonempty open set on which f is not identically $-\infty$ and is bounded from above.

Proof. It suffices to show that (ii) implies (i). Let $U \subset V$ be an open set such that

$$f(v) \le a$$

for all $v \in U$ for some positive constant a, and fix any $v_0 \in U$ such that $f(v_0) \in \mathbb{R}$. By the previous proposition f is continuous at v_0, and so in particular, it is finite in a neighborhood of v_0. By Remark 4.23 the function f is proper. Let now v_1 be any point in the interior of $\mathrm{dom}_e\, f$ and find a neighborhood U_1 of v_1 such that $U_1 \subset \mathrm{dom}_e\, f$. By the continuity of the function

$$t \in \mathbb{R} \mapsto v_0 + t(v_1 - v_0)$$

at $t = 1$ we may find $t > 1$ such that the point $v_t := v_0 + t(v_1 - v_0)$ is in U_1. The continuous invertible function $L : V \to V$, defined by

$$L(v) := \frac{1}{t} v_t + \left(1 - \frac{1}{t}\right) v, \quad v \in V,$$

maps U into an open set $L(U)$ and $L(v_0) = v_1 \in L(U)$. Moreover, by the convexity of f, for any $w \in L(U)$, letting $v := L^{-1}(w) \in U$, we have

$$f(w) = f\left(\frac{1}{t}v_t + \left(1 - \frac{1}{t}\right)v\right)$$

$$\leq \frac{1}{t}f(v_t) + \left(1 - \frac{1}{t}\right)f(v) \leq \frac{1}{t}f(v_t) + \left(1 - \frac{1}{t}\right)a,$$

which shows that f is bounded from above in a neighborhood of v_1. Using again the previous proposition, we conclude that f is continuous at v_1, and, given the arbitrariness of v_1 in the interior of $\mathrm{dom}_e\, f$, this concludes the proof.

Exercise 4.44. We construct an example of a convex function that is not continuous. Let V be an infinite-dimensional normed space and let $\{e_n\} \subset V$ be a sequence of linearly independent vectors of norm one. Let $f : V \to \mathbb{R}$ be the linear function satisfying $f(e_n) := n$ for all $n \in \mathbb{N}$ and $f :\equiv 0$ outside the span of $\{e_n\}$. Prove that f is linear but not continuous.

If in the previous theorem we assume in addition that f is lower semicontinuous and that V is a Banach space, then we may remove the assumption that f is bounded from above on an open set.

Corollary 4.45. *Let V be a Banach space and let $f : V \to (-\infty, \infty]$ be a lower semicontinuous convex function. Then f is continuous over the interior of $\mathrm{dom}_e\, f$.*

Proof. Let v_0 be in the interior of $\mathrm{dom}_e\, f$. By replacing f with $f(\cdot - v_0)$ we may assume, without loss of generality, that $v_0 = 0$, so that we may find $B(0, r) \subset \mathrm{dom}_e\, f$. Since f is lower semicontinuous and convex, the set

$$C := \{v \in V : f(v) \leq f(0) + 1\}$$

is closed and convex and $0 \in C$. We claim that

$$V = \bigcup_{n=1}^{\infty} nC.$$

Indeed, if $v \in V$, $v \neq 0$, then the function

$$g(t) := f(tv), \quad t \in \mathbb{R},$$

is convex and $\left(-\frac{r}{|v|}, \frac{r}{|v|}\right) \subset \mathrm{dom}_e\, g$. In particular, g is continuous in $\left(-\frac{r}{|v|}, \frac{r}{|v|}\right)$, and so there exists $0 < \delta < \frac{r}{|v|}$ such that

$$|f(tv) - f(0)| = |g(t) - g(0)| \leq 1$$

for all $|t| \leq \delta$. Hence $tv \in C$ for all $|t| \leq \delta$, which implies that $v \in nC$ for all $n \geq \frac{1}{\delta}$. This proves the claim.

By the Baire category theorem there exists $n_0 \in \mathbb{N}$ such that the set $n_0 C$ has nonempty interior. In turn, C has nonempty interior and thus we may apply the previous theorem to conclude that f is continuous over the interior of $\mathrm{dom}_e\, f$.

Exercise 4.46. Let $V := L^1\,([0,1])$, where the underlying measure is the Lebesgue measure. Show that the functional

$$I : L^1\,([0,1]) \to [0,\infty]$$

defined by

$$I\,(v) := \begin{cases} \displaystyle\int_0^1 |v\,(x)|^2 \; dx \text{ if } v \in L^2\,([0,1]), \\ \infty \qquad\qquad\qquad\quad \text{otherwise}, \end{cases}$$

is convex and lower semicontinuous, but the sets

$$\left\{ v \in L^1\,([0,1]) : \int_0^1 |u\,(x)|^2 \; dx \leq t \right\}, \quad t > 0,$$

have empty interiors.

The proof of the next result follows from Theorem 4.43 and adapting Step 1 of the proof of Theorem 4.36. We omit the details.

Theorem 4.47. *Let V be a normed space and let $f : V \to (-\infty, \infty]$ be a convex function. Then there exists a nonempty open set over which f is bounded from above if and only if f is locally Lipschitz in the interior of $\mathrm{dom}_e\, f$, which is nonempty.*

Next we study differentiability properties of convex functions. We begin by introducing the notion of subdifferential.

Definition 4.48. *Let V be a locally convex topological vector space, let $f : V \to [-\infty, \infty]$, and let $v_0 \in V$ be such that $f\,(v_0) \in \mathbb{R}$. The function f is said to be subdifferentiable at v_0 if there exists $v' \in V'$ such that*

$$f\,(v) \geq f\,(v_0) + \langle v', v - v_0 \rangle_{V',V} \quad \text{for all } v \in V.$$

The element v' is called a subgradient *of f at v_0, and the set of all subgradients at v_0 is called the* subdifferential *of f at v_0 and is denoted by $\partial f\,(v_0)$. Precisely,*

$$\partial f\,(v_0) = \{v' \in V : f\,(v) \geq f\,(v_0) + \langle v', v - v_0 \rangle_{V',V} \quad \text{for all } v \in V\}.$$

If f is not subdifferentiable at v_0, then $\partial f\,(v_0) := \emptyset$.

The one-sided directional derivative *of f at v_0 in the direction $v \in V$ is defined by*

$$\frac{\partial^+ f}{\partial v}\,(v_0) := \lim_{t \to 0^+} \frac{f\,(v_0 + tv) - f\,(v_0)}{t}$$

whenever the limit exists.

Note that

$$f(v_0) = \min_{v \in V} f(v) \quad \text{if and only if } 0 \in \partial f(v_0). \tag{4.22}$$

Remark 4.49. If V is a locally convex topological vector space, $f : V \to [-\infty, \infty]$ is convex, and $v_0 \in V$ is such that $f(v_0) \in \mathbb{R}$, then for every $v \in V$ the one-sided directional derivatives $\frac{\partial^+ f}{\partial v}(v_0)$ always exist in view of Proposition 4.34. It may be verified that

$$v \in V \mapsto \frac{\partial^+ f}{\partial v}(v_0)$$

is positively homogeneous of degree one, convex, and so subadditive. Moreover,

$$-\frac{\partial^+ f}{\partial(-v)}(v_0) \leq \frac{\partial^+ f}{\partial v}(v_0).$$

In addition, if f is subdifferentiable at v_0, then

$$\frac{\partial^+ f}{\partial v}(v_0) \geq \sup_{v' \in \partial f(v_0)} \langle v', v \rangle_{V',V} > -\infty$$

for all $v \in V$. Indeed, let $v \in V$ and let $v' \in \partial f(v_0)$. Then by definition of subdifferential for all $t > 0$ we have

$$\frac{f(v_0 + tv) - f(v_0)}{t} \geq \langle v', v \rangle_{V',V}. \tag{4.23}$$

Dividing by t and letting $t \to 0^+$ yields

$$\frac{\partial^+ f}{\partial v}(v_0) \geq \langle v', v \rangle_{V',V},$$

and given the arbitrariness of $v' \in \partial f(v_0)$ we get

$$\frac{\partial^+ f}{\partial v}(v_0) \geq \sup_{v' \in \partial f(v_0)} \langle v', v \rangle_{V',V}.$$

Exercise 4.50. Show that the convex function $f : \mathbb{R}^m \to (-\infty, \infty]$ defined by

$$f(z) := \begin{cases} -\left(1 - |z|^2\right)^{\frac{1}{2}} & \text{if } |z| \leq 1, \\ \infty & \text{otherwise}, \end{cases}$$

is differentiable, and so subdifferentiable (see Theorem 4.61 below) in the open unit ball $\{z \in \mathbb{R}^m : |z| < 1\}$ but it is not subdifferentiable at points z with $|z| = 1$.

We now study the existence of the subdifferential.

Theorem 4.51. *Let V be a locally convex topological vector space and let $f : V \to [-\infty, \infty]$ be a convex function. If there exists $v_0 \in V$ such that $f(v_0) \in \mathbb{R}$ and f is continuous at v_0, then $\partial f(v) \neq \emptyset$ for all v in the interior of $\mathrm{dom}_e f$. In particular, $\partial f(v_0) \neq \emptyset$.*

We begin with a preliminary result.

Lemma 4.52. *Let V be a locally convex topological vector space, let $f : V \to [-\infty, \infty]$ be convex, and let $v_0 \in V$ be such that $f(v_0) \in \mathbb{R}$ and f is continuous at v_0. If $C \subset V \times \mathbb{R}$ is a convex set that does not intersect the interior of $\mathrm{epi}\, f$ and such that $(v_0, s_0) \in C$ for some $s_0 \in \mathbb{R}$, then there exist $v' \in V'$ and $\alpha \in \mathbb{R}$ such that*

$$f(v) \geq \alpha + \langle v', v \rangle_{V',V} \quad \text{for all } v \in V,$$
$$t \leq \alpha + \langle v', v \rangle_{V',V} \quad \text{for all } (v, t) \in C.$$

Proof. Since f is convex, by Proposition 4.22 $\mathrm{epi}\, f$ is a convex set of $V \times \mathbb{R}$ and we claim that its interior is nonempty. Indeed, fix $t > f(v_0)$. Since f is continuous at v_0 there exists a neighborhood U of v_0 such that

$$f(v) < t \quad \text{for all } v \in U.$$

Hence the open set $U \times (t, \infty)$ is a subset of $\mathrm{epi}\, f$, which shows that the interior of $\mathrm{epi}\, f$ is nonempty. By the first geometric form of the Hahn–Banach theorem, with A being the interior of $\mathrm{epi}\, g$ and $E := C$, we may find a continuous linear functional $L : V \times \mathbb{R} \to \mathbb{R}$, $L \neq 0$, and a number $\alpha \in \mathbb{R}$ such that

$$L(v, t) \geq \alpha \quad \text{for all } (v, t) \in A \text{ and } L(v, t) \leq \alpha \quad \text{for all } (v, t) \in C.$$

Note that by the continuity of L it follows that the first inequality actually holds for all $(v, t) \in \mathrm{epi}\, f$. The functional L has the form $L(v, t) = \langle v', v \rangle_{V',V} + \alpha_0 t$ for some $v' \in V'$ and $\alpha_0 \in \mathbb{R}$. Hence

$$\langle v', v \rangle_{V',V} + \alpha_0 t \geq \alpha \quad \text{for all } (v, t) \in \mathrm{epi}\, f, \tag{4.24}$$
$$\langle v', v \rangle_{V',V} + \alpha_0 t \leq \alpha \quad \text{for all } (v, t) \in C.$$

Let $v \in \mathrm{dom}_e f$. Then for any $t \geq f(v)$,

$$\langle v', v \rangle_{V',V} + \alpha_0 t \geq \alpha,$$

and so letting $t \to \infty$, we obtain that $\alpha_0 \geq 0$. If $\alpha_0 = 0$, then since $(v_0, f(v_0)) \in \mathrm{epi}\, f$ and $(v_0, s_0) \in C$, taking $v = v_0$ in (4.24) we get $\langle v', v_0 \rangle_{V',V} = \alpha$. Again by (4.24),

$$\langle v', v - v_0 \rangle_{V',V} \geq 0 \quad \text{for all } v \in \mathrm{dom}_e f.$$

But since f is continuous at v_0 we may find a neighborhood of zero $U_1 \subset V$ such that $v_0 + U_1 \subset \mathrm{dom}_e f$. This implies that

$$\langle v', u \rangle_{V',V} \geq 0 \quad \text{for all } u \in U_1,$$

and since U_1 is absorbing, we have that

$$\langle v', u \rangle_{V',V} \geq 0 \quad \text{for all } u \in V,$$

which is a contradiction unless $v' = 0$. But since $L \neq 0$, this is not possible, and so $\alpha_0 > 0$, so that by (4.24),

$$f(v) \geq \frac{\alpha}{\alpha_0} - \frac{1}{\alpha_0} \langle v', v \rangle_{V',V} \quad \text{for all } v \in V,$$

$$t \leq \frac{\alpha}{\alpha_0} - \frac{1}{\alpha_0} \langle v', v \rangle_{V',V} \quad \text{for all } (v, t) \in C,$$

which gives the desired conclusion.

We are now ready to prove Theorem 4.51.

Proof (Theorem 4.51). By Theorem 4.43 the function f is proper and continuous in the interior of $\text{dom}_e f$. Hence it suffices to show that $\partial f(v_0) \neq \emptyset$. By the previous lemma with $C := \{(v_0, f(v_0))\}$, there exist $v' \in V'$ and $\alpha \in \mathbb{R}$ such that

$$f(v) \geq \alpha + \langle v', v \rangle_{V',V} \quad \text{for all } v \in V, \quad f(v_0) \leq \alpha + \langle v', v_0 \rangle_{V',V}.$$

Taking $v = v_0$ it follows that $f(v_0) = \alpha + \langle v', v_0 \rangle_{V',V}$, and so

$$f(v) \geq f(v_0) + \langle v', v - v_0 \rangle_{V',V} \quad \text{for all } v \in V.$$

Hence $\partial f(v_0) \neq \emptyset$.

As a corollary of the previous theorem we have the following.

Corollary 4.53. *Let* $f : \mathbb{R}^m \to (-\infty, \infty]$ *be a proper convex function. If* $z \in \text{ri}_{\text{aff}}(\text{dom}_e f)$, *then* $\partial f(z) \neq \emptyset$. *In particular, if* f *is real-valued, then* $\partial f(z) \neq \emptyset$ *for every* $z \in \mathbb{R}^m$.

Proof. Fix a point $z_0 \in \text{ri}_{\text{aff}}(\text{dom}_e f)$. If $\text{dom}_e f$ consists only of z_0, then

$$f(z_0) = \min_{z \in \mathbb{R}^m} f(z),$$

and so $0 \in \partial f(z_0)$ by (4.22).

If $\text{dom}_e f$ has more than one point, by Remark 4.1 the affine hull of its effective domain $\text{aff}(\text{dom}_e f)$ is, up to a translation, a subspace W of $\mathbb{R}^m$ of dimension $1 \leq \ell \leq m$. Since linear changes of variables preserve convexity, without loss of generality we may assume that $\text{aff}(\text{dom}_e f) = \mathbb{R}^\ell \times \{0\}$. Define

$$g(w) := f((w, 0)), \quad w \in \mathbb{R}^\ell. \tag{4.25}$$

Then the relative interior of $\mathrm{dom}_e\, g$ reduces simply to the interior of $\mathrm{dom}_e\, g$, and so, writing $z_0 = (w_0, 0)$, then w_0 belongs to the interior of $\mathrm{dom}_e\, g$. By Theorem 4.36 the function g is continuous in the interior of $\mathrm{dom}_e\, g$, and so by Theorem 4.51 we have that $\partial g\, (w_0)$ is nonempty. Hence there exists $\xi_0 \in \mathbb{R}^\ell$ such that

$$g\,(w) \geq g\,(w_0) + \xi_0 \cdot (w - w_0) \quad \text{for all } w \in \mathrm{dom}_e\, g.$$

Since $\mathrm{aff}\,(\mathrm{dom}_e\, f) = \mathbb{R}^\ell \times \{0\}$, by (4.25) the previous inequality may be written as

$$f\,(z) \geq f\,(z_0) + (\xi_0, 0) \cdot (z - z_0) \quad \text{for all } z \in \mathrm{aff}\,(\mathrm{dom}_e\, f).$$

On the other hand, if $z \notin \mathrm{aff}\,(\mathrm{dom}_e\, f)$, then $f\,(z) = \infty$, and so $(\xi_0, 0) \in \partial f\,(z_0)$.

Remark 4.54. It is actually possible to show that if $f\,(z_0) \in \mathbb{R}$ and f is not subdifferentiable at z_0, then

$$\frac{\partial^+ f}{\partial v}\,(z_0) = -\infty$$

for all $v \in -z_0 + \mathrm{ri}_{\mathrm{aff}}\,(\mathrm{dom}_e\, f)$. See Theorem 4.56 below.

Exercise 4.55. The set of points at which a convex function is subdifferentiable may be larger than $\mathrm{ri}_{\mathrm{aff}}\,(\mathrm{dom}_e\, f)$. Indeed, let $m = 2$ and consider the function

$$f\,(z) = f\,(z_1, z_2) = \begin{cases} \max\left\{1 - z_1^{\frac{1}{2}}, |z_1|\right\} & \text{if } z_1 \geq 0, \\ \infty & \text{otherwise.} \end{cases}$$

Prove that f is subdifferentiable everywhere in the half-plane $\{z_1 \geq 0\}$ except in the relative interior of the segment joining $(0, 1)$ and $(0, -1)$. Note that the set on which f is subdifferentiable is not convex.

The following theorem relates the subdifferential to one-sided directional derivatives.

Theorem 4.56. *Let V be a Banach space and let $f : V \to (-\infty, \infty]$ be a proper lower semicontinuous convex function. Assume that the interior of $\mathrm{dom}_e\, f$ is nonempty. If $v_0 \in \mathrm{dom}_e\, f$, then one of the following two alternatives holds:*

(i) $\partial f\,(v_0) = \emptyset$ and

$$\frac{\partial^+ f}{\partial v}\,(v_0) = -\infty$$

for all $v \in -v_0 + \overset{\circ}{\mathrm{dom}}_e\, f$;

(ii) $\partial f(v_0) \neq \emptyset$, the function $v \mapsto \frac{\partial^+ f}{\partial v}(v_0)$ is continuous on $-v_0 + \overset{\circ}{\text{dom}}_e f$, and

$$\frac{\partial^+ f}{\partial v}(v_0) = \max_{v' \in \partial f(v_0)} \langle v', v \rangle_{V', V}$$

for all v in $-v_0 + \overset{\circ}{\text{dom}}_e f$.

Proof. Let A denote the interior of $\text{dom}_e f$.

Step 1: We claim that if $v_0 \in A$ then

$$\frac{\partial^+ f}{\partial v}(v_0) = \max_{v' \in \partial f(v_0)} \langle v', v \rangle_{V', V}$$

for all $v \in V$. By Corollary 4.45 and Theorem 4.51, f is continuous and subdifferentiable in A, and so by Remark 4.49,

$$\frac{\partial^+ f}{\partial v}(v_0) \geq \sup_{v' \in \partial f(v_0)} \langle v', v \rangle_{V', V} > -\infty$$

for all $v \in V$.

To prove the reverse inequality we show that

$$p(v) := \frac{\partial^+ f}{\partial v}(v_0), \quad v \in V,$$

is continuous. Since $v_0 \in A$, by Corollary 4.45 and Theorem 4.47 there exist $r, M > 0$ such that

$$f(w) - f(v_0) \leq M \|w - v_0\|$$

for all $w \in B(v_0, r)$. Hence if $v \in V \setminus \{0\}$, then for $0 < t < r/\|v\|$ we have

$$\frac{f(v_0 + tv) - f(v_0)}{t} \leq M \|v\|,$$

and letting $t \to 0^+$ we conclude that

$$p(v) \leq M \|v\|.$$

Since $p : V \to \mathbb{R}$ is convex by Remark 4.49, it follows from Theorem 4.43 that it is actually continuous. Fix $v_1 \in V \setminus \{0\}$. By Lemma 4.52, with p and v_1 in place of f and v_0 and with $C := \{(v_1, p(v_1))\}$, we may find $v' \in V'$ and $\alpha \in \mathbb{R}$ such that

$$p(v) \geq \alpha + \langle v', v \rangle_{V', V} \quad \text{for all } v \in V, \quad p(v_1) \leq \alpha + \langle v', v_1 \rangle_{V', V}. \qquad (4.26)$$

Replacing v with tv, $t > 0$, in the first inequality and using the fact that p is positively homogeneous (see Remark 4.49), we obtain that

$$t(p(v) - \langle v', v \rangle_{V', V}) \geq \alpha \quad \text{for all } v \in V \text{ and } t > 0.$$

Letting $t \to 0^+$ we conclude that $\alpha \leq 0$, while if we divide by $t > 0$ and then let $t \to \infty$ we obtain

$$\frac{\partial^+ f}{\partial v}(v_0) = p(v) \geq \langle v', v \rangle_{V',V} \quad \text{for all } v \in V,$$

which implies in particular that $v' \in \partial f(v_0)$. It now follows from the second inequality in (4.26) that $\alpha = 0$ and

$$\frac{\partial^+ f}{\partial v_1}(v_0) = p(v_1) = \langle v', v_1 \rangle_{V',V},$$

and so the claim holds.

Step 2: Assume that $\frac{\partial^+ f}{\partial v}(v_0) = -\infty$ for all $v \in -v_0 + A$. We claim that $\partial f(v_0) = \emptyset$. Indeed, if not, fix $v \in -v_0 + A$ and let $v' \in \partial f(v_0)$. By letting $t \to 0^+$ in (4.23) we obtain a contradiction.

Step 3: Assume that $\frac{\partial^+ f}{\partial v_1}(v_0) > -\infty$ for some $v_1 \in -v_0 + A$. Consider the sequence of functions

$$g_n(v) := \frac{1}{n}\left[f\left(v_0 + \frac{1}{n} v \right) - f(v_0) \right], \quad v \in V.$$

Each g_n is convex, proper, lower semicontinuous, and continuous on $-v_0 + A$ by Corollary 4.45. By Remark 4.49 the function

$$g(v) := \inf_{n \in \mathbb{N}} g_n(v) = \lim_{n \to \infty} g_n(v) = \frac{\partial^+ f}{\partial v}(v_0), \quad v \in V,$$

is convex, and since $\mathrm{dom}_e\, g \supset \mathrm{dom}_e\, g_1$, we deduce that $-v_0 + A \subset \mathrm{dom}_e\, g$. Also $g(v_1) \in \mathbb{R}$. Since g is finite at some point in the interior of its domain, g is proper and in particular finite on $-v_0 + A$ (see Remark 4.23).

Since $-v_0 + A$ is a Baire space and g is the limit of a sequence of continuous functions on $-v_0 + A$, the set of points at which g is continuous is dense. By Corollary 4.45, g is continuous on $-v_0 + A$. Hence by Step 1, for every $v_1 \in -v_0 + A$ and $v \in V$,

$$\frac{\partial^+ g}{\partial v}(v_1) = \max_{v' \in \partial g(v_1)} \langle v', v \rangle_{V',V}. \tag{4.27}$$

Fix $v' \in \partial g(v_1)$ and $v \in V$. Since g is subadditive by Remark 4.49, we have

$$\frac{\partial^+ f}{\partial v}(v_0) = g(v) \geq g(v_1 + v) - g(v_1) \geq \langle v', v \rangle_{V',V},$$

and given the arbitrariness of $v \in V$ it follows that $v' \in \partial f(v_0)$, and in turn, $\partial g(v_1) \subset \partial f(v_0)$. This implies in particular that $\partial f(v_0)$ is nonempty, and so by Remark 4.49,

$$\sup_{v' \in \partial f(v_0)} \langle v', v_1 \rangle_{V',V} \leq \frac{\partial^+ f}{\partial v_1}(v_0).$$

By (4.27) there is $v_1' \in \partial g(v_1)$ such that

$$\frac{\partial^+ g}{\partial v_1}(v_1) = \langle v_1', v_1 \rangle_{V',V}.$$

Then $v_1' \in \partial f(v_0)$ and

$$\langle v_1', v_1 \rangle_{V',V} = \frac{\partial^+ g}{\partial v_1}(v_1) = \lim_{t \to 0^+} \frac{g(v_1 + tv_1) - g(v_1)}{t}$$

$$= g(v_1) \lim_{t \to 0^+} \frac{(1+t) - 1}{t} = g(v_1) = \frac{\partial^+ f}{\partial v_1}(v_0),$$

where we have used the fact that g is positively homogeneous. Hence

$$\max_{v' \in \partial f(v_0)} \langle v', v_1 \rangle_{V',V} = \frac{\partial^+ f}{\partial v_1}(v_0)$$

for all $v_1 \in -v_0 + A$.

Remark 4.57. If V is a Banach space and $f : V \to (-\infty, \infty]$ is a proper lower semicontinuous convex function, then considering the restriction $f :$ aff $(\mathrm{dom}_e\, f) \to (-\infty, \infty]$, the previous theorem still holds provided the interior of the effective domain in aff $(\mathrm{dom}_e\, f)$ is nonempty, i.e., $\mathrm{ri}_{\mathrm{aff}}(\mathrm{dom}_e\, f) \neq \emptyset$.

When V is only a locally convex topological vector space we can still prove a weaker form of the previous theorem:

Proposition 4.58. *Let V be a locally convex topological vector space and let $f : V \to (-\infty, \infty]$ be a proper convex function. If $v_0 \in V$ is such that $f(v_0) \in \mathbb{R}$ and f is continuous at v_0, then*

$$\frac{\partial^+ f}{\partial v}(v_0) = \max_{v' \in \partial f(v_0)} \langle v', v \rangle_{V',V}$$

for all $v \in V$.

Proof. By Theorem 4.51, f is subdifferentiable at v_0, and so by Remark 4.49,

$$\frac{\partial^+ f}{\partial v}(v_0) \geq \sup_{v' \in \partial f(v_0)} \langle v', v \rangle_{V',V} > -\infty$$

for all $v \in V$. To prove the reverse inequality, as in Step 1 of the previous theorem it suffices to show that the seminorm

$$p(v) := \frac{\partial^+ f}{\partial v}(v_0), \quad v \in V,$$

is continuous. Since

$$p(v) = \inf_{t>0} \frac{f(v_0 + tv) - f(v_0)}{t} \leq f(v_0 + v) - f(v_0)$$

and f is continuous at v_0, it follows that p is bounded from above in a neighborhood of v_0, and so it is continuous by Theorem 4.43.

Next we study the subdifferentiability of the sum of two convex functions.

Proposition 4.59. *Let V be a locally convex topological vector space and let f_1, $f_2 : V \to (-\infty, \infty]$ be two proper convex, lower semicontinuous functions. Assume that there exists*

$$v_0 \in \mathrm{dom}_e f_1 \cap \mathrm{dom}_e f_2$$

such that f_1 is continuous at v_0. Then for every $v \in V$,

$$\partial (f_1 + f_2)(v) = \partial f_1(v) + \partial f_2(v).$$

Proof. **Step 1:** Let $v \in V$. If $v_1' \in \partial f_1(v)$ and $v_2' \in \partial f_2(v)$, then

$$f_1(w) \geq f_1(v) + \langle v_1', w - v \rangle_{V',V} \quad \text{for all } w \in V,$$
$$f_2(w) \geq f_2(v) + \langle v_2', w - v \rangle_{V',V} \quad \text{for all } w \in V,$$

and so, adding the two inequalities, we conclude that $v_1' + v_2' \in \partial (f_1 + f_2)(v)$. Hence if $\partial f_1(v)$ and $\partial f_2(v)$ are nonempty, then $\partial (f_1 + f_2)(v)$ is nonempty and

$$\partial (f_1 + f_2)(v) \supset \partial f_1(v) + \partial f_2(v).$$

Step 2: Conversely, if $\partial (f_1 + f_2)(v)$ is nonempty, let $v' \in \partial (f_1 + f_2)(v)$. Then

$$(f_1 + f_2)(w) \geq (f_1 + f_2)(v) + \langle v', w - v \rangle_{V',V} \quad \text{for all } w \in V. \tag{4.28}$$

In particular, $f_1(v)$, $f_2(v) \in \mathbb{R}$. Define

$$g(w) := f_1(w) - f_1(v) + \langle v', w - v \rangle_{V',V}, \quad w \in V,$$

and

$$C := \{(w, t) \in V \times \mathbb{R} : t \leq f_2(v) - f_2(w)\}.$$

Then g is a convex function and C a convex set. We claim that g and C satisfy the hypotheses of Lemma 4.52. Indeed, since f_1 is proper and continuous at v_0 then so is g. Moreover, in view of (4.28), if $(w, t) \in \mathrm{epi}\, g \cap C$, then

$$g(w) = f_1(w) - f_1(v) + \langle v', w - v \rangle_{V',V} \leq t \leq f_2(v) - f_2(w)$$
$$\leq f_1(w) - f_1(v) + \langle v', w - v \rangle_{V',V},$$

which implies that $g(w) = t$ and $f_2(v) - f_2(w) = t$, that is, $(w, t) \in \partial(\text{epi } g) \cap \partial C$. Finally, $(v_0, f_2(v) - f_2(v_0)) \in C$. Hence we may apply Lemma 4.52 to find $w' \in V'$ and $\alpha \in \mathbb{R}$ such that

$$g(w) = f_1(w) - f_1(v) + \langle v', w - v \rangle_{V',V} \geq \alpha + \langle w', w \rangle_{V',V} \geq f_2(v) - f_2(w)$$

for all $w \in V$. Taking $w = v$ yields $\langle w', v \rangle_{V',V} = -\alpha$, so that

$$f_1(w) - f_1(v) + \langle v', w - v \rangle_{V',V} \geq \langle w', w - v \rangle_{V',V} \geq f_2(v) - f_2(w)$$

for all $w \in V$, and so $v' - w' \in \partial f_1(v)$ and $w' \in \partial f_2(v)$. Hence

$$v' = (v' - w') + w' \in \partial f_1(v) + \partial f_2(v),$$

which proves that if $\partial(f_1 + f_2)(v)$ is nonempty, then

$$\partial(f_1 + f_2)(v) \subset \partial f_1(v) + \partial f_2(v).$$

Step 3: To conclude the proof we observe that if either $\partial f_1(v)$ or $\partial f_2(v)$ is empty, then by Step 2 so must be $\partial(f_1 + f_2)(v)$. If both $\partial f_1(v)$ and $\partial f_2(v)$ are nonempty, then by Steps 1 and 2, $\partial(f_1 + f_2)(v)$ is also nonempty and

$$\partial(f_1 + f_2)(v) = \partial f_1(v) + \partial f_2(v).$$

This completes the proof.

We now turn to the relation between subdifferentiability and (Gâteaux) differentiability.

Definition 4.60. *Let V be a locally convex topological vector space. A function $f : V \to [-\infty, \infty]$ is Gâteaux differentiable at $v_0 \in V$ if $f(v_0) \in \mathbb{R}$ and there exists $v' \in V'$ such that for every $v \in V$,*

$$\lim_{t \to 0^+} \frac{f(v_0 + tv) - f(v_0)}{t} = \langle v', v \rangle_{V',V}.$$

The element v' is called the Gâteaux differential of f at v_0 and is denoted by $f'(v_0)$.

Theorem 4.61. *Let V be a locally convex topological vector space and let $f : V \to [-\infty, \infty]$ be a convex function. If f is Gâteaux differentiable at $v_0 \in V$, then it is subdifferentiable at v_0 and $\partial f(v_0) = \{f'(v_0)\}$. Conversely, if f is continuous and finite at $v_0 \in V$ and the subdifferential of f at v_0 is a singleton, then f is Gâteaux differentiable at v_0.*

Proof. Assume that f is Gâteaux differentiable at $v_0 \in V$, let $v \in V$, and define $g(t) := f(v_0 + tv)$. By Proposition 4.34 the difference quotient

$$t \mapsto \frac{g(t) - g(0)}{t - 0}$$

is nondecreasing in $\mathbb{R} \setminus \{t_0\}$. Hence

$$f(v_0 + v) - f(v_0) = \frac{g(1) - g(0)}{1 - 0} \geq \lim_{t \to 0^+} \frac{g(t) - g(0)}{t - 0}$$

$$= \lim_{t \to 0^+} \frac{f(v_0 + tv) - f(v_0)}{t} = \langle f'(v_0), v \rangle_{V', V},$$

which implies that $f'(v_0) \in \partial f(v_0)$. We claim that $\partial f(v_0) = \{f'(v_0)\}$. Indeed, if $v' \in \partial f(v_0)$, then for any $v \in V$ and $t > 0$,

$$\frac{f(v_0 + tv) - f(v_0)}{t} \geq \langle v', v \rangle_{V', V}.$$

By letting $t \to 0^+$ we obtain that

$$\langle f'(v_0), v \rangle_{V', V} \geq \langle v', v \rangle_{V', V} \text{ for all } v \in V,$$

which implies that $v' = f'(v_0)$.

Conversely, assume that f is continuous and finite at $v_0 \in V$ and the subdifferential of f at v_0 is a singleton $\{v'\}$. Then $f > -\infty$ by Remark 4.23, and so by Proposition 4.58 and the fact that $\partial f(v_0) = \{v'\}$ we deduce that

$$\frac{\partial^+ f}{\partial v}(v_0) = \langle v', v \rangle_{V', V}$$

for all $v \in V$, which shows that f is Gâteaux differentiable at v_0.

The next result shows that for smooth functions convexity is equivalent to the monotonicity of the Gâteaux differential.

Theorem 4.62. *Let E be a convex subset of a locally convex topological vector space V and let $f : V \to [-\infty, \infty]$ be Gâteaux differentiable in E. Then the following three conditions are equivalent:*

(i) $f : E \to \mathbb{R}$ is convex;
(ii) for all $v, w \in E$,

$$f(v) \geq f(w) + \langle f'(w), v - w \rangle_{V', V};$$

(iii) for all $v, w \in E$,

$$\langle f'(v) - f'(w), v - w \rangle_{V', V} \geq 0.$$

Proof. Assume that (i) holds. Since f is Gâteaux differentiable in E, by the previous theorem f is subdifferentiable at every $w \in E$ and $\partial f(w) = \{f'(w)\}$. Hence (ii) holds.

Assume next that (ii) holds. Then for all $v, w \in E$,

$$f(v) \geq f(w) + \langle f'(w), v - w \rangle_{V',V},$$
$$f(w) \geq f(v) - \langle f'(v), v - w \rangle_{V',V},$$

and by adding these inequalities we obtain that

$$0 \geq \langle f'(w) - f'(v), v - w \rangle_{V',V},$$

which gives (iii).

Finally, assume that (iii) holds and fix $v, w \in E$. Since f is Gâteaux differentiable in E, the function $g : [0,1] \to \mathbb{R}$, defined by

$$g(t) := f(tv + (1-t)w), \quad t \in [0,1],$$

is differentiable and

$$g'(t) = \langle f'(tv + (1-t)w), v - w \rangle_{V',V}.$$

If $s > t$, then

$$\begin{aligned} g'(s) - g'(t) &= \langle f'(sv + (1-s)w) - f'(tv + (1-t)w), v - w \rangle_{V',V} \\ &= \frac{1}{s-t} \langle f'(sv + (1-s)w) - f'(tv + (1-t)w), \\ &\quad (sv + (1-s)w) - (tv + (1-t)w) \rangle_{V',V} \geq 0. \end{aligned}$$

Hence g' is nondecreasing, and so g is convex. In particular,

$$f(tv + (1-t)w) = g(t) \leq (1-t)g(0) + tg(1) = (1-t)f(w) + tf(v),$$

which implies the convexity of f.

Remark 4.63. A similar result holds for strictly convex functions provided we require the inequalities (i) and (ii) to be strict when $v \neq w$.

As a consequence of the previous theorem we now specialize the result of Theorem 4.36 to obtain a p-Lipschitz condition for separately convex functions with algebraic growth.

Proposition 4.64. *Let $f : \mathbb{R}^m \to \mathbb{R}$ be a separately convex function such that*

$$|f(z)| \leq C(1 + |z|^p)$$

for some $C > 0$, $p \geq 1$, and all $z \in \mathbb{R}^m$. Then

$$|f(z) - f(w)| \leq C\left(1 + |z|^{p-1} + |w|^{p-1}\right)|z - w|$$

for all $z, w \in \mathbb{R}^m$.

Proof. **Step 1:** We first assume that $f \in C^\infty(\mathbb{R}^m)$. Let $z := (z_1, \ldots, z_m) \in \mathbb{R}^m$ be fixed and consider

$$g(t) := f(z_1, \ldots, z_{i-1}, t, z_{i+1}, \ldots, z_m), \quad t \in \mathbb{R}.$$

Since g is convex and smooth by Theorem 4.62 (see also Proposition 4.34), for all s and $t \in \mathbb{R}$ we have

$$g(t+s) - g(t) \geq g'(t) s.$$

Thus, with $s := 1 + |z|$, $t := z_i$,

$$g'(t) = \frac{\partial f}{\partial z_i}(z) \leq \frac{g(t+s) - g(t)}{s} \leq C \frac{(1+|z|^p)}{1+|z|} \leq C\left(1 + |z|^{p-1}\right).$$

Also, $g(t-s) - g(t) \geq -g'(t) s$, and so

$$-g'(t) \leq \frac{g(t-s) - g(t)}{s} \leq C \frac{(1+|z|^p)}{1+|z|} \leq C\left(1 + |z|^{p-1}\right).$$

Hence

$$\left|\frac{\partial f}{\partial z_i}(z)\right| \leq C\left(1 + |z|^{p-1}\right)$$

for every $i = 1, \ldots, m$.

Let $z, w \in \mathbb{R}^m$. By the mean value theorem there is $\theta \in (0,1)$ such that

$$
\begin{aligned}
|f(z) - f(w)| &= |\nabla f(\theta z + (1-\theta) w) \cdot (z - w)| \\
&\leq C\left(1 + |\theta z + (1-\theta) w|^{p-1}\right) |(z-w)| \\
&\leq C\left(1 + |z|^{p-1} + |w|^{p-1}\right) |z - w|,
\end{aligned}
$$

where we have used the fact that if $0 < q < \infty$ and $a, b \geq 0$, then

$$(a+b)^q \leq \max\left\{1, 2^{q-1}\right\}(a^q + b^q).$$

Step 2: To remove the smoothness hypothesis consider a standard mollifier φ_ε and let

$$f_\varepsilon(z) := \int_{\mathbb{R}^m} \varphi_\varepsilon(w) f(z-w)\, dw, \quad \varepsilon > 0.$$

The function f_ε is still separately convex and, in addition,

$$
\begin{aligned}
|f_\varepsilon(z)| &\leq \int_{\mathbb{R}^m} \varphi_\varepsilon(w) |f(z-w)|\, dw \leq C \int_{B(0,\varepsilon)} \varphi_\varepsilon(w)(1 + |z-w|^p)\, dw \\
&\leq C(1 + |z|^p) \int_{\mathbb{R}^m} \varphi_\varepsilon(w)\, dw = C(1 + |z|^p),
\end{aligned}
$$

where the constants are independent of ε. By the previous step,

$$|f_\varepsilon(z) - f_\varepsilon(w)| \leq C \left(1 + |z|^{p-1} + |w|^{p-1}\right)|z - w|,$$

and since $f_\varepsilon(z) \to f(z)$ pointwise as $\varepsilon \to 0^+$ (see Theorem 2.75) we obtain the desired result for f.

We conclude this subsection by proving that differentiable separately convex functions are of class C^1.

Theorem 4.65. *Let $B \subset \mathbb{R}^m$ be an open ball. If $f : B \to \mathbb{R}$ is separately convex and E is the set of points in B at which f is differentiable, then $\nabla f : E \to \mathbb{R}^m$ is continuous.*

Proof. Let $z_0 \in E$ and define

$$h(z) := f(z) - f(z_0) - \nabla f(z_0) \cdot (z - z_0), \quad z \in B.$$

Then the function h is separately convex and differentiable in E. By Theorem 4.36,

$$|\nabla f(z) - \nabla f(z_0)| = |\nabla h(z)| \leq \mathrm{Lip}\,(h; B(z_0, r)) \leq \sqrt{m}\frac{\mathrm{osc}\,(h; B(z_0, 2r))}{r}$$

for any $z \in E$ with $|z - z_0| < r$ and with $B(z_0, 2r) \subset B$. Since f is differentiable at z_0 we have that

$$\lim_{r \to 0^+} \sup_{z \in E, |z - z_0| < r} |\nabla f(z) - \nabla f(z_0)| \leq \lim_{r \to 0^+} \sqrt{m}\frac{\mathrm{osc}\,(h; B(z_0, 2r))}{r} = 0,$$

and thus $\nabla f(z) \to \nabla f(z_0)$ as $z \to z_0$, $z \in E$.

4.6 Recession Function

In this section we introduce and study some properties of the recession function of a convex function. This material will be used in Section 5.2.3.

Definition 4.66. *If $C \subset \mathbb{R}^m$ is a nonempty convex set, then its recession cone is defined by*

$$C^\infty := \{z \in \mathbb{R}^m : tz + w \in C \text{ for all } w \in C \text{ and } t \geq 0\}.$$

Proposition 4.67. *If $C \subset \mathbb{R}^m$ is nonempty and convex, then*

(i) *C^∞ is convex and $C^\infty = \{z \in \mathbb{R}^m : z + w \in C \text{ for all } w \in C\}$, i.e., $z + C \subset C$ if and only if $tz + C \subset C$ for all $t \geq 0$.*

If, in addition, C is closed, then

(ii) C^∞ is closed;
(iii) $z \in C^\infty$ if and only if $z = \lim_{n \to \infty} \theta_n w_n$ with $w_n \in C$ and $\theta_n \to 0^+$.

Proof. In part (i) the convexity of C^∞ and the inclusion

$$C^\infty \subset \{z \in \mathbb{R}^m : z + w \in C \text{ for all } w \in C\} =: D$$

are immediate. To prove the converse inclusion let $z \in D$. Then $z + w \in C$ for all $w \in C$. By induction, it follows that $w + nz \in C$ for all $n \in \mathbb{N}$. Therefore, if $t \in [n-1, n]$, then writing $t = \theta (n-1) + (1-\theta) n$ for some $\theta \in [0, 1]$, we have

$$w + tz = \theta (w + (n-1) z) + (1-\theta) (w + nz) \in C$$

by convexity of C, and so $z \in C^\infty$.

To prove (ii) let $z = \lim_{n \to \infty} z_n$, $z_n \in C^\infty$. In view of part (i) it suffices to prove that $w + z \in C$ for all $w \in C$. Suppose that $w + z \notin C$ for some $w \in C$. Since C is closed, there would exist $r > 0$ such that $B(w + z, r) \cap C = \emptyset$. However, $w + z_n \in C$ for every n and thus

$$0 < r \leq \text{dist}(w + z, C) \leq |(w + z) - (w + z_n)| \to 0,$$

and this is a contradiction.

Similarly, in part (iii), if $z = \lim_{n \to \infty} \theta_n w_n$, with $w_n \in C$ and $\theta_n \to 0^+$, and z were not in C^∞, then there would exist $w \in C$ and $r > 0$ such that $B(w + z, r) \cap C = \emptyset$. If n is sufficiently large we may assume that $\theta_n \in [0, 1]$, and so

$$(1 - \theta_n) w + \theta_n w_n \in C$$

by convexity of C, with

$$w + z = \lim_{n \to \infty} \{(1 - \theta_n) w + \theta_n w_n + \theta_n w\}.$$

Once again we would reach a contradiction because

$$0 < r \leq \text{dist}(w + z, C) \leq |\theta_n w| \to 0.$$

Conversely, if $z \in C^\infty$ and $w_0 \in C$, then $nz + w_0 = w_n \in C$ for all $n \in \mathbb{N}$, and so $\theta_n w_n \to z$ with $\theta_n := \frac{1}{n}$.

Remark 4.68. It can be shown that the notion of recession cone of an arbitrary set $E \subset \mathbb{R}^m$ introduced as

$$E^\infty := \{z \in \mathbb{R}^m : \text{there exist } w_n \in E, \, \theta_n \to 0^+ \text{ such that } \theta_n w_n \to z\}$$

inherits several properties of the recession cone of a convex set (see [RocWe98]). Note, however, that E^∞ defined in this way is always closed, which is in contrast to Definition 4.66 (see Theorem 4.67 (iii)).

We now introduce the concept of recession function.

Definition 4.69. *Let $f : \mathbb{R}^m \to (-\infty, \infty]$ be a proper convex function. The recession function of f is the function $f^\infty : \mathbb{R}^m \to [-\infty, \infty]$ defined by*

$$f^\infty(z) := \sup \{f(w + z) - f(w) : w \in \mathrm{dom}_e f\}, \quad z \in \mathbb{R}^m.$$

The next result relates the recession function of a convex function to the recession cone of its epigraph.

Theorem 4.70. *Let $f : \mathbb{R}^m \to (-\infty, \infty]$ be a proper convex function. The recession function f^∞ of f is a positively homogeneous proper convex function and*

$$\mathrm{epi}\, f^\infty = (\mathrm{epi}\, f)^\infty. \tag{4.29}$$

Moreover, if f is lower semicontinuous, then so is f^∞, and for every $w \in \mathrm{dom}_e f$ we have

$$f^\infty(z) = \sup_{t>0} \frac{f(w + tz) - f(w)}{t} = \lim_{t \to \infty} \frac{f(w + tz) - f(w)}{t}. \tag{4.30}$$

Proof. We begin by showing that

$$\mathrm{epi}\, f^\infty = (\mathrm{epi}\, f)^\infty. \tag{4.31}$$

Note that in view of Propositions 4.22 and 4.67(i) this entails that $\mathrm{epi}\, f^\infty$ is convex, and so f^∞ is convex. Convexity of f^∞ also follows from the fact that f^∞ is the supremum of a family of convex functions.

By Proposition 4.67(i) we have

$$
\begin{aligned}
(\mathrm{epi}\, f)^\infty &= \{(z, t) \in \mathbb{R}^m \times \mathbb{R} : (w, s) + (z, t) \in \mathrm{epi}\, f \\
&\qquad \text{for all } (w, s) \in \mathrm{epi}\, f\} \\
&= \{(z, t) \in \mathbb{R}^m \times \mathbb{R} : f(w + z) \leq s + t \\
&\qquad \text{for all } w \in \mathrm{dom}_e f \text{ and all } s \in \mathbb{R} \text{ such that } s \geq f(w)\} \\
&= \{(z, t) \in \mathbb{R}^m \times \mathbb{R} : f(w + z) \leq f(w) + t \\
&\qquad \text{for all } w \in \mathrm{dom}_e f\} \\
&= \{(z, t) \in \mathbb{R}^m \times \mathbb{R} : f^\infty(z) \leq t\} = \mathrm{epi}\, f^\infty.
\end{aligned}
$$

Since for a fixed $w \in \mathrm{dom}_e f$,

$$f^\infty(z) \geq f(w + z) - f(w)$$

for all $z \in \mathbb{R}^m$, we have that f^∞ never takes the value $-\infty$. On the other hand, f^∞ is not identically ∞ because $f^\infty(0) = 0$. Hence f^∞ is proper.

To prove that f^∞ is positively homogeneous let $z \in \mathbb{R}^m$ and $t > 0$. We first claim that

$$tf^\infty(z) \geq f^\infty(tz). \tag{4.32}$$

If $f^\infty(z) = \infty$ there is nothing to prove. Thus assume that $f^\infty(z) < \infty$. Then $(z, f^\infty(z)) \in \mathrm{epi}\, f^\infty$, and so by (4.29) and the definition of recession cone,

$$t\left(z, f^{\infty}(z)\right) + \operatorname{epi} f \subset \operatorname{epi} f.$$

In particular, $t\left(z, f^{\infty}(z)\right) + (w, f(w)) \in \operatorname{epi} f$ for all $w \in \operatorname{dom}_e f$, and thus

$$f(w + tz) \leq f(w) + t f^{\infty}(z), \tag{4.33}$$

or equivalently,

$$f(w + tz) - f(w) \leq t f^{\infty}(z)$$

for all $w \in \operatorname{dom}_e f$. Taking the supremum on the left-hand side over all $w \in \operatorname{dom}_e f$, it follows that

$$f^{\infty}(tz) \leq t f^{\infty}(z)$$

for all $z \in \mathbb{R}^m$ and $t > 0$. To prove the reverse inequality it suffices to replace z and t in the previous inequality with tz and $\frac{1}{t}$.

Assume next that f is lower semicontinuous. Note that by (4.31) and Proposition 4.67(ii), $\operatorname{epi} f^{\infty}$ is closed, and thus by Proposition 3.10, f^{∞} is lower semicontinuous.

To prove (4.30), in view of Proposition 4.34 it suffices to establish the equality

$$f^{\infty}(z) = \sup_{t>0} \frac{f(w + tz) - f(w)}{t}$$

for all $z \in \mathbb{R}^m$ and for any fixed $w \in \operatorname{dom}_e f$. Fix $w_0 \in \operatorname{dom}_e f$. We show first that

$$f^{\infty}(z) \geq \frac{f(w_0 + tz) - f(w_0)}{t}$$

for all $z \in \mathbb{R}^m$. Reasoning as in the proof of (4.32) we have that (4.33) holds for all $w \in \operatorname{dom}_e f$ and for every $t \geq 0$. In particular,

$$f(w_0 + tz) \leq f(w_0) + t f^{\infty}(z)$$

for all $t \geq 0$ and $z \in \mathbb{R}^m$, which gives

$$\sup_{t>0} \frac{f(w_0 + tz) - f(w_0)}{t} \leq f^{\infty}(z) \tag{4.34}$$

for all $z \in \mathbb{R}^m$. Conversely, if for $z \in \mathbb{R}^m$,

$$\sup_{t>0} \frac{f(w_0 + tz) - f(w_0)}{t} = \infty,$$

then there is nothing to prove; otherwise, let

$$s \geq \frac{f(w_0 + tz) - f(w_0)}{t}$$

for all $t > 0$. Then $st + f(w_0) \geq f(w_0 + tz)$ for all $t > 0$, and so

$$(w_0 + tz, st + f(w_0)) \in \operatorname{epi} f \tag{4.35}$$

for all $t > 0$. Write

$$(z, s) = \lim_{n \to \infty} \frac{1}{n} \left((w_0, f(w_0)) + n(z, s) \right).$$

By (4.35) and Proposition 4.67(iii), it follows that $(z, s) \in (\text{epi } f)^\infty$, and now invoking (4.31), we conclude that $f^\infty(z) \leq s$. Given the arbitrariness of s, we deduce that

$$f^\infty(z) \leq \sup_{t>0} \frac{f(w_0 + tz) - f(w_0)}{t}.$$

This, together with (4.34), yields (4.30).

Remark 4.71. For nonconvex functions it is also possible to define the recession function using formula (4.29), and where $(\text{epi } f)^\infty$ is given as in Remark 4.68 (see [RocWe98]). We call attention, however, to the fact that the recession function introduced in this way is always lower semicontinuous, in contrast to the case of a convex function (see Theorem 4.70).

Exercise 4.72. Prove that:

(i) If $1 \leq p < \infty$, then the recession function of $f(z) := |z|^p$ is

$$f^\infty(z) = \begin{cases} \infty & \text{if } z \neq 0, \\ 0 & \text{if } z = 0, \end{cases}$$

for $p > 1$, while $f^\infty(z) = f(z) = |z|$ if $p = 1$;

(ii) the recession function of $f(z) := \sqrt{|z|^2 + 1}$ is $f^\infty(z) = |z|$;

(iii) the recession function of

$$f(z) := Az \cdot z,$$

where A is a symmetric positive semidefinite matrix in $\mathbb{R}^{m \times m}$, is

$$f^\infty(z) = \begin{cases} \infty & \text{if } Az \neq 0, \\ 0 & \text{if } Az = 0. \end{cases}$$

The next result shows that the recession function is of interest only for convex functions that are not superlinear at infinity.

Theorem 4.73. *Let $f : \mathbb{R}^m \to (-\infty, \infty]$ be a proper convex lower semicontinuous function. Then*

$$\liminf_{|z| \to \infty} \frac{f(z)}{|z|} = \inf_{|z|=1} f^\infty(z).$$

Proof. By (4.29),

$$\inf_{|z|=1} f^\infty(z) = \inf_{|z|=1} \inf \{ t \in \mathbb{R} : (z, t) \in \text{epi } f^\infty \}$$

$$= \inf_{|z|=1} \inf \{ t \in \mathbb{R} : (z, t) \in (\text{epi } f)^\infty \}.$$

Since f is convex and lower semicontinuous, by Propositions 3.10 and 4.22, epi f is closed and convex, and by Theorem 4.67 (iii) we have

$$\inf_{|z|=1} f^\infty (z) = \inf_{|z|=1} \inf \{t \in \mathbb{R} : (z,t) \in (\text{epi } f)^\infty\}$$

$$= \inf_{|z|=1} \inf \{t \in \mathbb{R} : \text{there are } (w_n, t_n) \in \text{epi } f, \, \theta_n \to 0^+,$$

$$\text{such that } \theta_n (w_n, t_n) \to (z,t)\}$$

$$= \inf \{t \in \mathbb{R} : \text{there are } (w_n, t_n) \in \text{epi } f, \, \theta_n \to 0^+,$$

$$\text{such that } |\theta_n w_n| \to 1, \theta_n t_n \to t\}$$

$$= \inf \{t \in \mathbb{R} : \text{there are } w_n \in \mathbb{R}^m, \, \theta_n \to 0^+,$$

$$\text{such that } |\theta_n w_n| \to 1, \theta_n f(w_n) \to t\}$$

$$= \inf \left\{t \in \mathbb{R} : \text{there is } w_n \in \mathbb{R}^m, \text{ such that } |w_n| \to \infty, \frac{f(w_n)}{|w_n|} \to t\right\}$$

$$= \liminf_{|z| \to \infty} \frac{f(z)}{|z|}.$$

This concludes the proof.

Remark 4.74. The previous theorem continues to hold for nonconvex functions provided f^∞ is defined as in Remark 4.71.

4.7 Approximation of Convex Functions

In this section we show that every convex lower semicontinuous function f may be written as the supremum of a sequence of affine functions.

If V is a topological vector space, an *affine continuous* function $g : V \to \mathbb{R}$ is a function of the form

$$g(v) = \alpha + \langle v', v \rangle_{V',V},$$

where $v' \in V'$ and $\alpha \in \mathbb{R}$.

Proposition 4.75. *Let V be a locally convex topological vector space and let $f : V \to (-\infty, \infty]$ be a convex and lower semicontinuous function. Then*

(i) there exists an affine continuous function g such that $g \le f$;
(ii) $f(v) = \sup \{g(v) : g \text{ affine continuous}, g \le f\}$.

Proof. (i) If $f \equiv \infty$, then the result is immediate. So assume that there exists $v_0 \in \text{dom}_e f$, and fix $t_0 < f(v_0)$. By the second geometric form of the Hahn–Banach theorem, with $C := \text{epi } f$, $K := \{(v_0, t_0)\}$, there exist a continuous linear functional $L : V \times \mathbb{R} \to \mathbb{R}$ and two numbers $\alpha \in \mathbb{R}$ and $\varepsilon > 0$ such that

$$L(v, f(v)) \ge \alpha + \varepsilon \quad \text{for all } v \in \text{dom}_e f \text{ and } L(v_0, t_0) \le \alpha - \varepsilon.$$

The functional L has the form $L(v,t) = \langle v', v \rangle_{V',V} + \alpha_0 t$ for some $v' \in V'$ and $\alpha_0 \in \mathbb{R}$. Hence

$$\langle v', v \rangle_{V',V} + \alpha_0 f(v) \geq \alpha + \varepsilon \quad \text{for all } v \in \text{dom}_e\, f \text{ and } \langle v', v_0 \rangle_{V',V} + \alpha_0 t_0 \leq \alpha - \varepsilon.$$

Taking $v = v_0$ we get $\langle v', v_0 \rangle_{V',V} + \alpha_0 f(v_0) \geq \alpha + \varepsilon > \alpha - \varepsilon \geq \langle v', v_0 \rangle_{V',V} + \alpha_0 t_0$, and since $t_0 < f(v_0)$ it follows that $\alpha_0 > 0$ and thus

$$f(v) \geq g(v) := -\langle \alpha_0^{-1} v', v \rangle_{V',V} + \alpha_0^{-1}(\alpha + \varepsilon) \quad \text{for all } v \in \text{dom}_e\, f.$$

Hence $f \geq g$ in V.

(ii) If $f \equiv \infty$, then the result is immediate, so assume as before that there exists $v_0 \in \text{dom}_e\, f$. For any fixed $t_0 < f(v_0)$, we claim that there exists an affine function g below f with $g(v_0) \geq t_0$. Indeed, it suffices to observe that in the construction in part (i),

$$g(v_0) = -\langle \alpha_0^{-1} v', v_0 \rangle_{V',V} + \alpha_0^{-1}(\alpha + \varepsilon) \geq t_0.$$

This completes the proof.

Remark 4.76. If f takes the value $-\infty$, then (i) fails, and thus the right-hand side in (ii) is identically $-\infty$, while there exist functions $f : V \to \{-\infty, \infty\}$ that are convex and lower semicontinuous with $f \not\equiv -\infty$. As an example let $V = \mathbb{R}$ and define

$$f(z) := \begin{cases} \infty & \text{if } z > 0, \\ -\infty & \text{if } z \leq 0. \end{cases}$$

When more is known about the space V it may be possible to restrict the supremum in part (ii) of the previous proposition to a countable family of affine functions. For simplicity, here we address the case $V = \mathbb{R}^m$.

Proposition 4.77. *Let $f : \mathbb{R}^m \to (-\infty, \infty]$ be convex and lower semicontinuous. Then*

$$f(z) = \sup_{i \in \mathbb{N}} \{a_i + b_i \cdot z\} \tag{4.36}$$

for all $z \in \mathbb{R}^m$ and for some $a_i \in \mathbb{R}$, $b_i \in \mathbb{R}^m$. Moreover,

$$f^\infty(z) = \sup_{i \in \mathbb{N}} \{b_i \cdot z\} \tag{4.37}$$

for all $z \in \mathbb{R}^m$.

We begin with a preliminary result based on Lindelöf's theorem.

Proposition 4.78. *Let $A \subset \mathbb{R}^m$ be an open set, let $\mathcal{G} \subset C(A)$, and let*

$$f = \sup_{g \in \mathcal{G}} g. \tag{4.38}$$

Then there exists a countable subfamily $\mathcal{G}_1 \subset \mathcal{G}$ such that

$$f = \sup_{g \in \mathcal{G}_1} g.$$

Proof. The space $C(A)$ has a countable base of open sets when it is endowed with the topology compatible with the metric

$$d(g,h) := \max_{n \in \mathbb{N}} \frac{1}{2^n} \frac{\|g - h\|_{C(K_n)}}{1 + \|g - h\|_{C(K_n)}},$$

where

$$A = \bigcup_{n=1}^{\infty} K_n,$$

K_n is compact, nonempty, and K_n is contained in the interior of K_{n+1}. Let

$$\mathcal{A}_1 := \{B(g,r) : g \in \mathcal{G}, 0 < r \leq 1\}.$$

By Lindelöf's theorem, $\mathcal{A}_1$ admits a countable subfamily $\mathcal{D}_1$ such that

$$\mathcal{G} \subset \bigcup_{B(g,r) \in \mathcal{A}_1} B(g,r) = \bigcup_{B(g,r) \in \mathcal{D}_1} B(g,r).$$

Inductively, for every $k \in \mathbb{N}$ let

$$\mathcal{A}_k := \left\{ B(g,r) : g \in \mathcal{G}, 0 < r \leq \frac{1}{k} \right\},$$

and again by Lindelöf's theorem we may find a countable subfamily $\mathcal{D}_k \subset \mathcal{A}_k$ such that

$$\mathcal{G} \subset \bigcup_{B(g,r) \in \mathcal{A}_k} B(g,r) = \bigcup_{B(g,r) \in \mathcal{D}_k} B(g,r). \tag{4.39}$$

Write

$$\mathcal{D}_k = \left\{ B\left(g_j^{(k)}, r_j^{(k)}\right) : j \in \mathbb{N} \right\}.$$

We claim that

$$f = \sup_{j,k \in \mathbb{N}} \left\{ g_j^{(k)} \right\}. \tag{4.40}$$

Indeed, by (4.38),

$$f \geq \sup_{j,k \in \mathbb{N}} \left\{ g_j^{(k)} \right\}.$$

Conversely, fix $z \in A$ and let $t < f(z)$. By (4.38) we may find $g \in \mathcal{G}$ such that

$$t < g(z). \tag{4.41}$$

Let $0 < \varepsilon < \frac{1}{2}(g(z) - t)$, and let $n \in \mathbb{N}$ be such that $z \in K_n$ and let $k_1 \in \mathbb{N}$ be large enough that

$$2^n \leq \varepsilon k_1. \tag{4.42}$$

By (4.39) there exist $g_{j_1}^{(k_1)}$ and $0 < r_{j_1}^{(k_1)} \leq \frac{1}{k_1}$ such that $g \in B\left(g_{j_1}^{(k_1)}, r_{j_1}^{(k_1)}\right)$. In particular,

$$\frac{1}{2^n} \frac{\left\| g - g_{j_1}^{(k_1)} \right\|_{C(K_n)}}{1 + \left\| g - g_{j_1}^{(k_1)} \right\|_{C(K_n)}} \leq r_{j_1}^{(k_1)}.$$

Hence

$$\left| g(z) - g_{j_1}^{(k_1)}(z) \right| \leq \left\| g - g_{j_1}^{(k_1)} \right\|_{C(K_n)} \leq \frac{2^n r_{j_1}^{(k_1)}}{1 - 2^n r_{j_1}^{(k_1)}} < 2\varepsilon,$$

where we have used (4.42). By (4.41) and the fact that $0 < \varepsilon < \frac{1}{2}(g(z) - t)$, we deduce that

$$t < g(z) - 2\varepsilon \leq g_{j_1}^{(k_1)}(z) \leq \sup_{j,k \in \mathbb{N}} \left\{ g_j^{(k)} \right\}(z),$$

and given the arbitrariness of $t < f(z)$ we obtain (4.40).

Proof (Proposition 4.77). By Proposition 4.75(ii),

$$f(z) = \sup \{ g(z) : g \text{ affine continuous}, g \leq f \}$$

for all $z \in \mathbb{R}^m$, and (4.36) follows from Proposition 4.78. To prove (4.37) fix $w \in \mathrm{dom}_e f$ and let $z \in \mathbb{R}^m$. Then for every $i \in \mathbb{N}$ and by Theorem 4.70,

$$f^\infty(z) = \lim_{t \to \infty} \frac{f(w + tz) - f(w)}{t} \geq \lim_{t \to \infty} \frac{a_i + b_i \cdot (w + tz) - f(w)}{t} = b_i \cdot z.$$

Therefore

$$f^\infty(z) \geq \sup_{i \in \mathbb{N}} \{ b_i \cdot z \}$$

for all $z \in \mathbb{R}^m$. Conversely, for $t > 0$,

$$\frac{f(w + tz) - f(w)}{t} = \sup_{i \in \mathbb{N}} \frac{a_i + b_i \cdot (w + tz) - f(w)}{t}$$

$$\leq \sup_{i \in \mathbb{N}} b_i \cdot z + \sup_{i \in \mathbb{N}} \frac{a_i + b_i \cdot w - f(w)}{t} = \sup_{i \in \mathbb{N}} b_i \cdot z,$$

and this concludes the proof of (4.37).

In the case that f is real-valued it is possible to give an explicit characterization of the coefficients a_i and b_i in the previous proposition.

Let $f : \mathbb{R}^m \to \mathbb{R}$ be a convex function and let $\varphi \in C_c^1(\mathbb{R}^m)$ be any function with $\varphi \geq 0$ and $\int_{\mathbb{R}^m} \varphi(z)\, dz = 1$. Define

$$a_\varphi := \int_{\mathbb{R}^m} f(z) \left((m+1)\varphi(z) + \nabla\varphi(z) \cdot z \right) dz,$$

$$b_\varphi := -\int_{\mathbb{R}^m} f(z) \nabla\varphi(z)\, dz.$$

When necessary we will also write $a_\varphi(f)$ and $b_\varphi(f)$ to highlight the dependence on f.

Theorem 4.79 (De Giorgi). *Let f and φ be as above. Then*

(i) $f(z) \geq a_\varphi + b_\varphi \cdot z$ for all $z \in \mathbb{R}^m$;

(ii) $f(z) = \sup_{k \in \mathbb{N},\, q \in \mathbb{Q}^m} \{a_{\varphi_{k,q}} + b_{\varphi_{k,q}} \cdot z\}$ for all $z \in \mathbb{R}^m$, where

$$\varphi_{k,q}(z) := k^m \varphi(k(q-z)), \quad z \in \mathbb{R}^m. \tag{4.43}$$

Proof. (i) Assume first that $f \in C^1(\mathbb{R}^m)$. Then by Theorem 4.61, for any z, $\xi \in \mathbb{R}^m$ we have

$$f(z) \geq f(\xi) + \nabla f(\xi) \cdot (z - \xi).$$

Multiply the previous inequality by $\varphi(\xi)$ and integrate in ξ over $\mathbb{R}^m$ to obtain

$$f(z) \geq \int_{\mathbb{R}^m} (f(\xi) - \nabla f(\xi) \cdot \xi)\, \varphi(\xi)\, d\xi + z \cdot \int_{\mathbb{R}^m} \nabla f(\xi)\, \varphi(\xi)\, d\xi.$$

Integrating by parts now yields

$$f(z) \geq \int_{\mathbb{R}^m} f(\xi)\, ((m+1)\, \varphi(\xi) + \nabla \varphi(\xi) \cdot \xi)\, d\xi - z \cdot \int_{\mathbb{R}^m} f(\xi)\, \nabla \varphi(\xi)\, d\xi.$$

This proves (i) when $f \in C^1(\mathbb{R}^m)$. In the general case, let ψ_ε be a standard mollifier. Applying the previous inequality to the smooth convex function $f_\varepsilon := \psi_\varepsilon * f$ gives

$$f_\varepsilon(z) \geq \int_{\mathbb{R}^m} f_\varepsilon(\xi)\, ((m+1)\, \varphi(\xi) + \nabla \varphi(\xi) \cdot \xi)\, d\xi - z \cdot \int_{\mathbb{R}^m} f_\varepsilon(\xi)\, \nabla \varphi(\xi)\, d\xi.$$

Since φ has compact support, by Theorem 2.75 we may now let $\varepsilon \to 0^+$.

(ii) Let $k \in \mathbb{N}$, $q \in \mathbb{Q}^m$. By replacing the function φ with $\varphi_{k,q}$ in (i) we obtain

$$f(z) \geq a_{\varphi_{k,q}} + b_{\varphi_{k,q}} \cdot z \text{ for all } z \in \mathbb{R}^m$$

and hence $f \geq g$, where

$$g(z) := \sup_{k \in \mathbb{N},\, q \in \mathbb{Q}^m} \{a_{\varphi_{k,q}} + b_{\varphi_{k,q}} \cdot z\}.$$

Since g is everywhere finite and convex, it is continuous, and so is f (see Corollary 4.38). Hence, to prove (ii) it suffices to show that

$$f(q) = g(q) \text{ for all } q \in \mathbb{Q}^m,$$

since $\mathbb{Q}^m$ is dense in $\mathbb{R}^m$.

For $k \in \mathbb{N}$, $q \in \mathbb{Q}^m$, we have

$$a_{\varphi_{k,q}} + b_{\varphi_{k,q}} \cdot z$$

$$= \int_{\mathbb{R}^m} f(\xi)\, (m+1)\, k^m \varphi(k(q-\xi)) - k^{m+1} \nabla \varphi(k(q-\xi)) \cdot (\xi - z)\, d\xi$$

$$= \int_{\mathbb{R}^m} f\left(q - \frac{w}{k}\right) ((m+1)\, \varphi(w) - \nabla \varphi(w) \cdot (k(q-z) - w))\, dw,$$

where we have made the change of variables $w = k(q - \xi)$. Taking $z = q$ we obtain

$$a_{\varphi_{k,q}} + b_{\varphi_{k,q}} \cdot q = \int_{\mathbb{R}^m} f\left(q - \frac{w}{k}\right)\left((m+1)\,\varphi(w) + \nabla\varphi(w) \cdot w\right) dw.$$

Since the integrand is continuous and with compact support, we may let $k \to \infty$ to get

$$\lim_{k \to \infty} a_{\varphi_{k,q}} + b_{\varphi_{k,q}} \cdot q = f(q) \int_{\mathbb{R}^m} \left((m+1)\,\varphi(w) + \nabla\varphi(w) \cdot w\right) dw = f(q),$$

where we have used the facts that

$$\int_{\mathbb{R}^m} \varphi(w)\,dw = 1, \quad \int_{\mathbb{R}^m} \nabla\varphi(w) \cdot w\,dw = -m\int_{\mathbb{R}^m} \varphi(w)\,dw = -m.$$

Since

$$f(q) = \lim_{k \to \infty} a_{\varphi_{k,q}} + b_{\varphi_{k,q}} \cdot q \leq g(q),$$

(ii) follows.

A simple use of the approximation result given in Proposition 4.75 gives an extension of Jensen's inequality for real-valued functions to functions that may take the value ∞. We begin with the real-valued case.

Theorem 4.80 (Jensen's inequality). *Let V be a Banach space and let $f : V \to \mathbb{R}$ be bounded from above in a neighborhood of a point. Then f is convex if and only if given any probability measure μ on a measurable space $(X, \mathfrak{M})$, where X has at least two distinct elements, and any function $g \in L^1((X, \mathfrak{M}, \mu); V)$, then*

$$f\left(\int_X g\,d\mu\right) \leq \int_X f \circ g\,d\mu. \tag{4.44}$$

Proof. Assume that Jensen's inequality (4.44) holds, and let $(X, \mathfrak{M}, \mu)$ be as in the statement. Fix $z_1, z_2 \in V$ and $\theta \in (0, 1)$. Set

$$\mu := \theta\delta_{x_1} + (1 - \theta)\delta_{x_2},$$

where $x_1, x_2 \in X$, $x_1 \neq x_2$. Define

$$g(x) := \begin{cases} z_1 \text{ if } x = x_1, \\ z_2 \text{ otherwise.} \end{cases}$$

Then

$$f\left(\int_X g\,d\mu\right) \leq \int_X f \circ g\,d\mu$$

becomes

$$f \left(\theta g \left(x_1 \right) + \left(1 - \theta \right) g \left(x_2 \right) \right) \leq \theta \left(f \circ g \right) \left(x_1 \right) + \left(1 - \theta \right) \left(f \circ g \right) \left(x_2 \right),$$

that is,

$$f \left(\theta z_1 + \left(1 - \theta \right) z_2 \right) \leq \theta f \left(z_1 \right) + \left(1 - \theta \right) f \left(z_2 \right),$$

and so f is convex.

Conversely, assume that f is convex. Then by Proposition 4.42 and Theorem 4.51, for every $v_0 \in V$ there exists $v' \in V'$ such that

$$f \left(v \right) \geq f \left(v_0 \right) + \langle v', v - v_0 \rangle_{V',V} \quad \text{for all } v \in V.$$

Let $(X, \mathfrak{M}, \mu)$ and g be as in the statement and set

$$v_0 := \int_X g \, d\mu.$$

Then

$$f \left(g \left(w \right) \right) \geq f \left(\int_X g \, d\mu \right) + \left\langle v', g \left(w \right) - \int_X g \, d\mu \right\rangle_{V',V} \qquad \text{for all } w \in X.$$

Integrating with respect to μ yields

$$\int_X f \left(g \left(w \right) \right) \, d\mu \left(w \right) \geq f \left(\int_X g \, d\mu \right),$$

which is the desired inequality.

We now extend Jensen's inequality to the case that f may take the value ∞.

Corollary 4.81. *Let V be a Banach space and let $f : V \to (-\infty, \infty]$ be a convex, lower semicontinuous function. Given any probability measure μ on a measurable space $(X, \mathfrak{M})$ and any function $g \in L^1 \left((X, \mathfrak{M}, \mu) ; V \right)$, then*

$$f \left(\int_X g \, d\mu \right) \leq \int_X f \circ g \, d\mu. \tag{4.45}$$

Proof. By Proposition 4.75,

$$f \left(v \right) = \sup \left\{ h \left(v \right) : h \text{ affine continuous, } h \leq f \right\}.$$

Let h be any affine function such that $f \geq h$. If X has only one element, then (4.45) holds as equality. Otherwise, applying Jensen's inequality to h we have

$$\int_X f \circ g \, d\mu \geq \int_X h \circ g \, d\mu \geq h \left(\int_X g \, d\mu \right),$$

and taking the supremum over all affine functions h below f yields

$$\int_X f \circ g \, d\mu \geq f \left(\int_X g \, d\mu \right).$$

This concludes the proof.

Exercise 4.82. Let $f : \mathbb{R}^m \to (-\infty, \infty]$ be a lower semicontinuous function and let $Q \subset \mathbb{R}^N$ be the unit cube. Prove that f is convex if and only if

$$f(z) \le \int_Q f(z + v(x))\, dx$$

for all $z \in \mathbb{R}^m$ and all Q-periodic functions $v \in L^1_{\text{loc}}\left(\mathbb{R}^N; \mathbb{R}^m\right)$, with

$$\int_Q v(x)\, dx = 0,$$

for which the right-hand side of the previous inequality is well-defined.

4.8 Convex Envelopes

As we will see in relaxation problems, in the case of nonconvex integrands f one is interested in "convexifying" f. This brings us to various notions of convex envelopes.

Definition 4.83. *Let V be a vector space and let $f : V \to [-\infty, \infty]$. The convex envelope $\mathcal{C}f : V \to [-\infty, \infty]$ of f is defined by*

$$\mathcal{C}f(v) := \sup\{g(v) : g : V \to [-\infty, \infty]\ \text{convex},\ g \le f\}.$$

The following result relates the convex envelope of a function with the convex hull of its epigraph.

Theorem 4.84. *Let V be a vector space and let $f : V \to [-\infty, \infty]$. Then*

$$\text{dom}_e\, \mathcal{C}f = \text{co}\,(\text{dom}_e\, f)$$

and

$$\mathcal{C}f(v) = \inf\{t \in \mathbb{R} : (v, t) \in \text{co}\,(\text{epi}\, f)\}, \quad v \in V.$$

Proof. If $v \in \text{co}\,(\text{dom}_e\, f)$, then by Proposition 4.2, v can be written as a convex combination of elements of $\text{dom}_e\, f$, that is,

$$v = \sum_{i=1}^{n} \theta_i v_i$$

for some $n \in \mathbb{N}$, $\sum_{i=1}^{n} \theta_i = 1$, $\theta_i \ge 0$, $v_i \in \text{dom}_e\, f$. Since $\mathcal{C}f \le f$, it follows that $v_i \in \text{dom}_e\, \mathcal{C}f$. Since $\text{dom}_e\, \mathcal{C}f$ is a convex set, we have that $v \in \text{dom}_e\, \mathcal{C}f$, and so $\text{co}\,(\text{dom}_e\, f) \subset \text{dom}_e\, \mathcal{C}f$. To prove the converse inclusion, note that the function

$$g(v) := \begin{cases} -\infty & \text{if } v \in \text{co}\,(\text{dom}_e\, f), \\ \infty & \text{otherwise,} \end{cases}$$

is convex and below f, and so $g \leq \mathcal{C}f$, which implies that

$$\mathrm{co}\,(\mathrm{dom}_e\, f) = \mathrm{dom}_e\, g \supset \mathrm{dom}_e\, \mathcal{C}f.$$

To prove the second part of the theorem define

$$h(v) := \inf\{t \in \mathbb{R} : (v,t) \in \mathrm{co}\,(\mathrm{epi}\, f)\}, \quad v \in V.$$

Since $\mathrm{co}\,(\mathrm{epi}\, f)$ is a convex set, by Proposition 4.22 it follows that h is convex. Moreover, if $v \in \mathrm{dom}_e\, f$, then $(v, f(v)) \in \mathrm{epi}\, f \subset \mathrm{co}\,(\mathrm{epi}\, f)$, and so $h(v) \leq f(v)$. Hence by the definition of convex envelope we have that

$$h \leq \mathcal{C}f.$$

To prove the reverse inequality let $g : V \to [-\infty, \infty]$ be any convex function, with $g \leq f$. Then $\mathrm{epi}\, f \subset \mathrm{epi}\, g$, and since $\mathrm{epi}\, g$ is a convex set by Proposition 4.22, we have that

$$\mathrm{co}\,(\mathrm{epi}\, f) \subset \mathrm{epi}\, g.$$

Hence for every $v \in \mathrm{dom}_e\, g$,

$$g(v) = \inf\{t \in \mathbb{R} : (v,t) \in \mathrm{epi}\, g\} \tag{4.46}$$
$$\leq \inf\{t \in \mathbb{R} : (v,t) \in \mathrm{co}\,(\mathrm{epi}\, f)\} = h(v).$$

On the other hand, if $v \in \mathrm{dom}_e\, h$, then there exists $t \in \mathbb{R}$ such that $(v,t) \in \mathrm{co}\,(\mathrm{epi}\, f)$. In particular, we may write

$$v = \sum_{i=1}^{n} \theta_i v_i$$

for some $n \in \mathbb{N}$, $\sum_{i=1}^{n} \theta_i = 1$, $\theta_i \geq 0$, $v_i \in \mathrm{dom}_e\, f$. Hence $v \in \mathrm{co}\,(\mathrm{dom}_e\, f) = \mathrm{dom}_e\, \mathcal{C}f$ by the first part of the proof. Since $g(v) \leq \mathcal{C}f(v) < \infty$, it follows that $v \in \mathrm{dom}_e\, g$. Thus $\mathrm{dom}_e\, h \subset \mathrm{dom}_e\, g$, which implies that the inequality (4.46) holds everywhere. In turn, $g \leq h$, and given the arbitrariness of g we conclude that $\mathcal{C}f \leq h$. This proves the second part of the theorem.

Remark 4.85. It follows by Remark 4.3 that if V is a topological vector space and $\mathrm{dom}_e\, f$ is open, then $\mathrm{dom}_e\, \mathcal{C}f$ is open.

Another type of convex envelope is obtained by restricting the class of admissible functions below f to affine continuous functions.

Definition 4.86. *Given a dual pair (V, W) and a function $f : V \to [-\infty, \infty]$, the* polar *or* conjugate function *$f^* : W \to [-\infty, \infty]$ of f is defined by*

$$f^*(w) := \sup_{v \in V} \{\langle v, w \rangle_{V,W} - f(v)\}, \quad w \in W,$$

and the bipolar *or* biconjugate function *$f^{**} : V \to [-\infty, \infty]$ of f is defined by*

$$f^{**}(v) := \sup_{w \in W} \{\langle v, w \rangle_{V,W} - f^*(w)\}, \quad v \in V.$$

The spaces V, W are endowed with the topologies $\sigma(V, W)$ and $\sigma(W, V)$, respectively. Since f^* is the supremum of a family of continuous and convex functions, f^* is convex and lower semicontinuous. The same holds true for f^{**}, and $f^{**} \leq f$. Even when f is convex and lower semicontinuous it may happen that $f^{**} \neq f$. Indeed, if f takes the value $-\infty$ at some point and $f \not\equiv -\infty$ (see Proposition 4.26), then $f^* \equiv \infty$, and in turn, $f^{**} \equiv -\infty$.

Remark 4.87. If V is a topological vector space and $f : V \to [-\infty, \infty]$, then we may take W to be the topological dual of V, so that

$$f^*(v') = \sup_{v \in V} \{\langle v', v \rangle_{V', V} - f(v)\}, \quad v' \in V',$$

and

$$f^{**}(v) = \sup_{v' \in V'} \{\langle v', v \rangle_{V', V} - f^*(v')\}, \quad v \in V.$$

In what follows, whenever the space W is not specified, it is understood that $W := V'$.

Proposition 4.88. *Let (V, W) be a dual pair and $f : V \to [-\infty, \infty]$. Then*

$$f^* = f^{***}.$$

Proof. From the definition of f^{**} we have that

$$f^{**}(v) \geq \langle v, w \rangle_{V, W} - f^*(w)$$

for all $v \in V$ and $w \in W$. Therefore

$$f^*(w) \geq f^{***}(w) = \sup_{v \in V} \{\langle v, w \rangle_{V, W} - f^{**}(v)\}.$$

Conversely,

$$f^{***}(w) \geq \langle v, w \rangle_{V, W} - f^{**}(v) \geq \langle v, w \rangle_{V, W} - f(v)$$

for all $v \in V$ and $w \in W$, where we have used the fact that $f^{**} \leq f$. Taking the supremum over all $v \in V$, we conclude that

$$f^{***}(w) \geq f^*(w)$$

for all $w \in W$, and this completes the proof.

Exercise 4.89. Let $V = \mathbb{R}^m$. Prove that:

(i) If $1 \leq p < \infty$, then the polar function of $f(z) = \frac{|z|^p}{p}$ is

$$f^*(w) = \frac{|w|^{p'}}{p'} \quad w \in \mathbb{R}^m,$$

for $p > 1$ and

$$f^*(w) = \begin{cases} \infty \text{ if } |w| > 1, \\ 0 \text{ if } |w| \leq 1, \end{cases}$$

if $p = 1$.

(ii) The polar function of $f(z) := \sqrt{|z|^2 + 1}$ is

$$f^*(w) = \begin{cases} \infty & \text{if } |w| > 1, \\ -\sqrt{1 - |w|^2} & \text{if } |w| \le 1. \end{cases}$$

(iii) Let $m = 1$. The polar function of $f(z) := e^z$ is

$$f^*(w) = \begin{cases} w \log w - w & \text{if } w > 0, \\ 0 & \text{if } w = 0, \\ \infty & \text{if } w < 0. \end{cases}$$

Remark 4.90. Note that if V is a topological vector space and $f : V \to [-\infty, \infty]$, $f \not\equiv \infty$, is positively homogeneous of degree one, i.e.,

$$f(tv) = tf(v)$$

for all $v \in V$ and $t > 0$, then

$$f^* : V' \to \{0, \infty\}.$$

Indeed, fix $v' \in V'$. If there exists $v_0 \in V$ such that $\langle v', v_0 \rangle_{V',V} - f(v_0) > 0$, then for all $t > 0$,

$$\begin{aligned} f^*(v') &= \sup_{v \in V} \{\langle v', v \rangle_{V',V} - f(v)\} \\ &\ge \langle v', tv_0 \rangle_{V',V} - f(tv_0) = t(\langle v', v_0 \rangle_{V',V} - f(v_0)), \end{aligned}$$

and by letting $t \to \infty$ we obtain $f^*(v') = \infty$. On the other hand, if $\langle v', v \rangle_{V',V} - f(v) \le 0$ for all $v \in V$, then $f^*(v') \le 0$. Let $v \in V$ be such that $f(v) < \infty$. Then

$$0 \ge f^*(v') \ge t(\langle v', v \rangle_{V',V} - f(v)) \to 0$$

as $t \to 0^+$. Hence $f^*(v') = 0$.

From the definition of f^* we have that

$$f^*(v') \ge \langle v', v \rangle_{V',V} - f(v)$$

for all $v \in V$ and $v' \in V'$. The next result characterizes pairs $(v, v') \in V \times V'$ for which equality holds.

Theorem 4.91. *Let V be a topological vector space, let $f : V \to (-\infty, \infty]$ be a convex function not identically equal to ∞, and let $(v, v') \in V \times V'$. Then $v' \in \partial f(v)$ if and only if*

$$f(v) + f^*(v') = \langle v', v \rangle_{V',V}. \tag{4.47}$$

Proof. Fix $(v, v') \in V \times V'$. If $v' \in \partial f(v)$, then

$$f(w) \geq f(v) + \langle v', w - v \rangle_{V', V} \quad \text{for all } w \in V,$$

or equivalently,

$$\langle v', v \rangle_{V', V} - f(v) \geq f^*(v') = \sup_{w \in V} \{\langle v', w \rangle_{V', V} - f(w)\}.$$

Since the opposite inequality holds by definition of f^*, equality (4.47) follows.

Conversely, assume that (4.47) holds. In particular, $f(v) \in \mathbb{R}$. By definition of $f^*(v')$ we have that for all $w \in V$,

$$f(w) - \langle v', w \rangle_{V', V} \geq -f^*(v') = f(v) - \langle v', v \rangle_{V', V},$$

that is,

$$f(w) \geq f(v) + \langle v', w - v \rangle_{V', V} \quad \text{for all } w \in V,$$

which is equivalent to $v' \in \partial f(v)$.

We now relate the various types of convex envelopes.

Theorem 4.92. *Let V be a locally convex topological vector space and $f : V \to [-\infty, \infty]$, $f \not\equiv \infty$. Then*

(i) $f^{**}(v) = \sup\{g(v) : g \text{ affine continuous, } g \leq f\}$ *for all $v \in V$. In particular, if f^{**} takes the value $-\infty$, then $f^{**} \equiv -\infty$;*

(ii) $f^{**} \leq \operatorname{lsc}(\mathcal{C}f) \leq \mathcal{C}(\operatorname{lsc} f) \leq \mathcal{C}f \leq f$;

(iii) *if, in addition, there exists an affine continuous function below f, then $f^{**} = \operatorname{lsc}(\mathcal{C}f)$.*

Proof. **Step 1:** Set

$$\tilde{f}(v) := \sup\{g(v) : g \text{ affine continuous, } g \leq f\}, \quad v \in V.$$

We first prove that $\tilde{f} \equiv -\infty$, i.e., the family of admissible functions g in the definition of $\tilde{f}$ is empty, if and only if $f^* \equiv \infty$. Indeed, if there exist $v' \in V'$ and $\alpha \in \mathbb{R}$ such that

$$\langle v', v \rangle_{V', V} + \alpha \leq f(v)$$

for every $v \in V$, then, equivalently,

$$\langle v', v \rangle_{V', V} - f(v) \leq -\alpha$$

for every $v \in V$. Therefore

$$f^*(v') \leq -\alpha. \tag{4.48}$$

Conversely, if there exists $v' \in V'$ such that $f^*(v') < \infty$, then $f^*(v') \in \mathbb{R}$ since $f \not\equiv \infty$. In view of the definition of $f^*(v')$, it follows that

$$\langle v', v \rangle_{V', V} - f^*(v') \leq f(v)$$

for every $v \in V$ and thus

$$g(v) := \langle v', v \rangle_{V',V} - f^*(v') \leq \tilde{f}(v). \qquad (4.49)$$

Step 2: We prove (i). If $f^* \equiv \infty$, then $f^{**} \equiv -\infty$, and by Step 1, property (i) holds. Suppose now that there exists $v' \in V'$ such that $f^*(v') < \infty$. Taking the supremum in (4.49) over all such v' yields $f^{**}(v) \leq \tilde{f}(v)$.

Conversely, by Step 1 there is at least one admissible function $g(v) = \langle v', v \rangle_{V',V} + \alpha$ in the definition of $\tilde{f}$. As in (4.48) we obtain $f^*(v') \leq -\alpha$, and we deduce that

$$f^{**}(v) \geq \langle v', v \rangle_{V',V} - f^*(v') \geq \langle v', v \rangle_{V',V} + \alpha.$$

Taking the supremum over all such pairs (v', α), we conclude that $f^{**}(v) \geq \tilde{f}(v)$.

(ii) The last two inequalities are immediate. Since $\mathcal{C}f \leq f$, then $\mathrm{lsc}(\mathcal{C}f) \leq \mathrm{lsc}\, f$, and using the fact that $\mathrm{lsc}(\mathcal{C}f)$ is convex by Proposition 4.25, we obtain that $\mathrm{lsc}(\mathcal{C}f) \leq \mathcal{C}(\mathrm{lsc}\, f)$.

Since f^{**} is lower semicontinuous, convex, and below f, we have that $f^{**} \leq \mathrm{lsc}(\mathcal{C}f)$.

(iii) Since $\mathrm{lsc}(\mathcal{C}f)$ is convex, lower semicontinuous, and above an affine continuous function, then by Proposition 4.75, invoking (ii), it follows that

$$\mathrm{lsc}(\mathcal{C}f)(v) = \sup\{g(v) : g \text{ affine continuous}, g \leq \mathrm{lsc}(\mathcal{C}f)(v)\} \qquad (4.50)$$
$$= (\mathrm{lsc}(\mathcal{C}f))^{**}(v),$$

where in the last equality we used part (i). Since $\mathrm{lsc}(\mathcal{C}f) \leq f$ by (4.50) we conclude that $\mathrm{lsc}(\mathcal{C}f) \leq f^{**}$.

Remark 4.93. (i) Note that if there is no affine continuous function below f, then Theorem 4.92(iii) does not hold in general. Observe that Remark 4.76 exhibits an example in which $f^{**} \equiv -\infty \lneq \mathrm{lsc}(\mathcal{C}f) = f$.

(ii) From Theorem 4.92 it follows that if there exists an affine continuous function below f and if $\mathcal{C}f$ is lower semicontinuous, then $f^{**} = \mathrm{lsc}(\mathcal{C}f) = \mathcal{C}(\mathrm{lsc}\, f) = \mathcal{C}f$. In particular, if $f : V \to \mathbb{R}$ is bounded from above in an open set and if there exists an affine continuous function g such that $f \geq g$, then

$$-\infty < g \leq f^{**} \leq \mathcal{C}(\mathrm{lsc}\, f) \leq \mathcal{C}f \leq f < \infty,$$

and so $\mathcal{C}f : V \to \mathbb{R}$. By Theorem 4.43 it follows that $\mathcal{C}f$ is continuous, and so $f^{**} = \mathrm{lsc}(\mathcal{C}f) = \mathcal{C}(\mathrm{lsc}\, f) = \mathcal{C}f$.

(iii) Note that when V is a Euclidean space, say $V = \mathbb{R}^m$, and $f : \mathbb{R}^m \to \mathbb{R}$ is bounded from below by an affine function g, then $\mathcal{C}f$ is continuous by Corollary 4.38, and so by (ii) we have that $f^{**} = \mathrm{lsc}(\mathcal{C}f) = \mathcal{C}(\mathrm{lsc}\, f) = \mathcal{C}f$.

However, when f takes the value ∞, then by Theorem 4.40, $f^{**} = \mathrm{lsc}\,(\mathcal{C}f)$ agrees with $\mathcal{C}f$ except possibly on $\mathrm{rb}_{\mathrm{aff}}\,(\mathrm{dom}_e\,\mathcal{C}f)$, and in particular, it may happen that $f^{**} \lneq \mathcal{C}\,(\mathrm{lsc}\,f)$ on $\mathrm{rb}_{\mathrm{aff}}\,(\mathrm{dom}_e\,\mathcal{C}f)$ as shown by the following example.

Exercise 4.94. Let $m = 2$ and consider the function

$$f(z) = f(z_1, z_2) = \begin{cases} z_2 - z_1 e^{z_2} & \text{if } z_2 \geq 0 \text{ and } 0 < z_1 \leq z_2 e^{-z_2}, \\ 0 & \text{if } z_2 \geq 0 \text{ and } z_2 e^{-z_2} < z_1, \\ \infty & \text{otherwise.} \end{cases}$$

Prove that

$$f^{**}(z_1, z_2) = \begin{cases} 0 & \text{if } z_1 \geq 0 \text{ and } z_2 \geq 0, \\ \infty & \text{otherwise,} \end{cases}$$

while

$$\mathcal{C}\,(\mathrm{lsc}\,f)\,(z_1, z_2) = \begin{cases} z_2 & \text{if } z_1 = 0 \text{ and } z_2 \geq 0, \\ 0 & \text{if } z_1 > 0 \text{ and } z_2 \geq 0, \\ \infty & \text{otherwise.} \end{cases}$$

Note that $\mathcal{C}\,(\mathrm{lsc}\,f)$ is convex but not lower semicontinuous.

If we restrict our attention to the space $V = \mathbb{R}^m$, then additional qualitative properties may be obtained for convex envelopes.

We start with Carathéodory's theorem, which improves (4.1) in that it limits the number of terms in the convex combination to at most $m + 1$.

Theorem 4.95 (Carathéodory). *Let $E \subset \mathbb{R}^m$. Then*

$$\mathrm{co}E = \left\{ \sum_{i=1}^{m+1} \theta_i z_i : \sum_{i=1}^{m+1} \theta_i = 1,\ \theta_i \geq 0,\ z_i \in E,\ i = 1, \ldots, m+1 \right\}.$$

Theorem 4.96. *Let $f : \mathbb{R}^m \to (-\infty, \infty]$. Then for all $z \in \mathbb{R}^m$,*

$$\mathcal{C}f(z) = \inf \left\{ \sum_{i=1}^{m+1} \theta_i f(z_i) : \theta_i \in [0,1],\ z_i \in \mathbb{R}^m,\ i = 1, \ldots, m+1, \right.$$
$$\left. \sum_{i=1}^{m+1} \theta_i = 1,\ \sum_{i=1}^{m+1} \theta_i z_i = z \right\}.$$

The formula is also valid if one takes only the combinations such that

$$\mathrm{aff}\,(\{z_1, \ldots, z_{m+1}\}) = \mathbb{R}^m.$$

In the previous theorem if $\theta_i = 0$ we set $\theta_i f(z_i) := 0$ even if $f(z_i) = \infty$. As a corollary we obtain the following result.

Corollary 4.97. *Let* $f : \mathbb{R}^m \to (-\infty, \infty]$ *be bounded from below by an affine function* g. *Then*

$$\overline{\mathrm{dom}_e\, f^{**}} = \overline{\mathrm{dom}_e\, \mathcal{C}f} = \overline{\mathrm{co}\,(\mathrm{dom}_e\, f)}. \tag{4.51}$$

Moreover, for all $z \notin \mathrm{rb}_{\mathrm{aff}}\,(\mathrm{dom}_e\, \mathcal{C}f)$,

$$f^{**}(z) = \inf \left\{ \sum_{i=1}^{m+1} \theta_i f(z_i) : \theta_i \in [0,1],\, z_i \in \mathbb{R}^m,\, i = 1, \ldots, m+1, \right.$$

$$\left. \sum_{i=1}^{m+1} \theta_i = 1,\, \sum_{i=1}^{m+1} \theta_i z_i = z \right\},$$

while for $z \in \mathrm{rb}_{\mathrm{aff}}\,(\mathrm{dom}_e\, \mathcal{C}f)$,

$$f^{**}(z) = \lim_{\theta \to 1^-} \mathcal{C}f\,((1-\theta)z_0 + \theta z)$$

for any fixed $z_0 \in \mathrm{ri}_{\mathrm{aff}}\,(\mathrm{dom}_e\, \mathcal{C}f)$.

Proof. By Theorem 4.92(iv), $f^{**} = \mathrm{lsc}\,(\mathcal{C}f)$, and so by Proposition 3.12, for every $z \in \mathbb{R}^m$,

$$f^{**}(z) = \inf_{\{z_n\}} \left\{ \liminf_{n \to \infty} \mathcal{C}f(z_n) : \{z_n\} \subset \mathbb{R}^m,\, z_n \to z \right\},$$

which implies that

$$\mathrm{dom}_e\, \mathcal{C}f \subset \mathrm{dom}_e\, f^{**} \subset \overline{\mathrm{dom}_e\, \mathcal{C}f}.$$

Property (4.51) now follows from Proposition 4.9 and Theorem 4.84.

The second part of the theorem is a consequence of Theorems 4.96 and Theorem 4.40.

If f is superlinear at infinity, then we have the following result.

Theorem 4.98. *Let* $f : \mathbb{R}^m \to [0, \infty]$ *be such that*

$$\lim_{|z| \to \infty} \frac{f(z)}{|z|} = \infty. \tag{4.52}$$

Then

$$f^{**}(z) = \mathcal{C}\,(\mathrm{lsc}\, f)\,(z)$$

$$= \min \left\{ \sum_{i=1}^{m+1} \theta_i\,(\mathrm{lsc}\, f)\,(z_i) : \theta_i \in [0,1],\, z_i \in \mathbb{R}^m,\, i = 1, \ldots, m+1, \right.$$

$$\left. \sum_{i=1}^{m+1} \theta_i = 1,\, \sum_{i=1}^{m+1} \theta_i z_i = z \right\}.$$

Proof. **Step 1:** Assume first that f is lower semicontinuous. We claim that $\mathcal{C}f$ is lower semicontinuous. To see this, let $z^{(n)} \to z$. We need to show that

$$\liminf_{n\to\infty} \mathcal{C}f\left(z^{(n)}\right) \geq \mathcal{C}f\left(z\right).$$

Without loss of generality we may assume that

$$\liminf_{n\to\infty} \mathcal{C}f\left(z^{(n)}\right) = \lim_{n\to\infty} \mathcal{C}f\left(z^{(n)}\right) < \infty$$

and

$$C := \sup_n \mathcal{C}f\left(z^{(n)}\right) < \infty. \tag{4.53}$$

For each $n \in \mathbb{N}$, by Theorem 4.96 we may find $\left\{\left(\theta_i^{(n)}, z_i^{(n)}\right)\right\} \subset [0,1] \times \mathbb{R}^m$, $i = 1, \dots, m+1$, such that

$$\sum_{i=1}^{m+1} \theta_i^{(n)} = 1, \quad \sum_{i=1}^{m+1} \theta_i^{(n)} z_i^{(n)} = z^{(n)},$$

and

$$\sum_{i=1}^{m+1} \theta_i^{(n)} f\left(z_i^{(n)}\right) \leq \mathcal{C}f\left(z^{(n)}\right) + \frac{1}{n} \tag{4.54}$$

as $k \to \infty$. Upon extracting a subsequence if necessary, for each $i = 1, \dots, m+1$ we may assume that $\theta_i^{(n)} \to \theta_i$ and that either $\left|z_i^{(n)}\right| \to \infty$ or $z_i^{(n)} \to z_i$ as $n \to \infty$. Let

$$I := \left\{i = 1, \dots, m+1 : \left|z_i^{(n)}\right| \to \infty \text{ as } n \to \infty\right\}.$$

Next we show that if $j \in I$ then $\theta_j^{(n)} \left|z_j^{(n)}\right| \to 0$. Indeed, for every fixed $\varepsilon > 0$, by (4.52) there exists $L > 0$ such that

$$f(z) \geq \frac{(C+1)}{\varepsilon} |z| \quad \text{for all } |z| \geq L,$$

where C is the constant given in (4.53). Since $j \in I$ then $\left|z_j^{(n)}\right| \geq L$ for all n sufficiently large, and so by the previous inequality, together with (4.53), (4.54), and the fact that $f \geq 0$, we obtain

$$\theta_j^{(n)} \left|z_j^{(n)}\right| \leq \frac{\varepsilon}{C+1} \theta_j^{(n)} f\left(z_j^{(n)}\right) \leq \frac{\varepsilon}{C+1} \sum_{i=1}^{m+1} \theta_i^{(n)} f\left(z_i^{(n)}\right) \leq \varepsilon,$$

which shows that $\theta_j^{(n)} \left|z_j^{(n)}\right| \to 0$. Note also that since $\left|z_j^{(n)}\right| \to \infty$ as $n \to \infty$, this implies in particular that $\theta_j = 0$. Hence

$$z^{(n)} = \sum_{i=1}^{m+1} \theta_i^{(n)} z_i^{(n)} = \sum_{i \notin I} \theta_i^{(n)} z_i^{(n)} + o(1),$$

and so

$$\sum_{i \notin I} \theta_i^{(n)} z_i^{(n)} \to z = \sum_{i \notin I} \theta_i z_i$$

as $n \to \infty$. Since f is lower semicontinuous, from (4.54) and the fact that $f \geq 0$ we conclude that

$$\lim_{n \to \infty} \mathcal{C}f\left(z^{(n)}\right) \geq \liminf_{n \to \infty} \sum_{i \notin I} \theta_i^{(n)} f\left(z_i^{(n)}\right) \geq \sum_{i \notin I} \liminf_{n \to \infty} \theta_i^{(n)} f\left(z_i^{(n)}\right) \quad (4.55)$$

$$\geq \sum_{i \notin I} \theta_i f\left(z_i\right) \geq \mathcal{C}f\left(z\right).$$

This proves the claim. Hence $\mathrm{lsc}\,(\mathcal{C}f) = \mathcal{C}f$, and so using once more the fact that $f \geq 0$, by Theorem 4.92(iv) we have that $f^{**} = \mathcal{C}f$, and thus from Theorem 4.96 it follows that

$$f^{**}(z) = \inf\left\{ \sum_{i=1}^{m+1} \theta_i \, (\mathrm{lsc}\,f)\,(z_i) : \theta_i \in [0,1]\,, z_i \in \mathbb{R}^m, i = 1, \ldots, m+1, \right.$$

$$\left. \sum_{i=1}^{m+1} \theta_i = 1, \; \sum_{i=1}^{m+1} \theta_i z_i = z \right\}.$$

To see that the infimum is realized it suffices to consider the special sequence $z^{(n)} := z$ for all $n \in \mathbb{N}$ in (4.55).

Step 2: We claim that

$$\lim_{|z| \to \infty} \frac{\mathrm{lsc}\,f\,(z)}{|z|} = \infty.$$

Indeed, by (4.52) for every fixed $M > 0$ there exists $L > 0$ such that

$$f(z) \geq M\,|z| \quad \text{for all } |z| > L.$$

Since $f \geq 0$ we have that

$$\mathrm{lsc}\,f\,(z) \geq \begin{cases} M\,|z| & \text{for } |z| > L, \\ 0 & \text{for } |z| \leq L, \end{cases}$$

for all $z \in \mathbb{R}^m$, which proves the claim.

By applying the previous step to $\mathrm{lsc}\,f$ we conclude that

$$(\mathrm{lsc}\,f)^{**} = \mathcal{C}\,(\mathrm{lsc}\,f).$$

By Theorem 4.92(iv) we have that

$$f^{**} = (f^{**})^{**} = (\mathrm{lsc}\,(\mathcal{C}f))^{**} \leq (\mathcal{C}\,(\mathrm{lsc}\,f))^{**} \leq (\mathrm{lsc}\,f)^{**} \leq (\mathcal{C}f)^{**} \leq f^{**},$$

which concludes the proof.

Remark 4.99. (i) Note that under the assumptions of the previous theorem,

$$\lim_{|z| \to \infty} \frac{f^{**}(z)}{|z|} = \infty. \tag{4.56}$$

Indeed, fix $M > 0$ and let $L > 0$ be such that

$$\frac{f(z)}{|z|} \geq M$$

for all $|z| \geq L$. Since $f \geq 0$ we obtain that

$$f(z) \geq M |z| - ML$$

for all $z \in \mathbb{R}^m$. Hence $f^{**}(z) \geq M |z| - ML$ for all $z \in \mathbb{R}^m$, and so

$$\liminf_{|z| \to \infty} \frac{f^{**}(z)}{|z|} \geq M.$$

Given the arbitrariness of M we conclude (4.56).

(ii) If $f : \mathbb{R} \to [-\infty, \infty]$ is an increasing function, then f^{**} is also increasing. In particular, if $\gamma : [0, \infty) \to [0, \infty)$ is increasing, with

$$\lim_{z \to \infty} \frac{\gamma(z)}{z} = \infty, \tag{4.57}$$

then extend γ to $\mathbb{R}$ by defining $\gamma(z) := \gamma(0)$ if $z < 0$. The extended function γ is increasing, and so is γ^{**}. Using an argument similar to (i) we have that for every $M > 0$ there exists $L > 0$ such that

$$\gamma(z) \geq Mz - ML \tag{4.58}$$

for all $z \geq 0$, and since $\gamma(0) \geq 0$, the inequality (4.58) holds for all $z \in \mathbb{R}$. Hence, as in (i) we conclude that γ^{**} satisfies (4.57). Moreover, γ^{**} is continuous since γ is real-valued and nonnegative. Here we recall Remark 2.30.

(iii) The previous theorem continues to hold if we assume that f is bounded from below rather than nonnegative. Indeed, it suffices to apply the result to the function $g := f - \inf f$.

As a consequence of the previous theorem we have the following proposition:

Proposition 4.100. *Let $f : \mathbb{R}^m \to (-\infty, \infty]$ be bounded from below by an affine function. Let $\{\varphi_j\}$ be a decreasing sequence of upper semicontinuous functions $\varphi_j : \mathbb{R}^m \to [0, \infty]$ such that*

$$\lim_{j \to \infty} \varphi_j(z) = 0$$

for all $z \in \mathbb{R}^m$ and

$$\lim_{|z| \to \infty} \frac{\varphi_j(z)}{|z|} = \infty. \tag{4.59}$$

Then

$$\mathcal{C}(\text{lsc } f) = \inf_{j \in \mathbb{N}} (f + \varphi_j)^{**}.$$

Proof. By hypothesis there exist $(\alpha, \beta) \in \mathbb{R} \times \mathbb{R}^m$ such that

$$f(z) \geq \alpha + \beta \cdot z \tag{4.60}$$

for all $z \in \mathbb{R}^m$. Fix $j \in \mathbb{N}$. By (4.59), (4.60) it follows that

$$\lim_{|z| \to \infty} \frac{(f + \varphi_j)(z)}{|z|} = \infty.$$

By (4.59) we may find $M > 0$ such that $\varphi_j(z) \geq |\beta| |z|$ whenever $|z| \geq M$, which, together with (4.60), entails

$$(f + \varphi_j) \geq \alpha - |\beta| M.$$

By Theorem 4.98 and Remark 4.99(iii) we have

$$(f + \varphi_j)^{**} = \mathcal{C}(\text{lsc } (f + \varphi_j))$$

and thus

$$\mathcal{C}(\text{lsc} f) \leq \inf_{j \in \mathbb{N}} (f + \varphi_j)^{**}. \tag{4.61}$$

To prove the reverse inequality, note first that since φ_j is upper semicontinuous, then $\text{lsc}(f + \varphi_j) - \varphi_j$ is lower semicontinuous and below f. Therefore $\text{lsc}(f + \varphi_j) - \varphi_j \leq \text{lsc } f$, or equivalently,

$$\text{lsc} (f + \varphi_j) \leq \text{lsc } f + \varphi_j \quad \text{for each } j \in \mathbb{N}.$$

Hence for every $z \in \mathbb{R}^m$ and by the fact that the bipolar is lower semicontinuous,

$$\inf_{j \in \mathbb{N}} (f + \varphi_j)^{**}(z) = \lim_{j \to \infty} (f + \varphi_j)^{**}(z)$$

$$\leq \lim_{j \to \infty} \text{lsc} (f + \varphi_j)(z) \tag{4.62}$$

$$\leq \text{lsc } f(z) + \lim_{j \to \infty} \varphi_j(z) = \text{lsc } f(z).$$

Here we have used the fact that since the sequence $\{\varphi_j\}$ is decreasing, then so are the sequences $\{(f + \varphi_j)^{**}\}$ and $\{\text{lsc} (f + \varphi_j)\}$. In particular,

$$\inf_{j \in \mathbb{N}} (f + \varphi_j)^{**} = \lim_{j \to \infty} (f + \varphi_j)^{**}$$

is a convex function, and thus from (4.62) we deduce that

$$\inf_{j \in \mathbb{N}} (f + \varphi_j)^{**} \leq \mathcal{C}(\text{lsc} f).$$

This, together with (4.61), gives the result.

Remark 4.101. When condition (4.59) is violated one can conclude only that

$$f^{**} \leq \inf_{j \in \mathbb{N}} (f + \varphi_j)^{**} \leq \mathcal{C}\,(\mathrm{lsc}\,f)\,.$$

However, if $f : \mathbb{R}^m \to \mathbb{R}$ satisfies (4.60), then in view of Remark 4.93 we have that

$$f^{**} = \inf_{j \in \mathbb{N}} (f + \varphi_j)^{**} = \mathcal{C}\,(\mathrm{lsc}\,f)\,.$$

A result in the same spirit of the previous proposition is the following.

Proposition 4.102. *Let $\{f_j\}$ be an increasing sequence of lower semicontinuous functions $f_j : \mathbb{R}^m \to [0,\infty]$ such that*

$$f_j(z) \geq \gamma\,(|z|)$$

for all $j \in \mathbb{N}$ and $z \in \mathbb{R}^m$, where $\gamma : [0,\infty) \to [0,\infty)$ is a function such that

$$\lim_{s \to \infty} \frac{\gamma\,(s)}{s} = \infty.$$

Then

$$\left(\sup_j f_j \right)^{**} = \sup_j f_j^{**}.$$

Proof. Let $f := \sup_j f_j$. Since $f \geq f_j$ for all $j \in \mathbb{N}$ we have $f^{**} \geq \sup_j f_j^{**}$. To prove the reverse inequality, let $(\alpha,\beta) \in \mathbb{R} \times \mathbb{R}^m$ be such that

$$f^{**}(z) \geq g(z) := \alpha + \beta \cdot z \qquad (4.63)$$

for all $z \in \mathbb{R}^m$. By the superlinear growth of γ, we may find $L > 0$ such that

$$\gamma\,(|z|) \geq g(z) \quad \text{for all } |z| \geq L. \qquad (4.64)$$

We claim that for every fixed $\varepsilon > 0$ there exists j_0 such

$$f_j(z) \geq g(z) - \varepsilon$$

for all $j \geq j_0$ and $|z| \leq L$. Indeed, if not, then there would exist a sequence $\{z_k\} \subset \overline{B}\,(0,L)$ and a subsequence $\{f_{j_k}\}$ of $\{f_j\}$ such that

$$f_{j_k}(z_k) < g(z_k) - \varepsilon.$$

Without loss of generality we may assume that $z_k \to z_0$. Since the sequence $\{f_{j_k}\}$ is increasing and by lower semicontinuity,

$$f_{j_k}(z_0) \leq \liminf_{n \to \infty} f_{j_k}(z_{k+n}) \leq \liminf_{n \to \infty} f_{j_{k+n}}(z_{k+n})$$

$$\leq \lim_{n \to \infty} g(z_{k+n}) - \varepsilon = g(z_0) - \varepsilon.$$

Letting $k \to \infty$ in the previous inequality, we would get

$$f(z_0) = \lim_{k \to \infty} f_{j_k}(z_0) \leq g(z_0) - \varepsilon,$$

which is in contradiction with (4.63).

Hence the claim holds, and together with (4.64), it yields

$$f_j(z) \geq g(z) - \varepsilon$$

for all $j \geq j_0$ and all $z \in \mathbb{R}^m$. Thus for all $j \geq j_0$ and all $z \in \mathbb{R}^m$,

$$f_j^{**}(z) \geq g(z) - \varepsilon$$

and in turn $\sup_j f_j^{**} \geq g - \varepsilon$. By letting $\varepsilon \to 0^+$ we obtain $\sup_j f_j^{**} \geq g$, and taking the supremum over all the affine functions g that are below f gives the desired inequality in view of Theorem 4.92.

We conclude this chapter with some regularity results for the convex envelope Cf of a smooth function f.

Theorem 4.103. *Let $f : \mathbb{R}^m \to (-\infty, \infty]$ be a continuous function. Assume that f is differentiable in $\mathrm{dom}_e\, f$. Then its convex envelope Cf is C^1 in a neighborhood of each point $z_0 \in \mathbb{R}^m$ satisfying*

$$Cf(z_0) < \liminf_{|z| \to \infty} f(z).$$

Moreover, if ∇f is locally Hölder continuous with exponent $0 < \alpha < 1$ or locally Lipschitz in $\mathrm{dom}_e\, f$, then ∇Cf has the same (local) regularity in the open set

$$\left\{ w \in \mathbb{R}^m : Cf(w) < \liminf_{|z| \to \infty} f(z) \right\}.$$

Lemma 4.104. *Let $B \subset \mathbb{R}^m$ be any open ball. If $g : B \to \mathbb{R}$ is convex, and $f : B \to \mathbb{R}$ is differentiable at $z_0 \in B$, $g \leq f$, $f(z_0) = g(z_0)$, then g is differentiable at z_0 and $\nabla g(z_0) = \nabla f(z_0)$.*

Proof. Since $g \leq f$ and f is differentiable at z_0, we have

$$\limsup_{z \to z_0} \frac{g(z) - g(z_0) - \nabla f(z_0) \cdot (z - z_0)}{|z - z_0|} \tag{4.65}$$

$$\leq \lim_{z \to z_0} \frac{f(z) - f(z_0) - \nabla f(z_0) \cdot (z - z_0)}{|z - z_0|} = 0.$$

Conversely, using Remark 4.37(i), for $\varepsilon > 0$ sufficiently small we have

$$\inf_{B_\infty(z_0, \varepsilon)} \{g(z) - g(z_0) - \nabla f(z_0) \cdot (z - z_0)\}$$

$$\geq -(2^m - 1) \sup_{B_\infty(z_0, \varepsilon)} \{g(z) - g(z_0) - \nabla f(z_0) \cdot (z - z_0)\}$$

$$\geq -(2^m - 1) \sup_{B_\infty(z_0, \varepsilon)} \{f(z) - f(z_0) - \nabla f(z_0) \cdot (z - z_0)\} = o(\varepsilon),$$

and so

$$\liminf_{z \to z_0} \frac{g(z) - g(z_0) - \nabla f(z_0) \cdot (z - z_0)}{|z - z_0|} \geq 0.$$

This, together with (4.65), implies that g is differentiable at z_0 and $\nabla g(z_0) = \nabla f(z_0)$.

Proof (Theorem 4.103). **Step 1:** We claim that the set

$$A := \left\{ z \in \mathbb{R}^m : Cf(z) < \liminf_{|w| \to \infty} f(w) \right\}$$

is open. If A is nonempty, then

$$\liminf_{|z| \to \infty} f(z) > -\infty,$$

and since f is continuous, we deduce that f must be bounded from below by some constant c. By replacing f with $f - c$, without loss of generality, we may assume that $f \geq 0$.

Since f is continuous its effective domain is open, and so by Remark 4.85, $\mathrm{dom}_e\, Cf$ is open. By Corollary 4.38, Cf is continuous in $\mathrm{dom}_e\, Cf$. In particular, the set A is open.

Step 2: Next we claim that Cf is differentiable in A. Indeed, fix $z_0 \in A$. By Theorem 4.96, let $\left\{ \left(\theta_i^{(n)}, z_i^{(n)} \right) \right\} \subset [0,1] \times \mathbb{R}^m$, $i = 1, \dots, m+1$, $n \in \mathbb{N}$, be a minimizing sequence such that

$$\sum_{i=1}^{m+1} \theta_i^{(n)} = 1, \quad \sum_{i=1}^{m+1} \theta_i^{(n)} z_i^{(n)} = z_0,$$

and

$$\sum_{i=1}^{m+1} \theta_i^{(n)} f\left(z_i^{(n)} \right) \to Cf(z_0) \tag{4.66}$$

as $n \to \infty$.

Upon extracting a subsequence if necessary, for each $i = 1, \dots, m+1$ we may assume that $\theta_i^{(n)} \to \theta_i$, and that either $\left| z_i^{(n)} \right| \to \infty$ or $z_i^{(n)} \to z_i$ as $n \to \infty$.

Fix

$$Cf(z_0) < s < t < \liminf_{|w| \to \infty} f(w),$$

let $0 < \varepsilon_0 < 1$ be so small that

$$t(1 - \varepsilon_0) > s, \tag{4.67}$$

and find $L > 0$ such that

$$f(z) \geq t \quad \text{for all } |z| \geq L. \tag{4.68}$$

Define

$$I := \left\{ i = 1, \ldots, m+1 : \left| z_i^{(n)} \right| \geq L \text{ for all } n \text{ large} \right\},$$
$$J := \{ i = 1, \ldots, m+1 \} \setminus I.$$

We claim that there exists $i \in J$ such that $\theta_i \geq \frac{\varepsilon_0}{m+1}$. Indeed, if this is not the case, then $\theta_i < \frac{\varepsilon_0}{m+1}$ for all $i \in J$, and so

$$\sum_{i \in I} \theta_i = 1 - \sum_{i \in J} \theta_i \geq 1 - \varepsilon_0 > 0. \tag{4.69}$$

By (4.68), for any $i \in I$ we have that $f\left(z_i^{(n)} \right) \geq t$ for all n sufficiently large, and so, using the fact that $f \geq 0$,

$$\sum_{i=1}^{m+1} \theta_i^{(n)} f\left(z_i^{(n)} \right) \geq \sum_{i \in I} \theta_i^{(n)} f\left(z_i^{(n)} \right) \geq t \sum_{i \in I} \theta_i^{(n)}.$$

Letting $n \to \infty$, by (4.66) and (4.69) we get

$$s > \mathcal{C}f(z_0) \geq t \sum_{i \in I} \theta_i \geq t(1 - \varepsilon_0),$$

which contradicts (4.67) and proves the claim. Hence, without loss of generality, we may assume that $\theta_1^{(n)} \to \theta_1 \geq \frac{\varepsilon_0}{m+1}$, $z_1^{(n)} \to z_1 \in \overline{B(0, L)}$ as $n \to \infty$. Since $f \geq 0$ we have that

$$\theta_1^{(n)} f\left(z_1^{(n)} \right) \leq \sum_{i=1}^{m+1} \theta_i^{(n)} f\left(z_i^{(n)} \right),$$

and so letting $n \to \infty$ by (4.66) and using the continuity of f we get

$$\frac{\varepsilon_0}{m+1} f(z_1) \leq \theta_1 f(z_1) = \lim_{n \to \infty} \theta_1^{(n)} f\left(z_1^{(n)} \right) \leq \mathcal{C}f(z_0) < s. \tag{4.70}$$

This shows that $z_1 \in \operatorname{dom}_e f$, and thus f is differentiable at z_1 by assumption.

By the convexity of $\mathcal{C}f$ and since for any $h \in \mathbb{R}^m$,

$$z_0 + h = \theta_1^{(n)} \left(z_1^{(n)} + \frac{h}{\theta_1^{(n)}} \right) + \sum_{i=2}^{m+1} \theta_i^{(n)} z_i^{(n)},$$

for all n sufficiently large we obtain

$$\mathcal{C}f\left(z_0+h\right)-\mathcal{C}f\left(z_0\right) \le \theta_1^{(n)}\mathcal{C}f\left(z_1^{(n)}+\frac{h}{\theta_1^{(n)}}\right)$$

$$+\sum_{i=2}^{m+1}\theta_i^{(n)}\mathcal{C}f\left(z_i^{(n)}\right)-\mathcal{C}f\left(z_0\right)$$

$$\le \theta_1^{(n)}\left[f\left(z_1^{(n)}+\frac{h}{\theta_1^{(n)}}\right)-f\left(z_1^{(n)}\right)\right]$$

$$+\left[\sum_{i=1}^{m+1}\theta_i^{(n)}f\left(z_i^{(n)}\right)-\mathcal{C}f\left(z_0\right)\right],$$

where we have used the fact that since $z_1 \in \mathrm{dom}_e\, f$ and $\mathrm{dom}_e\, f$ is open, $\left\{z_1^{(n)}\right\} \subset \mathrm{dom}_e\, f$ for all n sufficiently large. Letting $n \to \infty$ in the previous inequality yields

$$\mathcal{C}f\left(z_0+h\right)-\mathcal{C}f\left(z_0\right) \le \theta_1\left[f\left(z_1+\frac{h}{\theta_1}\right)-f\left(z_1\right)\right] \qquad (4.71)$$

for all $h \in \mathbb{R}^m$. Since by assumption f is differentiable at z_1, it follows in particular that the right-hand side is finite for all h sufficiently small, say $|h| < r$. In turn, the nonnegative convex function $\mathcal{C}f\left(z_0+\cdot\right)$ is finite for the same values of h. Since the left-hand side is a convex function in the variable h, the previous lemma implies that $\mathcal{C}f\left(z_0+\cdot\right)$ is differentiable at 0 and $\nabla\left(\mathcal{C}f\right)\left(z_0\right) = \nabla f\left(z_1\right)$.

Thus we have shown that $\mathcal{C}f$ is differentiable in A, and by Theorem 4.65(i) it follows that $\nabla\left(\mathcal{C}f\right)$ is continuous on A.

Step 3: Finally, assume that ∇f is locally Hölder continuous with exponent $0 < \alpha < 1$ or locally Lipschitz in $\mathrm{dom}_e\, f$ and let U be an open set compactly contained in A. Find $U \subset\subset D \subset\subset A$. By the continuity of $\mathcal{C}f$ and the definition of the set A we may find s,

$$0 < s < \liminf_{|z|\to\infty} f\left(z\right),$$

such that $\mathcal{C}f\left(z\right) < s$ for all $z \in \overline{D}$. Fix

$$s < t < \liminf_{|z|\to\infty} f\left(z\right),$$

and let $\varepsilon_0 > 0$ and $L > 0$ be as in (4.67) and (4.68). By the previous step, for any $z \in \overline{D}$ we may find $z_1^{(z)} \in \mathrm{dom}_e\, f \cap \overline{B\left(0,L\right)}$ and $\frac{\varepsilon_0}{m+1} \le \theta_1 \le 1$ such that $\nabla\left(\mathcal{C}f\right)\left(z\right) = \nabla f\left(z_1^{(z)}\right)$ and (4.70) and (4.71) hold. We claim that there exists an open set U_1 compactly contained in $\mathrm{dom}_e\, f$ such that $z_1^{(z)} \in U_1$ for all $z \in \overline{D}$. Indeed, if not, then we may find a sequence $\{z_k\} \subset \overline{D}$ converging to some $z \in \overline{D}$ such that $z_1^{(z_k)} \to z_1 \in \overline{B\left(0,L\right)} \setminus \mathrm{dom}_e\, f$. But by (4.70),

$$\frac{\varepsilon_0}{m+1} f\left(z_1^{(z_k)}\right) \le Cf(z_k) < s.$$

Letting $k \to \infty$ and using the continuity of f we obtain a contradiction since $z_1 \notin \text{dom}_e\, f$. Hence the claim holds.

Let $U_1 \subset\subset U_2 \subset\subset \text{dom}_e\, f$ and by hypothesis let $C = C(U_2) > 0$ be such that

$$|\nabla f(z) - \nabla f(w)| \le C\, |z - w|^\alpha \tag{4.72}$$

for all z, $w \in U_2$. Let $r > 0$ be so small that $w + \frac{(m+1)h}{\varepsilon_0} \in U_2$ for all $|h| < r$ and all $w \in U_1$.

If $z \in \overline{U}$, then by what we just proved, $z_1^{(z)} \in U_1$, and so $z_1^{(z)} + \frac{h}{\theta_1^{(z)}} \in U_2$ for all $|h| < r$. By the mean value theorem and the fact that $\nabla(\mathcal{C}f)(z) = \nabla f\left(z_1^{(z)}\right)$ we obtain

$$\mathcal{C}f(z+h) - \mathcal{C}f(z) - \nabla(\mathcal{C}f)(z) \cdot h$$

$$\le \theta_1^{(z)}\left[f\left(z_1^{(z)} + \frac{h}{\theta_1^{(z)}}\right) - f\left(z_1^{(z)}\right) - \nabla f\left(z_1^{(z)}\right) \cdot \frac{h}{\theta_1^{(z)}}\right]$$

$$= \theta_1^{(z)}\left[\nabla f\left(w_1^{(z,h)}\right) \cdot \frac{h}{\theta_1^{(z)}} - \nabla f\left(z_1^{(z)}\right) \cdot \frac{h}{\theta_1^{(z)}}\right]$$

$$\le \theta_1^{(z)} C\left|\frac{h}{\theta_1^{(z)}}\right|^{1+\alpha} \le \frac{(m+1)^\alpha}{\varepsilon_0^\alpha} C\, |h|^{1+\alpha}$$

for some $w_1^{(z,h)}$ on the segment of endpoints $z_1^{(z)}$ and $z_1^{(z)} + \frac{h}{\theta_1^{(z)}}$, and where we have used (4.71), (4.72), and the fact that $\theta_1^{(z)} \in \left[\frac{\varepsilon_0}{m+1}, 1\right]$. Hence also by Theorem 4.62,

$$0 \le \mathcal{C}f(z+h) - \mathcal{C}f(z) - \nabla(\mathcal{C}f)(z) \cdot h \le \frac{(m+1)^\alpha}{\varepsilon_0^\alpha} C\, |h|^{1+\alpha}$$

for all $|h| < r$. By Remark 4.37(iii) applied to the convex function

$$g(h) := \mathcal{C}f(z+h) - \mathcal{C}f(z) - \nabla(\mathcal{C}f)(z) \cdot h$$

we obtain that

$$|\nabla(\mathcal{C}f)(z+h) - \nabla(\mathcal{C}f)(z)| \tag{4.73}$$

$$= |\nabla g(h)| \le \text{Lip}(g; B(0, 2\,|h|)) \le \frac{\text{osc}(g; B(0, 4\,|h|))}{2\,|h|}$$

$$\le 4^{1+\alpha}\frac{(m+1)^\alpha}{2\varepsilon_0^\alpha} C\, |h|^\alpha$$

for all $|h| < \frac{1}{4}r$.

Fix $\bar{z} \in \overline{U}$ and let $r^{(\bar{z})} > 0$ be so small that $B\left(\bar{z}, r^{(\bar{z})}\right) \subset D$ and $r^{(\bar{z})} < \frac{r}{8}$. We claim that $\nabla\left(\mathcal{C}f\right)$ is Hölder continuous with exponent $0 < \alpha < 1$ or Lipschitz in $B\left(\bar{z}, r^{(\bar{z})}\right)$. To see this, let z, $w \in B\left(\bar{z}, r^{(\bar{z})}\right)$ and write

$$w = z + h,$$

where $h := w - z$ is such that

$$|h| = |w - z| \le |w - \bar{z}| + |z - \bar{z}| < 2\bar{r} < \frac{r}{4}.$$

By (4.73),

$$\left|\nabla\mathcal{C}f\left(w\right) - \nabla\left(\mathcal{C}f\right)\left(z\right)\right| 4^{1+\alpha} \frac{(m+1)^{\alpha}}{2\varepsilon_0^{\alpha}} C\left|w - z\right|^{\alpha},$$

which proves the claim.

Since the family of balls $\left\{B\left(\bar{z}, r^{(\bar{z})}\right)\right\}_{\bar{z} \in \overline{U}}$ is an open cover for the compact set $\overline{U}$, we can find a finite number of balls that still cover $\overline{U}$.

Hence $\nabla\mathcal{C}f$ is locally Hölder continuous with exponent $0 < \alpha < 1$ or locally Lipschitz.

Example 4.105. The next two examples show the sharpness of the previous theorem.

(i) The function $f\left(z\right) = f\left(z_1, z_2\right) = \sqrt{\exp\left(-z_1^2\right) + z_2^2}$ shows that the condition $\mathcal{C}f\left(z\right) < \liminf_{|z| \to \infty} f\left(z\right)$ cannot be eliminated.

(ii) Note that in general one cannot go beyond the regularity stated in the previous theorem. Indeed, any smooth function $f : \mathbb{R} \to [0, \infty)$ such that $\mathcal{C}f = f$ outside $[-1, 1]$, $f^{-1}\left(\{0\}\right) = \{-1, 1\}$, and $f''\left(\pm 1\right) > 0$ shows that $\mathcal{C}f$ may not be of class C^2 even if f is.

4.9 Star-Shaped Sets

Given a vector space V, a set $E \subset V$ is said to be *star-shaped with respect to a set* $F \subset E$ if E is star-shaped with respect to each point of F, i.e., if $\theta v + (1 - \theta) w \in E$ for all $v \in F$, $w \in E$, and $\theta \in (0, 1)$.

If $V = \mathbb{R}^m$ and $E \subset \mathbb{R}^m$ is star-shaped with respect to a point $z_0 \in \mathbb{R}^m$, the function $\varphi : S^{m-1} \to [0, \infty]$, defined by

$$\varphi\left(z\right) := \sup\left\{s \ge 0 : z_0 + sz \in E\right\} \quad z \in S^{m-1},$$

is called the *radial function* of the set E. In the next theorem, we assume, without loss of generality, that $z_0 = 0$ and for any z_1, $z_2 \in S^{m-1}$ we define $d\left(z_1, z_2\right)$ to be the measure of the angle formed by the vectors z_1 and z_2.

Exercise 4.106. Consider the ball $B(0,r) \subset \mathbb{R}^m$, let $w \notin B(0,r)$, and define

$$C := \mathrm{co}\,(B(0,r) \cup \{w\}).$$

Let φ be the radial function of C. Prove that

$$\varphi(z) = \begin{cases} r & \begin{array}{l} \text{if } z \in S^{m-1} \\ \text{and } d(z_1, z) \geq \arccos \dfrac{r}{|w|}, \end{array} \\[2em] \dfrac{r}{\cos\left(\arccos\dfrac{r}{|w|} - d(z_1, z)\right)} & \begin{array}{l} \text{if } z \in S^{m-1} \\ \text{and } d(z_1, z) < \arccos \dfrac{r}{|w|}. \end{array} \end{cases}$$

Theorem 4.107. *Assume that $C \subset \mathbb{R}^m$ is a bounded closed set, star-shaped with respect to a ball $B(0, \varepsilon) \subset C$, and let $B(0, R) \supset C$. Then for all z_1, $z_2 \in S^{m-1}$,*

$$|\varphi(z_1) - \varphi(z_2)| \leq \left(R\sqrt{\left(\frac{R}{\varepsilon}\right)^2 - 1}\right) d(z_1, z_2). \tag{4.74}$$

We begin with a preliminary result that is of independent interest. Given an arbitrary nonempty set $E \subset \mathbb{R}^m$, a *convex component* of E is a maximal (in the sense of inclusion) convex subset of E. The *kernel* of E, denoted by $\ker E$, is the set

$$\ker E := \{z \in E : \theta z + (1 - \theta) w \in E \text{ for all } w \in E \text{ and } \theta \in (0,1)\}.$$

Lemma 4.108. *Given an arbitrary nonempty set $E \subset \mathbb{R}^m$, the kernel of E is the intersection of all convex components of E.*

Proof. Let $\{C_\alpha\}_{\alpha \in I}$ be the family of all convex components of E. Define

$$C_0 := \bigcap_{\alpha \in I} C_\alpha$$

and let $z \in C_0$. Since

$$E = \bigcup_{\alpha \in I} C_\alpha,$$

for all $w \in E$ there exists $\alpha \in I$ such that $w \in C_\alpha$, and so

$$\theta z + (1 - \theta) w \in C_\alpha \subset E$$

for all $\theta \in (0,1)$, which implies that $z \in \ker E$ and, in turn, that $C_0 \subset \ker E$.

Conversely, let $z \in \ker E$ and let C_α be an arbitrary convex component of E. Let

$$K_\alpha := \{\theta z + (1 - \theta) w : \theta \in [0,1],\ w \in C_\alpha\}.$$

Then $C_\alpha \subset K_\alpha$ and, since $z \in \ker E$, $K_\alpha \subset E$. Also, K_α is convex, since for any w_1, $w_2 \in C_\alpha$ and θ, θ_1, $\theta_2 \in (0,1)$ we have

$$
\theta(\theta_1 z + (1 - \theta_1) w_1) + (1 - \theta)(\theta_2 z + (1 - \theta_2) w_2)
$$
$$
= [\theta\theta_1 + (1 - \theta)\theta_2] z + (1 - [\theta\theta_1 + (1 - \theta)\theta_2])
$$
$$
\times \left[\frac{\theta(1 - \theta_1)}{1 - [\theta\theta_1 + (1 - \theta)\theta_2]} w_1 + \frac{(1 - \theta)(1 - \theta_2)}{1 - [\theta\theta_1 + (1 - \theta)\theta_2]} w_2 \right],
$$

which is still in K_α since C_α is convex. By the maximality of C_α it follows that $C_\alpha = K_\alpha$, and so $z \in C_\alpha$. Given the arbitrariness of C_α we have that $z \in C_0$, which shows that $\ker E \subset C_0$, and this completes the proof.

We are now ready to prove Theorem 4.107.

Proof (Theorem 4.107). **Step 1:** Assume first that C is convex. Let z_1, $z_2 \in S^{m-1}$ and assume that $\varphi(z_2) \le \varphi(z_1) := s_1$. The point $w_1 := s_1 z_1$ is in C, and so the set

$$
C_{w_1} := \mathrm{co}\,(B(0, \varepsilon) \cup \{w_1\})
$$

is contained in C. Hence, if φ_{w_1} denotes the radial function of C_{w_1}, then $\varphi_{w_1} \le \varphi$ and $\varphi_{w_1}(z_1) = \varphi(z_1)$. Moreover, by Exercise 4.106, for any $z \in S^{m-1}$,

$$
\varphi_{w_1}(z) = \begin{cases} \varepsilon & \text{if } \theta(z) \ge \theta_0, \\ \frac{\varepsilon}{\cos(\theta_0 - \theta(z))} & \text{if } \theta(z) < \theta_0, \end{cases}
$$

where

$$
\theta(z) := d(z_1, z), \quad \theta_0 := \arccos \frac{\varepsilon}{s_1}.
$$

If $d(z_1, z_2) > \theta_0$ let $z_3 \in S^{m-1}$ be a point between z_1 and z_2 such that $d(z_1, z_3) = \theta_0$, while if $d(z_1, z_2) \le \theta_0$ let $z_3 := z_2$. Then

$$
g(z_3) := \frac{\varphi_{w_1}(z_1) - \varphi_{w_1}(z_3)}{d(z_1, z_3)} = \frac{s_1 - \frac{\varepsilon}{\cos(\theta_0 - \theta(z_3))}}{\theta(z_3)}
$$
$$
= \frac{s_1 \cos(\theta_0 - \theta(z_3)) - \varepsilon}{\theta(z_3) \cos(\theta_0 - \theta(z_3))}.
$$

Note that $g(z_3)$ increases as z_3 tends to z_1. Moreover, by Taylor's formula, if $\theta(z_3)$ is small, then

$$
g(z_3) = \frac{\varepsilon \cos\theta(z_3) + s_1 \sqrt{1 - \left(\frac{\varepsilon}{s_1}\right)^2} \sin\theta(z_3) - \varepsilon}{\theta(z_3) \left[\frac{\varepsilon}{s_1} \cos\theta(z_3) - \sqrt{1 - \left(\frac{\varepsilon}{s_1}\right)^2} \sin\theta(z_3) \right]}
$$
$$
= \frac{s_1 \sqrt{1 - \left(\frac{\varepsilon}{s_1}\right)^2} (1 + o(1))}{\frac{\varepsilon}{s_1} (1 + o(1))}
$$
$$
= s_1 \sqrt{\left(\frac{s_1}{\varepsilon}\right)^2 - 1} (1 + o(1)).
$$

Hence

$$\frac{\varphi(z_1) - \varphi(z_2)}{d(z_1, z_2)} \leq \frac{\varphi_{w_1}(z_1) - \varphi_{w_1}(z_2)}{d(z_1, z_2)}$$

$$\leq \frac{\varphi_{w_1}(z_1) - \varphi_{w_1}(z_3)}{d(z_1, z_3)} = g(z_3)$$

$$\leq R\sqrt{\left(\frac{R}{\varepsilon}\right)^2 - 1}.$$

Step 2: Assume now that $C \subset \mathbb{R}^m$ is as in the statement of the problem and let $\{C_\alpha\}_{\alpha \in I}$ be the family of all convex components of C. Since C is star-shaped with respect to a ball $B(0, \varepsilon)$, we have that $B(0, \varepsilon) \subset \ker C$, and so in view of the previous lemma, we may apply the previous step to each convex component C_α of C.

We claim that

$$\varphi(z) = \max_{\alpha \in I} \varphi_\alpha(z), \tag{4.75}$$

where φ_α is the radial function of C_α. Indeed, since $C_\alpha \subset C$ for all $\alpha \in I$, we have that $\varphi_\alpha \leq \varphi$. Conversely, if $\varphi(z) = s$, then $w := sz \in C$, and so we may find $\alpha \in I$ such that $sz \in C_\alpha$. Thus $\varphi_\alpha(z) \geq s$, which, together with the fact that $\varphi \geq \varphi_\alpha$, shows that $\varphi(z) = \varphi_\alpha(z)$, so that (4.75) holds.

We are now ready to conclude the proof. For any $z_1, z_2 \in S^{m-1}$ assume that $\varphi(z_2) \leq \varphi(z_1)$ and let $\alpha \in I$ be such that $\varphi_\alpha(z_1) = \varphi(z_1)$. Then by (4.75),

$$\frac{\varphi(z_1) - \varphi(z_2)}{d(z_1, z_2)} = \frac{\varphi_\alpha(z_1) - \varphi(z_2)}{d(z_1, z_2)} \leq \frac{\varphi_\alpha(z_1) - \varphi_\alpha(z_2)}{d(z_1, z_2)} \leq R\sqrt{\left(\frac{R}{\varepsilon}\right)^2 - 1}.$$

This completes the proof.

Exercise 4.109. Let C be as in Exercise 4.106 and let $R := |w|$. Prove that for this set, the constant $R\sqrt{\left(\frac{R}{\varepsilon}\right)^2 - 1}$ in (4.74) is sharp.

Part III

Functionals Defined on L^p

Integrands $f = f(z)$

Mathematics is the science which
draws necessary conclusions.

Benjamin Peirce (1809–1880)

We study necessary and sufficient conditions for the sequential lower semi-continuity of functionals of the form

$$v \in L^p(E; \mathbb{R}^m) \mapsto \int_E f(v(x)) \, dx,$$

where $1 \leq p \leq \infty$ and

$$f : \mathbb{R}^m \to [-\infty, \infty].$$

We assume throughout this chapter that E is a Lebesgue measurable subset of $\mathbb{R}^N$. In view of Proposition 1.68 note that without loss of generality we may always assume that the domain of integration E is a Borel set.

We are interested in the following types of convergence:

- strong convergence in $L^p(E; \mathbb{R}^m)$ for $1 \leq p \leq \infty$;
- weak convergence in $L^p(E; \mathbb{R}^m)$ for $1 \leq p < \infty$;
- weak star convergence in $L^\infty(E; \mathbb{R}^m)$;
- weak star convergence in the sense of measures in $L^1(E; \mathbb{R}^m)$.

The reason why the last type of convergence is of interest is that for $1 < p \leq \infty$, bounded sequences in $L^p(E; \mathbb{R}^m)$ admit weakly convergent subsequences (respectively weakly star if $p = \infty$), but when $p = 1$, due to lack of reflexivity of the space $L^1(E; \mathbb{R}^m)$ one can conclude only that an energy-bounded sequence $\{v_n\} \subset L^1(E; \mathbb{R}^m)$ with

$$\sup_n \|v_n\|_{L^1} < \infty$$

admits a subsequence (not relabeled) such that $v_n \mathcal{L}^N \lfloor E \overset{*}{\rightharpoonup} \lambda$ in $\mathcal{M}(E; \mathbb{R}^m)$, i.e., if for all $u \in C_0(E)$,

$$\int_E u\, v_n\, dx \to \int_E u\, d\lambda$$

(see Remark 2.51). In particular, we will provide conditions under which the lower semicontinuity property

$$\liminf_{n\to\infty} \int_E f(v_n)\, dx \geq \int_E f\left(\frac{d\lambda}{d\mathcal{L}^N}\right) dx$$

holds whenever $v_n \mathcal{L}^N \lfloor E \xrightarrow{*} \lambda$ in $\mathcal{M}(E; \mathbb{R}^m)$ and λ admits the Radon–Nikodym decomposition

$$\lambda = \frac{d\lambda}{d\mathcal{L}^N}\, \mathcal{L}^N \lfloor E + \lambda_s,$$

with λ_s and $\mathcal{L}^N \lfloor E$ mutually singular.

The proofs in the next sections use heavily the fact that the Lebesgue measure is nonatomic (see Remark 1.161).

5.1 Well-Posedness

As mentioned in Section 1.1.1, measurability of functions with values in the Euclidean space $\mathbb{R}^m$ or in the extended real line $\overline{\mathbb{R}} := [-\infty, \infty]$ is understood when these target spaces are endowed with the Borel σ-algebras associated to the respective usual topologies. Therefore if $v \in L^p(E; \mathbb{R}^m)$ and $f : \mathbb{R}^m \to [-\infty, \infty]$, then measurability of $f \circ v : E \to [-\infty, \infty]$ is ensured provided f is a Borel function.

Theorem 5.1. *Let E be a Borel subset of $\mathbb{R}^N$ with finite measure, let $1 \leq p \leq \infty$, and let $f : \mathbb{R}^m \to [-\infty, \infty]$ be a Borel function. Then*

$$\int_E (f(v(x)))^- dx < \infty \tag{5.1}$$

for every $v \in L^p(E; \mathbb{R}^m)$ if and only if there exists a constant $C > 0$ such that

$$f(z) \geq -C\left(1 + |z|^p\right) \text{ for all } z \in \mathbb{R}^m \text{ if } 1 \leq p < \infty, \tag{5.2}$$

$$f \text{ is locally bounded from below if } p = \infty. \tag{5.3}$$

The proof uses the following lemma.

Lemma 5.2. *Let $L > 0$ and let $a_n \to \infty$. Then there exists $b_n \geq 0$ such that*

$$\sum_{n=1}^{\infty} b_n \leq L \quad \text{and} \quad \sum_{n=1}^{\infty} a_n b_n = \infty.$$

Proof. Without loss of generality we may assume that $L = 1$. For every $k \in \mathbb{N}$ define

$$M_k := \min\left\{n : a_l \geq 2^k \text{ for all } l \geq n\right\}.$$

Let $\{M_{k_j}\}$ be a strictly increasing subsequence of $\{M_k\}$, with $k_j \geq j$, and define

$$b_n := \begin{cases} 0 & \text{if } n < M_{k_1}, \\ \dfrac{1}{2^{k_j}} \dfrac{1}{M_{k_{j+1}} - M_{k_j}} & \text{if } M_{k_j} \leq n < M_{k_{j+1}}. \end{cases}$$

Then

$$\sum_{n=1}^{\infty} b_n = \sum_{j=1}^{\infty} \sum_{n=M_{k_j}}^{M_{k_{j+1}}-1} \frac{1}{2^{k_j}} \frac{1}{M_{k_{j+1}} - M_{k_j}} = \sum_{j=1}^{\infty} \frac{1}{2^{k_j}} \leq \sum_{j=1}^{\infty} \frac{1}{2^j} = 1$$

and

$$\sum_{n=1}^{\infty} a_n b_n = \sum_{j=1}^{\infty} \sum_{n=M_{k_j}}^{M_{k_{j+1}}-1} \frac{1}{2^{k_j}} \frac{a_n}{M_{k_{j+1}} - M_{k_j}}$$

$$\geq \sum_{j=1}^{\infty} \sum_{n=M_{k_j}}^{M_{k_{j+1}}-1} \frac{1}{M_{k_{j+1}} - M_{k_j}} = \infty.$$

This completes the proof.

We are now ready to prove Theorem 5.1.

Proof (Theorem 5.1). Note that (5.2) and (5.3) imply (5.1) for $1 \leq p < \infty$ and $p = \infty$, respectively.

Conversely, assume that (5.1) is satisfied. By considering constant functions, it follows that f cannot take the value $-\infty$.

Step 1: Suppose first that $1 \leq p < \infty$. Assume by contradiction that (5.2) does not hold. Then we can find a sequence $\{z_n\} \subset \mathbb{R}^m$ such that

$$\lim_{n \to \infty} \frac{f(z_n)}{1 + |z_n|^p} = -\infty.$$

By the previous lemma with $L := |E|$ and

$$a_n := -\frac{f(z_n)}{1 + |z_n|^p},$$

we can find $b_n \geq 0$ such that

$$\sum_{n=1}^{\infty} b_n \leq |E| \quad \text{and} \quad \sum_{n=1}^{\infty} \frac{f(z_n)}{1 + |z_n|^p} b_n = -\infty. \tag{5.4}$$

By Corollary 1.21 we may construct a countable family $\{E_n\}$ of mutually disjoint measurable subsets of E, with

$$|E_n| = \frac{b_n}{1 + |z_n|^p}.$$

Setting

$$v := \sum_{n=1}^{\infty} z_n \chi_{E_n},$$

it now follows from (5.4) that

$$\int_E |v|^p \, dx = \sum_{n=1}^{\infty} |z_n|^p \, |E_n| \leq \sum_{n=1}^{\infty} b_n \leq |E| < \infty,$$

while

$$\int_E (f(v))^- \, dx \geq -\sum_{n=1}^{\infty} f(z_n) \, |E_n| = -\sum_{n=1}^{\infty} \frac{f(z_n)}{1 + |z_n|^p} b_n = -\infty.$$

Step 2: If $p = \infty$, assume by contradiction that (5.3) does not hold. Then we can find a bounded sequence $\{z_n\} \subset \mathbb{R}^m$ such that

$$\lim_{n \to \infty} f(z_n) = -\infty.$$

By the previous lemma with $L := |E|$ and $a_n := -f(z_n)$, we can find $b_n \geq 0$ such that

$$\sum_{n=1}^{\infty} b_n \leq |E| \quad \text{and} \quad \sum_{n=1}^{\infty} f(z_n) b_n = -\infty. \tag{5.5}$$

By Corollary 1.21 we may construct a countable family $\{E_n\}$ of mutually disjoint measurable subsets of E, with $|E_n| = b_n$. Setting

$$v := \sum_{n=1}^{\infty} z_n \chi_{E_n},$$

it now follows from (5.5) and the fact that $\{z_n\}$ is bounded that $v \in L^\infty(E; \mathbb{R}^m)$ and

$$\int_E (f(v))^- \, dx \geq -\sum_{n=1}^{\infty} f(z_n) \, |E_n| = -\sum_{n=1}^{\infty} f(z_n) b_n = -\infty.$$

This concludes the proof.

Remark 5.3. Applying the previous proposition to f and $-f$, it follows that if $f : \mathbb{R}^m \to [-\infty, \infty]$ is a Borel function, then $f \circ v \in L^1(E)$ for every $v \in L^p(E; \mathbb{R}^m)$ if and only if f does not take the values $-\infty$ and ∞ and, moreover, there exists a constant $C > 0$ such that

$$|f(z)| \le C(1 + |z|^p) \text{ for all } z \in \mathbb{R}^m \text{ if } 1 \le p < \infty,$$
$$f \text{ is locally bounded if } p = \infty.$$

Exercise 5.4. Prove that the previous proposition still holds if $L^p(E; \mathbb{R}^m)$ is replaced by the space $L^p((X, \mathfrak{M}, \mu); \mathbb{R}^m)$, where $(X, \mathfrak{M}, \mu)$ is a measure space with μ finite and nonatomic. What can you conclude if the measure μ has atoms (recall Proposition 1.22)?

Exercise 5.5. Let E be a Borel subset of $\mathbb{R}^N$ with finite measure, let $1 \le p \le \infty$, and let $f : \mathbb{R}^m \to [-\infty, \infty]$ be a Borel function. Prove that the integral $\int_E f(v(x)) \, dx$ is well-defined for every $v \in L^p(E; \mathbb{R}^m)$ (that is, the integrals $\int_E (f(v(x)))^- \, dx$ and $\int_E (f(v(x)))^+ \, dx$ are not both infinite) if and only if f or $-f$ (or both) satisfies condition (5.2) if $1 \le p < \infty$ and (5.3) if $p = \infty$.

Next we address the case that E has infinite measure.

Theorem 5.6. *Let E be a Borel subset of $\mathbb{R}^N$ with infinite measure, let $1 \le p \le \infty$, and let $f : \mathbb{R}^m \to [-\infty, \infty]$ be a Borel function. Then*

$$\int_E (f(v(x)))^- \, dx < \infty \tag{5.6}$$

for every $v \in L^p(E; \mathbb{R}^m)$ if and only if there exists a constant $C > 0$ such that

$$f(z) \ge -C|z|^p \text{ for all } z \in \mathbb{R}^m \text{ if } 1 \le p < \infty, \tag{5.7}$$
$$f(z) \ge 0 \text{ for all } z \in \mathbb{R}^m \text{ if } p = \infty. \tag{5.8}$$

Proof. Note that (5.7) and (5.8) imply (5.6) for $1 \le p < \infty$ and $p = \infty$, respectively.

Conversely, assume that (5.6) is satisfied. For $p = \infty$ and taking constant functions, we have that f cannot take negative values.

Consider now the case $1 \le p < \infty$. First note that $f > -\infty$. Indeed, if $f(z_0) = -\infty$ for some $z_0 \in \mathbb{R}^m$, using the fact that the Lebesgue measure is nonatomic, choose a measurable subset $E_1 \subset E$, with $0 < |E_1| < \infty$, and define

$$v(x) := \begin{cases} z_0 & \text{if } x \in E_1, \\ 0 & \text{otherwise.} \end{cases}$$

Then $v \in L^p(E; \mathbb{R}^m)$ and still

$$\int_E (f(v))^- \, dx \ge (f(z_0))^- |E_1| = \infty,$$

which violates (5.6).

Assume by contradiction that (5.7) does not hold. Then we can find a sequence $\{z_n\} \subset \mathbb{R}^m$ such that

$$\lim_{n \to \infty} \frac{f(z_n)}{|z_n|^p} = -\infty,$$

and thus

$$\lim_{n \to \infty} \frac{|z_n|^p}{f(z_n)} = 0.$$

Extract a subsequence $\{z_{n_k}\}$ such that

$$\sum_{k=1}^{\infty} \frac{|z_{n_k}|^p}{-f(z_{n_k})} < \infty.$$

By Proposition 1.20 we may construct a countable family $\{E_k\}$ of mutually disjoint measurable subsets of E, with

$$|E_k| = -\frac{1}{f(z_{n_k})}.$$

Defining

$$v := \sum_{k=1}^{\infty} z_{n_k} \chi_{E_k},$$

it now follows that

$$\int_E |v|^p \, dx = \sum_{k=1}^{\infty} \frac{|z_{n_k}|^p}{-f(z_{n_k})} < \infty,$$

while

$$\int_E (f(v))^- \, dx \geq -\sum_{k=1}^{\infty} f(z_{n_k}) |E_k| = \infty.$$

Hence we have reached a contradiction, and so (5.7) holds.

Exercise 5.7. Prove that the previous proposition still holds if $L^p(E; \mathbb{R}^m)$ is replaced by the space $L^p((X, \mathfrak{M}, \mu); \mathbb{R}^m)$, where $(X, \mathfrak{M}, \mu)$ is a measure space with μ nonatomic. What can you conclude if the measure μ has atoms (recall Proposition 1.22)?

Exercise 5.8. Let E be a Borel subset of $\mathbb{R}^N$ with infinite measure, let $1 \leq p \leq \infty$, and let $f : \mathbb{R}^m \to [-\infty, \infty]$ be a Borel function. Prove that the integral $\int_E f(v(x)) \, dx$ is well-defined for every $v \in L^p(E; \mathbb{R}^m)$ (that is, the integrals $\int_E (f(v(x)))^- \, dx$ and $\int_E (f(v(x)))^+ \, dx$ are not both infinite) if and only if f or $-f$ (or both) satisfies condition (5.7) if $1 \leq p < \infty$ and (5.8) if $p = \infty$.

5.2 Sequential Lower Semicontinuity

This section is dedicated to finding necessary and sufficient conditions for sequential lower semicontinuity with respect to strong convergence in L^p, $1 \leq p \leq \infty$, weak convergence in L^p, $1 \leq p < \infty$, weak star convergence in L^∞, and weak star convergence in the sense of measures.

5.2.1 Strong Convergence in L^p

From now on we will assume that the integrand f satisfies the appropriate growth conditions from below that are necessary and sufficient to guarantee that $(f \circ v)^- \in L^1(E)$ for all $L^p(E; \mathbb{R}^m)$, so that the functional

$$v \in L^p(E; \mathbb{R}^m) \mapsto \int_E f(v(x))\, dx$$

is well-defined (see Theorems 5.1 and 5.6).

Theorem 5.9. *Let E be a Borel subset of $\mathbb{R}^N$ with finite measure, let $1 \leq p \leq \infty$, and let $f : \mathbb{R}^m \to (-\infty, \infty]$ be a Borel function. Assume that there exists $C > 0$ such that*

$$f(z) \geq -C(1 + |z|^p) \text{ for all } z \in \mathbb{R}^m \text{ if } 1 \leq p < \infty, \tag{5.9}$$

$$f \text{ is locally bounded from below if } p = \infty. \tag{5.10}$$

Then the functional

$$v \in L^p(E; \mathbb{R}^m) \mapsto \int_E f(v(x))\, dx$$

is sequentially lower semicontinuous with respect to strong convergence in $L^p(E; \mathbb{R}^m)$ if and only if f is lower semicontinuous.

Proof. Assume that

$$v \in L^p(E; \mathbb{R}^m) \mapsto \int_E f(v)\, dx$$

is sequentially lower semicontinuous with respect to strong convergence in $L^p(E; \mathbb{R}^m)$. Let $\{z_n\} \subset \mathbb{R}^m$ be such that $z_n \to z$, and set $v_n(x) := z_n$, $v(x) := z$. Then $v_n \to v$ in $L^p(E; \mathbb{R}^m)$, and so

$$f(z) = \frac{1}{|E|} \int_E f(v)\, dx \leq \liminf_{n \to \infty} \frac{1}{|E|} \int_E f(v_n)\, dx = \liminf_{n \to \infty} f(z_n),$$

i.e., f is lower semicontinuous.

Conversely, suppose that f is lower semicontinuous and let $v_n \to v$ in $L^p(E; \mathbb{R}^m)$. Consider first the case $1 \leq p < \infty$.

Without loss of generality we may assume that

$$\liminf_{n \to \infty} \int_E f(v_n) \, dx = \lim_{n \to \infty} \int_E f(v_n) \, dx,$$

and, by extracting a further subsequence if necessary (see Theorem 2.20), that $v_n \to v$ pointwise $\mathcal{L}^N$ a.e. in E. By (5.9) we can apply Fatou's lemma to conclude that

$$\liminf_{n \to \infty} \int_E f(v_n) \, dx + \int_E [C + C |v|^p] \, dx$$

$$= \liminf_{n \to \infty} \int_E [f(v_n) + C + C |v_n|^p] \, dx$$

$$\geq \int_E \liminf_{n \to \infty} [f(v_n) + C + C |v_n|^p] \, dx$$

$$\geq \int_E f(v) \, dx + \int_E [C + C |v|^p] \, dx,$$

and thus, since E has finite measure,

$$\liminf_{n \to \infty} \int_E f(v_n) \, dx \geq \int_E f(v) \, dx. \tag{5.11}$$

The case $p = \infty$ is similar. Let

$$M := \sup_n \|v_n\|_{L^\infty} < \infty.$$

By (5.10) there exists a constant $C > 0$ such that

$$f(z) \geq -C \text{ for all } z \in \mathbb{R}^m \text{ with } |z| \leq M.$$

Hence, again by Fatou's lemma, we have that

$$\liminf_{n \to \infty} \int_E f(v_n) \, dx + C |E| = \liminf_{n \to \infty} \int_E [f(v_n) + C] \, dx$$

$$\geq \int_E \liminf_{n \to \infty} [f(v_n) + C] \, dx \geq \int_E f(v) \, dx + C |E|,$$

and so (5.11) holds.

Remark 5.10. Applying the previous proposition to f and $-f$ it follows that if $f : \mathbb{R}^m \to \mathbb{R}$ is a Borel function satisfying (5.9) and (5.10), then the functional

$$v \in L^p (E; \mathbb{R}^m) \mapsto \int_E f(v(x)) \, dx$$

is continuous with respect to strong convergence in $L^p (E; \mathbb{R}^m)$ if and only if f is continuous.

Exercise 5.11. Prove that the previous proposition still holds if $L^p(E; \mathbb{R}^m)$ is replaced by the space $L^p((X, \mathfrak{M}, \mu); \mathbb{R}^m)$, where $(X, \mathfrak{M}, \mu)$ is a measure space with μ finite.

Next we consider the case that E has infinite measure.

Theorem 5.12. *Let E be a Borel subset of $\mathbb{R}^N$ with infinite measure, let $1 \leq p \leq \infty$, and let $f : \mathbb{R}^m \to (-\infty, \infty]$ be a Borel function. Assume that there exists $C > 0$ such that*

$$f(z) \geq -C|z|^p \text{ for all } z \in \mathbb{R}^m \text{ if } 1 \leq p < \infty,$$
$$f(z) \geq 0 \text{ for all } z \in \mathbb{R}^m \text{ if } p = \infty.$$

Then the functional

$$v \in L^p(E; \mathbb{R}^m) \mapsto \int_E f(v(x))\, dx$$

is sequentially lower semicontinuous with respect to strong convergence in $L^p(E; \mathbb{R}^m)$ if and only if f is lower semicontinuous.

Proof. Without loss of generality we may assume that there exists $v_0 \in L^p(E; \mathbb{R}^m)$ such that

$$\int_E f(v_0(x))\, dx < \infty. \tag{5.12}$$

The proof of the sufficiency follows an argument entirely similar to that of the previous theorem. As for the necessity, consider a Borel subset $E_1 \subset E$ with finite measure, and let $z_n \to z$ in $\mathbb{R}^m$. Define

$$v_n(x) := \begin{cases} z_n & \text{if } x \in E_1, \\ v_0(x) & \text{if } x \in E \setminus E_1. \end{cases}$$

Then $v_n \to v$ in $L^p(E; \mathbb{R}^m)$, where

$$v(x) := \begin{cases} z & \text{if } x \in E_1, \\ v_0(x) & \text{if } x \in E \setminus E_1. \end{cases}$$

Hence

$$|E_1| f(z) + \int_{E \setminus E_1} f(v_0)\, dx = \int_E f(v)\, dx \leq \liminf_{n \to \infty} \int_E f(v_n)\, dx$$

$$= |E_1| \liminf_{n \to \infty} f(z_n) + \int_{E \setminus E_1} f(v_0)\, dx,$$

and by (5.12) we conclude that

$$f(z) \leq \liminf_{n \to \infty} f(z_n).$$

Thus the proof is complete.

Exercise 5.13. Prove that the previous proposition still holds if $L^p(E; \mathbb{R}^m)$ is replaced by $L^p((X, \mathfrak{M}, \mu); \mathbb{R}^m)$, where $(X, \mathfrak{M}, \mu)$ is a measure space that admits at least one measurable set of positive finite measure. What happens if $\mu : \mathfrak{M} \to \{0, \infty\}$?

5.2.2 Weak Convergence and Weak Star Convergence in L^p

Since functionals that are sequentially lower semicontinuous with respect to weak (respectively weak star when $p = \infty$) convergence in $L^p(E; \mathbb{R}^m)$ are sequentially lower semicontinuous also with respect to strong convergence in $L^p(E; \mathbb{R}^m)$, without loss of generality we may assume in what follows that the integrand f satisfies all the necessary conditions for strong convergence in $L^p(E; \mathbb{R}^m)$.

Theorem 5.14. *Let E be a Borel subset of $\mathbb{R}^N$ with finite measure, let $1 \leq p \leq \infty$, and let $f : \mathbb{R}^m \to (-\infty, \infty]$ be a lower semicontinuous function. Assume that there exists $C > 0$ such that*

$$f(z) \geq -C(1 + |z|^p) \quad \text{for all } z \in \mathbb{R}^m \text{ if } 1 \leq p < \infty,$$
$$f \text{ is locally bounded from below if } p = \infty.$$

Then the functional

$$v \in L^p(E; \mathbb{R}^m) \mapsto \int_E f(v(x))\, dx$$

is sequentially lower semicontinuous with respect to weak convergence in $L^p(E; \mathbb{R}^m)$ (weak star if $p = \infty$) if and only if f is convex.

Proof (Sufficiency). We consider only the case $1 \leq p < \infty$, since the case $p = \infty$ can be treated in an entirely similar way. The proof of the sufficiency part is based on the *the blowup method*. Assume that f is convex.

Step 1: Suppose first that f is nonnegative. Let $\{v_n\} \subset L^p(E; \mathbb{R}^m)$ be a sequence weakly converging to some $v \in L^p(E; \mathbb{R}^m)$. Without loss of generality we may assume that

$$\liminf_{n \to \infty} \int_E f(v_n)\, dx = \lim_{n \to \infty} \int_E f(v_n)\, dx < \infty.$$

By Remark 2.51, passing to a subsequence if necessary, there exists a (positive) Radon measure μ such that

$$f(v_n(\cdot))\, \mathcal{L}^N \lfloor E \overset{*}{\rightharpoonup} \mu \quad \text{in } \mathcal{M}(E; \mathbb{R}) \tag{5.13}$$

as $n \to \infty$. We claim that

$$\frac{d\mu}{d\mathcal{L}^N}(x_0) \geq f(v(x_0)) \quad \text{for } \mathcal{L}^N \text{ a.e. } x_0 \in E. \tag{5.14}$$

If (5.14) holds, then the conclusion of the theorem follows. Indeed, since by the Radon–Nikodym and Lebesgue decomposition theorems

$$\mu = \frac{d\mu}{d\mathcal{L}^N}\, \mathcal{L}^N \lfloor E + \mu_s,$$

where $\mu_s \geq 0$, by Proposition 1.203(i) with $X = E$ we have

$$\lim_{n\to\infty} \int_E f(v_n)\, dx \geq \mu(E) \geq \int_E \frac{d\mu}{d\mathcal{L}^N}\, dx \geq \int_E f(v)\, dx.$$

Thus, to conclude the proof of the theorem, it suffices to prove (5.14) for $\mathcal{L}^N$ a.e. $x_0 \in E$.

Take $x_0 \in E$ a point of Lebesgue density one for E such that, in view of Besicovitch's derivation theorem and Corollary 1.159,

$$\frac{d\mu}{d\mathcal{L}^N}(x_0) = \lim_{\varepsilon\to 0^+} \frac{\mu(Q(x_0,\varepsilon)\cap E)}{\varepsilon^N} < \infty, \tag{5.15}$$

$$\lim_{\varepsilon\to 0^+} \frac{1}{\varepsilon^N} \int_{Q(x_0,\varepsilon)\cap E} |v(x) - v(x_0)|\, dx = 0. \tag{5.16}$$

Since f is convex and lower semicontinuous, by Proposition 4.77 we may write

$$f(z) = \sup_{i\in\mathbb{N}} \{a_i + b_i \cdot z\}, \tag{5.17}$$

for some $a_i \in \mathbb{R}$, $b_i \in \mathbb{R}^m$. By Proposition 1.15 we may choose $\varepsilon_k \searrow 0$ such that $\mu(\partial Q(x_0,\varepsilon_k)\cap E) = 0$. Using Proposition 1.203(iii), with $X := E$ and the (relatively) open set $A := Q(x_0,\varepsilon_k)\cap E$, by (5.15) for any fixed $i \in \mathbb{N}$ we have

$$\frac{d\mu}{d\mathcal{L}^N}(x_0) = \lim_{k\to\infty} \frac{\mu(Q(x_0,\varepsilon_k)\cap E)}{\varepsilon_k^N} = \lim_{k\to\infty} \lim_{n\to\infty} \frac{1}{\varepsilon_k^N} \int_{Q(x_0,\varepsilon_k)\cap E} f(v_n)\, dx$$

$$\geq \liminf_{k\to\infty} \liminf_{n\to\infty} \frac{1}{\varepsilon_k^N} \int_{Q(x_0,\varepsilon_k)\cap E} (a_i + b_i \cdot v_n)\, dx \tag{5.18}$$

$$= \liminf_{k\to\infty} \frac{1}{\varepsilon_k^N} \int_{Q(x_0,\varepsilon_k)\cap E} (a_i + b_i \cdot v)\, dx,$$

where we have used the fact that $v_n \rightharpoonup v$ in $L^p(E;\mathbb{R}^m)$. Since x_0 is a point of density one for E, we have

$$\lim_{k\to\infty} \frac{1}{\varepsilon_k^N} \int_{Q(x_0,\varepsilon_k)\cap E} a_i\, dx = a_i,$$

and by (5.16) we obtain

$$\lim_{k\to\infty} \frac{1}{\varepsilon_k^N} \int_{Q(x_0,\varepsilon_k)\cap E} b_i \cdot (v(x) - v(x_0))\, dx = 0.$$

Hence from (5.18) it follows that

$$\frac{d\mu}{d\mathcal{L}^N}(x_0) \geq a_i + b_i \cdot v(x_0),$$

and taking the supremum over all i and using (5.17) we conclude (5.14).

Step 2: Since $f(z) \geq a_1 + b_1 \cdot z$ for all $z \in \mathbb{R}^m$ (see (5.17)), we have that the function $f(z) - (a_1 + b_1 \cdot z)$ is convex, nonnegative, and lower semicontinuous. By Step 1,

$$\liminf_{n \to \infty} \int_E f(v_n)\, dx - a_1 |E| - \int_E b_1 \cdot v\, dx$$

$$= \liminf_{n \to \infty} \int_E (f(v_n) - (a_1 + b_1 \cdot v_n))\, dx$$

$$\geq \int_E f(v)\, dx - a_1 |E| - \int_E b_1 \cdot v\, dx.$$

Since E has finite measure and $v \in L^p(E; \mathbb{R}^m)$, the result follows.

Proof (Necessity). Let $z_1, z_2 \in \mathbb{R}^m$, $y \in S^{N-1}$, $\theta \in (0,1)$, and for $n \in \mathbb{N}$ and $x \in \mathbb{R}^m$ define

$$v_n(x) := z_2 + \chi(nx \cdot y)(z_1 - z_2),$$

where χ is the characteristic function of the interval $[0, \theta]$ in $[0,1]$ extended periodically to $\mathbb{R}$ with period one. In view of Example 2.86 we have that $v_n \overset{*}{\rightharpoonup} z_2 + \theta(z_1 - z_2)$ in $L^\infty(E; \mathbb{R}^d)$, hence weakly in $L^p(E; \mathbb{R}^m)$. Thus

$$f(z_2 + \theta(z_1 - z_2)) = \frac{1}{|E|} \int_E f(z_2 + \theta(z_1 - z_2))\, dx$$

$$\leq \liminf_{n \to \infty} \frac{1}{|E|} \int_E f(z_2 + \chi(nx \cdot y)(z_1 - z_2))\, dx$$

$$= \liminf_{n \to \infty} \frac{1}{|E|} \int_E \chi(nx \cdot y) f(z_1) + (1 - \chi(nx \cdot y)) f(z_2)\, dx$$

$$= \theta f(z_1) + (1 - \theta) f(z_2).$$

Therefore f is convex.

Exercise 5.15. Show that the sufficiency part of the previous proof still holds with the Lebesgue measure replaced by a (positive) finite Radon measure[1]. Prove also that the necessity part of the theorem continues to hold for $L^p((X, \mathfrak{M}, \nu); \mathbb{R}^m)$, where $(X, \mathfrak{M}, \nu)$ is a measure space with μ finite and nonatomic (recall Proposition 2.87).

In order to prove the analogous result for the case in which E has infinite measure we present first an auxiliary result.

---

[1] More generally, using an argument different from the one offered here, the sufficiency part may be extended to $L^p((X, \mathfrak{M}, \nu); \mathbb{R}^m)$, where $(X, \mathfrak{M}, \nu)$ is a measure space with ν finite (see [Bu89]).

Proposition 5.16. *Let* $f : \mathbb{R}^m \to (-\infty, \infty]$ *be a convex, lower semicontinuous function such that*

$$f(z) \geq -\frac{1}{p}|z|^p \text{ for all } z \in \mathbb{R}^m \text{ and for some } 1 \leq p < \infty.$$

Then there exist a constant $a \geq 0$ *and a vector* $b \in \mathbb{R}^m$ *such that*

$$f(z) \geq a + b \cdot z \geq -\frac{1}{p}|z|^p \text{ for all } z \in \mathbb{R}^m. \tag{5.19}$$

Moreover, $|b| \leq 1$ *for* $p = 1$, *while for* $p > 1$,

$$|b|^{p'} \leq \left[\frac{1}{(p')^{1/p}} + \frac{1}{p^{1/p}(p-1)^{1/p'}} \right]^{p'} a. \tag{5.20}$$

Proof. It suffices to consider the case that f is not identically ∞. The existence of some $a \in \mathbb{R}$ and a vector $b \in \mathbb{R}^m$ satisfying (5.19) follows from Theorem 4.19 applied to the convex sets epi f and

$$A := \left\{ (z,t) \in \mathbb{R}^m \times \mathbb{R} : t < -\frac{1}{p}|z|^p \right\}.$$

To see this, let $(\beta, \gamma) \in (\mathbb{R}^m \times \mathbb{R}) \setminus \{(0,0)\}$ and $s \in \mathbb{R}$ be such that

$$\beta \cdot z + \gamma t \geq s \text{ for every } (z,t) \in \text{epi } f,$$
$$\beta \cdot z + \gamma t \leq s \text{ for every } (z,t) \in A.$$

We claim that $\gamma > 0$. If $z \in \text{dom}_e f$, then letting $t \to \infty$ in the first inequality we obtain that $\gamma \geq 0$. Assume by contradiction that $\gamma = 0$. Then $\beta \neq 0$, and so, taking $z_n := n\beta$ and $t_n < -\frac{1}{p}|z_n|^p$, $n \in \mathbb{N}$, we have that $(z_n, t_n) \in A$. It follows that

$$n|\beta|^2 \leq s$$

for all $n \in \mathbb{N}$, which is a contradiction. Hence $\gamma > 0$, and so

$$\frac{\beta}{\gamma} \cdot z + f(z) \geq \frac{s}{\gamma} \quad \text{for all } z \in \text{dom}_e f,$$

$$\frac{s}{\gamma} \geq \frac{\beta}{\gamma} \cdot z - \frac{1}{p}|z|^p \quad \text{for all } z \in \mathbb{R}^m,$$

and in turn,

$$f(z) \geq \frac{s}{\gamma} - \frac{\beta}{\gamma} \cdot z \geq -\frac{1}{p}|z|^p$$

for all $z \in \mathbb{R}^m$. It suffices to define $a := \frac{s}{\gamma}$, $b := -\frac{\beta}{\gamma}$ to obtain (5.19).

Taking $z = 0$ in (5.19) we obtain that $a \geq 0$. Assume that $b \neq 0$ and in (5.19) set

$$z := -\mu \frac{b}{|b|}$$

with $\mu > 0$. We get

$$a - \mu |b| \geq -\frac{\mu^p}{p}$$

for all $\mu > 0$, or, equivalently,

$$|b| \leq \inf_{\mu > 0} \left(\frac{a}{\mu} + \frac{\mu^{p-1}}{p} \right) = \begin{cases} 1 & \text{if } p = 1, \\ a^{1/p'} \left[\frac{1}{(p')^{1/p}} + \frac{1}{p^{1/p}(p-1)^{1/p'}} \right] & \text{if } p > 1. \end{cases}$$

This concludes the proof.

We are now ready to treat the case that E has infinite measure.

Theorem 5.17. *Let E be a Borel subset of $\mathbb{R}^N$ with infinite measure, let $1 \leq p \leq \infty$, and let $f : \mathbb{R}^m \to (-\infty, \infty]$ be a lower semicontinuous function. Assume that there exists $C > 0$ such that*

$$f(z) \geq -C|z|^p \text{ for all } z \in \mathbb{R}^m \text{ if } 1 \leq p < \infty, \tag{5.21}$$
$$f(z) \geq 0 \text{ for all } z \in \mathbb{R}^m \text{ if } p = \infty. \tag{5.22}$$

Suppose that there exists $v_0 \in L^p(E; \mathbb{R}^m)$ such that

$$\int_E f(v_0(x)) \, dx < \infty.$$

Then the functional

$$v \in L^p(E; \mathbb{R}^m) \mapsto \int_E f(v(x)) \, dx$$

is sequentially lower semicontinuous with respect to weak convergence in $L^p(E; \mathbb{R}^m)$ (weak star if $p = \infty$) if and only if

(i) f is convex;
(ii) $f \geq 0$ if $1 < p \leq \infty$;
(iii) $f(0) = 0$ if $1 \leq p < \infty$;
(iv) $\min f = 0$ if $p = \infty$.

Proof. We begin by proving the necessity part. Using an argument similar to that of Theorem 5.12 it can be shown that sequential lower semicontinuity of the functional with respect to weak convergence in $L^p(E; \mathbb{R}^m)$ (weak star if $p = \infty$) yields sequential lower semicontinuity in $L^p(E_1; \mathbb{R}^m)$ for any Borel subset $E_1 \subset E$ with finite measure. Therefore, by the previous theorem we conclude that f is convex.

In view of (5.22), we need to prove (ii) only for $1 < p < \infty$. By Proposition 5.16 there exist a constant $a \geq 0$ and a vector $b \in \mathbb{R}^m$ such that

$$f(z) \geq a + b \cdot z \geq -C |z|^p \text{ for all } z \in \mathbb{R}^m.$$

If $a = 0$, then by (5.20), $b = 0$; hence $f \geq 0$. If $a > 0$, let $\varepsilon > 0$ be such that $a - \varepsilon |b| > 0$. Since

$$|\{x \in E : |v_0(x)| > \varepsilon\}| \leq \frac{1}{\varepsilon^p} \int_E |v_0(x)|^p \, dx < \infty,$$

the set

$$E_\varepsilon := \{x \in E : |v_0(x)| \leq \varepsilon\}$$

has infinite measure. Therefore

$$\infty > \int_{E_\varepsilon} f(v_0) \, dx \geq \int_{E_\varepsilon} (a + b \cdot v_0) \, dx \geq (a - \varepsilon |b|) |E_\varepsilon| = \infty,$$

and we have reached a contradiction.

We now prove (iii), i.e., that $f(0) = 0$ for all $1 \leq p < \infty$. By Proposition 4.77 we may write

$$f(z) = \sup_{i \in \mathbb{N}} \{a_i + b_i \cdot z\}$$

for all $z \in \mathbb{R}^m$. If $a_i > 0$ for some $i \in \mathbb{N}$, then reasoning as above but now with $a_i - \varepsilon |b_i| > 0$, we obtain that

$$\infty > \int_{E_\varepsilon} (f(v_0))^+ \, dx \geq \int_{E_\varepsilon} (a_i + b_i \cdot v_0)^+ \, dx \geq (a_i - \varepsilon |b_i|) |E_\varepsilon| = \infty,$$

which is a contradiction. Hence $a_i \leq 0$ for all $i \in \mathbb{N}$, and so $f(0) = \sup_{i \in \mathbb{N}} \{a_i\} \leq 0$, which, together with either (5.21) or (5.22), yields $f(0) = 0$.

Finally we prove (iv). Let $R := \|v_0\|_{L^\infty(E;\mathbb{R}^m)}$. Since f is lower semicontinuous, by the Weierstrass theorem there exists $z_0 \in \overline{B(0, R)}$ such that

$$\min_{\overline{B(0,R)}} f = f(z_0) \geq 0.$$

If $f(z_0) > 0$, then

$$\infty > \int_E f(v_0) \, dx \geq f(z_0) |E| = \infty,$$

which is a contradiction. Hence $f(z_0) = 0$, and since f is nonnegative, it follows that

$$\min_{\mathbb{R}^m} f = f(z_0) = 0.$$

This concludes the proof of the necessity part of the theorem.

Conversely, in the cases in which $1 < p \leq \infty$ we have $f \geq 0$, and so the same argument as in the proof of the previous theorem yields sequential lower semicontinuity. If $p = 1$ and if $f \geq 0$, then similar reasoning will assert sequential lower semicontinuity. If $p = 1$ and if f takes negative values, then using once more Proposition 5.16, and since $a \geq 0$, we have that $g(z) = f(z) - b \cdot z$ is a nonnegative, convex, lower semicontinuous function. Hence if $v_n \rightharpoonup v$ in L^1, then

$$\liminf_{n \to \infty} \int_E f(v_n) \, dx - \int_E b \cdot v \, dx$$

$$= \liminf_{n \to \infty} \int_E g(v_n) \, dx \geq \int_E g(v) \, dx$$

$$= \int_E f(v) \, dx - \int_E b \cdot v \, dx,$$

and thus

$$\liminf_{n \to \infty} \int_E f(v_n) \, dx \geq \int_E f(v) \, dx,$$

which gives the desired result.

Exercise 5.18. For which measures does the necessity part of the theorem continue to hold? For which measures does the sufficiency part of the theorem continue to hold?

5.2.3 Weak Star Convergence in the Sense of Measures

As we remarked in the beginning of this chapter, due to lack of reflexivity of the space $L^1(E;\mathbb{R}^m)$, if $\{v_n\} \subset L^1(E;\mathbb{R}^m)$ is such that

$$\sup_n \|v_n\|_{L^1} < \infty,$$

one can conclude only that $\{v_n\}$ admits a subsequence (not relabeled) such that $v_n \mathcal{L}^N \lfloor E \overset{*}{\rightharpoonup} \lambda$ in $\mathcal{M}(E;\mathbb{R}^m)$. Thus we will also study sequential lower semicontinuity under this natural notion of convergence. In particular, we will address necessary and sufficient conditions under which the lower semicontinuity property

$$\liminf_{n \to \infty} \int_E f(v_n) \, dx \geq \int_E f\left(\frac{d\lambda}{d\mathcal{L}^N}\right) dx$$

holds, whenever $v_n \mathcal{L}^N \lfloor E \overset{*}{\rightharpoonup} \lambda$ in $\mathcal{M}(E;\mathbb{R}^m)$ and λ admits the Radon–Nikodym decomposition

$$\lambda = \frac{d\lambda}{d\mathcal{L}^N} \mathcal{L}^N \lfloor E + \lambda_s,$$

with λ_s and $\mathcal{L}^N \lfloor E$ mutually singular.

Since functionals that are sequentially lower semicontinuous with respect to weak star convergence in the sense of measures are in particular sequentially lower semicontinuous also with respect to weak convergence in $L^1(E;\mathbb{R}^m)$, without loss of generality we may assume in what follows that the integrand f satisfies all the necessary conditions for weak convergence in $L^1(E;\mathbb{R}^m)$.

We will study the following cases:

- The Borel set E has finite measure. We divide this case into two subcases:
  - There exists a compact set $K \subset E$ such that

$$|E \setminus K| = 0. \qquad (5.23)$$

  - For all compact sets $K \subset E$,

$$|E \setminus K| > 0. \qquad (5.24)$$

- The Borel set E has infinite measure.

Note that (5.24) is satisfied if, e.g., E is open.

When (5.23) holds, then for every $v \in L^1(E; \mathbb{R}^m)$ we have that

$$\int_E f(v(x)) \, dx = \int_K f(v(x)) \, dx,$$

and thus there is no loss of generality in considering the functional

$$v \in L^1(K; \mathbb{R}^m) \mapsto \int_K f(v(x)) \, dx.$$

This case will be treated in the next theorem.

Theorem 5.19. *Let E be a compact subset of $\mathbb{R}^N$ and let $f : \mathbb{R}^m \to (-\infty, \infty]$ be a convex, lower semicontinuous function. Then the functional*

$$v \in L^1(E; \mathbb{R}^m) \mapsto \int_E f(v(x)) \, dx$$

is sequentially lower semicontinuous with respect to weak star convergence in the sense of measures.

More generally, for any sequence $\{v_n\} \subset L^1(E; \mathbb{R}^m)$ such that

$$v_n \mathcal{L}^N \lfloor E \overset{*}{\rightharpoonup} \lambda \quad in \ \mathcal{M}(E; \mathbb{R}^m)$$

we have

$$\liminf_{n \to \infty} \int_E f(v_n) \, dx \geq \int_E f\left(\frac{d\lambda}{d\mathcal{L}^N}\right) dx + \int_E f^\infty\left(\frac{d\lambda}{d\|\lambda_s\|}\right) d\|\lambda_s\|,$$

where f^∞ is the recession function of f.

Proof. Without loss of generality we may assume that f is not identically ∞, since otherwise there is nothing to prove.

Step 1: Assume that f is nonnegative. Let $\{v_n\} \subset L^1(E; \mathbb{R}^m)$ be such that $v_n \mathcal{L}^N \lfloor E \overset{*}{\rightharpoonup} \lambda$ in $\mathcal{M}(E; \mathbb{R}^m)$.

As in the sufficiency proof of Theorem 5.14, we use the blowup method. Let the Radon measure μ be as in (5.13). In view of Proposition A.57 we have that

$$\sup_n \int_E |v_n|\, dx < \infty,$$

and so by extracting a further subsequence if necessary, we may assume that

$$|v_n|\, \mathcal{L}^N \lfloor E \overset{*}{\rightharpoonup} \nu \quad \text{in } \mathcal{M}(E; \mathbb{R})$$

for some (positive) Radon measure ν (see Remark 2.51).

We claim that for $\mathcal{L}^N$ a.e. $x_0 \in E$,

$$\frac{d\mu}{d\mathcal{L}^N}(x_0) \geq f\left(\frac{d\lambda}{d\mathcal{L}^N}(x_0)\right), \tag{5.25}$$

and for $\|\lambda_s\|$ a.e. $x_0 \in E$,

$$\frac{d\mu}{d\|\lambda_s\|}(x_0) \geq f^\infty\left(\frac{d\lambda}{d\|\lambda_s\|}(x_0)\right). \tag{5.26}$$

If (5.25) and (5.26) hold, then the conclusion of the theorem follows. Indeed, since by Corollary 1.116,

$$\mu = \frac{d\mu}{d\mathcal{L}^N}\, \mathcal{L}^N \lfloor E + \frac{d\mu}{d\|\lambda_s\|}\, \|\lambda_s\| + \mu_s,$$

where μ_s is a (positive) Radon measure singular with respect to $\mathcal{L}^N \lfloor E + \|\lambda_s\|$, by Proposition 1.203(i) with $X = E$ we have

$$\lim_{n\to\infty} \int_E f(v_n)\, dx \geq \mu(E) \geq \int_E \frac{d\mu}{d\mathcal{L}^N}\, dx + \int_E \frac{d\mu}{d\|\lambda_s\|}\, d\|\lambda_s\|$$

$$\geq \int_E f\left(\frac{d\lambda}{d\mathcal{L}^N}\right) dx + \int_E f^\infty\left(\frac{d\lambda}{d\|\lambda_s\|}\right) d\|\lambda_s\|.$$

Step 2: We prove (5.25). Take $x_0 \in E$ a point of Lebesgue density one of E with respect to $\mathcal{L}^N$ such that

$$\frac{d\mu}{d\mathcal{L}^N}(x_0) = \lim_{\varepsilon\to 0^+} \frac{\mu(Q(x_0,\varepsilon) \cap E)}{\varepsilon^N} < \infty, \tag{5.27}$$

$$\frac{d\lambda}{d\mathcal{L}^N}(x_0) = \lim_{\varepsilon\to 0^+} \frac{\lambda(Q(x_0,\varepsilon) \cap E)}{\varepsilon^N}, \tag{5.28}$$

where we have used Theorem 1.155.

Since f is convex and lower semicontinuous, by Proposition 4.77 we may write

$$f(z) = \sup_{i\in\mathbb{N}} \{a_i + b_i \cdot z\}. \tag{5.29}$$

By Proposition 1.15 we may choose $\varepsilon_k \searrow 0$ such that $\mu(\partial Q(x_0,\varepsilon_k) \cap E) = 0$ and $\nu(\partial Q(x_0,\varepsilon_k) \cap E) = 0$. Invoking Corollary 1.204 with $X := E$ and the (relatively) open set $A := Q(x_0,\varepsilon_k) \cap E$, for all $i, k \in \mathbb{N}$ we have

$$\lim_{n\to\infty} \int_{Q(x_0,\varepsilon_k)\cap E} b_i \cdot v_n \, dx = \int_{Q(x_0,\varepsilon_k)\cap E} b_i \cdot d\lambda, \qquad (5.30)$$

while by Proposition 1.203,

$$\frac{d\mu}{d\mathcal{L}^N}(x_0) = \lim_{k\to\infty} \frac{\mu(Q(x_0,\varepsilon_k)\cap E)}{\varepsilon_k^N} = \lim_{k\to\infty}\lim_{n\to\infty} \frac{1}{\varepsilon_k^N} \int_{Q(x_0,\varepsilon_k)\cap E} f(v_n) \, dx$$

$$\geq \liminf_{k\to\infty}\liminf_{n\to\infty} \frac{1}{\varepsilon_k^N} \int_{Q(x_0,\varepsilon_k)\cap E} (a_i + b_i \cdot v_n) \, dx \qquad (5.31)$$

$$= a_i + \liminf_{k\to\infty} \frac{1}{\varepsilon_k^N} \int_{Q(x_0,\varepsilon_k)\cap E} b_i \cdot d\lambda,$$

where in the last identity we used the fact that $x_0 \in E$ is a point of density one of E. Using (5.28) we have that

$$\liminf_{k\to\infty} \frac{1}{\varepsilon_k^N} \int_{Q(x_0,\varepsilon_k)\cap E} b_i \cdot d\lambda = b_i \cdot \frac{d\lambda}{d\mathcal{L}^N}(x_0),$$

which, together with (5.31), yields

$$\frac{d\mu}{d\mathcal{L}^N}(x_0) \geq a_i + b_i \cdot \frac{d\lambda}{d\mathcal{L}^N}(x_0).$$

The inequality (5.25) now follows from (5.29).

Step 3: We prove (5.26). Take $x_0 \in E$ such that (see Theorem 1.155)

$$\frac{d\mu}{d\|\lambda_s\|}(x_0) = \lim_{\varepsilon\to 0^+} \frac{\mu(Q(x_0,\varepsilon)\cap E)}{\|\lambda_s\|(Q(x_0,\varepsilon)\cap E)} < \infty, \qquad (5.32)$$

$$\frac{d\lambda}{d\|\lambda_s\|}(x_0) = \lim_{\varepsilon\to 0^+} \frac{\lambda(Q(x_0,\varepsilon)\cap E)}{\|\lambda_s\|(Q(x_0,\varepsilon)\cap E)}, \qquad (5.33)$$

$$\lim_{\varepsilon\to 0^+} \frac{\mathcal{L}^N(Q(x_0,\varepsilon)\cap E)}{\|\lambda_s\|(Q(x_0,\varepsilon)\cap E)} = 0. \qquad (5.34)$$

Choosing $\varepsilon_k \searrow 0$ such that $\mu(\partial Q(x_0,\varepsilon_k)\cap E) = 0$ and $\nu(\partial Q(x_0,\varepsilon_k)\cap E) = 0$ and reasoning as in the previous step we obtain

$$\frac{d\mu}{d\|\lambda_s\|}(x_0) = \lim_{k\to\infty} \frac{\mu(Q(x_0,\varepsilon_k)\cap E)}{\|\lambda_s\|(Q(x_0,\varepsilon_k)\cap E)}$$

$$= \lim_{k\to\infty}\lim_{n\to\infty} \frac{1}{\|\lambda_s\|(Q(x_0,\varepsilon_k)\cap E)} \int_{Q(x_0,\varepsilon_k)\cap E} f(v_n) \, dx$$

$$\geq \liminf_{k\to\infty}\liminf_{n\to\infty} \frac{1}{\|\lambda_s\|(Q(x_0,\varepsilon_k)\cap E)} \int_{Q(x_0,\varepsilon_k)\cap E} (a_i + b_i \cdot v_n) \, dx$$

$$\qquad\qquad (5.35)$$

$$= \liminf_{k\to\infty} \frac{1}{\|\lambda_s\|(Q(x_0,\varepsilon_k)\cap E)} \int_{Q(x_0,\varepsilon_k)\cap E} b_i \cdot d\lambda,$$

where in the last equality we have used (5.34) and (5.30). By (5.33) we have that

$$\liminf_{k \to \infty} \frac{1}{\|\lambda_s\| (Q(x_0, \varepsilon_k) \cap E)} \int_{Q(x_0, \varepsilon_k) \cap E} b_i \cdot d\lambda(x) \tag{5.36}$$

$$= b_i \cdot \lim_{k \to \infty} \frac{\lambda(Q(x_0, \varepsilon_k) \cap E)}{\|\lambda_s\| (Q(x_0, \varepsilon_k) \cap E)} = b_i \cdot \frac{d\lambda}{d \|\lambda_s\|}(x_0),$$

which, together with (5.35), yields

$$\frac{d\mu}{d \|\lambda_s\|}(x_0) \geq b_i \cdot \frac{d\lambda}{d \|\lambda_s\|}(x_0).$$

Taking the supremum over all i, it follows from (5.29) and Proposition 4.77 that

$$\frac{d\mu}{d \|\lambda_s\|}(x_0) \geq f^\infty \left(\frac{d\lambda}{d \|\lambda_s\|}(x_0) \right).$$

Step 4: Here we remove the additional assumption that $f \geq 0$. By Proposition 4.75 there exist $a \in \mathbb{R}$ and $b \in \mathbb{R}^m$ such that $f(z) \geq a + b \cdot z$ (see (5.17)); we have that the function $g(z) := f(z) - (a + b \cdot z)$ is convex, nonnegative, and lower semicontinuous. Using (for the first time) the fact that E is compact we have that $\mathcal{M}(E)$ is the dual of $C_0(E) = C_b(E)$, and so

$$\lim_{n \to \infty} \int_E b \cdot v_n \, dx = \int_E b \cdot d\lambda.$$

Hence, by Steps 1–3 applied to g,

$$\liminf_{n \to \infty} \int_E f(v_n) \, dx - a |E| - \int_E b \cdot d\lambda = \liminf_{n \to \infty} \int_E g(v_n) \, dx$$

$$\geq \int_E g \left(\frac{d\lambda}{d\mathcal{L}^N} \right) dx + \int_E g^\infty \left(\frac{d\lambda}{d \|\lambda_s\|} \right)$$

$$= \int_E f \left(\frac{d\lambda}{d\mathcal{L}^N} \right) dx - a |E| - \int_E b \cdot \frac{d\lambda}{d\mathcal{L}^N} \, dx$$

$$+ \int_E f^\infty \left(\frac{d\lambda}{d \|\lambda_s\|} \right) d \|\lambda_s\| - \int_E b \cdot \frac{d\lambda}{d \|\lambda_s\|} \, d \|\lambda_s\|$$

$$= \int_E f \left(\frac{d\lambda}{d\mathcal{L}^N} \right) dx + \int_E f^\infty \left(\frac{d\lambda}{d \|\lambda_s\|} \right) d \|\lambda_s\| - a |E| - \int_E b \cdot d\lambda,$$

where we have used the fact that by Theorem 4.70, for every $w \in \mathrm{dom}_e \, f$ we have

$$g^\infty(z) = \lim_{t \to \infty} \frac{g(w + tz) - g(w)}{t}$$

$$= \lim_{t \to \infty} \frac{f(w + tz) - b \cdot (tz) - f(w)}{t} = f^\infty(z) - b \cdot z.$$

Since E has finite measure and $\int_E b \cdot d\lambda$ is well-defined and finite, the result follows.

Exercise 5.20. Show that the previous theorem still holds if the Lebesgue measure is replaced by a (positive) nonatomic finite Radon measure.

We now turn to the case in which (5.24) holds.

Theorem 5.21. *Let E be a Borel subset of $\mathbb{R}^N$ with finite measure for which there is no compact set $K \subset E$ such that $\mathcal{L}^N (E \setminus K) = 0$, and let $f : \mathbb{R}^m \to (-\infty, \infty]$ be a convex, lower semicontinuous function. Then the functional*

$$v \in L^1 (E; \mathbb{R}^m) \mapsto \int_E f (v(x))\, dx$$

is sequentially lower semicontinuous with respect to weak star convergence in the sense of measures if and only if

$$\liminf_{|z| \to \infty} \frac{f(z)}{|z|} \geq 0. \tag{5.37}$$

More generally, if $f : \mathbb{R}^m \to (-\infty, \infty]$ is a convex, lower semicontinuous function satisfying (5.37), then for any sequence $\{v_n\} \subset L^1(E; \mathbb{R}^m)$ such that

$$v_n \mathcal{L}^N \lfloor E \overset{*}{\rightharpoonup} \lambda \quad in \ \mathcal{M}(E; \mathbb{R}^m),$$

we have

$$\liminf_{n \to \infty} \int_E f(v_n)\, dx \geq \int_E f \left(\frac{d\lambda}{d\mathcal{L}^N} \right) dx + \int_E f^\infty \left(\frac{d\lambda}{d\|\lambda_s\|} \right) d\|\lambda_s\|.$$

Remark 5.22. We note that in view of Theorem 4.73, condition (5.37) is equivalent to $f^\infty \geq 0$.

Proof (Sufficiency). Without loss of generality we may assume that f is not identically ∞, since otherwise there is nothing to prove. By condition (5.37) for $\varepsilon > 0$, then there exists $M > 0$ such that

$$\frac{f(z)}{|z|} \geq -\varepsilon$$

for all $z \in \mathbb{R}^m$ with $|z| > M$. Since f is lower semicontinuous, by the Weierstrass theorem $f(\cdot) + \varepsilon |\cdot|$ is bounded from below in $\overline{B}(0, M)$, and so there exists $a_\varepsilon \in \mathbb{R}$ such that

$$g_\varepsilon (z) := f(z) + \varepsilon |z| + a_\varepsilon \geq 0$$

for all $z \in \mathbb{R}^m$.

Let $\{v_n\} \subset L^1 (E; \mathbb{R}^m)$ be such that $v_n \mathcal{L}^N \lfloor E \overset{*}{\rightharpoonup} \lambda$ in $\mathcal{M}(E; \mathbb{R}^m)$. By Proposition A.57 we have

$$\sup_n \|v_n\|_{L^1} =: C < \infty.$$

As we already remarked in the proof of the previous theorem, the hypothesis that E is compact is not needed for nonnegative integrands. Hence we may apply Steps 1–3 of the previous theorem to g_ε to obtain

$$\liminf_{n\to\infty} \int_E f(v_n)\, dx + \varepsilon C + a_\varepsilon |E| \geq \liminf_{n\to\infty} \int_E g_\varepsilon(v_n)\, dx$$

$$\geq \int_E g_\varepsilon \left(\frac{d\lambda}{d\mathcal{L}^N}\right) dx + \int_E g_\varepsilon^\infty \left(\frac{d\lambda}{d\|\lambda_s\|}\right)$$

$$= \int_E f\left(\frac{d\lambda}{d\mathcal{L}^N}\right) dx + \varepsilon \int_E \left|\frac{d\lambda}{d\mathcal{L}^N}\right| dx + a_\varepsilon |E|$$

$$+ \int_E f^\infty \left(\frac{d\lambda}{d\|\lambda_s\|}\right) d\|\lambda_s\| + \varepsilon \int_E \left|\frac{d\lambda}{d\|\lambda_s\|}\right| d\|\lambda_s\|,$$

where we have used the fact that by Theorem 4.70, for every $w \in \mathrm{dom}_e\, f$ we have

$$g^\infty(z) = \lim_{t\to\infty} \frac{g(w+tz) - g(w)}{t}$$

$$= \lim_{t\to\infty} \frac{f(w+tz) + \varepsilon |w + tz| - f(w) - \varepsilon |w|}{t} = f^\infty(z) + \varepsilon |z|.$$

Hence

$$\liminf_{n\to\infty} \int_E f(v_n)\, dx + \varepsilon C \geq \int_E f\left(\frac{d\lambda}{d\mathcal{L}^N}\right) dx + \varepsilon \int_E \left|\frac{d\lambda}{d\mathcal{L}^N}\right| dx$$

$$+ \int_E f^\infty \left(\frac{d\lambda}{d\|\lambda_s\|}\right) d\|\lambda_s\| + \varepsilon \int_E \left|\frac{d\lambda}{d\|\lambda_s\|}\right| d\|\lambda_s\|,$$

and the result now follows by letting $\varepsilon \to 0$.

Proof (Necessity). **Step 1:** We claim that there exists a sequence $\{K_j\} \subset E$ of pairwise disjoint compact sets with positive Lebesgue measure such that for any compact $K \subset E$ we have that $K_j \subset E \setminus K$ for all j sufficiently large.

Indeed, assume first that E is unbounded in the measure-theoretic sense, that is

$$\left|E \setminus \overline{B(0,r)}\right| > 0 \text{ for all } r > 0. \tag{5.38}$$

By the inner regularity of the Lebesgue measure there exists a compact set $K_1 \subset E$ with $|K_1| > 0$. Let $r_1 \geq 1$ be so large that $K_1 \subset \overline{B(0,r_1)}$. Since $\left|E \setminus \overline{B(0,r_1)}\right| > 0$, again by the inner regularity of the Lebesgue measure we may find a compact set $K_2 \subset E \setminus \overline{B(0,r_1)}$ with $|K_2| > 0$. Find $r_2 > r_1 + 1$ such that $K_2 \subset \overline{B(0,r_2)}$. Inductively we can construct a sequence of pairwise disjoint compact sets $\{K_j\}$ and a sequence of radii $\{r_j\}$ such that $|K_j| > 0$, $K_j \subset E \setminus \overline{B(0, r_{j-1})}$, and $r_j > r_{j-1} + 1$. The sequence $\{K_j\}$ satisfies the claim.

Next assume that (5.38) fails, so that there exists $R > 0$ such that $\left|E \setminus \overline{B(0,R)}\right| = 0$. Define

$$C := \left\{ x \in \overline{E} : \left| E \cap \overline{B\left(x,r\right)} \right| > 0 \text{ for all } r > 0 \right\} \cap \overline{B\left(0,R\right)}.$$

We show that C is compact. Since C is bounded it suffices to prove that C is closed. Thus, let $\{x_l\} \subset C$ be such that $x_l \to x$. Then $x \in \overline{E} \cap \overline{B\left(0,R\right)}$. Fix $r > 0$ and let l be so large that $|x_l - x| < \frac{r}{2}$. Since $x_l \in C$ we have that $\left| E \cap \overline{B\left(x_l, \frac{r}{2}\right)} \right| > 0$, and so

$$\left| E \cap \overline{B\left(x,r\right)} \right| \geq \left| E \cap \overline{B\left(x_l, \frac{r}{2}\right)} \right| > 0,$$

where we have used the fact that $\overline{B\left(x,r\right)} \supset \overline{B\left(x_l, \frac{r}{2}\right)}$. Hence $x \in C$.

Next we prove that $|E \setminus C| = 0$. Indeed,

$$E \setminus C \subset E_1 \cup \left(E \setminus \overline{B\left(0,R\right)} \right),$$

where

$$E_1 := \left\{ x \in E \cap \overline{B\left(0,R\right)} : \left| E \cap \overline{B\left(x,r\right)} \right| = 0 \text{ for some } r > 0 \right\},$$

and since $\left| E \setminus \overline{B\left(0,R\right)} \right| = 0$, it suffices to show that $|E_1| = 0$. This follows from the Vitali–Besicovitch covering theorem, since the family of closed balls

$$\left\{ \overline{B\left(x,r\right)} : x \in E_1, \left| E \cap \overline{B\left(x,r\right)} \right| = 0 \right\}$$

is a fine Morse cover for E_1. Hence we have shown that $|E \setminus C| = 0$. In view of the hypothesis on E we cannot have that $E \supset C$, and so there exists a point $x_1 \in C \setminus E$. By the definition of the set C we have that $\left| E \cap \overline{B\left(x_1,r\right)} \right| > 0$ for all $r > 0$.

We are now ready to construct the sequence $\{K_j\}$ even in this case. By the inner regularity of the Lebesgue measure, there exists a compact set $K_1 \subset E$ with $|K_1| > 0$. Since $x_1 \in \overline{E} \setminus E$ we have that $\operatorname{dist}\left(x_1, K_1\right) > 0$. Let $r_1 < 1$ be so small that $K_1 \subset E \setminus \overline{B\left(x_1, r_1\right)}$. Since $\left| E \cap \overline{B\left(x_1, r_1\right)} \right| > 0$, again by the inner regularity of the Lebesgue measure, there exists a compact set $K_2 \subset E \cap \overline{B\left(x_1, r_1\right)}$ with $|K_2| > 0$. Find $0 < r_2 < \min\left\{r_1, \frac{1}{2}\right\}$ such that $K_2 \subset E \setminus \overline{B\left(x_1, r_2\right)}$. Inductively we can construct a sequence of pairwise disjoint compact sets $\{K_j\} \subset E$ and a sequence of radii $\{r_j\}$ such that $|K_j| > 0$, $K_j \subset \overline{B\left(x_1, r_{j-1}\right)} \setminus \overline{B\left(x_1, r_j\right)}$, and $r_j < \min\left\{r_{j-1}, \frac{1}{j}\right\}$. Moreover, since $x_1 \in \overline{E} \setminus E$, for any compact $K \subset E$ we have that $\operatorname{dist}\left(x_1, K\right) > 0$, and so there is $j_0 \in \mathbb{N}$ such that $K_j \subset E \setminus K$ for all $j \geq j_0$. Hence the claim is satisfied also in this case.

Step 2: We prove that (5.37) holds. Suppose, without loss of generality, that f is not identically ∞ and let $z_0 \in \operatorname{dom}_e f$. Replacing f with the integrands

$$g(z) := f(z - z_0) - f(z_0), \quad z \in \mathbb{R}^m,$$

and using the fact that E has finite measure, we have that the functional

$$v \in L^1(E; \mathbb{R}^m) \mapsto \int_E g(v(x)) \, dx$$

is still sequentially lower semicontinuous with respect to weak star convergence in the sense of measures. Moreover, (5.37) holds for g if and only if it holds for f. Thus, without loss of generality, we may assume that $f(0) = 0$. Assume by contradiction that

$$\liminf_{|z| \to \infty} \frac{f(z)}{|z|} < 0.$$

Then we may find a sequence $\{z_n\} \subset \mathbb{R}^m$ such that $|z_n| \nearrow \infty$ and

$$f(z_n) \le -c|z_n| \quad \text{for all } n \in \mathbb{N}, \tag{5.39}$$

for some constant $c > 0$. Let $\{K_j\} \subset E$ be the sequence constructed in Step 1 and select a subsequence $\{z_{n_j}\}$ such that

$$|K_j| > \frac{1}{|z_{n_j}|}.$$

By Proposition 1.20 find a Borel set $E_j \subset K_j$ such that

$$|E_j| = \frac{1}{|z_{n_j}|}.$$

Define

$$v_j := z_{n_j} \chi_{E_j}.$$

We claim that $v_j \mathcal{L}^N \lfloor E \overset{*}{\rightharpoonup} 0$ in $\mathcal{M}(E; \mathbb{R}^m)$. Indeed, fix $\varphi \in C_0(E)$. Given $\varepsilon > 0$ find a compact set $K \subset E$ such that

$$|\varphi(x)| \le \varepsilon \quad \text{if } x \in E \setminus K. \tag{5.40}$$

If j is sufficiently large, then $E_j \subset E \setminus K$, and so

$$\left| \int_E \varphi v_j \, dx \right| = |z_{n_j}| \left| \int_{E_j} \varphi \, dx \right| \le \varepsilon |z_{n_j}| |E_j| = \varepsilon. \tag{5.41}$$

On the other hand, by (5.39) we have

$$\int_E f(v_j) \, dx = f(z_{n_j}) \frac{1}{|z_{n_j}|} \le -c.$$

Hence

$$\liminf_{j \to \infty} \int_E f(v_j) \, dx \le -c < \int_E f(0) \, dx = 0,$$

which is a contradiction and proves the theorem.

Remark 5.23. Note that if f is lower semicontinuous, then condition (5.37) is stronger than (5.2) with $p = 1$, although it still does not guarantee that f is bounded from below, as confirmed by the function $f : \mathbb{R} \to (-\infty, \infty]$ defined by

$$f(z) := \begin{cases} -\log z & \text{if } z > 0, \\ \infty & \text{if } z \leq 0. \end{cases}$$

Exercise 5.24. Show that the sufficiency part of the previous theorem still holds with the Lebesgue measure replaced by a (positive) finite Radon measure, while the necessity part requires in addition the measure to be nonatomic. What happens in the case of measures with atoms?

We consider next the case that E has infinite measure. In view of Theorems 5.6 and 5.17, it suffices to consider the case that $f(0) = 0$ and $f(z) \geq -C\,|z|$ for all $z \in \mathbb{R}^m$.

Theorem 5.25. *Let E be a Borel subset of $\mathbb{R}^N$ with infinite measure, and let $f : \mathbb{R}^m \to (-\infty, \infty]$ be a convex, lower semicontinuous function such that $f(0) = 0$. Assume that there exists $C > 0$ such that*

$$f(z) \geq -C\,|z| \quad \text{for all } z \in \mathbb{R}^m.$$

Then the functional

$$v \in L^1(E; \mathbb{R}^m) \mapsto \int_E f(v(x))\, dx$$

is sequentially lower semicontinuous with respect to weak star convergence in the sense of measures if and only if $f \geq 0$.

More generally, if $f : \mathbb{R}^m \to [0, \infty]$ is a convex, lower semicontinuous function such that $f(0) = 0$, then for any sequence $\{v_n\} \subset L^1(E; \mathbb{R}^m)$ such that $v_n \mathcal{L}^N \lfloor E \overset{*}{\rightharpoonup} \lambda$ in $\mathcal{M}(E; \mathbb{R}^m)$ we have

$$\liminf_{n \to \infty} \int_E f(v_n)\, dx \geq \int_E f\left(\frac{d\lambda}{d\mathcal{L}^N}\right) dx + \int_E f^\infty \left(\frac{d\lambda}{d\|\lambda_s\|}\right) d\|\lambda_s\|.$$

Proof. Since $f \geq 0$, the proof of the sufficiency is identical to Steps 1–3 of the proof of Theorem 5.19.

To prove the necessity we proceed somewhat as in Step 2 of the proof of the necessity part of the previous theorem. Assume by contradiction that there exists $z \in \mathbb{R}^m$ such that $f(z) < 0$. Since E has infinite measure,

$$\lim_{R \to \infty} |E \cap B(0, R)| = |E| = \infty,$$

and so there is $R_1 > 1$ such that $|E \cap B(0, R_1)| > \frac{1}{|z|}$. By Proposition 1.20 there exists a Borel set $E_1 \subset E \cap B(0, R_1)$ such that $|E_1| = \frac{1}{|z|}$. Since $|E \setminus B(0, R_1)| = \infty$ we may find $R_2 > R_1 + 1$ such that

$$|E \cap (B(0, R_2) \setminus B(0, R_1))| > \frac{1}{|z|}$$

and then use Proposition 1.20 to obtain a Borel set

$$E_2 \subset E \cap (B(0, R_2) \setminus B(0, R_1))$$

such that $|E_2| = \frac{1}{|z|}$. Recursively, we may construct a sequence $R_n \nearrow \infty$ and Borel sets $\{E_n\} \subset E$ such that

$$E_n \subset B(0, R_n) \setminus \bigcup_{j=1}^{n-1} B(0, R_j) \tag{5.42}$$

and

$$|E_n| = \frac{1}{|z|}.$$

Note that by (5.42) and the fact that $R_n \nearrow \infty$ for any compact set $K \subset E$ we have that $E_n \subset E \setminus K$ for all n sufficiently large. Therefore, if we define

$$v_n := z\chi_{E_n},$$

then we may proceed as in (5.40) and (5.41) to conclude that $v_n \mathcal{L}^N \lfloor E \overset{*}{\rightharpoonup} 0$ in $\mathcal{M}(E; \mathbb{R}^m)$. On the other hand, since $f(0) = 0$ we have

$$\int_E f(v_n) \, dx = f(z)\frac{1}{|z|} < 0.$$

Hence

$$\liminf_{n\to\infty} \int_E f(v_n) \, dx = f(z)\frac{1}{|z|} < \int_E f(0) \, dx = 0,$$

contradicting the sequential lower semicontinuity. This concludes the proof of the theorem.

Exercise 5.26. Prove that the sufficiency part of the previous theorem still holds with the Lebesgue measure replaced by a (positive) Radon measure, while the necessity part holds if in addition, the measure is nonatomic. What can you conclude if the measure has atoms?

5.2.4 Weak Star Convergence in $\left(C_b\left(\overline{E}; \mathbb{R}^m\right)\right)'$

In the previous theorems we identified $L^1(E; \mathbb{R}^m)$ with a subspace of

$$\left(C_0(E; \mathbb{R}^m)\right)' = \mathcal{M}(E; \mathbb{R}^m),$$

as is usual in the literature. However, when E is not compact this choice carries disadvantages. For example, linear functionals of the form

$$v \in L^1(E; \mathbb{R}^m) \mapsto \int_E (a + b \cdot v(x))\, dx$$

are not sequentially lower continuous with respect to weak star convergence in $(C_0(E))'$ unless $b = 0$ as required by (5.2).

Moreover, another drawback of weak star convergence in $(C_0(E; \mathbb{R}^m))'$ is that, unless the set is closed, it misses measures that concentrate at the boundary (see Prohorov's theorem). As an example, let $N = 1$, $E = (0, 1)$, and

$$u_n := n\chi_{(0, \frac{1}{n})}.$$

Then $u_n \mathcal{L}^1 \lfloor (0, 1) \overset{*}{\rightharpoonup} 0$ in $\mathcal{M}((0, 1); \mathbb{R})$, while $u_n \mathcal{L}^1 \overset{*}{\rightharpoonup} \delta_0$ in $\mathcal{M}(\mathbb{R}; \mathbb{R})$.

To overcome this, a possibility is to consider $L^1(E; \mathbb{R}^m)$ as a subspace of $(C_b(\overline{E}; \mathbb{R}^m))'$.

The structure of the proof of the previous theorem used strongly Radon–Nikodym type arguments that continue to hold if E is bounded, since in this case $C_b(\overline{E}; \mathbb{R}^m) = C_0(\overline{E}; \mathbb{R}^m)$, and thus $(C_b(\overline{E}; \mathbb{R}^m))'$ is still the set of signed Radon measures. If E is unbounded, then the dual of $C_b(\overline{E}; \mathbb{R}^m)$ is the space of regular finitely additive signed measures $\mathrm{rba}(\overline{E}; \mathbb{R}^m)$, where the Radon–Nikodym theorem may fail (see Theorem 1.118).

Theorem 5.27. *Let E be a bounded Borel subset of $\mathbb{R}^N$, and let $f : \mathbb{R}^m \to (-\infty, \infty]$ be convex and lower semicontinuous. Then the functional*

$$v \in L^1(E; \mathbb{R}^m) \mapsto \int_E f(v(x))\, dx$$

is sequentially lower semicontinuous with respect to weak star convergence in $\mathcal{M}(\overline{E}; \mathbb{R}^m)$.

More generally, for any sequence $\{v_n\} \subset L^1(E; \mathbb{R}^m)$ such that

$$\chi_E v_n \mathcal{L}^N \lfloor \overline{E} \overset{*}{\rightharpoonup} \lambda \quad \text{in } \mathcal{M}(\overline{E}; \mathbb{R}^m)$$

we have

$$\liminf_{n \to \infty} \int_E f(v_n)\, dx \geq \int_E f\left(\frac{d\lambda}{d\mathcal{L}^N}\right) dx + \int_{\partial E \setminus E} f^\infty\left(\frac{d\lambda}{d\mathcal{L}^N}\right) dx \qquad (5.43)$$

$$+ \int_{\overline{E}} f^\infty\left(\frac{d\lambda}{d\|\lambda_s\|}\right) d\|\lambda_s\|.$$

Proof. **Step 1:** Assume that $f \geq 0$ and let $\{v_n\} \subset L^1(E; \mathbb{R}^m)$ be such that $\chi_E v_n \mathcal{L}^N \lfloor \overline{E} \overset{*}{\rightharpoonup} \lambda$ in $\mathcal{M}(\overline{E}; \mathbb{R}^m)$. The argument follows closely that of Step 1 of the proof of Theorem 5.19 with the following changes:

1. (5.13) is replaced by

$$\chi_E f(v_n(\cdot))\, \mathcal{L}^N \lfloor \overline{E} \overset{*}{\rightharpoonup} \mu$$

as $n \to \infty$ in $\mathcal{M}(\overline{E}; \mathbb{R}^m)$.

2. Inequalities (5.25) and (5.26) are complemented by

$$\frac{d\mu}{d\mathcal{L}^N}(x_0) \geq f^\infty \left(\frac{d\lambda}{d\mathcal{L}^N}(x_0) \right), \tag{5.44}$$

for $\mathcal{L}^N$ a.e. $x_0 \in \partial E \setminus E$. The proof of (5.44) follows closely that of (5.25), namely take $x_0 \in \partial E \setminus E$ a point of density one for $\partial E \setminus E$ with respect to $\mathcal{L}^N$ such that

$$\frac{d\mu}{d\mathcal{L}^N}(x_0) = \lim_{\varepsilon \to 0^+} \frac{\mu(Q(x_0,\varepsilon) \cap \overline{E})}{|Q(x_0,\varepsilon) \cap \overline{E}|} < \infty, \tag{5.45}$$

$$\frac{d\lambda}{d\mathcal{L}^N}(x_0) = \lim_{\varepsilon \to 0^+} \frac{\lambda(Q(x_0,\varepsilon) \cap \overline{E})}{|Q(x_0,\varepsilon) \cap \overline{E}|}, \tag{5.46}$$

where we have used Theorem 1.155. Note that

$$\frac{|Q(x_0,\varepsilon) \cap E|}{\varepsilon^N} = 1 - \frac{|Q(x_0,\varepsilon) \cap (\mathbb{R}^N \setminus E)|}{\varepsilon^N} \tag{5.47}$$

$$\leq 1 - \frac{|Q(x_0,\varepsilon) \cap (\partial E \setminus E)|}{\varepsilon^N} \to 0$$

as $\varepsilon \to 0^+$.

By Proposition 1.15 we may choose $\varepsilon_k \searrow 0$ such that $\mu(\partial Q(x_0,\varepsilon_k) \cap \overline{E}) = 0$ and $\nu(\partial Q(x_0,\varepsilon_k) \cap \overline{E}) = 0$. Invoking Corollary 1.204 with $X := \overline{E}$ and the (relatively) open set $A := Q(x_0,\varepsilon_k) \cap \overline{E}$, for all i, $k \in \mathbb{N}$ we have

$$\lim_{n \to \infty} \int_{Q(x_0,\varepsilon_k) \cap E} b_i \cdot v_n \, dx = \lim_{n \to \infty} \int_{Q(x_0,\varepsilon_k) \cap \overline{E}} b_i \cdot d\lambda_n \tag{5.48}$$

$$= \int_{Q(x_0,\varepsilon_k) \cap \overline{E}} b_i \cdot d\lambda,$$

where $\lambda_n := \chi_E v_n \mathcal{L}^N \lfloor \overline{E}$. As in (5.31) we have

$$\frac{d\mu}{d\mathcal{L}^N}(x_0) = \lim_{k \to \infty} \frac{\mu(Q(x_0,\varepsilon_k) \cap \overline{E})}{\varepsilon^N}$$

$$\geq \liminf_{k \to \infty} \liminf_{n \to \infty} \frac{1}{\varepsilon^N} \int_{Q(x_0,\varepsilon_k) \cap E} (a_i + b_i \cdot v_n(x)) \, dx$$

$$= \liminf_{k \to \infty} \frac{1}{\varepsilon_k^N} \int_{Q(x_0,\varepsilon_k) \cap \overline{E}} b_i \cdot d\lambda(x),$$

where in the last identity we used (5.47) and (5.48). By (5.46) we have that

$$\liminf_{k \to \infty} \frac{1}{\varepsilon_k^N} \int_{Q(x_0,\varepsilon_k) \cap \overline{E}} b_i \cdot d\lambda(x) = b_i \cdot \frac{d\lambda}{d\mathcal{L}^N}(x_0),$$

which yields

$$\frac{d\mu}{d\mathcal{L}^N}(x_0) \geq b_i \cdot \frac{d\lambda}{d\mathcal{L}^N}(x_0).$$

By taking the supremum over all i we obtain (5.44).

3. The inequality (5.26) continues to hold for $\|\lambda_s\|$ a.e. $x_0 \in \overline{E}$. To see this, replace E by $\overline{E}$ in (5.32)–(5.34) and in (5.36), and replace (5.35) by

$$
\begin{aligned}
\frac{d\mu}{d\|\lambda_s\|}(x_0) &= \lim_{k\to\infty} \frac{\mu(Q(x_0,\varepsilon_k) \cap \overline{E})}{\|\lambda_s\|\,(Q(x_0,\varepsilon_k) \cap \overline{E})} \\
&= \lim_{k\to\infty} \lim_{n\to\infty} \frac{1}{\|\lambda_s\|\,(Q(x_0,\varepsilon_k) \cap \overline{E})} \int_{Q(x_0,\varepsilon_k)\cap E} f(v_n)\,dx \\
&\geq \liminf_{k\to\infty} \liminf_{n\to\infty} \frac{1}{\|\lambda_s\|\,(Q(x_0,\varepsilon_k) \cap \overline{E})} \int_{Q(x_0,\varepsilon_k)\cap E} (a_i + b_i \cdot v_n)\,dx \\
&= \liminf_{k\to\infty} \frac{1}{\|\lambda_s\|\,(Q(x_0,\varepsilon_k) \cap \overline{E})} \int_{Q(x_0,\varepsilon_k)\cap \overline{E}} b_i \cdot d\lambda,
\end{aligned}
$$

where in the last inequality we have used (5.48) and (5.34) (with E replaced by $\overline{E}$).

Step 2: Here we remove the additional assumption that $f \geq 0$. Since f is convex and lower semicontinuous, there exist $a \in \mathbb{R}$ and $b \in \mathbb{R}^m$ such that

$$
f(z) \geq a + b \cdot z
$$

for all $z \in \mathbb{R}^m$. Setting $g(z) := f(z) - a - b \cdot z$, since g is a nonnegative, lower semicontinuous convex integrand, by Step 1, for any sequence $\{v_n\} \subset L^1(E;\mathbb{R}^m)$ such that $\chi_E v_n \mathcal{L}^N \lfloor \overline{E} \overset{*}{\rightharpoonup} \lambda$ in $\mathcal{M}(\overline{E};\mathbb{R}^m)$ we have

$$
\begin{aligned}
\liminf_{n\to\infty} \int_E f(v_n)\,dx - a\,|E| - \int_{\overline{E}} b \cdot d\lambda &= \liminf_{n\to\infty} \int_E g(v_n)\,dx \\
&\geq \int_E g\left(\frac{d\lambda}{d\mathcal{L}^N}\right)dx + \int_{\partial E \setminus E} g^\infty\left(\frac{d\lambda}{d\mathcal{L}^N}\right)dx + \int_{\overline{E}} g^\infty\left(\frac{d\lambda}{d\|\lambda_s\|}\right)d\|\lambda_s\| \\
&= \int_E f\left(\frac{d\lambda}{d\mathcal{L}^N}\right)dx - a\,|E| - \int_E b \cdot \frac{d\lambda}{d\mathcal{L}^N}\,dx + \int_{\partial E \setminus E} f^\infty\left(\frac{d\lambda}{d\mathcal{L}^N}\right)dx \\
&\quad - \int_{\partial E \setminus E} b \cdot \frac{d\lambda}{d\mathcal{L}^N}\,dx + \int_{\overline{E}} f^\infty\left(\frac{d\lambda}{d\|\lambda_s\|}\right)d\|\lambda_s\| - \int_{\overline{E}} b \cdot \frac{d\lambda}{d\|\lambda_s\|}\,d\|\lambda_s\|,
\end{aligned}
$$

where we have used the facts that since $1 \in C_b(\overline{E})$, we have

$$
\lim_{n\to\infty} \int_E b \cdot v_n\,dx = \int_{\overline{E}} b \cdot d\lambda,
$$

and that by Theorem 4.70, for every $w \in \mathrm{dom}_e\, f$,

$$
\begin{aligned}
g^\infty(z) &= \lim_{t\to\infty} \frac{g(w+tz) - g(w)}{t} \\
&= \lim_{t\to\infty} \frac{f(w+tz) - b \cdot (tz) - f(w)}{t} = f^\infty(z) - b \cdot z.
\end{aligned}
$$

It now suffices to observe that

$$\int_{\overline{E}} b \cdot d\lambda = \int_E b \cdot \frac{d\lambda}{d\mathcal{L}^N}\, dx + \int_{\partial E \setminus E} b \cdot \frac{d\lambda}{d\mathcal{L}^N}\, dx + \int_{\overline{E}} b \cdot \frac{d\lambda}{d\,\|\lambda_s\|}\, d\,\|\lambda_s\|,$$

and so we obtain (5.43).

Exercise 5.28. Prove that the previous theorem still holds with the Lebesgue measure replaced by a (positive) Radon measure.

5.3 Integral Representation

In the previous sections we have studied necessary and sufficient conditions on the integrand f for the functional

$$v \in L^p(E; \mathbb{R}^m) \mapsto \int_E f(v(x))\, dx$$

to be well-defined and lower semicontinuous with respect to various types of convergence. As we will explain in more detail in the next section, when these conditions are violated one looks for a relaxed or effective energy. The next natural question is whether this relaxed energy still has an integral form for some new integrand h and if so, what is the relation between h and the original integrand f. As a first step in this direction, in this section we give conditions under which the relaxation of an abstract functional admits an integral representation. The main result is the following theorem.

Theorem 5.29. *Let $\Omega \subset \mathbb{R}^N$ be an open set, $1 \le p < \infty$, and let*

$$I : L^p(\Omega; \mathbb{R}^m) \times \mathcal{B}(\Omega) \to [0, \infty]$$

satisfy the following properties:

(I_1) $I(v; \cdot)$ *is additive, that is,*

$$I(v; B_1 \cup B_2) = I(v; B_1) + I(v; B_2)$$

for all $v \in L^p(\Omega; \mathbb{R}^m)$ and B_1, $B_2 \in \mathcal{B}(\Omega)$ such that $B_1 \cap B_2 = \emptyset$;
(I_2) $I(v; \cdot)$ *is local, that is,*

$$I(v; B) = I(w; B)$$

for all v, $w \in L^p(\Omega; \mathbb{R}^m)$ such that $v = w$ $\mathcal{L}^N$ a.e. on $B \in \mathcal{B}(\Omega)$;
(I_3) *there exist $v_0 \in L^p(\Omega; \mathbb{R}^m)$ and a finite Radon measure μ absolutely continuous with respect to the Lebesgue measure such that*

$$I(v_0; B) \le \mu(B)$$

for all $B \in \mathcal{B}(\Omega)$;

(I_4) $I(v; Q(x_0, \varepsilon)) = I(v(x_0 + (\cdot - y_0)); Q(y_0, \varepsilon))$ *for all* $v \in L^p(\Omega; \mathbb{R}^m)$, *and all cubes* $Q(x_0, \varepsilon), Q(y_0, \varepsilon) \subset \Omega$.

For every $B \in \mathcal{B}(\Omega)$ *let* τ *be either the weak or the strong topology in* $L^p(B; \mathbb{R}^m)$ *and let* $\mathcal{E}_p(\cdot; B)$ *be the greatest functional below* $I(\cdot, B)$ *that is sequentially lower semicontinuous with respect to* τ. *Then there exists a Borel function* $h : \mathbb{R}^m \to [0, \infty]$ *(depending on* τ*) such that*

$$\mathcal{E}_p(v; B) = \int_B h(v)\, dx \qquad (5.49)$$

for every $B \in \mathcal{B}(\Omega)$ *and* $v \in L^p(\Omega; \mathbb{R}^m)$.

In particular, if I is sequentially lower semicontinuous with respect to the strong topology, then I admits an integral representation as in (5.49).

We begin with an auxiliary result on *Yosida transforms*.

Lemma 5.30. *Let* (V, d) *be a metric space, let* $\Psi : V \to [0, \infty]$ *be a lower semicontinuous function, and let* $g : [0, \infty) \to [0, \infty)$ *be an increasing function such that* $g(0) = 0$ *and* $g(s) > 0$ *for* $s > 0$. *For every* $k \in \mathbb{N}$ *and* $v \in V$ *define*

$$\Psi_k(v) := \inf \{\Psi(w) + kg(d(w, v)) : w \in V\}.$$

Then

$$\Psi = \sup_{k \in \mathbb{N}} \Psi_k.$$

Proof. Let $v \in V$. Then for every $k \in \mathbb{N}$ we have

$$\Psi_k(v) \leq \Psi(v) + kg(d(v, v)) = \Psi(v),$$

where we have used the fact that $g(0) = 0$. Taking the supremum over all $k \in \mathbb{N}$ implies that

$$\sup_{k \in \mathbb{N}} \Psi_k(v) \leq \Psi(v).$$

To prove the reverse inequality, let $t < \Psi(v)$. Since Ψ is lower semicontinuous there exists $r > 0$ such that

$$t < \inf \{\Psi(w) : w \in B(v, r)\}.$$

Since $g(r) > 0$ we may find $k_0 \in \mathbb{N}$ such that $k_0 g(r) > t$. Then for every $w \in B(v, r)$ we have

$$\Psi(w) + k_0 g(d(w, v)) \geq \Psi(w) > t,$$

while for all $w \in V \setminus B(v, r)$ we obtain

$$\Psi(w) + k_0 g(d(w, v)) \geq k_0 g(r) > t,$$

where we have used the fact that g is increasing. Combining the two previous inequalities yields

$$\sup_{k \in \mathbb{N}} \Psi_k (v) \geq \Psi_{k_0} (v) = \inf \{ \Psi (w) + k_0 g \left(d \left(w, v \right) \right) : w \in V \} \geq t.$$

It now suffices to let $t \nearrow \Psi (v)$.

We now turn to the proof of Theorem 5.29.

Proof (Theorem 5.29). **Step 1:** We claim that $\mathcal{E}_p$ satisfies conditions (I_1)–(I_4). In view of Remark A.85, for every countable ordinal α we define a functional $\mathcal{H}_\alpha$ as follows. For $v \in L^p \left(\Omega; \mathbb{R}^m \right)$ and $B \in \mathcal{B}(\Omega)$ set

$$\mathcal{H}_0 \left(v; B \right) := I \left(v; B \right),$$

$$\mathcal{H}_{\beta+1} \left(v; B \right) := \inf \left\{ \liminf_{n \to \infty} \mathcal{H}_\beta \left(v_n; B \right) : \{ v_n \} \subset L^p(\Omega; \mathbb{R}^m), \right.$$

$$\left. v_n \overset{\tau}{\to} v \text{ in } L^p(B; \mathbb{R}^m) \right\} \text{ if } \beta \text{ is not a limit ordinal,}$$

$$\mathcal{H}_\beta \left(v; B \right) := \inf \{ \mathcal{H}_\alpha \left(v; B \right) : \alpha < \beta, \alpha \text{ not a limit ordinal} \}$$

if β is a limit ordinal. We recall that by Proposition 3.17,

$$\mathcal{E}_p(v; B) = \inf \{ \mathcal{H}_\beta \left(v; B \right) : \beta \text{ countable ordinal} \}, \tag{5.50}$$

and thus it suffices to show that each functional $\mathcal{H}_\beta$ satisfies conditions (I_1)–(I_4). Properties (I_2)–(I_4) hold. We prove (I_1) using Remark A.83. By assumption, $\mathcal{H}_0$ satisfies (I_1). Fix a countable ordinal β that is not a limit ordinal and assume that $\mathcal{H}_\beta$ satisfies (I_1). We claim that $\mathcal{H}_{\beta+1}$ has the same property. Let $B_1, B_2 \in \mathcal{B}(\Omega)$ be such that $B_1 \cap B_2 = \emptyset$. Fix $\varepsilon > 0$ and find two sequences $\{ v_n \} \subset L^p \left(\Omega; \mathbb{R}^m \right)$, $\{ w_n \} \subset L^p \left(\Omega; \mathbb{R}^m \right)$ τ-converging to v in $L^p \left(B_1; \mathbb{R}^m \right)$ and $L^p \left(B_2; \mathbb{R}^m \right)$, respectively, such that

$$\lim_{n \to \infty} \mathcal{H}_\beta(v_n; B_1) \leq \mathcal{H}_{\beta+1}(v; B_1) + \varepsilon,$$

$$\lim_{n \to \infty} \mathcal{H}_\beta(w_n; B_2) \leq \mathcal{H}_{\beta+1}(v; B_2) + \varepsilon.$$

Define

$$z_n \left(x \right) := \begin{cases} v_n \left(x \right) & \text{for } x \in B_1, \\ w_n \left(x \right) & \text{for } x \in \Omega \setminus B_1. \end{cases}$$

Then $z_n \overset{\tau}{\to} v$ in $L^p \left(B_1 \cup B_2; \mathbb{R}^m \right)$ and hence by (I_1) for $\mathcal{H}_\beta$,

$$\mathcal{H}_{\beta+1}(v; B_1 \cup B_2) \leq \lim_{n \to \infty} \mathcal{H}_\beta \left(z_n; B_1 \cup B_2 \right)$$

$$= \lim_{n \to \infty} \mathcal{H}_\beta \left(v_n; B_1 \right) + \lim_{n \to \infty} \mathcal{H}_\beta \left(w_n; B_2 \right)$$

$$\leq \mathcal{H}_{\beta+1}(v; B_1) + \mathcal{H}_{\beta+1}(v; B_2) + 2\varepsilon.$$

By letting $\varepsilon \to 0^+$ we obtain

$$\mathcal{H}_{\beta+1}(v; B_1 \cup B_2) \leq \mathcal{H}_{\beta+1}(v; B_1) + \mathcal{H}_{\beta+1}(v; B_2).$$

To prove the opposite inequality, fix $\varepsilon > 0$ and let $\{v_n\} \subset L^p(\Omega; \mathbb{R}^m)$ be a sequence τ converging in $L^p(B_1 \cup B_2; \mathbb{R}^m)$ to v and such that

$$\lim_{n \to \infty} \mathcal{H}_\beta(v_n; B_1 \cup B_2) \leq \mathcal{H}_{\beta+1}(v; B_1 \cup B_2) + \varepsilon.$$

Then, by the definition of $\mathcal{H}_{\beta+1}$ and (I_1) for $\mathcal{H}_\beta$,

$$\mathcal{H}_{\beta+1}(v; B_1) + \mathcal{H}_{\beta+1}(v; B_2) \leq \liminf_{n \to \infty} \mathcal{H}_\beta(v_n; B_1) + \liminf_{n \to \infty} \mathcal{H}_\beta(v_n; B_2)$$
$$\leq \lim_{n \to \infty} \mathcal{H}_\beta(v_n; B_1 \cup B_2) \leq \mathcal{H}_{\beta+1}(v; B_1 \cup B_2) + \varepsilon.$$

By letting $\varepsilon \to 0^+$ we have proved the claim.

Finally, if β is a limit ordinal and $\mathcal{H}_\alpha$ satisfies condition (I_1) for each $\alpha < \beta$, α not a limit ordinal, then so does

$$\mathcal{H}_\beta := \inf\{\mathcal{H}_\alpha : \alpha < \beta, \, \alpha \text{ not a limit ordinal}\}. \qquad (5.51)$$

To see this, note that for any $\alpha < \beta$, α not a limit ordinal, we have that

$$\mathcal{H}_\alpha(v; B_1 \cup B_2) = \mathcal{H}_\alpha(v; B_1) + \mathcal{H}_\alpha(v; B_2),$$

and so taking the infimum over all such α yields

$$\mathcal{H}_\beta(v; B_1 \cup B_2) \geq \mathcal{H}_\beta(v; B_1) + \mathcal{H}_\beta(v; B_2).$$

To prove the reverse inequality fix $\varepsilon > 0$ and by (5.51) find a nonlimit ordinal α with $\alpha < \beta$ such that

$$\mathcal{H}_\alpha(v; B_1 \cup B_2) \leq \mathcal{H}_\beta(v; B_1 \cup B_2) + \varepsilon.$$

Since property (I_1) holds for $\mathcal{H}_\alpha$ and using again (5.51), we get

$$\mathcal{H}_\beta(v; B_1) + \mathcal{H}_\beta(v; B_2) \leq \mathcal{H}_\alpha(v; B_1) + \mathcal{H}_\alpha(v; B_2)$$
$$= \mathcal{H}_\alpha(v; B_1 \cup B_2) \leq \mathcal{H}_\beta(v; B_1 \cup B_2) + \varepsilon.$$

It now suffices to let $\varepsilon \to 0^+$.

Step 2: We claim that $\mathcal{E}_p(v; \cdot)$ is a measure, absolutely continuous with respect to the Lebesgue measure. By (I_1) and Proposition 1.9 it is enough to prove that

$$\mathcal{E}_p(v; B_n) \to \mathcal{E}_p(v; B) \quad \text{whenever } B_n \nearrow B.$$

Since $\mathcal{E}_p \geq 0$ we have

$$\limsup_{n \to \infty} \mathcal{E}_p(v; B_n) \leq \mathcal{E}_p(v; B).$$

Setting

$$v_n(x) := \begin{cases} v(x) & \text{if } x \in B_n, \\ v_0(x) & \text{if } x \in \Omega \setminus B_n, \end{cases}$$

since $v_n \to v$ in L^p by (I_1)–(I_3) for $\mathcal{E}_p$ (see Step 1) and Proposition 1.7(ii) (applied to the finite measure μ) we have that

$$\mathcal{E}_p(v; B) \leq \liminf_{n \to \infty} \mathcal{E}_p(v_n; B) = \liminf_{n \to \infty} \left[\mathcal{E}_p(v; B_n) + \mathcal{E}_p(v_0; B \setminus B_n) \right]$$

$$\leq \liminf_{n \to \infty} \mathcal{E}_p(v; B_n) + \limsup_{n \to \infty} \mu(B \setminus B_n) = \liminf_{n \to \infty} \mathcal{E}_p(v; B_n).$$

Thus $\mathcal{E}_p(v; \cdot)$ is a measure. To prove that it is absolutely continuous with respect to the Lebesgue measure, let B be a Borel set of zero measure. By (I_2) and (I_3) for $\mathcal{E}_p$ we have

$$0 \leq \mathcal{E}_p(v; B) = \mathcal{E}_p(v_0; B) \leq \mu(B) = 0.$$

At this point, for every $v \in L^p(\Omega; \mathbb{R}^m)$ we are in a position to apply the Radon–Nikodym theorem to represent $\mathcal{E}_p(v; \cdot)$ as

$$\mathcal{E}_p(v; B) = \int_B h_v(x) \, dx, \quad B \in \mathcal{B}(\Omega),$$

where the Radon–Nikodym derivative $h_v = \frac{d\mathcal{E}_p(v;\cdot)}{d\mathcal{L}^N}$ is a nonnegative measurable function. However, since in general there is no assurance that $\mathcal{E}_p(v; \cdot)$ is a Radon measure, we cannot apply Besicovitch's derivation theorem to obtain an explicit formula of h_v in terms of $\mathcal{E}_p(v; \cdot)$. This is the reason why we need to consider the Yosida transforms, and this leads us to Step 3.

Step 3: For every $k \in \mathbb{N}$, $B \in \mathcal{B}(\Omega)$, and $v \in L^p(\Omega; \mathbb{R}^m)$ we introduce the *Yosida transform*

$$\mathcal{Y}^k(v; B) := \inf \left\{ \mathcal{E}_p(w; B) + k \int_B |w - v|^p \, dx : w \in L^p(\Omega; \mathbb{R}^m) \right\}. \quad (5.52)$$

Then $\mathcal{Y}^k$ still satisfies (I_4) and by Lemma 5.30 (with $g(s) := s^{\frac{1}{p}}$) for all $B \in \mathcal{B}(\Omega)$ and $v \in L^p(\Omega; \mathbb{R}^m)$ we have

$$\mathcal{E}_p(v; B) = \sup_{k \in \mathbb{N}} \mathcal{Y}^k(v; B). \quad (5.53)$$

Moreover, taking $w = v_0$ in the definition of $\mathcal{Y}^k$ yields

$$0 \leq \mathcal{Y}^k(v; B) \leq \mu(B) + k \int_B |v - v_0|^p \, dx \quad (5.54)$$

for all $k \in \mathbb{N}$, $B \in \mathcal{B}(\Omega)$, and $v \in L^p(\Omega; \mathbb{R}^m)$. Note that this implies in particular that in the definition of $\mathcal{Y}^k(v; B)$ it is enough to consider those $w \in L^p(\Omega; \mathbb{R}^m)$ for which

$$\mathcal{E}_p(w; B) + k \int_B |w - v|^p \, dx \le \mu(B) + k \int_B |v - v_0|^p \, dx,$$

so that, since $\mathcal{E}_p \ge 0$ and $k \ge 1$,

$$\int_B |w|^p \, dx \le C \left(\mu(B) + \int_B |v|^p \, dx + \int_B |v_0|^p \, dx \right). \tag{5.55}$$

Fix $k \in \mathbb{N}$, $B \in \mathcal{B}(\Omega)$, and u, v, $w \in L^p(\Omega; \mathbb{R}^m)$, with

$$\int_B |w|^p \, dx \le C \left(\mu(B) + \int_B |v|^p \, dx + \int_B |u|^p + \int_B |v_0|^p \, dx \right).$$

Using the inequality

$$\left| |z_1|^p - |z_2|^p \right| \le p \left(\max\left\{ |z_1|^{p-1}, |z_2|^{p-1} \right\} \right) |z_1 - z_2|,$$

which holds for all z_1, $z_2 \in \mathbb{R}^m$, we obtain

$$\begin{aligned}
\mathcal{Y}^k(v; B) \le & \mathcal{E}_p(w; B) + k \int_B |w - v|^p \, dx \\
\le & \mathcal{E}_p(w; B) + k \int_B |w - u|^p \, dx \\
& + kp \int_B \left(\max\left\{ |w - v|^{p-1}, |w - u|^{p-1} \right\} \right) |u - v| \, dx \\
\le & \mathcal{E}_p(w; B) + k \int_B |w - u|^p \, dx \\
& + Ck \left(\mu(B) + \int_B (|u|^p + |v_0|^p + |v|^p) \, dx \right)^{1/p'} \left(\int_B |v - u|^p \, dx \right)^{1/p},
\end{aligned}$$

where we have used Hölder's inequality and (5.55). Taking the infimum over all admissible $w \in L^p(\Omega; \mathbb{R}^m)$ yields

$$\begin{aligned}
\mathcal{Y}^k(v; B) \le & \mathcal{Y}^k(u; B) \\
& + Ck \left(\mu(B) + \int_B (|u|^p + |v_0|^p + |v|^p) \, dx \right)^{1/p'} \left(\int_B |v - u|^p \, dx \right)^{1/p},
\end{aligned}$$

and in turn, by interchanging u and v, we get

$$\begin{aligned}
& |\mathcal{Y}^k(v; B) - \mathcal{Y}^k(u; B)| \\
& \le Ck \left(\mu(B) + \int_B (|u|^p + |v_0|^p + |v|^p) \, dx \right)^{1/p'} \left(\int_B |v - u|^p \, dx \right)^{1/p},
\end{aligned}$$

$$\tag{5.56}$$

for all $k \in \mathbb{N}$, $B \in \mathcal{B}(\Omega)$ and u, $v \in L^p(\Omega; \mathbb{R}^m)$.

Substep 3 a: We claim that for every $v \in L^p(\Omega; \mathbb{R}^m)$ the set function $\mathcal{Y}^k(v; \cdot)$ is a Radon measure absolutely continuous with respect to the Lebesgue measure. In view of (5.54) and by Proposition 1.60 it suffices to prove that $\mathcal{Y}^k(v; \cdot)$ is countably additive. Consider a sequence of pairwise disjoint sets $\{B_n\} \subset \mathcal{B}(\Omega)$, and let

$$B := \bigcup_{n=1}^{\infty} B_n.$$

Given $u \in L^p(\Omega; \mathbb{R}^m)$, by Step 2 we have

$$\mathcal{E}_p(u; B) + k \int_B |u - v|^p \, dx = \sum_{n=1}^{\infty} \left[\mathcal{E}_p(u; B_n) + k \int_{B_n} |u - v|^p \, dx \right]$$

$$\geq \sum_{n=1}^{\infty} \mathcal{Y}^k(v; B_n).$$

Taking the infimum over all such u, by (5.52) we deduce that

$$\mathcal{Y}^k(v; B) \geq \sum_{n=1}^{\infty} \mathcal{Y}^k(v; B_n).$$

Conversely, fix $\varepsilon > 0$ and for every $n \in \mathbb{N}$ choose $u_n \in L^p(\Omega; \mathbb{R}^m)$ such that (see (5.52))

$$\mathcal{E}_p(u_n; B_n) + k \int_{B_n} |u_n - v|^p \, dx \leq \frac{\varepsilon}{2^n} + \mathcal{Y}^k(v; B_n). \tag{5.57}$$

Fix $l \in \mathbb{N}$ and define

$$w_l(x) := \begin{cases} u_n(x) & \text{if } x \in B_n, 1 \leq n \leq l, \\ v_0(x) & \text{if } x \in \Omega \setminus \bigcup_{n=1}^{l} B_n. \end{cases}$$

By (5.52), the fact that $\mathcal{E}_p(w_l; \cdot)$ is a measure, (I_2), (I_3) for $\mathcal{E}_p$ (see Step 1), Proposition 1.7(ii) (applied to the finite measure μ), and (5.57), in this order, we obtain

$$\mathcal{Y}^k(v; B) \leq \liminf_{l \to \infty} \left[\mathcal{E}_p(w_l; B) + k \int_B |w_l - v|^p \, dx \right]$$

$$\leq \limsup_{l \to \infty} \left[\mu \left(v_0; B \setminus \bigcup_{n=1}^{l} B_n \right) + k \int_{B \setminus \bigcup_{n=1}^{l} B_n} |v_0 - v|^p \, dx \right]$$

$$+ \liminf_{l \to \infty} \sum_{n=1}^{l} \left[\mathcal{E}_p(u_n; B_n) + k \int_{B_n} |u_n - v|^p \, dx \right]$$

$$\leq \liminf_{l \to \infty} \sum_{n=1}^{l} \left[\mathcal{Y}^k(v; B_n) + \frac{\varepsilon}{2^n} \right] = \varepsilon + \sum_{n=1}^{\infty} \mathcal{Y}^k(v; B_n).$$

It suffices now to let $\varepsilon \to 0^+$.

Substep 3b: Fix a cube $Q(y_0, r_0) \subset \Omega$ and for every $z \in \mathbb{R}^m$ set

$$f_k(z) := \liminf_{r \to 0^+} \frac{1}{r^N} \mathcal{Y}^k(z; Q(y_0, r)). \tag{5.58}$$

We claim that

$$\mathcal{Y}^k(v; B) = \int_B f_k(v) \, dx \tag{5.59}$$

for every $v \in L^p(\Omega; \mathbb{R}^m)$ and $B \in \mathcal{B}(\Omega)$.

Since $\mathcal{Y}^k(v; \cdot)$ is a Radon measure absolutely continuous with respect to the Lebesgue measure, by the Radon–Nikodym theorem we may write

$$\mathcal{Y}^k(v; B) = \int_B h_{k,v}(x) \, dx,$$

where $h_{k,v} = \frac{d\mathcal{Y}^k(v; \cdot)}{d\mathcal{L}^N}$. Take $x_0 \in \Omega$ such that, by Besicovitch's derivation theorem and Corollary 1.159,

$$h_{k,v}(x_0) = \lim_{\varepsilon \to 0^+} \frac{\mathcal{Y}^k(v; Q(x_0, \varepsilon))}{\varepsilon^N} < \infty, \tag{5.60}$$

$$\lim_{\varepsilon \to 0^+} \frac{1}{\varepsilon^N} \int_{Q(x_0, \varepsilon)} |v(x) - v(x_0)|^p \, dx = 0, \tag{5.61}$$

$$\lim_{\varepsilon \to 0^+} \frac{\mu(v; Q(x_0, \varepsilon))}{\varepsilon^N} < \infty, \tag{5.62}$$

$$\lim_{\varepsilon \to 0^+} \frac{1}{\varepsilon^N} \int_{Q(x_0, \varepsilon)} |v_0(x) - v_0(x_0)|^p \, dx = 0. \tag{5.63}$$

Since by (5.56), for every $\varepsilon > 0$ so small that $Q(x_0, \varepsilon) \subset \Omega$ we have

$$|\mathcal{Y}^k(v; Q(x_0, \varepsilon)) - \mathcal{Y}^k(v(x_0); Q(x_0, \varepsilon))|$$

$$\leq Ck \left(\mu(Q(x_0, \varepsilon)) + \int_{Q(x_0, \varepsilon)} (|v(x_0)|^p + |v_0|^p + |v|^p) \, dx \right)^{1/p'}$$

$$\times \left(\int_{Q(x_0, \varepsilon)} |v(x) - v(x_0)|^p \, dx \right)^{1/p},$$

it follows from (5.58), (5.60)–(5.63) that

$$h_{k,v}(x_0) = \lim_{\varepsilon \to 0^+} \frac{\mathcal{Y}^k(v; Q(x_0, \varepsilon))}{\varepsilon^N} = \lim_{\varepsilon \to 0^+} \frac{\mathcal{Y}^k(v(x_0); Q(x_0, \varepsilon))}{\varepsilon^N}$$

$$= \lim_{\varepsilon \to 0^+} \frac{\mathcal{Y}^k(v(x_0); Q(y_0, \varepsilon))}{\varepsilon^N} = f_k(v(x_0)),$$

where we have used (I_4) for $\mathcal{Y}^k$, and this confirms (5.59).

Since for $k \leq n$ we have $\mathcal{Y}^k(v; B) \leq \mathcal{Y}^n(v; B)$, it follows that

$$f_k(z) \leq f_n(z)$$

for all $z \in \mathbb{R}^m$. Define

$$h(z) := \sup_{k \in \mathbb{N}} f_k(z).$$

By (5.53), (5.59), and the Lebesgue monotone convergence theorem, we have

$$\mathcal{E}_p(v; B) = \sup_{k \in N} \mathcal{Y}^k(v; B) = \sup_{k \in N} \int_B f_k(v) \, dx$$
$$= \lim_{k \to \infty} \int_B f_k(v) \, dx = \int_B h(v) \, dx$$

for every $v \in L^p(\Omega; \mathbb{R}^m)$ and $B \in \mathcal{B}(\Omega)$.

In the case that $p = \infty$, an integral representation analogous to the one provided in the previous proposition still holds, where now the relevant topologies τ are either the weak star topology or the Mackey topology in $L^p(B; \mathbb{R}^m)$.

Theorem 5.31. *Let $\Omega \subset \mathbb{R}^N$ be an open set and let*

$$I : L^\infty(\Omega; \mathbb{R}^m) \times \mathcal{B}(\Omega) \to [0, \infty]$$

satisfy properties (I_1)–(I_4) of the previous theorem. For every $B \in \mathcal{B}(\Omega)$ let τ be either the weak star or the Mackey topology in $L^\infty(B; \mathbb{R}^m)$ and let $\mathcal{E}_\infty(\cdot; B)$ be the greatest functional below $I(\cdot, B)$ that is sequentially lower semicontinuous with respect to τ. Then there exists a Borel function $h : \mathbb{R}^m \to [0, \infty]$ (depending on τ) such that

$$\mathcal{E}_\infty(v; B) = \int_B h(v) \, dx$$

for every $B \in \mathcal{B}(\Omega)$ and $v \in L^\infty(\Omega; \mathbb{R}^m)$.
If, in addition, I satisfies the growth condition

(I_5) *for every $M > 0$ there exists a nonnegative function $\gamma_M \in L^1(\Omega)$ such that for all $B \in \mathcal{B}(\Omega)$,*

$$I(v; B) \leq \int_B \gamma_M(x) \, dx$$

for all $v \in L^\infty(\Omega; \mathbb{R}^m)$ with $|v(x)| \leq M$ for $\mathcal{L}^N$ a.e. $x \in B$,

then τ can also be taken as the strong topology in $L^\infty(B; \mathbb{R}^m)$.

Proof. Exactly as in Step 1 of the previous proof it can be shown that $\mathcal{E}_\infty$ satisfies properties (I_1)–(I_4). Fix $M \geq \|v_0\|_\infty$ and define a new functional

$$v \in L^1(\Omega; \mathbb{R}^m) \mapsto \mathcal{T}_M(v; B) := \mathcal{E}_\infty(\tau_M \circ v; B),$$

where τ_M is given by

$$\tau_M(z) := \begin{cases} z & \text{if } |z| \leq M, \\ \dfrac{z}{|z|}M & \text{if } |z| > M. \end{cases}$$

Then $\mathcal{T}_M$ satisfies (I_1)–(I_4), where to assert (I_3) we use the fact that $\tau_M \circ v_0 = v_0$. We claim that $\mathcal{T}_M$ is sequentially lower semicontinuous with respect to strong convergence in L^1. Indeed, let $\{v_n\} \subset L^1(\Omega; \mathbb{R}^m)$ be any sequence converging strongly in $L^1(\Omega; \mathbb{R}^m)$ to some $v \in L^1(\Omega; \mathbb{R}^m)$. Without loss of generality we may assume that

$$\liminf_{n \to \infty} \mathcal{T}_M(v_n; B) = \lim_{n \to \infty} \mathcal{T}_M(v_n; B)$$

and then extract a subsequence (not relabeled) of $\{v_n\}$ that converges to v pointwise $\mathcal{L}^N$ a.e. in Ω. Then $\{\tau_M \circ v_n\}$ converges to $\tau_M \circ v$ with respect to the Mackey topology, and so by (I_4) we have

$$\liminf_{n \to \infty} \mathcal{T}_M(v_n; B) = \lim_{n \to \infty} \mathcal{E}_\infty(\tau_M \circ v_n; B) \geq \mathcal{E}_\infty(\tau_M \circ v; B) = \mathcal{T}_M(v; B).$$

Since the functional $\mathcal{T}_M$ satisfies all the hypotheses of the previous theorem we may find a Borel function $h_M : \mathbb{R}^m \to [0, \infty]$ such that

$$\mathcal{T}_M(v; B) = \int_B h_M(v) \, dx$$

for all $v \in L^1(\Omega; \mathbb{R}^m)$ and $B \in \mathcal{B}(\Omega)$. This implies that

$$\mathcal{E}_\infty(v; B) = \int_B h_M(v) \, dx$$

for all $v \in L^\infty(\Omega; \mathbb{R}^m)$ with $\|v\|_{L^\infty(\Omega;\mathbb{R}^m)} \leq M$ and $B \in \mathcal{B}(\Omega)$. We now repeat the same reasoning for some $L > M$ to conclude that

$$\int_B h_L(v) \, dx = \int_B h_M(v) \, dx$$

for all $v \in L^\infty(\Omega; \mathbb{R}^m)$, with $\|v\|_{L^\infty(\Omega;\mathbb{R}^m)} \leq M$, and for all $B \in \mathcal{B}(\Omega)$, which yields

$$h_L(z) = h_M(z)$$

for all $z \in \mathbb{R}^m$ with $|z| \leq M$. Take a sequence $M_k \nearrow \infty$ and for each $z \in \mathbb{R}^m$ define

$$h(z) := h_{M_k}(z)$$

if $|z| \leq M_k$. We obtain that

$$\mathcal{E}_p(v; B) = \int_B h(v) \, dx$$

for all $v \in L^\infty(\Omega; \mathbb{R}^m)$ and all $B \in \mathcal{B}(\Omega)$.

Suppose now that (I_5) holds. To show that $\mathcal{T}_M$ is sequentially lower semi-continuous with respect to L^1 convergence, fix $B \in \mathcal{B}(\Omega)$ and consider a sequence $\{v_n\} \subset L^1(\Omega; \mathbb{R}^m)$ converging to v in $L^1(B; \mathbb{R}^m)$. For any fixed $\varepsilon > 0$ find a compact set $K \subset B$ such that

$$\int_{B \setminus K} \gamma_M(x) \, dx < \varepsilon, \tag{5.64}$$

and a number $\delta > 0$ such that

$$\int_E \gamma_M(x) \, dx < \varepsilon \tag{5.65}$$

for all $E \subset B$ with $|E| < \delta$.

By Egoroff's theorem there exists a set $E \subset K$ such that $|E| < \delta$ and v_n converges uniformly to v in $K \setminus E$. Define

$$w_n := \begin{cases} v_n & \text{in } K \setminus E, \\ v & \text{otherwise.} \end{cases}$$

Then the sequence $\{\tau_M \circ w_n\}$ converges in $L^\infty(B; \mathbb{R}^m)$ to $\tau_M \circ v \in L^\infty(B; \mathbb{R}^m)$, and by (I_1), (I_2), and (I_4),

$$\begin{aligned}
\liminf_{n \to \infty} \mathcal{T}_M(v_n; B) &= \liminf_{n \to \infty} \mathcal{E}_\infty(\tau_M \circ v_n; B) \\
&\geq \liminf_{n \to \infty} \mathcal{E}_\infty(\tau_M \circ v_n; K \setminus E) \\
&= \liminf_{n \to \infty} \mathcal{E}_\infty(\tau_M \circ w_n; K \setminus E) \\
&\geq \mathcal{E}_\infty(\tau_M \circ v; K \setminus E) \tag{5.66} \\
&\geq \mathcal{E}_\infty(\tau_M \circ v; B) - \int_{B \setminus E} \gamma_M \, dx \\
&\geq \mathcal{T}_M(v; B) - 2\varepsilon,
\end{aligned}$$

where we have used (5.64) and (5.65). It suffices to let $\varepsilon \to 0^+$.

5.4 Relaxation

In this section we consider functionals of the form

$$v \in L^p(E; \mathbb{R}^m) \mapsto I(v) := \int_E f(v) \, dx,$$

where E is a Borel subset of $\mathbb{R}^N$, $1 \leq p \leq \infty$, and $f : \mathbb{R}^m \to (-\infty, \infty]$ is a Borel function, $f \not\equiv \infty$.

As described in Chapter 3, if the functional I is not sequentially lower semicontinuous with respect to a given topology τ, then it is of interest to characterize the *relaxed energy*

$$\mathcal{E} : L^p (E; \mathbb{R}^m) \to [-\infty, \infty]$$

of I, that is, the greatest functional $\mathcal{E}$ below I that is sequentially lower semicontinuous with respect to the topology τ. We will consider the following types of convergence:

- weak convergence in $L^p (E; \mathbb{R}^m)$ for $1 \leq p < \infty$;
- weak star convergence in $L^\infty (E; \mathbb{R}^m)$;
- weak star convergence in the sense of measures in $L^1 (E; \mathbb{R}^m)$.

By Propositions 3.16 and 3.18, when I satisfies a suitable coercivity condition, then $\mathcal{E}(v)$ coincides with

$$\mathcal{I}(v) := \inf \left\{ \liminf_{n \to \infty} I(v_n) : \{v_n\} \subset L^p (E; \mathbb{R}^m), v_n \xrightarrow{\tau} v \right\}.$$

5.4.1 Weak Convergence and Weak Star Convergence in L^p, $1 \leq p \leq \infty$

In this subsection we characterize the *relaxed energy*

$$\mathcal{E}_p : L^p (E; \mathbb{R}^m) \to [-\infty, \infty]$$

of I with respect to weak convergence in L^p, $1 \leq p < \infty$ (respectively weak star convergence if $p = \infty$), that is, the functional $\mathcal{E}_p$ is the greatest functional below I that is sequentially lower semicontinuous with respect to weak (respectively weak star if $p = \infty$) convergence in $L^p (E; \mathbb{R}^m)$.

For every $v \in L^p(E; \mathbb{R}^m)$ define

$$\mathcal{I}_p(v) := \inf \Big\{ \liminf_{n \to \infty} I(v_n) : \{v_n\} \subset L^p(E; \mathbb{R}^m),$$

$$v_n \rightharpoonup v \ (\xrightarrow{*} \text{ if } p = \infty) \text{ in } L^p(E; \mathbb{R}^m) \Big\}.$$

Since $\mathcal{E}_p \leq I$, for every $v \in L^p(E; \mathbb{R}^m)$ we have

$$\mathcal{E}_p(v) \leq \mathcal{I}_p(v).$$

Theorem 5.32. *Let E be a Borel subset of $\mathbb{R}^N$ with finite measure, $1 \leq p \leq \infty$, and let $f : \mathbb{R}^m \to (-\infty, \infty]$ be a Borel function, $f \not\equiv \infty$, such that*

$$f(z) \geq a + b \cdot z \text{ for all } z \in \mathbb{R}^m, \tag{5.67}$$

where $a \in \mathbb{R}$, $b \in \mathbb{R}^m$. Then for every $v \in L^p(E; \mathbb{R}^m)$ we have

$$\mathcal{E}_p(v) = \int_E f^{**}(v)\, dx. \tag{5.68}$$

Moreover, for every $v \in L^p(E; \mathbb{R}^m)$ for which $\mathcal{I}_p(v) < \infty$ we have

$$\mathcal{I}_p(v) = \int_E \mathcal{C}\,(\mathrm{lsc}\ f)\,(v)\, dx, \tag{5.69}$$

where the function $\mathcal{C}\,(\mathrm{lsc}\ f)$ is the convex envelope of the lower semicontinuous envelope of f.

Remark 5.33. (i) We recall that by Propositions 3.16, 3.18 we have

$$\mathcal{E}_p(v) = \mathcal{I}_p(v) = \int_E f^{**}(v)\, dx$$

for every $v \in L^p(E; \mathbb{R}^m)$ provided f satisfies the coercivity condition

$$f(z) \geq C\gamma\,(|z|) - \frac{1}{C} \text{ for all } z \in \mathbb{R}^m,$$

where $C > 0$ and where for $t \geq 0$,

$$\gamma(t) := t^p \quad \text{if } 1 < p < \infty, \tag{5.70}$$

$$\gamma(t) := \begin{cases} 0 & \text{if } 0 \leq t < R, \\ \infty & \text{if } t \geq R, \end{cases} \quad \text{if } p = \infty, \tag{5.71}$$

for some $R > 0$, and $\gamma : [0, \infty) \to [0, \infty]$ is an increasing function with

$$\lim_{t \to \infty} \frac{\gamma(t)}{t} = \infty$$

if $p = 1$.

(ii) In view of Remark 4.93, if $f : \mathbb{R}^m \to \mathbb{R}$ satisfies (5.67), then $f^{**} = \mathrm{lsc}\,(\mathcal{C}f) = \mathcal{C}\,(\mathrm{lsc}\ f) = \mathcal{C}f$, and so by the previous theorem, for every $v \in L^p(E; \mathbb{R}^m)$ for which $\mathcal{I}_p(v) < \infty$ we have

$$\mathcal{E}_p(v) = \mathcal{I}_p(v) = \int_E f^{**}(v)\, dx.$$

We now turn to the proof of Theorem 5.32

Proof. Since the functional

$$v \in L^p(E; \mathbb{R}^m) \mapsto \int_E (a + b \cdot v)\, dx$$

is continuous with respect to weak (respectively weak star if $p = \infty$) convergence, replacing $f(z)$ by $f(z) - (a + b \cdot z)$, we may assume, without loss of generality, that $f \geq 0$.

Hence f^{**} is convex, lower semicontinuous, and nonnegative. By Theorem 5.14 the functional

$$v \in L^p(E; \mathbb{R}^m) \mapsto \int_E f^{**}(v) \, dx$$

is sequentially weakly (respectively weakly star if $p = \infty$) lower semicontinuous in $L^p(E; \mathbb{R}^m)$, and since $f^{**} \leq f$ we deduce that for every $v \in L^p(E; \mathbb{R}^m)$,

$$\mathcal{E}_p(v) \geq \int_E f^{**}(v) \, dx.$$

To prove the reverse inequality, let Ω be an open subset of $\mathbb{R}^N$ with finite measure that contains E, and for every Borel set $B \in \mathcal{B}(\Omega)$ we define the functional

$$v \in L^p(\Omega; \mathbb{R}^m) \mapsto I(v; B) := \int_B f(v) \, dx.$$

Note that I satisfies (I_1), (I_2), (I_4), while to prove (I_3) it suffices to set $v_0 \equiv z_0$, where $z_0 \in \mathbb{R}^m$ is such that $f(z_0) \in \mathbb{R}$.

By Theorems 5.29 and 5.31 there exists a Borel function $h : \mathbb{R}^m \to [0, \infty]$ such that

$$\mathcal{E}_p(v; B) = \int_B h(v) \, dx$$

for every $B \in \mathcal{B}(\Omega)$ and $v \in L^p(\Omega; \mathbb{R}^m)$.

Note that for every fixed $z \in \mathbb{R}^m$, taking $v(x) :\equiv z$ and using the fact that $\mathcal{E}_p \leq I$, we obtain that

$$h(z) |\Omega| \leq f(z) |\Omega|,$$

which, since $|\Omega| < \infty$, implies that $h \leq f$.

Suppose now that $v \in L^p(E; \mathbb{R}^m)$. Consider the extension

$$\overline{v}(x) := \begin{cases} v(x) & \text{if } x \in E, \\ v_0(x) & \text{if } x \in \Omega \setminus E. \end{cases}$$

Then

$$\mathcal{E}_p(v) = \mathcal{E}_p(\overline{v}; E) = \int_E h(v) \, dx.$$

Since $\mathcal{E}_p$ is sequentially lower semicontinuous and below I and since $h \geq 0$, by Theorems 5.9 and 5.14 we conclude that h must be lower semicontinuous, convex, and since $h \leq f$ then $h \leq f^{**}$.

This completes the proof of (5.68).

To prove (5.69) fix $v \in L^p(E; \mathbb{R}^m)$ and $\varepsilon > 0$. By the definition of $\mathcal{I}_p(v)$ we may find $\{v_n\} \subset L^p(E; \mathbb{R}^m)$, $v_n \rightharpoonup v$ ($\overset{*}{\rightharpoonup}$ if $p = \infty$) in $L^p(E; \mathbb{R}^m)$ such that

$$\lim_{n \to \infty} I\left(v_n; E\right) \leq \mathcal{I}_p(v) + \varepsilon. \qquad (5.72)$$

If $p = 1$, then by Theorem 2.29 let $\gamma : [0, \infty) \to [0, \infty)$ be an increasing continuous function with

$$\lim_{t \to \infty} \frac{\gamma(t)}{t} = \infty$$

and such that

$$\sup_{n} \int_E \gamma\left(|v_n|\right) \, dx =: c < \infty. \qquad (5.73)$$

If $1 < p < \infty$ (respectively $p = \infty$), then (5.73) continues to hold, provided we define γ as in (5.70) (respectively in (5.71) for an appropriate $R > 0$).

Although the sequence $\{v_n\}$, and hence γ, depends on ε, for simplicity of notation we do not indicate this dependence. For $k \in \mathbb{N}$ and all $z \in \mathbb{R}^m$ define

$$f_k(z) := f(z) + \frac{1}{k}\gamma\left(|z|\right),$$

with corresponding energy $I^{(k)}$ and envelopes $\mathcal{E}_p^{(k)}$ and $\mathcal{I}_p^{(k)}$. By Remark 5.33(i) and (5.68) we have

$$\mathcal{E}_p^{(k)}(v) = \mathcal{I}_p^{(k)}(v) = \int_E \left(f_k\right)^{**}(v) \, dx.$$

Since the sequence $\{f_k\}$ is decreasing, the same is true for $\left\{(f_k)^{**}\right\}$. By Proposition 4.100,

$$\mathcal{C}\left(\operatorname{lsc} f\right) = \inf_{k \in \mathbb{N}} \left(f_k\right)^{**}.$$

We claim that

$$\mathcal{I}_p(v) = \inf_{k \in \mathbb{N}} \mathcal{I}_p^{(k)}(v).$$

Since $f \leq f_k$ we have $\mathcal{I}_p(v) \leq \inf_{k \in \mathbb{N}} \mathcal{I}_p^{(k)}(v)$. On the other hand, by (5.73) and (5.72),

$$\mathcal{I}_p^{(k)}(v) \leq \lim_{n \to \infty} I_k\left(v_n; E\right) \leq \lim_{n \to \infty} I\left(v_n; E\right) + \frac{c}{k} \leq \mathcal{I}_p(v) + \varepsilon + \frac{c}{k}.$$

Hence

$$\mathcal{I}_p(v) \leq \inf_{k \in \mathbb{N}} \mathcal{I}_p^{(k)}(v) \leq \mathcal{I}_p(v) + \varepsilon,$$

and given the arbitrariness of ε, the claim is proved.

If now $\mathcal{I}_p(v) < \infty$, then by the Lebesgue monotone convergence theorem,

$$\mathcal{I}_p(v) = \inf_{k \in \mathbb{N}} \mathcal{I}_p^{(k)}(v) = \lim_{k \to \infty} \mathcal{I}_p^{(k)}(v) = \lim_{k \to \infty} \int_E \left(f_k\right)^{**}(v) \, dx$$

$$= \int_E \mathcal{C}\left(\operatorname{lsc} f\right)(v) \, dx.$$

Thus (5.69) holds.

Remark 5.34. In the case that E has infinite measure the previous theorem still holds provided the constant a in (5.67) is zero, and we assume that there exists $v_0 \in L^1(E; \mathbb{R}^m)$ such that

$$\int_E f(v_0) \, dx < \infty.$$

When $\mathcal{I}_\infty(v) = \infty$ the inequality $\mathcal{I}_\infty(v) = \int_E \mathcal{C}(\operatorname{lsc} f)(v) \, dx$ may fail.

Exercise 5.35. Let $N = m = 2$ and consider the function

$$f(z) = f(z_1, z_2) = \begin{cases} \infty & \text{if } z_1 \leq 0, \\ \frac{1}{z_1} - e^{z_2^2} & \text{if } 0 < z_1 \leq e^{-z_2^2}, \\ 0 & \text{if } z_1 > e^{-z_2^2}. \end{cases}$$

Prove that

$$\mathcal{C}(\operatorname{lsc} f)(z_1, z_2) = \begin{cases} \infty & \text{if } z_1 \leq 0, \\ 0 & \text{if } z_1 > 0, \end{cases}$$

while for any integer $k \in \mathbb{N}$,

$$(f_k)^{**}(z_1, z_2) = \begin{cases} \infty & \text{if } z_1 \leq 0 \text{ or } z_1 \geq k \text{ or } |z_2| \geq k, \\ \frac{1}{z_1} - e^{k^2} & \text{if } 0 < z_1 \leq e^{-k^2} \text{ and } |z_2| < k, \\ 0 & \text{if } e^{-k^2} < z_1 < k \text{ and } |z_2| < k. \end{cases}$$

Let $E := (0,1) \times (-1,1)$ and $v(x,y) := (x,0)$. Prove that

$$\mathcal{I}_\infty(v) \neq \int_E \mathcal{C}(\operatorname{lsc} f)(v) \, dx.$$

5.4.2 Weak Star Convergence in the Sense of Measures

In this subsection we consider functionals of the form

$$v \in L^1(\Omega; \mathbb{R}^m) \mapsto I(v) := \int_\Omega f(v) \, dx,$$

where Ω is an open subset of $\mathbb{R}^N$ and $f : \mathbb{R}^m \to (-\infty, \infty]$ is a Borel function, $f \not\equiv \infty$. For any $\lambda \in \mathcal{M}(\Omega; \mathbb{R}^m)$ we define the functional

$$H(\lambda) := \begin{cases} I(v) & \text{if } \lambda = v \, \mathcal{L}^N \lfloor \Omega, \\ \infty & \text{otherwise,} \end{cases}$$

and we characterize the *relaxed energy*

$$\mathcal{H} : \mathcal{M}(\Omega; \mathbb{R}^m) \to [-\infty, \infty]$$

of H with respect to weak star convergence in $\mathcal{M}(\Omega; \mathbb{R}^m)$, that is, the functional $\mathcal{H}$ is the greatest functional below H that is sequentially lower semicontinuous with respect to weak star convergence in $\mathcal{M}(\Omega; \mathbb{R}^m)$.

For every $\lambda \in \mathcal{M}(\Omega; \mathbb{R}^m)$ define

$$\mathcal{I}(\lambda) := \inf \left\{ \liminf_{n \to \infty} I(v_n) : \{v_n\} \subset L^1(\Omega; \mathbb{R}^m), \right.$$
$$\left. v_n \, \mathcal{L}^N \lfloor \Omega \overset{*}{\rightharpoonup} \lambda \text{ in } \mathcal{M}(\Omega; \mathbb{R}^m) \right\}.$$

Since $\mathcal{H} \leq H$, for every $\lambda \in \mathcal{M}(\Omega; \mathbb{R}^m)$ we have

$$\mathcal{H}(\lambda) \leq \mathcal{I}(\lambda).$$

Theorem 5.36. *Let Ω be an open subset of $\mathbb{R}^N$, and let $f : \mathbb{R}^m \to [0, \infty]$ be a Borel function. Assume that there exists $v_0 \in L^1(\Omega; \mathbb{R}^m)$ such that*

$$\int_\Omega f(v_0) \, dx < \infty.$$

Then for every $\lambda \in \mathcal{M}(\Omega; \mathbb{R}^m)$ we have

$$\mathcal{H}(\lambda) = \int_\Omega f^{**}\left(\frac{d\lambda}{d\mathcal{L}^N}\right) dx + \int_\Omega (f^{**})^\infty \left(\frac{d\lambda}{d\|\lambda_s\|}\right) d\|\lambda_s\|. \qquad (5.74)$$

Moreover, for every $\lambda \in \mathcal{M}(\Omega; \mathbb{R}^m)$ for which $\mathcal{I}(\lambda) < \infty$ we have

$$\mathcal{I}(\lambda) = \int_\Omega g\left(\frac{d\lambda}{d\mathcal{L}^N}\right) dx + \int_\Omega h\left(\frac{d\lambda}{d\|\lambda_s\|}\right) d\|\lambda_s\|, \qquad (5.75)$$

where

$$g = \inf_k \left(f + \frac{1}{k}|\cdot|\right)^{**}, \quad h = \inf_k \left(\left(f + \frac{1}{k}|\cdot|\right)^{**}\right)^\infty.$$

Remark 5.37. We recall that by Proposition 3.16 we have

$$\mathcal{I}(\lambda) = \mathcal{H}(\lambda) = \int_\Omega f^{**}\left(\frac{d\lambda}{d\mathcal{L}^N}\right) dx + \int_\Omega (f^{**})^\infty \left(\frac{d\lambda}{d\|\lambda_s\|}\right) d\|\lambda_s\|$$

for every $\lambda \in \mathcal{M}(\Omega; \mathbb{R}^m)$ provided f satisfies the coercivity condition

$$f(z) \geq C |z| \text{ for all } z \in \mathbb{R}^m,$$

where $C > 0$.

We are now ready to prove Theorem 5.36.

Proof. Since f^{**} is convex, lower semicontinuous, and nonnegative, and $f^{**} \leq f$, by Theorems 5.19 and 5.25 we deduce that for every $\lambda \in \mathcal{M}(\Omega; \mathbb{R}^m)$,

$$\mathcal{H}(\lambda) \geq \int_\Omega f^{**}\left(\frac{d\lambda}{d\mathcal{L}^N}\right) dx + \int_\Omega (f^{**})^\infty \left(\frac{d\lambda}{d\|\lambda_s\|}\right) d\|\lambda_s\|. \qquad (5.76)$$

To prove the reverse inequality it suffices to consider the case that the right-hand side in (5.76) is finite. We start by showing that if $\lambda = v \, \mathcal{L}^N \lfloor \Omega$ for some $L^1(\Omega; \mathbb{R}^m)$, then

$$\mathcal{H}(\lambda) = \int_\Omega f^{**}(v) \, dx.$$

Indeed, setting $\mathcal{E}_1$ to be the greatest functional below H that is sequentially lower semicontinuous with respect to L^1 weak convergence, then by Theorem 5.32 it follows that

$$\mathcal{H}(\lambda) \leq \mathcal{E}_1(v) = \int_\Omega f^{**}(v) \, dx,$$

and equality follows by virtue of (5.76).

Consider now an arbitrary $\lambda \in \mathcal{M}(\Omega; \mathbb{R}^m)$ and a family of standard mollifiers $\{\varphi_\varepsilon\}$. Fix $\delta > 0$, let

$$\Omega_\delta := \{x \in \Omega : \operatorname{dist}(x, \partial\Omega) > \delta\},$$

and define

$$\lambda_{\varepsilon,\delta} := v_{\varepsilon,\delta} \, \mathcal{L}^N \lfloor \Omega,$$

where

$$v_{\varepsilon,\delta} := \begin{cases} \varphi_\varepsilon * \lambda & \text{in } \Omega_\delta, \\ v_0 & \text{on } \Omega \setminus \Omega_\delta. \end{cases}$$

By Remark 2.80 it can be shown that for any $\psi \in C_0(\Omega)$,

$$\lim_{\delta \to 0^+} \lim_{\varepsilon \to 0^+} \int_\Omega \psi \, d\lambda_{\varepsilon,\delta} = \int_\Omega \psi \, d\lambda.$$

Moreover, since $f^{**} \circ \frac{d\lambda}{d\mathcal{L}^N} \in L^1(\Omega)$, it follows that

$$\lim_{\delta \to 0^+} \lim_{\varepsilon \to 0^+} \int_{\Omega_\delta} \varphi_\varepsilon * f^{**}\left(\frac{d\lambda}{d\mathcal{L}^N}\right) dx = \int_\Omega f^{**}\left(\frac{d\lambda}{d\mathcal{L}^N}\right) dx.$$

Using a diagonalization argument we may find a sequence $\varepsilon_n \to 0$ such that

$$\lambda_n \overset{*}{\rightharpoonup} \lambda \text{ in } \mathcal{M}(\Omega; \mathbb{R}^m),$$

$$\lim_{n \to \infty} \int_{\Omega_{\delta_n}} \varphi_{\varepsilon_n} * f^{**}\left(\frac{d\lambda}{d\mathcal{L}^N}\right) dx = \int_\Omega f^{**}\left(\frac{d\lambda}{d\mathcal{L}^N}\right) dx, \qquad (5.77)$$

where $\delta_n := \frac{1}{n}$ and $\lambda_n := \lambda_{\varepsilon_n, \delta_n}$. Hence, with $v_n := v_{\varepsilon_n, \delta_n}$, we have

$$\mathcal{H}(\lambda; \Omega) \leq \liminf_{n \to \infty} \mathcal{H}(\lambda_n; \Omega) = \liminf_{n \to \infty} \int_\Omega f^{**}(v_n) \, dx$$

$$\leq \limsup_{n \to \infty} \int_{\Omega \setminus \Omega_{\delta_n}} f^{**}(v_0) \, dx + \liminf_{n \to \infty} \int_{\Omega_{\delta_n}} f^{**}(\varphi_{\varepsilon_n} * \lambda) \, dx$$

$$\leq \limsup_{n \to \infty} \int_{\Omega_{\delta_n}} f^{**}\left(\varphi_{\varepsilon_n} * \frac{d\lambda}{d\mathcal{L}^N}\right) dx$$

$$+ \liminf_{n \to \infty} \int_{\Omega_{\delta_n}} (f^{**})^\infty (\varphi_{\varepsilon_n} * \lambda_s) \, dx,$$

where in the last inequality we have used the facts that $f^{**} \circ v_0 \in L^1(\Omega)$ and (see, e.g., (4.33) with $t = 1$)

$$f^{**}(a + b) \leq f^{**}(a) + (f^{**})^{\infty}(b).$$

By Jensen's inequality,

$$\int_{\Omega_{\delta_n}} f^{**} \left(\varphi_{\varepsilon_n} * \frac{d\lambda}{d\mathcal{L}^N} \right) dx = \int_{\Omega_{\delta_n}} f^{**} \left(\int_{\Omega} \varphi_{\varepsilon_n}(x - y) \frac{d\lambda}{d\mathcal{L}^N}(y) \, dy \right) dx$$

$$\leq \int_{\Omega_{\delta_n}} \int_{\Omega} \varphi_{\varepsilon_n}(x - y) f^{**} \left(\frac{d\lambda}{d\mathcal{L}^N}(y) \right) dy dx$$

$$= \int_{\Omega_{\delta_n}} \varphi_{\varepsilon_n} * f^{**} \left(\frac{d\lambda}{d\mathcal{L}^N} \right) dx,$$

and by (5.77) we deduce that

$$\limsup_{n \to \infty} \int_{\Omega_{\delta_n}} f^{**} \left(\varphi_{\varepsilon_n} * \frac{d\lambda}{d\mathcal{L}^N} \right) dx \leq \int_{\Omega} f^{**} \left(\frac{d\lambda}{d\mathcal{L}^N} \right) dx.$$

Furthermore,

$$\int_{\Omega_{\delta_n}} (f^{**})^{\infty} (\varphi_{\varepsilon_n} * \lambda_s) \, dx = \int_{\Omega_{\delta_n}} (f^{**})^{\infty} \left(\int_{\Omega} \varphi_{\varepsilon_n}(x - y) \, d\lambda_s(y) \right) dx$$

$$= \int_{\Omega_{\delta_n}} (f^{**})^{\infty} \left(\int_{\Omega} \varphi_{\varepsilon_n}(x - y) \|\lambda_s\|(\Omega) \frac{d\lambda}{d\|\lambda_s\|}(y) \, d\frac{\|\lambda_s\|}{\|\lambda_s\|(\Omega)}(y) \right) dx$$

$$\leq \int_{\Omega_{\delta_n}} \int_{\Omega} \varphi_{\varepsilon_n}(x - y) (f^{**})^{\infty} \left(\frac{d\lambda}{d\|\lambda_s\|}(y) \right) d\|\lambda_s\|(y) \, dx,$$

where we used Jensen's inequality and the homogeneity of $(f^{**})^{\infty}$. By Fubini's theorem and using the fact that $\int_{\mathbb{R}^N} \varphi_{\varepsilon_n} \, dx = 1$ we deduce that

$$\int_{\Omega_{\delta_n}} (f^{**})^{\infty} (\varphi_{\varepsilon_n} * \lambda_s) \, dx \leq \int_{\Omega} (f^{**})^{\infty} \left(\frac{d\lambda}{d\|\lambda_s\|}(y) \right) d\|\lambda_s\|(y).$$

We have proved that

$$\mathcal{H}(\lambda) \leq \int_{\Omega} f^{**} \left(\frac{d\lambda}{d\mathcal{L}^N} \right) dx + \int_{\Omega} (f^{**})^{\infty} \left(\frac{d\lambda}{d\|\lambda_s\|} \right) d\|\lambda_s\|.$$

To prove (5.75), for $k \in \mathbb{N}$ and all $z \in \mathbb{R}^m$ define

$$f_k(z) := f(z) + \frac{1}{k} |z|$$

with corresponding energy $I^{(k)}$ and envelopes $\mathcal{I}^{(k)}$ and $\mathcal{H}^{(k)}$. By Remark 5.37(i) and (5.74) we have

$$\mathcal{I}^{(k)}(\lambda) = \mathcal{H}^{(k)}(\lambda) = \int_{\Omega} f_k^{**}\left(\frac{d\lambda}{d\mathcal{L}^N}\right) dx + \int_{\Omega} (f_k^{**})^{\infty}\left(\frac{d\lambda}{d\|\lambda_s\|}\right) d\|\lambda_s\|.$$

Since the sequence $\{f_k\}$ is decreasing, the same is true for $\{(f_k)^{**}\}$. We claim that

$$\mathcal{I}(\lambda) = \inf_{k \in \mathbb{N}} \mathcal{I}^{(k)}(\lambda).$$

Since $f \le f_k$ we have $\mathcal{I}(\lambda) \le \inf_{k \in \mathbb{N}} \mathcal{I}^{(k)}(\lambda)$. On the other hand, by the definition of $\mathcal{I}(\lambda)$ we may find $\{v_n\} \subset L^1(\Omega; \mathbb{R}^m)$, $v_n \mathcal{L}^N \lfloor \Omega \overset{*}{\rightharpoonup} \lambda$ in $\mathcal{M}(\Omega; \mathbb{R}^m)$, such that

$$\lim_{n \to \infty} I(v_n) \le \mathcal{I}(\lambda) + \varepsilon.$$

Without loss of generality we may assume that

$$\sup_n \int_{\Omega} |v_n|\, dx =: c < \infty,$$

and so

$$\mathcal{I}^{(k)}(\lambda) \le \lim_{n \to \infty} I_k(v_n) \le \lim_{n \to \infty} I(v_n) + \frac{c}{k} \le \mathcal{I}(\lambda) + \varepsilon + \frac{c}{k}.$$

Hence

$$\mathcal{I}(\lambda) \le \inf_{k \in \mathbb{N}} \mathcal{I}^{(k)}(\lambda) \le \mathcal{I}(\lambda) + \varepsilon,$$

and given the arbitrariness of ε, the claim is established.

If now $\mathcal{I}(\lambda) < \infty$, then by the Lebesgue monotone convergence theorem,

$$\mathcal{I}(\lambda) = \inf_{k \in \mathbb{N}} \mathcal{I}^{(k)}(\lambda) = \lim_{k \to \infty} \mathcal{I}^{(k)}(\lambda)$$

$$= \lim_{k \to \infty} \left(\int_{\Omega} f_k^{**}\left(\frac{d\lambda}{d\mathcal{L}^N}\right) dx + \int_{\Omega} (f_k^{**})^{\infty}\left(\frac{d\lambda}{d\|\lambda_s\|}\right) d\|\lambda_s\|\right)$$

$$= \int_{\Omega} g\left(\frac{d\lambda}{d\mathcal{L}^N}\right) dx + \int_{\Omega} h\left(\frac{d\lambda}{d\|\lambda_s\|}\right) d\|\lambda_s\|.$$

This proves (5.75).

5.5 Minimization

We conclude this chapter with a minimization problem. We begin with the case that f is real-valued.

Theorem 5.38. *Let E be a Borel subset of $\mathbb{R}^N$ with finite measure, let $1 \le p \le \infty$, and let $f : \mathbb{R}^m \to \mathbb{R}$ be a Borel function bounded from below by an affine function. If $z_0 \in \mathbb{R}^m$, then*

$$\inf\left\{\int_E f(v)\, dx : v \in L^p(E; \mathbb{R}^m),\, \frac{1}{|E|}\int_E v\, dx = z_0\right\} = f^{**}(z_0)\,|E|,$$

and the infimum is attained if and only if

$$z_0 \in \mathrm{co} M_{z_0},$$

where

$$M_{z_0} := \{z \in \mathbb{R}^m : f(z) = f^{**}(z_0) + \beta \cdot (z - z_0) \text{ for all } \beta \in \partial f^{**}(z_0)\}.$$

Lemma 5.39. *Let E be a Borel subset of $\mathbb{R}^N$, let μ be a (positive) finite Radon measure on E, and let $v \in L^1((E, \mathcal{B}(E), \mu); \mathbb{R}^m)$. Then*

$$\frac{1}{\mu(E)} \int_E v \, d\mu \in \mathrm{co}\{v(x) : x \in E, x \text{ is a Lebesgue point of } v\}.$$

Proof. The proof is by induction on m. Let

$$z_0 := \frac{1}{\mu(E)} \int_E v \, d\mu, \quad G := \{v(x) : x \in E, x \text{ a Lebesgue point of } v\}.$$

For $m = 1$ it is not difficult to show that $\mathrm{co}\, G$ is the (possibly infinite) interval of endpoints $\mathrm{essinf}_E v$ and $\mathrm{esssup}_E v$. If $z_0 \notin \mathrm{co}\, G$, then either $z_0 \geq \mathrm{esssup}_E v$ or $z_0 \leq \mathrm{essinf}_E v$. Assume that $z_0 \geq \mathrm{esssup}_E v$. Then $z_0 - v(x) \geq 0$ for μ a.e. $x \in E$, and so since

$$\frac{1}{\mu(E)} \int_E (z_0 - v(x)) \, d\mu(x) = 0,$$

we deduce that $v(x) = z_0$ for μ a.e. $x \in E$. In turn, $G = \{z_0\}$, which contradicts the fact that $z_0 \notin \mathrm{co}\, G$. The case $z_0 \leq \mathrm{essinf}_E v$ is treated in an analogous way.

Assume that the result is true for functions with values in $\mathbb{R}^{m-1}$ and let $v \in L^1((E, \mathcal{B}(E), \mu); \mathbb{R}^m)$. If $z_0 \notin \mathrm{co}\, G$, then by Theorem 4.19 (with $C_1 = \{z_0\}$ and $C_2 = \mathrm{co}\, G$) we may find a half-space

$$H = \{z \in \mathbb{R}^m : b \cdot (z - z_0) \geq 0\}$$

through z_0 containing $\mathrm{co}\, G$, where $b \in \mathbb{R}^m$, $b \neq 0$. Then from the definition of z_0 and G and since $G \subset H$,

$$0 = \int_E b \cdot (v(x) - z_0) \, d\mu(x) = \int_{\{y \in E: \, b \cdot (v(y) - z_0) \geq 0\}} b \cdot (v(x) - z_0) \, d\mu(x)$$

$$+ \int_{\{y \in E: \, b \cdot (v(y) - z_0) < 0\}} b \cdot (v(x) - z_0) \, d\mu(x).$$

Hence

$$\int_E |b \cdot (v(x) - z_0)| \, d\mu(x) = 2 \int_{\{y \in E: \, b \cdot (v(y) - z_0) < 0\}} |b \cdot (v(x) - z_0)| \, d\mu(x) = 0,$$

since $v(x) \in H$ for μ a.e. $x \in E$. This implies that

$$v(x) \in \{z \in \mathbb{R}^m : b \cdot (z - z_0) = 0\}$$

for μ a.e. $x \in E$, and thus the function v takes values on an $(m-1)$-dimensional hyperplane. By the induction hypothesis we have that $z_0 \in \operatorname{co} G$, which is a contradiction.

Proof (Theorem 5.38). We begin by observing that by Theorem 4.96,

$$\inf \left\{ \int_E f(s) \, dx : s \in L^p(E; \mathbb{R}^m), \, s \text{ simple}, \, \frac{1}{|E|} \int_E s \, dx = z_0 \right\}$$

$$= \inf \left\{ \sum_{i=1}^n \theta_i f(z_i) : n \in \mathbb{N}, \, \theta_i \in [0,1], \, z_i \in \mathbb{R}^m, \, i = 1, \ldots, n, \right.$$

$$\left. \sum_{i=1}^n \theta_i = 1, \, \sum_{i=1}^n \theta_i z_i = z_0 \right\} = Cf(z_0) |E|,$$

and so

$$\inf \left\{ \int_E f(v) \, dx : v \in L^p(E; \mathbb{R}^m), \, \frac{1}{|E|} \int_E v \, dx = z_0 \right\} \leq Cf(z_0) |E|.$$

By Remark 4.93 and the fact that f is real-valued, the right-hand side of the previous inequality coincides with $f^{**}(z_0) |E|$.

Conversely, by Jensen's inequality,

$$\int_E f(v) \, dx \geq \int_E f^{**}(v) \, dx \geq f^{**}(z_0) |E|, \qquad (5.78)$$

and so

$$\inf \left\{ \int_E f(v) \, dx : v \in L^p(E; \mathbb{R}^m), \, \frac{1}{|E|} \int_E v \, dx = z_0 \right\} \geq f^{**}(z_0) |E|.$$

Suppose now that the infimum is attained at some function $v \in L^p(E; \mathbb{R}^m)$ with

$$\frac{1}{|E|} \int_E v(x) \, dx = z_0. \qquad (5.79)$$

By (5.78) we have

$$\int_E f(v) \, dx = \int_E f^{**}(v) \, dx = f^{**}(z_0) |E|, \qquad (5.80)$$

and so $f(v(x)) = f^{**}(v(x))$ for $\mathcal{L}^N$ a.e. $x \in E$.

Therefore, given any $\beta \in \partial f^{**}(z_0)$ (recall Theorem 4.53), it follows that

$$f(v(x)) = f^{**}(v(x)) \geq f^{**}(z_0) + \beta \cdot (v(x) - z_0),$$

and the inequality must be an equality for $\mathcal{L}^N$ a.e. $x \in E$ or else, in view of (5.79), (5.80) would be violated. We deduce that

$$v(x) \in \{z \in \mathbb{R}^m : f(z) = f^{**}(z) = f^{**}(z_0) + \beta \cdot (z - z_0)\}$$

for $\mathcal{L}^N$ a.e. $x \in E$, which, together with (5.79) and Lemma 5.39, yields

$$z_0 \in \mathrm{co}\{z \in \mathbb{R}^m : f(z) = f^{**}(z) = f^{**}(z_0) + \beta \cdot (z - z_0)\}.$$

Conversely, assume that $z_0 \in \mathrm{co}\, M_{z_0}$ and write

$$z_0 = \sum_{i=1}^{m+1} \theta_i z_i, \tag{5.81}$$

where $\theta_i \in [0, 1]$, $z_i \in M_{z_0}$, $i = 1, \ldots, m+1$, and $\sum_{i=1}^{m+1} \theta_i = 1$. By Corollary 1.21 we may find a partition of E into measurable subsets E_i such that

$$|E_i| = \theta_i |E|,$$

$i = 1, \ldots, m+1$, and define

$$v := \sum_{i=1}^{m+1} \chi_{E_i} z_i.$$

By (5.81), v is admissible, and for a fixed $\beta \in \partial f^{**}(z_0)$ we have

$$\int_E f(v)\, dx = \sum_{i=1}^{m+1} |E_i| f(z_i) = \sum_{i=1}^{m+1} |E_i| (f^{**}(z_0) + \beta \cdot (z_i - z_0))$$
$$= |E| f^{**}(z_0),$$

where we have used (5.81) and the fact that $z_i \in M_{z_0}$.

Next we consider the case in which f takes the value ∞.

Theorem 5.40. *Let E be a Borel subset of $\mathbb{R}^N$ with finite measure, let $1 \leq p \leq \infty$, and let $f : \mathbb{R}^m \to (-\infty, \infty]$ be a Borel function bounded from below by an affine function. If $z_0 \in \mathbb{R}^m$, then*

$$\inf\left\{\int_E f(v)\, dx : v \in L^p(E; \mathbb{R}^m),\, \frac{1}{|E|}\int_E v\, dx = z_0\right\} = \mathcal{C}f(z_0)|E|. \tag{5.82}$$

Moreover, if $f(z_0) > \mathcal{C}f(z_0)$, then the infimum is attained if and only if there exists an affine space H of dimension ℓ, $\ell = 1, \ldots, m$, such that

$$z_0 \in \mathrm{co}N_{z_0},$$

where

$$N_{z_0} := \{z \in H : f(z) = \mathcal{C}f(z_0) + \beta \cdot (z - z_0) \text{ for all } \beta \in \partial(f|_H)^{**}(z_0)\}.$$

Proof. **Step 1:** As in the previous proof we have that

$$\inf\left\{\int_E f(v)\,dx : v \in L^p\left(E;\mathbb{R}^m\right),\ \frac{1}{|E|}\int_E v\,dx = z_0\right\} \le \mathcal{C}f\left(z_0\right)|E|.$$

In particular, if for *every* $v \in L^p\left(E;\mathbb{R}^m\right)$ with average z_0 we have

$$\int_E f(v)\,dx = \infty,$$

then the previous inequality implies that $\mathcal{C}f\left(z_0\right) = \infty$ and (5.82) holds. Hence in what follows we always take for granted the existence of admissible functions $v \in L^p\left(E;\mathbb{R}^m\right)$ with finite energy.

Step 2: Assume that $z_0 \in \mathrm{ri}_{\mathrm{aff}}\left(\mathrm{co}\left(\mathrm{dom}_e f\right)\right)$. In this case we take $H := \mathbb{R}^m$. By Corollary 4.97,

$$f^{**}\left(z_0\right) = \mathcal{C}f\left(z_0\right) = \inf\left\{\sum_{i=1}^{m+1}\theta_i f\left(z_i\right) : \theta_i \in [0,1],\ z_i \in \mathbb{R}^m,\ i = 1,\ldots,m+1,\right.$$
$$\left.\sum_{i=1}^{m+1}\theta_i = 1,\ \sum_{i=1}^{m+1}\theta_i z_i = z_0\right\},$$

and so we can proceed as in the first part of the proof of the previous theorem to conclude that (5.82) holds.

Moreover, by Theorem 4.53 we have that f^{**} is subdifferentiable at z_0, and so, as in the second part of proof of the previous theorem, we can show that the infimum is attained if and only if $z_0 \in \mathrm{co}\, N_{z_0}$.

Step 3: Assume that $z_0 \in \overline{\mathrm{co}\left(\mathrm{dom}_e f\right)} \setminus \mathrm{ri}_{\mathrm{aff}}\left(\mathrm{co}\left(\mathrm{dom}_e f\right)\right)$, and by Corollary 4.20 find a half-space

$$H^+ = \{z \in \mathbb{R}^m : b \cdot (z - z_0) \ge 0\}$$

through z_0 containing $\overline{\mathrm{co}\left(\mathrm{dom}_e f\right)}$, $b \neq 0$. Note that $z_0 \in H := \partial H^+$.

Given any admissible test function $v \in L^p\left(E;\mathbb{R}^m\right)$ such that

$$\int_E f(v)\,dx < \infty,$$

we have that $v(x) \in \mathrm{dom}_e f$ for $\mathcal{L}^N$ a.e. $x \in E$, and so, reasoning as in the proof of Lemma 5.39, we conclude that $v(x) \in H$ for $\mathcal{L}^N$ a.e. $x \in E$. With an abuse of notation we may identify H with $\mathbb{R}^{m-1}$ and denote by $f_1 : \mathbb{R}^{m-1} \to (-\infty,\infty]$ the restriction of f to H. Then

$$\inf\left\{\int_E f(v)\,dx : v \in L^p\left(E;\mathbb{R}^m\right),\ \frac{1}{|E|}\int_E v\,dx = z_0\right\} \tag{5.83}$$
$$= \inf\left\{\int_E f_1(v)\,dx : v \in L^p\left(E;\mathbb{R}^{m-1}\right),\ \frac{1}{|E|}\int_E v\,dx = z_0\right\}.$$

If $z_0 \in \mathrm{ri_{aff}}\left(\mathrm{co}\left(\mathrm{dom}_e\, f_1\right)\right)$, then applying the previous step to f_1 yields

$$\inf\left\{\int_E f_1(v)\,dx : v \in L^p\left(E; \mathbb{R}^{m-1}\right),\ \frac{1}{|E|}\int_E v\,dx = z_0\right\} = \mathcal{C}f_1(z_0)\,|E|\,,$$
(5.84)

and the infimum is realized if and only if

$$z_0 \in \mathrm{co}\left\{z \in \mathbb{R}^{m-1} : f_1(z) = f_1^{**}(z_0) + \beta\cdot(z - z_0) \text{ for all } \beta \in \partial f_1^{**}(z_0)\right\}.$$

Since f_1 is the restriction of f to H, by Theorem 4.96 we have $\mathcal{C}f(z_0) \le \mathcal{C}f_1(z_0)$, and so by Step 1, (5.83), and (5.84) we conclude that

$$\mathcal{C}f(z_0) = \mathcal{C}f_1(z_0)\,,$$

and in turn (5.82) follows.

If $z_0 \in \overline{\mathrm{co}\left(\mathrm{dom}_e\, f_1\right)} \setminus \mathrm{ri_{aff}}\left(\mathrm{co}\left(\mathrm{dom}_e\, f_1\right)\right)$, then we repeat the first part of this step to lower the dimension to $\mathbb{R}^{m-2}$.

Continuing in this way we either prove (5.82) or reduce to dimension $m = 1$, that is,

$$\inf\left\{\int_E f(v)\,dx : v \in L^p\left(E; \mathbb{R}^m\right),\ \frac{1}{|E|}\int_E v\,dx = z_0\right\}$$
$$= \inf\left\{\int_E f_{m-1}(v)\,dx : v \in L^p\left(E; \mathbb{R}\right),\ \frac{1}{|E|}\int_E v\,dx = z_0\right\}.$$

Note that the convex set $\mathrm{co}\left(\mathrm{dom}_e\, f_{m-1}\right) \subset \mathbb{R}$ is an interval I. If z_0 belongs to the interior of I, then we use Step 2 to obtain (5.82), while if z_0 is one of the endpoints of I, then as in the first part of this step we conclude that the only admissible function $v \in L^p\left(E; \mathbb{R}\right)$ such that

$$\int_E f_{m-1}(v)\,dx < \infty$$

is the function $v(x) \equiv z_0$, and so

$$\inf\left\{\int_E f_{m-1}(v)\,dx : v \in L^p\left(E; \mathbb{R}\right),\ \frac{1}{|E|}\int_E v\,dx = z_0\right\} = f(z_0)\,|E|\,,$$

and (5.82) follows once again by Step 1.

Step 4: If $z_0 \notin \overline{\mathrm{co}\left(\mathrm{dom}_e\, f\right)}$, then by Corollary 4.97 it follows that $\mathcal{C}f(z_0) = f^{**}(z_0) = \infty$, so that by (5.78),

$$\inf\left\{\int_E f(v)\,dx : v \in L^p\left(E; \mathbb{R}^m\right),\ \frac{1}{|E|}\int_E v\,dx = z_0\right\} = \infty,$$

and (5.82) follows.

6

Integrands $f = f(x, z)$

In this chapter we study well-posedness, lower semicontinuity, and relaxation for functionals

$$v \in L^p(E; \mathbb{R}^m) \mapsto \int_E f(x, v(x))\, dx.$$

Unless otherwise specified, in what follows, $Cf(x, \cdot)$, $f^*(x, \cdot)$, $f^{**}(x, \cdot)$, and lsc $f(x, \cdot)$ stand for the convex envelope, polar, bipolar, and lower semicontinuous envelope, respectively, of $f(x, \cdot)$.

Several results in the following chapters rely heavily on multifunctions. Thus we begin with a preliminary section that provides an overview of their definition and main properties.

6.1 Multifunctions

6.1.1 Measurable Selections

In order to introduce the concept of multifunction we exhibit a simple example in which a measurable selection criterion is needed.

Consider a bounded continuous function $f : \mathbb{R}^d \times K \to \mathbb{R}$, with $K \subset \mathbb{R}^m$ compact, and let $u : \Omega \subset \mathbb{R}^N \to \mathbb{R}^d$ be a measurable function, where Ω is open. Define

$$\bar{f}(x) := \min_{y \in K} f(u(x), y) \quad x \in \Omega.$$

Aumann's measurable selection theorem (that relies on the notion of multifunctions) will entail the existence of a measurable minimizing function $\bar{y} : \Omega \to K$ such that

$$\bar{f}(x) = f(u(x), \bar{y}(x))$$

for every $x \in \Omega$.

Definition 6.1. *Given two nonempty sets* X, Y, *a* multifunction *or* correspondence *from* X *to* Y *is a map from* X *to the family of nonempty subsets of* Y, *namely*

$$\Gamma : X \to \mathcal{P}(Y) \setminus \{\emptyset\}.$$

The graph *of a multifunction* Γ *is the set*

$$\mathrm{Gr}\,\Gamma := \{(x, y) \in X \times Y : y \in \Gamma(x)\}.$$

A function $u : X \to Y$ *is said to be a* selection *of the multifunction* Γ *if* $u(x) \in \Gamma(x)$ *for every* $x \in X$.

We now study measurability properties of multifunctions.

Definition 6.2. *Let* $(X, \mathfrak{M})$ *be a measurable space, let* Y *be a topological space, and let* $\Gamma : X \to \mathcal{P}(Y) \setminus \{\emptyset\}$ *be a multifunction.*

(i) Γ *is* measurable *if for every closed set* $C \subset Y$ *the set*

$$\Gamma^-(C) := \{x \in X : \Gamma(x) \cap C \neq \emptyset\}$$

belongs to $\mathfrak{M}$;

(ii) Γ *is* weakly measurable *if for every open set* $A \subset Y$ *the set*

$$\Gamma^-(A) := \{x \in X : \Gamma(x) \cap A \neq \emptyset\}$$

belongs to $\mathfrak{M}$.

The relation between these two notions of measurability is explained in the next proposition.

Proposition 6.3. *Let $(X, \mathfrak{M})$ be a measurable space and let Y be a metric space. If $\Gamma : X \to \mathcal{P}(Y) \setminus \{\emptyset\}$ is measurable, then it is weakly measurable.*

Proof. Since Y is a metric space, every open set $A \subset Y$ may be written as a countable union of closed sets,

$$A = \bigcup_n C_n.$$

Hence

$$\Gamma^-(A) = \{x \in X : \Gamma(x) \cap A \neq \emptyset\} = \bigcup_n \{x \in X : \Gamma(x) \cap C_n \neq \emptyset\} \in \mathfrak{M},$$

and the desired result follows.

Remark 6.4. If $Y = \mathbb{R}^m$ and Γ is closed-valued, then weak measurability implies measurability. Indeed, let $K \subset \mathbb{R}^m$ be a compact set and for each $n \in \mathbb{N}$ define the open set

$$A_n := \left\{ z \in \mathbb{R}^m : \operatorname{dist}(z, K) < \frac{1}{n} \right\}.$$

Then $\overline{A_n}$ is compact and $A_n \supset \overline{A_{n+1}}$. Moreover, if $x \in X$, then $\Gamma(x) \cap A_n \neq \emptyset$ for all $n \in \mathbb{N}$ if and only if $\Gamma(x) \cap \overline{A_n} \neq \emptyset$ for all $n \in \mathbb{N}$, and since $\Gamma(x)$ is closed, the latter is equivalent by the compactness of K to

$$\Gamma(x) \cap K = \bigcap_{n=1}^{\infty} \left(\Gamma(x) \cap \overline{A_n} \right) \neq \emptyset.$$

Therefore, by weak measurability,

$$\Gamma^-(K) = \bigcap_{n=1}^{\infty} \Gamma^-(A_n) \in \mathfrak{M}.$$

Finally, if $C \subset \mathbb{R}^m$ is a closed set, write

$$C = \bigcup_{n=1}^{\infty} \left(C \cap \overline{B(0,n)} \right).$$

Then

$$\Gamma^-(C) = \bigcup_{n=1}^{\infty} \Gamma^-(C \cap \overline{B(0,n)}) \in \mathfrak{M}$$

since $C \cap \overline{B(0,n)}$ is compact.

We now prove a selection theorem for closed-valued multifunctions.

Theorem 6.5. *Let $(X, \mathfrak{M})$ be a measurable space, let Y be a complete separable metric space, and let Γ be a multifunction from X to the nonempty closed subsets of Y. Then the following properties are equivalent:*

(i) Γ is weakly measurable;
(ii) there exists a sequence of measurable selections $u_n : X \to Y$, $n \in \mathbb{N}$, such that $\{u_n(x)\}_n$ is dense in $\Gamma(x)$ for every $x \in X$;
(iii) for every $y \in Y$ the map

$$x \in X \mapsto \operatorname{dist}(y, \Gamma(x))$$

is measurable.

The proof requires the following auxiliary result, which is of interest in itself.

Lemma 6.6. *Let $(X, \mathfrak{M})$ be a measurable space and let Y be a metric space. Then the pointwise limit of a sequence of measurable functions $u_n : X \to Y$, $n \in \mathbb{N}$, is measurable.*

Proof. Let $\{u_n\}$ be a sequence of measurable functions $u_n : X \to Y$, $n \in \mathbb{N}$, and let $u : X \to Y$ be its pointwise limit. In view of Remark 1.70 it suffices to show that $u^{-1}(C) \in \mathfrak{M}$ for any closed set $C \subset Y$. To see this, we write

$$C = \bigcap_{k=1}^{\infty} A_k,$$

where

$$A_k := \left\{ y \in C : \operatorname{dist}(y, C) < \frac{1}{k} \right\}$$

is an open set. The proof will be completed once we show that

$$u^{-1}(C) = \bigcap_{k=1}^{\infty} \bigcup_{l=1}^{\infty} \bigcap_{n=l}^{\infty} u_n^{-1}(A_k). \tag{6.1}$$

Indeed, $\mathfrak{M}$ is closed under countable unions and intersections and $u_n^{-1}(A_k) \in \mathfrak{M}$ for all $n, k \in \mathbb{N}$. To prove (6.1) let $x \in u^{-1}(C)$. Since $u_n(x) \to u(x) \in C$ as $n \to \infty$, for any fixed $k \in \mathbb{N}$ we have that $u(x) \in A_k$, and so there exists $l \in \mathbb{N}$ such that $u_n(x) \in A_k$ for all $n \geq l$. Hence $x \in \bigcap_{n=l}^{\infty} u_n^{-1}(A_k)$, which implies that

$$x \in \bigcup_{l=1}^{\infty} \bigcap_{n=l}^{\infty} u_n^{-1}(A_k)$$

for all $k \in \mathbb{N}$. It follows that $x \in \bigcap_{k=1}^{\infty} \bigcup_{l=1}^{\infty} \bigcap_{n=l}^{\infty} u_n^{-1}(A_k)$, which shows that

$$u^{-1}(C) \subset \bigcap_{k=1}^{\infty} \bigcup_{l=1}^{\infty} \bigcap_{n=l}^{\infty} u_n^{-1}(A_k).$$

Conversely, assume that $x \in \bigcap_{k=1}^{\infty} \bigcup_{l=1}^{\infty} \bigcap_{n=l}^{\infty} u_n^{-1}(A_k)$. Then for all $k \in \mathbb{N}$ there exists $l \in \mathbb{N}$ such that $x \in \bigcap_{n=l}^{\infty} u_n^{-1}(A_k)$, or equivalently $u_n(x) \in A_k$ for all $n \geq l$. This implies that $u(x) = \lim_{n \to \infty} u_n(x) \in \overline{A_k}$ for all $k \in \mathbb{N}$, that is,

$$\mathrm{dist}(u(x), C) \leq \frac{1}{k}$$

for all $k \in \mathbb{N}$. Letting $k \to \infty$ and using the fact that C is closed yields $u(x) \in C$, and so

$$u^{-1}(C) \supset \bigcap_{k=1}^{\infty} \bigcup_{l=1}^{\infty} \bigcap_{n=l}^{\infty} u_n^{-1}(A_k),$$

which completes the proof.

We now turn to the proof of Theorem 6.5

Proof (Theorem 6.5). **Step 1:** Assume that $\Gamma^-(A) \in \mathfrak{M}$ for every open set $A \subset Y$. We show that there exists a measurable selection of Γ. Let $\{y_n\}_n$ be a countable dense set in Y. For every $x \in X$ let n_x be the smallest positive integer such that $\Gamma(x) \cap B(y_{n_x}, 1) \neq \emptyset$. Note that n_x exists since $\Gamma(x)$ is nonempty and $\{y_n\}_n$ is a dense set in Y. Define $v_0(x) := y_{n_x}$. Then by definition of n_x, for every n we have that

$$v_0^{-1}(\{y_n\}) = \Gamma^-(B(y_n, 1)) \setminus \bigcup_{j<n} \Gamma^-(B(y_j, 1)).$$

We claim that v_0 is measurable. Indeed, $v_0 : X \to \{y_n\}_n$, and so for any open set $A \subset Y$,

$$v_0^{-1}(A) = \bigcup_{y_n \in A} v_0^{-1}(\{y_n\}) \tag{6.2}$$

$$= \bigcup_{y_n \in A} \left(\Gamma^-(B(y_n, 1)) \setminus \bigcup_{j<n} \Gamma^-(B(y_j, 1)) \right) \in \mathfrak{M},$$

where we have used the facts that $\Gamma^-(B(y_n, 1)) \in \mathfrak{M}$ for every n and that $\mathfrak{M}$ is a σ-algebra.

Assume next that $v_0, \ldots, v_k : X \to \{y_n\}_n$ measurable functions have been chosen such that

$$\mathrm{dist}(v_i(x), \Gamma(x)) < \frac{1}{2^i}, \quad d(v_i(x), v_{i+1}(x)) \leq \frac{1}{2^{i-1}} \tag{6.3}$$

for all $x \in X$ and for all $i = 1, \ldots, k$. Since $v_k : X \to \{y_n\}_n$, the sets $E_j^{(k)} := v_k^{-1}(\{y_j\})$, $j \in \mathbb{N}$, form a partition of X. Moreover, if $x \in E_j^{(k)}$, then by $(6.3)_1$, with $i = k$,

$$\Gamma(x) \cap B\left(y_j, \frac{1}{2^k}\right) \neq \emptyset,$$

and so, again by the density of $\{y_n\}_n$ in Y, there exists a smallest positive integer n_x such that

$$\Gamma(x) \cap B\left(y_j, \frac{1}{2^k}\right) \cap B\left(y_{n_x}, \frac{1}{2^{k+1}}\right) \neq \emptyset. \tag{6.4}$$

Define $v_{k+1}(x) := y_{n_x}$ for all $x \in E_j^{(k)}$. Reasoning as in (6.2) we have that $v_{k+1} : X \to \{y_n\}_n$ is measurable, and by (6.4) for all $x \in E_j^{(k)}$,

$$\text{dist}\left(v_{k+1}(x), \Gamma(x)\right) < \frac{1}{2^{k+1}}$$

and

$$d\left(v_k(x), v_{k+1}(x)\right) = d\left(y_j, y_{n_x}\right) \leq \frac{1}{2^k} + \frac{1}{2^{k+1}} \leq \frac{1}{2^{k-1}}.$$

Hence we have constructed a sequence $\{v_j\}_{j \in \mathbb{N}}$ of measurable functions satisfying (6.3). It follows from (6.3)$_2$ that for each $x \in X$ the sequence $\{v_j(x)\}_{j \in \mathbb{N}}$ is a Cauchy sequence in the complete metric space Y. Thus $v_j(x) \to v(x)$ as $j \to \infty$ for all $x \in X$, and since $\Gamma(x)$ is closed, by (6.3)$_1$ we have that $v(x) \in \Gamma(x)$ for all $x \in X$. By the previous lemma the function v is measurable.

Step 2: We prove that (i) implies (ii). Fix $i, n \in \mathbb{N}$ and for all $x \in X$ define

$$\Gamma_{i,n}(x) := \begin{cases} \Gamma(x) \cap B\left(y_n, \frac{1}{2^i}\right) & \text{if } \Gamma(x) \cap B\left(y_n, \frac{1}{2^i}\right) \neq \emptyset, \\ \Gamma(x) & \text{otherwise.} \end{cases}$$

Then $\Gamma_{i,n} : X \to \mathcal{P}(Y) \setminus \{\emptyset\}$ (but $\Gamma_{i,n}$ is not closed-valued in general). We claim that for every open set $A \subset Y$ the set $(\Gamma_{i,n})^-(A)$ belongs to $\mathfrak{M}$. Indeed, we have

$$(\Gamma_{i,n})^-(A) = \left\{ x \in X : \Gamma(x) \cap B\left(y_n, \frac{1}{2^i}\right) \cap A \neq \emptyset \right\}$$

$$\cup \left\{ x \in X : \Gamma(x) \cap B\left(y_n, \frac{1}{2^i}\right) = \emptyset, \, \Gamma(x) \cap A \neq \emptyset \right\}$$

$$= \Gamma^- \left(B\left(y_n, \frac{1}{2^i}\right) \cap A \right)$$

$$\cup \left(\left(X \setminus \Gamma^- \left(B\left(y_n, \frac{1}{2^i}\right) \right) \right) \cap \Gamma^-(A) \right) \in \mathfrak{M}.$$

Define next the multifunction $\overline{\Gamma}_{i,n} : X \to \{C \subset Y, C \text{ closed}, C \neq \emptyset\}$ as

$$\overline{\Gamma}_{i,n}(x) := \overline{\Gamma_{i,n}(x)}.$$

Note that for every $x \in X$ and every open set $A \subset Y$ we have that $\overline{\Gamma_{i,n}(x)} \cap A \neq \emptyset$ if and only if $\Gamma_{i,n}(x) \cap A \neq \emptyset$, and so

$$\left(\overline{\Gamma}_{i,n}\right)^{-}(A) = \left\{x \in X : \overline{\Gamma}_{i,n}(x) \cap A \neq \emptyset\right\}$$
$$= \left\{x \in X : \Gamma_{i,n}(x) \cap A \neq \emptyset\right\} = \left(\Gamma_{i,n}\right)^{-}(A) \in \mathfrak{M}.$$

Hence by Step 1 there exists a measurable selection $u_{i,n} : X \to Y$ of $\overline{\Gamma}_{i,n}$. Since $\overline{\Gamma}_{i,n}(x) \subset \Gamma(x)$ for all $x \in X$ it follows that $u_{i,n}$ is a measurable selection of Γ. It remains to show that for each $x \in X$ the family $\{u_{i,n}(x)\}_{i,n \in \mathbb{N}}$ is dense in $\Gamma(x)$. To see this, fix $x \in X$ and $y \in \Gamma(x)$. For every $i \in \mathbb{N}$, by the density of $\{y_n\}_n$ in Y, we may find y_n such that

$$d(y, y_n) \leq \frac{1}{2^{i+1}}.$$

Hence $\Gamma(x) \cap B\left(y_n, \frac{1}{2^i}\right) \neq \emptyset$, and so by the definition of $\Gamma_{i,n}(x)$ we have that

$$u_{i,n}(x) \in \overline{\Gamma}_{i,n}(x) = \overline{\Gamma(x) \cap B\left(y_n, \frac{1}{2^i}\right)}.$$

It follows that

$$d(y, u_{i,n}(x)) \leq d(y, y_n) + d(y_n, u_{i,n}(x)) \leq \frac{1}{2^{i+1}} + \frac{1}{2^i},$$

which, given the arbitrariness of $i \in \mathbb{N}$, implies that the family $\{u_{i,n}(x)\}_{i,n \in \mathbb{N}}$ is dense in $\Gamma(x)$.

Step 3: We show that (ii) implies (i). Assume that there exists a sequence of measurable selections $u_n : X \to Y$ such that $\{u_n(x)\}_n$ is dense in $\Gamma(x)$ for every $x \in X$. Then for every open set $A \subset Y$ and for every n we have that $u_n^{-1}(A) \in \mathfrak{M}$, and so

$$\Gamma^{-}(A) = \{x \in X : \Gamma(x) \cap A \neq \emptyset\} = \left\{x \in X : \overline{\bigcup_n \{u_n(x)\}} \cap A \neq \emptyset\right\}$$

$$= \left\{x \in X : \left(\bigcup_n \{u_n(x)\}\right) \cap A \neq \emptyset\right\} = \bigcup_n u_n^{-1}(A) \in \mathfrak{M}.$$

Step 4: We prove that (i) is equivalent to (iii). Since Y is separable, every open set $A \subset Y$ is the countable union of countably many open balls, say

$$A = \bigcup_n B(y_n, r_n).$$

Since

$$\Gamma^{-}(A) = \bigcup_n \Gamma^{-}(B(y_n, r_n)),$$

it follows that (i) holds if and only if $\Gamma^{-}(B(y, r)) \in \mathfrak{M}$ for all $y \in Y$ and $r > 0$. On the other hand,

$$\Gamma^{-}(B(y, r)) = \{x \in X : \Gamma(x) \cap B(y, r) \neq \emptyset\}$$
$$= \{x \in X : \operatorname{dist}(y, \Gamma(x)) < r\},$$

and so (i) is equivalent to the measurability of the function

$$x \in X \mapsto \operatorname{dist}(y, \Gamma(x))$$

for all $y \in Y$, which is (iii). This concludes the proof.

Any of the previous properties implies the measurability of the graph of Γ. Indeed, the following result holds.

Corollary 6.7. *Let $(X, \mathfrak{M})$ be a measurable space, let Y be a complete separable metric space, and let Γ be a multifunction from X to the nonempty closed subsets of Y. If Γ satisfies any of the conditions (i)–(iii) of the previous theorem, then $\operatorname{Gr}\Gamma$ belongs to $\mathfrak{M} \otimes \mathcal{B}(Y)$.*

Proof. Since Γ is closed-valued,

$$\operatorname{Gr}\Gamma = \{(x, y) \in X \times Y : y \in \Gamma(x)\}$$
$$= \{(x, y) \in X \times Y : \operatorname{dist}(y, \Gamma(x)) = 0\},$$

and thus it suffices to show that the mapping

$$(x, y) \in X \times Y \mapsto \psi(x, y) := \operatorname{dist}(y, \Gamma(x))$$

is $\mathfrak{M} \otimes \mathcal{B}(Y)$-measurable. Let $\{y_n\}_n$ be a countable dense set in Y. For every $y \in Y$ and $k \in \mathbb{N}$ let $n_{y,k}$ be the smallest positive integer such that $y \in B(y_{n_{y,k}}, \frac{1}{k})$. Note that $n_{y,k}$ exists since $\{y_n\}_n$ is dense in Y. Define

$$\psi_k(x, y) := \psi\left(x, y_{n_{y,k}}\right), \quad (x, y) \in X \times Y.$$

If $k \to \infty$, then $y_{n_{y,k}} \to y$ for all $y \in Y$, and so

$$\psi_k(x, y) = \psi\left(x, y_{n_{y,k}}\right) = \operatorname{dist}\left(y_{n_{y,k}}, \Gamma(x)\right) \to \operatorname{dist}(y, \Gamma(x)),$$

where we have used the continuity of the distance function. Since the pointwise limit of $\mathfrak{M} \otimes \mathcal{B}(Y)$-measurable functions is still $\mathfrak{M} \otimes \mathcal{B}(Y)$-measurable, it remains to prove that each function ψ_k is $\mathfrak{M} \otimes \mathcal{B}(Y)$-measurable. This follows from the fact that on the set

$$X \times \left(B\left(y_n, \frac{1}{k}\right) \setminus \bigcup_{j < n} B\left(y_j, \frac{1}{k}\right)\right)$$

the function ψ_k coincides with the function

$$(x, y) \in X \times Y \mapsto \varphi_n(x) := \operatorname{dist}(y_n, \Gamma(x)),$$

and so for any open set $A \subset \mathbb{R}$,

$$\psi_k^{-1}(A) = \{(x,y) \in X \times Y : \psi_k(x,y) \in A\}$$

$$= \bigcup_{n=1}^{\infty} \left\{ (x,y) \in X \times \left(B\left(y_n, \frac{1}{k}\right) \setminus \bigcup_{\substack{j \in \mathbb{N}, \\ j < n}} B\left(y_j, \frac{1}{k}\right) \right) : \psi_k(x,y) \in A \right\}$$

$$= \bigcup_{n=1}^{\infty} \left\{ (x,y) \in X \times \left(B\left(y_n, \frac{1}{k}\right) \setminus \bigcup_{\substack{j \in \mathbb{N}, \\ j < n}} B\left(y_j, \frac{1}{k}\right) \right) : \varphi_n(x) \in A \right\}$$

$$= \bigcup_{n=1}^{\infty} \varphi_n^{-1}(A) \times \left(B\left(y_n, \frac{1}{k}\right) \setminus \bigcup_{\substack{j \in \mathbb{N}, \\ j < n}} B\left(y_j, \frac{1}{k}\right) \right) \in \mathfrak{M} \otimes \mathcal{B}(Y),$$

where we have used the fact that $\varphi_n^{-1}(A) \in \mathfrak{M}$ by (iii). This completes the proof.

We now turn to the case that Γ is not necessarily closed-valued. We begin by showing that $\mathbb{N}^{\mathbb{N}} := \{f : \mathbb{N} \to \mathbb{N}\}$ is a complete separable metric space.

Proposition 6.8. *The function $d : \mathbb{N}^{\mathbb{N}} \times \mathbb{N}^{\mathbb{N}} \to [0, \infty)$ defined as*

$$d(f,g) := \begin{cases} 0 & \text{if } f = g, \\ \frac{1}{k} & \text{if } f \neq g \text{ and } k = \min\{j \in \mathbb{N} : f(j) \neq g(j)\}, \end{cases} \tag{6.5}$$

is a metric and $\left(\mathbb{N}^{\mathbb{N}}, d\right)$ is a complete separable metric space.

Proof. To prove that d is a metric it suffices to prove the triangle inequality. We will actually prove more, namely that

$$d(f,g) \leq \max\{d(f,h), d(h,g)\}$$

for all f, g, $h \in \mathbb{N}^{\mathbb{N}}$. Let

$$k_1 = \min\{j \in \mathbb{N} : f(j) \neq h(j)\}, \quad k_2 = \min\{j \in \mathbb{N} : g(j) \neq h(j)\},$$

with, say, $k_1 \leq k_2$. Then $g(j) = f(j) = h(j)$ for all $j = 1, \dots, k_1$, and so

$$\min\{j \in \mathbb{N} : f(j) \neq g(j)\} \geq k_1,$$

which implies that

$$d(f,g) \leq \frac{1}{k_1} = \max\{d(f,h), d(h,g)\}.$$

We now show that $\left(\mathbb{N}^{\mathbb{N}}, d\right)$ is a complete metric space. Let $\{f_n\} \subset \mathbb{N}^{\mathbb{N}}$ be a Cauchy sequence. Then for all $k \in \mathbb{N}$ there exists $n_k \in \mathbb{N}$ such that

$$d(f_n, f_l) < \frac{1}{k} \tag{6.6}$$

for all $l, n \geq n_k$. Without loss of generality we may assume that $n_k > n_{k-1}$. Taking $l = n_k$ it follows from (6.5) and (6.6) that $f_n(j) = f_{n_k}(j)$ for all $j = 1, \ldots, k$ and all $n \geq n_k$. Since the sequence $\{n_k\}_k$ is increasing, for all $k \in \mathbb{N}$ we have that $f_{n_l}(k) = f_{n_k}(k)$ for all $l > k$. For all $k \in \mathbb{N}$ define $f(k) := f_{n_k}(k)$. Then

$$d(f_n, f) < \frac{1}{k}$$

for all $k \in \mathbb{N}$ and $n \geq n_k$. Hence $\{f_n\}$ converges to f, and so $(\mathbb{N}^\mathbb{N}, d)$ is a complete metric space.

To show that $(\mathbb{N}^\mathbb{N}, d)$ is separable, for all $n \in \mathbb{N}$ and all $m_1, \ldots, m_n \in \mathbb{N}$ define,

$$f_{m_1,\ldots,m_n}(j) := \begin{cases} m_j & \text{if } 1 \leq j \leq n, \\ 0 & \text{otherwise.} \end{cases}$$

Note that for each fixed $n \in \mathbb{N}$ the family

$$\{f_{m_1,\ldots,m_n} : m_1, \ldots, m_n \in \mathbb{N}\}$$

is countable, and so is

$$\{f_{m_1,\ldots,m_n} : m_1, \ldots, m_n \in \mathbb{N}, \, n \in \mathbb{N}\}.$$

We show that this family is dense in $(\mathbb{N}^\mathbb{N}, d)$. Fix $f \in \mathbb{N}^\mathbb{N}$ and $k \in \mathbb{N}$. Then $f(j) = f_{f(1),\ldots,f(k)}(j)$ for all $j = 1, \ldots, k$, and so

$$d\left(f, f_{f(1),\ldots,f(k)}\right) \leq \frac{1}{k}.$$

Hence $(\mathbb{N}^\mathbb{N}, d)$ is separable.

Note that for every $f \in \mathbb{N}^\mathbb{N}$ and $r > 0$,

$$B_{\mathbb{N}^\mathbb{N}}(f, r) = \left\{ g \in \mathbb{N}^\mathbb{N} : g(j) = f(j) \text{ for all } j = 1, \ldots, \left[\frac{1}{r}\right] \right\}. \tag{6.7}$$

In what follows, if $(X, \mathfrak{M}, \mu)$ is a measure space, we denote by $\mathfrak{M}_\mu$ the σ-algebra of all μ^*-measurable subsets, where μ^* is the outer measure that extends μ given in Corollary 1.38. We recall that $\mathfrak{M}_\mu \supset \mathfrak{M}$.

Theorem 6.9. *Let* $(X, \mathfrak{M}, \mu)$ *be a measure space and let* Y *be a complete separable metric space. Let* $\mathcal{R}$ *be the class of all rectangles* $F \times C$, *where* $F \in \mathfrak{M}$ *and* $C \subset Y$ *is closed. Given any Suslin-*$\mathcal{R}$ *set* $E \subset X \times Y$ *on* X, *there exists a sequence* $\{u_n\}$ *of* $\mathfrak{M}_\mu$-*measurable functions* $u_n : \pi_X(E) \to Y$, $n \in \mathbb{N}$, *such that* $(x, u_n(x)) \in E$ *for all* $x \in \pi_X(E)$ *and all* $n \in \mathbb{N}$. *Moreover, for all* $(x, y) \in E$ *and* $\varepsilon > 0$ *there exists* $n \in \mathbb{N}$ *such that* $d_Y(y, u_n(x)) \leq \varepsilon$.

Proof. Since E is a Suslin-$\mathcal{R}$ set there exists a Suslin scheme $\mathcal{E} : \mathcal{I} \to \mathcal{R}$ such that

$$E = A(\mathcal{E}) = \bigcup_{f \in \mathbb{N}^{\mathbb{N}}} \bigcap_{k=1}^{\infty} \left(F_{f|_k} \times C_{f|_k} \right),$$

where $F_{f|_k} \in \mathfrak{M}$ and $C_{f|_k} \subset Y$ is closed. The set E is the projection onto $X \times Y$ of the set $E^* \subset X \times Y \times \mathbb{N}^{\mathbb{N}}$ defined by

$$E^* := \bigcup_{f \in \mathbb{N}^{\mathbb{N}}} \bigcap_{k=1}^{\infty} \left(F_{f|_k} \times C_{f|_k} \times I_{f|_k} \right) = \bigcup_{f \in \mathbb{N}^{\mathbb{N}}} \left(F_f \times C_f \times \{f\} \right), \qquad (6.8)$$

where (see (6.7))

$$I_{f|_k} := \{ g \in \mathbb{N}^{\mathbb{N}} : g|_k = f|_k \} = B_{\mathbb{N}^{\mathbb{N}}} \left(f, \frac{1}{k} \right)$$

and

$$F_f := \bigcap_{k=1}^{\infty} F_{f|_k}, \quad C_f := \bigcap_{k=1}^{\infty} C_{f|_k}.$$

Let $\mathcal{R}^*$ be the class of all rectangles $F \times C$, where $F \in \mathfrak{M}$ and $C \subset Y \times \mathbb{N}^{\mathbb{N}}$ is closed. Rewriting E^* as

$$E^* = \bigcup_{f \in \mathbb{N}^{\mathbb{N}}} \bigcap_{k=1}^{\infty} \left(F_{f|_k} \times K_{f|_k} \right),$$

where $K_{f|_k} := C_{f|_k} \times \{f\}$ is closed in $Y \times \mathbb{N}^{\mathbb{N}}$, we see that E^* is a Suslin-$\mathcal{R}^*$ set. Consider the multifunction

$$\Gamma : \pi_X(E) \to \mathcal{P}\left(Y \times \mathbb{N}^{\mathbb{N}} \right) \setminus \{\emptyset\}$$

defined by

$$\Gamma(x) := \left\{ (y, f) \in Y \times \mathbb{N}^{\mathbb{N}} : (x, y, f) \in E^* \right\}.$$

Note that $\pi_X(E)$ belongs to $\mathfrak{M}_\mu$ by Theorem 1.135. We claim that $\Gamma(x)$ is closed in $Y \times \mathbb{N}^{\mathbb{N}}$ for all $x \in \pi_X(E)$, or equivalently that $\left(Y \times \mathbb{N}^{\mathbb{N}} \right) \setminus \Gamma(x)$ is open. Fix $x \in \pi_X(E)$ and $(y, f) \in \left(Y \times \mathbb{N}^{\mathbb{N}} \right) \setminus \Gamma(x)$. Then $(x, y, f) \notin E^*$ by definition of Γ, and so by (6.8) there exists $k \in \mathbb{N}$ such that

$$(x, y, f) \notin F_{f|_k} \times C_{f|_k} \times I_{f|_k},$$

that is, since $f \in I_{f|_k}$, $(x, y) \notin F_{f|_k} \times C_{f|_k}$. There are now two cases. If $x \notin F_{f|_k}$, then $Y \times I_{f|_k}$ is a neighborhood of (y, f) and $(x, z, g) \notin F_{f|_k} \times Y \times I_{f|_k}$ for all $(z, g) \in Y \times I_{f|_k}$. Hence

$$\Gamma(x) \cap \left(Y \times I_{f|_k} \right) = \emptyset.$$

If $y \notin C_{f|_k}$, then, since $C_{f|_k}$ is closed in Y, there exists $B_Y(y, r) \subset Y$ such that $B_Y(y, r) \subset Y \setminus C_{f|_k}$. Then $B_Y(y, r) \times I_{f|_k}$ is a neighborhood of (y, f) and $(x, z, g) \notin F_{f|_k} \times C_{f|_k} \times I_{f|_k}$ for all $(z, g) \in B_Y(y, r) \times I_{f|_k}$. Hence

$$\Gamma(x) \cap \left(B_Y(y, r) \times I_{f|_k}\right) = \emptyset.$$

Thus we have proved that

$$\Gamma : \pi_X(E) \to \left\{C \subset Y \times \mathbb{N}^{\mathbb{N}}, \ C \text{ closed, } C \neq \emptyset\right\}.$$

Next we claim that $\Gamma^-(U) \in \mathfrak{M}_\mu$ for all open sets $U \subset Y \times \mathbb{N}^{\mathbb{N}}$. Indeed, fix an open set $U \subset Y \times \mathbb{N}^{\mathbb{N}}$ and note that

$$\Gamma^-(U) = \{x \in \pi_X(E) : \Gamma(x) \cap U \neq \emptyset\} = \tilde{\pi}_X(E^* \cap (X \times U)),$$

where $\tilde{\pi}_X : X \times Y \times \mathbb{N}^{\mathbb{N}} \to X$ is the usual projection. Since $Y \times \mathbb{N}^{\mathbb{N}}$ is a metric space, we can write U as a countable union of closed balls, and so $X \times U$ is a Suslin-$\mathcal{R}^*$ set. By Theorem 1.133, $E^* \cap (X \times U)$ is a Suslin-$\mathcal{R}^*$ set, and so by Theorem 1.135 the set $\Gamma^-(U) = \tilde{\pi}_X(E^* \cap (X \times U))$ belongs to $\mathfrak{M}_\mu$. Hence we are now in a position to apply Theorem 6.5 to find a sequence $\{v_n\}$ of $\mathfrak{M}_\mu$-measurable selections $v_n : \pi_X(E) \to Y \times \mathbb{N}^{\mathbb{N}}$ of Γ such that $\{v_n(x)\}_n$ is dense in $\Gamma(x)$ for every $x \in X$. For all $n \in \mathbb{N}$ define

$$u_n(x) := \pi_Y(v_n(x)), \quad x \in \pi_X(E),$$

where $\pi_Y : Y \times \mathbb{N}^{\mathbb{N}} \to Y$ is the projection onto Y. If $x \in \pi_X(E)$, then $v_n(x) \in \Gamma(x)$, and so $(x, v_n(x)) \in E^*$. Hence $(x, u_n(x)) \in E$, since E is the projection of E^* onto $X \times Y$.

Moreover, if $A \subset Y$ is open, then

$$
\begin{aligned}
u_n^{-1}(A) &= \{x \in \pi_X(E) : \pi_Y(v_n(x)) \in A\} \\
&= \{x \in \pi_X(E) : v_n(x) \in A \times \mathbb{N}^{\mathbb{N}}\} \\
&= v_n^{-1}\left(A \times \mathbb{N}^{\mathbb{N}}\right) \in \mathfrak{M}_\mu,
\end{aligned}
$$

which shows that u_n is $\mathfrak{M}_\mu$-measurable for all $n \in \mathbb{N}$.

Finally, for all $(x, y) \in E$ and $\varepsilon > 0$ we have that $(x, y, f) \in E^*$ for any fixed $f \in \mathbb{N}^{\mathbb{N}}$. Hence $(y, f) \in \Gamma(x)$, and since $\{v_n(x)\}_n$ is dense in $\Gamma(x)$, there exists $n \in \mathbb{N}$ such that

$$d_{Y \times \mathbb{N}^{\mathbb{N}}}((y, f), v_n(x)) \leq \varepsilon,$$

and so in particular,

$$d_Y(y, u_n(x)) = d_Y(y, \pi_Y(v_n(x))) \leq d_{Y \times \mathbb{N}^{\mathbb{N}}}((y, f), v_n(x)) \leq \varepsilon.$$

This completes the proof.

As a corollary of the previous theorem we obtain the following result.

Theorem 6.10 (Aumann measurable selection). *Let* $(X, \mathfrak{M}, \mu)$ *be a measure space, let* Y *be a complete separable metric space, and let*

$$\Gamma : X \to \mathcal{P}(Y) \setminus \{\emptyset\}$$

be a multifunction. Assume that the graph of Γ *belongs to* $\mathfrak{M} \otimes \mathcal{B}(Y)$ *(or more generally to* $\mathfrak{M}_\mu \otimes \mathcal{B}(Y)$*). Then there exists a sequence* $\{u_n\}$ *of* $\mathfrak{M}_\mu$*-measurable selections* $u_n : X \to Y$ *such that* $\{u_n(x)\}$ *is dense in* $\overline{\Gamma(x)}$ *for every* $x \in X$.

Proof. As in the proof of the projection theorem we have that $\operatorname{Gr}\Gamma$ is a Suslin-$\mathcal{R}$ set. Hence by the previous theorem there exists a sequence $\{u_n\}$ of $\mathfrak{M}_\mu$-measurable functions $u_n : X \to Y$ such that $(x, u_n(x)) \in \operatorname{Gr}\Gamma$ for all $x \in X$ and $n \in \mathbb{N}$, where we have used the fact that $\pi_X(\operatorname{Gr}\Gamma) = X$. It follows from the definition of $\operatorname{Gr}\Gamma$ that $u_n(x) \in \Gamma(x)$ for all $x \in X$. Hence u_n is a measurable selection of Γ. Finally, if $z \in \overline{\Gamma(x)}$ and $\varepsilon > 0$, then we may find $y \in \Gamma(x)$ such that $d_Y(y, z) \leq \varepsilon$. By the properties of the sequence $\{u_n\}$ we may find $n \in \mathbb{N}$ such that $d_Y(y, u_n(x)) \leq \varepsilon$. Hence $d_Y(z, u_n(x)) \leq 2\varepsilon$, and the proof is complete.

Remark 6.11. In particular, if Γ is closed-valued and $\mathfrak{M}_\mu = \mathfrak{M}$, then by the previous theorem and Corollary 6.7 the condition that $\operatorname{Gr}\Gamma \in \mathfrak{M} \otimes \mathcal{B}(Y)$ is equivalent to any of the conditions (i)–(iii) in Theorem 6.5.

To apply Aumann selection theorem to the example at the beginning of this subsection it suffices to define $\Gamma : \Omega \to \mathcal{P}(K) \setminus \{\emptyset\}$ as

$$\Gamma(x) := \left\{ y \in K : \bar{f}(x) = f(u(x), y) \right\},$$

and to consider the restriction to Ω of the σ-algebra of Lebesgue measurable sets of $\mathbb{R}^N$. Recall that this σ-algebra is the completion of the σ-algebra of Borel sets with respect to the Lebesgue measure.

Exercise 6.12. Let $(X, \mathfrak{M})$ be a measurable space and let Y be a topological space.

(i) Show that if

$$\Gamma(x) := \{u(x)\}, \quad x \in X, \tag{6.9}$$

for some function $u : X \to Y$, then Γ is measurable if and only if u is measurable.

(ii) Let $Y = \mathbb{R}$ and let

$$\Gamma(x) := (-\infty, u(x)], \quad x \in X. \tag{6.10}$$

Prove that Γ is measurable if and only if u is measurable.

Next we introduce the concept of the supremum of a family of multifunctions.

Theorem 6.13 (Essential supremum). *Let* $(X, \mathfrak{M}, \mu)$ *be a measure space, with* μ *a (positive) σ-finite measure, and let* Y *be a separable metric space. Let* $\mathcal{F} = \{\Gamma_\alpha\}_{\alpha \in J}$ *be a nonempty family of closed-valued weakly measurable multifunctions. Then there exists a countable set* $I_0 \subset J$ *such that the closed-valued multifunction* Γ_0 *defined by*

$$\Gamma_0(x) := \overline{\bigcup_{\alpha \in I_0} \Gamma_\alpha(x)}, \quad x \in X, \tag{6.11}$$

is weakly measurable and satisfies the following properties:

(i) for every $\alpha \in J$ *we have*

$$\Gamma_\alpha(x) \subset \Gamma_0(x)$$

for μ a.e. $x \in X$;
(ii) if Γ' *is a closed-valued weakly measurable multifunction such that*

$$\Gamma_\alpha(x) \subset \Gamma'(x)$$

for μ a.e. $x \in X$ *and every* $\alpha \in J$, *then*

$$\Gamma_0(x) \subset \Gamma'(x)$$

for μ a.e. $x \in X$.

Proof. Let $\{y_k\} \subset Y$ be a dense set and consider the countable family of balls $\{B\left(y_k, \frac{1}{l}\right) : k, l \in \mathbb{N}\}$. By Remark 1.107(ii), for every $k, l \in \mathbb{N}$, there exists a countable set $I_{k,l} \subset J$ such that the set

$$\bigcup_{\alpha \in I_{k,l}} (\Gamma_\alpha)^- \left(B\left(y_k, \frac{1}{l}\right)\right)$$

is the essential union of the family of sets $\left\{(\Gamma_\alpha)^- \left(B\left(y_k, \frac{1}{l}\right)\right)\right\}_{\alpha \in J}$. This implies, in particular, that for every $\beta \in J$, up to a set of μ measure zero,

$$(\Gamma_\beta)^- \left(B\left(y_k, \frac{1}{l}\right)\right) \subset \bigcup_{\alpha \in I_{k,l}} (\Gamma_\alpha)^- \left(B\left(y_k, \frac{1}{l}\right)\right). \tag{6.12}$$

Set

$$I_0 := \bigcup_{k,l \in \mathbb{N}} I_{k,l}$$

and define Γ_0 as in (6.11). In order to show that Γ_0 is a weakly measurable multifunction, fix an open set $A \subset Y$ and let $x \in X$ be such that $\Gamma_0(x) \cap A \neq \emptyset$. Since $\Gamma_0(x)$ is closed and A is open, it follows that

$$\left(\bigcup_{\alpha \in I_0} \Gamma_\alpha(x)\right) \cap A \neq \emptyset,$$

and this shows that

$$(\Gamma_0)^- (A) = \bigcup_{\alpha \in I_0} (\Gamma_\alpha)^- (A) \in \mathfrak{M}.$$

Property (ii) follows from the definition of Γ_0 (see (6.11)). To prove (i) fix $\beta \in J$. Using (6.12) it follows that for every $k, l \in \mathbb{N}$, up to a set of μ measure zero,

$$(\Gamma_\beta)^- \left(B \left(y_k, \frac{1}{l} \right) \right) \subset (\Gamma_0)^- \left(B \left(y_k, \frac{1}{l} \right) \right). \tag{6.13}$$

Hence the set

$$E_\beta := \bigcup_{k,l \in \mathbb{N}} \left((\Gamma_\beta)^- \left(B \left(y_k, \frac{1}{l} \right) \right) \setminus (\Gamma_0)^- \left(B \left(y_k, \frac{1}{l} \right) \right) \right)$$

has measure zero. We claim that if $x \notin E_\beta$, then $\Gamma_\beta (x) \subset \Gamma_0 (x)$. Indeed, fix one such x and let $y \in \Gamma_\beta (x)$. Since the sequence $\{y_k\}$ is dense in Y, we can find subsequences $\{y_{k_x}\} \subset \{y_k\}$ and $\{l_{k_x}\} \subset \mathbb{N}$ such that $l_{k_x} \to \infty$ and $y \in B \left(y_{k_x}, \frac{1}{l_{k_x}} \right)$. We deduce that

$$x \in (\Gamma_\beta)^- \left(B \left(y_{k_x}, \frac{1}{l_{k_x}} \right) \right) \subset (\Gamma_0)^- \left(B \left(y_k, \frac{1}{l} \right) \right),$$

where we have used (6.13) and the fact that $x \notin E_\beta$. Therefore we may find $z_{k_x} \in B \left(y_{k_x}, \frac{1}{l_{k_x}} \right) \cap \Gamma_0 (x)$, and since $z_{k_x} \to y$ and $\Gamma_0 (x)$ is closed, we conclude that $y \in \Gamma_0 (x)$. This concludes the proof.

The multifunction Γ_0 is called the *essential supremum* of the family $\mathcal{F}$ with respect to the measure μ. We omit the reference to the dependence on the measure μ when it is clear from the context.

Remark 6.14. (i) Under the hypotheses of Theorem 1.108, the essential supremum u_0 defined in (1.49) satisfies

$$u_0 (x) = \sup_{y \in \widetilde{\Gamma}_0(x)} y$$

for μ a.e. $x \in X$, where $\widetilde{\Gamma}_0$ is the essential supremum of the family of multifunctions $\widetilde{\Gamma}_\alpha : X \to \mathcal{P} ([-\infty, \infty]) \setminus \{\emptyset\}$ given by

$$\widetilde{\Gamma}_\alpha (x) := \{u_\alpha (x)\}.$$

Indeed, setting

$$v (x) := \sup_{y \in \widetilde{\Gamma}_0(x)} y,$$

then $v(x) \geq u_\alpha(x)$ for every $\alpha \in J$ and for μ a.e. $x \in X$ because $\widetilde{\Gamma}_0(x) \supset \{u_\alpha(x)\}$, and so $v(x) \geq u_0(x)$ for μ a.e. $x \in X$ by property (ii) of Definition 1.106. Conversely, define $\Gamma'(x) := [-\infty, u_0(x)]$. Then $\Gamma'(x) \supset \{u_\alpha(x)\}$ for every $\alpha \in J$ and for μ a.e. $x \in X$ by property (i) of Definition 1.106, and hence $\Gamma'(x) \supset \widetilde{\Gamma}_0(x)$ for μ a.e. $x \in X$. In particular, it follows that $v(x) \leq u_0(x)$ for μ a.e. $x \in X$.

(ii) Under the hypotheses of the previous theorem, if $\mathcal{F}$ is a convex set of the space of measurable functions, then $\widetilde{\Gamma}_0(x)$ is a convex set for μ a.e. $x \in X$. Indeed, by (6.11) there exists a sequence $\{v_n\} \subset \mathcal{F}$ such that

$$\widetilde{\Gamma}_0(x) = \overline{\bigcup_n \{v_n(x)\}}$$

for μ a.e. $x \in X$. To prove that $\widetilde{\Gamma}_0(x)$ is a convex set for μ a.e. $x \in X$, we consider $\theta \in [0, 1] \cap \mathbb{Q}$ and we show that

$$\theta \widetilde{\Gamma}_0(x) + (1 - \theta) \widetilde{\Gamma}_0(x) \subset \widetilde{\Gamma}_0(x) \tag{6.14}$$

for all $x \in X \setminus X_\theta$, where $\mu(X_\theta) = 0$. Setting

$$X_0 := \bigcup_{\theta \in [0,1] \cap \mathbb{Q}} X_\theta,$$

it follows that $\mu(X_0) = 0$, and due to the closedness of $\widetilde{\Gamma}_0(x)$ we conclude that (6.14) holds for every $\theta \in [0, 1]$ and for all $x \in X \setminus X_0$.
To establish (6.14), define

$$X_\theta := \bigcup_{n,l} \left\{ x \in X : \theta v_n(x) + (1 - \theta) v_l(x) \notin \widetilde{\Gamma}_0(x) \right\}.$$

Since $\mathcal{F}$ is convex, $\theta v_n + (1 - \theta) v_l \in \mathcal{F}$ for every $n, l \in \mathbb{N}$, and thus $\mu(X_\theta) = 0$. Fix $x \in X \setminus X_\theta$, let $y_1, y_2 \in \widetilde{\Gamma}_0(x)$, and find $\{v_{n_1}\}, \{v_{n_2}\} \subset \{v_n\}$ such that $v_{n_i}(x) \to y_i$, $i = 1, 2$. Since $x \in X \setminus X_\theta$, we have that $\theta v_{n_1}(x) + (1 - \theta) v_{n_2}(x) \in \widetilde{\Gamma}_0(x)$, and so $\theta y_1 + (1 - \theta) y_2 \in \overline{\widetilde{\Gamma}_0(x)} = \widetilde{\Gamma}_0(x)$.

(iii) Under the hypotheses of the previous theorem, if ν is a (positive) σ-finite measure such that μ is absolutely continuous with respect to ν, then

$$\Gamma_\nu(x) = \Gamma_\mu(x)$$

for μ a.e. $x \in X$, where Γ_ν and Γ_μ are the essential suprema of the family of multifunctions $\mathcal{F}$ with respect to ν and μ, respectively. Indeed, for every $\alpha \in J$ we have

$$\Gamma_\alpha(x) \subset \Gamma_\nu(x)$$

for ν a.e. $x \in X$, therefore for μ a.e. $x \in X$, and so by the definition of essential supremum with respect to μ, we conclude that

$$\Gamma_\mu(x) \subset \Gamma_\nu(x)$$

for μ a.e. $x \in X$. Conversely, find a countable set $I_0 \subset J$ such that

$$\Gamma_\nu (x) := \overline{\bigcup_{\alpha \in I_0} \Gamma_\alpha (x)}, \quad x \in X. \tag{6.15}$$

For every $\alpha \in I_0$ there exists $N_\alpha \in \mathfrak{M}$ with $\mu (N_\alpha) = 0$ such that

$$\Gamma_\alpha (x) \subset \Gamma_\mu (x)$$

for all $x \in X \setminus N_\alpha$. Hence setting

$$N_0 := \bigcup_{\alpha \in I_0} N_\alpha,$$

it follows that $\mu (N_0) = 0$ and for all $x \in X \setminus N_0$,

$$\bigcup_{\alpha \in I_0} \Gamma_\alpha (x) \subset \Gamma_\mu (x).$$

Since $\Gamma_\mu (x)$ is closed, in view of (6.15) we have $\Gamma_\nu (x) \subset \Gamma_\mu (x)$ for all $x \in X \setminus N_0$.

6.1.2 Continuous Selections

We study next the existence of continuous selections.

Definition 6.15. *Let X and Y be topological spaces. A multifunction Γ from X to the nonempty subsets of Y is said to be* lower semicontinuous *if for every open set $A \subset Y$ the set*

$$\Gamma^- (A) := \{x \in X : \Gamma (x) \cap A \neq \emptyset\}$$

is open in X.

Exercise 6.16. Let X, Y be topological spaces.

(i) Let $v : Y \to X$ be onto and define

$$\Gamma : X \to \mathcal{P} (Y) \setminus \{\emptyset\}$$

as

$$\Gamma (x) := v^{-1} (\{x\}), \quad x \in X.$$

Prove that Γ is lower semicontinuous if and only if v is open, i.e., it maps open sets into open sets. In this case a continuous selection $u : X \to Y$ is a continuous "inverse", in the sense that $u (x) \in v^{-1} (\{x\})$ for all $x \in X$.

(ii) Let $Y := [-\infty, \infty]$, and let $w : X \to [-\infty, \infty)$ and $v : X \to (-\infty, \infty]$ be such that $w \leq v$. Define

$$\Gamma : X \to \mathcal{P}(Y) \setminus \{\emptyset\}$$

as

$$\Gamma(x) := \{y \in \mathbb{R} : w(x) \leq y \leq v(x)\}, \quad x \in X.$$

Show that Γ is lower semicontinuous if and only if w is upper semicontinuous and v lower semicontinuous. In this case a continuous selection $u : X \to Y$ is a continuous function such that $w \leq u \leq v$.

It can be verified that lower semicontinuity of Γ is equivalent to lower semicontinuity of u when Γ is of the form (6.10), while in the case (6.9) lower semicontinuity of Γ implies continuity of u. This follows from the theorem below.

Theorem 6.17 (Michael continuous selection theorem). *Let X be a metric space and let Y be a Banach space. If the multifunction*

$$\Gamma : X \to \{C \subset Y : C \text{ convex, nonempty, closed}\}$$

is lower semicontinuous, then for every $x_0 \in X$ and every $y_0 \in \Gamma(x_0)$ there exists a continuous function $u : X \to Y$ such that $u(x_0) = y_0$ and $u(x) \in \Gamma(x)$ for every $x \in X$.

In order to prove the previous theorem we need some auxiliary results that are of interest in themselves. We begin by showing that given any cover in an open space there exists a locally finite partition of unity subordinated to it.

Theorem 6.18 (Partition of unity). *Let X be a metric space and let $\{U_\alpha\}_{\alpha \in I}$ be an open cover of X. Then there exists a locally finite partition of unity subordinated to it.*

Proof. **Step 1:** We assume first that I is finite, say $I = \{1, \ldots, n\}$. For each $i = 1, \ldots, n$ define the continuous function

$$u_i(x) := \text{dist}(x, X \setminus U_i), \quad x \in X,$$

and set $u := \sum_{i=1}^{n} u_i$. Note that $u > 0$, since $U_1, \ldots, U_n$ is a cover of X and $u_i > 0$ in U_i, $i = 1, \ldots, n$. Next, for each $i = 1, \ldots, n$ we define the continuous function

$$v_i(x) := \max\left\{u_i(x) - \frac{1}{n+1}u(x), 0\right\}, \quad x \in X.$$

We claim that $\text{supp}\, v_i \subset U_i$ for each $i = 1, \ldots, n$. Indeed,

$$\text{supp}\, v_i = \overline{\left\{x \in X : u_i(x) > \frac{1}{n+1}u(x)\right\}} \subset \left\{x \in X : u_i(x) \geq \frac{1}{n+1}u(x)\right\}.$$

Since $u > 0$, the closed set $\left\{ x \in X : u_i(x) \geq \frac{1}{n+1} u(x) \right\}$ is contained in $U_i = \{ x \in X : u_i(x) > 0 \}$.

Next we show that $\sum_{i=1}^n v_i > 0$. For all $x \in X$ we have

$$\sum_{i=1}^n v_i(x) \geq \sum_{i=1}^n \left(u_i(x) - \frac{1}{n+1} u(x) \right) = u(x) - \frac{n}{n+1} u(x) = \frac{1}{n+1} u(x) > 0.$$

It now suffices to define for each $i = 1, \ldots, n$,

$$\varphi_i(x) := \frac{v_i(x)}{\sum_{j=1}^n v_j(x)}, \quad x \in X.$$

Step 2: Assume next that $I = \mathbb{N}$. For each $i \in \mathbb{N}$ define the continuous function $u_i : X \to \left[0, \frac{1}{2^i} \right]$ by

$$u_i(x) := \min \left\{ \operatorname{dist}(x, X \setminus U_i), \frac{1}{2^i} \right\}, \quad x \in X.$$

Then $u_i > 0$ in U_i, $u_i = 0$ outside U_i, and $u_i \leq \frac{1}{2^i}$. Hence the function $u := \sum_{i=1}^\infty \frac{1}{2^{i-1}} u_i$ is continuous and $u > 0$ since $\{U_i\}_{i \in \mathbb{N}}$ is a cover of X. Next, for $i \in \mathbb{N}$ we define the continuous function $v_i : X \to [0, 1]$ by

$$v_i(x) := \max \left\{ u_i(x) - \frac{1}{3} u(x), 0 \right\}, \quad x \in X.$$

As in the previous step we have that $\operatorname{supp} v_i \subset U_i$. We claim that $\{v_i\}_{i \in \mathbb{N}}$ is locally finite. For any fixed $x \in X$, since u is positive and continuous there exist $\varepsilon, r > 0$ such that $u(y) > \varepsilon$ for all $y \in B(x, r)$. Let $i_0 \in \mathbb{N}$ be so large that $\frac{1}{2^{i_0}} < \frac{\varepsilon}{3}$. From the definition of v_i and the fact that $u_i \leq \frac{1}{2^i}$ it follows that $v_i(y) = 0$ for all $y \in B(x, r)$ and $i \geq i_0$.

Next we show that $\sum_{i=1}^\infty v_i > 0$. Fix $x \in X$. Since $u_i(x) > 0$ for some $i \in \mathbb{N}$ and $u_n \leq \frac{1}{2^n} < \frac{1}{2^i}$ for all $n \geq i$, it follows that there exists $i_0 \in \mathbb{N}$ such that

$$u_{i_0}(x) = \sup_{i \in \mathbb{N}} u_i(x),$$

and so

$$u(x) = \sum_{i=1}^\infty \frac{1}{2^{i-1}} u_i(x) \leq u_{i_0}(x) \sum_{i=1}^\infty \frac{1}{2^{i-1}} = 2 u_{i_0}(x).$$

Hence

$$v_{i_0}(x) \geq u_{i_0}(x) - \frac{1}{3} u(x) \geq u_{i_0}(x) - \frac{2}{3} u_{i_0}(x) = \frac{1}{3} u_{i_0}(x) > 0.$$

It now suffices to define for each $i \in \mathbb{N}$,

$$\varphi_i(x) := \frac{v_i(x)}{\sum\limits_{j=1}^{\infty} v_j(x)}, \quad x \in X.$$

Step 3: Finally, if $\{U_\alpha\}_{\alpha \in I}$ is an arbitrary open cover of X, note that for any $x \in X$,

$$X = \bigcup_{n=1}^{\infty} \overline{B(x, n)}.$$

Since each closed ball $\overline{B(x, n)}$ is compact it may be covered by a finite number of open sets in the cover. Thus we may select a countable subcover of X and then apply Step 2 (defining $\varphi_\alpha \equiv 0$ if U_α does not belong to the countable subcover).

Remark 6.19. (i) Note that as a consequence of Step 3 in the above proof, all but at most a countable number of φ_α are identically zero.

(ii) Given $x_0 \in X$ it is possible to construct the locally finite partition of unity such that $\varphi_\alpha(x_0) = 1$ for some α and thus $\varphi_\beta(x_0) = 0$ for all $\beta \neq \alpha$. To see this, choose $\alpha \in I$ such that $x_0 \in U_\alpha$ and let $r > 0$ be such that $\overline{B(x_0, r)} \subset U_\alpha$. It suffices to apply the previous theorem to the open cover $\left\{ \hat{U}_\beta \right\}_{\beta \in I}$ of X, where

$$\hat{U}_\beta := \begin{cases} U_\beta \setminus \overline{B(x_0, r)} & \text{if } \beta \neq \alpha, \\ U_\alpha & \text{if } \beta = \alpha. \end{cases}$$

As a consequence of the previous result we have the following.

Lemma 6.20. *Let X be a metric space, let Y be a locally convex topological vector space, and let*

$$\Gamma : X \to \{ C \subset Y : C \text{ convex, nonempty} \}$$

be a lower semicontinuous multifunction. If $x_0 \in X$, $y_0 \in \Gamma(x_0)$, and if $A \subset Y$ is a convex, balanced neighborhood of the origin, then there exists a continuous function $u : X \to Y$ such that $u(x_0) = y_0$ and $u(x) \in (\Gamma(x) + A)$ for all $x \in X$.

Proof. For every $x \in X$ choose $y_x \in \Gamma(x)$, with $y_{x_0} := y_0$. Since Γ is lower semicontinuous it follows that $\Gamma^-(y_x + A)$ is open in X and thus the family $\{\Gamma^-(y_x + A)\}_{x \in X}$ is an open cover of X. By Remark 6.19 there exists a locally finite partition of unity $\{\varphi_x\}_{x \in X}$ subordinated to $\{\Gamma^-(y_x + A)\}_{x \in X}$ such that $\varphi_{x_0}(x_0) = 1$ and $\varphi_x(x_0) = 0$ for all $x \neq x_0$. Note that if $\varphi_x(z) > 0$, then $z \in \Gamma^-(y_x + A)$, and so $y_x \in \Gamma(z) + A$, where we have used the fact that A is balanced. Since $\Gamma(z)$ and A are convex sets, so is their sum $\Gamma(z) + A$, and since $\{\varphi_x\}_{x \in X}$ is locally finite, the convex combination

$$u(z) := \sum_{x \in X} \varphi_x(z) y_x$$

belongs to $\Gamma(z) + A$. The function u is continuous and

$$u(x_0) = \sum_{x \in X} \varphi_x(x_0) y_x = \varphi_{x_0}(x_0) y_{x_0} = y_0.$$

This completes the proof.

Lemma 6.21. *Let X be a metric space, let Y be a topological vector space, and let*

$$\Gamma : X \to \mathcal{P}(Y) \setminus \{\emptyset\}$$

be a lower semicontinuous multifunction. Let $A \subset Y$ be a neighborhood of the origin and let $u : X \to Y$ be a continuous function such that

$$\Gamma(x) \cap (u(x) + A) \neq \emptyset$$

for all $x \in X$. Then the multifunction $\Gamma' : X \to \mathcal{P}(Y) \setminus \{\emptyset\}$, defined by

$$\Gamma'(x) := \Gamma(x) \cap (u(x) + A), \quad x \in X,$$

is lower semicontinuous.

Proof. Fix an open set $W \subset Y$. To prove that Γ' is lower semicontinuous we need to show that the set $(\Gamma')^-(W)$ is open in X. Let $x_0 \in (\Gamma')^-(W)$. If $y_0 \in \Gamma'(x_0) \cap W$, then in particular, by the definition of Γ', y_0 belongs to the open set $(u(x_0) + A) \cap W$, and thus we may find a balanced neighborhood $D \subset Y$ of the origin such that

$$y_0 + D + D \subset (u(x_0) + A) \cap W, \tag{6.16}$$

where we have used the continuity of addition. Since u is continuous, the set $U := u^{-1}(u(x_0) + D)$ is a neighborhood of x_0 in X. Also, since $y_0 \in \Gamma(x_0) \cap (y_0 + D)$, we have that x_0 belongs to the set $\Gamma^-(y_0 + D)$, which is open since Γ is lower semicontinuous.

We claim that the neighborhood $U_0 := U \cap \Gamma^-(y_0 + D)$ of x_0 is contained in $(\Gamma')^-(W)$. Given the arbitrariness of $x_0 \in (\Gamma')^-(W)$ this will prove that $(\Gamma')^-(W)$ is an open set in X, and in turn that Γ' is lower semicontinuous.

To prove the claim we begin by showing that

$$y_0 + D \subset u(x) + A \tag{6.17}$$

for all $x \in U$. To see this, let $x \in U$ and $y \in D$. By the definition of U and the fact that D is balanced we have that $(u(x_0) - u(x)) \in D$, while by (6.16) we may find $z \in A$ such that $u(x_0) + z \in W$ and

$$y_0 + (u(x_0) - u(x)) + y = u(x_0) + z.$$

Hence
$$y_0 + y = u(x) + z \in u(x) + A$$
and so (6.17) holds.

If $x \in U_0 = U \cap \Gamma^-(y_0 + D)$, then there exists $y \in \Gamma(x) \cap (y_0 + D)$, and so by (6.16) we have that $y = y + 0 \in y_0 + D + D \subset W$. On the other hand, by (6.17),

$$\Gamma'(x) \cap W = \Gamma(x) \cap (u(x) + A) \cap W \supset \Gamma(x) \cap (y_0 + D) \cap W \neq \emptyset,$$

since $y \in \Gamma(x) \cap (y_0 + D) \cap W$. Hence $\Gamma'(x) \cap W \neq \emptyset$ for all $x \in U_0$, which shows that $U_0 \subset (\Gamma')^-(W)$.

We now turn to the proof of Michael's continuous selection theorem.

Proof (Michael's continuous selection theorem). By Lemma 6.20 there exists a continuous function $u_1 : X \to Y$ such that $u_1(x_0) = y_0$ and

$$u_1(x) \in \left(\Gamma(x) + B\left(0, \frac{1}{2}\right) \right)$$

for all $x \in X$. Inductively, assume that continuous functions $u_1, \dots, u_\ell : X \to Y$ have been constructed such that $u_n(x_0) = y_0$,

$$u_n(x) \in \left(\Gamma(x) + B\left(0, \frac{1}{2^n}\right) \right), \quad u_n(x) \in \left(u_{n-1}(x) + B\left(0, \frac{1}{2^{n-2}}\right) \right)$$
(6.18)

for all $x \in X$ and all $n = 2, \dots, \ell$. Define the multifunction

$$\Gamma_\ell(x) := \Gamma(x) \cap \left(u_\ell(x) + B\left(0, \frac{1}{2^\ell}\right) \right). \tag{6.19}$$

Note that for all $x \in X$ the set $\Gamma_\ell(x)$ is nonempty by (6.18) and convex, since it is the intersection of two convex sets. By Lemma 6.21, Γ_ℓ is lower semicontinuous, and so we may apply Lemma 6.20 to Γ_ℓ to find a continuous function $u_{\ell+1} : X \to Y$ such that for all $x \in X$,

$$u_{\ell+1}(x) \in \left(\Gamma_\ell(x) + B\left(0, \frac{1}{2^{\ell+1}}\right) \right).$$

Then, by (6.19), $u_{\ell+1}(x) \in (\Gamma(x) + B(0, \frac{1}{2^{\ell+1}}))$, while

$$u_{\ell+1}(x) \in \left(u_\ell(x) + B\left(0, \frac{1}{2^\ell}\right) + B\left(0, \frac{1}{2^{\ell+1}}\right) \right) \subset u_\ell(x) + B\left(0, \frac{1}{2^{\ell-1}}\right)$$

for all $x \in X$.

Hence we have constructed a sequence $\{u_n\}$ of continuous functions satisfying (6.18) for all positive integers $n \geq 2$. In particular, by (6.18)$_2$,

$$\|u_{n+\ell}(x) - u_n(x)\|_Y \leq \frac{1}{2^{n-2}}$$

for all $x \in X$ and for all $n \geq 2$ and $\ell \in \mathbb{N}$. Since Y is a Banach space it follows that $C(X;Y)$ is also a Banach space and thus $\{u_n\}$ is a Cauchy sequence in $C(X;Y)$. Thus $\{u_n\}$ converges uniformly to a continuous function $u : X \to Y$ such that $u(x_0) = y_0$. Since $\Gamma(x)$ is a closed set, it follows from $(6.18)_1$ that $u(x) \in \Gamma(x)$ for all $x \in X$.

Remark 6.22. Note that the fact that X is a metric space has been used in Lemma 6.20 only to guarantee the existence of a locally finite partition of unity subordinated to an arbitrary cover of X. Thus Michael's continuous selection theorem continues to hold when X is a topological vector space for which a locally finite partition of unity subordinated to an arbitrary cover of X exists, namely for paracompact spaces. We refer to [Mi56] and [W04] for more details.

As a corollary of the previous theorem we have the following result (see also Example 6.16 above).

Corollary 6.23. *Let X, Y be Banach spaces and let $L : Y \to X$ be linear, continuous, and onto. Then there exists a continuous function $u : X \to Y$ such that $u(x) \in L^{-1}(\{x\})$ for all $x \in X$.*

Proof. Define
$$\Gamma : X \to \mathcal{P}(Y) \setminus \{\emptyset\}$$
as
$$\Gamma(x) := L^{-1}(\{x\}), \quad x \in X.$$
By the open mapping theorem the function L sends open sets into open sets and thus (see Example 6.16 above) Γ is lower semicontinuous. Since L is linear, continuous, and onto, the set $\Gamma(x)$ is convex, closed, and nonempty for all $x \in X$. Hence we may apply the previous theorem to obtain a continuous selection $u : X \to Y$.

6.2 Integrands

In the remainder of this chapter, $E \subset \mathbb{R}^N$ is a Lebesgue measurable set, and as usual, $\mathcal{B}(E)$ is the class of Borel subsets of E.

6.2.1 Equivalent Integrands

Consider a function
$$f : E \times \mathbb{R}^m \to [-\infty, \infty].$$
We say that $f : E \times \mathbb{R}^m \to [-\infty, \infty]$ is $\mathcal{L}^N \times \mathcal{B}$ *measurable* if it is measurable with respect to the σ-algebra generated by products of Lebesgue measurable subsets of E and Borel subsets of $\mathbb{R}^m$. Then for every measurable function $v : E \to \mathbb{R}^m$ the function $f(\cdot, v(\cdot))$ is Lebesgue measurable.

Proposition 6.24. *Let $E \subset \mathbb{R}^N$ be a Lebesgue measurable set and let f, g : $E \times \mathbb{R}^m \to [-\infty, \infty]$ be $\mathcal{L}^N \times \mathcal{B}$ measurable functions. Assume that for every $M > 0$ there exists a nonnegative function $\gamma_M \in L^1(E)$ such that*

$$f(x, z), g(x, z) \geq -\gamma_M(x)$$

for $\mathcal{L}^N$ a.e. $x \in E$ and all $z \in \mathbb{R}^m$ with $|z| \leq M$. Then the following two conditions are equivalent:

(i) for every $v \in L^\infty(E; \mathbb{R}^m)$ and every Lebesgue measurable set $G \subset E$,

$$\int_G f(x, v(x)) \, dx \leq \int_G g(x, v(x)) \, dx;$$

(ii) $f(x, z) \leq g(x, z)$ for $\mathcal{L}^N$ a.e. $x \in E$ and all $z \in \mathbb{R}^m$.

Proof. Obviously (ii) implies (i). The proof of the converse implication is established in two steps.

Step 1: Assume that E has finite measure. For $k \in \mathbb{N}$ define $f_k := \inf\{f, k\}$, $g_k := \inf\{g, k\}$. It is enough to show that

$$f_k(x, z) \leq g_k(x, z) \text{ for } \mathcal{L}^N \text{ a.e. } x \in E \text{ and all } z \in \mathbb{R}^m \text{ with } |z| \leq k.$$

For $n \in \mathbb{N}$ define

$$S_{n,k} := \left\{ (x, z) \in E \times \mathbb{R}^m : |z| \leq k, \, f_k(x, z) > g_k(x, z) + \frac{1}{n} \right\},$$

$$S_{n,k}(x) := \{z \in \mathbb{R}^m : (x, z) \in S_{n,k}\},$$
$$\Pi_{n,k} := \{x \in E : S_{n,k}(x) \neq \emptyset\}.$$

Since the set $S_{n,k}$ is $\mathcal{L}^N \times \mathcal{B}$ measurable, we have that $\Pi_{n,k}$ is Lebesgue measurable by the projection theorem. We claim that $|\Pi_{n,k}| = 0$. Indeed, assume by contradiction that $|\Pi_{n,k}| > 0$. Then by Aumann's selection theorem we may find measurable functions $v_{n,k} : \Pi_{n,k} \to \mathbb{R}^m$ such that $v_{n,k}(x) \in S_{n,k}(x)$ for every $x \in \Pi_{n,k}$. Hence $|v_{n,k}(x)| \leq k$. Extend $v_{n,k}$ to be zero on $E \setminus \Pi_{n,k}$. Then $v_{n,k} \in L^\infty(E; \mathbb{R}^m)$, and so by hypothesis,

$$\int_{\Pi_{n,k}} f(x, v_{n,k}(x)) \, dx \leq \int_{\Pi_{n,k}} g(x, v_{n,k}(x)) \, dx. \tag{6.20}$$

By definition of $S_{n,k}$ and f_k,

$$k \geq f_k(x, v_{n,k}(x)) > g_k(x, v_{n,k}(x)) + \frac{1}{n} \tag{6.21}$$

for all $x \in \Pi_{n,k}$, and so, since $k > g_k(x, v_{n,k}(x)) = \inf\{g(x, v_{n,k}(x)), k\}$, we have that

$$g\left(x, v_{n,k}\left(x\right)\right) + \frac{1}{n} = g_k\left(x, v_{n,k}\left(x\right)\right) + \frac{1}{n} < f_k\left(x, v_{n,k}\left(x\right)\right) \leq f\left(x, v_{n,k}\left(x\right)\right),$$

and upon integration over $\Pi_{n,k}$,

$$\int_{\Pi_{n,k}} g\left(x, v_{n,k}\left(x\right)\right)\,dx + \frac{1}{n}\left|\Pi_{n,k}\right| \leq \int_{\Pi_{n,k}} f\left(x, v_{n,k}\left(x\right)\right)\,dx$$

$$\leq \int_{\Pi_{n,k}} g\left(x, v_{n,k}\left(x\right)\right)\,dx$$

by (6.20). Since E is bounded, it follows by (6.21) that the right-hand side of the previous inequality is finite, and thus $\left|\Pi_{n,k}\right| = 0$. Define

$$N_k := \bigcup_{n=1}^{\infty} \Pi_{n,k}.$$

Then $\left|N_k\right| = 0$ and hence

$$f_k\left(x, z\right) \leq g_k\left(x, z\right) \text{ for all } x \in E \setminus N_k \text{ and all } z \in \mathbb{R}^m \text{ with } \left|z\right| \leq k.$$

Step 2: In the case that E has infinite measure it suffices to apply Step 1 to $E_j := E \cap B\left(0, j\right)$.

Remark 6.25. Since every Lebesgue measurable set $G \subset E$ may be written as the union of a Borel set and a set of Lebesgue measure zero, property (i) may be equivalently stated for Borel sets $G \subset E$.

Similarly, we can prove the following result.

Proposition 6.26. *Let $E \subset \mathbb{R}^N$ be a Lebesgue measurable set and let $f, g : E \times \mathbb{R}^m \to \left[-\infty, \infty\right]$ be $\mathcal{L}^N \times \mathcal{B}$ measurable functions such that*

$$f\left(x, v\left(x\right)\right) = g\left(x, v\left(x\right)\right) \text{ for } \mathcal{L}^N \text{ a.e. } x \in E$$

for every $v \in C_b\left(E; \mathbb{R}^m\right)$. Then

$$f\left(x, z\right) = g\left(x, z\right) \text{ for } \mathcal{L}^N \text{ a.e. } x \in E \text{ and all } z \in \mathbb{R}^m.$$

Proof. Let

$$S := \left\{\left(x, z\right) \in E \times \mathbb{R}^m : f\left(x, z\right) \neq g\left(x, z\right)\right\},$$
$$S\left(x\right) := \left\{z \in \mathbb{R}^m : \left(x, z\right) \in S\right\},$$
$$\Pi := \left\{x \in E : S\left(x\right) \neq \emptyset\right\}.$$

Since the set S is $\mathcal{L}^N \times \mathcal{B}$ measurable, by the projection theorem we have that Π is a Lebesgue measurable set, and so by Aumann's selection theorem we may find measurable functions $w : \Pi \to \mathbb{R}^m$ such that $w\left(x\right) \in S\left(x\right)$ for every $x \in \Pi$. Assume by contradiction that $\mathcal{L}^N\left(\Pi\right) > 0$. Then by Lusin's theorem there exists a compact set $K \subset \Pi$ such that $\mathcal{L}^N\left(K\right) > 0$ and w is continuous on K. By the Tietze extension theorem we may extend w outside K as a function $v \in C_b\left(E; \mathbb{R}^m\right)$. Hence $\left(x, v\left(x\right)\right) \in S$ for all $x \in K$, but this contradicts the hypothesis.

Motivated by the previous proposition, in what follows we say that two functions $f, g : E \times \mathbb{R}^m \to [-\infty, \infty]$ are *equivalent integrands* (in the L^p sense) if for every $v \in C_b (E; \mathbb{R}^m)$ we have

$$g(x, v(x)) = f(x, v(x)) \text{ for } \mathcal{L}^N \text{ a.e. } x \in E.$$

Note that for equivalent integrands,

$$\widetilde{\int_E} f(x, v(x)) \, dx = \int_E g(x, v(x)) \, dx$$

whenever the integrals are defined. Since Lebesgue integration does not distinguish g from f, for simplicity, and in analogy with Lebesgue spaces, we identify equivalent integrands.

6.2.2 Normal and Carathéodory Integrands

In this subsection we introduce the notion of normal and Carathéodory integrands.

Definition 6.27. *Let $E \subset \mathbb{R}^N$ be a Lebesgue measurable set and let $B \subset \mathbb{R}^m$ be a Borel set. A function $f : E \times B \to [-\infty, \infty]$ is said to be a normal integrand if:*

(i) for $\mathcal{L}^N$ a.e. $x \in E$ the function $f(x, \cdot)$ is lower semicontinuous on B;
(ii) there exists a Borel function $g : E \times B \to [-\infty, \infty]$ such that

$$f(x, \cdot) = g(x, \cdot)$$

for $\mathcal{L}^N$ a.e. $x \in E$.

The next result can be regarded as a "uniform" Lusin theorem.

Theorem 6.28. *Let $E \subset \mathbb{R}^N$ be a Lebesgue measurable set, let $B \subset \mathbb{R}^m$ be a Borel set, and let $f : E \times B \to [-\infty, \infty]$ be such that for $\mathcal{L}^N$ a.e. $x \in E$ the function $f(x, \cdot)$ is lower semicontinuous. Then the following conditions are equivalent:*

(i) f is a normal integrand.
(ii) For every $\varepsilon > 0$ there exists a closed set $C_\varepsilon \subset E$, with $\mathcal{L}^N (E \setminus C_\varepsilon) \leq \varepsilon$, such that the restriction of f to $C_\varepsilon \times B$ is lower semicontinuous.

Proof. **Step 1:** We prove that (i) implies (ii). By eventually modifying f on a subset of E of Lebesgue measure zero, we may assume that $f(x, \cdot)$ is lower semicontinuous on B for all $x \in E$ and that f is a Borel function on $E \times B$. Moreover, after composing f with an isomorphism from $[-\infty, \infty]$ to $[0, 1]$, we may assume that $f : E \times B \to [0, 1]$. Let $\mathcal{A}$ be the countable family obtained by taking all finite unions of open balls with rational radii centered at $z \in \mathbb{Q}^m$ and

$$\mathcal{F} := \{ r\chi_A : r \in \mathbb{Q} \cap [0,1],\ A \in \mathcal{A} \}.$$

Then $\mathcal{F}$ is a countable family of lower semicontinuous functions. For any lower semicontinuous $g : B \to [0,1]$ we have

$$g(z) = \sup \{ \psi(z) : \psi \in \mathcal{F} \text{ and } \psi \leq g \} \tag{6.22}$$

for all $z \in B$. Indeed, the inequality $\geq$ is immediate. To prove the reverse inequality fix $z_0 \in B$. If $g(z_0) = 0$, then there is nothing to prove. Thus assume that $g(z_0) > 0$ and let $r \in \mathbb{Q}$ be such that

$$g(z_0) > r > 0.$$

Since g is lower semicontinuous, there exists $\varepsilon > 0$ such that

$$g(z) > r$$

for all $z \in B(z_0, \varepsilon)$. Consider a ball $B(z_1, \varepsilon_1) \in \mathcal{A}$ such that

$$z_0 \in B(z_1, \varepsilon_1) \subset B(z_0, \varepsilon).$$

Then

$$\sup \{ \psi(z_0) : \psi \in \mathcal{F} \text{ and } \psi \leq g \} \geq r\chi_{B(z_1,\varepsilon_1)}(z_0) = r.$$

It suffices to let $r \nearrow g(z_0)$ to obtain (6.22).

Since $\mathcal{F}$ is countable, we may write $\mathcal{F} = \{ r_n \chi_{A_n} \}$, and set

$$G_n := \{ x \in E : f(x, \cdot) \geq r_n \chi_{A_n}(\cdot) \},$$
$$F_n := \{ (x,z) \in E \times B : f(x,z) < r_n \chi_{A_n}(z) \}.$$

Using the facts that f is a Borel function and $r_n \chi_{A_n}$ is a lower semicontinuous function, we have that the set F_n is $\mathcal{L}^N \times \mathcal{B}$ measurable. The set G_n is the complement of the projection of F_n and thus it is Lebesgue measurable by the projection theorem.

Since $f(x, \cdot)$ is lower semicontinuous on B, for every $x \in E$ in view of (6.22) we may write

$$f(x,z) = \sup_n \chi_{G_n}(x)\, r_n \chi_{A_n}(z) \tag{6.23}$$

for all $(x,z) \in E \times B$. By Remark 1.95 applied to each function χ_{G_n}, for every fixed $\varepsilon > 0$ we may find a closed set $C_n \subset E$ such that

$$\mathcal{L}^N (E \setminus C_n) \leq \frac{\varepsilon}{2^{n+1}}$$

and the restriction of χ_{G_n} to C_n is continuous. Let

$$C_\varepsilon := \bigcap_{n=1}^{\infty} C_n.$$

Then C_ε is closed and $\mathcal{L}^N (E \setminus C_\varepsilon) \le \varepsilon$.

Since each function $\chi_{G_n} (x) \, r_n \chi_{A_n} (z)$ restricted to $C_\varepsilon \times B$ is lower semi-continuous, it follows that f, which is their pointwise supremum, is also lower semicontinuous on $C_\varepsilon \times B$ by Proposition 3.5.

Step 2: We prove that (ii) implies (i). Taking $\varepsilon := \frac{1}{k}$, $k \in \mathbb{N}$, let $C_k \subset E$ be the closed set given in (ii) and let

$$F := \bigcup_{k=1}^{\infty} C_k.$$

Then F is a Borel set, with $\mathcal{L}^N (E \setminus F) = 0$, and $f : F \times B \to [-\infty, \infty]$ is a Borel function. Setting

$$g(x, z) := \begin{cases} f(x, z) & \text{if } x \in F, \\ \infty & \text{if } x \notin F, \end{cases}$$

we have that g is a Borel function such that $f(x, \cdot) = g(x, \cdot)$ for $\mathcal{L}^N$ a.e. $x \in E$. Hence f is a normal integrand. $\blacksquare$

Remark 6.29. (i) Theorem 6.28 still holds if the Lebesgue measure $\mathcal{L}^N$ is replaced by a (positive) Radon measure on a measurable set E.

(ii) It follows from the previous theorem that if f is a normal integrand, then for every $\varepsilon > 0$ and every compact set $K \subset E$ there exists a compact set $K_\varepsilon \subset K$ such that $\mathcal{L}^N (K \setminus K_\varepsilon) \le \varepsilon$ for which the restriction of f to $K_\varepsilon \times B$ is lower semicontinuous. To see this, it suffices to apply Theorem 6.28 to f restricted to $K \times B$.

In the special case that f is nonnegative we obtain a stronger result.

Corollary 6.30. *Let $E \subset \mathbb{R}^N$ be a Lebesgue measurable set, let $B \subset \mathbb{R}^m$ be a Borel set, and let $f : E \times B \to [0, \infty]$ be a normal integrand. Then there exists a sequence $\{(E_n, \varphi_n)\}$, where $E_n \subset E$ has finite Lebesgue measure and $\varphi_n \in C_c(\mathbb{R}^m)$, such that*

$$f(x, z) = \sup_n \chi_{E_n} (x) \varphi_n (z)$$

for every $(x, z) \in E \times B$.

Proof. We proceed as in the previous theorem, with the exception that $[-\infty, \infty]$ need not be replaced by $[0, 1]$ via an isomorphism because f is bounded from below. Accordingly, the family $\mathcal{F}$ now is

$$\mathcal{F} := \{r\chi_A : r \in \mathbb{Q} \cap [0, \infty), \, A \in \mathcal{A}\}.$$

Then, as before, (6.23) holds, that is,

$$f(x, z) = \sup_n \chi_{G_n} (x) \, r_n \chi_{A_n} (z) \tag{6.24}$$

for every $(x, z) \in E \times B$. Since $A_n \subset \mathbb{R}^m$ is open, for each $k \in \mathbb{N}$ the function

$$\psi_{n,k}(z) := \inf \{1, k \, \mathrm{dist}(z, \mathbb{R}^m \setminus A_n)\}$$

is continuous and

$$\chi_{A_n}(z) = \sup_k \psi_{n,k}(z).$$

Thus for every $(x, z) \in E \times B$,

$$f(x, z) = \sup_{n,k} \chi_{G_n}(x) \, r_n \psi_{n,k}(z).$$

Let $\{F_j\}$ be an increasing sequence of Lebesgue sets with finite measure such that

$$\bigcup_{j=1}^{\infty} F_j = E,$$

and let $\{\psi_j\} \subset C_c(\mathbb{R}^m)$ be an increasing sequence of nonnegative functions such that for every $z \in \mathbb{R}^m$,

$$\sup_j \psi_j(z) = 1.$$

Then for every $(x, z) \in E \times B$,

$$f(x, z) = \sup_{n,k,j} \chi_{G_n \cap F_j}(x) \, r_n \psi_{n,k}(z) \, \psi_j(z).$$

This completes the proof.

In the next result we present conditions that are equivalent to normality and may be easier to verify in applications.

We will also give an alternative proof to Theorem 6.28. For simplicity we consider only the case $B := \mathbb{R}^m$.

Proposition 6.31. *Let $E \subset \mathbb{R}^N$ be a Lebesgue measurable set and let $f : E \times \mathbb{R}^m \to [-\infty, \infty]$ be such that for $\mathcal{L}^N$ a.e. $x \in E$ the function $f(x, \cdot)$ is lower semicontinuous. Then the following conditions are equivalent:*

(i) f is a normal integrand;
(ii) f is $\mathcal{L}^N \times \mathcal{B}$ measurable;
(iii) the set

$$E_\infty := \{x \in E : f(x, \cdot) \text{ is lower semicontinuous and } \not\equiv \infty\}$$

is Lebesgue measurable and the multifunction

$$\Gamma : E_\infty \to \{C \subset \mathbb{R}^{m+1} : C \text{ nonempty, closed}\}$$

defined as

$$\Gamma(x) := \mathrm{epi}\, f(x, \cdot) = \{(z, t) \in \mathbb{R}^m \times \mathbb{R} : f(x, z) \leq t\}, \quad x \in E_\infty,$$

is measurable.

The proof is hinged on the following result, which is of interest in itself.

Lemma 6.32. *Let $G \subset \mathbb{R}^N$ be a Borel set endowed with the σ-algebra of Lebesgue measurable sets and let*

$$\Gamma : G \to \{C \subset \mathbb{R}^m : C \text{ nonempty, closed}\}$$

be a multifunction. Then the following conditions are equivalent:

(i) Γ is measurable;
(ii) there exists a multifunction

$$\Gamma' : G \to \{C \subset \mathbb{R}^m : C \text{ nonempty, closed}\}$$

such that $\mathrm{Gr}\,\Gamma'$ is a Borel set of $G \times \mathbb{R}^m$, and $\Gamma(x) = \Gamma'(x)$ for $\mathcal{L}^N$ a.e. $x \in G$;
(iii) for every $\varepsilon > 0$ there exists a closed set $C_\varepsilon \subset G$, with $\mathcal{L}^N(G \setminus C_\varepsilon) \leq \varepsilon$, such that the graph of Γ restricted to C_ε, that is,

$$\{(x, z) \in C_\varepsilon \times \mathbb{R}^m : z \in \Gamma(x)\},$$

is closed.

Proof. **Step 1:** We first prove that (i) implies (iii) under the additional assumption that $\mathcal{L}^N(G) < \infty$. Let $\{C_n\}$ be the countable family of all closed subsets of $\mathbb{R}^m$ that are complementary to open balls with center in $\mathbb{Q}^m$ and positive rational radius. Write $\mathbb{R}^m \setminus C_n = B_m(z_n, r_n)$. We claim that for all $x \in G$,

$$\Gamma(x) = \bigcap_{n:\, C_n \supset \Gamma(x)} C_n.$$

Indeed, if $x \in G$, then the set $\mathbb{R}^m \setminus \Gamma(x)$ is open, and so it can be written as the union of all open balls with center in $\mathbb{Q}^m$ and positive rational radius that are contained in it. De Morgan's law now establishes the claim. Let $G_n := \Gamma^-(B_m(z_n, r_n))$ and set

$$F_n := G \setminus G_n = \{x \in G : \Gamma(x) \subset C_n\}.$$

Then G_n and F_n are Lebesgue measurable by Proposition 6.3 and

$$\mathrm{Gr}\,\Gamma = \{(x, z) \in G \times \mathbb{R}^m : z \in \Gamma(x)\} \tag{6.25}$$

$$= \bigcap_{n=1}^{\infty} [(G_n \times \mathbb{R}^m) \cup (F_n \times C_n)].$$

To see the latter, fix $(x, z) \in \mathrm{Gr}\,\Gamma$ and $n \in \mathbb{N}$. If $\Gamma(x) \subset C_n$, then $x \in F_n$, and so $(x, z) \in F_n \times C_n$, while if $\Gamma(x)$ is not contained in C_n, then $x \in G_n$, and so

$$\operatorname{Gr}\Gamma \subset \bigcap_{n=1}^{\infty}\left[(G_n \times \mathbb{R}^m) \cup (F_n \times C_n)\right].$$

Conversely, let $(x,z) \in \bigcap_{n=1}^{\infty}\left[(G_n \times \mathbb{R}^m) \cup (F_n \times C_n)\right]$ and assume by contradiction that $z \notin \Gamma(x)$. Since $\Gamma(x)$ is closed we have that $\operatorname{dist}(z, \Gamma(x)) > 0$, and so by the density of the rationals in the reals we may find $n \in \mathbb{N}$ such that $z \in B_m(z_n, r_n)$ and $B_m(z_n, r_n) \cap \Gamma(x) = \emptyset$. This last condition implies that $x \notin G_n$; therefore $(x,z) \in (F_n \times C_n)$, which is a contradiction since $z \in B_m(z_n, r_n)$. Thus (6.25) holds.

For every $\varepsilon > 0$ and $n \in \mathbb{N}$ we may find compact sets $K_n \subset G_n$ and $K_n' \subset F_n$ such that

$$\mathcal{L}^N\left(G \setminus (K_n \cup K_n')\right) \le \frac{\varepsilon}{2^n}.$$

Define

$$K_\varepsilon := \bigcap_{n=1}^{\infty} K_n \cup K_n'.$$

Then K_ε is compact,

$$\mathcal{L}^N(G \setminus K_\varepsilon) = \mathcal{L}^N\left(G \setminus \left(\bigcap_{n=1}^{\infty} K_n \cup K_n'\right)\right)$$

$$= \mathcal{L}^N\left(\bigcup_{n=1}^{\infty}(G \setminus (K_n \cup K_n'))\right) \le \varepsilon,$$

and, by (6.25),

$$\{(x,z) \in K_\varepsilon \times \mathbb{R}^m : z \in \Gamma(x)\} = \bigcap_{n=1}^{\infty}\left[(K_n \times \mathbb{R}^m) \cup (K_n' \times C_n)\right],$$

which is a closed set.

Step 2: To prove that (i) implies (iii) in the general case, we write

$$G = \bigcup_{k=1}^{\infty} G^{(k)}, \quad G^{(k)} := \{x \in G : k - 1 \le |x| < k\},$$

and apply the previous step to Γ restricted to each $G^{(k)}$ to find compact sets $K^{(k)} \subset G^{(k)}$, with

$$\mathcal{L}^N\left(G^{(k)} \setminus K^{(k)}\right) \le \frac{\varepsilon}{2^k},$$

such that the set

$$\left\{(x,z) \in K^{(k)} \times \mathbb{R}^m : z \in \Gamma(x)\right\}$$

is closed. Let

$$C_\varepsilon := \bigcup_{k=1}^{\infty} K^{(k)}.$$

410 6 Integrands $f = f(x, z)$

Then only finitely many $K^{(k)}$ intersect any bounded region, and so C_ε is closed. We claim that the set

$$\{(x, z) \in C_\varepsilon \times \mathbb{R}^m : z \in \Gamma(x)\}$$

is closed. Indeed, let $\{(x_j, z_j)\} \subset C_\varepsilon \times \mathbb{R}^m$ be such that $z_j \in \Gamma(x_j)$ for all $j \in \mathbb{N}$ and $(x_j, z_j) \to (x, z)$ as $j \to \infty$, for some $(x, z) \in C_\varepsilon \times \mathbb{R}^m$. Then the sequence $\{x_j\}$ must be bounded by some positive integer ℓ, and so

$$(x_j, z_j) \in \bigcup_{k=1}^{\ell} \left\{(x, z) \in K^{(k)} \times \mathbb{R}^m : z \in \Gamma(x)\right\}$$

for all $j \in \mathbb{N}$. Since it is a finite union of closed sets, this set is closed, and thus it must contain (x, z), that is, $z \in \Gamma(x)$. Thus the claim holds.

Step 3: Next we show that (iii) implies (ii). Taking $\varepsilon := \frac{1}{k}$, $k \in \mathbb{N}$, let $C_k \subset G$ be the closed set given in (iii) and let

$$F := \bigcup_{k=1}^{\infty} C_k.$$

Then F is a Borel set and $\mathcal{L}^N(G \setminus F) = 0$. Moreover, the set

$$\{(x, z) \in F \times \mathbb{R}^m : z \in \Gamma(x)\} = \bigcup_{k=1}^{\infty} \{(x, z) \in C_k \times \mathbb{R}^m : z \in \Gamma(x)\}$$

is a Borel set since it is the union of closed sets. Define

$$\Gamma' : G \to \{C \subset \mathbb{R}^m : C \text{ nonempty, closed}\}$$

as

$$\Gamma'(x) := \begin{cases} \Gamma(x) & \text{if } x \in F, \\ \mathbb{R}^m & \text{if } x \in G \setminus F. \end{cases}$$

Then $\Gamma(x) = \Gamma'(x)$ for all $x \in F$ and

$$\mathrm{Gr}\, \Gamma' = \{(x, z) \in F \times \mathbb{R}^m : z \in \Gamma(x)\} \cup ((G \setminus F) \times \mathbb{R}^m),$$

which is a Borel set.

Step 4: Finally, we prove that (ii) implies (i). The equivalence of the fact that $\mathrm{Gr}\, \Gamma'$ is a Borel set with the weak measurability of the multifunction Γ' is a consequence of Remark 6.11 together with the completeness of the Lebesgue measure on the σ-algebra of Lebesgue measurable subsets. In turn, by Remark 6.4, Γ' is measurable. Let $C \subset \mathbb{R}^m$ be any closed set. Then $(\Gamma')^-(C)$ is Lebesgue measurable. Since $\Gamma(x) = \Gamma'(x)$ for $\mathcal{L}^N$ a.e. $x \in G$, we deduce that $\Gamma^-(C)$ differs from $(\Gamma')^-(C)$ by at most a set of Lebesgue measure zero, and so, again by the completeness of the Lebesgue measure, it is Lebesgue measurable.

We now turn to the proof of Proposition 6.31

Proof (Proposition 6.31). Let $E_1 \subset E$ be the set of points $x \in E$ such that the function $f(x, \cdot)$ is lower semicontinuous. By hypothesis $\mathcal{L}^N (E \setminus E_1) = 0$.

Step 1: We prove that (ii) implies (iii). Since f is $\mathcal{L}^N \times \mathcal{B}$ measurable, then so is the function

$$h : E_1 \times \mathbb{R}^m \times \mathbb{R} \to [-\infty, \infty]$$

defined by

$$h(x, z, t) := f(x, z) - t.$$

In turn, the set

$$F := \{(x, z, t) \in E_1 \times \mathbb{R}^m \times \mathbb{R} : h(x, z, t) \leq 0\}$$

is $\mathcal{L}^N \times \mathcal{B}(\mathbb{R}^m \times \mathbb{R})$ measurable. By the projection theorem, the projection of F onto $\mathbb{R}^N$ is Lebesgue measurable. Since $\pi_{\mathbb{R}^N}(F) = E_\infty$, we have proved that the set E_∞ is Lebesgue measurable. It remains to show that Γ is measurable. Note that if $x \in E_\infty$, then the function $f(x, \cdot)$ is lower semicontinuous, and so the set epi $f(x, \cdot)$ is closed in $\mathbb{R}^m \times \mathbb{R}$ by Proposition 3.4. Thus Γ is well-defined. Moreover,

$$\mathrm{Gr}\, \Gamma = \{(x, z, t) \in E_\infty \times \mathbb{R}^m \times \mathbb{R} : (z, t) \in \Gamma(x)\} = F,$$

and so it is $\mathcal{L}^N \times \mathcal{B}(\mathbb{R}^m \times \mathbb{R})$ measurable. Thus, by Remark 6.11 and the completeness of the Lebesgue measure, Γ is weakly measurable, and actually measurable by Remark 6.4. Thus (iii) holds.

Step 2: We show that (iii) implies (i). By Proposition 1.68 we may find a Borel set $G \subset E_\infty$ such that $\mathcal{L}^N (E_\infty \setminus G) = 0$. Since the restriction of Γ to G is still measurable, by Lemma 6.32, for every $\varepsilon > 0$ there exists a closed set $C_\varepsilon \subset G$, with $\mathcal{L}^N (G \setminus C_\varepsilon) \leq \varepsilon$, such that the graph of Γ restricted to C_ε, that is,

$$\{(x, z, t) \in C_\varepsilon \times \mathbb{R}^m \times \mathbb{R} : (z, t) \in \Gamma(x)\}$$
$$= \{(x, z, t) \in C_\varepsilon \times \mathbb{R}^m \times \mathbb{R} : f(x, z) \leq t\} = \mathrm{epi}\, f|_{C_\varepsilon \times \mathbb{R}^m},$$

is closed. It follows by Proposition 3.4 that the function

$$f : C_\varepsilon \times \mathbb{R}^m \to [-\infty, \infty]$$

is lower semicontinuous. On the other hand, if $x \in E_1 \setminus E_\infty$, then $f(x, z) = \infty$ for all $z \in \mathbb{R}^m$. By the inner regularity of the Lebesgue measure we may find a closed set $C'_\varepsilon \subset E_1 \setminus E_\infty$ such that $\mathcal{L}^N ((E_1 \setminus E_\infty) \setminus C'_\varepsilon) \leq \varepsilon$. Since C'_ε is closed, the function

$$f : (C_\varepsilon \cup C'_\varepsilon) \times \mathbb{R}^m \to [-\infty, \infty]$$

is still lower semicontinuous and $\mathcal{L}^N (E \setminus (C_\varepsilon \cup C'_\varepsilon)) \leq 2\varepsilon$. It now follows from Theorem 6.28 that f is a normal integrand.

Step 3: Finally, we prove that (i) implies (ii). Assume that f is a normal integrand. Then there exist a measurable set $E_2 \subset E$, with $\mathcal{L}^N (E \setminus E_2) = 0$, and a Borel function $g : E \times \mathbb{R}^m \to [-\infty, \infty]$ such that

$$f (x, \cdot) = g (x, \cdot)$$

for all $x \in E_2$. This implies that $f : E_2 \times \mathbb{R}^m \to [-\infty, \infty]$ is $\mathcal{L}^N \times \mathcal{B}$ measurable. Since $\mathcal{L}^N (E \setminus E_2) = 0$, by the completeness of the Lebesgue measure it follows that $f : E \times \mathbb{R}^m \to [-\infty, \infty]$ is $\mathcal{L}^N \times \mathcal{B}$ measurable.

Next we introduce the notion of Carathéodory function.

Definition 6.33. *Let $E \subset \mathbb{R}^N$ be a Lebesgue measurable set and let $B \subset \mathbb{R}^m$ be a Borel set. A function $f : E \times B \to [-\infty, \infty]$ is said to be a* Carathéodory *function if*

(i) for $\mathcal{L}^N$ a.e. $x \in E$ the function $f (x, \cdot)$ is continuous on B;
(ii) for all $z \in B$ the function $f (\cdot, z)$ is measurable on E.

The relation between normal integrands and Carathéodory functions is the subject of the next theorem.

Theorem 6.34. *Let $E \subset \mathbb{R}^N$ be a Lebesgue measurable set, let $B \subset \mathbb{R}^m$ be a Borel set, and let $f : E \times B \to [-\infty, \infty]$ be a Carathéodory function. Then f is a normal integrand.*

Proof. By eventually modifying f on a subset of E of Lebesgue measure zero, we may assume that $f (x, \cdot)$ is continuous on B for all $x \in E$ and that for all $z \in B$ the function $f (\cdot, z)$ is measurable on E. Moreover, after composing f with an isomorphism from $[-\infty, \infty]$ to $[0, 1]$, we may assume that $f : E \times B \to [0, 1]$. Let $\mathcal{A}$ and $\mathcal{F} = \{r_n \chi_{A_n}\}$ be as in Theorem 6.28 and let $\{z_k\}$ be dense in B. For each $k, n \in \mathbb{N}$, define

$$G_{k,n} := \{x \in E : f (x, z_k) \geq r_n \chi_{A_n} (z_k)\}.$$

Since $f (\cdot, z_k)$ is measurable on E it follows that the set $G_{k,n}$ is Lebesgue measurable. In turn, the set

$$G_n := \bigcap_{k=1}^{\infty} G_{k,n} = \{x \in E : f (x, z_k) \geq r_n \chi_{A_n} (z_k) \text{ for all } k \in \mathbb{N}\}$$

is Lebesgue measurable. We claim that

$$G_n = \{x \in E : f (x, z) \geq r_n \chi_{A_n} (z) \text{ for all } z \in B\}.$$

Indeed, if $x \in G_n$ and $z \in B$, since $\{z_k\}$ is dense in B we may find a subsequence $\{z_{k_j}\}$ of $\{z_k\}$ that converges to z. Since $f\left(x, z_{k_j}\right) \geq r_n \chi_{A_n}\left(z_{k_j}\right)$ for all $j \in \mathbb{N}$, using the continuity of $f\left(x, \cdot\right)$ and the lower semicontinuity of $r_n \chi_{A_n}$ we conclude that $f\left(x, z\right) \geq r_n \chi_{A_n}\left(z\right)$, and so the claim holds.

Since $f\left(x, \cdot\right)$ is continuous on B for every $x \in E$, in view of (6.22) we may write

$$f\left(x, z\right) = \sup_n \chi_{G_n}\left(x\right) r_n \chi_{A_n}\left(z\right)$$

for all $(x, z) \in E \times B$. By Proposition 1.68, for all $n \in \mathbb{N}$ we may find Borel sets $H_n \subset G_n$ such that $\mathcal{L}^N\left(H_n\right) = \mathcal{L}^N\left(G_n\right)$. Define the function

$$g\left(x, z\right) = \sup_n \chi_{H_n}\left(x\right) r_n \chi_{A_n}\left(z\right)$$

$x \in E$, $z \in B$. The function g is a Borel function since it is the supremum of Borel functions. Moreover,

$$f\left(x, \cdot\right) = g\left(x, \cdot\right)$$

for $\mathcal{L}^N$ a.e. $x \in E$. Since $f\left(x, \cdot\right)$ is continuous on B for every $x \in E$ it follows that f is a normal integrand.

The analogous statement of Theorem 6.28 for Carathéodory functions is Scorza-Dragoni's theorem below.

Theorem 6.35 (Scorza-Dragoni). *Let $E \subset \mathbb{R}^N$ be a Lebesgue measurable set, let $B \subset \mathbb{R}^m$ be a Borel set, and let $f : E \times B \to [-\infty, \infty]$ be a Carathéodory function. Then for every $\varepsilon > 0$ there exists a closed set $C_\varepsilon \subset E$, with $\mathcal{L}^N\left(E \setminus C_\varepsilon\right) \leq \varepsilon$, such that the restriction of f to $C_\varepsilon \times B$ is continuous.*

Proof. Since f and $-f$ are normal integrands in view of the previous theorem, the result follows by applying Theorem 6.28 to both f and $-f$.

6.2.3 Convex Integrands

In this subsection we extend the results of Section 4.7 to functions f that depend also on x. We begin with the case that f is real-valued.

Theorem 6.36. *Let $E \subset \mathbb{R}^N$ be a Lebesgue measurable set and let $f : E \times \mathbb{R}^m \to \mathbb{R}$ be an $\mathcal{L}^N \times \mathcal{B}$ measurable function such that $f\left(x, \cdot\right)$ is convex in $\mathbb{R}^m$ for $\mathcal{L}^N$ a.e. $x \in E$. Then there exist measurable functions $a_i : E \to \mathbb{R}$ and $b_i : E \to \mathbb{R}^m$ such that*

$$f\left(x, z\right) = \sup_{i \in \mathbb{N}}\{a_i\left(x\right) + b_i\left(x\right) \cdot z\}$$

for $\mathcal{L}^N$ a.e. $x \in E$ and for all $z \in \mathbb{R}^m$.

Moreover, if f is nonnegative, then the functions a_i and b_i may be taken to be bounded.

Proof. By De Giorgi's theorem, for $\mathcal{L}^N$ a.e. $x \in E$ and for all $z \in \mathbb{R}^m$ we may write

$$f(x, z) = \sup_{i \in \mathbb{N}} \{a_i(x) + b_i(x) \cdot z\},$$

where

$$a_i(x) := \int_{\mathbb{R}^m} f(x, z)((m+1)\varphi_i(z) + \nabla\varphi_i(z) \cdot z) \, dz, \qquad (6.26)$$

$$b_i(x) := -\int_{\mathbb{R}^m} f(x, z) \nabla\varphi_i(z) \, dz,$$

and the functions φ_i are of the form

$$\varphi_i(z) := k_i^m \varphi(k_i(q_i - z)), \quad z \in \mathbb{R}^m,$$

for $k_i \in \mathbb{N}$, $q_i \in \mathbb{Q}^m$, and some $\varphi \in C_c^1(\mathbb{R}^m)$ (see (4.43)).

The measurability of a_i and b_i follows from Exercise 6.38 below.

To prove the second part of the theorem, assume that $f \geq 0$. By the first part of the theorem, and since f is nonnegative, we may write

$$f(x, z) = \sup_{i \in \mathbb{N}} \{a_i(x) + b_i(x) \cdot z\}^+$$

for $\mathcal{L}^N$ a.e. $x \in E$ and for all $z \in \mathbb{R}^m$. For $k \in \mathbb{N}_0$ define $\sigma_0 :\equiv 0$ and

$$\sigma_k(s) := \begin{cases} 1 & s \leq k-1, \\ -s+k & k-1 < s < k, \\ 0 & s > k, \end{cases}$$

and let

$$\phi_{i,k}(x) := \sigma_k(|a_i(x)| + |b_i(x)|).$$

Since $0 \leq \phi_{i,k} \leq 1$, it follows that

$$(a_i(x) + b_i(x) \cdot z)^+ = \sup_{k \in \mathbb{N}} \{\phi_{i,k}(x) a_i(x) + \phi_{i,k}(x) b_i(x) \cdot z\},$$

for $\mathcal{L}^N$ a.e. $x \in E$ and for all $z \in \mathbb{R}^m$. Note that $\phi_{i,k} a_i \in L^\infty(E)$ and $\phi_{i,k} b_i \in L^\infty(E; \mathbb{R}^m)$.

Remark 6.37. Note that if f is continuous in the x variable, then it follows from formulas (6.26) that the functions a_i and b_i are continuous, and, in turn, so are the functions $\phi_{i,k}$.

Exercise 6.38. Under the hypotheses of Theorem 6.36, prove that the functions a_i and b_i defined in (6.26) are measurable.

Next we show that the statement of the previous theorem continues to hold when f takes the value ∞, although there is no explicit representation of the approximating functions.

Theorem 6.39. *Let $E \subset \mathbb{R}^N$ be a Lebesgue measurable set and let $f : E \times \mathbb{R}^m \to (-\infty, \infty]$ be a normal integrand such that $f(x, \cdot)$ is convex in $\mathbb{R}^m$ for $\mathcal{L}^N$ a.e. $x \in E$. Then there exist measurable functions $a_i : E \to \mathbb{R}$ and $b_i : E \to \mathbb{R}^m$ such that*

$$f(x, z) = \sup_{i \in \mathbb{N}} \{a_i(x) + b_i(x) \cdot z\}$$

for $\mathcal{L}^N$ a.e. $x \in E$ and for all $z \in \mathbb{R}^m$.

Moreover, if f is nonnegative, then the functions a_i and b_i may be taken to be bounded.

Proof. Without loss of generality we may assume that $f(x, \cdot)$ is convex and lower semicontinuous in $\mathbb{R}^m$ for all $x \in E$.

Consider the multifunction

$$\Gamma : E \to \{C \subset \mathbb{R} \times \mathbb{R}^m, \ C \text{ closed, nonempty}\}$$

defined by

$$\Gamma(x) := \{(a, b) \in \mathbb{R} \times \mathbb{R}^m : a + b \cdot z \leq f(x, z) \text{ for all } z \in \mathbb{R}^m\}$$

for $x \in E$. Note that $\Gamma(x) \neq \emptyset$ for all for $x \in E$ by Proposition 4.75, and next we prove that the graph of Γ belongs to $\mathfrak{M}(E) \otimes \mathcal{B}(\mathbb{R} \times \mathbb{R}^m)$, where $\mathfrak{M}(E)$ is the σ-algebra of Lebesgue measurable subsets of E. Let $\mathcal{A}$ be the countable family of all finite unions of open balls of $\mathbb{R}^m$ with rational radii centered at $z \in \mathbb{Q}^m$. For $A \in \mathcal{A}$, $x \in E$, and $(a, b) \in \mathbb{R} \times \mathbb{R}^m$ set

$$f_A(x) := \inf_{z \in A} f(x, z), \quad g_A(a, b) := \inf_{z \in A} (a + b \cdot z).$$

Then $g_A : \mathbb{R} \times \mathbb{R}^m \to \mathbb{R}$ is upper semicontinuous and f_A is measurable. To check the latter statement we observe that the set

$$f_A^{-1}((-\infty, t)) = \left\{x \in E : \inf_{z \in A} f(x, z) < t\right\}$$
$$= \{x \in E : f(x, z) < t \text{ for some } z \in A\}$$

is the projection onto E of the set

$$f^{-1}((-\infty, t)) \cap (E \times A) = \{(x, z) \in E \times A : f(x, z) < t\},$$

so the measurability of the set $f_A^{-1}((-\infty, t))$ follows from the projection theorem.

In particular, we have that the function

$$(x, a, b) \in E \times \mathbb{R} \times \mathbb{R}^m \mapsto f_A(x) - g_A(a, b) \tag{6.27}$$

is $\mathfrak{M}(E) \otimes \mathcal{B}(\mathbb{R} \times \mathbb{R}^m)$ measurable. We claim that the graph of Γ coincides with

$$G := \bigcap_{A \in \mathcal{A}} \{(x, a, b) \in E \times \mathbb{R} \times \mathbb{R}^m : f_A(x) \geq g_A(a, b)\}.$$

Indeed, if $(a, b) \in \Gamma(x)$ and if $A \in \mathcal{A}$, then for any $w \in A$,

$$\inf_{z \in A} (a + b \cdot z) \leq a + b \cdot w \leq f(x, w).$$

Therefore, taking the infimum over all such w, we conclude that $g_A(a, b) \leq f_A(x)$. Conversely, if $(x, a, b) \in G$ and if there exists $z_0 \in \mathbb{R}^m$ such that

$$a + b \cdot z_0 > f(x, z_0),$$

then choose $\varepsilon > 0$ so small that

$$a + b \cdot z_0 > f(x, z_0) + \varepsilon.$$

Let $A \in \mathcal{A}$ be a neighborhood of z_0 for which

$$\inf_{z \in A} (a + b \cdot z) \geq a + b \cdot z_0 - \varepsilon.$$

Then

$$g_A(a, b) = \inf_{z \in A} (a + b \cdot z) > f(x, z_0) \geq \inf_{z \in A} f(x, z) = f_A(x),$$

and this contradicts the fact that $(x, a, b) \in G$.

The claim together with (6.27) ensures that the graph of Γ belongs to $\mathfrak{M}(E) \otimes \mathcal{B}(\mathbb{R} \times \mathbb{R}^m)$.

By Aumann's measurable selection theorem there exists a sequence of measurable functions $a_i : E \to \mathbb{R}$ and $b_i : E \to \mathbb{R}^m$ such that

$$(a_i(x), b_i(x)) \in \Gamma(x)$$

for every $x \in E$ and $\{(a_i(x), b_i(x))\}$ is dense in $\Gamma(x)$. We claim that for all $x \in E$ and $z \in \mathbb{R}^m$,

$$f(x, z) = \sup_{i \in \mathbb{N}} \{a_i(x) + b_i(x) \cdot z\}. \tag{6.28}$$

By the definition of Γ we have that

$$f(x, z) \geq \sup_{i \in \mathbb{N}} \{a_i(x) + b_i(x) \cdot z\}.$$

To prove the reverse inequality, fix $(x, z) \in E \times \mathbb{R}^m$ and let

$$f(x, z) > t.$$

By Proposition 4.75 (ii) we may find $(a, b) \in \Gamma(x)$ such that

$$a + b \cdot z > t.$$

Since $\{(a_i(x), b_i(x))\}_{i \in \mathbb{N}}$ is dense in $\Gamma(x)$, for every $\varepsilon > 0$ there exists $i \in \mathbb{N}$ such that

$$|a - a_i(x)| + |b - b_i(x)| \leq \frac{\varepsilon}{1 + |z|}.$$

Hence we deduce that

$$\sup_{j \in \mathbb{N}} \{a_j(x) + b_j(x) \cdot z\} \geq a_i(x) + b_i(x) \cdot z > t - 2\varepsilon,$$

and by letting first $\varepsilon \to 0^+$ and then $t \nearrow f(x, z)$, we obtain (6.28).

This concludes the first part of the proof. The second part follows exactly as in the previous theorem. We omit the details.

Next we consider the case in which the function f is lower semicontinuous in x uniformly with respect to z.

Proposition 6.40. *Let $\Omega \subset \mathbb{R}^N$ be an open set and let $f : \Omega \times \mathbb{R}^m \to [0, \infty]$ be such that $f(x, \cdot)$ is convex and lower semicontinuous in $\mathbb{R}^m$ for all $x \in E$. Assume that for every $x_0 \in \Omega$ and $\varepsilon > 0$ there exists $\delta > 0$ such that*

$$f(x, z) \geq (1 - \varepsilon) f(x_0, z) \tag{6.29}$$

for all $x \in \Omega$, with $|x - x_0| \leq \delta$, and for all $z \in \mathbb{R}^m$. Then

$$f(x, z) = \sup_{i \in \mathbb{N}} \left\{ \varphi_i(x) \left(a_i + b_i \cdot z\right)^+ \right\}$$

for all $(x, z) \in \Omega \times \mathbb{R}^m$, where $\varphi_i \in C_c^\infty(\Omega)$, $\varphi_i \geq 0$, $a_i \in \mathbb{R}$, and $b_i \in \mathbb{R}^m$, $i \in \mathbb{N}$.

Proof. Let $\mathcal{G}$ be the class of all functions $g : \Omega \times \mathbb{R}^m \to [0, \infty)$ of the form

$$g(x, z) = \varphi(x) \left(a + b \cdot z\right)^+, \quad (x, z) \in \Omega \times \mathbb{R}^m,$$

where $\varphi \in C_c^\infty(\Omega)$, $\varphi \geq 0$, $a \in \mathbb{R}$, and $b \in \mathbb{R}^m$, and such that $g(x, z) \leq f(x, z)$ for all $(x, z) \in \Omega \times \mathbb{R}^m$. Note that $\mathcal{G} \neq \emptyset$ because $0 \in \mathcal{G}$.

We claim that

$$f(x, z) = \sup_{g \in \mathcal{G}} g(x, z) \quad \text{for all } (x, z) \in \Omega \times \mathbb{R}^m. \tag{6.30}$$

By definition of $\mathcal{G}$, it follows that

$$f \geq \sup_{g \in \mathcal{G}} g.$$

Conversely, fix $x_0 \in \Omega$, $\varepsilon > 0$, and let δ be such that (6.29) is satisfied. Consider a cutoff function $\varphi \in C_c^\infty(\Omega)$, with $0 \leq \varphi \leq 1$, $\varphi \equiv 1$ on $B(x_0, \delta/2)$, $\varphi \equiv 0$ outside $B(x_0, \delta)$. Using Proposition 4.77 we can write

$$f(x_0, z) = \sup_{i \in \mathbb{N}} (a_i + b_i \cdot z)^+,$$

for some $a_i \in \mathbb{R}$ and $b_i \in \mathbb{R}^m$. Consider

$$f_i^\varepsilon(x, z) := (1 - \varepsilon)\varphi(x)(a_i + b_i \cdot z)^+.$$

By (6.29) it follows that $f_i^\varepsilon \in \mathcal{G}$, and we get

$$(1 - \varepsilon)f(x_0, z) = \sup_{i \in \mathbb{N}} f_i^\varepsilon(x_0, z) \le \sup_{g \in \mathcal{G}} g(x_0, z);$$

hence the claim (6.30) follows by letting $\varepsilon \to 0^+$.

By Proposition 4.78 and (6.30) there exists a sequence f_i in $\mathcal{G}$ such that $f(x, z) = \sup_{i \in \mathbb{N}} f_i(x, z)$ for all (x, z) in $\Omega \times \mathbb{R}^m$.

Remark 6.41. (i) By further specializing the class $\mathcal{G}$, we may also assume the functions φ_j to have the form

$$\varphi_j(x) = \varphi_j^1(x_1)\,\varphi_j^2(x_2)\ldots\varphi_j^N(x_N),$$

where $\varphi_j^i \in C_c^\infty(\Omega)$, $\varphi_j^i \ge 0$, $i = 1, \ldots, N$, $j \in \mathbb{N}$.

(ii) In the case in which Ω is replaced by a measurable set E, the statement of the proposition still holds, provided the functions φ_i are required only to be continuous and bounded, and the proof is similar with obvious adaptations.

We are now ready to prove the general case.

Proposition 6.42. *Let $E \subset \mathbb{R}^N$ be a Lebesgue measurable set and let $f : E \times \mathbb{R}^m \to (-\infty, \infty]$ be a lower semicontinuous function such that $f(x, \cdot)$ is convex in $\mathbb{R}^m$ for all $x \in E$. Assume that one of the following two conditions is satisfied:*

(i) there exists a continuous function $v_0 : E \to \mathbb{R}^m$ with

$$(f(\cdot, v_0(\cdot)))^+ \in L^\infty(E); \tag{6.31}$$

(ii) the set E is closed and there exists a function $\gamma : [0, \infty) \to [0, \infty)$, with

$$\lim_{s \to \infty} \frac{\gamma(s)}{s} = \infty$$

such that

$$f(x, z) \ge \gamma(|z|)$$

for all $x \in E$ and $z \in \mathbb{R}^m$.

Then there exist two sequences of continuous functions

$$a_i : E \to \mathbb{R}, \qquad b_i : E \to \mathbb{R}^m,$$

such that

$$f(x, z) = \sup_{i \in \mathbb{N}} \{a_i(x) + b_i(x) \cdot z\}$$

for all $x \in E$ and $z \in \mathbb{R}^m$.

Moreover, if f is nonnegative, then the functions a_i and b_i may be taken to be bounded and condition (i) may be weakened to

$$(f(\cdot, v_0(\cdot)))^+ \in L^\infty_{\mathrm{loc}}(E). \tag{6.32}$$

Proof. **Step 1**: Assume first that condition (i) is satisfied and define

$$g(x, z) := f(x, z + v_0(x)) - C_0,$$

where $C_0 := \left\| (f(\cdot, v_0(\cdot)))^+ \right\|_{L^\infty(E;\mathbb{R})}$. Then g satisfies the same hypotheses of f and

$$-\infty < g(x, 0) = f(x, v_0(x)) - C_0 \le 0. \tag{6.33}$$

Consider the multifunction

$$\Gamma : E \to \{C \subset \mathbb{R} \times \mathbb{R}^m : C \ne \emptyset, \text{ convex, closed}\}$$

defined by

$$\Gamma(x) := \{(a, b) \in \mathbb{R} \times \mathbb{R}^m : g(x, z) \ge a + b \cdot z \quad \text{for all } z \in \mathbb{R}^m\}.$$

By Proposition 4.75 we have that $\Gamma(x)$ is well-defined. We claim that Γ is lower semicontinuous. Fix an open set $A \subset \mathbb{R} \times \mathbb{R}^m$. We need to show that

$$\Gamma^-(A) := \{x \in E : \Gamma(x) \cap A \ne \emptyset\}$$

is relatively open in E. Assume that $\Gamma^-(A) \ne \emptyset$ and fix $x_0 \in \Gamma^-(A)$, $(a_0, b_0) \in A$, and $\varepsilon > 0$ such that

$$g(x_0, z) \ge a_0 + b_0 \cdot z \quad \text{for all } z \in \mathbb{R}^m, \tag{6.34}$$

$$\{(a, b) \in \mathbb{R} \times \mathbb{R}^m : |a - a_0| + |b - b_0| < 4\varepsilon\} \subset A. \tag{6.35}$$

We claim that there exists an open ball B in $\mathbb{R}^N$ containing x_0 such that

$$g(x, z) \ge a_0 + b_0 \cdot z - \varepsilon(1 + |z|) \quad \text{for all } x \in B \cap E \text{ and } z \in \mathbb{R}^m. \tag{6.36}$$

Assume that (6.36) is false. Then there exist a sequence $\{x_n\} \subset E$ converging to x_0 and a sequence $\{z_n\} \subset \mathbb{R}^m$ such that

$$g(x_n, z_n) < a_0 + b_0 \cdot z_n - \varepsilon(1 + |z_n|) \quad \text{for all } n \in \mathbb{N}. \tag{6.37}$$

If $\sup_n |z_n| < \infty$, then we can assume that z_n converges to some $z_0 \in \mathbb{R}^m$, and due to the lower semicontinuity of g, letting $n \to \infty$ in (6.37) we get

$$g(x_0, z_0) \leq a_0 + b_0 \cdot z_0 - \varepsilon(1 + |z_0|),$$

which contradicts (6.34). Therefore $\sup_n |z_n| = \infty$, and without loss of generality we can assume that $|z_n| \to \infty$, with

$$w_n := \frac{z_n}{|z_n|} \to w_0$$

for some $w_0 \in \mathbb{R}^m$ as $n \to \infty$. Fix $\gamma > 0$ and let n be so large that $|z_n|\gamma > 1$. By the convexity of $g(x, \cdot)$ and (6.37),

$$g\left(x_n, \frac{w_n}{\gamma}\right) - a_0 - b_0 \cdot \frac{w_n}{\gamma}$$

$$\leq \frac{1}{\gamma|z_n|} [g(x_n, z_n) - a_0 - b_0 \cdot z_n] + \left(1 - \frac{1}{\gamma|z_n|}\right)(g(x_n, 0) - a_0)$$

$$\leq -\frac{\varepsilon(1 + |z_n|)}{\gamma|z_n|} - a_0 \left(1 - \frac{1}{\gamma|z_n|}\right),$$

where we have used the fact that $g(x_n, 0) \leq 0$ by (6.33). Letting $n \to \infty$ and using the lower semicontinuity of g and (6.34) gives

$$0 \leq g\left(x_0, \frac{w_0}{\gamma}\right) - a_0 - b_0 \cdot \frac{w_0}{\gamma} \leq -\frac{\varepsilon}{\gamma} - a_0 < 0,$$

provided γ is taken sufficiently small. We arrive at a contradiction, and this proves (6.36).

Next we claim that $B \cap E \subset \Gamma^-(A)$. Fix $x \in B \cap E$ and define

$$h(z) := g(x, z) - (a_0 + b_0 \cdot z), \quad z \in \mathbb{R}^m.$$

By (6.36) and Theorem 4.19 applied to the convex sets epi h and

$$C := \{(z, t) \in \mathbb{R}^m \times \mathbb{R} : t + 2\varepsilon + \varepsilon|z| \leq 0\}$$

we may find $(b', \alpha_0) \in (\mathbb{R}^m \times \mathbb{R}) \setminus \{(0, 0)\}$ and a number $\alpha \in \mathbb{R}$ such that

$$b' \cdot z + \alpha_0 t \geq \alpha \quad \text{for all } (z, t) \in \text{epi } h, \tag{6.38}$$
$$b' \cdot z + \alpha_0 t \leq \alpha \quad \text{for all } (z, t) \in C.$$

Let $z \in \text{dom}_e h$. Then for any $t \geq h(z)$,

$$b' \cdot z + \alpha_0 t \geq \alpha,$$

and so, letting $t \to \infty$ we obtain that $\alpha_0 \geq 0$. If $\alpha_0 = 0$, then, since for every $s > 0$ the point $(sb', -2\varepsilon - \varepsilon s|b'|)$ is in C, we obtain

$$s\,|b'|^2 \leq \alpha$$

for all $s > 0$, which can hold only if $b' = 0$. Since this would contradict the fact that $(b', \alpha_0) \neq (0, 0)$, we have that $\alpha_0 > 0$, and so from (6.38) we obtain

$$h(z) \geq \frac{\alpha}{\alpha_0} - \frac{b'}{\alpha_0} \cdot z \geq -2\varepsilon - \varepsilon|z| \quad \text{for all } z \in \mathbb{R}^m.$$

Setting $a := \frac{\alpha}{\alpha_0}$ and $b := -\frac{b'}{\alpha_0}$, we have proved that

$$g(x, z) - a_0 - b_0 \cdot z \geq a + b \cdot z \geq -2\varepsilon - \varepsilon|z| \quad \text{for all } z \in \mathbb{R}^m.$$

Consequently, $a \geq -2\varepsilon$ and $|b| \leq \varepsilon$. Let $\omega := \inf\{a, 2\varepsilon\}$. Then

$$g(x, z) \geq a_0 + \omega + (b_0 + b) \cdot z \quad \text{for all } z \in \mathbb{R}^m,$$

and thus by (6.35) it follows that $x \in \Gamma^-(A)$, and the claim is proved.

Next we show that for every $(x, z) \in E \times \mathbb{R}^m$,

$$g(x, z) = \sup \{a(x) + b(x) \cdot z : a \in C(E), b \in C(E; \mathbb{R}^m), \quad\quad (6.39)$$
$$g(y, \xi) \geq a(y) + b(y) \cdot \xi \text{ for all } y \in E \text{ and } \xi \in \mathbb{R}^m\}.$$

One inequality is immediate. To prove the reverse inequality fix $(x_0, z_0) \in E \times \mathbb{R}^m$ and let

$$g(x_0, z_0) > t.$$

Since

$$g(x_0, z_0) = \sup \{a + b \cdot z_0 : (a, b) \in \Gamma(x_0)\},$$

there exists $(a_0, b_0) \in \Gamma(x_0)$ such that

$$a_0 + b_0 \cdot z_0 > t.$$

Since Γ is lower semicontinuous, by the Micheal continuous selection theorem we may find $a \in C(E)$, $b \in C(E; \mathbb{R}^m)$ such that $(a(x), b(x)) \in \Gamma(x)$ for every $x \in E$ and

$$(a(x_0), b(x_0)) = (a_0, b_0).$$

Therefore the supremum in the right-hand side of (6.39) at the point (x_0, z_0) is bigger than t, and letting $t \nearrow g(x_0, z_0)$ we conclude (6.39).

By Proposition 4.78 we get the desired result for the function g, say,

$$g(x, z) = \sup_{i \in \mathbb{N}} \{\tilde{a}_i(x) + b_i(x) \cdot z\}$$

for all $x \in E$ and $z \in \mathbb{R}^m$, and some $\tilde{a}_i \in C(E)$, $b_i \in C(E; \mathbb{R}^m)$. We conclude that

$$f(x, z) = \sup_{i \in \mathbb{N}} \{a_i(x) + b_i(x) \cdot z\}$$

for all $x \in E$ and $z \in \mathbb{R}^m$, where

$$a_i(x) := \tilde{a}_i(x) - b_i(x) \cdot v_0(x) + C_0.$$

We recall that v_0 is a continuous function.

Finally, if f is nonnegative and (6.32) holds, for each $n \in \mathbb{N}$ let $\psi_n \in C_c^\infty(\mathbb{R}^N)$, $0 \le \psi_n \le 1$, be a cutoff function such that $\psi_n(x) = 1$ for $|x| \le n$ and $\psi_n(x) = 0$ for $|x| \ge n + 1$. Define

$$f_n(x, z) := \psi_n(x) f(x, z), \quad (x, z) \in E \times \mathbb{R}^m.$$

Then f satisfies the same hypotheses of f and also (6.31). Thus, we can apply the first part of the proof to f_n to find two sequences of continuous functions

$$a_{i,n} : E \to \mathbb{R}, \qquad b_{i,n} : E \to \mathbb{R}^m,$$

such that

$$f_n(x, z) = \sup_{i \in \mathbb{N}} \{a_{i,n}(x) + b_{i,n}(x) \cdot z\} \tag{6.40}$$

for all $x \in E$, $z \in \mathbb{R}^m$, and $n \in \mathbb{N}$. Since $f \ge 0$, we have that $f = \sup_{n \in \mathbb{N}} f_n$, and so by (6.40) we obtain

$$f(x, z) = \sup_{i,n \in \mathbb{N}} \{a_{i,n}(x) + b_{i,n}(x) \cdot z\}$$

for all $x \in E$, $z \in \mathbb{R}^m$. As in the final part of the proof of Theorem 6.36 we may take the functions $a_{i,n}$ and $b_{i,n}$ to be bounded. Note that the truncation function $\sigma_k(s)$ is Lipschitz continuous. This completes the proof of the proposition in the case that (i) holds.

Step 2: Next we assume that condition (ii) holds, and without loss of generality, we suppose that $f \ge 1$ and $\gamma \ge 1$ (if not, carry out the proof with $f + 1$ and $\gamma + 1$ in place of f and γ, respectively).

Define the Yosida transform

$$f_n(x, z) := \inf \{f(y, z) + n |x - y| : y \in E\},$$

$x \in E$, $z \in \mathbb{R}^m$. For each $n \in \mathbb{N}$,

$$f(x, z) \ge f_{n+1}(x, z) \ge f_n(x, z) \ge \gamma(|z|)$$

for all $x \in E$, $z \in \mathbb{R}^m$. We claim that

$$f(x, z) = \sup_n f_n(x, z), \tag{6.41}$$

$x \in E$, $z \in \mathbb{R}^m$. This follows from the previous inequality if $\sup_n f_n(x, z) = \infty$, so fix $(x, z) \in E \times \mathbb{R}^m$ such that $\sup_n f_n(x, z) < \infty$. By the definition of $f_n(x, z)$ we may find a sequence $\{x_n\} \subset E$ such that for each $n \in \mathbb{N}$,

$$f_n(x, z) \ge f(x_n, z) + n |x - x_n| - \frac{1}{2^n}.$$

Since $\sup_n f_n(x, z) < \infty$, it follows that $x_n \to x$, and thus, using the lower semicontinuity of f, we obtain that

$$\lim_{n \to \infty} f_n(x, z) \geq \liminf_{n \to \infty} \left(f(x_n, z) + n \,|x - x_n| - \frac{1}{2^n} \right)$$
$$\geq \liminf_{n \to \infty} f(x_n, z) \geq f(x, z),$$

and the claim is proved.

Next we show that for each $n \in \mathbb{N}$ the function $f_n(x, \cdot)$ is lower semicontinuous. Indeed, let $z_k \to z$. Without loss of generality we may assume that

$$\liminf_{k \to \infty} f_n(x, z_k) = \lim_{k \to \infty} f_n(x, z_k) < \infty$$

(since otherwise there is nothing to prove), and so there exists a bounded sequence $\{x_k\} \subset E$ such that

$$f_n(x, z_k) \geq f(x_k, z_k) + n \,|x - x_k| - \frac{1}{2^k}.$$

By extracting a subsequence if necessary, and since E is closed, we may assume that $x_k \to x_0 \in E$. Letting $k \to \infty$ in the previous inequality, and using once again the lower semicontinuity of f, we get

$$\liminf_{k \to \infty} f_n(x, z_k) \geq \liminf_{k \to \infty} \left(f(x_k, z_k) + n \,|x - x_k| - \frac{1}{2^k} \right)$$
$$\geq f(x_0, z) + n \,|x - x_0| \geq f_n(x, z).$$

Thus we are in a position to apply Proposition 4.102, and in view of the convexity of $f(x, \cdot)$ we have

$$f(x, z) = \sup_n f_n^{**}(x, z). \tag{6.42}$$

We claim that condition (6.29) is satisfied. Indeed, fix $\varepsilon > 0$ and let $0 < \delta \leq \varepsilon/n$. If $|x - x_0| \leq \delta$ and $x \in E$, since $f_n^{**} \geq 1$, then

$$\varepsilon f_n^{**}(x_0, z) \geq n \,|x - x_0|,$$

and using the fact that

$$f_n^{**}(x, z) + n \,|x - x_0| \geq f_n^{**}(x_0, z) \tag{6.43}$$

for every $x \in E$ and $z \in \mathbb{R}^m$, we conclude that

$$f_n^{**}(x, z) \geq f_n^{**}(x_0, z) - n \,|x - x_0| \geq (1 - \varepsilon) f_n^{**}(x_0, z)$$

for all $|x - x_0| \leq \delta$ and $z \in \mathbb{R}^m$.

We may now apply Remark 6.41(ii) to approximate each function $f_n^{**}(x, z)$.

If f is nonnegative, then as in the final part of the proof of Theorem 6.36 we may take the functions a_i and b_i to be bounded. Note that the truncation function $\sigma_k(s)$ is Lipschitz continuous.

The convex envelope of an $\mathcal{L}^N \times \mathcal{B}$ measurable integrand is normal. This property follows from the proposition below.

Proposition 6.43. *Let $E \subset \mathbb{R}^N$ be a Lebesgue measurable set and let $f :$ $E \times \mathbb{R}^m \to [-\infty, \infty]$ be $\mathcal{L}^N \times \mathcal{B}$ measurable. Then f^*, and therefore also f^{**}, is a normal integrand.*

Proof. Since f is $\mathcal{L}^N \times \mathcal{B}$ measurable, then so is the function

$$h : E \times \mathbb{R}^m \times \mathbb{R} \to [-\infty, \infty]$$

defined by

$$h(x, z, t) := f(x, z) - t.$$

In turn, the set

$$F := \{(x, z, t) \in E \times \mathbb{R}^m \times \mathbb{R} : h(x, z, t) \leq 0\}$$

is $\mathcal{L}^N \times \mathcal{B}(\mathbb{R}^m \times \mathbb{R})$ measurable. By the projection theorem, the projection of F onto $\mathbb{R}^N$, denoted by E_∞, is Lebesgue measurable.

Define the multifunction

$$\Gamma : E_\infty \to \mathcal{P}(\mathbb{R}^{m+1}) \setminus \{\emptyset\}$$

as

$$\Gamma(x) := \operatorname{epi} f(x, \cdot) = \{(z, t) \in \mathbb{R}^m \times \mathbb{R} : f(x, z) \leq t\}, \quad x \in E_\infty.$$

Since

$$\begin{aligned}
\operatorname{Gr} \Gamma &= \{(x, z, t) \in E_\infty \times \mathbb{R}^m \times \mathbb{R} : (z, t) \in \Gamma(x)\} \\
&= \{(x, z, t) \in E_\infty \times \mathbb{R}^m \times \mathbb{R} : f(x, z) \leq t\} = F,
\end{aligned}$$

it follows that $\operatorname{Gr} \Gamma$ is $\mathcal{L}^N \times \mathcal{B}(\mathbb{R}^m \times \mathbb{R})$ measurable, and so by Aumann's measurable selection theorem there exists a sequence of Lebesgue measurable functions $\{(v_n, \psi_n)\}$, $v_n : E_\infty \to \mathbb{R}^m$, $\psi_n : E_\infty \to \mathbb{R}$, $n \in \mathbb{N}$, such that for each $x \in E_\infty$ the set $\{(v_n(x), \psi_n(x))\}_{n \in \mathbb{N}}$ is dense in $\overline{\operatorname{epi} f(x, \cdot)}$.

We claim that for all $(x, w) \in E_\infty \times \mathbb{R}^m$,

$$f^*(x, w) = \sup_{(z, t) \in \operatorname{epi} f(x, \cdot)} \{z \cdot w - t\}.$$

Indeed, if $(x, w) \in E_\infty \times \mathbb{R}^m$, then $f(x, \cdot) \not\equiv \infty$, and so $\operatorname{epi} f(x, \cdot)$ is nonempty. Let $(z', t) \in \operatorname{epi} f(x, \cdot)$. Then

$$f^*(x, w) = \sup_{z \in \mathbb{R}^m} \{z \cdot w - f(x, z)\} \geq z' \cdot w - f(x, z_0) \geq z' \cdot w - t$$

and so

$$f^* (x, w) \geq \sup_{(z,t) \in \text{epi } f(x, \cdot)} \{z \cdot w - t\}.$$

The reverse inequality follows from the facts that

$$f^* (x, w) = \sup \{z \cdot w - f (x, z) : z \in \text{dom}_e \, f (x, \cdot)\}$$

and that $(z, f (x, z)) \in \text{epi } f (x, \cdot)$ for all $z \in \text{dom}_e \, f (x, \cdot)$.

Thus the claim holds, and so, since for each $x \in E_\infty$, the set

$$\{(v_n (x), \psi_n (x))\}$$

is dense in $\overline{\text{epi } f (x, \cdot)}$, we have that for all $(x, w) \in E_\infty \times \mathbb{R}^m$,

$$f^* (x, w) = \sup_{n \in \mathbb{N}} \{v_n (x) \cdot w - \psi_n (x)\}.$$

For each $n \in \mathbb{N}$ define the Carathéodory function

$$g_n (x, w) := v_n (x) \cdot w - \psi_n (x), \quad (x, w) \in E_\infty \times \mathbb{R}^m.$$

Then we have proved that for all $(x, w) \in E_\infty \times \mathbb{R}^m$,

$$f^* (x, w) = \sup_{n \in \mathbb{N}} g_n (x, w).$$

Since Carathéodory functions are normal integrands, it follows that $f^* : E_\infty \times \mathbb{R}^m \to [-\infty, \infty]$ is a normal integrand. On the other hand, if $x \notin E_\infty$, then $f (x, \cdot) \equiv \infty$, and so

$$f^* (x, \cdot) \equiv -\infty.$$

Since the set $E \setminus E_\infty$ is measurable, it follows that $f^* : E \times \mathbb{R}^m \to [-\infty, \infty]$ is a normal integrand.

The next proposition will be used in the proof of Theorems 6.49 and 7.13.

Proposition 6.44. *Let $E \subset \mathbb{R}^N$ be a Lebesgue measurable set with finite measure and let $s : E \times \mathbb{R}^m \to [0, \infty]$ be such that for $\mathcal{L}^N$ a.e. $x \in E$ the function $s (x, \cdot)$ is lower semicontinuous on $\mathbb{R}^m$. Then there exists a normal integrand $g : E \times \mathbb{R}^m \to [0, \infty]$ such that*

(i) $s (x, z) \leq g (x, z)$ for $\mathcal{L}^N$ a.e. $x \in E$ and for all $z \in \mathbb{R}^m$;

(ii) if $h : E \times \mathbb{R}^m \to [0, \infty]$ is $\mathcal{L}^N \times \mathcal{B}$ measurable and $s (x, z) \leq h (x, z)$ for $\mathcal{L}^N$ a.e. $x \in E$ and for all $z \in \mathbb{R}^m$, then $g (x, z) \leq h (x, z)$ for $\mathcal{L}^N$ a.e. $x \in E$ and for all $z \in \mathbb{R}^m$;

(iii) for every $v \in L^1 (E; \mathbb{R}^m)$ and for every Borel set $G \subset E$,

$$\int_G g (x, v (x)) \, dx = \inf \Big\{ \int_G \psi (x) \, dx \colon \psi : G \to [0, \infty] \text{ measurable},$$

$$\psi (x) \geq s (x, v (x)) \text{ for } \mathcal{L}^N \text{ a.e. } x \in G \Big\}.$$

Moreover, if $m = d + l$ with d, $l \in \mathbb{N}$, and if for $\mathcal{L}^N$ a.e. $x \in E$ and for all $u \in \mathbb{R}^d$ the function $s(x, u, \cdot)$ is convex in $\mathbb{R}^l$, then the same holds for g.

Proof. By Proposition 1.68 the set E is the union of a Borel set with a set of Lebesgue measure zero. Therefore, without loss of generality we may assume that E is a Borel set.

Using the same notation as in the proof of Corollary 6.30, by (6.24) we can write

$$s(x, z) = \sup_n \chi_{G_n}(x) r_n \chi_{A_n}(z)$$

for every $(x, z) \in E \times \mathbb{R}^m$, where

$$G_n := \{x \in E : s(x, \cdot) \geq r_n \chi_{A_n}(\cdot)\}.$$

Since the Lebesgue outer measure is a Radon outer measure, it follows by Remark 1.51 that for every $n \in \mathbb{N}$ there exists a Borel set $B_n \supset G_n$ such that

$$\mathcal{L}_o^N(G_n) = \mathcal{L}^N(B_n). \tag{6.44}$$

Note that since E is Borel and $G_n \subset E$, we can assume that $B_n \subset E$. Define

$$g(x, z) := \sup_n \chi_{B_n}(x) r_n \chi_{A_n}(z) \tag{6.45}$$

for every $(x, z) \in E \times \mathbb{R}^m$. The function g is a Borel function with $s \leq g$, and this proves (i).

To verify (ii) let $h : E \times \mathbb{R}^m \to [0, \infty]$ be an $\mathcal{L}^N \times \mathcal{B}$ measurable function such that $s(x, z) \leq h(x, z)$ for $\mathcal{L}^N$ a.e. $x \in E$ and for all $z \in \mathbb{R}^m$, and let $v : E \to \mathbb{R}^m$ be a Lebesgue measurable function.

For $n \in \mathbb{N}$ set

$$F_n := \{x \in E : h(x, v(x)) \geq \chi_{B_n}(x) r_n \chi_{A_n}(v(x))\}.$$

The set F_n is Lebesgue measurable and

$$\mathcal{L}^N(G_n \setminus F_n) = 0. \tag{6.46}$$

Indeed, let E_h be a Lebesgue measurable set in E, with $\mathcal{L}^N(E \setminus E_h) = 0$, and such that $s \leq h$ in $E_h \times \mathbb{R}^m$. If $x \in G_n \cap E_h$, then, since $B_n \supset G_n$,

$$h(x, v(x)) \geq s(x, v(x)) \geq r_n \chi_{A_n}(v(x)) = \chi_{B_n}(x) r_n \chi_{A_n}(v(x)),$$

and thus $x \in F_n$. This proves (6.46). In particular, by (6.44),

$$\mathcal{L}^N(B_n \setminus F_n) = 0,$$

and so

$$h(x, v(x)) \geq \chi_{B_n}(x) r_n \chi_{A_n}(v(x))$$

for $\mathcal{L}^N$ a.e. $x \in E$ and for all $n \in \mathbb{N}$. Consequently, by (6.45),

$$h\left(x, v\left(x\right)\right) \geq g\left(x, v\left(x\right)\right)$$

for $\mathcal{L}^N$ a.e. $x \in E$, and so

$$\int_G h\left(x, v\left(x\right)\right) \, dx \geq \int_G g\left(x, v\left(x\right)\right) \, dx$$

for every Lebesgue measurable set $G \subset E$ and for all Lebesgue measurable functions $v : E \to \mathbb{R}^m$. By Proposition 6.24 we obtain (ii).

Now we prove (iii). Fix $v \in L^1\left(E; \mathbb{R}^m\right)$ and a Borel set $G \subset E$. Setting $\psi_1\left(x\right) := g\left(x, v\left(x\right)\right)$, it follows that

$$\int_G g\left(x, v\left(x\right)\right) \, dx = \int_G \psi_1\left(x\right) \, dx$$
$$\geq \inf \left\{ \int_G \psi\left(x\right) \, dx \colon \psi : G \to [0, \infty] \text{ measurable,} \right.$$
$$\left. \psi\left(x\right) \geq s\left(x, v\left(x\right)\right) \text{ for } \mathcal{L}^N \text{ a.e. } x \in G \right\}.$$

Conversely, if ψ is admissible for the right-hand side of the above inequality, then an argument similar to that used to establish (ii), where now $\psi\left(x\right)$ replaces $h\left(x, v\left(x\right)\right)$, yields

$$\psi\left(x\right) \geq g\left(x, v\left(x\right)\right)$$

for $\mathcal{L}^N$ a.e. $x \in G$.

Finally, assume that $m = d + l$ with $d, l \in \mathbb{N}$, and that for $\mathcal{L}^N$ a.e. $x \in E$, and for all $u \in \mathbb{R}^d$ the function $s\left(x, u, \cdot\right)$ is convex in $\mathbb{R}^l$. For every Borel set $G \subset E$ and for every Lebesgue measurable function $u : G \to \mathbb{R}^d$, $v : G \to \mathbb{R}^l$ define

$$\int_G^* s\left(x, u\left(x\right), v\left(x\right)\right) \, dx := \inf \left\{ \int_G \psi\left(x\right) \, dx \colon \psi : G \to [0, \infty] \text{ measurable,} \right.$$
$$\left. \psi\left(x\right) \geq s\left(x, u\left(x\right), v\left(x\right)\right) \text{ for } \mathcal{L}^N \text{ a.e. } x \in G \right\}.$$

Note that the map

$$v \mapsto \int_G^* s\left(x, u\left(x\right), v\left(x\right)\right) \, dx$$

is convex, and by (iii) so is

$$v \mapsto \int_G g\left(x, u\left(x\right), v\left(x\right)\right) \, dx.$$

Hence for any fixed $\theta \in [0, 1] \cap \mathbb{Q}$ and $v, w \in L^\infty\left(E; \mathbb{R}^l\right)$ we have

$$\int_G g\left(x, u\left(x\right), \left(1 - \theta\right) v\left(x\right) + \theta w\left(x\right)\right) dx$$

$$\leq \left(1 - \theta\right) \int_G g\left(x, u\left(x\right), v\left(x\right)\right) dx + \theta \int_G g\left(x, u\left(x\right), w\left(x\right)\right) dx.$$

Since this is true for all $G \in \mathcal{B}\left(E\right)$, $u \in L^\infty\left(E; \mathbb{R}^d\right)$, and $v, w \in L^\infty\left(E; \mathbb{R}^l\right)$, it follows from Proposition 6.24 that there exists a set $N_\theta \subset E$, with $\left|N_\theta\right| = 0$, such that

$$g\left(x, u, \left(1 - \theta\right) \zeta + \theta w\right) \leq \left(1 - \theta\right) g\left(x, u, \zeta\right) + \theta g\left(x, u, w\right) \qquad (6.47)$$

for all $x \in E \setminus N_\theta$, $u \in \mathbb{R}^d$, and $\zeta, w \in \mathbb{R}^l$. Let

$$N_0 := \bigcup_{\theta \in [0,1] \cap \mathbb{Q}} N_\theta.$$

Then $\left|N_0\right| = 0$ and (6.47) holds for all $\theta \in [0, 1] \cap \mathbb{Q}$, $x \in E \setminus N_0$, $u \in \mathbb{R}^d$, and all $\zeta, w \in \mathbb{R}^l$. Now fix $\theta \in [0, 1]$ and let $\theta_n \in [0, 1] \cap \mathbb{Q}$ be such that $\theta_n \to \theta$. By (6.47),

$$g\left(x, u, \left(1 - \theta_n\right) \zeta + \theta_n w\right) \leq \left(1 - \theta_n\right) g\left(x, u, \zeta\right) + \theta_n g\left(x, u, w\right)$$

for all $n \in \mathbb{N}$, $x \in E \setminus N_0$, $u \in \mathbb{R}^d$, and all $\zeta, w \in \mathbb{R}^l$. By letting $n \to \infty$ and using the fact that $g\left(x, u, \cdot\right)$ is lower semicontinuous in $\mathbb{R}^l$ for $\mathcal{L}^N$ a.e. $x \in E$, we establish that for $\mathcal{L}^N$ a.e. $x \in E$ the function $g\left(x, u, \cdot\right)$ is convex. This concludes the proof.

6.3 Well-Posedness

Throughout this section we assume that the set E has finite measure. Let $f : E \times \mathbb{R}^m \to [-\infty, \infty]$ be an $\mathcal{L}^N \times \mathcal{B}$ measurable function. For every measurable function $v : E \to \mathbb{R}^m$ consider the superposition operator

$$v \mapsto f\left(\cdot, v\left(\cdot\right)\right).$$

In this section we study necessary and sufficient conditions for the superposition operator to map $L^p\left(E; \mathbb{R}^m\right)$ into $L^1\left(E\right)$. We begin with the case $1 \leq p < \infty$.

6.3.1 Well-Posedness, $1 \leq p < \infty$

In this subsection we characterize the class of $\mathcal{L}^N \times \mathcal{B}$ measurable integrands f for which

$$\int_E \left(f\left(x, v\left(x\right)\right)\right)^- dx < \infty$$

for every $v \in L^p (E; \mathbb{R}^m)$. The main result of this subsection is given by the following theorem. Note that although its proof its somewhat involved, the main novelty with respect to similar results available in the literature (see, e.g., [Bu89], [Kr76]) is that no regularity assumptions (e.g., continuity or lower semicontinuity) are made on $f(x, \cdot)$.

Theorem 6.45. *Let* $E \subset \mathbb{R}^N$ *be a Lebesgue measurable set with finite measure, let* $1 \leq p < \infty$, *and let* $f : E \times \mathbb{R}^m \to [-\infty, \infty]$ *be* $\mathcal{L}^N \times \mathcal{B}$ *measurable. Then*

$$\int_E (f(x, v(x)))^- \, dx < \infty \qquad (6.48)$$

for every $v \in L^p (E; \mathbb{R}^m)$ *if and only if there exist a nonnegative function* $\gamma \in L^1 (E)$ *and a constant* $C > 0$ *such that*

$$f(x, z) \geq -C |z|^p - \gamma(x) \text{ for } \mathcal{L}^N \text{ a.e. } x \in E \text{ and for all } z \in \mathbb{R}^m. \qquad (6.49)$$

Proof. If (6.49) holds, then

$$(f(x, z))^- \leq C |z|^p + \gamma(x)$$

for $\mathcal{L}^N$ a.e. $x \in E$ and for all $z \in \mathbb{R}^m$, and therefore (6.48) is satisfied.

To prove the converse implication, we observe that replacing f by $-f^-$, we may assume without loss of generality that $f \leq 0$.

Step 1: Fix a Borel function $g : E \to [1, \infty]$ and for every $v \in L^p (E; \mathbb{R}^m)$ define the function $g_v : E \to [-\infty, 0]$ as

$$g_v (x) := \begin{cases} f(x, v(x)) & \text{if } |v(x)| \leq g(x), \\ 0 & \text{otherwise.} \end{cases} \qquad (6.50)$$

Let $J(g) : E \to [-\infty, 0]$ be the essential infimum of the family

$$\{g_v : v \in L^p (E; \mathbb{R}^m)\}.$$

By Remark 1.107(i) it follows that if g_1, $g_2 : E \to [1, \infty]$ are two measurable functions such that $g_1 = g_2 \ \mathcal{L}^N$ a.e. in some Borel set $B \in \mathcal{B}(E)$, then

$$J(g_1) = J(g_2) \ \mathcal{L}^N \text{ a.e. in } B. \qquad (6.51)$$

Step 2: For each $x \in E$ and $n \in \mathbb{N}$ define

$$h(x, n) := J(n)(x).$$

We claim that there exist a function $\gamma \in L^1 (E)$ and a constant $C > 0$ such that

$$h(x, n) \geq -Cn^p - \gamma(x) \text{ for } \mathcal{L}^N \text{ a.e. } x \in E \text{ and for all } n \in \mathbb{N}. \qquad (6.52)$$

Assuming that the claim holds, in the remainder of this step we prove that (6.49) is satisfied. For any $v \in L^p(E; \mathbb{R}^m)$ construct a function $g \in L^p(E; \mathbb{N})$ such that

$$|v(x)| \le g(x) \le |v(x)| + 1 \tag{6.53}$$

for $\mathcal{L}^N$ a.e. $x \in E$.

We show that

$$h(x, g(x)) = J(g)(x) \tag{6.54}$$

for $\mathcal{L}^N$ a.e. $x \in E$. Indeed, since g is integer-valued, for each $n \in \mathbb{N}$ let

$$E_n := \{x \in E : g(x) = n\}.$$

By the locality property (6.51), for all $n \in \mathbb{N}$ and for $\mathcal{L}^N$ a.e. $x \in E_n$ we have

$$h(x, g(x)) = h(x, n) = J(n)(x) = J(g)(x),$$

which proves (6.54).

In turn, also by (6.50), (6.53), and (6.54), for every Borel set B and for $\mathcal{L}^N$ a.e. $x \in B$ we have

$$f(x, v(x)) = g_v(x) \ge J(g)(x) = h(x, g(x)),$$

and so by (6.52) and (6.53),

$$\int_B f(x, v(x))\, dx \ge \int_B h(x, g(x))\, dx \ge -\int_B (C(g(x))^p + \gamma(x))\, dx$$

$$\ge -\int_B (C(|v(x)| + 1)^p + \gamma(x))\, dx$$

$$\ge -\int_B \left[C\left(2^{p-1}|v(x)|^p + 2^{p-1}\right) + \gamma(x) \right] dx.$$

Since this inequality holds for all $v \in L^p(E; \mathbb{R}^m)$ and for every Borel set B, we may apply Proposition 6.24 to deduce

$$f(x, z) \ge -C\left(2^{p-1}|z|^p + 2^{p-1}\right) - \gamma(x)$$

for $\mathcal{L}^N$ a.e. $x \in E$ and for all $z \in \mathbb{R}^m$.

The remainder of the proof is devoted to proving claim (6.52).

Step 3: Let $g \in L^p(E; [1, \infty))$ and let $t \in \mathbb{R}$ be any number such that $t > \int_E J(g)(x)\, dx$. We claim that there exists $v \in L^p(E; \mathbb{R}^m)$ such that

$$\int_E f(x, v(x))\, dx \le t, \quad \|v\|_{L^p(E;\mathbb{R}^m)} \le 2\|g\|_{L^p(E)}.$$

Since $t > \int_E J(g)(x)\, dx$, by Proposition 1.92 we may find a nonpositive simple function $w \in L^1(E)$ such that $t > \int_E w(x)\, dx$ and $w \ge J(g)$ $\mathcal{L}^N$ a.e. in E with strict inequality whenever $J(g) < 0$.

By Theorem 1.108 there exists a sequence $\{v_n\} \subset L^p(E; \mathbb{R}^m)$ such that

$$J(g)(x) = \inf_{n \in \mathbb{N}} g_{v_n}(x) \tag{6.55}$$

for $\mathcal{L}^N$ a.e. in E. Let $\tilde{E}$ be the set of points of density one of E at which (6.55) holds. Let E_0 be the set of all points $x \in \tilde{E}$ such that $w(x) < 0$, x is a Lebesgue point of w, a p-Lebesgue point for g and all v_n, and a point of approximate continuity for all g_{v_n}, where all the functions in question have been extended by zero outside E.

Let λ be the measure defined for $B \in \mathcal{B}(E)$ by

$$\lambda(B) := \int_B w(x)\, dx. \tag{6.56}$$

Then $\lambda(E \setminus E_0) = 0$.

If $x \in E_0$, then $J(g)(x) \leq w(x) < 0$, and so $J(g)(x) < w(x)$. In particular, there exists $n_x \in \mathbb{N}$ such that

$$0 > w(x) \geq g_{v_{n_x}}(x), \tag{6.57}$$

and since $g_{v_{n_x}}(x) < 0$, it follows from (6.50) that

$$g_{v_{n_x}}(x) = f(x, v_{n_x}(x)),$$

and in turn, we deduce that

$$g(x) \geq |v_{n_x}(x)|. \tag{6.58}$$

Fix $\varepsilon > 0$ so small that

$$(1 - \varepsilon)\lambda(E) < t, \tag{6.59}$$

and find $r_x > 0$ such that for all $0 < r < r_x$ we have

$$\|v_{n_x}\|_{L^p(B(x,r) \cap E; \mathbb{R}^m)} < 2\|g\|_{L^p(B(x,r) \cap E)}, \tag{6.60}$$

$$\frac{\left|\{y \in B(x,r) \cap E : \left|g_{v_{n_x}}(y) - g_{v_{n_x}}(x)\right| \geq \varepsilon\}\right|}{|B(x,r) \cap E|} \leq \varepsilon, \tag{6.61}$$

$$\int_{B(x,r) \cap E} w(y)\, dy \geq |B(x,r) \cap E|\,(w(x) - \varepsilon). \tag{6.62}$$

Note that to obtain (6.60) we have used the fact that x is a p-Lebesgue point for g and v_{n_x}, (6.58), and the fact that $g(x) \geq 1$ (this is the first and only time that we use the fact that g is bounded away from zero).

We observe that for all $0 < r < r_x$,

$$\int_{B(x,r) \cap E} g_{v_{n_x}}(y)\, dy \leq (1 - \varepsilon)|B(x,r) \cap E|\,(g_{v_{n_x}}(x) + \varepsilon). \tag{6.63}$$

Indeed, since $g_{v_{n_x}} \leq 0$ if $g_{v_{n_x}}(x) + \varepsilon \geq 0$ then the inequality holds. If $g_{v_{n_x}}(x) + \varepsilon < 0$, then

$$\int_{B(x,r)\cap E} g_{v_{n_x}}(y)\, dy \leq \int_{B(x,r)\cap E\cap\{|g_{v_{n_x}}(\cdot) - g_{v_{n_x}}(x)| \leq \varepsilon\}} g_{v_{n_x}}(y)\, dy$$

$$\leq (g_{v_{n_x}}(x) + \varepsilon)\left|\{y \in B(x,r)\cap E : |g_{v_{n_x}}(y) - g_{v_{n_x}}(x)| \leq \varepsilon\}\right|$$

$$\leq (g_{v_{n_x}}(x) + \varepsilon)(1 - \varepsilon)|B(x,r)\cap E|,$$

where we have used (6.61).

Set $\mathcal{G} := \{B(x,r) : x \in E_0, \, 0 < r < r_x\}$. By the Vitali–Besicovitch covering theorem there exist disjoint balls $B(x_1, r_1), \ldots, B(x_k, r_k)$ such that

$$\lambda\left(E_0 \setminus \bigcup_{i=1}^{k} B(x_i, r_i)\right) \geq -\varepsilon. \tag{6.64}$$

Define
$$v(x) := \begin{cases} v_{n_{x_i}}(x) & \text{if } x \in B(x_i, r_i)\cap E, \, i = 1, \ldots, k, \\ 0 & \text{otherwise.} \end{cases}$$

By (6.60) we have

$$\|v\|_{L^p(E;\mathbb{R}^m)}^p \leq \sum_{i=1}^{k} \|v_{n_{x_i}}\|_{L^p(B(x_i, r_i)\cap E;\mathbb{R}^m)}^p \leq 2^p \|g\|_{L^p(E)}^p.$$

Moreover, by (6.50), (6.56), (6.57), (6.58), (6.62), (6.63), and (6.64), since $\lambda(E) = \lambda(E_0) < t$ and $f \leq 0$, we have

$$\int_E f(x, v)\, dx \leq \sum_{i=1}^{k} \int_{B(x_i, r_i)\cap E} f(x, v_{n_{x_i}}(x))\, dx$$

$$\leq \sum_{i=1}^{k} \int_{B(x_i, r_i)\cap E} g_{v_{n_{x_i}}}(y)\, dy$$

$$\leq (1 - \varepsilon) \sum_{i=1}^{k} |B(x_i, r_i)\cap E| \left(g_{v_{n_{x_i}}}(x_i) + \varepsilon\right)$$

$$\leq (1 - \varepsilon) \sum_{i=1}^{k} |B(x_i, r_i)\cap E| (w(x_i) + \varepsilon)$$

$$\leq (1 - \varepsilon) \sum_{i=1}^{k} \int_{B(x_i, r_i)\cap E} (w(y) + 2\varepsilon)\, dy$$

$$\leq (1 - \varepsilon) \lambda\left(\bigcup_{i=1}^{k} B(x_i, r_i)\cap E\right) + 2\varepsilon |E|$$

$$\leq (1 - \varepsilon) \lambda(E) + \varepsilon + 2\varepsilon |E| \leq t,$$

provided ε is sufficiently small and where we have used (6.59).

Step 4: Let $g \in L^p\left(E; [1, \infty)\right)$. We claim that

$$\int_E J\left(g\right)(x)\, dx > -\infty.$$

Indeed, assume by contradiction that $\int_E J\left(g\right)(x)\, dx = -\infty$. By Corollary 1.90 we may find a sequence of disjoint sets E_n such that

$$\int_{E_n} J\left(g\right)(x)\, dx = -\infty.$$

Applying Step 3 to each E_n we may find functions $v_n \in L^p\left(E_n; \mathbb{R}^m\right)$ such that

$$\int_{E_n} f\left(x, v_n\right) dx \leq -1, \quad \|v_n\|_{L^p(E_n; \mathbb{R}^m)} \leq 2C \|g\|_{L^p(E_n)}.$$

Defining

$$v := \sum_{n=1}^{\infty} v_n \chi_{E_n},$$

then $v \in L^p\left(E; \mathbb{R}^m\right)$ and

$$-\infty < \int_E f\left(x, v\right) dx \leq \sum_{n=1}^{\infty} \int_{E_n} f\left(x, v_n\right) dx \leq \sum_{n=1}^{\infty} -1 = -\infty,$$

which is a contradiction.

Step 5: We claim that there exists an integer $k \in \mathbb{N}$ such that the function $\psi_k : E \to \mathbb{N} \cup \{\infty\}$ defined by

$$\psi_k\left(x\right) := \sup\left\{n \in \mathbb{N} : h\left(x, n\right) \leq -kn^p\right\}, \quad x \in E, \qquad (6.65)$$

belongs to $L^p\left(E\right)$. Indeed, assume by contradiction that $\int_E \left(\psi_k\right)^p dx = \infty$ for all $k \in \mathbb{N}$. By Corollary 1.90 there exist pairwise disjoint Borel sets $B_k \in \mathcal{B}\left(E\right)$ such that

$$\int_{B_k} \left(\psi_k\right)^p dx = \infty \text{ for all } k \in \mathbb{N}. \qquad (6.66)$$

By the definition of ψ_k we may find measurable functions $w_k : B_k \to \mathbb{N}$ such that

$$h\left(x, w_k\left(x\right)\right) \leq -k\left(w_k\left(x\right)\right)^p \text{ for all } x \in B_k \text{ and } \int_{B_k} \left(w_k\right)^p dx \geq \frac{1}{k^2}. \qquad (6.67)$$

Indeed, set

$$C_k := \left\{x \in B_k : \psi_k\left(x\right) = \infty\right\}. \qquad (6.68)$$

If $|C_k| = 0$, we can simply take $w_k := \psi_k$, while if $|C_k| > 0$, then since (6.66) continues to hold on C_k, without loss of generality we can replace B_k with C_k. For each $x \in C_k$ we define

$$w_k(x) := \min\left\{ n \in \mathbb{N} : h(x,n) \le -kn^p, \, n \ge \frac{1}{k^{\frac{2}{p}} |C_k|^{\frac{1}{p}}} \right\}.$$

Note that in view of (6.65) and (6.68), w_k is measurable and well-defined, and that (6.67) holds.

Find Borel sets $E_k \subset B_k$ such that

$$\int_{E_k} (w_k)^p \, dx = \frac{1}{k^2}$$

and define

$$w(x) := \begin{cases} w_k(x) & \text{if } x \in E_k, k \in \mathbb{N}, \\ 1 & \text{otherwise.} \end{cases}$$

Then $w \in L^p(E)$, since

$$\int_E (w)^p \, dx = \sum_{k=1}^{\infty} \int_{E_k} (w_k)^p \, dx + \left| E \setminus \bigcup_{k=1}^{\infty} E_k \right| \le \sum_{k=1}^{\infty} \frac{1}{k^2} + |E| < \infty,$$

while (6.67) and the fact that $h \le 0$ yields

$$\int_E J(w)(x) \, dx = \int_E h(x, w(x)) \, dx \le \sum_{k=1}^{\infty} \int_{E_k} h(x, w_k(x))$$

$$\le -\sum_{k=1}^{\infty} \int_{E_k} k (w_k(x))^p = -\sum_{k=1}^{\infty} \frac{1}{k} = -\infty,$$

which contradicts Step 4. Note that we have used the fact that $w(E) \subset \mathbb{N}$, so that $h(x, w(x)) := J(w)(x)$ for $\mathcal{L}^N$ a.e. $x \in E$. Hence the claim holds and there exists $k \in \mathbb{N}$ such that $\psi_k \in L^p(E)$.

By the definition of ψ_k we have that

$$h(x,n) > -kn^p \tag{6.69}$$

for all $x \in E$ and all integers $n > \psi_k(x)$. Since $1 \le \psi_k(x) < \infty$ for $\mathcal{L}^N$ a.e. $x \in E$, we may find a measurable function $\psi : E \to \mathbb{N}$ that is everywhere less than or equal to ψ_k and such that

$$h(x, \psi(x)) = \min\{h(x,n) : n \le \psi_k(x)\} \quad \text{for } \mathcal{L}^N \text{ a.e. } x \in E.$$

Set $\gamma(x) := -h(x, \psi(x))$. Since $0 \le \psi \le \psi_k \in L^p(E)$, by Step 4 we deduce that $\gamma \in L^1(E)$ (where again we have used the fact that $h(x, \psi(x)) = J(\psi)(x)$), which, together with (6.69), yields (6.52).

By applying the previous theorem to f and $-f$ we obtain necessary and sufficient conditions for a superposition operator $v \mapsto f(\cdot, v(\cdot))$ to map $L^p(E; \mathbb{R}^m)$ into $L^1(E)$. Note that, in contrast to classical results in the literature (see Lemma 17.6 in [Kr76]), no continuity assumptions are made on f in the z variable.

Corollary 6.46 (Integral operators). *Let $E \subset \mathbb{R}^N$ be a Lebesgue measurable set with finite measure, let $1 \leq p < \infty$, and let $f : E \times \mathbb{R}^m \to [-\infty, \infty]$ be $\mathcal{L}^N \times \mathcal{B}$ measurable. Then the superposition operator $v \mapsto f(\cdot, v(\cdot))$ maps $L^p(E; \mathbb{R}^m)$ into $L^1(E)$ if and only if there exist a nonnegative function $\gamma \in L^1(E)$ and a constant $C > 0$ such that*

$$|f(x, z)| \leq C |z|^p + \gamma(x) \text{ for } \mathcal{L}^N \text{ a.e. } x \in E \text{ and for all } z \in \mathbb{R}^m.$$

6.3.2 Well-Posedness, $p = \infty$

The proof for the case $p = \infty$ follows from the results in the previous subsection.

Theorem 6.47. *Let $E \subset \mathbb{R}^N$ be a Lebesgue measurable set with finite measure, and let $f : E \times \mathbb{R}^m \to [-\infty, \infty]$ be $\mathcal{L}^N \times \mathcal{B}$ measurable. Then*

$$\int_E (f(x, v(x)))^- \, dx < \infty$$

for every $v \in L^\infty(E; \mathbb{R}^m)$ if and only if for every $M > 0$ there exists a nonnegative function $\gamma_M \in L^1(E)$ such that

$$f(x, z) \geq -\gamma_M(x) \tag{6.70}$$

for $\mathcal{L}^N$ a.e. $x \in E$ and all $z \in \mathbb{R}^m$ with $|z| \leq M$.

Proof. Fix $M > 0$ and define the functions

$$\tau_M(z) := \begin{cases} z & \text{if } |z| \leq M, \\ \dfrac{z}{|z|} M & \text{if } |z| > M, \end{cases} \tag{6.71}$$

and

$$f_M(x, z) := f(x, \tau_M(z)). \tag{6.72}$$

We claim that the functional

$$v \in L^1(E; \mathbb{R}^m) \mapsto \int_E f_M(x, v(x)) \, dx \tag{6.73}$$

is well-defined and does not take the value $-\infty$. Indeed, this follows from the fact that if $v \in L^1(E; \mathbb{R}^m)$, then $\tau_M \circ v \in L^\infty(E; \mathbb{R}^m)$ and

$$\int_E f_M(x, v(x)) \, dx = \int_E f(x, (\tau_M \circ v)(x)) \, dx.$$

By Theorem 6.45 there exist a function $a_M \in L^1(E)$ and a constant $C_M > 0$ such that

$$f_M(x, z) \geq -C_M |z| - a_M(x) \text{ for } \mathcal{L}^N \text{ a.e. } x \in E \text{ and for all } z \in \mathbb{R}^m.$$

In particular,

$$f(x, z) \geq -C_M M - a_M(x)$$

for $\mathcal{L}^N$ a.e. $x \in E$ and for all $z \in \mathbb{R}^m$ with $|z| \leq M$.

As in the previous subsection, by applying the previous theorem to f and $-f$ we can characterize the class of operators $v \mapsto f(\cdot, v(\cdot))$ that map $L^\infty(E; \mathbb{R}^m)$ into $L^1(E)$.

Corollary 6.48 (Integral operators). *Let $E \subset \mathbb{R}^N$ be a Lebesgue measurable set with finite measure, and let $f : E \times \mathbb{R}^m \to [-\infty, \infty]$ be $\mathcal{L}^N \times \mathcal{B}$ measurable. Then the superposition operator $v \mapsto f(\cdot, v(\cdot))$ maps $L^\infty(E; \mathbb{R}^m)$ into $L^1(E)$ if and only if for every $M > 0$ there exists a nonnegative function $\gamma_M \in L^1(E)$ such that*

$$|f(x, z)| \leq \gamma_M(x)$$

for $\mathcal{L}^N$ a.e. $x \in E$ and all $z \in \mathbb{R}^m$ with $|z| \leq M$.

6.4 Sequential Lower Semicontinuity

In this section we study necessary and sufficient conditions for the sequential lower semicontinuity of functionals of the form

$$v \in L^p(E; \mathbb{R}^m) \mapsto \int_E f(x, v(x))\, dx, \tag{6.74}$$

where $1 \leq p \leq \infty$ and

$$f : E \times \mathbb{R}^m \to [-\infty, \infty]$$

is $\mathcal{L}^N \times \mathcal{B}$ measurable.

We are interested in the following types of convergence:

- strong convergence in $L^p(E; \mathbb{R}^m)$ for $1 \leq p \leq \infty$;
- weak convergence in $L^p(E; \mathbb{R}^m)$ for $1 \leq p < \infty$;
- weak star convergence in $L^\infty(E; \mathbb{R}^m)$;
- weak star convergence in the sense of measures in $L^1(E; \mathbb{R}^m)$.

From now on we will assume that the integrand f satisfies the appropriate growth conditions from below that are necessary and sufficient to guarantee that $(f \circ v)^- \in L^1(E)$ for all $v \in L^p(E; \mathbb{R}^m)$, so that the functional

$$v \in L^p(E; \mathbb{R}^m) \mapsto \int_E f(x, v(x))\, dx$$

is well-defined (see Theorems 6.45 and 6.47).

6.4.1 Strong Convergence in L^p, $1 \leq p < \infty$

In this subsection we study necessary and sufficient conditions for the sequential lower semicontinuity of the functional (6.74) with respect to strong convergence in L^p.

Theorem 6.49. *Let $E \subset \mathbb{R}^N$ be a Lebesgue measurable set, let $1 \leq p < \infty$, and let $f : E \times \mathbb{R}^m \to (-\infty, \infty]$ be $\mathcal{L}^N \times \mathcal{B}$ measurable. Assume that there exist a nonnegative function $\gamma \in L^1(E)$ and a constant $C > 0$ such that*

$$f(x,z) \geq -C|z|^p - \gamma(x) \text{ for } \mathcal{L}^N \text{ a.e. } x \in E \text{ and for all } z \in \mathbb{R}^m. \quad (6.75)$$

Then the functional

$$v \in L^p(E; \mathbb{R}^m) \mapsto \int_E f(x, v(x))\, dx$$

is sequentially lower semicontinuous with respect to strong convergence in $L^p(E; \mathbb{R}^m)$ if and only if (up to equivalent integrands) $f(x, \cdot)$ is lower semicontinuous in $\mathbb{R}^m$ for $\mathcal{L}^N$ a.e. $x \in E$.

Proof. Without loss of generality we may assume that there exists $v_0 \in L^p(E; \mathbb{R}^m)$ such that

$$\int_E f(x, v_0(x))\, dx < \infty. \quad (6.76)$$

The proof of the sufficiency part is very similar to that of Theorem 5.9, and therefore we omit it.

We divide the proof of the necessity part into four steps.

Step 1: We claim that for every $B \in \mathcal{B}(E)$ the functional

$$v \in L^p(E; \mathbb{R}^m) \mapsto \int_B f(x, v(x))\, dx$$

is sequentially lower semicontinuous with respect to strong convergence in $L^p(E; \mathbb{R}^m)$. Indeed, let $\{v_n\} \subset L^p(E; \mathbb{R}^m)$ be a sequence strongly converging to some $v \in L^p(E; \mathbb{R}^m)$. Define

$$w_n(x) := \begin{cases} v_n(x) & \text{if } x \in B, \\ v_0(x) & \text{if } x \in E \setminus B, \end{cases} \qquad w(x) := \begin{cases} v(x) & \text{if } x \in B, \\ v_0(x) & \text{if } x \in E \setminus B. \end{cases}$$

Then $w_n \to w$ in $L^p(E; \mathbb{R}^m)$ and thus

$$\int_B f(x, v)\, dx + \int_{E \setminus B} f(x, v_0)\, dx = \int_E f(x, w)\, dx$$

$$\leq \liminf_{n \to \infty} \int_E f(x, w_n)\, dx$$

$$= \liminf_{n \to \infty} \int_B f(x, v_n)\, dx + \int_{E \setminus B} f(x, v_0)\, dx.$$

Hence

$$\int_B f(x, v)\, dx \leq \liminf_{n \to \infty} \int_B f(x, v_n)\, dx,$$

where we have used the fact that $f(\cdot, v_0(\cdot)) \in L^1(E)$ by (6.75) and (6.76).

Step 2: Assume that E has finite measure and that $f(x, z) \geq 0$ for $\mathcal{L}^N$ a.e. $x \in E$ and all $z \in \mathbb{R}^m$. Fix *any* countable set Y of $L^p(E; \mathbb{R}^m)$ and for $(x, z) \in E \times \mathbb{R}^m$ define

$$h_Y(x, z) := \begin{cases} f(x, w(x)) & \text{if there is } w \in Y \text{ such that } w(x) = z, \\ \infty & \text{otherwise.} \end{cases} \quad (6.77)$$

By Exercise 6.50, h_Y is $\mathcal{L}^N \times \mathcal{B}$ measurable. Hence by Proposition 6.44(i) and (ii), where the roles of the functions s and h there are played here by the functions

$$s_Y(x, z) := \liminf_{w \to z} h_Y(x, w), \quad (x, z) \in E \times \mathbb{R}^m, \quad (6.78)$$

and h_Y, respectively, there exists a normal integrand $g_Y : E \times \mathbb{R}^m \to [0, \infty]$ such that

$$s_Y(x, z) \leq g_Y(x, z) \leq h_Y(x, z) \quad (6.79)$$

for $\mathcal{L}^N$ a.e. $x \in E$ and all $z \in \mathbb{R}^m$. By (6.77), (6.78), and (6.79), we have that

$$f(x, w(x)) \geq g_Y(x, w(x)) \quad (6.80)$$

for $\mathcal{L}^N$ a.e. $x \in E$ and all $w \in Y$.

Step 3: For every $v \in L^p(E; \mathbb{R}^m)$ let $B \subset E$ be any Borel set such that

$$B \subset \{x \in E : g_Y(x, v(x)) \in \mathbb{R}\}.$$

We claim that for every $\varepsilon > 0$ there exists $w \in L^p(E; \mathbb{R}^m)$, with $\|w\|_{L^p(E;\mathbb{R}^m)} \leq \varepsilon$, and a Borel set $B_1 \subset E$, with $|B \setminus B_1| \leq \varepsilon$, such that

$$g_Y(x, v(x)) \geq f(x, v(x) + w(x)) - \varepsilon \text{ for } \mathcal{L}^N \text{ a.e. } x \in B_1. \quad (6.81)$$

Let B_0 be the set of all points of B for which (6.79) holds for all $z \in \mathbb{R}^m$ and that are points of density one for E, p-Lebesgue points for v and for all functions $w \in Y$, and points of approximate continuity for $g_Y(x, w(x))$ and $f(x, w(x))$ for all $w \in Y$, where all the functions in question have been extended by zero outside of E. Note that $|B \setminus B_0| = 0$.

Fix $\delta > 0$. For every $x \in B_0$ we have $g_Y(x, v(x)) < \infty$, and by (6.77), (6.78), and (6.79), there exists $v_x \in Y$ such that

$$|v(x) - v_x(x)| < \delta, \quad g_Y(x, v(x)) \geq f(x, v_x(x)) - \delta.$$

Since $x \in B_0$, we may find $r_x > 0$ such that for all $0 < r < r_x$ we have

$$\|v - v_x\|_{L^p(B(x,r) \cap E; \mathbb{R}^m)} < 2\delta |B(x, r) \cap E|^{1/p}, \quad (6.82)$$

$$|B(x, r) \cap \{y \in E : g_Y(y, v(y)) \geq f(y, v_x(y)) - 2\delta\}| \quad (6.83)$$
$$\geq (1 - \delta)|B(x, r) \cap E|.$$

Set $\mathcal{G} := \{B(x,r) : x \in B_0, \, 0 < r < r_x\}$. By the Vitali-Besicovitch covering theorem there exist disjoint balls $B(x_1, r_1), \ldots, B(x_k, r_k)$ such that

$$\left| B_0 \setminus \bigcup_{i=1}^k B(x_i, r_i) \right| \leq \delta. \tag{6.84}$$

Define

$$w(x) := \begin{cases} v_{x_i}(x) - v(x) & \text{if } x \in B(x_i, r_i) \cap E, \, i = 1, \ldots, k, \\ 0 & \text{otherwise.} \end{cases}$$

By (6.82) we have

$$\|w\|_{L^p(E;\mathbb{R}^m)}^p \leq \sum_{i=1}^k \|v_{x_i} - v\|_{L^p(B(x_i,r_i) \cap E;\mathbb{R}^m)}^p$$

$$\leq 2^p \delta^p \sum_{i=1}^k |B(x_i, r_i) \cap E| \leq 2^p \delta^p |E|.$$

Let

$$B_1 := \{x \in E : g_Y(x, v(x)) \geq f(x, v(x) + w(x)) - 2\delta\}.$$

Since $f(x, v(x) + w(x)) = f(x, v_{x_i}(x))$ for $\mathcal{L}^N$ a.e. $x \in B(x_i, r_i) \cap E$, by (6.83) we have

$$|(B(x_i, r_i) \cap E) \setminus B_1| \leq \delta |B(x_i, r_i) \cap E|$$

and by (6.84),

$$|B \setminus B_1| = |B_0 \setminus B_1| \leq \left| B_0 \setminus \bigcup_{i=1}^k B(x_i, r_i) \right| + \sum_{i=1}^k |(B(x_i, r_i) \cap E) \setminus B_1|$$

$$\leq \delta + \sum_{i=1}^k |B(x_i, r_i) \cap E| \delta \leq \delta(1 + |E|).$$

Hence (6.81) holds.

Step 4: We claim that

$$g_Y(x, z) \geq f(x, z) \tag{6.85}$$

for $\mathcal{L}^N$ a.e. $x \in E$ and all $z \in \mathbb{R}^m$. In view of Proposition 6.24 it suffices to show that

$$\int_B g_Y(x, v) \, dx \geq \int_B f(x, v) \, dx \tag{6.86}$$

for any fixed Borel set $B \subset E$ and $v \in L^p(E; \mathbb{R}^m)$. If the left-hand side of (6.86) is infinite there is nothing to prove. Thus, since $g_Y \geq 0$, it suffices to consider the case that $g_Y(\cdot, v(\cdot)) \in L^1(B)$. Define

$$T := \{x \in B : g_Y(x, v(x)) \in \mathbb{R}\}.$$

Then $|B \setminus T| = 0$. By Step 3 (see (6.81)), for every $n \in \mathbb{N}$ there exists $w_n \in L^p(E; \mathbb{R}^m)$, with $\|w_n\|_{L^p(E; \mathbb{R}^m)} \leq \frac{1}{2^n}$, and a Borel set $B_n \subset E$, with $|T \setminus B_n| \leq \frac{1}{2^n}$, such that

$$g_Y(x, v(x)) \geq f(x, v(x) + w_n(x)) - \frac{1}{2^n} \text{ for } \mathcal{L}^N \text{ a.e. } x \in B_n.$$

For any fixed integer $k \in \mathbb{N}$ set

$$C_k := \bigcap_{n > k} B_n.$$

Then $v + w_n$ converges to v in $L^p(E; \mathbb{R}^m)$ and

$$g_Y(x, v(x)) \geq f(x, v(x) + w_n(x)) - \frac{1}{2^k}$$

for $\mathcal{L}^N$ a.e. $x \in C_k$ for all $n > k$. Hence by Step 1,

$$\int_B g_Y(x, v) \, dx \geq \int_{C_k} g_Y(x, v) \, dx \geq \liminf_{n \to \infty} \int_{C_k} \left[f(x, v + w_n) - \frac{1}{2^k} \right] dx$$

$$\geq \int_{C_k} f(x, v) \, dx - \frac{1}{2^k} |C_k| \geq \int_{C_k} f(x, v) \, dx - \frac{1}{2^k} |E|,$$

where we have used the fact that $g_Y \geq 0$. Since $|B \setminus T| = 0$ we have

$$|B \setminus C_k| = |T \setminus C_k| \leq \sum_{n > k} |T \setminus B_n| \leq \sum_{n > k} \frac{1}{2^n} = \frac{1}{2^k} \to 0$$

as $k \to \infty$, and so letting $k \to \infty$ and using the Lebesgue dominated convergence theorem we conclude (6.86).

Step 5: For each $k \in \mathbb{N}$ define the functional

$$v \in L^p(E; \mathbb{R}^m) \mapsto I_k(v; E) := \int_E f_k(x, v(x)) \, dx,$$

where

$$f_k(x, z) := \Gamma_k(f(x, z)), \quad \Gamma_k(s) := \begin{cases} k \text{ if } s > k, \\ s \text{ if } 0 \leq s \leq k. \end{cases}$$

We claim that for every $B \in \mathcal{B}(E)$ the functional $I_k(\cdot; B)$ is sequentially lower semicontinuous with respect to strong convergence in $L^p(E; \mathbb{R}^m)$. Indeed, let $\{v_n\} \subset L^p(E; \mathbb{R}^m)$ be a sequence strongly converging to some $v \in L^p(E; \mathbb{R}^m)$ and set

$$Y := \{v_n : n \in \mathbb{N}\} \cup \{v\}.$$

By (6.80) and (6.85) we have that

$$f\left(x, v_n\left(x\right)\right) = g_Y\left(x, v_n\left(x\right)\right)$$

for $\mathcal{L}^N$ a.e. $x \in E$ and all $n \in \mathbb{N}$, and hence

$$f_k\left(x, v_n\left(x\right)\right) = g_{Y,k}\left(x, v_n\left(x\right)\right)$$

for $\mathcal{L}^N$ a.e. $x \in E$ and all $n \in \mathbb{N}$ and where $g_{Y,k} := \Gamma_k \circ g_Y$. Since the function $g_{Y,k}\left(x, \cdot\right)$ is lower semicontinuous, by Fatou's lemma it follows that

$$\liminf_{n \to \infty} \int_B f_k\left(x, v_n\left(x\right)\right)\, dx = \liminf_{n \to \infty} \int_B g_{Y,k}\left(x, v_n\left(x\right)\right)\, dx$$

$$\geq \int_B g_{Y,k}\left(x, v\left(x\right)\right)\, dx = \int_B f_k\left(x, v\left(x\right)\right)\, dx,$$

and the claim is proved.

Step 6: Consider $\mathcal{B}\left(E\right)$ as a subset of $L^1\left(E\right)$. Since $L^1\left(E\right)$ is separable we may find a countable set $\mathcal{S}\left(E\right)$ of $\mathcal{B}\left(E\right)$ that is dense in $\mathcal{B}\left(E\right)$.

Since $I_k\left(\cdot; B\right)$ is sequentially lower semicontinuous with respect to strong convergence in $L^p\left(E; \mathbb{R}^m\right)$, by Proposition 3.7 we may find a countable set Y_0 of $L^p\left(E; \mathbb{R}^m\right)$ such that for every $v \in L^p\left(E; \mathbb{R}^m\right)$, $B \in \mathcal{S}\left(E\right)$, and $k \in \mathbb{N}$,

$$I_k\left(v; B\right) = \liminf_{w \in Y_0, w \to v} I_k\left(w; B\right). \tag{6.87}$$

Hence for every $v \in L^p\left(E; \mathbb{R}^m\right)$, $B \in \mathcal{S}\left(E\right)$, and $k \in \mathbb{N}$, by (6.87), (6.80), and (6.85), and again using Fatou's lemma we have

$$\int_B f_k\left(x, v\left(x\right)\right)\, dx = \liminf_{w \in Y_0, w \to v} \int_B f_k\left(x, w\left(x\right)\right)\, dx$$

$$= \liminf_{w \in Y_0, w \to v} \int_B g_{Y_0,k}\left(x, w\left(x\right)\right)\, dx$$

$$\geq \int_B g_{Y_0,k}\left(x, v\left(x\right)\right)\, dx.$$

Since $0 \leq f_k$, $g_{Y_0,k} \leq k$ and $\mathcal{S}\left(E\right)$ is dense in $\mathcal{B}\left(E\right)$, the previous inequality holds for every $B \in \mathcal{B}\left(E\right)$ and thus by Proposition 6.24 we have that

$$f_k\left(x, z\right) \geq g_{Y_0,k}\left(x, z\right)$$

for $\mathcal{L}^N$ a.e. $x \in E$ and all $z \in \mathbb{R}^m$ and $k \in \mathbb{N}$. By letting $k \to \infty$ we conclude that

$$f\left(x, z\right) \geq g_{Y_0}\left(x, z\right)$$

for $\mathcal{L}^N$ a.e. $x \in E$ and all $z \in \mathbb{R}^m$, which, together with (6.85), establishes the necessity part of the theorem in the case that E has finite measure and $f \geq 0$.

Step 7: In the case that E has finite measure but f can take negative values, by (6.75) we have that the function

$$g(x, z) := f(x, z) + C|z|^p + \gamma(x)$$

is nonnegative, $\mathcal{L}^N \times \mathcal{B}$ measurable, and the functional

$$H(v; B) := \int_B f(x, v) + C|v|^p + \gamma(x)\, dx$$

is sequentially lower semicontinuous with respect to strong convergence in $L^p(E; \mathbb{R}^m)$. Hence, by Steps 1–6, $f(x, \cdot) + C|\cdot|^p + \gamma(x)$ is lower semicontinuous, and in turn, $f(x, \cdot)$ is lower semicontinuous.

Step 8: Finally, if E has infinite measure, then by Step 1 we can apply Steps 2–7 to the set $E \cap B(0, j)$ for each $j \in \mathbb{N}$. This completes the proof of the necessity part.

Exercise 6.50. Prove that the function h_Y defined in Step 2 of the previous proof is $\mathcal{L}^N \times \mathcal{B}$ measurable.

As in the previous subsection, by applying the previous theorem to f and $-f$ we can characterize the class of all continuous operators $v \mapsto f(\cdot, v(\cdot))$ that map $L^\infty(E; \mathbb{R}^m)$ into $L^1(E)$.

Corollary 6.51 (Continuity of integral operators). *Let $E \subset \mathbb{R}^N$ be a Lebesgue measurable set, let $1 \leq p < \infty$, and let $f : E \times \mathbb{R}^m \to [-\infty, \infty]$ be $\mathcal{L}^N \times \mathcal{B}$ measurable. Assume that there exist a nonnegative function $\gamma \in L^1(E)$ and a constant $C > 0$ such that*

$$|f(x, z)| \leq C|z|^p + \gamma(x) \text{ for } \mathcal{L}^N \text{ a.e. } x \in E \text{ and for all } z \in \mathbb{R}^m.$$

Then the functional

$$v \in L^p(E; \mathbb{R}^m) \mapsto \int_E f(x, v(x))\, dx$$

is continuous with respect to strong convergence in $L^p(E; \mathbb{R}^m)$ if and only if (up to equivalent integrands) $f(x, \cdot)$ is continuous in $\mathbb{R}^m$ for $\mathcal{L}^N$ a.e. $x \in E$.

6.4.2 Strong Convergence in L^∞

When $p = \infty$ we have an analogous result.

Theorem 6.52. *Let $E \subset \mathbb{R}^N$ be a Lebesgue measurable set, and let $f : E \times \mathbb{R}^m \to (-\infty, \infty]$ be $\mathcal{L}^N \times \mathcal{B}$ measurable. Assume that for every $M > 0$ there exists a nonnegative function $\gamma_M \in L^1(E)$ such that*

$$f(x, z) \geq -\gamma_M(x) \tag{6.88}$$

for $\mathcal{L}^N$ a.e. $x \in E$ and all $z \in \mathbb{R}^m$ with $|z| \leq M$. Then the functional

$$v \in L^{\infty}(E; \mathbb{R}^m) \mapsto \int_E f(x, v(x)) \, dx$$

is sequentially lower semicontinuous with respect to strong convergence in $L^{\infty}(E; \mathbb{R}^m)$ if and only if (up to equivalent integrands) $f(x, \cdot)$ is lower semicontinuous in $\mathbb{R}^m$ for $\mathcal{L}^N$ a.e. $x \in E$.

Proof. Without loss of generality we may assume that there exists $v_0 \in L^{\infty}(E; \mathbb{R}^m)$ such that

$$\int_E f(x, v_0(x)) \, dx < \infty.$$

The sufficiency part of the theorem follows from Fatou's lemma. To prove the necessity part, we start by observing that as in Step 1 of the proof of the previous theorem, we may assume, without loss of generality, that for every $B \in \mathcal{B}(E)$ the functional

$$v \in L^{\infty}(E; \mathbb{R}^m) \mapsto \int_B f(x, v(x)) \, dx$$

is sequentially lower semicontinuous with respect to strong convergence in $L^{\infty}(E; \mathbb{R}^m)$. Also, by Step 8 it is enough to consider the case that E has finite measure.

Fix $M > 0$ and consider the function f_M defined in (6.72) and the corresponding functional defined in (6.73). By (6.88) the function f_M satisfies (6.75) with $p = 1$. We claim that the functional

$$v \in L^1(E; \mathbb{R}^m) \mapsto \int_E f_M(x, v(x)) \, dx \qquad (6.89)$$

is sequentially lower semicontinuous with respect to strong convergence in $L^1(E; \mathbb{R}^m)$. Indeed, let $\{v_n\} \subset L^1(E; \mathbb{R}^m)$ be any a sequence converging strongly in $L^1(E; \mathbb{R}^m)$ to some $v \in L^1(E; \mathbb{R}^m)$. Without loss of generality we may assume that

$$\liminf_{n \to \infty} \int_E f_M(x, v_n(x)) \, dx = \lim_{n \to \infty} \int_E f_M(x, v_n(x)) \, dx$$

and then extract a subsequence (not relabeled) of $\{v_n\}$ that converges to v pointwise $\mathcal{L}^N$ a.e. in E.

Fix $\varepsilon > 0$. By Egoroff's theorem there exists a set $B_\varepsilon \subset E$ such that $|B_\varepsilon| < \varepsilon$ and v_n converges uniformly to v in $E \setminus B_\varepsilon$. Define

$$w_n := \begin{cases} v_n & \text{on } E \setminus B_\varepsilon, \\ v & \text{on } B_\varepsilon. \end{cases}$$

Then the sequence $\{\tau_M \circ w_n\}$ converges in $L^{\infty}(E; \mathbb{R}^m)$ to $\tau_M \circ v$, where τ_M is the function defined as in (6.71). By hypothesis it follows that

$$\liminf_{n \to \infty} \int_E [f_M(x, v_n) + \gamma_M] \, dx \geq \liminf_{n \to \infty} \int_{E \setminus B_\varepsilon} [f_M(x, v_n) + \gamma_M(x)] \, dx$$

$$= \liminf_{n \to \infty} \int_{E \setminus B_\varepsilon} [f(x, \tau_M \circ w_n) + \gamma_M(x)] \, dx$$

$$\geq \int_{E \setminus B_\varepsilon} [f(x, \tau_M \circ v) + \gamma_M(x)] \, dx$$

$$= \int_{E \setminus B_\varepsilon} [f_M(x, v) + \gamma_M(x)] \, dx.$$

Letting $\varepsilon \to 0^+$ we conclude that

$$\liminf_{n \to \infty} \int_E f_M(x, v_n) \, dx \geq \int_E f_M(x, v) \, dx$$

and the claim is proved. Since the functional (6.89) satisfies all the hypotheses of the previous theorem, we have that $f_M(x, \cdot)$ is lower semicontinuous in $\mathbb{R}^m$ for $\mathcal{L}^N$ a.e. $x \in E$.

Choose $M_k := k$ and let $N_k \subset E$ with $|N_k| = 0$ be such that $f_{M_k}(x, \cdot)$ is lower semicontinuous in $\mathbb{R}^m$ for all $x \in E \setminus N_k$. Define

$$N_0 := \bigcup_{k=1}^{\infty} N_k.$$

Then $|N_0| = 0$. We claim that $f(x, \cdot)$ is lower semicontinuous in $\mathbb{R}^m$ for all $x \in E \setminus N_0$. Indeed, fix $x \in E \setminus N_0$ and let $\{z_n\} \subset \mathbb{R}^m$ be such that $z_n \to z$ and take $M_k > |z_n|$ for all n. Then

$$\liminf_{n \to \infty} f(x, z_n) = \liminf_{n \to \infty} f_{M_k}(x, z_n) \geq f_{M_k}(x, z) = f(x, z).$$

This completes the proof.

The analogue of Corollary 6.51 is the following result.

Corollary 6.53 (Continuity of integral operators). *Let $E \subset \mathbb{R}^N$ be a Lebesgue measurable set and let $f : E \times \mathbb{R}^m \to [-\infty, \infty]$ be $\mathcal{L}^N \times \mathcal{B}$ measurable. Assume that for every $M > 0$ there exists a nonnegative function $\gamma_M \in L^1(E)$ such that*

$$|f(x, z)| \leq \gamma_M(x)$$

for $\mathcal{L}^N$ a.e. $x \in E$ and all $z \in \mathbb{R}^m$ with $|z| \leq M$. Then the functional

$$v \in L^\infty(E; \mathbb{R}^m) \mapsto \int_E f(x, v(x)) \, dx$$

is continuous with respect to strong convergence in $L^\infty(E; \mathbb{R}^m)$ if and only if (up to equivalent integrands) $f(x, \cdot)$ is continuous in $\mathbb{R}^m$ for $\mathcal{L}^N$ a.e. $x \in E$.

6.4.3 Weak Convergence in L^p, $1 \leq p < \infty$

We study necessary and sufficient conditions for sequential lower semicontinuity with respect to weak convergence in $L^p(E; \mathbb{R}^m)$, $1 \leq p < \infty$, and weak star convergence in $L^\infty(E; \mathbb{R}^m)$. Since functionals that are sequentially lower semicontinuous with respect to weak convergence in $L^p(E; \mathbb{R}^m)$ (respectively weak star when $p = \infty$) are also sequentially lower semicontinuous with respect to strong convergence in $L^p(E; \mathbb{R}^m)$, without loss of generality, in what follows we may assume that the integrand f satisfies all the necessary conditions for strong convergence in $L^p(E; \mathbb{R}^m)$ obtained in the previous subsections.

Theorem 6.54. *Let $E \subset \mathbb{R}^N$ be a Lebesgue measurable set, let $1 \leq p < \infty$, and let $f : E \times \mathbb{R}^m \to (-\infty, \infty]$ be $\mathcal{L}^N \times \mathcal{B}$ measurable. Assume that $f(x, \cdot)$ is lower semicontinuous in $\mathbb{R}^m$ for $\mathcal{L}^N$ a.e. $x \in E$ and that there exist a nonnegative function $\gamma \in L^1(E)$ and a constant $C > 0$ such that*

$$f(x, z) \geq -C|z|^p - \gamma(x) \text{ for } \mathcal{L}^N \text{ a.e. } x \in E \text{ and for all } z \in \mathbb{R}^m.$$

Then the functional

$$v \in L^p(E; \mathbb{R}^m) \mapsto \int_E f(x, v(x)) \, dx$$

is sequentially lower semicontinuous with respect to weak convergence in $L^p(E; \mathbb{R}^m)$ if and only if

(i) $f(x, \cdot)$ is convex in $\mathbb{R}^m$ for $\mathcal{L}^N$ a.e. $x \in E$;
(ii) there exist two functions $a \in L^1(E)$ and $b \in L^{p'}(E; \mathbb{R}^m)$ such that

$$f(x, z) \geq a(x) + b(x) \cdot z$$

for $\mathcal{L}^N$ a.e. $x \in E$ and all $z \in \mathbb{R}^m$.

Proof (Sufficiency). Without loss of generality we may assume that there exists $v_0 \in L^p(E; \mathbb{R}^m)$ such that

$$\int_E f(x, v_0(x)) \, dx < \infty.$$

The proof is very similar to the sufficiency proof of Theorem 5.14 and we only indicate the main changes. As in Step 2 we can assume, without loss of generality, that $f(x, z) \geq 0$ for $\mathcal{L}^N$ a.e. $x \in E$ and all $z \in \mathbb{R}^m$. Using the blowup method as in Step 1 of the sufficiency proof of Theorem 5.14 we need to show only that

$$\frac{d\mu}{d\mathcal{L}^N}(x_0) = \lim_{\varepsilon \to 0^+} \frac{\mu(Q(x_0, \varepsilon) \cap E)}{\varepsilon^N} \geq f(x_0, v(x_0)) \qquad \text{for } \mathcal{L}^N \text{ a.e. } x_0 \in E.$$

By Theorem 6.39 there exist two sequences of bounded measurable functions

$$a_i : E \to \mathbb{R}, \quad b_i : E \to \mathbb{R}^m,$$

such that

$$f(x, z) = \sup_{i \in \mathbb{N}} \{a_i(x) + b_i(x) \cdot z\}$$

for $\mathcal{L}^N$ a.e. $x \in E$ and all $z \in \mathbb{R}^m$.

Fix a point $x_0 \in E$ of density one for E that satisfies (5.15) and is a Lebesgue point of all the L^1_{loc} functions $a_i(\cdot)\chi_E(\cdot)$ and $(b_i(\cdot) \cdot v(\cdot))\chi_E(\cdot)$. Then as in (5.18) we can conclude that

$$
\begin{aligned}
\frac{d\mu}{d\mathcal{L}^N}(x_0) &= \lim_{k \to \infty} \frac{\mu(Q(x_0, \varepsilon_k) \cap E)}{\varepsilon_k^N} \\
&= \lim_{k \to \infty} \lim_{n \to \infty} \frac{1}{\varepsilon_k^N} \int_{Q(x_0, \varepsilon_k) \cap E} f(x, v_n(x))\, dx \\
&\geq \liminf_{k \to \infty} \liminf_{n \to \infty} \frac{1}{\varepsilon_k^N} \int_{Q(x_0, \varepsilon_k) \cap E} (a_i(x) + b_i(x) \cdot v_n(x))\, dx \\
&= \liminf_{k \to \infty} \frac{1}{\varepsilon_k^N} \int_{Q(x_0, \varepsilon_k) \cap E} (a_i(x) + b_i(x) \cdot v(x))\, dx \\
&= a_i(x_0) + b_i(x_0) \cdot v(x_0),
\end{aligned}
$$

where we have used the fact that $v_n \rightharpoonup v$ in $L^p(E; \mathbb{R}^m)$. By taking the supremum over all i we conclude that

$$\frac{d\mu}{d\mathcal{L}^N}(x_0) \geq f(x_0, v(x_0))$$

as desired.

Proof (Necessity). Without loss of generality, we may assume that there exists $v_0 \in L^p(E; \mathbb{R}^m)$ such that

$$\int_E f(x, v_0(x))\, dx < \infty, \tag{6.90}$$

and, replacing $f(x, z)$ with $f(x, z - v_0(x))$, that $v_0 = 0$. As in the proof of Step 1 of Theorem 6.49, we can show that for every $B \in \mathcal{B}(E)$ the functional

$$v \in L^p(E; \mathbb{R}^m) \mapsto \int_B f(x, v(x))\, dx \tag{6.91}$$

is sequentially lower semicontinuous with respect to weak convergence in $L^p(E; \mathbb{R}^m)$.

Step 1: We claim that $f(x, \cdot)$ is convex in $\mathbb{R}^m$ for $\mathcal{L}^N$ a.e. $x \in E$. We consider first the case that $|E| < \infty$. Fix $\theta \in [0, 1] \cap \mathbb{Q}$ and let $v, w \in L^\infty(E; \mathbb{R}^m)$. Let $\{h_n\}$ be a sequence of functions in $L^\infty(E; \{0, 1\})$ such that (see Example 2.86)

$$h_n \overset{*}{\rightharpoonup} \theta \quad \text{in } L^\infty(E).$$

For $x \in E$ define

$$u_n(x) := (1 - h_n(x)) v(x) + h_n(x) w(x).$$

Then $u_n \overset{*}{\rightharpoonup} u$ in $L^\infty(E; \mathbb{R}^m)$, where

$$u(x) := (1 - \theta) v(x) + \theta w(x),$$

and so by the sequential lower semicontinuity of the energy (6.91), for every $B \in \mathcal{B}(E)$ we have

$$\int_B f(x, (1 - \theta) v(x) + \theta w(x))\, dx \leq \liminf_{n \to \infty} \int_B f(x, u_n(x))\, dx$$

$$= \liminf_{n \to \infty} \int_B (1 - h_n(x)) f(x, v(x)) + h_n(x) f(x, w(x))\, dx$$

$$= (1 - \theta) \int_B f(x, v(x))\, dx + \theta \int_B f(x, w(x))\, dx.$$

Since this is true for all $B \in \mathcal{B}(E)$ and $v,\, w \in L^\infty(E; \mathbb{R}^m)$, it follows from Proposition 6.24 that there exists a set $N_\theta \subset E$, with $|N_\theta| = 0$, such that

$$f(x, (1 - \theta) z + \theta w) \leq (1 - \theta) f(x, z) + \theta f(x, w) \qquad (6.92)$$

for all $x \in E \setminus N_\theta$ and all $z,\, w \in \mathbb{R}^m$. Let

$$N_0 := \bigcup_{\theta \in [0,1] \cap \mathbb{Q}} N_\theta.$$

Then $|N_0| = 0$ and (6.92) holds for all $\theta \in [0,1] \cap \mathbb{Q}$, $x \in E \setminus N_0$, and all z, $w \in \mathbb{R}^m$. Now fix $\theta \in [0,1]$ and let $\theta_n \in [0,1] \cap \mathbb{Q}$ be such that $\theta_n \to \theta$. By (6.92),

$$f(x, (1 - \theta_n) z + \theta_n w) \leq (1 - \theta_n) f(x, z) + \theta_n f(x, w)$$

for all $n \in \mathbb{N}$, $x \in E \setminus N_0$, and all $z,\, w \in \mathbb{R}^m$. By letting $n \to \infty$ and using the fact that $f(x, \cdot)$ is lower semicontinuous in $\mathbb{R}^m$ for $\mathcal{L}^N$ a.e. $x \in E$ we prove the claim in the case $|E| < \infty$. If $|E| = \infty$ and in view of the sequential lower semicontinuity of the energy (6.91) it suffices to apply the first part of this step to $|E \cap B(0, j)|$ for each $j \in \mathbb{N}$.

Step 2: Replacing $f(x, z)$ with $\frac{1}{pC}(f(x, z) - \gamma(x))$, we may assume, without loss of generality, that

$$f(x, z) \geq -\frac{1}{p}|z|^p \quad \text{for } \mathcal{L}^N \text{ a.e. } x \in E \text{ and for all } z \in \mathbb{R}^m.$$

Thus for $\mathcal{L}^N$ a.e. $x \in E$ we may apply Proposition 5.16 to define two functions $a : E \to [0, \infty)$, $b : E \to \mathbb{R}^m$ such that

$$f(x, z) \geq a(x) + b(x) \cdot z \geq -\frac{1}{p} |z|^p \text{ for } \mathcal{L}^N \text{ a.e. } x \in E \text{ and all } z \in \mathbb{R}^m. \quad (6.93)$$

By Aumann's selection theorem we may assume that a and b are measurable. Moreover, if $p = 1$, then $|b(x)| \leq 1$, while for $p > 1$,

$$|b(x)|^{p'} \leq C(p) a(x) \quad (6.94)$$

for some constant $C(p) > 0$. It remains to show that $a \in L^1(E)$, which by (6.94) will entail $b \in L^{p'}(E; \mathbb{R}^m)$. Taking $z = 0$ in (6.93), it follows that

$$f(x, 0) \geq a(x) \geq 0,$$

which, by (6.90), implies that $a \in L^1(E)$.

Remark 6.55. Note that the proof of the sufficiency still holds under L^∞ weak star convergence for nonnegative $\mathcal{L}^N \times \mathcal{B}$ measurable integrands $f : E \times \mathbb{R}^m \to [0, \infty]$ such that $f(x, \cdot)$ is lower semicontinuous and convex in $\mathbb{R}^m$ for $\mathcal{L}^N$ a.e. $x \in E$.

6.4.4 Weak Star Convergence in L^∞

The case $p = \infty$ follows from the results in the previous subsections.

Theorem 6.56. *Let $E \subset \mathbb{R}^N$ be a Lebesgue measurable set, and let $f : E \times \mathbb{R}^m \to (-\infty, \infty]$ be $\mathcal{L}^N \times \mathcal{B}$ measurable. Assume that $f(x, \cdot)$ is lower semicontinuous in $\mathbb{R}^m$ for $\mathcal{L}^N$ a.e. $x \in E$ and that for every $M > 0$ there exists a nonnegative function $a_M \in L^1(E)$ such that*

$$f(x, z) \geq -a_M(x)$$

for $\mathcal{L}^N$ a.e. $x \in E$ and all $z \in \mathbb{R}^m$ with $|z| \leq M$. Then the functional

$$v \in L^\infty(E; \mathbb{R}^m) \mapsto \int_E f(x, v(x)) \, dx$$

is sequentially lower semicontinuous with respect to weak star convergence in $L^\infty(E; \mathbb{R}^m)$ if and only if $f(x, \cdot)$ is convex in $\mathbb{R}^m$ for $\mathcal{L}^N$ a.e. $x \in E$.

Proof. To prove sufficiency, let $v_n \overset{*}{\rightharpoonup} v$ in $L^\infty(E; \mathbb{R}^m)$. Then

$$\sup_n \|v_n\|_{L^\infty(\Omega; \mathbb{R}^m)} \leq M$$

for some $M > 0$, and so

$$f(x, v_n(x)) \geq -a_M(x)$$

for $\mathcal{L}^N$ a.e. $x \in E$ and all $n \in \mathbb{N}$. Define

$$f_M(x, z) := \max\{f(x, z), -a_M(x)\}.$$

Note that

$$f(x, v_n(x)) = f_M(x, v_n(x)) \quad \text{and} \quad f(x, v(x)) = f_M(x, v(x))$$

for $\mathcal{L}^N$ a.e. $x \in E$, all $n \in \mathbb{N}$, and

$$f_M(x, z) + a_M(x)$$

is a nonnegative $\mathcal{L}^N \times \mathcal{B}$ measurable function, convex and lower semicontinuous in the z variable. By Remark 6.55 we have

$$\liminf_{n \to \infty} \int_E f(x, v_n)\, dx = \liminf_{n \to \infty} \int_E [f_M(x, v_n) + a_M(x)]\, dx - \int_E a_M(x)\, dx$$

$$\geq \int_E [f_M(x, v) + a_M(x)]\, dx - \int_E a_M(x)\, dx$$

$$= \int_E f(x, v)\, dx.$$

The proof of the necessity condition is entirely identical to that of Theorem 6.54.

6.4.5 Weak Star Convergence in the Sense of Measures

In this subsection we consider only the case in which the domain of integration is an open subset $\Omega \subset \mathbb{R}^N$. As we remarked in Section 5.2.3, due to lack of reflexivity of the space $L^1(\Omega; \mathbb{R}^m)$, if $\{v_n\} \subset L^1(\Omega; \mathbb{R}^m)$ is such that

$$\sup_n \|v_n\|_{L^1} < \infty,$$

then one can conclude only that $\{v_n\}$ admits a subsequence (not relabeled) such that $v_n \mathcal{L}^N \lfloor \Omega \overset{*}{\rightharpoonup} \lambda$ in $\mathcal{M}(\Omega; \mathbb{R}^m)$, i.e., if for all $u \in C_0(\Omega)$,

$$\int_\Omega u v_n\, dx \to \int_\Omega u\, d\lambda.$$

This subsection is devoted to the study of sequential lower semicontinuity under this natural notion of convergence. In particular, we will address necessary and sufficient conditions under which the lower semicontinuity property

$$\liminf_{n \to \infty} \int_\Omega f(x, v_n)\, dx \geq \int_\Omega f\left(x, \frac{d\lambda}{d\mathcal{L}^N}\right) dx$$

holds whenever $v_n \mathcal{L}^N \lfloor \Omega \overset{*}{\rightharpoonup} \lambda$ in $\mathcal{M}(\Omega; \mathbb{R}^m)$ and λ admits the Radon–Nikodym decomposition

$$\lambda = \frac{d\lambda}{d\mathcal{L}^N} \, \mathcal{L}^N \lfloor \Omega + \lambda_s,$$

with λ_s and $\mathcal{L}^N$ mutually singular.

Since functionals that are sequentially lower semicontinuous with respect to this type of convergence are also sequentially lower semicontinuous with respect to weak convergence in $L^1 (\Omega; \mathbb{R}^m)$, without loss of generality we may assume in what follows that the integrand f satisfies all the necessary conditions for weak convergence in $L^1 (\Omega; \mathbb{R}^m)$ (see the previous subsection).

In what follows, to each function $\psi : \Omega \to \mathbb{R}^m$ we associate the multifunction $\Gamma_\psi : \Omega \to \mathcal{P}(\mathbb{R}^m) \setminus \{\emptyset\}$ defined by

$$\Gamma_\psi (x) := \{\psi(x)\}, \quad x \in \Omega.$$

Theorem 6.57. *Let $\Omega \subset \mathbb{R}^N$ be an open set and let $f : \Omega \times \mathbb{R}^m \to [-\infty, \infty]$ be Borel measurable. Assume that $f(x, \cdot)$ is convex and lower semicontinuous in $\mathbb{R}^m$ for $\mathcal{L}^N$ a.e. $x \in \Omega$ and that there exist a function $a \in L^1(\Omega)$ and a constant $C > 0$ such that*

$$f(x, z) \geq a(x) + C|z| \tag{6.95}$$

for $\mathcal{L}^N$ a.e. $x \in \Omega$ and all $z \in \mathbb{R}^m$. Then the functional

$$v \in L^1 (\Omega; \mathbb{R}^m) \mapsto \int_\Omega f(x, v(x)) \, dx$$

is sequentially lower semicontinuous with respect to weak star convergence in the sense of measures if and only if

$$f(x, z) = \tilde{f}(x, z) \tag{6.96}$$

for $\mathcal{L}^N$ a.e. $x \in \Omega$ and all $z \in \mathbb{R}^m$, where

$$\tilde{f}(x, z) := \sup_{\xi \in \Gamma(x)} [\xi \cdot z - f^*(x, \xi)], \quad (x, z) \in (\Omega, \mathbb{R}^m), \tag{6.97}$$

and Γ is the essential supremum of the family of multifunctions $\mathcal{F} := \{\Gamma_\psi : \psi \in \mathcal{X}\}$ (with respect to the Lebesgue measure), where

$$\mathcal{X} := \{\psi \in C_0(\Omega; \mathbb{R}^m) : f^*(\cdot, \psi(\cdot)) \in L^1(\Omega)\},$$

and $f^*(x, \cdot)$ is the polar function of $f(x, \cdot)$.

More generally, if (6.96) holds, then for any sequence $\{v_n\} \subset L^1(\Omega; \mathbb{R}^m)$ such that $v_n \mathcal{L}^N \lfloor \Omega \overset{*}{\rightharpoonup} \lambda$ in $\mathcal{M}(\Omega; \mathbb{R}^m)$ we have

$$\liminf_{n \to \infty} \int_\Omega f(x, v_n) \, dx \geq \int_\Omega f\left(x, \frac{d\lambda}{d\mathcal{L}^N}\right) dx + \int_\Omega f_s\left(x, \frac{d\lambda}{d\|\lambda_s\|}\right) d\|\lambda_s\|, \tag{6.98}$$

where

$$f_s(x, z) := \sup\{\psi(x) \cdot z : \psi \in \mathcal{X}\} \tag{6.99}$$

for all $x \in \Omega$ and all $z \in \mathbb{R}^m$.

Remark 6.58. Often in the literature

$$\liminf_{n\to\infty} \int_\Omega f(x, v_n) \, dx \geq \int_\Omega f\left(x, \frac{d\lambda}{d\mathcal{L}^N}\right) dx + \int_\Omega f^\infty\left(x, \frac{d\lambda}{d\|\lambda_s\|}\right) d\|\lambda_s\|$$

is written in place of (6.98). A word of caution is needed here: although

$$\int_\Omega f(x, v(x)) \, dx = \int_\Omega \tilde{f}(x, v(x)) \, dx$$

for every $v \in L^1(\Omega; \mathbb{R}^m)$, it may happen that

$$f^\infty(x, \cdot) \neq f_s(x, \cdot) \qquad (6.100)$$

on a set of Lebesgue measure zero E for which $\|\lambda_s\|(E) > 0$ for some measure λ.

See the next exercise.

Exercise 6.59. An example of (6.100) is given by

$$f(x, z) := \begin{cases} |z| & \text{if } |z|\sqrt{|x|} \leq 1, \\ 2|z| - \dfrac{1}{\sqrt{|x|}} & \text{if } |z|\sqrt{|x|} > 1, \end{cases}$$

for $x \in (-1, 1)$ and $z \in \mathbb{R}$. Prove that $f_s \neq f^\infty$.

To prove Theorem 6.57 we need some preliminary results.

Proposition 6.60. *Let $\Omega \subset \mathbb{R}^N$ be an open set, let $g : \Omega \times \mathbb{R}^m \to (-\infty, \infty]$ be Borel measurable, and let $\nu : \mathfrak{M}(\Omega) \to [0, \infty)$ be a positive finite Radon measure. Assume that there exist two functions $a \in L^1(\Omega, \nu)$ and $b \in L^1(\Omega, \nu; \mathbb{R}^m)$ such that*

$$g(x, z) \geq a(x) + b(x) \cdot z \qquad (6.101)$$

for ν a.e. $x \in \Omega$ and all $z \in \mathbb{R}^m$. Suppose further that $g(x, \cdot)$ is convex and lower semicontinuous in $\mathbb{R}^m$ for ν a.e. $x \in \Omega$, and let

$$\mathcal{X}_{g,\nu} := \left\{\psi \in C_0(\Omega) : g(\cdot, \psi(\cdot)) \in L^1(\Omega, \nu)\right\}. \qquad (6.102)$$

If $\mathcal{X}_{g,\nu} \neq \emptyset$, then

$$\inf_{\psi \in \mathcal{X}_{g,\nu}} \int_\Omega g(x, \psi(x)) \, d\nu = \int_\Omega \hat{g}(x) \, d\nu, \qquad (6.103)$$

where

$$\hat{g}(x) := \inf_{\xi \in \Gamma_\nu(x)} g(x, \xi)$$

and Γ_ν is the essential supremum of the family of multifunctions $\mathcal{F} := \{\Gamma_\psi : \psi \in \mathcal{X}_{g,\nu}\}$ with respect to ν. In addition,

$$\hat{g}(x) = \inf_{n \in \mathbb{N}} g(x, \psi_n(x)) \tag{6.104}$$

for ν a.e. $x \in \Omega$ whenever

$$\Gamma_\nu(x) = \overline{\{\psi_n(x)\}}$$

for ν a.e. $x \in \Omega$.

Proof. Note that the convexity of $g(x, \cdot)$ together with (6.101) implies that the family $\mathcal{X}_{g,\nu}$ is convex. Hence by Remark 6.14, $\Gamma_\nu(x)$ is convex and there exists a sequence $\{\psi_n\} \subset \mathcal{X}_{g,\nu}$ such that

$$\Gamma_\nu(x) = \overline{\{\psi_n(x)\}} \tag{6.105}$$

for ν a.e. $x \in \Omega$. Define

$$N_0 := \Omega \setminus \bigcup_{n=1}^{\infty} \{x \in \Omega : g(x, \psi_n(x)) \in \mathbb{R}\}.$$

By (6.102) the set N_0 has ν measure zero, and by Proposition 4.41 and (6.105),

$$\hat{g}(x) = \inf_{n \in \mathbb{N}} g(x, \psi_n(x)) \tag{6.106}$$

for $\mathcal{L}^N$ a.e. $x \in \Omega$. It follows that the function $\hat{g}$ is measurable. Moreover, for any $\psi \in \mathcal{X}_{g,\nu}$ we have that $\psi(x) \in \Gamma_\nu(x)$ for ν a.e. $x \in \Omega$, and so

$$g(x, \psi(x)) \geq \hat{g}(x)$$

for ν a.e. $x \in \Omega$. By (6.102) it follows that $(\hat{g})^+ \in L^1(\Omega, \nu)$; thus

$$\inf_{\psi \in \mathcal{X}_{g,\nu}} \int_\Omega g(x, \psi(x)) \, d\nu \geq \int_\Omega \hat{g}(x) \, d\nu,$$

where the right-hand side is well-defined. To prove the reverse inequality let $t \in \mathbb{R}$ be such that

$$t > \int_\Omega \hat{g}(x) \, d\nu.$$

Then

$$\int_\Omega (\hat{g})^-(x) \, d\nu > \int_\Omega (\hat{g})^+(x) \, d\nu - t,$$

and so by applying Proposition 1.92 to the function $(\hat{g})^-$ we may find a simple function $s \in L^1(\Omega, \nu)$ such that $0 \leq s \leq (\hat{g})^-$ in Ω with

$$\int_\Omega s(x) \, d\nu > \int_\Omega (\hat{g})^+(x) \, d\nu - t.$$

Then the function $\alpha := (\hat{g})^+ - s$ belongs to $L^1(\Omega, \nu)$, $\alpha(x) \geq \hat{g}(x)$ for ν a.e. $x \in \Omega$, and

$$t > \int_{\Omega} \alpha(x) \, d\nu. \tag{6.107}$$

By the Vitali–Carathéodory theorem we may assume that α is lower semicontinuous.

Fix $w_0 \in \mathcal{X}_{g,\nu}$. By (6.102), for any fixed $\varepsilon > 0$ there exists a compact set $K \subset \Omega \setminus N_0$ such that

$$\int_{\Omega \setminus K} |g(x, w_0(x))| + |\alpha(x)| \, d\nu < \varepsilon, \tag{6.108}$$

and a number $\delta > 0$ such that

$$\int_{E} |g(x, w_0(x))| + |\alpha(x)| \, d\nu < \varepsilon \tag{6.109}$$

for all $E \subset \Omega$ with $\nu(E) < \delta$. By Lusin's theorem for each $n \in \mathbb{N}$ there exists a compact set $K^{(n)} \subset K$ with $\nu(K \setminus K^{(n)}) < \frac{\delta}{2^n}$ such that $g(\cdot, \psi_n(\cdot))$ is continuous on $K^{(n)}$. Let

$$K_\varepsilon := \bigcap_{n=1}^{\infty} K^{(n)}.$$

Then $\nu(K \setminus K_\varepsilon) < \delta$ and $g(\cdot, \psi_n(\cdot))$ is continuous on K_ε for all $n \in \mathbb{N}$. Let $\eta > 0$ be small enough that $\eta \nu(K) < \varepsilon$, and for every $n \in \mathbb{N}$ set

$$E_n := \{x \in K_\varepsilon : g(x, \psi_n(x)) < \alpha(x) + \eta\}. \tag{6.110}$$

Since each function $g(\cdot, \psi_n(\cdot)) - \alpha(\cdot)$ is upper semicontinuous in K_ε, the set E_n is relatively open in K_ε, and by (6.106) and the fact that $\alpha \geq \hat{g}$, we have

$$K_\varepsilon = \bigcup_{n=1}^{\infty} E_n.$$

By compactness there exists $\ell \in \mathbb{N}$ such that

$$K_\varepsilon = \bigcup_{n=1}^{\ell} E_n.$$

Let $A_\varepsilon \subset\subset \Omega$ be an open set such that $A_\varepsilon \supset K_\varepsilon$ and

$$\int_{A_\varepsilon \setminus K_\varepsilon} (g(x, \psi_n(x)))^+ \, d\nu < \frac{\varepsilon}{\ell} \tag{6.111}$$

for all $n = 1, \ldots, \ell$. Since the sets E_n are relatively open in K_ε we may find open sets $A_n \subset A_\varepsilon$ such that $A_n \cap K_\varepsilon = E_n$. Construct a partition of unity $\varphi_0, \ldots, \varphi_\ell \in C(\overline{\Omega})$ with

$$\sum_{n=0}^{\ell} \varphi_n \equiv 1 \text{ on } \overline{A_\varepsilon}$$

such that $\operatorname{supp} \varphi_n \subset A_n$ for all $n = 1, \ldots, \ell$, and $\operatorname{supp} \varphi_0 \subset \Omega \setminus K_\varepsilon$. Since $\varphi_0 = 1$ on ∂A_ε we may define φ_0 continuously to be identically equal to one in $\Omega \setminus \overline{A_\varepsilon}$. Hence now

$$\sum_{n=0}^{\ell} \varphi_n \equiv 1 \text{ in } \Omega$$

and

$$\psi := \varphi_0 w_0 + \sum_{n=1}^{\ell} \varphi_n \psi_n \in \mathcal{X}_{g,\nu}$$

by the convexity of $g(x, \cdot)$, (6.101), and (6.102). Moreover, again by the convexity of $g(x, \cdot)$ we have

$$
\begin{aligned}
\int_\Omega g(x, \psi(x))\, d\nu &\leq \int_\Omega \varphi_0(x) g(x, w_0(x))\, d\nu + \sum_{n=1}^{\ell} \int_\Omega \varphi_n(x) g(x, \psi_n(x))\, d\nu \\
&\leq \int_{\Omega \setminus K_\varepsilon} |g(x, w_0(x))|\, d\nu + \sum_{n=1}^{\ell} \int_{K_\varepsilon} \varphi_n(x) g(x, \psi_n(x))\, d\nu \\
&\quad + \sum_{n=1}^{\ell} \int_{A_\varepsilon \setminus K_\varepsilon} (g(x, \psi_n(x)))^+\, d\nu \\
&\leq 2\varepsilon + \int_{K_\varepsilon} (\alpha(x) + \eta)\, d\nu + \varepsilon \leq t + 6\varepsilon,
\end{aligned}
$$

where we have used (6.107), (6.108), (6.109), (6.110), and (6.111). Hence

$$\inf_{w \in \mathcal{X}_{g,\nu}} \int_\Omega g(x, w(x))\, d\nu \leq t + 6\varepsilon.$$

By letting first $\varepsilon \to 0^+$ and then $t \searrow \int_\Omega \hat{g}(x)\, d\nu$ we conclude that

$$\inf_{\psi \in \mathcal{X}_{g,\nu}} \int_\Omega g(x, \psi(x))\, d\nu \leq \int_\Omega \hat{g}(x)\, d\nu.$$

This completes the proof.

Remark 6.61. The analogue of (6.103) holds for supremum in place of infimum provided there exist two functions $a \in L^1(\Omega, \nu)$ and $b \in L^1(\Omega, \nu; \mathbb{R}^m)$ such that

$$g(x, z) \leq a(x) + b(x) \cdot z$$

for ν a.e. $x \in \Omega$ and all $z \in \mathbb{R}^m$, and that $g(x, \cdot)$ is concave and upper semicontinuous in $\mathbb{R}^m$ for ν a.e. $x \in \Omega$. In this case,

$$\sup_{\psi \in \mathcal{X}_{g,\nu}} \int_\Omega g(x, \psi(x))\, d\nu = \int_\Omega \check{g}(x)\, d\nu,$$

where

$$\breve{g}(x) := \sup_{\xi \in \Gamma_\nu(x)} g(x, \xi)$$

and Γ_ν is still the essential supremum of the family of multifunctions $\mathcal{F} := \{\Gamma_\psi : \psi \in \mathcal{X}_{g,\nu}\}$ with respect to ν. In addition,

$$\breve{g}(x) = \sup_{n \in \mathbb{N}} g(x, \psi_n(x)) \qquad (6.112)$$

for ν a.e. $x \in \Omega$ whenever

$$\Gamma_\nu(x) = \overline{\{\psi_n(x)\}}$$

for ν a.e. $x \in \Omega$.

The next exercise illustrates the fact that in the previous theorem $\mathcal{X}_{g,\nu}$ cannot be taken to be the entire space $C_0(\Omega)$.

Exercise 6.62. Let $\Omega = \mathbb{R}$, $m = 1$.

(i) Construct a compact set $K \subset \mathbb{R}$ with empty interior and such that $|K| > 0$.

(ii) Define

$$g(x, z) := \begin{cases} z & \text{if } x \in K, \\ I_{\{0\}}(z) & \text{otherwise,} \end{cases}$$

where $I_{\{0\}}$ is the indicator function of $\{0\}$, namely

$$I_{\{0\}}(z) = \begin{cases} 0 & \text{if } z = 0, \\ \infty & \text{otherwise.} \end{cases}$$

Prove that

$$\inf_{\psi \in C_0(\mathbb{R})} \int_\Omega g(x, \psi(x)) \, d\nu = 0.$$

(iii) Prove that the essential supremum of the family of multifunctions $\mathcal{F} := \{\Gamma_\psi : \psi \in C_0(\mathbb{R})\}$ is the multifunction

$$\Gamma_\nu(x) \equiv \mathbb{R}$$

for $x \in \mathbb{R}$.

(iv) Prove that (6.103) fails.

Proof (Necessity of Theorem 6.57). Without loss of generality we may assume that there exists $v_0 \in L^1(\Omega; \mathbb{R}^m)$ such that

$$\int_\Omega f(x, v_0(x)) \, dx < \infty. \qquad (6.113)$$

Since $L^1(\Omega; \mathbb{R}^m)$ may be identified with a subspace of $\mathcal{M}(\Omega; \mathbb{R}^m)$ that is the dual of the separable space $C_0(\Omega; \mathbb{R}^m)$, by (6.95) we are in a position to apply Proposition 3.16 to conclude that the functional

$$I(v) := \int_\Omega f(x, v) \, dx, \quad v \in L^1(\Omega; \mathbb{R}^m),$$

is lower semicontinuous with respect to weak star convergence.

Consider now the duality pair $(L^1(\Omega; \mathbb{R}^m), C_0(\Omega; \mathbb{R}^m))$ under the duality

$$\langle u, \phi \rangle_{L^1(\Omega;\mathbb{R}^m), C_0(\Omega;\mathbb{R}^m)} := \int_\Omega u \cdot \phi \, dx$$

for $u \in L^1(\Omega; \mathbb{R}^m)$ and $C_0(\Omega; \mathbb{R}^m)$. Then the functional I is convex and lower semicontinuous with respect to the topology $\sigma(L^1, C_0)$. It follows from Proposition 4.92, taking as V the topological vector space

$$\left(L^1(\Omega; \mathbb{R}^m), \sigma(L^1, C_0)\right),$$

that $I = I^{**}$.

Since by Theorem A.72 the topological dual of $(L^1(\Omega; \mathbb{R}^m), \sigma(L^1, C_0))$ is $C_0(\Omega; \mathbb{R}^m)$, if $v \in L^1(\Omega; \mathbb{R}^m)$, then

$$I(v) = I^{**}(v) = \sup_{\psi \in C_0(\Omega;\mathbb{R}^m)} \left[\int_\Omega \psi \cdot v \, dx - I^*(\psi) \right]. \tag{6.114}$$

Step 1: We claim that for $\psi \in C_0(\Omega; \mathbb{R}^m)$,

$$I^*(\psi) = \int_\Omega f^*(x, \psi(x)) \, dx. \tag{6.115}$$

Note that by (6.95) and (6.113) the function

$$\gamma(x) := -f(x, v_0(x)) \tag{6.116}$$

belongs to $L^1(\Omega)$ and since

$$f^*(x, \xi) \geq \xi \cdot v_0(x) + \gamma(x) \tag{6.117}$$

for $\mathcal{L}^N$ a.e. $x \in \Omega$ and all $\xi \in \mathbb{R}^m$, by Proposition 6.43 the right-hand side of (6.115) is well-defined.

By Definition 4.86 we have

$$I^*(\psi) = \sup_{v \in L^1(\Omega;\mathbb{R}^m)} \int_\Omega (\psi \cdot v - f(x, v)) \, dx. \tag{6.118}$$

Since

$$f^*(x, \xi) = \sup_{w \in \mathbb{R}^m} [\xi \cdot w - f(x, w)],$$

then

$$I^* (\psi) \leq \int_\Omega f^* (x, \psi(x)) \, dx.$$

To prove the reverse inequality, let $t \in \mathbb{R}$ be such that

$$\int_\Omega f^* (x, \psi(x)) \, dx > t$$

and fix $\varepsilon > 0$. Let

$$E_\infty := \{x \in \Omega : f^* (x, \psi(x)) = \infty\}$$

and consider first the case $|E_\infty| > 0$. Let $K \subset E_\infty$ be any compact set with positive measure, and using Aumann's selection theorem and the definition of f^* select a measurable function $w : K \to \mathbb{R}^m$ such that

$$\psi(x) \cdot w(x) - f(x, w(x)) \geq \frac{2}{|K|} \left(t + \int_\Omega |\psi \cdot v_0 + \gamma| \, dy \right) \qquad (6.119)$$

for $\mathcal{L}^N$ a.e. $x \in K$. By Lusin's theorem there exists a compact set $K_\varepsilon \subset K$ with $|K \setminus K_\varepsilon| < \frac{|K|}{2}$ such that w is continuous on K_ε. Define

$$v_\varepsilon (x) := \begin{cases} w(x) & \text{if } x \in K_\varepsilon, \\ v_0(x) & \text{if } x \in \Omega \setminus K_\varepsilon. \end{cases} \qquad (6.120)$$

Then $v_\varepsilon \in L^1(\Omega; \mathbb{R}^m)$, and by (6.116) and (6.119),

$$\int_\Omega (\psi \cdot v_\varepsilon - f(x, v_\varepsilon)) \, dx \geq \frac{2|K_\varepsilon|}{|K|} \left(t + \int_\Omega |\psi \cdot v_0 + \gamma| \, dy \right)$$
$$- \int_{\Omega \setminus K_\varepsilon} |\psi \cdot v_0 + \gamma| \, dx \geq t,$$

and so

$$I^* (\psi) = \sup_{\varpi \in L^1(\Omega; \mathbb{R}^m)} \left[\int_\Omega \psi \cdot \varpi \, dx - \int_\Omega f(x, \varpi) \, dx \right] \geq t.$$

Given the arbitrariness of t we conclude that

$$I^* (\psi) \geq \int_\Omega f^* (x, \psi(x)) \, dx. \qquad (6.121)$$

If $|E_\infty| = 0$, then find a compact set $K \subset \Omega$ such that

$$\int_K f^* (x, \psi(x)) \, dx > t, \qquad \int_{\Omega \setminus K} |\psi \cdot v_0 + \gamma| \, dx < \varepsilon, \qquad (6.122)$$

and a number $\delta = \delta(\varepsilon) > 0$ such that

$$\int_E |\psi \cdot v_0 + \gamma| \, dx < \varepsilon \qquad (6.123)$$

for all $E \subset \Omega$ with $|E| < \delta$. The function δ may be assumed to be increasing with $\varepsilon > 0$.

As before, let $w : K \to \mathbb{R}^m$ be a measurable function such that

$$f^*(x, \psi(x)) \le \psi(x) \cdot w(x) - f(x, w(x)) + \frac{\varepsilon}{|K|} \qquad (6.124)$$

for $\mathcal{L}^N$ a.e. $x \in K$, and by Lusin's theorem find a compact set $K_\varepsilon \subset K$ with $|K \setminus K_\varepsilon| < \delta$ such that w is continuous on K_ε. Choose K_ε in such a way that if $\varepsilon < \varepsilon'$, then $K_\varepsilon \supset K_{\varepsilon'}$. Define v_ε as in (6.120). Then $v_\varepsilon \in L^1(\Omega; \mathbb{R}^m)$, and by (6.116) and (6.124),

$$\int_\Omega (\psi \cdot v_\varepsilon - f(x, v_\varepsilon)) \, dx$$

$$\ge \int_{K_\varepsilon} f^*(x, \psi) \, dx - \varepsilon - \int_{\Omega \setminus K_\varepsilon} |\psi \cdot v_0 + \gamma| \, dx$$

$$= \int_{K_\varepsilon} (f^*(x, \psi))^+ \, dx - \int_{K_\varepsilon} (f^*(x, \psi))^- \, dx$$

$$\quad - \varepsilon - \int_{\Omega \setminus K_\varepsilon} |\psi \cdot v_0 + \gamma| \, dx$$

$$\ge \int_{K_\varepsilon} (f^*(x, \psi))^+ \, dx - \int_K (f^*(x, \psi))^- \, dx$$

$$\quad - \varepsilon - \int_{\Omega \setminus K} |\psi \cdot v_0 + \gamma| \, dx$$

$$\ge \int_{K_\varepsilon} (f^*(x, \psi))^+ \, dx - \int_K (f^*(x, \psi))^- \, dx - 3\varepsilon,$$

where we have used (6.122) and (6.123).

By (6.118) we deduce that

$$I^*(\psi) \ge \int_{K_\varepsilon} (f^*(x, \psi))^+ \, dx - \int_K (f^*(x, \psi))^- \, dx - 3\varepsilon,$$

and so letting $\varepsilon \to 0$ and using the Lebesgue monotone convergence theorem we obtain

$$I^*(\psi) \ge \int_K f^*(x, \psi) \, dx \ge t$$

by (6.122), and thus (6.121) follows.

Step 2: By (6.114) and Step 1 we have that for every $v \in L^1(\Omega; \mathbb{R}^m)$,

$$I(v) = \sup_{\psi \in C_0(\Omega; \mathbb{R}^m)} \int_\Omega \psi(x) \cdot v(x) - f^*(x, \psi(x)) \, dx.$$

Fix $v \in L^1(\Omega; \mathbb{R}^m)$ and for $x \in \Omega$ and $\xi \in \mathbb{R}^m$ define

$$g_v(x, \xi) := \xi \cdot v(x) - f^*(x, \xi).$$

Note that for $\psi \in C_0(\Omega; \mathbb{R}^m)$ we have that $g_v(\cdot, \psi(\cdot)) \in L^1(\Omega)$ if and only if $\psi \in \mathcal{X}$. Hence we invoke Remark 6.61 to deduce that

$$I(v) = \sup_{\psi \in \mathcal{X}} \int_\Omega g_v(x, \psi(x))\, dx = \int_\Omega \bar{h}_v(x)\, dx,$$

where

$$\bar{h}_v(x) := \sup_{\xi \in \Gamma(x)} [\xi \cdot v(x) - f^*(x, \xi)] = \tilde{f}(x, v(x)).$$

Thus we have shown that

$$\int_\Omega f(x, v(x))\, dx = \int_\Omega \tilde{f}(x, v(x))\, dx \qquad (6.125)$$

for all $v \in L^1(\Omega; \mathbb{R}^m)$. Since $f \geq \tilde{f}$ this implies in particular that $f(x, v(x)) = \tilde{f}(x, v(x))$ for $\mathcal{L}^N$ a.e. $x \in \Omega$, whenever $f(\cdot, v(\cdot)) \in L^1(\Omega)$.

Fix $w \in L^\infty(\Omega; \mathbb{R}^m)$ and $B \in \mathcal{B}(\Omega)$ bounded, and define

$$v(x) := \begin{cases} w(x) & \text{if } x \in B, \\ v_0(x) & \text{if } x \in \Omega \setminus B. \end{cases}$$

By (6.125),

$$\int_B f(x, w(x))\, dx + \int_{\Omega \setminus B} f(x, v_0(x))\, dx$$
$$= \int_B \tilde{f}(x, v(x))\, dx + \int_{\Omega \setminus B} \tilde{f}(x, v_0(x))\, dx.$$

Since $f(\cdot, v_0(\cdot)) \in L^1(\Omega; \mathbb{R}^m)$ it follows

$$\int_{\Omega \setminus B} f(x, v_0(x))\, dx = \int_{\Omega \setminus B} \tilde{f}(x, v_0(x))\, dx < \infty,$$

and so

$$\int_B f(x, w(x))\, dx = \int_B \tilde{f}(x, v(x))\, dx$$

for all $w \in L^\infty(\Omega; \mathbb{R}^m)$ and $B \in \mathcal{B}(\Omega)$ bounded. We may now apply Theorem 6.24 to conclude that

$$f(x, z) = \tilde{f}(x, z)$$

for $\mathcal{L}^N$ a.e. $x \in \Omega$ and all $z \in \mathbb{R}^m$.

Proof (Sufficiency of Theorem 6.57). Let $\{v_n\} \subset L^1(\Omega; \mathbb{R}^m)$ be such that $v_n \mathcal{L}^N \lfloor \Omega \overset{*}{\rightharpoonup} \lambda$ in $\mathcal{M}(\Omega; \mathbb{R}^m)$ and set $\nu := \mathcal{L}^N \lfloor \Omega + |\lambda_s|$. Let Γ_ν be the essential supremum of the family of multifunctions $\mathcal{F} := \{\Gamma_\psi : \psi \in \mathcal{X}\}$ with respect to ν.

Step 1: We claim that

$$f_s(x, z) = \sup\{\xi \cdot z : \xi \in \Gamma_\nu(x)\} \tag{6.126}$$

for ν a.e. $x \in \Omega$ and all $z \in \mathbb{R}^m$. By Remark 6.14(ii) there exists a sequence $\{b_i\} \subset \mathcal{X}$ such that

$$\Gamma_\nu(x) = \overline{\{b_i(x)\}} \tag{6.127}$$

for ν a.e. $x \in \Omega$. Let $\{\varphi_n\}$ be a countable dense subset of $\mathcal{X}$. Then

$$\Gamma_\nu(x) = \overline{\{b_i(x)\} \cup \{\psi_n(x)\}}$$

for ν a.e. $x \in \Omega$. Therefore, without loss of generality, in (6.127) we may assume that sequence $\{b_i\}$ is dense in $\mathcal{X}$, and thus for every $x \in \Omega$ and all $z \in \mathbb{R}^m$,

$$f_s(x, z) = \sup\{\psi(x) \cdot z : \psi \in \mathcal{X}\} = \sup_{i \in \mathbb{N}}\{b_i(x) \cdot z\},$$

and the claim (6.126) follows from (6.127).

Step 2: We claim that

$$f(x, z) = \sup_{i \in \mathbb{N}}\{b_i(x) \cdot z - f^*(x, b_i(x))\} \tag{6.128}$$

for $\mathcal{L}^N$ a.e. $x \in \Omega$ and all $z \in \mathbb{R}^m$.

The equality (6.128) follows from (6.96), the fact that

$$\Gamma_\nu(x) = \Gamma(x)$$

for $\mathcal{L}^N$ a.e. $x \in \Omega$ (see Remark 6.14(iii)), and (6.112).

Step 3: The remainder of the proof follows closely that of Theorem 5.19. We indicate only the main changes. Define

$$a_i(x) := -f^*(x, b_i(x)).$$

By replacing $f(x, z)$ with $f(x, z) - a(x)$ we can assume, without loss of generality, that $f \geq 0$.

We choose a point $x_0 \in \Omega$ that satisfies (6.128), (5.27)–(5.28),

$$\lim_{\varepsilon \to 0^+} \frac{\|\lambda\|(Q(x_0, \varepsilon))}{\varepsilon^N} < \infty, \tag{6.129}$$

and is a Lebesgue point for all functions a_i, $i \in \mathbb{N}$. Choosing $\varepsilon_k \searrow 0$ such that $\mu(\partial Q(x_0, \varepsilon_k)) = 0$ and $\|\lambda\|(\partial Q(x_0, \varepsilon_k)) = 0$, in place of (5.31) we have

$$\frac{d\mu}{d\mathcal{L}^N}(x_0) = \lim_{k\to\infty}\lim_{n\to\infty}\frac{1}{\varepsilon_k^N}\int_{Q(x_0,\varepsilon_k)} f(x,v_n(x))\,dx$$

$$\geq \limsup_{k\to\infty}\limsup_{n\to\infty}\frac{1}{\varepsilon_k^N}\int_{Q(x_0,\varepsilon_k)}[a_i(x)+b_i(x)\cdot v_n(x)]\,dx$$

$$= a_i(x_0) + \limsup_{k\to\infty}\frac{1}{\varepsilon_k^N}\int_{Q(x_0,\varepsilon_k)} b_i(x)\cdot d\lambda(x),$$

where we have used Proposition 1.203(iii) and the fact that x_0 is a Lebesgue point for a_i. Write

$$\frac{1}{\varepsilon_k^N}\int_{Q(x_0,\varepsilon_k)} b_i(x)\cdot d\lambda(x) = \frac{1}{\varepsilon_k^N}\int_{Q(x_0,\varepsilon_k)}(b_i(x)-b_i(x_0))\cdot d\lambda(x)$$

$$+ b_i(x_0)\cdot\frac{\lambda(Q(x_0,\varepsilon_k))}{\varepsilon_k^N}$$

and use the continuity of b_i and (6.129) to conclude that

$$\lim_{k\to\infty}\frac{1}{\varepsilon_k^N}\int_{Q(x_0,\varepsilon_k)} b_i(x)\cdot d\lambda(x) = b_i(x_0)\cdot\lim_{k\to\infty}\frac{\lambda(Q(x_0,\varepsilon_k))}{\varepsilon_k^N}$$

$$= b_i(x_0)\cdot\frac{d\lambda}{d\mathcal{L}^N}(x_0)$$

by (5.28). Hence

$$\frac{d\mu}{d\mathcal{L}^N}(x_0)\geq a_i(x_0)+b_i(x_0)\cdot\frac{d\lambda}{d\mathcal{L}^N}(x_0),$$

and taking the supremum over all $i\in\mathbb{N}$, by (6.128) we get

$$\frac{d\mu}{d\mathcal{L}^N}(x_0)\geq f\left(x_0,\frac{d\lambda}{d\mathcal{L}^N}(x_0)\right).$$

It remains to show that for $\|\lambda_s\|$ a.e. $x_0\in E$,

$$\frac{d\mu}{d\|\lambda_s\|}(x_0)\geq f_s\left(x_0,\frac{d\lambda}{d\|\lambda_s\|}(x_0)\right).$$

Take $x_0\in\Omega$ that satisfies (6.126), (5.32)–(5.34), and

$$\lim_{\varepsilon\to 0^+}\frac{1}{\|\lambda_s\|(Q(x_0,\varepsilon))}\int_{Q(x_0,\varepsilon)}|a_i|\,dx = 0, \tag{6.130}$$

for all $i\in\mathbb{N}$.

Choosing $\varepsilon_k\searrow 0$ such that $\mu(\partial Q(x_0,\varepsilon_k))=0$ and $\|\lambda\|(\partial Q(x_0,\varepsilon_k))=0$, then

$$\frac{d\mu}{d\|\lambda_s\|}(x_0)$$

$$= \lim_{k \to \infty} \frac{\mu(Q(x_0, \varepsilon_k))}{\|\lambda_s\|(Q(x_0, \varepsilon_k))}$$

$$= \lim_{k \to \infty} \lim_{n \to \infty} \frac{1}{\|\lambda_s\|(Q(x_0, \varepsilon_k))} \int_{Q(x_0, \varepsilon_k)} f(x, v_n(x)) \, dx$$

$$\geq \limsup_{k \to \infty} \limsup_{n \to \infty} \frac{1}{\|\lambda_s\|(Q(x_0, \varepsilon_k))} \int_{Q(x_0, \varepsilon_k)} [a_i(x) + b_i(x) \cdot v_n(x)] \, dx$$

$$\tag{6.131}$$

$$= \limsup_{k \to \infty} \frac{1}{\|\lambda_s\|(Q(x_0, \varepsilon_k))} \int_{Q(x_0, \varepsilon_k)} b_i(x) \cdot d\lambda(x),$$

where in the last equality we have used (6.130). Write

$$\frac{1}{\|\lambda_s\|(Q(x_0, \varepsilon_k))} \int_{Q(x_0, \varepsilon_k)} b_i(x) \cdot d\lambda(x)$$

$$= \frac{1}{\|\lambda_s\|(Q(x_0, \varepsilon_k))} \int_{Q(x_0, \varepsilon_k)} (b_i(x) - b_i(x_0)) \cdot d\lambda(x)$$

$$+ b_i(x_0) \cdot \frac{\lambda(Q(x_0, \varepsilon_k))}{\|\lambda_s\|(Q(x_0, \varepsilon_k))}$$

and use the continuity of b_i and (5.33) to conclude that

$$\lim_{k \to \infty} \frac{1}{\|\lambda_s\|(Q(x_0, \varepsilon_k))} \int_{Q(x_0, \varepsilon_k)} b_i(x) \cdot d\lambda(x)$$

$$= b_i(x_0) \cdot \lim_{k \to \infty} \frac{\lambda(Q(x_0, \varepsilon_k))}{\|\lambda_s\|(Q(x_0, \varepsilon_k))} = b_i(x_0) \cdot \frac{d\lambda}{d\|\lambda_s\|}(x_0).$$

Hence by (6.131) we obtain

$$\frac{d\mu}{d\|\lambda_s\|}(x_0) \geq b_i(x_0) \cdot \frac{d\lambda}{d\|\lambda_s\|}(x_0),$$

and taking the supremum over all i it follows by (6.126) that

$$\frac{d\mu}{d\|\lambda_s\|}(x_0) \geq f_s\left(x_0, \frac{d\lambda}{d\|\lambda_s\|}(x_0)\right).$$

This completes the proof.

Remark 6.63. Condition (6.95) has been used only in the necessity part of the proof to conclude that the sequentially lower semicontinuous functional I is actually lower semicontinuous with respect to weak star convergence. In the sufficiency part of the proof we need only a much weaker condition, namely that there exist functions $a \in L^1(\Omega)$ and $b \in C_0(\Omega; \mathbb{R}^m)$ such that

$$f(x, z) \geq a(x) + b(x) \cdot z$$

for $\mathcal{L}^N$ a.e. $x \in \Omega$ and all $z \in \mathbb{R}^m$. Indeed, since $b \in C_0(\Omega)$, it suffices to replace $f(x, z)$ with $f(x, z) - (a(x) + b(x) \cdot z)$ in Step 3 of the sufficiency proof.

Note that under the hypotheses of Theorem 6.57, since $f(x, \cdot)$ is convex and lower semicontinuous in $\mathbb{R}^m$ for $\mathcal{L}^N$ a.e. $x \in \Omega$ and (6.95) holds, by Theorem 4.92(iii) we have that $f(x, \cdot) = f^{**}(x, \cdot)$ for $\mathcal{L}^N$ a.e. $x \in \Omega$. Hence by (6.97) and the definition of $f^{**}(x, \cdot)$ the inequality

$$f(x, \cdot) = f^{**}(x, \cdot) \geq \tilde{f}(x, \cdot)$$

is satisfied for $\mathcal{L}^N$ a.e. $x \in \Omega$. To our knowledge, it is still an open problem to characterize the class of integrands f for which the opposite inequality holds. By the sufficiency and the necessity parts and by (6.128), sequential lower semicontinuity holds if and only if f can be written (up to equivalent integrands) as

$$f(x, z) = \sup_{i \in \mathbb{N}} \{a_i(x) + b_i(x) \cdot z\}, \quad (x, z) \in \Omega \times \mathbb{R}^m, \tag{6.132}$$

for some $a_i \in L^1_{\text{loc}}(\Omega)$ and $b_i \in C(\Omega; \mathbb{R}^m)$ (with at least one $a_i \in L^1(\Omega)$ and $b_i \in C_0(\Omega; \mathbb{R}^m)$). In particular, if $f \geq 0$, then we can set $a_1 :\equiv 0$ and $b_1 :\equiv 0$. Note that by Proposition 4.77, if (6.132) holds, then

$$f^{\infty}(x, z) = \sup_{i \in \mathbb{N}} \{b_i(x) \cdot z\}, \quad (x, z) \in \Omega \times \mathbb{R}^m,$$

and so the recession function f^{∞} must be lower semicontinuous in (x, z) (up to equivalent integrands). Next we present two easily verifiable conditions under which (6.132) holds.

Corollary 6.64. *Let* $f : \Omega \times \mathbb{R}^m \to [0, \infty]$ *be Borel measurable. Assume that* $f(x, \cdot)$ *is convex and lower semicontinuous in* $\mathbb{R}^m$ *for* $\mathcal{L}^N$ *a.e.* $x \in \Omega$. *Then* (6.96) *holds if either of the two following conditions is satisfied:*

(i) f *is lower semicontinuous in* $\Omega \times \mathbb{R}^m$ *and there exists a continuous function* $v_0 : \Omega \to \mathbb{R}^m$ *with* $f(\cdot, v_0(\cdot)) \in L^{\infty}_{\text{loc}}(\Omega; \mathbb{R})$;

(ii) $f : \Omega \times \mathbb{R}^m \to [0, \infty)$ *is locally bounded, and* $\nabla_z f$ *exists for all* $x \in \Omega$ *and for* $\mathcal{L}^m$ *a.e.* $z \in \mathbb{R}^m$, *with* $\nabla_z f(\cdot, z)$ *continuous.*

Proof. By Proposition 6.42 it follows that condition (i) implies (6.132), while if (ii) is satisfied, then by De Giorgi's theorem 6.36, for $\mathcal{L}^N$ a.e. $x \in \Omega$ and all $z \in \mathbb{R}^m$ we may write

$$f(x, z) = \sup_{i \in \mathbb{N}} \{a_i(x) + b_i(x) \cdot z\},$$

where

$$a_i(x) := \int_{\mathbb{R}^m} f(x, z) \left((m+1) \varphi_i(z) + \nabla \varphi_i(z) \cdot z \right) dz,$$

$$b_i(x) := -\int_{\mathbb{R}^m} f(x, z) \nabla \varphi_i(z) \, dz,$$

and $\varphi_i \in C_c^1(\mathbb{R}^m)$ are nonnegative and $\int_{\mathbb{R}^m} \varphi_i(z) \, dz = 1$. Since f is locally bounded, $a_i \in L_{loc}^1(\Omega)$. Integrating by parts, we have that

$$b_i(x) := \int_{\mathbb{R}^m} \nabla_z f(x, z) \varphi_i(z) \, dz,$$

and so the continuity of b_i follows by the Lebesgue dominated convergence theorem using also Theorem 4.36.

6.5 Integral Representation in L^p

Throughout this section we assume that $E \subset \mathbb{R}^N$ is a Lebesgue measurable set, and we find sufficient conditions for a functional

$$\mathcal{E} : L^p(E; \mathbb{R}^m) \to (-\infty, \infty]$$

to have an integral representation of the type

$$\mathcal{E}(v) = \int_E h(x, v(x)) \, dx$$

for all $v \in L^p(E; \mathbb{R}^m)$ and for some integrand h. We start with the case that $1 \le p < \infty$.

Theorem 6.65. *Let $E \subset \mathbb{R}^N$ be a Lebesgue measurable set, let $1 \le p < \infty$, and let*

$$I : L^p(E; \mathbb{R}^m) \times \mathcal{B}(E) \to (-\infty, \infty]$$

satisfy the following properties:

(I_1) $I(v; \cdot)$ *is additive, that is,*

$$I(v; B_1 \cup B_2) = I(v; B_1) + I(v; B_2)$$

for all $v \in L^p(E; \mathbb{R}^m)$ and all B_1, $B_2 \in \mathcal{B}(E)$ such that $B_1 \cap B_2 = \emptyset$;

(I_2) $I(\cdot; B)$ *is local, that is, for all $B \in \mathcal{B}(E)$,*

$$I(v; B) = I(w; B)$$

for all $v, w \in L^p(E; \mathbb{R}^m)$ such that $v = w$ $\mathcal{L}^N$ a.e. on $B \in \mathcal{B}(E)$.

For every $B \in \mathcal{B}(E)$ let τ be either the weak or the strong topology in $L^p(B; \mathbb{R}^m)$ and let $\mathcal{E}_p(\cdot; B)$ be the greatest functional below $I(\cdot, B)$ that is sequentially lower semicontinuous with respect to τ. Assume that

(I_3) $\mathcal{E}_p$ *is proper, that is,* $\mathcal{E}_p > -\infty$, *and there exists* $v_0 \in L^p(E; \mathbb{R}^m)$ *such that*

$$\mathcal{E}_p(v_0; B) \in \mathbb{R}$$

for every $B \in \mathcal{B}(E)$.

Then there exists a normal integrand $h : E \times \mathbb{R}^m \to (-\infty, \infty]$ that satisfies the condition

$$h(x, z) \geq -C |z|^p - a(x)$$

for $\mathcal{L}^N$ a.e. $x \in E$ and for all $z \in \mathbb{R}^m$, for some nonnegative function $a \in L^1(E)$ and some constant $C > 0$, and such that

$$\mathcal{E}_p(v; B) = \mathcal{E}_p(v_0; B) + \int_B h(x, v(x)) \, dx$$

for all $v \in L^p(E; \mathbb{R}^m)$ and $B \in \mathcal{B}(E)$.

Proof. **Step 1:** As in Step 1 of the proof of Theorem 5.29 we can show that $\mathcal{E}_p$ satisfies conditions (I_1) and (I_2). Note that here we use heavily the fact that, by (I_3), the functional $\mathcal{E}_p$ and, in view of (5.50), all the functionals $\mathcal{H}_\beta$ defined in Step 1 of Theorem 5.29 never take the value $-\infty$, so that $\mathcal{H}_\beta(v; B_1) + \mathcal{H}_\beta(v; B_2)$ and $\mathcal{E}_p(v; B_1) + \mathcal{E}_p(v; B_2)$ are always well-defined.

Step 2: By replacing the functional $\mathcal{E}_p(\cdot; B)$ with $\mathcal{E}_p(\cdot + v_0; B) - \mathcal{E}_p(v_0; B)$, where v_0 is the function given in (I_3), in the remainder of the proof we assume that

$$\mathcal{E}_p(0; B) = 0 \text{ for each } B \in \mathcal{B}(E). \tag{6.133}$$

We claim that for every fixed $\varepsilon > 0$ there exists $\delta = \delta(\varepsilon) > 0$ such that

$$\mathcal{E}_p(v; B) + \frac{\varepsilon}{\delta} \int_B |v|^p \, dx \geq -\varepsilon \tag{6.134}$$

for every $v \in L^p(E; \mathbb{R}^m)$ and $B \in \mathcal{B}(E)$. Since $\mathcal{E}_p(\cdot; E)$ is sequentially lower semicontinuous with respect to strong convergence in $L^p(E; \mathbb{R}^m)$, by (6.133) this implies that for every fixed $\varepsilon > 0$ there exists $\delta = \delta(\varepsilon) > 0$ such that

$$\mathcal{E}_p(v; E) > -\varepsilon \text{ for all } v \in L^p(E; \mathbb{R}^m) \text{ with } \int_E |v|^p \, dx \leq \delta. \tag{6.135}$$

For $B \in \mathcal{B}(E)$ and $v \in L^p(E; \mathbb{R}^m)$ we may find $k \in \mathbb{N}$ and $\eta \in [0, 1)$ such that

$$\int_B |v|^p \, dx = (k + \eta) \delta \tag{6.136}$$

and, by Propositions 1.89 and 1.20, $k + 1$ pairwise disjoint subsets of B such that

$$B = \bigcup_{i=i}^{k+1} B_i, \quad \int_{B_i} |v|^p \, dx \leq \delta \text{ for all } i = 1, \ldots, k+1.$$

Indeed, let $B_1 \subset B$, $B_1 \in \mathcal{B}(E)$, be such that

$$\int_{B_1} |v|^p \, dx = \frac{k + \eta}{k + 1} \delta.$$

Recursively, having found disjoint Borel sets $B_1, \dots, B_i \subset B$ with $i \leq k$ and

$$\int_{B_i} |v|^p \, dx = \frac{k + \eta}{k + 1} \delta,$$

and since

$$\int_{B \setminus \bigcup_{j=1}^{i} B_j} |v|^p \, dx = (k + \eta) \delta - \frac{k + \eta}{k + 1} \delta i \geq \frac{k + \eta}{k + 1} \delta (k + 1 - i),$$

we may find $B_{i+1} \subset B \setminus \bigcup_{j=1}^{i} B_j$, $B_{i+1} \in \mathcal{B}(E)$, such that

$$\int_{B_{i+1}} |v|^p \, dx = \frac{k + \eta}{k + 1} \delta.$$

By Step 1, (6.133), and (6.135) we have

$$\mathcal{E}_p(v; B_i) = \mathcal{E}_p(v \chi_{B_i}; E) > -\varepsilon \text{ for all } i = 1, \dots, k + 1,$$

which implies that, again by Step 1,

$$\mathcal{E}_p(v; B) = \sum_{i=i}^{k+1} \mathcal{E}_p(v; B_i) \geq -(k + 1) \varepsilon.$$

In turn, by (6.136) this yields

$$\mathcal{E}_p(v; B) + \frac{\varepsilon}{\delta} \int_B |v|^p \, dx \geq -(k + 1) \varepsilon + (k + \eta) \varepsilon \geq -\varepsilon$$

for all $\varepsilon > 0$, $B \in \mathcal{B}(E)$, and $v \in L^p(E; \mathbb{R}^m)$, and the claim is proved.

Next we show that $\mathcal{E}_p(v; \cdot)$ is a measure, absolutely continuous with respect to the Lebesgue measure. By Proposition 1.168 and (I_1), to prove that $\mathcal{E}_p(v; \cdot)$ is a signed measure it is enough to verify that

$$\mathcal{E}_p(v; B_n) \to \mathcal{E}_p(v; B) \text{ whenever } B_n \nearrow B, \ B_n, B \in \mathcal{B}(E).$$

By (I_1), (I_2), and (6.133),

$$\mathcal{E}_p(v; B) = \mathcal{E}_p(v \chi_B; E) \leq \liminf_{n \to \infty} \mathcal{E}_p(v \chi_{B_n}; E) = \liminf_{n \to \infty} \mathcal{E}_p(v; B_n).$$

To obtain the opposite inequality, by (6.134) for every $\varepsilon > 0$ we have that

$$\mathcal{E}_p(v;B) + \frac{\varepsilon}{\delta}\int_B |v|^p \, dx = \mathcal{E}_p(v;B_n) + \frac{\varepsilon}{\delta}\int_{B_n} |v|^p \, dx$$

$$+ \mathcal{E}_p(v;B\setminus B_n) + \frac{\varepsilon}{\delta}\int_{B\setminus B_n} |v|^p \, dx$$

$$\geq \mathcal{E}_p(v;B_n) + \frac{\varepsilon}{\delta}\int_{B_n} |v|^p \, dx - \varepsilon.$$

Hence

$$\mathcal{E}_p(v;B) + \frac{\varepsilon}{\delta}\int_B |v|^p \, dx \geq \limsup_{n\to\infty} \mathcal{E}(v;B_n) + \frac{\varepsilon}{\delta}\int_B |v|^p \, dx - \varepsilon,$$

or, equivalently,

$$\mathcal{E}_p(v;B) \geq \limsup_{n\to\infty} \mathcal{E}_p(v;B_n) - \varepsilon,$$

and we may now let $\varepsilon \to 0^+$.

Thus $\mathcal{E}_p(v;\cdot)$ is a measure, and by (I_2) and (6.133) it is absolutely continuous with respect to the Lebesgue measure.

Step 3: For every $\varepsilon > 0$, $B \in \mathcal{B}(E)$, and $v \in L^p(E;\mathbb{R}^m)$ we define

$$\Phi_\varepsilon(v;B) := -\mathcal{E}_p(v;B) - \frac{\varepsilon}{\delta}\int_B |v|^p \, dx,$$

where $\delta = \delta(\varepsilon) > 0$ is the constant given in the previous step. For every $\varepsilon > 0$ and $B \in \mathcal{B}(E)$ we also set

$$\mu_\varepsilon(B) := \sup\{\Phi_\varepsilon(v;B') : B' \in \mathcal{B}(E),\, B' \subset B,\, v \in L^p(E;\mathbb{R}^m)\}.$$

We claim that μ_ε is a (positive) Borel measure. Note that $\mu_\varepsilon \geq 0$ by (6.133), while $\mu_\varepsilon \leq \varepsilon$ by (6.134).

If $B_1,\, B_2 \in \mathcal{B}(E)$, with $B_1 \subset B_2$, then every Borel subset of B_1 is a Borel subset of B_2, and so $\mu_\varepsilon(B_1) \leq \mu_\varepsilon(B_2)$, that is, μ_ε is an increasing set function.

If $B_1,\, B_2 \in \mathcal{B}(E)$, with $B_1 \cap B_2 = \emptyset$, then for any $B' \in \mathcal{B}(E)$, with $B' \subset B_1 \cup B_2$, and any $v \in L^p(E;\mathbb{R}^m)$ we have

$$\Phi_\varepsilon(v;B') = \Phi_\varepsilon(v;B' \cap B_1) + \Phi_\varepsilon(v;B' \cap B_2)$$
$$\leq \mu_\varepsilon(B_1) + \mu_\varepsilon(B_2),$$

where we have used the fact that $\Phi_\varepsilon(v;\cdot)$ is a signed measure. Taking the supremum over all admissible B' and v gives

$$\mu_\varepsilon(B_1 \cup B_2) \leq \mu_\varepsilon(B_1) + \mu_\varepsilon(B_2). \tag{6.137}$$

To prove the reverse inequality fix real numbers $t_i < \mu_\varepsilon(B_i)$, $i = 1, 2$, and find $B_i' \in \mathcal{B}(E)$, $B_i' \subset B_i$, and $v_i \in L^p(E;\mathbb{R}^m)$ such that

$$t_i < \Phi_\varepsilon (v_i; B_i'),$$

$i = 1, 2$. Define

$$v(x) := \begin{cases} v_1(x) \text{ if } x \in B_1, \\ v_2(x) \text{ if } x \in E \setminus B_2. \end{cases}$$

Then $v \in L^p(E; \mathbb{R}^m)$, and so

$$t_1 + t_2 < \Phi_\varepsilon(v_1; B_1') + \Phi_\varepsilon(v_2; B_2') = \Phi_\varepsilon(v; B_1') + \Phi_\varepsilon(v; B_2')$$
$$= \Phi_\varepsilon(v; B_1' \cup B_2') \le \mu_\varepsilon(B_1 \cup B_2),$$

where we have used the locality of $\Phi_\varepsilon(\cdot; B_i')$, $i = 1, 2$, and again the fact that $\Phi_\varepsilon(v; \cdot)$ is a signed measure. Letting $t_i \nearrow \mu_\varepsilon(B_i)$, $i = 1, 2$, gives, also by (6.137),

$$\mu_\varepsilon(B_1 \cup B_2) = \mu_\varepsilon(B_1) + \mu_\varepsilon(B_2).$$

Since $\mu_\varepsilon(\emptyset) = 0$ it follows that μ_ε is a finitely additive measure. Hence by Proposition 1.9, to prove that μ_ε is a measure it is enough to verify that

$$\mu_\varepsilon(B_n) \to \mu_\varepsilon(B) \text{ whenever } B_n \nearrow B, \ B_n, B \in \mathcal{B}(E).$$

Using the fact that μ_ε is an increasing set function, we have that

$$\sup_{n \in \mathbb{N}} \mu_\varepsilon(B_n) \le \mu_\varepsilon(B).$$

To prove the reverse inequality fix a real number $t < \mu_\varepsilon(B)$ and find $B' \in \mathcal{B}(E)$, with $B' \subset B_i$, and $v \in L^p(E; \mathbb{R}^m)$ such that

$$t < \Phi_\varepsilon(v; B').$$

Since $\Phi_\varepsilon(v; \cdot)$ is a signed measure, we have that

$$t < \Phi_\varepsilon(v; B') = \lim_{n \to \infty} \Phi_\varepsilon(v; B' \cap B_n) \le \sup_{n \in \mathbb{N}} \mu_\varepsilon(B_n),$$

and so letting $t \nearrow \mu_\varepsilon(B)$ we obtain

$$\mu_\varepsilon(B) \le \sup_{n \in \mathbb{N}} \mu_\varepsilon(B_n).$$

Hence μ_ε is a (positive) finite Borel measure. Moreover, if $B \in \mathcal{B}(E)$ has Lebesgue measure zero, then $\mu_\varepsilon(B) = 0$ by (6.133). Hence μ_ε is absolutely continuous with respect to the Lebesgue measure and thus by the Radon–Nikodym theorem there exists a function $a_\varepsilon \in L^1(E)$ such that

$$\mu_\varepsilon(B) = \int_B a_\varepsilon(x) \, dx$$

for all $B \in \mathcal{B}(E)$. By the definition of μ_ε and $\Phi_\varepsilon(v; B)$ and by (6.134) we obtain that

$$\Phi_\varepsilon(v; B) = -\mathcal{E}_p(v; B) - \frac{\varepsilon}{\delta} \int_B |v|^p \, dx \le \mu_\varepsilon(B) = \int_B a_\varepsilon(x) \, dx$$

for all $B \in \mathcal{B}(E)$ and $v \in L^p(E; \mathbb{R}^m)$, or equivalently,

$$\mathcal{E}_p(v; B) \ge -\int_B \left(a_\varepsilon(x) + \frac{\varepsilon}{\delta} |v|^p \right) dx$$

for all $B \in \mathcal{B}(E)$ and $v \in L^p(E; \mathbb{R}^m)$. Take $\varepsilon = 1$ and let $a := a_1$ and $C := \frac{1}{\delta(1)}$. Then

$$\mathcal{E}_p(v; B) \ge -\int_B \left(a(x) + C |v|^p \right) dx$$

for all $B \in \mathcal{B}(E)$ and $v \in L^p(E; \mathbb{R}^m)$.

As in the proof of Theorem 5.29, since $\mathcal{E}_p(v; \cdot)$ may not be a Radon measure, next we consider the Yosida transforms. In what follows we replace the functional $\mathcal{E}_p(v; B)$ with

$$\mathcal{E}_p(v; B) + \int_B \left(a(x) + C |v|^p \right) dx$$

and thus the new functional, still denoted by $\mathcal{E}_p$, is nonnegative and

$$\mathcal{E}_p(0; B) = \int_B a(x) \, dx$$

for all $B \in \mathcal{B}(E)$.

Step 4: For every $k \in \mathbb{N}$, $B \in \mathcal{B}(E)$, and $v \in L^p(E; \mathbb{R}^m)$ define

$$\mathcal{Y}^k(v; B) := \inf \left\{ \mathcal{E}_p(u; B) + k \int_B |w(x) - v(x)|^p \, dx : w \in L^p(E; \mathbb{R}^m) \right\}.$$

Exactly as in Substep 3a of the proof of Theorem 5.29, for $v \in L^p(E; \mathbb{R}^m)$ we have that $\mathcal{Y}^k(v; \cdot)$ is a (positive) finite Radon measure absolutely continuous with respect to the Lebesgue measure, that

$$\mathcal{E}_p(v; B) = \sup_k \mathcal{Y}^k(v; B) \qquad (6.138)$$

for every $B \in \mathcal{B}(E)$, that

$$\mathcal{Y}^k(v; B) \le \mathcal{E}_p(0; B) + k \int_B |v|^p \, dx = \int_B \left(a + k |v|^p \right) dx \qquad (6.139)$$

for all $k \in \mathbb{N}$, $B \in \mathcal{B}(E)$, and that

$$|\mathcal{Y}^k(v; B) - \mathcal{Y}^k(u; B)|$$
$$\le Ck \left(\int_B \left(a + |u|^p + |v|^p \right) dx \right)^{1/p'} \left(\int_B |v - u|^p \, dx \right)^{1/p}. \qquad (6.140)$$

for all $k \in \mathbb{N}$, $B \in \mathcal{B}(E)$, and $u \in L^p(E; \mathbb{R}^m)$.

Let

$$\mathfrak{M} := \{B \in \mathcal{B}(E) : |B| < \infty\}.$$

By (6.139), for every $z \in \mathbb{R}^m$ the set function

$$B \in \mathfrak{M} \mapsto \mathcal{Y}^k(z; B)$$

is nonnegative, countably additive, and bounded from above by a measure that is finite on $\mathfrak{M}$ and absolutely continuous with respect to the Lebesgue measure. By Corollary 1.38 we may extend $\mathcal{Y}^k(z; \cdot)$ to a measure $\mu_{k,z}$ defined on $\mathcal{B}(E)$. Since $\mu_{k,z}(B) = \mathcal{Y}^k(z; B)$ for all $B \in \mathfrak{M}$, it follows that $\mu_{k,z}$ is still absolutely continuous with respect to the Lebesgue measure and it is σ-finite. Thus by the Radon–Nikodym theorem we may find a nonnegative measurable function $f_{k,z} : E \to [0, \infty]$ such that

$$\mu_{k,z}(B) = \int_B f_{k,z}(x)\, dx$$

for all $B \in \mathcal{B}(E)$, and so

$$\mathcal{Y}^k(z; B) = \int_B f_{k,z}(x)\, dx \tag{6.141}$$

for all $B \in \mathfrak{M}$.

In particular, if $j \in \mathbb{N}$ and $z, w \in \mathbb{Q}^m$ are such that $|z|, |w| \le j$, then by (6.140) and Young inequality,

$$|\mathcal{Y}^k(z; B) - \mathcal{Y}^k(w; B)| \le Ck \left(\int_B (a + 2j^p)\, dx \right)^{1/p'} |B|^{1/p} |z - w|$$

$$\le Ck \left[\int_B (a + 2j^p)\, dx + |B| \right] |z - w|$$

for all $B \in \mathfrak{M}$.

Thus, also by (6.139), (6.141), and Proposition 1.87 there exists a measurable set $N_1 \subset E$, with $|N_1| = 0$, such that

$$0 \le f_{k,z}(x) \le a(x) + k|z|^p \tag{6.142}$$

for every $x \in E \setminus N_1$ and every $z \in \mathbb{Q}^m$, and

$$|f_{k,z}(x) - f_{k,w}(x)| \le Ck(a(x) + 2j^p + 1)|z - w| \tag{6.143}$$

for every $x \in E \setminus N_1$, $j \in \mathbb{N}$, and every $z, w \in \mathbb{Q}^m$, with $|z|, |w| \le j$.

Fix $x \in E \setminus N_1$. Since the function $z \mapsto f_{k,z}(x)$ is locally Lipschitz continuous on $\mathbb{Q}^m$, using the density of $\mathbb{Q}^m$ we may uniquely extend it to a locally Lipschitz continuous function on $\mathbb{R}^m$ that still satisfies (6.142) and (6.143). For every $x \in E$ and $z \in \mathbb{R}^m$ define

$$f_k(x,z) := \begin{cases} f_{k,z}(x) & x \in E \setminus N_1, \\ 0 & x \in N_1. \end{cases}$$

Then by (6.142) and (6.143) the function $f_k(x,z)$ is a Carathéodory function and

$$0 \leq f_k(x,z) \leq a(x) + k|z|^p$$

for every $x \in E$ and $z \in \mathbb{R}^m$.

Since $\mathcal{Y}^k(v; \cdot)$ is a measure, it follows from (6.141) that

$$\mathcal{Y}^k(s;B) = \int_B f_k(x,s(x)) \, dx$$

for all $B \in \mathcal{B}(E)$ and for every simple function s that takes values on $\mathbb{Q}^m$ and such that

$$|\{x \in E : s(x) \neq 0\}| < \infty.$$

Using the continuity with respect to strong convergence in $L^p(E;\mathbb{R}^m)$ of both sides of the previous identity (see (6.140), (6.142), (6.143), and Corollary 6.51) we conclude that for every $v \in L^p(E;\mathbb{R}^m)$,

$$\mathcal{Y}^k(v;B) = \int_B f_k(x,v(x)) \, dx.$$

Since for $k \leq n$ we have $\mathcal{Y}^k(v;B) \leq \mathcal{Y}^n(v;B)$, we may find a set N_2 with $|N_2| = 0$ such that

$$f_k(x,z) \leq f_n(x,z)$$

for all $x \in E \setminus N_2$ and $z \in \mathbb{R}^m$. Define

$$h(x,z) := \begin{cases} \sup_{k \in \mathbb{N}} f_k(x,z) & x \in E \setminus N_2, \\ 0 & x \in N_2. \end{cases}$$

By Theorem 6.34, each f_k is a normal integrand, and thus h, being the supremum of an increasing sequence of normal integrands, is still normal. Hence by (6.138) and the Lebesgue monotone convergence theorem we have

$$\mathcal{E}_p(v;B) = \int_B h(x,v(x)) \, dx$$

for every $v \in L^p(E;\mathbb{R}^m)$ and $B \in \mathcal{B}(E)$.

Exercise 6.66. Let $\Omega = (0,1)$, let $m = 1$, and for every $v \in L^1(\Omega)$ and $x \in \Omega$ let

$$T(v)(x) := \begin{cases} 1 \text{ if } |\{y \in \Omega : v(y) = v(x)\}| = 0, \\ 0 \text{ otherwise.} \end{cases}$$

For every $v \in L^1(\Omega)$ and $B \in \mathcal{B}(\Omega)$ define

$$I(v;B) := \int_B T(v)(x) \, dx.$$

(i) For every $v \in L^1(\Omega)$ and $B \in \mathcal{B}(\Omega)$ define

$$N(v; B) := \{t \in \mathbb{R} : |\{y \in \Omega : v(x) = t\}| \neq 0\}$$

and prove that $N(v; B)$ is at most countable and that if $w \in L^p(\Omega)$ is such that $v = w \, \mathcal{L}^1$ a.e. on B, then $N(v; B) = N(w; B)$.

(ii) For every $v \in L^1(\Omega)$, $B \in \mathcal{B}(\Omega)$, and $x \in \Omega$ let

$$T^B(v)(x) := \begin{cases} 1 \text{ if } v(x) \notin N(v; B), \\ 0 \text{ otherwise.} \end{cases}$$

Prove that $T(v)(x) = T^B(v)(x)$ for $\mathcal{L}^1$ a.e. $x \in B$.

(iii) Prove that the functional I satisfies conditions (I_1), (I_2), and (I_3).

(iv) Prove that for every $B \in \mathcal{B}(\Omega)$ and $c \in \mathbb{R}$,

$$I(v_c; B) = |B|,$$

where $v_c(x) := x + c$ for $x \in \Omega$.

(v) Prove that there is no integrand $h : \Omega \times \mathbb{R} \to [0, \infty]$ such that

$$I(v; B) = \int_B h(x, v(x)) \, dx$$

for every $v \in L^1(\Omega)$ and $B \in \mathcal{B}(\Omega)$.

Next we address the case that $p = \infty$.

Theorem 6.67. *Let $E \subset \mathbb{R}^N$ be a Lebesgue measurable set and let*

$$I : L^\infty(E; \mathbb{R}^m) \times \mathcal{B}(E) \to (-\infty, \infty]$$

satisfy properties (I_1)–(I_2) of the previous theorem. For every $B \in \mathcal{B}(E)$ let τ be either the weak star or the Mackey topology in $L^\infty(B; \mathbb{R}^m)$ and let $\mathcal{E}_\infty(\cdot; B)$ be the greatest functional below $I(\cdot, B)$ that is sequentially lower semicontinuous with respect to τ. Assume that $\mathcal{E}_\infty$ satisfies assumption (I_3) of the previous theorem. Then there exists a normal integrand $h : E \times \mathbb{R}^m \to (-\infty, \infty]$ (depending on τ) that satisfies condition (6.70) such that

$$\mathcal{E}_\infty(v; B) = \mathcal{E}_\infty(v_0; B) + \int_B h(x, v(x)) \, dx$$

for all $v \in L^\infty(E; \mathbb{R}^m)$ and $B \in \mathcal{B}(E)$.

If, in addition, I satisfies the growth condition

$(I_5)'$ *for every $M > 0$ there exists $\gamma_M \in L^1(E)$ such that for all $B \in \mathcal{B}(E)$,*

$$|I(v; B)| \leq \int_B \gamma_M(x) \, dx$$

for all $v \in L^\infty(E; \mathbb{R}^m)$ with $|v(x)| \leq M$ for $\mathcal{L}^N$ a.e. $x \in B$, *then τ can also be taken as the strong topology in $L^\infty(B;\mathbb{R}^m)$.*

Proof. The proof is almost identical to that of Theorem 5.31. The only slight difference concerns the case in which τ is the strong topology. Note that in this case, by $(I_5)'$ the functional

$$v \in L^1(E;\mathbb{R}^m) \mapsto T_M(v;B) + \int_B \gamma_M \, dx$$

is nonnegative, and so applying to this functional the reasoning in (5.66), once again we deduce that T_M is sequentially lower semicontinuous with respect to strong convergence in L^1.

6.6 Relaxation in L^p

In this section we treat relaxation properties of functionals of the form

$$v \in L^p(E;\mathbb{R}^m) \mapsto I(v) := \int_E f(x,v)\,dx,$$

where E is a Borel subset of $\mathbb{R}^N$, $1 \le p \le \infty$, and $f : E \times \mathbb{R}^m \to (-\infty,\infty]$ is an $\mathcal{L}^N \times \mathcal{B}$ measurable function.

As in Section 5.4 we consider the *relaxed energy*

$$\mathcal{E} : L^p(E;\mathbb{R}^m) \to [-\infty,\infty]$$

of I, that is, the greatest functional $\mathcal{E}$ below I that is sequentially lower semicontinuous with respect to the topology τ. We will address the following types of convergence:

- weak convergence in $L^p(E;\mathbb{R}^m)$ for $1 \le p < \infty$;
- weak star convergence in $L^\infty(E;\mathbb{R}^m)$;
- weak star convergence in the sense of measures $L^1(E;\mathbb{R}^m)$.

By Propositions 3.16 and 3.18, when I satisfies a suitable coercivity condition we have that $\mathcal{E}(v)$ coincides with

$$\mathcal{I}(v) := \inf\left\{\liminf_{n\to\infty} I(v_n;E) : \{v_n\} \subset L^p(E;\mathbb{R}^m),\, v_n \xrightarrow{\tau} v\right\}$$

for all $v \in L^p(E;\mathbb{R}^m)$.

6.6.1 Weak Convergence and Weak Star Convergence in L^p, $1 \le p \le \infty$

In this subsection we characterize the *relaxed energy*

$$\mathcal{E}_p : L^p(E; \mathbb{R}^m) \to [-\infty, \infty]$$

of I with respect to weak convergence in L^p, $1 \leq p \leq \infty$, that is, the functional $\mathcal{E}_p$ is the greatest functional below I that is sequentially lower semicontinuous with respect to weak (respectively weak star if $p = \infty$) convergence in $L^p(E; \mathbb{R}^m)$.

For every $v \in L^p(E; \mathbb{R}^m)$ define

$$\mathcal{I}_p(v) := \inf \left\{ \liminf_{n \to \infty} I(v_n) : \{v_n\} \subset L^p(E; \mathbb{R}^m), \right.$$

$$\left. v_n \rightharpoonup v \; (\overset{*}{\rightharpoonup} \text{ if } p = \infty) \text{ in } L^p(E; \mathbb{R}^m) \right\}.$$

Since $\mathcal{E}_p \leq I$, for every $v \in L^p(E; \mathbb{R}^m)$ we have

$$\mathcal{E}_p(v) \leq \mathcal{I}_p(v). \tag{6.144}$$

Note that the functional $\mathcal{I}_p$ may fail to be sequentially lower semicontinuous, and thus in general we may have strict inequality in (6.144). Indeed, we have the following theorem.

Theorem 6.68. *Let E be a Borel subset of $\mathbb{R}^N$, let $1 \leq p \leq \infty$, and let $f : E \times \mathbb{R}^m \to (-\infty, \infty]$ be an $\mathcal{L}^N \times \mathcal{B}$ measurable function satisfying*

$$f(x, z) \geq a(x) + b(x) \cdot z \tag{6.145}$$

for $\mathcal{L}^N$ a.e. $x \in E$ and for all $z \in \mathbb{R}^m$, where $a \in L^1(E)$, $b \in L^{p'}(E; \mathbb{R}^m)$, and $C > 0$. Then for every $v \in L^p(E; \mathbb{R}^m)$ we have

$$\mathcal{E}_p(v) = \int_E f^{**}(x, v(x)) \, dx,$$

where for every fixed $x \in E$ the function $f^{**}(x, \cdot)$ is the bipolar of $f(x, \cdot)$. In addition, for every $v \in L^p(E; \mathbb{R}^m)$ for which $\mathcal{I}_p(v) < \infty$ we have

$$\mathcal{I}_p(v) = \int_E \mathcal{C}(\operatorname{lsc} f)(x, v(x)) \, dx,$$

where for every fixed $x \in E$ the function $\mathcal{C}(\operatorname{lsc} f)(x, \cdot)$ is the convex envelope of the lower semicontinuous envelope of $f(x, \cdot)$.

Remark 6.69. (i) We recall that by Propositions 3.16, 3.18, we have

$$\mathcal{E}_p(v) = \mathcal{I}_p(v) = \int_E f^{**}(x, v) \, dx$$

for every $v \in L^p(E; \mathbb{R}^m)$ provided the function f in Theorem 6.68 satisfies the additional coercivity condition

$$f(x, z) \geq C\gamma(|z|) - \psi(x)$$

for $\mathcal{L}^N$ a.e. $x \in E$ and for all $z \in \mathbb{R}^m$, where $C > 0$, $\psi \in L^1(E)$, and where for $t \geq 0$,

$$\gamma(t) := t^p \quad \text{if } 1 < p < \infty,$$

$$\gamma(t) := \begin{cases} 0 & \text{if } 0 \leq t < R, \\ \infty & \text{if } t \geq R, \end{cases} \quad \text{if } p = \infty,$$

for some $R > 0$, and $\gamma : [0, \infty) \to [0, \infty]$ is an increasing function with

$$\lim_{t \to \infty} \frac{\gamma(t)}{t} = \infty$$

if $p = 1$.

(ii) In view of Remark 4.93, if the function f in Theorem 6.68 is real-valued, that is, $f : E \times \mathbb{R}^m \to \mathbb{R}$, then for $\mathcal{L}^N$ a.e. $x \in E$,

$$f^{**}(x, \cdot) = \mathrm{lsc}\,(\mathcal{C}f)(x, \cdot) = \mathcal{C}\,(\mathrm{lsc}\,f)(x, \cdot) = \mathcal{C}f(x, \cdot),$$

and so by the previous theorem, for every $v \in L^p(E; \mathbb{R}^m)$ for which $\mathcal{I}_p(v) < \infty$ we have

$$\mathcal{E}_p(v) = \mathcal{I}_p(v) = \int_E f^{**}(x, v)\, dx.$$

Proof. Without loss of generality we may assume that there exists $v_0 \in L^p(E; \mathbb{R}^m)$ such that

$$\int_E f(x, v_0(x))\, dx < \infty. \tag{6.146}$$

Since the functional

$$v \in L^p(E; \mathbb{R}^m) \mapsto \int_E (a(x) + b(x) \cdot v(x))\, dx$$

is continuous with respect to weak (respectively weak star if $p = \infty$) convergence, replacing $f(x, z)$ by $f(x, z) - (a(x) + b(x) \cdot z)$, we may assume without loss of generality that $f \geq 0$. For every Borel set $B \in \mathcal{B}(E)$ we define the functional

$$v \in L^p(E; \mathbb{R}^m) \mapsto I(v; B) := \int_B f(x, v)\, dx.$$

Since I satisfies the hypotheses (I_1)–(I_2) in Theorems 6.65, 6.67 and $\mathcal{E}_p$ satisfies hypothesis (I_3), using these results we deduce the existence of a normal function $h : E \times \mathbb{R}^m \to [0, \infty]$ such that

$$\mathcal{E}_p(v; B) = \mathcal{E}_p(v_0; B) + \int_B h(x, v)\, dx \tag{6.147}$$

for every $B \in \mathcal{B}(E)$ and $v \in L^p(E; \mathbb{R}^m)$.

We claim that $\mathcal{E}_p(v_0; \cdot)$ is a Radon measure. By Step 1 of Theorem 6.65 and by (6.146), $\mathcal{E}_p(v_0; \cdot)$ is nonnegative, finite, and finitely additive, and so in view of Propositions 1.60 and 1.9 it suffices to prove that

$$\mathcal{E}_p \left(v_0; \bigcup_{n=1}^{\infty} E_n \right) = \lim_{n \to \infty} \mathcal{E}_p(v_0; E_n)$$

for every increasing sequence $\{E_n\}$ of Lebesgue measurable sets $E_n \subset E$. Consider one such sequence $\{E_n\}$. Then

$$\mathcal{E}_p \left(v_0; \bigcup_{n=1}^{\infty} E_n \right) = \mathcal{E}_p (v_0; E_k) + \mathcal{E}_p \left(v_0; \left(\bigcup_{n=1}^{\infty} E_n \right) \setminus E_k \right)$$
$$\geq \mathcal{E}_p (v_0; E_k).$$

Therefore

$$\mathcal{E}_p \left(v_0; \bigcup_{n=1}^{\infty} E_n \right) \geq \limsup_{k \to \infty} \mathcal{E}_p(v_0; E_k).$$

Conversely,

$$\mathcal{E}_p \left(v_0; \bigcup_{n=1}^{\infty} E_n \right) = \mathcal{E}_p (v_0; E_k) + \mathcal{E}_p \left(v_0; \left(\bigcup_{n=1}^{\infty} E_n \right) \setminus E_k \right)$$
$$\leq \mathcal{E}_p (v_0; E_k) + \int_{\left(\bigcup_{n=1}^{\infty} E_n \right) \setminus E_k} f(x, v_0(x)) \, dx.$$

By (6.146) we have

$$\mathcal{E}_p \left(v_0; \bigcup_{n=1}^{\infty} E_n \right) \leq \liminf_{k \to \infty} \mathcal{E}_p(v_0; E_k).$$

Having concluded that $\mathcal{E}_p(v_0; \cdot)$ is a Radon measure, we observe that since

$$\mathcal{E}_p(v_0; B) \leq \int_B f(x, v_0(x)) \, dx$$

for every Borel set $B \subset E$, by the Radon–Nikodym theorem there exists a function $\psi \in L^1(E)$ such that

$$\mathcal{E}_p(v_0; B) = \int_B \psi(x) \, dx$$

for every Borel set $B \subset E$.

Hence (6.147) reduces to

$$\mathcal{E}_p(v; B) = \int_B (\psi(x) + h(x, v)) \, dx$$

for every $B \in \mathcal{B}(E)$ and $v \in L^p(E; \mathbb{R}^m)$. Observe that the function $\tilde{h} : E \times \mathbb{R}^m \to [0, \infty]$ defined by

$$\tilde{h}(x, z) := \psi(x) + h(x, z)$$

for $x \in E$ and $z \in \mathbb{R}^m$ is a normal integrand, and that the functional

$$v \in L^p(E; \mathbb{R}^m) \mapsto \int_E \tilde{h}(x, v) \, dx$$

satisfies the hypotheses of Theorem 6.54. Therefore $\tilde{h}(x, \cdot)$ is convex in $\mathbb{R}^m$ for $\mathcal{L}^N$ a.e. $x \in E$.

Moreover, in view of Proposition 6.43, f^{**} is a nonnegative normal integrand. By Theorems 6.54 and 6.56, for every $B \in \mathcal{B}(E)$ the functional

$$v \in L^p(E; \mathbb{R}^m) \mapsto \int_B f^{**}(x, v) \, dx$$

is sequentially weakly (respectively weakly star if $p = \infty$) lower semicontinuous in $L^p(E; \mathbb{R}^m)$, and since $f^{**} \leq f$, we deduce that for every $v \in L^p(E; \mathbb{R}^m)$,

$$\mathcal{E}_p(v, B) = \int_B \tilde{h}(x, v) \, dx \geq \int_B f^{**}(x, v) \, dx.$$

Hence for every Borel subset $B \subset E$ and for every $v \in L^\infty(E; \mathbb{R}^m)$,

$$\int_B f^{**}(x, v) \, dx \leq \int_B \tilde{h}(x, v) \, dx \leq \int_B f(x, v) \, dx,$$

and thus by Proposition 6.24,

$$f^{**}(x, z) \leq \tilde{h}(x, z) \leq f(x, z)$$

for $\mathcal{L}^N$ a.e. $x \in E$ and for all $z \in \mathbb{R}^m$. Since $\tilde{h}(x, \cdot)$ is convex in $\mathbb{R}^m$ for $\mathcal{L}^N$ a.e. $x \in E$, it follows that

$$f^{**}(x, z) = \tilde{h}(x, z)$$

for $\mathcal{L}^N$ a.e. $x \in E$ and for all $z \in \mathbb{R}^m$.

The last statement in the theorem can be proved following an argument entirely identical to that used in the proof of Theorem 5.32 starting from (5.72). We observe that $\mathcal{C}(\text{lsc } f)$ is a Borel function, since it can be written as

$$\mathcal{C}(\text{lsc } f)(x, z) = \inf_{k \in \mathbb{N}} (f_k)^{**}(x, z)$$

and the functions $(f_k)^{**}$ are normal integrands by Proposition 6.43.

6.6.2 Weak Star Convergence in the Sense of Measures in L^1

In this subsection we consider only the case in which the domain of integration is an open set $\Omega \subset \mathbb{R}^N$. We study the relaxation of functionals of the type

$$v \in L^1(\Omega; \mathbb{R}^m) \mapsto I(v) := \int_\Omega f(x, v(x)) \, dx, \qquad (6.148)$$

where

$$f : \Omega \times \mathbb{R}^m \to (-\infty, \infty],$$

with respect to weak star convergence in the sense of measures.

For any $\lambda \in \mathcal{M}(\Omega; \mathbb{R}^m)$ we define the functional

$$H(\lambda) := \begin{cases} I(v) & \text{if } \lambda = v \, \mathcal{L}^N \lfloor \Omega \text{ for some } v \in L^1(\Omega; \mathbb{R}^m), \\ \infty & \text{otherwise,} \end{cases}$$

and we characterize the *relaxed energy*

$$\mathcal{H} : \mathcal{M}(\Omega; \mathbb{R}^m) \to [-\infty, \infty]$$

of H with respect to weak star convergence in $\mathcal{M}(\Omega; \mathbb{R}^m)$, that is, the functional $\mathcal{H}$ is the greatest functional below H that is sequentially lower semicontinuous with respect to weak star convergence in $\mathcal{M}(\Omega; \mathbb{R}^m)$.

For every $\lambda \in \mathcal{M}(\Omega; \mathbb{R}^m)$ define

$$\mathcal{I}(\lambda) := \inf \Big\{ \liminf_{n \to \infty} I(v_n) : \{v_n\} \subset L^1(\Omega; \mathbb{R}^m), \qquad (6.149)$$

$$v_n \, \mathcal{L}^N \lfloor \Omega \overset{*}{\rightharpoonup} \lambda \text{ in } \mathcal{M}(\Omega; \mathbb{R}^m) \Big\}.$$

Since $\mathcal{H} \leq H$, for every $\lambda \in \mathcal{M}(\Omega; \mathbb{R}^m)$ we have

$$\mathcal{H}(\lambda) \leq \mathcal{I}(\lambda).$$

Theorem 6.70. *Let $\Omega \subset \mathbb{R}^N$ be an open set and let $f : \Omega \times \mathbb{R}^m \to (-\infty, \infty]$ be Borel measurable. Assume that there exist a function $a \in L^1(\Omega)$ and a constant $C > 0$ such that*

$$f(x, z) \geq a(x) + C|z| \qquad (6.150)$$

for $\mathcal{L}^N$ a.e. $x \in \Omega$ and all $z \in \mathbb{R}^m$. Then for every $\lambda \in \mathcal{M}(\Omega; \mathbb{R}^m)$,

$$\mathcal{H}(\lambda) = \mathcal{I}(\lambda) = \int_\Omega \tilde{f}\left(x, \frac{d\lambda}{d\mathcal{L}^N}\right) dx + \int_\Omega f_s\left(x, \frac{d\lambda}{d\|\lambda_s\|}\right) d\|\lambda_s\|,$$

where for $x \in \Omega$ and $z \in \mathbb{R}^m$,

$$\tilde{f}(x, z) := \sup \{\xi \cdot z - f^*(x, \xi) : \xi \in \Gamma(x)\}, \qquad (6.151)$$

$$f_s(x, z) := \sup \{\psi(x) \cdot z : \psi \in \mathcal{X}\},$$

Γ is the essential supremum of the family of multifunctions $\{\Gamma_\psi : \psi \in \mathcal{X}\}$, with

$$\mathcal{X} := \left\{ \psi \in C_0\left(\Omega; \mathbb{R}^m\right) : f^*\left(\cdot, \psi\left(\cdot\right)\right) \in L^1\left(\Omega\right) \right\},$$

and for every fixed $x \in \Omega$ the function $f^*\left(x, \cdot\right)$ is the polar function of $f\left(x, \cdot\right)$.

Proof. Without loss of generality we may assume that there exists $v_0 \in L^1(\Omega; \mathbb{R}^m)$ such that

$$\int_\Omega f\left(x, v_0\left(x\right)\right) dx < \infty.$$

Step 1: We show that, without loss of generality, we may assume that $f\left(x, \cdot\right) = f^{**}\left(x, \cdot\right)$, and hence in what follows $f\left(x, \cdot\right)$ is convex and lower semicontinuous. Applying Proposition 6.43 we deduce that f^{**} is a normal integrand.

For any $\lambda \in \mathcal{M}\left(\Omega; \mathbb{R}^m\right)$ we define the functional

$$J\left(\lambda\right) := \begin{cases} \int_\Omega f^{**}\left(x, v\left(x\right)\right) dx & \text{if } \lambda = v \ \mathcal{L}^N \lfloor \Omega \text{ for some } v \in L^1\left(\Omega; \mathbb{R}^m\right), \\ \infty & \text{otherwise,} \end{cases}$$

and let $\mathcal{J}$ be the greatest functional below J that is sequentially lower semicontinuous with respect to weak star convergence in $\mathcal{M}\left(\Omega; \mathbb{R}^m\right)$.

Since $\mathcal{J} \le H$, we have

$$\mathcal{J} \le \mathcal{H}.$$

Moreover, if $\mathcal{E}_1$ is the greatest functional below I that is sequentially lower semicontinuous with respect to weak convergence in $L^1\left(\Omega; \mathbb{R}^m\right)$, then by Theorem 6.68,

$$\mathcal{H}\left(v \ \mathcal{L}^N \lfloor \Omega\right) \le \mathcal{E}_1\left(v\right) = \int_\Omega f^{**}\left(x, v\left(x\right)\right) dx$$

for every $v \in L^1\left(\Omega; \mathbb{R}^m\right)$. Therefore

$$\mathcal{H}\left(\lambda\right) \le J\left(\lambda\right)$$

for all $\lambda \in \mathcal{M}\left(\Omega; \mathbb{R}^m\right)$, and by relaxing with respect to weak star convergence in $\mathcal{M}\left(\Omega; \mathbb{R}^m\right)$, we conclude that

$$\mathcal{H}\left(\lambda\right) \le \mathcal{J}\left(\lambda\right).$$

Thus we have proved that $\mathcal{J} = \mathcal{H}$. In turn, the integrand f may be replaced by the normal integrand f^{**}, and with an abuse of notation, in the sequel we write f in place of f^{**}. Therefore below, when we write f^* this should be read as f^{***}, but in view of Proposition 4.88, f^* and f^{***} coincide; hence there is no ambiguity.

Step 2: We claim that the functional $\mathcal{I}$ defined in (6.149) is convex. Let $\lambda_1, \lambda_2 \in \mathcal{M}(\Omega; \mathbb{R}^m)$ and let $\theta \in (0, 1)$. It suffices to consider the case in which $\mathcal{I}\left(\lambda_1\right)$ and $\mathcal{I}\left(\lambda_2\right)$ are both finite. Hence for any fixed $\varepsilon > 0$ we may find $\{v_n\}$, $\{w_n\} \subset L^1(\Omega; \mathbb{R}^m)$ such that $v_n \, \mathcal{L}^N \lfloor \Omega \overset{*}{\rightharpoonup} \lambda_1$, $w_n \, \mathcal{L}^N \lfloor \Omega \overset{*}{\rightharpoonup} \lambda_2$ in $\mathcal{M}(\Omega; \mathbb{R}^m)$, and

$$\lim_{n \to \infty} \int_\Omega f(x, v_n(x)) \, dx \leq \mathcal{I}(\lambda_1) + \varepsilon,$$

$$\lim_{n \to \infty} \int_\Omega f(x, w_n(x)) \, dx \leq \mathcal{I}(\lambda_2) + \varepsilon.$$

Let $\{h_k\}$ be a sequence of functions in $L^\infty(\Omega; \{0, 1\})$ such that $h_k \overset{*}{\rightharpoonup} \theta$ in $L^\infty(\Omega)$ (see Example 2.86). For $x \in \Omega$ define

$$u_{n,k}(x) := (1 - h_k(x)) v_n(x) + h_k(x) w_n(x).$$

Since

$$\lim_{n \to \infty} \lim_{k \to \infty} \int_\Omega f(x, u_{n,k}) \, dx$$

$$= \lim_{n \to \infty} \lim_{k \to \infty} \int_\Omega (1 - h_k) f(x, v_n) + h_k f(x, w_n) \, dx$$

$$= (1 - \theta) \lim_{n \to \infty} \int_\Omega f(x, v_n) \, dx + \theta \lim_{n \to \infty} \int_\Omega f(x, w_n) \, dx$$

$$\leq (1 - \theta) \mathcal{I}(\lambda_1) + \theta \mathcal{I}(\lambda_2) + \varepsilon,$$

and since

$$\lim_{n \to \infty} \lim_{k \to \infty} \int_\Omega u_{n,k} \psi \, dx = \langle (1 - \theta) \lambda_1 + \theta \lambda_2, \psi \rangle_{\mathcal{M}, C_0}$$

for a countable dense family of functions ψ in $C_0(\Omega)$, using a diagonalization argument we may find an increasing sequence $\{k_n\}$ of positive integers such that

$$u_{n,k_n} \, \mathcal{L}^N \lfloor \Omega \overset{*}{\rightharpoonup} (1 - \theta) \lambda_1 + \theta \lambda_2$$

and

$$\lim_{n \to \infty} \int_\Omega f(x, u_{n,k_n}) \, dx \leq (1 - \theta) \mathcal{I}(\lambda_1) + \theta \mathcal{I}(\lambda_2) + \varepsilon.$$

We deduce that

$$\mathcal{I}((1 - \theta) \lambda_1 + \theta \lambda_2) \leq (1 - \theta) \mathcal{I}(\lambda_1) + \theta \mathcal{I}(\lambda_2) + \varepsilon,$$

and the convexity of $\mathcal{I}$ follows by letting $\varepsilon \to 0$.

Step 3: In view of (6.150), for every $\lambda \in \mathcal{M}(\Omega; \mathbb{R}^m)$,

$$H(\lambda) \geq \int_\Omega a \, dx + C \|\lambda\|,$$

which shows that H is coercive. By Proposition 3.16 we conclude that $\mathcal{H} = \mathcal{I}$ and that $\mathcal{H}$ is lower semicontinuous with respect to weak star convergence in $\mathcal{M}(\Omega; \mathbb{R}^m)$.

Consider now the duality pair $(\mathcal{M}(\Omega; \mathbb{R}^m), C_0(\Omega; \mathbb{R}^m))$ under the duality

$$\langle u, \phi \rangle_{\mathcal{M}(\Omega; \mathbb{R}^m), C_0(\Omega; \mathbb{R}^m)} := \int_\Omega \phi \cdot d\lambda$$

for $\lambda \in \mathcal{M}(\Omega; \mathbb{R}^m)$ and $C_0(\Omega; \mathbb{R}^m)$. Then the functional $\mathcal{H}$ is convex and lower semicontinuous with respect to the topology $\sigma(\mathcal{M}, C_0)$. It follows from Proposition 4.92, taking as V the topological vector space

$$(\mathcal{M}(\Omega; \mathbb{R}^m), \sigma(L^1, C_0)),$$

that

$$\mathcal{H} = \mathcal{H}^{**} \leq H^{**}.$$

Conversely, by Theorem 4.92, H^{**} is sequentially lower semicontinuous and below H, and therefore $\mathcal{H}^{**} \geq H^{**}$. Since by Theorem A.72 the topological dual of $(\mathcal{M}(\Omega; \mathbb{R}^m), \sigma(L^1, C_0))$ is $C_0(\Omega; \mathbb{R}^m)$, for each $\lambda \in \mathcal{M}(\Omega; \mathbb{R}^m)$,

$$\mathcal{H}(\lambda) = H^{**}(\lambda)$$

$$= \sup_{\psi \in C_0(\Omega; \mathbb{R}^m)} \left[\int_\Omega \psi(x) \cdot d\lambda(x) - H^*(\psi) \right] \qquad (6.152)$$

$$= \sup_{\psi \in C_0(\Omega; \mathbb{R}^m)} \left[\int_\Omega \psi(x) \cdot d\lambda(x) - \int_\Omega f^*(x, \psi(x)) \, dx \right],$$

where in the last identity we have used Step 1 of the necessity part of the proof of Theorem 6.57.

Fix $\lambda \in \mathcal{M}(\Omega; \mathbb{R}^m)$ and let $E \subset \Omega$ be a Borel set such that

$$\mathcal{L}^N(E) = 0 = \|\lambda_s\|(\Omega \setminus E). \qquad (6.153)$$

Set $\nu := \mathcal{L}^N \lfloor \Omega + \|\lambda_s\|$. Then $\lambda \ll \nu$ and

$$\frac{d\lambda}{d\nu}(x) = \begin{cases} \dfrac{d\lambda}{d\mathcal{L}^N}(x) & \text{if } x \in \Omega \setminus E, \\ \dfrac{d\lambda}{d\|\lambda_s\|}(x) & \text{if } x \in E, \end{cases}$$

with $\frac{d\lambda}{d\nu} \in L^1(\Omega, \nu)$. Define

$$h(x, z) := \begin{cases} f^*(x, z) & \text{if } x \in \Omega \setminus E, \\ 0 & \text{if } x \in E. \end{cases} \qquad (6.154)$$

Then for $\psi \in C_0(\Omega; \mathbb{R}^m)$,

$$\int_\Omega \psi(x) \cdot d\lambda(x) - \int_\Omega f^*(x, \psi(x)) \, dx$$

$$= \int_\Omega \left[\psi(x) \cdot \frac{d\lambda}{d\nu}(x) - h(x, \psi(x)) \right] d\nu.$$

Hence by (6.152),

$$\mathcal{H}(\lambda) = \sup_{\psi \in C_0(\Omega; \mathbb{R}^m)} \int_\Omega \left[\psi(x) \cdot \frac{d\lambda}{d\nu}(x) - h(x, \psi(x)) \right] d\nu.$$

We show that

$$\mathcal{H}(\lambda) = \sup_{\psi \in \mathcal{X}} \int_\Omega \left[\psi(x) \cdot \frac{d\lambda}{d\nu}(x) - h(x, \psi(x)) \right] d\nu.$$

Indeed, if $\psi \notin \mathcal{X}$, then since by (6.117),

$$\int_\Omega f^*(x, \psi(x)) \, dx > -\infty,$$

we have

$$\int_\Omega h(x, \psi(x)) \, d\nu = \int_\Omega f^*(x, \psi(x)) \, dx = \infty,$$

and thus

$$\int_\Omega \left[\psi(x) \cdot \frac{d\lambda}{d\nu}(x) - h(x, \psi(x)) \right] d\nu = -\infty.$$

Define

$$g(x, \xi) := \xi \cdot \frac{d\lambda}{d\nu}(x) - h(x, \xi),$$

and note that by (6.153) and (6.154), if $\psi \in C_0(\Omega; \mathbb{R}^m)$, then

$$\int_\Omega g(x, \psi(x)) \, d\nu = -\int_\Omega f^*(x, \psi(x)) \, dx + \int_\Omega \psi(x) \cdot \frac{d\lambda}{d\nu}(x) \, d\nu,$$

and so $g(\cdot, \psi(\cdot)) \in L^1(\Omega, \nu)$ if and only if $\psi \in \mathcal{X}$. Hence we may apply Remark 6.61 to deduce that

$$\mathcal{H}(\lambda) = \int_\Omega \breve{g}(x) \, d\nu,$$

where

$$\breve{g}(x) := \sup_{\xi \in \Gamma_\nu(x)} \left(\xi \cdot \frac{d\lambda}{d\nu}(x) - h(x, \xi) \right),$$

where Γ_ν is the essential supremum of the family of multifunctions $\mathcal{F} := \{\Gamma_\psi : \psi \in \mathcal{X}\}$ with respect to ν.

Since $\mathcal{L}^N \lfloor \Omega \ll \nu$, by Remark 6.14(iii),

$$\Gamma_\nu(x) = \Gamma(x)$$

for $\mathcal{L}^N$ a.e. $x \in \Omega$, and so for $\mathcal{L}^N$ a.e. $x \in \Omega \setminus E$ we have

$$\breve{g}(x) = \sup_{\xi \in \Gamma(x)} \left(\xi \cdot \frac{d\lambda}{d\mathcal{L}^N}(x) - f^*(x, \xi) \right) = \tilde{f}\left(x, \frac{d\lambda}{d\mathcal{L}^N}(x)\right).$$

On the other hand, by Step 1 of the sufficiency proof of Theorem 6.57, for $\|\lambda_s\|$ a.e. $x \in E$,

$$\check{g}(x) = \sup_{\xi \in \Gamma_\nu(x)} \left(\xi \cdot \frac{d\lambda}{d\|\lambda_s\|}(x) \right) = f_s\left(x, \frac{d\lambda}{d\|\lambda_s\|}(x) \right).$$

This concludes the proof.

Remark 6.71. (i) Without the growth condition (6.150) for nonnegative integrands f it is possible to prove that

$$\mathcal{H}(\lambda) = \int_\Omega h\left(x, \frac{d\lambda}{d\mathcal{L}^N} \right) dx + \int_\Omega h^\infty\left(x, \frac{d\lambda}{d\|\lambda_s\|} \right) d\|\lambda_s\|$$

for each $\lambda \in \mathcal{M}(\Omega; \mathbb{R}^m)$, where $h : \Omega \times \mathbb{R}^m \to [0, \infty]$ is a normal integrand, with $h(x, \cdot)$ convex and h^∞ lower semicontinuous in $\Omega \times \mathbb{R}^m$. We refer to [AmBu88] for the precise statement, which holds without any measurability hypotheses on f (provided the integral in (6.148) is replaced by the Lebesgue upper integral). However, in this case we are not aware of any explicit formula relating h to f.

(ii) If in place of the growth condition (6.150) we assume that there exist functions $a \in L^1(\Omega)$ and $b \in C_0(\Omega; \mathbb{R}^m)$ such that

$$f(x, z) \geq a(x) + b(x) \cdot z$$

for $\mathcal{L}^N$ a.e. $x \in \Omega$ and all $z \in \mathbb{R}^m$, then the previous theorem can be applied to show that for every $\lambda \in \mathcal{M}(\Omega; \mathbb{R}^m)$,

$$\mathcal{E}(\lambda) = \int_\Omega \tilde{f}\left(x, \frac{d\lambda}{d\mathcal{L}^N} \right) dx + \int_\Omega f_s\left(x, \frac{d\lambda}{d\|\lambda_s\|} \right) d\|\lambda_s\|,$$

where $\mathcal{E}$ is the greatest functional below H that is lower semicontinuous with respect to weak star convergence in $\mathcal{M}(\Omega; \mathbb{R}^m)$ (we refer to [BouVa88] for more details). Note that in general we can conclude only that $\mathcal{E} \leq \mathcal{H}$.

Exercise 6.72. Let $\Omega \subset \mathbb{R}^N$ be an open set, let $a : \Omega \to [0, \infty]$ be a locally integrable function bounded away from zero, and let

$$f(x, z) := a(x) |z|$$

for $x \in \Omega$ and $z \in \mathbb{R}^m$. Prove that for every $\lambda \in \mathcal{M}(\Omega; \mathbb{R}^m)$,

$$\mathcal{H}(\lambda) = \int_\Omega \hat{a}(x) d\|\lambda\|,$$

where $\hat{a}$ is the lower semicontinuous envelope of the function

$$\bar{a}(x) := \limsup_{\varepsilon \to 0^+} \frac{1}{\varepsilon^N} \int_{Q(x,\varepsilon)} a(y) dy, \quad x \in \Omega.$$

Exercise 6.73. Let $\Omega \subset \mathbb{R}^N$ be an open set, let $a : \Omega \to [0, \infty]$ be a measurable function bounded away from zero, and let

$$f(x, z) := \frac{1}{2} a(x) |z|^2$$

for $x \in \Omega$ and $z \in \mathbb{R}^m$. Let Ω' be the largest open set on which $\frac{1}{a}$ is integrable. Prove that for every $\lambda \in \mathcal{M}(\Omega; \mathbb{R}^m)$,

$$\mathcal{H}(\lambda) = \begin{cases} \frac{1}{2} \int_\Omega a(x) \left| \frac{d\lambda}{d\mathcal{L}^N} \right|^2 dx & \text{if } \|\lambda_s\|(\Omega') = 0, \\ \infty & \text{if } \|\lambda_s\|(\Omega') > 0. \end{cases}$$

Integrands $f = f(x, u, z)$

In this chapter we consider functionals of the type

$$(u, v) \in L^q\left(E; \mathbb{R}^d\right) \times L^p\left(E; \mathbb{R}^m\right) \mapsto I(u, v) := \int_E f(x, u(x), v(x)) \, dx, \quad (7.1)$$

where $1 \leq p, q \leq \infty$, E is a Lebesgue measurable subset of $\mathbb{R}^N$, and $f : E \times \mathbb{R}^d \times \mathbb{R}^m \to [-\infty, \infty]$ is an $\mathcal{L}^N \times \mathcal{B}$ measurable function that is measurable with respect to the σ-algebra generated by products of Lebesgue measurable subsets of E and Borel subsets of $\mathbb{R}^d \times \mathbb{R}^m$.

In what follows, normal and Carathéodory integrands should be understood in the sense of Definitions 6.27, 6.33 with respect to the variables $(x, (u, z))$.

7.1 Convex Integrands

In this section we prove approximation results for normal integrands under some convexity assumptions. We begin with the case that f is real-valued.

Theorem 7.1. *Let E be a Lebesgue measurable subset of $\mathbb{R}^N$ and let $f : E \times \mathbb{R}^d \times \mathbb{R}^m \to \mathbb{R}$ be a Carathéodory function such that $f(x, \cdot, \cdot) \in L^\infty_{\text{loc}}\left(\mathbb{R}^d \times \mathbb{R}^m\right)$*

for $\mathcal{L}^N$ a.e. $x \in E$ and $f(x, u, \cdot)$ is convex in $\mathbb{R}^m$ for $\mathcal{L}^N$ a.e. $x \in E$ and all $u \in \mathbb{R}^d$. Then there exist two sequences of Carathéodory functions

$$a_i : E \times \mathbb{R}^d \to \mathbb{R}, \quad b_i : E \times \mathbb{R}^d \to \mathbb{R}^m,$$

such that

$$f(x, u, z) = \sup_{i \in \mathbb{N}} \{a_i(x, u) + b_i(x, u) \cdot z\}$$

for $\mathcal{L}^N$ a.e. $x \in E$ and all $u \in \mathbb{R}^d$ and $z \in \mathbb{R}^m$.

Moreover, if f is nonnegative, then the functions a_i and b_i may be taken to be bounded.

Proof. By De Giorgi's theorem (Theorem 4.79) for $\mathcal{L}^N$ a.e. $x \in E$ and all $u \in \mathbb{R}^d$ and $z \in \mathbb{R}^m$ we may write

$$f(x, u, z) = \sup_{i \in \mathbb{N}} \{a_i(x, u) + b_i(x, u) \cdot z\},$$

where

$$a_i(x, u) := \int_{\mathbb{R}^m} f(x, u, z) \left((m + 1)\varphi_i(z) + \nabla\varphi_i(z) \cdot z\right) dz,$$

$$b_i(x, u) := -\int_{\mathbb{R}^m} f(x, u, z) \nabla\varphi_i(z) \, dz,$$

and the functions φ_i are of the form

$$\varphi_i(z) := k_i^m \varphi(k_i(q_i - z)), \quad z \in \mathbb{R}^m,$$

for $k_i \in \mathbb{N}$, $q_i \in \mathbb{Q}^m$, and some $\varphi \in C_c^1(\mathbb{R}^m)$ (see (4.43)).

Since $f(x, \cdot, \cdot) \in L_{\text{loc}}^\infty(\mathbb{R}^d \times \mathbb{R}^m)$ for $\mathcal{L}^N$ a.e. $x \in E$, it follows by the Lebesgue dominated convergence theorem that the functions a_i and b_i are Carathéodory.

If, in addition, f is nonnegative, then for $k \in \mathbb{N}_0$ define $\sigma_0 :\equiv 0$ and

$$\sigma_k(s) := \begin{cases} 1 & 0 \le s \le k - 1, \\ -s + k & k - 1 < s < k, \\ 0 & s > k, \end{cases}$$

if $k \ge 1$. We claim that

$$(a_i(x, u) + b_i(x, u) \cdot z)^+$$
$$= \sup_{k \in \mathbb{N}_0} \sigma_k(|a_i(x, u)| + |b_i(x, u)|)(a_i(x, u) + b_i(x, u) \cdot z)$$

for $\mathcal{L}^N$ a.e. $x \in E$ and all $u \in \mathbb{R}^d$ and $z \in \mathbb{R}^m$. Indeed, if $a_i(x, u) + b_i(x, u) \cdot z \le 0$, then since $\sigma_k \ge 0$, the supremum on the right-hand side is reached for $k = 0$. If $a_i(x, u) + b_i(x, u) \cdot z > 0$, then since $0 \le \sigma_k \le 1$ for all $k \ge 0$, it follows that

$$a_i\left(x,u\right) + b_i\left(x,u\right) \cdot z$$
$$\geq \sigma_k\left(\left|a_i\left(x,u\right)\right| + \left|b_i\left(x,u\right)\right|\right)\left(a_i\left(x,u\right) + b_i\left(x,u\right) \cdot z\right).$$

It suffices to take $k \in \mathbb{N}$ so large that $\left|a_i\left(x,u\right)\right| + \left|b_i\left(x,u\right)\right| \leq k - 1$, so that

$$\sigma_k\left(\left|a_i\left(x,u\right)\right| + \left|b_i\left(x,u\right)\right|\right) = 1.$$

Then the functions

$$a_{i,k}\left(x,u\right) := \sigma_k\left(\left|a_i\left(x,u\right)\right| + \left|b_i\left(x,u\right)\right|\right)a_i\left(x,u\right),$$
$$b_{i,k}\left(x,u\right) := \sigma_k\left(\left|a_i\left(x,u\right)\right| + \left|b_i\left(x,u\right)\right|\right)b_i\left(x,u\right),$$

where $x \in E$ and $u \in \mathbb{R}^d$, are bounded Carathéodory functions.

A similar result holds if f is allowed to take the value ∞, but in this case there is no explicit formula for the functions a_i and b_i.

Theorem 7.2. *Let E be a Lebesgue measurable subset of $\mathbb{R}^N$ and let $f : E \times \mathbb{R}^d \times \mathbb{R}^m \to (-\infty, \infty]$ be a normal integrand such that $f\left(x,u,\cdot\right)$ is convex in $\mathbb{R}^m$ for $\mathcal{L}^N$ a.e. $x \in E$ and all $u \in \mathbb{R}^d$. Assume also that f satisfies one of the following two conditions:*

(i) there exists a continuous function $v_0 : \mathbb{R}^d \to \mathbb{R}^m$ such that

$$\left(f(x,\cdot,v_0(\cdot))\right)^+ \in L^\infty(\mathbb{R}^d)$$

for $\mathcal{L}^N$ a.e. $x \in E$;
(ii) there exists a function $\gamma : [0,\infty) \to [0,\infty)$, with

$$\lim_{s \to \infty} \frac{\gamma\left(s\right)}{s} = \infty,$$

such that

$$f\left(x,u,z\right) \geq \gamma\left(\left|z\right|\right)$$

for $\mathcal{L}^N$ a.e. $x \in E$ and for all $u \in \mathbb{R}^d$ and $z \in \mathbb{R}^m$.

Then there exist two sequences of Carathéodory functions

$$a_i : E \times \mathbb{R}^d \to \mathbb{R}, \quad b_i : E \times \mathbb{R}^d \to \mathbb{R}^m,$$

such that

$$f(x,u,z) = \sup_{i \in \mathbb{N}}\left\{a_i(x,u) + b_i(x,u) \cdot z\right\} \tag{7.2}$$

for $\mathcal{L}^N$ a.e. $x \in E$ and all $u \in \mathbb{R}^d$ and $z \in \mathbb{R}^m$.
Moreover, if f is nonnegative, then the functions a_i and b_i may be taken to be bounded.

Proof. Assume first that (i) holds. Then for $\mathcal{L}^N$ a.e. $x \in E$,

$$\operatorname{esssup}\left(f(x, \cdot, v_0(\cdot))\right)^+ < \infty,$$

and due to the lower semicontinuity of the function

$$u \in \mathbb{R}^d \mapsto \left(f(x, u, v_0(u))\right)^+, \tag{7.3}$$

we conclude that for $\mathcal{L}^N$ a.e. $x \in E$,

$$\sup_{u \in \mathbb{R}^d} \left(f(x, u, v_0(u))\right)^+ < \infty. \tag{7.4}$$

Define $g : E \to [0, \infty]$ as

$$g(x) := \sup_{u \in \mathbb{Q}^d} \left(f(x, u, v_0(u))\right)^+, \quad x \in E.$$

Then g is measurable and by (7.4) real-valued for $\mathcal{L}^N$ a.e. $x \in E$. By applying Lusin's theorem to g and Theorem 6.28 to the normal integrand f, we may find a sequence of increasing compact sets $\{K_n\} \subset E$, with

$$\left| E \setminus \bigcup_{n=1}^{\infty} K_n \right| = 0, \tag{7.5}$$

such that $g : K_n \to [0, \infty)$ is continuous and $f : K_n \times \mathbb{R}^d \times \mathbb{R}^m \to (-\infty, \infty]$ is lower semicontinuous for each $n \in \mathbb{N}$. Hence for each fixed $n \in \mathbb{N}$ the function g is bounded on K_n by some constant $M_n > 0$. Using once again the lower semicontinuity of the function (7.3) we obtain

$$0 \leq \left(f(x, u, v_0(u))\right)^+ \leq M_n$$

for all $(x, u) \in K_n \times \mathbb{R}^d$. Thus we may apply Theorem 6.42(i) (where the role of x in that theorem is now played by the variables (x, u), and E there is $K_n \times \mathbb{R}^d$ here) to write

$$f(x, u, z) = \sup_{i \in \mathbb{N}} \left\{ a_{i,n}(x, u) + b_{i,n}(x, u) \cdot z \right\} \tag{7.6}$$

for all $x \in K_n$, $u \in \mathbb{R}^d$, and $z \in \mathbb{R}^m$, where $a_i \in C\left(K_n \times \mathbb{R}^d\right)$ and $b_i \in C\left(K_n \times \mathbb{R}^d; \mathbb{R}^m\right)$.

Moreover, since $K_n \subset K_{n+1}$, by replacing $a_{i,n}$ and $b_{i,n}$ with $a_{i,n+1}|_{K_n \times \mathbb{R}^d}$ and $b_{i,n+1}|_{K_n \times \mathbb{R}^d}$, respectively, we can assume that

$$a_{i,n}(x, u) = a_{i,n+1}(x, u), \quad b_{i,n}(x, u) = b_{i,n+1}(x, u)$$

for all $x \in K_n$ and $u \in \mathbb{R}^d$. Hence the functions

$$a_i : \bigcup_{n=1}^{\infty} K_n \to \mathbb{R}, \quad b_i : \bigcup_{n=1}^{\infty} K_n \to \mathbb{R}^m$$

defined by

$$a_i(x, u) := a_{i,n}(x, u) \text{ if } x \in K_n \text{ and } u \in \mathbb{R}^d,$$

$$b_i(x, u) := b_{i,n}(x, u) \text{ if } x \in K_n \text{ and } u \in \mathbb{R}^d,$$

are Carathéodory and, in view of (7.5), satisfy (7.2).

When condition (ii) is satisfied, then one can proceed as in the first part of the theorem to find an increasing sequence of compact sets $\{K_n\} \subset E$ such that (7.5) holds and $f : K_n \times \mathbb{R}^d \times \mathbb{R}^m \to (-\infty, \infty]$ is lower semicontinuous for each $n \in \mathbb{N}$. Thus we may apply Theorem 6.42(ii) (where the role of x in that theorem is now played by the variables (x, u), and the closed set E there is $K_n \times \mathbb{R}^d$ here) and then continue as in (7.6).

The second part of the proof follows as in the proof of the previous theorem. We omit the details.

7.2 Well-Posedness

In this section we characterize the class of $\mathcal{L}^N \times \mathcal{B}$ measurable integrands f for which

$$\int_E (f(x, u(x), v(x)))^- dx < \infty$$

for every $u \in L^q(E; \mathbb{R}^d)$ and $v \in L^p(E; \mathbb{R}^m)$.

Theorem 7.3. *Let $E \subset \mathbb{R}^N$ be a Lebesgue measurable set with finite measure, $1 \le p, q \le \infty$, and let $f : E \times \mathbb{R}^d \times \mathbb{R}^m \to [-\infty, \infty]$ be $\mathcal{L}^N \times \mathcal{B}$ measurable. Then*

$$\int_E (f(x, u(x), v(x)))^- dx < \infty \tag{7.7}$$

for every $u \in L^q(E; \mathbb{R}^d)$ and $v \in L^p(E; \mathbb{R}^m)$ if and only if

(i) when $1 \le p, q < \infty$, there exist a nonnegative function $\omega \in L^1(E)$ and a constant $C > 0$ such that

$$f(x, u, z) \ge -C(|u|^q + |z|^p) - \omega(x)$$

for $\mathcal{L}^N$ a.e. $x \in E$ and for all $(u, z) \in \mathbb{R}^d \times \mathbb{R}^m$;

(ii) when $p = q = \infty$, for every $M > 0$ there exists a nonnegative function $\omega_M \in L^1(E)$ such that

$$f(x, u, z) \ge -\omega_M(x)$$

for $\mathcal{L}^N$ a.e. $x \in E$ and for all $(u, z) \in \mathbb{R}^d \times \mathbb{R}^m$ with $|u|, |z| \le M$;

(iii) when $1 \le p < \infty$ and $q = \infty$, for every $M > 0$ there exist a nonnegative function $\omega_M \in L^1(E)$ and a constant $C_M > 0$ such that

$$f(x, u, z) \ge -C_M |z|^p - \omega_M(x)$$

for $\mathcal{L}^N$ a.e. $x \in E$ and for all $(u, z) \in \mathbb{R}^d \times \mathbb{R}^m$ with $|u| \le M$;

(iv) when $p = \infty$ and $1 \leq q < \infty$, for every $M > 0$ there exist a nonnegative function $\omega_M \in L^1(E)$ and a constant $C_M > 0$ such that

$$f(x, u, z) \geq -C_M |u|^q - \omega_M(x)$$

for $\mathcal{L}^N$ a.e. $x \in E$ and for all $(u, z) \in \mathbb{R}^d \times \mathbb{R}^m$ with $|z| \leq M$.

Proof. **Step 1:** We prove (i). Without loss of generality, we may assume that $p \geq q$. Define the integrand

$$g(x, w, z) := f\left(x, w |w|^{\frac{p}{q}-1}, z\right).$$

Note that if $w \in L^p(E; \mathbb{R}^d)$, then $w |w|^{\frac{p}{q}-1} \in L^q(E; \mathbb{R}^d)$, and hence by (7.7) we have that

$$\int_E (g(x, w(x), v(x)))^- \, dx < \infty$$

for every $(w, v) \in L^p(E; \mathbb{R}^{d+m})$. By Theorem 6.45, this implies that there exist a nonnegative function $\omega \in L^1(E)$ and a constant $C > 0$ such that

$$f\left(x, w |w|^{\frac{p}{q}-1}, z\right) \geq -C(|w|^p + |z|^p) - \omega(x)$$

for $\mathcal{L}^N$ a.e. $x \in E$ and for all $(w, z) \in \mathbb{R}^d \times \mathbb{R}^m$, or, equivalently,

$$f(x, u, z) \geq -C(|u|^q + |z|^p) - \omega(x)$$

for $\mathcal{L}^N$ a.e. $x \in E$ and for all $(u, z) \in \mathbb{R}^d \times \mathbb{R}^m$.

Step 2: Case (ii) follows from Theorem 6.47, where the variable z there is now replaced by the pair (u, z).

Step 3: We treat only the case (iii), since the case (iv) may be obtained from (iii) by interchanging u and z.

As in the proof of Theorem 6.47, fix $M > 0$ and define the functions

$$\tau_M(u) := \begin{cases} u & \text{if } |u| \leq M, \\ \dfrac{u}{|u|} M & \text{if } |u| > M, \end{cases}$$

and

$$f_M(x, u, z) := f(x, \tau_M(u), z).$$

Note that if $u \in L^p(E; \mathbb{R}^d)$, then $\tau_M(u) \in L^q(E; \mathbb{R}^d)$. Since the functional

$$(u, v) \in L^p(E; \mathbb{R}^{d+m}) \mapsto \int_E f_M(x, u(x), v(x)) \, dx$$

is well-defined and does not take the value $-\infty$, by Theorem 6.45 there exist a function $a_M \in L^1(E)$ and a constant $C_M > 0$ such that if $u \in \mathbb{R}^d$, $|u| \leq M$, then

$$f(x, u, z) = f_M(x, u, z) \geq -C_M(|u|^p + |z|^p) - a_M(x)$$
$$\geq -C_M |z|^p - (C_M M^p + a_M(x))$$

for $\mathcal{L}^N$ a.e. $x \in E$ and for all $z \in \mathbb{R}^m$.

7.3 Sequential Lower Semicontinuity

In this section we study necessary and sufficient conditions for the sequential lower semicontinuity of functionals of the form

$$(u,v) \in L^q\left(E; \mathbb{R}^d\right) \times L^p\left(E; \mathbb{R}^m\right) \mapsto \int_E f\left(x, u\left(x\right), v\left(x\right)\right) \, dx,$$

where $1 \leq p, q \leq \infty$ and

$$f : E \times \mathbb{R}^d \times \mathbb{R}^m \to [-\infty, \infty]$$

is $\mathcal{L}^N \times \mathcal{B}$ measurable.

We are interested in the following types of convergence:

- strong convergence in $L^q\left(E; \mathbb{R}^d\right) \times L^p\left(E; \mathbb{R}^m\right)$ for $1 \leq p, q \leq \infty$;
- strong–weak convergence in $L^q\left(E; \mathbb{R}^d\right) \times L^p\left(E; \mathbb{R}^m\right)$ for $1 \leq p, q < \infty$.

From now on we will assume that the integrand f satisfies the appropriate growth conditions from below that are necessary and sufficient to guarantee that $(f \circ (u,v))^- \in L^1\left(E\right)$ for all $(u,v) \in L^q\left(E; \mathbb{R}^d\right) \times L^p\left(E; \mathbb{R}^m\right)$, so that the functional

$$(u,v) \in L^q\left(E; \mathbb{R}^d\right) \times L^p\left(E; \mathbb{R}^m\right) \mapsto \int_E f\left(x, u\left(x\right), v\left(x\right)\right) \, dx$$

is well-defined (see Theorem 7.3).

7.3.1 Strong–Strong Convergence

In this subsection we study necessary and sufficient conditions for the sequential lower semicontinuity of the functional (6.74) with respect to strong convergence $L^q\left(E; \mathbb{R}^d\right) \times L^p\left(E; \mathbb{R}^m\right)$ for $1 \leq p, q \leq \infty$. The next theorem follows from Theorem 7.4.

Theorem 7.4. *Let $E \subset \mathbb{R}^N$ be a Lebesgue measurable set with finite measure, $1 \leq p, q \leq \infty$, and let $f : E \times \mathbb{R}^d \times \mathbb{R}^m \to (-\infty, \infty]$ be an $\mathcal{L}^N \times \mathcal{B}$ measurable function that satisfies conditions (i)–(iv) of Theorem 7.3. Then the functional*

$$(u,v) \in L^q\left(E; \mathbb{R}^d\right) \times L^p\left(E; \mathbb{R}^m\right) \mapsto \int_E f\left(x, u\left(x\right), v\left(x\right)\right) \, dx \qquad (7.8)$$

is sequentially lower semicontinuous with respect to strong convergence in $L^q\left(E; \mathbb{R}^d\right) \times L^p\left(E; \mathbb{R}^m\right)$ if and only if (up to equivalent integrands) $f\left(x, \cdot, \cdot\right)$ is lower semicontinuous in $\mathbb{R}^d \times \mathbb{R}^m$ for $\mathcal{L}^N$ a.e. $x \in E$.

Proof. Suppose first that the functional (7.8) is sequentially lower semicontinuous with respect to strong convergence in $L^q(E; \mathbb{R}^d) \times L^p(E; \mathbb{R}^m)$. Without loss of generality we may assume that $q \geq p$. Since E has finite measure, we have that $L^q(E; \mathbb{R}^d)$ is continuously embedded in $L^p(E; \mathbb{R}^d)$, and so by hypothesis it follows that the functional

$$(u, v) \in L^q(E; \mathbb{R}^{d+m}) \mapsto \int_E f(x, u(x), v(x))\, dx$$

is sequentially lower semicontinuous with respect to strong convergence in $L^q(E; \mathbb{R}^{d+m})$. The result now follows from Theorems 6.49 and 6.52.

The converse follows by applying Fatou's lemma to the modified integrand obtained by adding to f the lower bound provided by (i)–(iv).

7.3.2 Strong–Weak Convergence $1 \leq p,\, q < \infty$

We study necessary and sufficient conditions for sequential lower semicontinuity of the functional (7.8) with respect to strong convergence in L^q, $1 \leq q < \infty$, and weak convergence in L^p, $1 \leq p < \infty$. Since functionals that are sequentially lower semicontinuous with respect to this kind of convergence are also sequentially lower semicontinuous with respect to strong convergence, without loss of generality, in what follows we may assume that the integrand f satisfies all the necessary conditions for strong convergence obtained in the previous subsections.

Theorem 7.5. *Let $E \subset \mathbb{R}^N$ be a Lebesgue measurable set with finite measure, $1 \leq p,\, q < \infty$, and let $f : E \times \mathbb{R}^d \times \mathbb{R}^m \to (-\infty, \infty]$ be an $\mathcal{L}^N \times \mathcal{B}$ measurable function such that*

$$f(x, u, z) \geq -C(|u|^q + |z|^p) - \omega(x) \tag{7.9}$$

for $\mathcal{L}^N$ a.e. $x \in E$ and for all $(u, z) \in \mathbb{R}^d \times \mathbb{R}^m$, for some $\omega \in L^1(E)$ and $C > 0$. Assume that $f(x, \cdot, \cdot)$ is lower semicontinuous in $\mathbb{R}^d \times \mathbb{R}^m$ for $\mathcal{L}^N$ a.e. $x \in E$. Then the functional

$$(u, v) \in L^q(E; \mathbb{R}^d) \times L^p(E; \mathbb{R}^m) \mapsto \int_E f(x, u(x), v(x))\, dx$$

is sequentially lower semicontinuous with respect to strong convergence in $L^q(E; \mathbb{R}^d)$ and to weak convergence in $L^p(E; \mathbb{R}^m)$ if and only if (up to equivalent integrands)

(i) $f(x, u, \cdot)$ is convex in $\mathbb{R}^m$ for $\mathcal{L}^N$ a.e. $x \in E$ and for all $u \in \mathbb{R}^d$;
(ii) for $\mathcal{L}^N$ a.e. $x \in E$ and all $(u, z) \in \mathbb{R}^d \times \mathbb{R}^m$,

$$f(x, u, z) \geq \alpha(x) + \beta(x, u) \cdot z - C|u|^q,$$

where $\alpha \in L^1(E)$, $\beta : E \times \mathbb{R}^d \to \mathbb{R}^m$ is an $\mathcal{L}^N \times \mathcal{B}$ measurable function, and $C > 0$;

(iii) if $p = 1$, then there exists a constant $M > 0$ such that

$$|\beta(x, u)| \le M$$

for $\mathcal{L}^N$ a.e. $x \in E$ and all $u \in \mathbb{R}^d$, while if $p > 1$, and if $\{u_n\} \subset L^q(E; \mathbb{R}^d)$ and $\{v_n\} \subset L^p(E; \mathbb{R}^m)$ are any two sequences, with u_n strongly convergent in $L^q(E; \mathbb{R}^d)$, and v_n weakly convergent in $L^p(E; \mathbb{R}^m)$, and such that

$$\sup_n \int_E f(x, u_n(x), v_n(x)) \, dx < \infty,$$

then the sequence $\left\{ |\beta(\cdot, u_n(\cdot))|^{p'} \right\}$ is equi-integrable.

Proof (Sufficiency). Let $\{u_n\} \subset L^q(E; \mathbb{R}^d)$ be a sequence converging strongly to $u \in L^q(E; \mathbb{R}^d)$ and let $\{v_n\} \subset L^p(E; \mathbb{R}^m)$ be a sequence weakly converging to $v \in L^p(E; \mathbb{R}^m)$. We assume that

$$\liminf_{n \to \infty} \int_E f(x, u_n, v_n) \, dx < \infty$$

or else there is nothing to prove.

Step 1: Suppose that f is bounded from below by a constant that, without loss of generality, we take to be zero. Let $\eta > 0$ and consider the perturbation of f,

$$f_\eta(x, u, z) := f(x, u, z) + \gamma_\eta(|z|),$$

where for $s \ge 0$,

$$\gamma_\eta(s) := \begin{cases} \eta s^p & \text{if } p > 1, \\ \eta \gamma(s) & \text{if } p = 1, \end{cases}$$

with $\gamma : [0, \infty) \to [0, \infty)$ an increasing continuous convex function provided by Theorem 2.29 and Remark 4.99 for the $L^1(E; \mathbb{R}^m)$ weakly converging sequence $\{v_n\}$, satisfying

$$\lim_{s \to \infty} \frac{\gamma(s)}{s} = \infty \quad \text{and} \quad \sup_n \int_E \gamma(|v_n|) \, dx < \infty.$$

We use the blowup method. Extracting a subsequence if necessary, we may assume that

$$\liminf_{n \to \infty} \int_E f_\eta(x, u_n, v_n) \, dx = \lim_{n \to \infty} \int_E f_\eta(x, u_n, v_n) \, dx < \infty,$$

that $u_n(x) \to u(x)$ for $\mathcal{L}^N$ a.e. $x \in E$, and that there exists a (positive) Radon measure μ such that

$$f_\eta(x, u_n(x), v_n(x)) \, \mathcal{L}^N \lfloor E \overset{*}{\rightharpoonup} \mu$$

as $n \to \infty$, weakly star in the sense of measures. We claim that

$$\lim_{n \to \infty} \int_E f_\eta(x, u_n, v_n) \, dx \geq \int_E f_\eta(x, u, v) \, dx. \tag{7.10}$$

As in the proof of Theorem 6.54, to prove (7.10) it is enough to establish that

$$\frac{d\mu}{d\mathcal{L}^N}(x_0) = \lim_{\varepsilon \to 0^+} \frac{\mu(Q(x_0, \varepsilon) \cap E)}{\varepsilon^N} \geq f_\eta(x_0, u(x_0), v(x_0))$$

for $\mathcal{L}^N$ a.e. $x_0 \in E$.

By Theorem 7.2 there exist two sequences of bounded Carathéodory functions

$$a_i : E \times \mathbb{R}^d \to \mathbb{R}, \quad b_i : E \times \mathbb{R}^d \to \mathbb{R}^m,$$

such that

$$f_\eta(x, u, z) = \sup_{i \in \mathbb{N}} \{a_i(x, u) + b_i(x, u) \cdot z\}$$

for $\mathcal{L}^N$ a.e. $x \in E$ and all $(u, z) \in \mathbb{R}^d \times \mathbb{R}^m$.

Take $x_0 \in E$ a point of density one of E such that

$$\frac{d\mu}{d\mathcal{L}^N}(x_0) = \lim_{\varepsilon \to 0^+} \frac{\mu(Q(x_0, \varepsilon) \cap E)}{\varepsilon^N} < \infty$$

and such that x_0 is a Lebesgue point of all the functions $a_i(\cdot, u(\cdot)) \chi_E(\cdot)$ and $b_i(\cdot, u(\cdot)) \cdot v(\cdot) \chi_E(\cdot)$. Choosing $\varepsilon_k \searrow 0$ such that $\mu(\partial Q(x_0, \varepsilon_k) \cap E) = 0$, for any fixed $i \in \mathbb{N}$ we have

$$\frac{d\mu}{d\mathcal{L}^N}(x_0) = \lim_{k \to \infty} \frac{\mu(Q(x_0, \varepsilon_k) \cap E)}{\varepsilon_k^N}$$

$$= \lim_{k \to \infty} \lim_{n \to \infty} \frac{1}{\varepsilon_k^N} \int_{Q(x_0, \varepsilon_k) \cap E} f_\eta(x, u_n, v_n) \, dx$$

$$\geq \limsup_{k \to \infty} \limsup_{n \to \infty} \frac{1}{\varepsilon_k^N} \int_{Q(x_0, \varepsilon_k) \cap E} [a_i(x, u_n) + b_i(x, u_n) \cdot v_n] \, dx.$$

By Proposition 2.61 we have $b_i(x, u_n) \cdot v_n \rightharpoonup b_i(x, u) \cdot v$ weakly in $L^1(E)$, while by the Lebesgue dominated convergence theorem,

$$\lim_{n \to \infty} \int_{Q(x_0, \varepsilon_k) \cap E} a_i(x, u_n) \, dx = \int_{Q(x_0, \varepsilon_k) \cap E} a_i(x, u) \, dx.$$

Hence

$$\frac{d\mu}{d\mathcal{L}^N}(x_0) \geq \limsup_{k \to \infty} \limsup_{n \to \infty} \frac{1}{\varepsilon_k^N} \int_{Q(x_0, \varepsilon_k) \cap E} [a_i(x, u_n) + b_i(x, u_n) \cdot v_n] \, dx$$

$$= \limsup_{k \to \infty} \frac{1}{\varepsilon_k^N} \int_{Q(x_0, \varepsilon_k) \cap E} [a_i(x, u) + b_i(x, u) \cdot v] \, dx$$

$$= a_i(x_0, u(x_0)) + b_i(x_0, u(x_0)) \cdot v(x_0),$$

where in the last equality we have used the fact that x_0 is a Lebesgue point of $a_i\left(\cdot, u\left(\cdot\right)\right)\chi_E\left(\cdot\right)$ and $b_i(\cdot, u\left(\cdot\right))\cdot v\left(\cdot\right)\chi_E\left(\cdot\right)$. Taking the supremum over all i we have that

$$\frac{d\mu}{d\mathcal{L}^N}(x_0) \geq f_\eta(x_0, u(x_0), v(x_0)),$$

and so (7.10) holds. Since

$$\sup_n \int_E \gamma\left(|v_n|\right) dx =: M < \infty,$$

from (7.10) we have that

$$\liminf_{n\to\infty} \int_E f(x, u_n, v_n)\, dx + \eta M \geq \lim_{n\to\infty} \int_E f_\eta(x, u_n, v_n)\, dx$$

$$\geq \int_E f_\eta(x, u, v)\, dx \geq \int_E f(x, u, v)\, dx,$$

where we used the fact that $f_\eta \geq f$. It suffices now to let $\eta \to 0^+$.

Step 2: In the general case, without loss of generality we may assume that

$$\sup_n \int_E f\left(x, u_n, v_n\right) dx < \infty.$$

We claim that $\{f^-\left(x, u_n, v_n\right)\}$ is weakly compact in $L^1\left(E\right)$. Indeed, boundedness of $\{f^-\left(x, u_n, v_n\right)\}$ in $L^1\left(E\right)$ follows from the fact that by (7.9),

$$0 \leq f^-\left(x, u_n, v_n\right) \leq C\left(|u_n|^q + |v_n|^p\right) + \omega\left(x\right)$$

for $\mathcal{L}^N$ a.e. $x \in E$ and for all $n \in \mathbb{N}$. This inequality entails the equi-integrability of $\{f^-\left(x, u_n, v_n\right)\}$ when $p = 1$ (due to the strong convergence of $\{u_n\}$ in L^q and the equi-integrability of $\{v_n\}$), and for $p > 1$ it suffices to note that in view of (ii) and Hölder's inequality,

$$\int_E f^-\left(x, u_n, v_n\right) dx \leq \int_E [|\alpha| + C\,|u_n|^q]\, dx$$

$$+ \left(\int_E |\beta\left(x, u_n\right)|^{p'}\, dx\right)^{1/p'} \left(\int_E |v_n|^p\, dx\right)^{1/p},$$

where the quantity of the right-hand side is uniformly small for small sets due to the boundedness of $\{v_n\}$ in L^p, the strong convergence of $\{u_n\}$ in L^q, and the equi-integrability of $\left\{|\beta\left(\cdot, u_n\left(\cdot\right)\right)|^{p'}\right\}$.

Hence for any fixed $\varepsilon > 0$ there exists $M > 0$ such that

$$\int_{\{x\in E:\, f^-(x, u_n, v_n)\geq M\}} f^-\left(x, u_n, v_n\right) dx \leq \varepsilon, \tag{7.11}$$

and so, with

$$f_M(x, u, z) := \max\{f(x, u, z), -M\},$$

we have

$$\int_E f(x, u_n, v_n)\, dx + \varepsilon$$

$$= \int_{\{x \in E:\, f(x, u_n, v_n) \geq -M\}} f_M(x, u_n, v_n)\, dx$$

$$\quad - \int_{\{x \in E:\, f(x, u_n, v_n) < -M\}} f^-(x, u_n, v_n)\, dx + \varepsilon$$

$$\geq \int_E f_M(x, u_n, v_n)\, dx + M\,|\{x \in E:\, f(x, u_n, v_n) < -M\}|$$

$$\quad - \int_{\{x \in E:\, f(x, u_n, v_n) < -M\}} f^-(x, u_n, v_n)\, dx + \varepsilon$$

$$\geq \int_E f_M(x, u_n, v_n)\, dx - \int_{\{x \in E:\, f^-(x, u_n, v_n) > M\}} f^-(x, u_n, v_n)\, dx + \varepsilon$$

$$\geq \int_E f_M(x, u_n, v_n)\, dx,$$

where we used (7.11) in the last inequality. Since f_M satisfies conditions (i)–(iii) and is bounded from below, by Step 1 it follows that

$$\liminf_{n \to \infty} \int_E f(x, u_n, v_n)\, dx + \varepsilon \geq \liminf_{n \to \infty} \int_E f_M(x, u_n, v_n)\, dx$$

$$\geq \int_E f_M(x, u, v)\, dx \geq \int_E f(x, u, v)\, dx,$$

and by letting $\varepsilon \to 0^+$ we obtain the desired result.

Remark 7.6. From Step 1 in the previous proof it follows that if f is bounded from below, then the sufficiency part of the previous theorem continues to hold if we assume only that $\{u_n\}$ is a sequence of measurable functions such that $u_n \to u$ in measure to some measurable function u.

The necessity proof of Theorem 7.5 relies on the following auxiliary lemma.

Lemma 7.7. Let $g : \mathbb{R}^m \to (-\infty, \infty]$ be a proper convex, lower semicontinuous function such that

$$g(z) \geq -\frac{1}{p}|z|^p \quad \text{for all } z \in \mathbb{R}^m \text{ and for some } 1 < p < \infty.$$

Let D_g be the set of vectors $\beta \in \mathbb{R}^m$ for which there exists $\alpha \geq 0$ such that

$$g(z) \geq \alpha + \beta \cdot z \geq -\frac{1}{p}|z|^p \quad \text{for all } z \in \mathbb{R}^m. \tag{7.12}$$

Then D_g is nonempty, convex, and compact and there exists a unique $\beta_0 \in D_g$ such that

$$|\beta_0| = \min\{|\beta| : \beta \in D_g\}.$$

Moreover, if $\beta_0 \neq 0$, then there exists $z_0 \in \mathbb{R}^m \setminus \{0\}$ such that

$$g(z_0) \leq -\frac{1}{p}|z_0||\beta_0| \quad and \quad |z_0|^p \geq |\beta_0|^{p'}. \tag{7.13}$$

Proof. **Step 1:** The fact that D_g is nonempty has been proved in Proposition 5.16. We recall that if (α, β) satisfies (7.12), then $\alpha \geq 0$ and

$$|\beta|^{p'} \leq C(p)\alpha. \tag{7.14}$$

Then D_g is convex and closed. To prove compactness, fix $w_0 \in \mathrm{dom}_e\, g$. Then by Young's inequality (see Exercise 4.24),

$$0 \leq \alpha \leq g(w_0) - \beta \cdot w_0 \leq g(w_0) + \frac{\varepsilon}{p'}|\beta|^{p'} + \frac{1}{\varepsilon^{p-1}p}|w_0|^p$$

$$\leq g(w_0) + \frac{\varepsilon}{p'}C(p)\alpha + \frac{1}{\varepsilon^{p-1}p}|w_0|^p.$$

Therefore, choosing $\varepsilon = \varepsilon(p) > 0$ so small that $\frac{\varepsilon}{p'}C(p) < 1$, it follows that

$$0 \leq \alpha \leq C_1(p)(g(w_0) + |w_0|^p) \tag{7.15}$$

for some constant $C_1(p)$. This, together with (7.14), implies that admissible pairs (α, β) satisfying (7.12) remain on a bounded set of $\mathbb{R} \times \mathbb{R}^m$, and the closedness of D_g entails its compactness.

Since D_g is convex we deduce that there exists a unique $\beta_0 \in D_g$ such that

$$|\beta_0| = \min\{|\beta| : \beta \in D_g\}.$$

Step 2: We consider the case $\beta_0 \neq 0$. We claim that $\beta \in D_g$ if and only if

$$\psi(\beta) := \frac{1}{p'}|\beta|^{p'} + g^*(\beta) \leq 0, \tag{7.16}$$

where g^* is the conjugate (or polar) function of g. Indeed, if $\beta \in D_g$, then for some $\alpha \geq 0$,

$$g(z) \geq \alpha + \beta \cdot z \geq -\frac{1}{p}|z|^p \quad \text{for all } z \in \mathbb{R}^m.$$

Since

$$h(z) := \alpha + \beta \cdot z + \frac{1}{p}|z|^p \geq 0 \quad \text{for all } z \in \mathbb{R}^m,$$

then necessarily

$$h\left(z_{\min}\right) = h\left(-|\beta|^{(2-p)/(p-1)}\beta\right) = \alpha - \frac{1}{p'}|\beta|^{p'} \geq 0,$$

and, in turn,

$$\beta \cdot z - g\left(z\right) \leq -\alpha \leq -\frac{1}{p'}|\beta|^{p'} \quad \text{for all } z \in \mathbb{R}^m.$$

Hence

$$g^*\left(\beta\right) := \sup_{z \in \mathbb{R}^m}\{\beta \cdot z - g\left(z\right)\} \leq -\frac{1}{p'}|\beta|^{p'}.$$

Conversely, if β satisfies (7.16), then

$$\beta \cdot z - g\left(z\right) \leq -\frac{1}{p'}|\beta|^{p'} \quad \text{for all } z \in \mathbb{R}^m,$$

and so (7.12) holds with

$$\alpha := \frac{1}{p'}|\beta|^{p'}.$$

Step 3: Let

$$C := \{\beta \in \mathbb{R}^m : \psi\left(\beta\right) \leq 0\}.$$

By the previous step we may write

$$|\beta_0| = \min\{|\beta| : \beta \in C\}. \tag{7.17}$$

If $\beta \in C$, then by convexity,

$$\psi\left(\beta_0 + t\left(\beta - \beta_0\right)\right) \leq 0$$

for all $0 < t \leq 1$, and thus $\beta_0 + t\left(\beta - \beta_0\right) \in C$ for all $0 < t \leq 1$. Setting

$$\varphi\left(t\right) := |\beta_0 + t\left(\beta - \beta_0\right)|^2, \quad t \in [0, 1],$$

and using the fact that φ has a minimum at $t = 0$, we have that $\varphi'\left(0\right) \geq 0$, i.e.,

$$-\beta_0 \cdot \left(\beta_0 - \beta\right) \geq 0 \tag{7.18}$$

for all $\beta \in C$.

Step 4: Consider the case that ψ is not subdifferentiable at β_0 or $\psi\left(\beta_0\right) < 0$.

Substep 4a: We claim that

$$-\beta_0 \cdot \left(\beta_0 - \beta\right) \geq 0 \tag{7.19}$$

for all $\beta \in \mathrm{dom}_e\, g^*$.

Indeed, if ψ is not subdifferentiable at β_0, then by Remark 4.54,

$$\frac{\partial^+ \psi}{\partial\left(\beta - \beta_0\right)}\left(\beta_0\right) = \lim_{t \to 0^+}\frac{\psi\left(\beta_0 + t\left(\beta - \beta_0\right)\right) - \psi\left(\beta_0\right)}{t} = -\infty$$

for all $\beta \in \text{ri}_{\text{aff}} \, \text{dom}_e \, \psi$. Hence if $\beta \in \text{ri}_{\text{aff}} \, \text{dom}_e \, \psi$, then $\psi \, (\beta_0 + t \, (\beta - \beta_0)) \leq 0$ for all $t > 0$ sufficiently small, and so $\beta_0 + t \, (\beta - \beta_0) \in C$. By Step 3 we obtain

$$-\beta_0 \cdot (\beta_0 - \beta) \geq 0$$

for all $\beta \in \text{ri}_{\text{aff}} \, \text{dom}_e \, \psi = \text{ri}_{\text{aff}} \, \text{dom}_e \, g^*$. In view of Proposition 4.7,

$$\overline{\text{dom}_e \, g^*} = \overline{\text{ri}_{\text{aff}} \, \text{dom}_e \, g^*},$$

and so (7.19) holds.

Next we consider the case

$$\psi \, (\beta_0) < 0.$$

If there were $\beta \in \text{dom} \, g^*$ such that

$$|\beta| < |\beta_0| \, ,$$

then for $\theta \in (0,1)$ we would have $|\beta_\theta| < |\beta_0|$, where

$$\beta_\theta := \theta \beta + (1 - \theta) \, \beta_0 \in \text{dom} \, g^*,$$

and by convexity,

$$\frac{1}{p'} \, |\beta_\theta|^{p'} + g^* \, (\beta_\theta)$$

$$\leq \theta \left(\frac{1}{p'} \, |\beta|^{p'} + g^* \, (\beta) \right) + (1 - \theta) \left(\frac{1}{p'} \, |\beta_0|^{p'} + g^* \, (\beta_0) \right) < 0$$

for θ sufficiently close to 0. This contradicts (7.17), and so, reasoning as in Step 3 but with $\text{dom} \, g^*$ in place of C, we conclude that (7.19) holds.

Substep 4b: We claim that

$$z_0 := -t\beta_0 + w_0$$

satisfies (7.13), where $w_0 \in \text{dom}_e \, g$ and $t > 0$ is sufficiently large. Indeed, by Proposition 4.92,

$$\begin{aligned}
g \, (z_0) = g^{**} \, (z_0) &= \sup \{ z_0 \cdot \beta - g^* \, (\beta) : \ \beta \in \text{dom} \, g^* \} \\
&= \sup \{ w_0 \cdot \beta - g^* \, (\beta) - t\beta_0 \cdot \beta : \ \beta \in \text{dom} \, g^* \} \\
&\leq \sup \{ w_0 \cdot \beta - g^* \, (\beta) : \ \beta \in \text{dom} \, g^* \} - t \, |\beta_0|^2 \\
&= g^{**} \, (w_0) - t \, |\beta_0|^2 = g \, (w_0) - t \, |\beta_0|^2 \, ,
\end{aligned}$$

where we have used (7.19). Since $p > 1$ if t is large enough, then

$$g \, (w_0) - t \, |\beta_0|^2 \leq -\frac{1}{p} \, |-t\beta_0 + w_0| \, |\beta_0| = -\frac{1}{p} \, |z_0| \, |\beta_0| \, ,$$

and $|z_0|^p \geq |\beta_0|^{p'}$.

Step 5: It remains to consider the case that

$$\psi(\beta_0) = 0 \text{ and } \partial \psi(\beta_0) \neq \emptyset. \tag{7.20}$$

Substep 5a: We show that if ψ does not attain a minimum at β_0, then

$$-\beta_0 \cdot z \leq 0 \text{ for all } z \in \mathbb{R}^m \text{ satisfying } z \cdot \xi \leq 0 \text{ for all } \xi \in \partial \psi(\beta_0). \tag{7.21}$$

Let

$$\varphi(z) := \frac{\partial^+ \psi}{\partial z}(\beta_0), \quad z \in \mathbb{R}^m.$$

We begin by showing that

$$-\beta_0 \cdot z \leq 0 \text{ for all } z \in \mathbb{R}^m \text{ such that } \varphi(z) < 0. \tag{7.22}$$

Indeed, if

$$0 > \varphi(z) = \frac{\partial^+ \psi}{\partial z}(\beta_0) = \lim_{t \to 0^+} \frac{\psi(\beta_0 + tz)}{t},$$

then $\psi(\beta_0 + tz) < 0$ for all $t > 0$ sufficiently small, and so $\beta_0 + tz \in C$ for those values of t. It now follows from (7.18) that $-\beta_0 \cdot tz \leq 0$, and so (7.22) holds.

Next we prove that

$$D_1 := \overline{\{z \in \mathbb{R}^m : \varphi(z) < 0\}} = \{z \in \mathbb{R}^m : \operatorname{lsc} \varphi(z) \leq 0\} =: D_2. \tag{7.23}$$

Indeed, if $z \in D_1$, then there exists a sequence $\{z_n\}$ converging to z and such that $\varphi(z_n) < 0$ for all $n \in \mathbb{N}$, which implies that

$$\operatorname{lsc} \varphi(z) \leq \liminf_{n \to \infty} \operatorname{lsc} \varphi(z_n) \leq \liminf_{n \to \infty} \varphi(z_n) \leq 0,$$

and thus $z \in D_2$.

Conversely, let $z \in D_2$. There are now two cases. If there exists $z_n \to z$ such that $\operatorname{lsc} \varphi(z_n) < 0$ for all n, then by a simple diagonalization argument we get $z \in D_1$. If not, then there exists $r > 0$ such that for all $w \in B(z, r)$ we have

$$\operatorname{lsc} \varphi(w) \geq 0.$$

Therefore z is a local minimum of the convex function $\operatorname{lsc} \varphi$, and so it must be its absolute minimum. We conclude that $\operatorname{lsc} \varphi \geq 0$ for all $w \in \mathbb{R}^m$. On the other hand, since by assumption ψ does not attain a minimum at β_0, there must exist $\beta \in \mathbb{R}^m$ such that $\psi(\beta) < 0$. Hence

$$\operatorname{lsc} \varphi(\beta - \beta_0) \leq \frac{\partial^+ \psi}{\partial(\beta - \beta_0)}(\beta_0) \leq \psi(\beta) < 0$$

by Proposition 4.34, which is a contradiction. This shows that $D_1 = D_2$.

Next we claim that

$$\varphi^* (\xi) = 0 \text{ if and only if } \xi \in \partial\psi (\beta_0). \tag{7.24}$$

Note that since φ is positively homogeneous and $\varphi \not\equiv \infty$, by Remark 4.90 φ^* takes only the values 0 and ∞.

If $\varphi^* (\xi) = 0$, then

$$\varphi (z) \geq z \cdot \xi \tag{7.25}$$

for all $z \in \mathbb{R}^m$, and by Proposition 4.34,

$$\psi (\beta_0 + z) - \psi (\beta_0) \geq \inf_{t>0} \frac{\psi (\beta_0 + tz) - \psi (\beta_0)}{t} = \frac{\partial^+ \psi}{\partial z} (\beta_0) \geq z \cdot \xi$$

for all $z \in \mathbb{R}^m$. But this implies that $\xi \in \partial\psi (\beta_0)$. Conversely, if $\xi \in \partial\psi (\beta_0)$, then

$$\varphi (z) = \inf_{t>0} \frac{\psi (\beta_0 + tz) - \psi (\beta_0)}{t} \geq z \cdot \xi,$$

and so $\varphi^* (\xi) \leq 0$. This proves (7.24) (since φ^* takes values only in $\{0, \infty\}$).

Since $\partial\psi (\beta_0) \neq \emptyset$ this implies in particular that $\varphi^* (\xi) \leq 0$ for some $\xi \in \mathbb{R}^n$, and só by (7.25) we are in a position to apply Theorem 4.92 to conclude that $\varphi^{**} = \operatorname{lsc}\varphi$. In turn, by (7.23) we have

$$\overline{\{z \in \mathbb{R}^m : \varphi (z) < 0\}} = \{z \in \mathbb{R}^m : \varphi^{**} (z) \leq 0\}. \tag{7.26}$$

Using once more the fact that φ^* takes only the values 0 and ∞, for every $z \in \mathbb{R}^m$ we have

$$\varphi^{**} (z) = \sup \{z \cdot \xi : \varphi^* (\xi) = 0\} = \sup \{z \cdot \xi : \xi \in \partial\psi (\beta_0)\}, \tag{7.27}$$

where we have used (7.24).

It now follows from (7.26) and (7.27) that

$$\overline{\{z \in \mathbb{R}^m : \varphi (z) < 0\}} = \{z \in \mathbb{R}^m : z \cdot \xi \leq 0 \text{ for all } \xi \in \partial\psi (\beta_0)\},$$

and in turn, from (7.22) we obtain (7.21).

Substep 5b: We claim that either there exists $t \geq 0$ such that

$$-t\beta_0 - \beta_0 |\beta_0|^{p'-2} \in \partial g^* (\beta_0) \tag{7.28}$$

or for every $w \in \partial g^* (\beta_0)$ and for every $t \geq 0$,

$$-t\beta_0 + w \in \partial g^* (\beta_0). \tag{7.29}$$

If ψ attains a minimum at β_0, then by (4.22), $0 \in \partial\psi (\beta_0)$, and so, since $\beta_0 \neq 0$, by Proposition 4.59 we have that $-\beta_0 |\beta_0|^{p'-2} \in \partial g^* (\beta_0)$ and this proves (7.28) with $t = 0$.

Suppose now that ψ does not attain a minimum at β_0. Then by (7.21), $-\beta_0 \in (K^*)^*$, where K is the convex cone generated by $\partial\psi(\beta_0)$ and K^* is the polar of K.

By Proposition 4.13 we have that $(K^*)^* = \overline{K}$, and so we are in a position to apply Theorem 4.15 to conclude that $-\beta_0$ belongs to the set

$$\{t\beta : t > 0, \, \beta \in \partial\psi(\beta_0)\} \cup \{\beta : t\beta + \partial\psi(\beta_0) \subset \partial\psi(\beta_0) \text{ for all } t \geq 0\}.$$

Therefore either there exists $t > 0$ such that

$$-t\beta_0 \in \partial\psi(\beta_0),$$

and so, by Proposition 4.59, (7.28) holds, or for any $w \in \partial\psi(\beta_0)$ we have

$$-t\beta_0 + w \in \partial\psi(\beta_0) \quad \text{for all } t \geq 0,$$

and so, again by Proposition 4.59, (7.29) is satisfied with w_0 any arbitrary element of $\partial\psi(\beta_0)$.

Substep 5c: Suppose first that (7.28) holds. We claim that

$$z_0 := -t\beta_0 - \beta_0 |\beta_0|^{p'-2}$$

satisfies (7.13). Since

$$|z_0| = |\beta_0| \left(t + |\beta_0|^{p'-2} \right) \geq |\beta_0|^{p'-1} = |\beta_0|^{\frac{1}{p-1}}$$

we have that $|z_0|^p \geq |\beta_0|^{p'}$. Moreover,

$$z_0 \cdot \beta_0 = -|z_0| \, |\beta_0| = -t |\beta_0|^2 - |\beta_0|^{p'}. \tag{7.30}$$

Since $z_0 \in \partial g^*(\beta_0)$, by Theorem 4.91, (7.20), and (7.30), we have

$$g(z_0) = z_0 \cdot \beta_0 - g^*(\beta_0) = z_0 \cdot \beta_0 + \frac{1}{p'} |\beta_0|^{p'}$$

$$= -|z_0| \, |\beta_0| + \frac{1}{p'} \left(|z_0| \, |\beta_0| - t |\beta_0|^2 \right) \leq -\frac{p'-1}{p'} |z_0| \, |\beta_0|.$$

Next consider the case in which (7.29) is satisfied and let

$$z_0 := -t\beta_0 - \beta_0 |\beta_0|^{p'-2} + w_0,$$

where $t > 0$ is to be chosen below. If t is large enough, then $|z_0|^p \geq |\beta_0|^{p'}$. Moreover, by (7.20) we get

$$g(z_0) = z_0 \cdot \beta_0 + \frac{1}{p'} |\beta_0|^{p'} = -t |\beta_0|^2 - \frac{1}{p} |\beta_0|^{p'} + w_0 \cdot \beta_0.$$

Since $p > 1$ and

$$-\frac{1}{p} |z_0| \, |\beta_0| = -\frac{1}{p} t |\beta_0|^2 \left(\left| \frac{\beta_0}{|\beta_0|} + \frac{1}{t |\beta_0|} \left(\beta_0 |\beta_0|^{p'-2} - w_0 \right) \right| \right)$$

for t large enough, we have $g(z_0) \leq -\frac{1}{p} |z_0| \, |\beta_0|$.

Proof (Necessity of Theorem 7.5). **Step 1:** To prove (i) we adapt the argument used in Theorem 6.54 to arrive at the inequality

$$\int_B f\left(x, u\left(x\right), \left(1 - \theta\right) v\left(x\right) + \theta w\left(x\right)\right) \, dx$$

$$\leq \left(1 - \theta\right) \int_B f\left(x, u\left(x\right), v\left(x\right)\right) \, dx + \theta \int_B f\left(x, u\left(x\right), w\left(x\right)\right) \, dx,$$

which holds for all $B \in \mathcal{B}\left(E\right)$ and $\left(u, v, w\right) \in L^\infty\left(E; \mathbb{R}^{d+2m}\right)$. The rest of the proof is now entirely similar.

Step 2: By Proposition 5.16 applied to

$$g\left(x, u, z\right) := \frac{1}{pC}\left(f\left(x, u, z\right) + C\left|u\right|^q + \omega\left(x\right)\right)$$

(note that the functional

$$\left(u, v\right) \in L^q\left(E; \mathbb{R}^d\right) \times L^p\left(E; \mathbb{R}^m\right) \mapsto \int_E g\left(x, u\left(x\right), v\left(x\right)\right) \, dx$$

is still sequentially lower semicontinuous with respect to strong convergence in $L^q\left(E; \mathbb{R}^d\right)$ and to weak convergence in $L^p\left(E; \mathbb{R}^m\right)$) we obtain

$$g\left(x, u, z\right) \geq \alpha\left(x, u\right) + \beta\left(x, u\right) \cdot z \geq -\frac{1}{p}\left|z\right|^p \tag{7.31}$$

for $\mathcal{L}^N$ a.e. $x \in E$ and all $\left(u, z\right) \in \mathbb{R}^d \times \mathbb{R}^m$, where $\alpha\left(x, u\right) \geq 0$ and $\beta\left(x, u\right)$ is the unique vector in $\mathbb{R}^m$ that is a solution of (see (7.17))

$$\left|\beta\left(x, u\right)\right| = \min\left\{\left|w\right| : \frac{1}{p'}\left|w\right|^{p'} + g^*\left(x, u, w\right) \leq 0\right\}.$$

Since g^* is $\mathcal{L}^N \times \mathcal{B}$ measurable by Proposition 6.43, it follows that β is still $\mathcal{L}^N \times \mathcal{B}$ measurable.

Hence

$$g\left(x, u, z\right) \geq \beta\left(x, u\right) \cdot z$$

for $\mathcal{L}^N$ a.e. $x \in E$ and all $\left(u, z\right) \in \mathbb{R}^d \times \mathbb{R}^m$, and this yields (ii).

Step 3: If $p = 1$, then by Proposition 5.16, $\left|\beta\left(x, u\right)\right| \leq 1$, and so there is nothing to prove.

If $p > 1$, let $\{u_n\}$ converge to u strongly in $L^q\left(E; \mathbb{R}^d\right)$, $v_n \rightharpoonup v$ weakly in $L^p\left(E; \mathbb{R}^m\right)$, and assume that

$$\sup_n \int_E g\left(x, u_n\left(x\right), v_n\left(x\right)\right) \, dx < \infty.$$

In view of (7.31) we have

$$\sup_n \int_E |g(x, u_n(x), v_n(x))| \, dx \le L < \infty \qquad (7.32)$$

for some $L > 0$. By lower semicontinuity, (7.31), and by taking L larger if necessary, we can assume, without loss of generality, that

$$\int_E |g(x, u(x), v(x))| \, dx \le L < \infty. \qquad (7.33)$$

We claim that $\left\{ |\beta(\cdot, u_n(\cdot))|^{p'} \right\}$ is equi-integrable.

We begin by showing that $\left\{ |\beta(\cdot, u_n(\cdot))|^{p'} \right\}$ is bounded in $L^1(E)$. By (7.14) and (7.15) we have

$$|\beta(x, u)|^{p'} \le C(p) \alpha(x, u) \le C(p) C_1(p) (g(x, u, z) + |z|^p) \qquad (7.34)$$

for $\mathcal{L}^N$ a.e. $x \in E$ and all $(u, z) \in \mathbb{R}^d \times \mathbb{R}^m$. Hence from (7.32) and the weak convergence of $\{v_n\}$ in L^p it follows that

$$\sup_n \int_E |\beta(x, u_n(x))|^{p'} \, dx \le C(p) C_1(p) L + \sup_n \int_E |v_n(x)|^p \, dx < \infty.$$

Assume by contradiction that $\left\{ |\beta(\cdot, u_n(\cdot))|^{p'} \right\}$ is not equi-integrable. Then we may find $\delta > 0$ and measurable sets $B_n \subset E$, with $|B_n| \to 0$, such that

$$\int_{B_n} |\beta(x, u_n(x))|^{p'} \, dx \ge \delta. \qquad (7.35)$$

Without loss of generality we may assume that $\beta(x, u_n(x)) \ne 0$ on B_n. By the previous lemma, for any n and any $x \in B_n$ the set

$$D_n(x) := \left\{ z \in \mathbb{R}^m : g(x, u_n(x), z) \le -\frac{1}{p} |\beta(x, u_n(x))| \, |z|, \right.$$

$$\left. |z|^p \ge |\beta(x, u_n(x))|^{p'} \right\}$$

is nonempty and closed, and thus by Aumann's selection theorem we may find a measurable function $\xi_n : E \to \mathbb{R}^m$ such that for $\mathcal{L}^N$ a.e. $x \in B_n$,

$$g(x, u_n(x), \xi_n(x)) \le -\frac{1}{p} |\beta(x, u_n(x))| \, |\xi_n(x)|, \qquad (7.36)$$

$$|\xi_n(x)|^p \ge |\beta(x, u_n(x))|^{p'}.$$

For $k, n \in \mathbb{N}$ define

$$B_{nk}^- := \left\{ x \in B_n : |\xi_n(x)|^p \le k |\beta(x, u_n(x))|^{p'} \right\}, \quad B_{nk}^+ := B_n \setminus B_{nk}^-. \qquad (7.37)$$

There are now three cases.

Case 1: If

$$\liminf_{k\to\infty}\liminf_{n\to\infty}\left[\int_{B_{nk}^-}|\xi_n(x)|^p\,dx+k\int_{B_{nk}^+}|\beta(x,u_n(x))|^{p'}\,dx\right]<\infty,$$

then without loss of generality, we may assume that

$$\int_{B_{nk}^-}|\xi_n(x)|^p\,dx+k\int_{B_{nk}^+}|\beta(x,u_n(x))|^{p'}\,dx\le M\ \text{for all }k,\,n\in\mathbb{N},$$

for some constant $M>0$. Fix $k\in\mathbb{N}$ so that $M/k<\delta/2$. Then by the previous inequality and (7.35),

$$\int_{B_{nk}^-}|\xi_n(x)|^p\,dx\le M,\quad\int_{B_{nk}^-}|\beta(x,u_n(x))|^{p'}\,dx\ge\frac{\delta}{2}\ \text{for all }n\in\mathbb{N}.\quad(7.38)$$

For $x\in E$ and $n\in\mathbb{N}$ define

$$U_n(x):=\left(1-\chi_{B_{nk}^-}(x)\right)u(x)+\chi_{B_{nk}^-}(x)u_n(x),$$
$$V_n(x):=\left(1-\chi_{B_{nk}^-}(x)\right)v(x)+\chi_{B_{nk}^-}(x)\xi_n(x).$$

Then $U_n\to u$ in $L^q\left(E;\mathbb{R}^d\right)$. By (7.38) and since $\left|B_{nk}^-\right|\to0$ as $n\to\infty$ we have that $V_n\in L^p\left(E;\mathbb{R}^m\right)$ and the V_n converge weakly to v in $L^p\left(E;\mathbb{R}^m\right)$. Indeed, we have that

$$\sup_n\int_E|V_n|^p\,dx<\infty,$$

and for every Borel set $B\subset E$,

$$\int_B|V_n-v|\,dx\le\int_{B_{nk}^-}(|\xi_n|+|v|)\,dx$$
$$\le C\left|B_{nk}^-\right|^{\frac{1}{p'}}\left(M^{\frac{1}{p}}+\|v\|_{L^p}\right)\to0$$

as $n\to\infty$.

By the definition of U_n, V_n, (7.36), and (7.38), we have

$$\int_E g(x,U_n(x),V_n(x))\,dx-\int_E g(x,u(x),v(x))\,dx$$
$$=\int_{B_{nk}^-}[g(x,u_n(x),\xi_n(x))-g(x,u(x),v(x))]\,dx$$
$$\le-\frac{1}{p}\int_{B_{nk}^-}|\beta(x,u_n(x))|\,|\xi_n(x)|\,dx-\int_{B_{nk}^-}g(x,u(x),v(x))\,dx$$
$$\le-\frac{1}{p}\int_{B_{nk}^-}|\beta(x,u_n(x))|^{p'}\,dx-\int_{B_{nk}^-}g(x,u(x),v(x))\,dx$$
$$\le-\frac{1}{p}\frac{\delta}{2}-\int_{B_{nk}^-}g(x,u(x),v(x))\,dx\to-\frac{1}{p}\frac{\delta}{2}$$

as $n \to \infty$, which contradicts the sequential lower semicontinuity of the functional. Note that here we used the fact that in view of (7.33), $g\left(\cdot, u\left(\cdot\right), v\left(\cdot\right)\right) \in L^1\left(E\right)$.

Case 2: If

$$\liminf_{k \to \infty} \liminf_{n \to \infty} \left[\int_{B_{nk}^-} |\xi_n\left(x\right)|^p \, dx + k \int_{B_{nk}^+} |\beta\left(x, u_n\left(x\right)\right)|^{p'} \, dx \right] = \infty$$

and

$$\liminf_{k \to \infty} \liminf_{n \to \infty} \int_{B_{nk}^+} |\beta\left(x, u_n\left(x\right)\right)|^{p'} \, dx = 0,$$

then

$$\limsup_{k \to \infty} \limsup_{n \to \infty} \int_{B_{nk}^-} |\beta\left(x, u_n\left(x\right)\right)|^{p'} \, dx \geq \delta,$$

and in turn,

$$\limsup_{k \to \infty} \limsup_{n \to \infty} k \int_{B_{nk}^-} |\beta\left(x, u_n\left(x\right)\right)|^{p'} \, dx = \infty.$$

Fix $k \in \mathbb{N}$ and a subsequence of $\{u_n\}$ (not relabeled) so that

$$k \int_{B_{nk}^-} |\beta\left(x, u_n\left(x\right)\right)|^{p'} \, dx \geq 1 \text{ for all } n \in \mathbb{N},$$

and by Proposition 1.20 find measurable sets $E_{nk} \subset B_{nk}^-$ such that

$$k \int_{E_{nk}} |\beta\left(x, u_n\left(x\right)\right)|^{p'} \, dx = 1 \text{ for all } n \in \mathbb{N},$$

and so by (7.37),

$$\int_{E_{nk}} |\xi_n\left(x\right)|^p \, dx \leq 1 \text{ for all } n \in \mathbb{N}.$$

These last two relations are analogous to (7.38) and thus lead to a contradiction as in Case 1.

Case 3: Finally, if

$$\liminf_{k \to \infty} \liminf_{n \to \infty} \left[\int_{B_{nk}^-} |\xi_n\left(x\right)|^p \, dx + k \int_{B_{nk}^+} |\beta\left(x, u_n\left(x\right)\right)|^{p'} \, dx \right] = \infty,$$

$$\liminf_{k \to \infty} \liminf_{n \to \infty} \int_{B_{nk}^+} |\beta\left(x, u_n\left(x\right)\right)|^{p'} \, dx =: T \in (0, \infty)$$

(note that the sequence $\left\{ |\beta\left(\cdot, u_n\left(\cdot\right)\right)|^{p'} \right\}$ is bounded in $L^1\left(E\right)$), then fix $k \in \mathbb{N}$ so large that

$$\frac{T}{2}\frac{1}{p}k^{1/p} > 1 + 2L$$

and

$$\liminf_{n\to\infty} \int_{B_{nk}^+} |\beta(x, u_n(x))|^{p'} dx > \frac{T}{2}.$$

Select $n_0 = n_0(k) \in \mathbb{N}$ so large that for all $n \geq n_0$,

$$\int_{B_{nk}^+} |\beta(x, u_n(x))|^{p'} dx > \frac{T}{2},$$

and so

$$\frac{1}{p}k^{1/p} \int_{B_{nk}^+} |\beta(x, u_n(x))|^{p'} dx > \frac{T}{2}\frac{1}{p}k^{1/p} > 1 + 2L$$

for all $n \geq n_0$. By Proposition 1.20 we may find $E_{nk} \subset B_{nk}^+$ measurable such that

$$\frac{1}{p}k^{1/p} \int_{E_{nk}} |\beta(x, u_n(x))|^{p'} dx = 1 + 2L \text{ for all } n \in \mathbb{N} \text{ large.} \quad (7.39)$$

Without loss of generality we may assume that $\beta(x, u_n) \neq 0$ on E_{nk}. Construct measurable functions $\alpha_n : E_{nk} \to [0, \infty)$ such that

$$|\alpha_n(x) \xi_n(x)|^p = k|\beta(x, u_n(x))|^{p'}. \quad (7.40)$$

Note that in view of (7.37), $0 \leq \alpha_n(x) \leq 1$ for $\mathcal{L}^N$ a.e. $x \in E_{nk}$ and for all $n \in \mathbb{N}$ large. For $x \in E$ and for all $n \in \mathbb{N}$ large define

$$U_n := (1 - \chi_{E_{nk}})u + \chi_{E_{nk}}u_n,$$
$$V_n := (1 - \chi_{E_{nk}})v + \chi_{E_{nk}}(\alpha_n \xi_n + (1 - \alpha_n)v_n).$$

By (7.39) and (7.40),

$$\int_E |V_n|^p dx \leq C\left(\int_E (|v|^p + |v_n|^p) dx + \int_{E_{nk}} |\alpha_n \xi_n|^p dx\right)$$
$$= C\left(\int_E (|v|^p + |v_n|^p) dx + \int_{E_{nk}} k|\beta(x, u_n(x))|^{p'} dx\right)$$
$$\leq C\left(\int_E (|v|^p + |v_n|^p) dx + \frac{1}{p}k^{1-1/p}\right),$$

and this shows that the sequence $\{V_n\}$ is bounded in $L^p(E; \mathbb{R}^m)$ (recall that k has been fixed). On the other hand, by Hölder's inequality,

$$\int_E |V_n - v| dx \leq \int_{E_{nk}} |v| dx + \int_{E_{nk}} |\alpha_n \xi_n + (1 - \alpha_n)v_n| dx$$
$$\leq \int_{E_{nk}} |v| dx + C|E_{nk}|^{1/p'}.$$

Since $|E_{nk}| \to 0$ as $n \to \infty$, it follows that $\{V_n\}$ converges weakly to v in $L^p(E; \mathbb{R}^m)$. Moreover, by the convexity of $g(x, u, \cdot)$,

$$\int_E g(x, U_n, V_n)\, dx - \int_E g(x, u, v)\, dx$$

$$= \int_{E_{nk}} g(x, u_n, \alpha_n \xi_n + (1 - \alpha_n) v_n)\, dx - \int_{E_{nk}} g(x, u, v)\, dx$$

$$\leq \int_{E_{nk}} \alpha_n g(x, u_n, \xi_n)\, dx + \int_{E_{nk}} (1 - \alpha_n) g(x, u_n, v_n)\, dx$$

$$- \int_{E_{nk}} g(x, u, v)\, dx$$

$$\leq -\frac{1}{p} \int_{E_{nk}} |\beta(x, u_n)|\, |\alpha_n \xi_n|\, dx + 2L$$

$$= -\frac{1}{p} k^{1/p} \int_{E_{nk}} |\beta(x, u_n)|^{p'}\, dx + 2L = -1,$$

where in the last inequality we used (7.36), (7.32), and (7.33), and the latter two equalities used (7.40) and (7.39) in this order.

This contradicts the sequential lower semicontinuity of the functional.

The proof of part (iii) in the previous theorem can be significantly simplified under an additional, mildly restrictive assumption.

Corollary 7.8. *Assume that in addition to the hypotheses of the previous theorem*

$$\int_E f(x, u(x), v_0(x))\, dx < \infty \tag{7.41}$$

for some $v_0 \in L^p(E; \mathbb{R}^m)$ and all $u \in L^q(E; \mathbb{R}^d)$. Then condition (iii) can be replaced by the condition that

$$|\beta(x, u)|^{p'} \leq C_1 |u|^q + b_1(x) \tag{7.42}$$

for $\mathcal{L}^N$ a.e. $x \in E$ and all $u \in \mathbb{R}^d$, for some constant $C_1 > 0$, and some function $b_1 \in L^1(E)$.

Proof. In the sufficiency part of the previous proof all that was required from β was that $\left\{ |\beta(\cdot, u_n(\cdot))|^{p'} \right\}$ be equi-integrable whenever $\{u_n\}$ converges strongly in $L^q(E; \mathbb{R}^d)$, which here is satisfied in view of (7.42).

Conversely, in the necessity part of the proof, by (7.34) and (7.41),

$$\int_E |\beta(x, u(x))|^{p'}\, dx$$

$$\leq C(p) C_1(p) \int_E (g(x, u(x), v_0(x)) + |v_0(x)|^p)\, dx < \infty$$

for all $u \in L^q(E; \mathbb{R}^d)$. The conclusion now follows from Corollary 6.46.

As corollaries of Theorem 7.5 we have the following:

Corollary 7.9. *Let $E \subset \mathbb{R}^N$ be a Lebesgue measurable set with finite measure and let $f : E \times \mathbb{R}^d \times \mathbb{R}^m \to [0, \infty]$ be an $\mathcal{L}^N \times \mathcal{B}$ measurable function. Assume that for $\mathcal{L}^N$ a.e. $x \in E$ the function $f(x, \cdot, \cdot)$ is lower semicontinuous in $\mathbb{R}^d \times \mathbb{R}^m$ and for $\mathcal{L}^N$ a.e. $x \in E$ and for all $u \in \mathbb{R}^d$ the function $f(x, u, \cdot)$ is convex. Then for any sequence $\{u_n\}$ of measurable functions such that $u_n \to u$ in measure for some function u, and for any sequence $\{v_n\} \subset L^1(E; \mathbb{R}^m)$ such that $v_n \overset{b}{\rightharpoonup} v$ (in the biting sense) for some function $v \in L^1(E; \mathbb{R}^m)$ we have*

$$\int_E f(x, u(x), v(x)) \, dx \leq \liminf_{n \to \infty} \int_E f(x, u_n(x), v_n(x)) \, dx.$$

Proof. Since $v_n \overset{b}{\rightharpoonup} v$ there exists a decreasing sequence $\{E_k\} \subset E$ of measurable sets, with $|E_k| \to 0$ as $k \to \infty$, such that

$$v_n \rightharpoonup v \text{ weakly in } L^1(E \setminus E_k; \mathbb{R}^m) \text{ for every } k.$$

Applying Theorem 7.5 and Remark 7.6 to the function $\chi_{E \setminus E_k}(x) f(x, u, z)$ we conclude that

$$\int_{E \setminus E_k} f(x, u, v) \, dx \leq \liminf_{n \to \infty} \int_{E \setminus E_k} f(x, u_n, v_n) \, dx$$

$$\leq \liminf_{n \to \infty} \int_E f(x, u_n, v_n) \, dx.$$

The result now follows by Lebesgue's monotone convergence theorem.

From the previous corollary we may obtain the following result.

Corollary 7.10. *Let $E \subset \mathbb{R}^N$ be a Lebesgue measurable set with finite measure, let $\{u_n\}$ be a sequence of nonnegative, measurable functions such that $u_n \to u$ in measure for some function u, and let $\{v_n\} \subset L^1(E; \mathbb{R}^m)$ be a sequence of nonnegative functions bounded in $L^1(E; \mathbb{R}^m)$ and such that*

$$\int_F v \, dx \leq \liminf_{n \to \infty} \int_F v_n \, dx$$

for every measurable set $F \subset E$ and for some nonnegative function $v \in L^1(E; \mathbb{R}^m)$. Then

$$\int_E uv \, dx \leq \liminf_{n \to \infty} \int_E u_n v_n \, dx. \tag{7.43}$$

Proof. Without loss of generality we may assume that

$$\liminf_{n \to \infty} \int_E u_n v_n \, dx = \lim_{n \to \infty} \int_E u_n v_n \, dx < \infty.$$

and that $v_n \overset{b}{\rightharpoonup} w$ for some function $w \in L^1(E; \mathbb{R}^m)$. Let $\{E_k\} \subset E$ be a decreasing sequence of measurable sets, with $|E_k| \to 0$ as $k \to \infty$, such that

$$v_n \rightharpoonup w \text{ weakly in } L^1(E \setminus E_k; \mathbb{R}^m) \text{ for every } k.$$

Then by hypothesis,

$$\int_{F \setminus E_k} v \, dx \leq \liminf_{n \to \infty} \int_{F \setminus E_k} v_n \, dx = \int_{F \setminus E_k} w \, dx$$

for every measurable set $F \subset E$, and so letting $k \to \infty$ we deduce that

$$\int_F v \, dx \leq \int_F w \, dx.$$

By Proposition 1.87 we conclude that $v(x) \leq w(x)$ for $\mathcal{L}^N$ a.e. $x \in E$. Setting $f(x, u, z) := (uz)^+$ by Corollary 7.9 we have

$$\int_E uv \, dx \leq \int_E uw \, dx \leq \liminf_{n \to \infty} \int_E u_n v_n \, dx.$$

This concludes the proof.

In the next two exercises we show that in general, we cannot weaken any further the modes of convergence of $\{u_n\}$ and $\{v_n\}$.

Exercise 7.11. (i) For $\theta \in (0, 1)$ let χ be the characteristic function of the interval $(0, \theta)$ extended periodically to $\mathbb{R}$ with period 1. Fix $\nu \in S^{N-1}$, $0 < a < b$, and define

$$u_n(x) := \chi(n\nu \cdot x), \quad v_n(x) := u_n(x)\,a + (1 - u_n(x))\,b.$$

Prove that $u_n \overset{*}{\rightharpoonup} \theta$, $v_n \overset{*}{\rightharpoonup} \theta a + (1 - \theta)b$ in $L^\infty(E)$ but (7.43) fails.
(ii) Define

$$v_n(x) := \begin{cases} n \text{ if } x \in \left[0, \frac{1}{n^2}\right] \cup \left[\frac{1}{n}, \frac{1}{n} + \frac{1}{n^2}\right] \cup \ldots \cup \left[\frac{n-1}{n}, \frac{n-1}{n} + \frac{1}{n^2}\right], \\ 0 \text{ otherwise.} \end{cases}$$

Prove that $v_n \overset{*}{\rightharpoonup} 1$, that

$$u_n := \chi_{(0,1) \setminus (\text{supp}\, v_n)} \to 1 \text{ strongly in } L^1(0, 1),$$

and that (7.43) fails. Note that $\{v_n\}$ does not satisfy the hypotheses of Corollary 7.10 if

$$B := (0, 1) \setminus \bigcup_{i=0}^{\infty} (\text{supp}\, v_i)$$

for some subsequence $\{v_i\}$ such that $\sum \frac{1}{i} = \theta < 1$, then $\mathcal{L}^1(B) = 1 - \sum \frac{1}{i} = 1 - \theta > 0$, and so

$$\mathcal{L}^1(B) = \int_B v \, dx > \liminf_{n \to \infty} \int_B v_n \, dx = 0.$$

Exercise 7.12. State and prove an analogue of Theorem 7.5 in the following cases:

(i) $p = q = \infty$;
(ii) $1 \leq p < \infty$ and $q = \infty$;
(iii) $1 \leq q < \infty$ and $p = \infty$.

7.4 Relaxation

Consider the *relaxed energy*

$$\mathcal{E}_{q,p} : L^q\left(E; \mathbb{R}^d\right) \times L^p\left(E; \mathbb{R}^m\right) \to [-\infty, \infty]$$

of the functional I defined in (7.1), that is, the functional $\mathcal{E}_{q,p}$ is the greatest functional that is sequentially lower semicontinuous with respect to strong convergence in $L^q\left(E; \mathbb{R}^d\right)$ and weak convergence in $L^p\left(E; \mathbb{R}^m\right)$ and less than or equal to I.

Theorem 7.13. *Let $E \subset \mathbb{R}^N$ be a Lebesgue measurable set with finite measure, let $1 \leq p, q < \infty$, and let $f : E \times \mathbb{R}^d \times \mathbb{R}^m \to (-\infty, \infty]$ be an $\mathcal{L}^N \times \mathcal{B}$ measurable function such that*

$$f(x, u, z) \geq \alpha(x, u) + \beta(x, u) \cdot z$$

for $\mathcal{L}^N$ a.e. $x \in E$ and for all $(u, z) \in \mathbb{R}^d \times \mathbb{R}^m$, where

$$\alpha : E \times \mathbb{R}^d \to \mathbb{R}, \quad \beta : E \times \mathbb{R}^d \to \mathbb{R}^m$$

are Carathéodory functions, with

$$|\alpha(x, u)| \leq a(x) + C|u|^q, \quad |\beta(x, u)| \leq b(x) + C|u|^{\frac{q}{p'}}$$

for $\mathcal{L}^N$ a.e. $x \in E$ and for all $u \in \mathbb{R}^d$, and for some $a \in L^1(E)$, $b \in L^{p'}(E)$, and $C > 0$. Then

$$\mathcal{E}_{q,p}((u, v); B) = \int_B \mathrm{S} f(x, u(x), v(x)) \, dx,$$

where for every fixed $x \in E$

$$\mathrm{S} f(x, u, z) := \sup\{g(u, z) : g(\cdot, \cdot) \leq f(x, \cdot, \cdot),$$

$$g \text{ is lower semicontinuous,}$$

$$g(w, \cdot) \text{ is convex in } \mathbb{R}^m \text{ for all } w \in \mathbb{R}^d\}$$

for all $(u, z) \in \mathbb{R}^d \times \mathbb{R}^m$.

Proof. Without loss of generality we may assume that there exists $(u_0, v_0) \in L^q(E; \mathbb{R}^d) \times L^p(E; \mathbb{R}^m)$ such that

$$\int_B f(x, u_0(x), v_0(x)) \, dx < \infty.$$

We claim that the functional

$$(u, v) \in L^q\left(E; \mathbb{R}^d\right) \times L^p\left(E; \mathbb{R}^m\right) \mapsto \int_E \left(\alpha(x, u) + \beta(x, u) \cdot v\right) dx$$

is continuous with respect to strong convergence in $L^q\left(E; \mathbb{R}^d\right)$ and weak convergence in $L^p\left(E; \mathbb{R}^m\right)$.

By Corollary 6.51 the functional

$$u \in L^q\left(E; \mathbb{R}^d\right) \mapsto \int_E \alpha(x, u) \, dx$$

is continuous with respect to strong convergence in $L^q\left(E; \mathbb{R}^d\right)$, and so, in view of (2.47), it remains to establish the continuity of

$$u \in L^q\left(E; \mathbb{R}^d\right) \mapsto \beta(x, u) \in L^{p'}\left(E; \mathbb{R}^m\right). \tag{7.44}$$

By Corollary 6.51 the functional

$$u \in L^q\left(E; \mathbb{R}^d\right) \mapsto \int_E |\beta(x, u)|^{p'} \, dx$$

is continuous with respect to strong convergence in $L^q\left(E; \mathbb{R}^d\right)$, and so the continuity of (7.44) follows from the Vitali convergence theorem.

Since

$$S(f(x, u, z) - \alpha(x, u) - \beta(x, u) \cdot z)$$
$$= S\left(f(x, u, z)\right) - \alpha(x, u) - \beta(x, u) \cdot z,$$

replacing $f(x, u, z)$ by $f(x, u, z) - (\alpha(x, u) + \beta(x, u) \cdot z)$, we may assume without loss of generality that $f \geq 0$.

Let Ω be an open subset of $\mathbb{R}^N$ with finite measure that contains E, and for every Borel set $B \in \mathcal{B}(\Omega)$ we define the functional

$$(u, v) \in L^q\left(E; \mathbb{R}^d\right) \times L^p\left(E; \mathbb{R}^m\right) \mapsto I(u, v; B) := \int_B f(x, u, v) \, dx,$$

where we have extended f to $(\Omega \setminus E) \times \mathbb{R}^d \times \mathbb{R}^m$ and u_0, v_0 to $(\Omega \setminus E)$ by zero.

As in the proof of Theorem 6.65, it can be shown that $\mathcal{E}_{q,p}$ satisfies (I_1)–(I_3) (where here τ is the strong–weak topology in $L^q\left(E; \mathbb{R}^d\right) \times L^p\left(E; \mathbb{R}^m\right)$). Hence we deduce the existence of a normal function

$$h : \Omega \times \mathbb{R}^d \times \mathbb{R}^m \to (-\infty, \infty]$$

such that

$$\mathcal{E}_{q,p}(u, v; B) = \mathcal{E}_{q,p}(u_0, v_0; B) + \int_B h(x, u, v) \, dx \qquad (7.45)$$

for every $B \in \mathcal{B}(\Omega)$, $u \in L^q(E; \mathbb{R}^d)$, and $v \in L^p(\Omega; \mathbb{R}^m)$.

As in the proof of Theorem 6.68 we can show that $\mathcal{E}_{q,p}(u_0, v_0; \cdot)$ is a Radon measure and that there exists a function $\psi \in L^1(\Omega)$ such that

$$\mathcal{E}_{q,p}(u_0, v_0; B) = \int_B \psi(x) \, dx$$

for every Borel set $B \subset \Omega$.

Hence (7.45) reduces to

$$\mathcal{E}_{q,p}(u, v; B) = \int_B (\psi(x) + h(x, u, v)) \, dx$$

for every $B \in \mathcal{B}(\Omega)$, $u \in L^q(E; \mathbb{R}^d)$, and $v \in L^p(\Omega; \mathbb{R}^m)$. Observe that the function $\tilde{h} : \Omega \times \mathbb{R}^d \times \mathbb{R}^m \to [0, \infty]$ defined by

$$\tilde{h}(x, u, z) := \psi(x) + h(x, u, z)$$

for $x \in \Omega$, $u \in \mathbb{R}^d$, and $z \in \mathbb{R}^m$ is a normal integrand, and that the functional

$$(u, v) \in L^q(E; \mathbb{R}^d) \times L^p(E; \mathbb{R}^m) \mapsto \int_\Omega \tilde{h}(x, u, v) \, dx$$

satisfies the hypotheses of Theorem 7.5. Therefore $\tilde{h}(x, u, \cdot)$ is convex in $\mathbb{R}^m$ for $\mathcal{L}^N$ a.e. $x \in \Omega$ and for all $u \in \mathbb{R}^d$. Moreover, since

$$\mathcal{E}_{q,p}(u, v; B) = \int_B \tilde{h}(x, u, v) \, dx \le \int_B f(x, u, v) \, dx$$

for every $B \in \mathcal{B}(\Omega)$, $u \in L^q(E; \mathbb{R}^d)$, and $v \in L^p(\Omega; \mathbb{R}^m)$, by Proposition 6.24,

$$\tilde{h}(x, u, z) \le f(x, u, z)$$

for $\mathcal{L}^N$ a.e. $x \in \Omega$ and for all $(u, z) \in \mathbb{R}^d \times \mathbb{R}^m$.

In particular, for $\mathcal{L}^N$ a.e. $x \in \Omega$ the function $\tilde{h}(x, \cdot, \cdot)$ is admissible in the definition of $S f(x, \cdot, \cdot)$, and so

$$\tilde{h}(x, u, z) \le S f(x, u, z) \qquad (7.46)$$

for $\mathcal{L}^N$ a.e. $x \in \Omega$ and for all $(u, z) \in \mathbb{R}^d \times \mathbb{R}^m$.

To prove the reverse inequality, we apply Proposition 6.44 with $S f$ in place of s to find a normal integrand $g : \Omega \times \mathbb{R}^d \times \mathbb{R}^m \to [0, \infty]$ satisfying properties (i)–(iii) and such that for $\mathcal{L}^N$ a.e. $x \in \Omega$ and for all $u \in \mathbb{R}^d$, the function $g(x, u, \cdot)$ is convex in $\mathbb{R}^m$. Since $\mathrm{S}\,f \leq f$ and f is $\mathcal{L}^N \times \mathcal{B}$ measurable, by (ii) (with $h := f$) we deduce that $g(x, u, z) \leq f(x, u, z)$ for $\mathcal{L}^N$ a.e. $x \in E$ and for all $(u, z) \in \mathbb{R}^d \times \mathbb{R}^m$. It follows from the definition of $\mathrm{S}\,f$ that $g(x, u, z) \leq \mathrm{S}\,f(x, u, z)$ for $\mathcal{L}^N$ a.e. $x \in E$ and for all $(u, z) \in \mathbb{R}^d \times \mathbb{R}^m$, which together with (i), yields $g(x, u, z) = \mathrm{S}\,f(x, u, z)$ for $\mathcal{L}^N$ a.e. $x \in E$ and for all $(u, z) \in \mathbb{R}^d \times \mathbb{R}^m$.

Thus $\mathrm{S}\,f$ satisfies all the hypotheses of Theorem 7.5, and so for every $B \in \mathcal{B}(\Omega)$ the functional

$$(u, v) \in L^q(E; \mathbb{R}^d) \times L^p(E; \mathbb{R}^m) \mapsto \int_B \mathrm{S}\,f(x, u, v)\, dx$$

is sequentially weakly lower semicontinuous, and since $\mathrm{S}\,f \leq f$ we deduce that for every $(u, v) \in L^q(E; \mathbb{R}^d) \times L^p(E; \mathbb{R}^m)$,

$$\mathcal{E}_p(u, v, B) = \int_B \tilde{h}(x, u, v)\, dx \geq \int_B \mathrm{S}\,f(x, u, v)\, dx.$$

Using once more Proposition 6.24, we conclude that

$$\mathrm{S}\,f(x, u, v) \leq \tilde{h}(x, u, z)$$

for $\mathcal{L}^N$ a.e. $x \in \Omega$ and for all $(u, z) \in \mathbb{R}^d \times \mathbb{R}^m$. This, together with (7.46), completes the proof.

We observe that by Theorem 4.92(iii),

$$\mathrm{S}\,f(x, u, z) \leq f^{**}(x, u, z), \tag{7.47}$$

where $f^{**}(x, u, z)$ is the polar of $f(x, u, \cdot)$ evaluated at z, and the strict inequality is possible.

Exercise 7.14. Let $d = m = 1$ and consider the function $f : \mathbb{R} \times \mathbb{R} \to [0, \infty)$ defined by

$$f(u, z) := (|z| + 1)^{|u|}$$

for $(u, z) \in \mathbb{R}$. Prove that for $(u, z) \in \mathbb{R}$,

$$f^{**}(u, z) = \begin{cases} (|z| + 1)^{|u|} & \text{if } |u| \geq 1, \\ 1 & \text{if } |u| < 1, \end{cases}$$

while

$$\mathrm{S}\,f(u, z) = \begin{cases} (|z| + 1)^{|u|} & \text{if } |u| > 1, \\ 1 & \text{if } |u| \leq 1. \end{cases}$$

Prove also that f^{**} and $\mathrm{S}\,f$ are not equivalent integrands.

However, under some relatively mild coercivity hypotheses on the integrand f it can be shown that $f^{**} = \mathrm{S}\,f$.

Proposition 7.15. *Let $E \subset \mathbb{R}^N$ be a Lebesgue measurable set with finite measure, and let $f : E \times \mathbb{R}^d \times \mathbb{R}^m \to [0, \infty]$ be an $\mathcal{L}^N \times \mathcal{B}$ measurable function such that*

$$f(x, u, z) \geq \gamma(|z|) \tag{7.48}$$

for $\mathcal{L}^N$ a.e. $x \in E$ and for all $(u, z) \in \mathbb{R}^d \times \mathbb{R}^m$, where $\gamma : [0, \infty) \to [0, \infty)$ satisfies

$$\lim_{s \to \infty} \frac{\gamma(s)}{s} = \infty.$$

Assume that $f(x, \cdot, \cdot)$ is lower semicontinuous in $\mathbb{R}^d \times \mathbb{R}^m$ for $\mathcal{L}^N$ a.e. $x \in E$. Then

$$f^{**}(x, u, z) = \mathrm{S}\, f(x, u, z)$$

for $\mathcal{L}^N$ a.e. $x \in E$ and for all $(u, z) \in \mathbb{R}^d \times \mathbb{R}^m$.

Proof. In view of (7.47) it remains to show that $f^{**} \leq \mathrm{S}\, f$. By the definition of $\mathrm{S}\, f$ this will be a consequence of the lower semicontinuity of $f^{**}(x, \cdot, \cdot)$ for $\mathcal{L}^N$ a.e. $x \in E$, which will be established next. The proof follows closely that of Proposition 6.42.

Fix $x \in E$ for which (7.48) holds and $f(x, \cdot, \cdot)$ is lower semicontinuous. We define the Yosida transforms

$$f_n(x, u, z) := \inf \left\{ f(x, y, z) + n\,|u - y| : y \in \mathbb{R}^d \right\}$$

$u \in \mathbb{R}^d$, $z \in \mathbb{R}^m$. As in (6.41), (6.42), and (6.43), we can show that for all u, $u_0 \in \mathbb{R}^d$, $z \in \mathbb{R}^m$,

$$f(x, u, z) = \sup_{n \in \mathbb{N}} f_n(x, u, z), \quad f^{**}(x, u, z) = \sup_n f_n^{**}(x, u, z),$$

and

$$f_n^{**}(x, u, z) + n\,|u - u_0| \geq f_n^{**}(x, u_0, z). \tag{7.49}$$

Therefore, to prove the lower semicontinuity of $f^{**}(x, \cdot, \cdot)$ it suffices to prove it for each f_n^{**}. Hence fix $n \in \mathbb{N}$ and let $(u_k, z_k) \to (u_0, z_0)$ as $k \to \infty$. Using the lower semicontinuity of $f_n^{**}(x, u_0, \cdot)$ together with (7.49), we have

$$f_n^{**}(x, u_0, z_0) \leq \liminf_{k \to \infty} f_n^{**}(x, u_0, z_k)$$

$$\leq \liminf_{k \to \infty} \left(n\,|u_k - u_0| + f_n^{**}(x, u_k, z_k) \right)$$

$$= \liminf_{k \to \infty} f_n^{**}(x, u_k, z_k),$$

and this concludes the proof.

Exercise 7.16. State and prove an analogue of Theorem 7.13 in the following cases:

(i) $p = q = \infty$;

(ii) $1 \leq p < \infty$ and $q = \infty$;

(iii) $1 \leq q < \infty$ and $p = \infty$.

8

Young Measures

Often in applications we deal with energies of the type

$$v \mapsto \int_E f\left(x, v\left(x\right)\right) dx, \tag{8.1}$$

where $f\left(x, \cdot\right)$ is nonconvex. The study of equilibria leads to the search for minimizers of the energy, possibly subject to constraints, and the nonconvexity of the energy density may prevent the existence of solutions. One way to "resolve" this issue is to relax the energy and, under appropriate hypotheses, use the direct method to obtain a minimizer for

$$v \mapsto \int_E f^{**}\left(x, v\left(x\right)\right) dx.$$

A major setback in this approach has to do with the fact that the relaxed energy density f^{**} may be far below f, and in this way, qualitative properties of the equilibrium energy may be lost.

A compromise between the nonexistence of the minimizers of the original problem and working with the convexified energy can be reached using the notion of Young measures. Here minimizers are obtained for an energy with energy density the original f and where the class of admissible fields is enlarged to include certain probability measures that are intrinsically related to minimizing sequences for (8.1).

8.1 The Fundamental Theorem for Young Measures

Let $E \subset \mathbb{R}^N$ be a Lebesgue measurable set with finite measure. By the Riesz representation theorem in L^p and in C_0 (see Theorems 2.112 and 1.200), the dual of the space $L^1(E; C_0(\mathbb{R}^m))$ is $L_w^\infty(E; \mathcal{M}(\mathbb{R}^m; \mathbb{R}))$. We recall that a function $\lambda : E \to \mathcal{M}(\mathbb{R}^m; \mathbb{R})$ belongs to $L_w^\infty(E; \mathcal{M}(\mathbb{R}^m; \mathbb{R}))$ if and only if for all $\varphi \in C_0(\mathbb{R}^m)$ the map

$$x \in E \mapsto \int_{\mathbb{R}^m} \varphi(v)\, d\lambda_x(v) \text{ is measurable}$$

and

$$\|\lambda\|_{L_w^\infty(E; \mathcal{M}(\mathbb{R}^m; \mathbb{R}))} := \operatorname*{esssup}_{x \in E} \|\lambda_x\|(\mathbb{R}^m) < \infty,$$

where for simplicity we write λ_x in place of $\lambda(x)$ and $\|\lambda_x\|(\mathbb{R}^m)$ is the total variation of the signed Radon measure λ_x.

In what follows, $\Pr(\mathbb{R}^m)$ is the family of all (positive) Radon measures that are also probability measures, that is, the family of all (positive) Radon measures with measure one.

Definition 8.1. *A* Young measure *$\nu : E \to \mathcal{M}(\mathbb{R}^m; \mathbb{R})$ is an element of the space $L_w^\infty(E; \mathcal{M}(\mathbb{R}^m; \mathbb{R}))$ such that $\nu_x \in \Pr(\mathbb{R}^m)$ for $\mathcal{L}^N$ a.e. $x \in E$.*

Theorem 8.2. *Let $E \subset \mathbb{R}^N$ be a Lebesgue measurable set with finite measure and let $\{v_n\}$ be a sequence of measurable functions, with $v_n : E \to \mathbb{R}^m$. Then there exist a subsequence $\{v_{n_k}\}$ of $\{v_n\}$ and a map $\nu \in L_w^\infty(E; \mathcal{M}(\mathbb{R}^m; \mathbb{R}))$ such that*

(i) $\nu_x \geq 0$, $\nu_x(\mathbb{R}^m) \leq 1$ for $\mathcal{L}^N$ a.e. $x \in E$;

(ii) $\varphi(v_{n_k}) \overset{*}{\rightharpoonup} \langle \nu_x, \varphi \rangle_{\mathcal{M}, C_0} = \int_{\mathbb{R}^m} \varphi(z)\, d\nu_x(z)$ in $L^\infty(E)$ for all $\varphi \in C_0(\mathbb{R}^m)$;

(iii) for any normal integrand $f : E \times \mathbb{R}^m \to (-\infty, \infty]$ bounded from below,

$$\liminf_{k \to \infty} \int_E f(x, v_{n_k}(x))\, dx \geq \int_E \overline{f}(x)\, dx,$$

where for $x \in E$,

$$\overline{f}(x) := \int_{\mathbb{R}^m} f(x, z)\, d\nu_x(z).$$

Moreover, the map $\nu \in L_w^\infty(E; \mathcal{M}(\mathbb{R}^m; \mathbb{R}))$ is a Young measure if and only if the subsequence $\{v_{n_k}\}$ satisfies

$$\lim_{t \to \infty} \limsup_{k \to \infty} |\{x \in E : |v_{n_k}(x)| > t\}| = 0. \tag{8.2}$$

Proof. Define $\nu_n : E \to \mathcal{M}(\mathbb{R}^m; \mathbb{R})$ as $\nu_n(x) := \delta_{v_n(x)}$ for $x \in E$. Then

$$\|\nu_n(x)\|(\mathbb{R}^m) = \int_{\mathbb{R}^m} d\delta_{v_n(x)} = 1,$$

and for any $\varphi \in C_0(\mathbb{R}^m)$ the map

$$x \in E \mapsto \int_{\mathbb{R}^m} \varphi(z) \, d\delta_{v_n(x)}(z) = \varphi(v_n(x))$$

is measurable. Thus $\nu_n \in L_w^\infty(E; \mathcal{M}(\mathbb{R}^m; \mathbb{R}))$ and

$$\|\nu_n\|_{L_w^\infty(E;\mathcal{M}(\mathbb{R}^m;\mathbb{R}))} = 1$$

for all n. Since $L_w^\infty(E; \mathcal{M}(\mathbb{R}^m; \mathbb{R}))$ is the dual of $L^1(E; C_0(\mathbb{R}^m))$, by Corollary A.55 there exists a subsequence $\{v_{n_k}\}$ of $\{v_n\}$ and a map $\nu \in L_w^\infty(E; \mathcal{M}(\mathbb{R}^m; \mathbb{R}))$ such that

$$\nu_{n_k} \overset{*}{\rightharpoonup} \nu \text{ in } L_w^\infty(E; \mathcal{M}(\mathbb{R}^m; \mathbb{R})),$$

i.e., for all $\psi \in L^1(E; C_0(\mathbb{R}^m))$ we have

$$\int_E \psi(x, v_{n_k}(x)) \, dx = \int_E \langle \nu_{n_k}(x), \psi(x, \cdot) \rangle_{\mathcal{M},C_0} \, dx \to \int_E \langle \nu_x, \psi(x, \cdot) \rangle_{\mathcal{M},C_0} \, dx \tag{8.3}$$

$$= \int_E \int_{\mathbb{R}^m} \psi(x, z) \, d\nu_x(z) \, dx.$$

In particular, if $h \in L^1(E)$ and $\varphi \in C_0(\mathbb{R}^m)$, then the function

$$\psi(x, z) := h(x) \, \varphi(z),$$

where $x \in E$ and $z \in \mathbb{R}^m$, belongs to $L^1(E; C_0(\mathbb{R}^m))$ by Theorem 2.108, since the function $|h(\cdot)| \, \|\varphi\|_\infty$ is in $L^1(E)$, and thus

$$\int_E h(x) \varphi(v_{n_k}(x)) \, dx \to \int_E h(x) \int_{\mathbb{R}^m} \varphi(z) \, d\nu_x(z) \, dx. \tag{8.4}$$

This proves (ii). Moreover, taking h, $\varphi \geq 0$ we obtain that $\nu_x \geq 0$ for $\mathcal{L}^N$ a.e. $x \in E$, and the lower semicontinuity of the norm implies that

$$\|\nu\|_{L_w^\infty(E;\mathcal{M}(\mathbb{R}^m;\mathbb{R}))} \leq 1,$$

so that (i) holds.

To prove (iii) note first that since $|E| < \infty$, without loss of generality we may assume that $f \geq 0$. We use Corollary 6.30 to find a sequence $\{(E_j, \varphi_j)\}$, where $E_j \subset E$ has finite Lebesgue measure and $\varphi_j \in C_c(\mathbb{R}^m)$, such that

$$f(x, z) = \sup_j \chi_{E_j}(x) \varphi_j(z)$$

for every $(x, z) \in E \times \mathbb{R}^m$. For each $j \in \mathbb{N}$ let

$$f_j(x, z) = \sup_{i \leq j} \chi_{E_i}(x) \varphi_i(z)$$

for every $(x, z) \in E \times \mathbb{R}^m$. We claim that $f_j \in L^1(E; C_0(\mathbb{R}^m))$. Indeed, this follows from Theorem 2.108 and the facts that for fixed $x \in E$ the function $f_j(x, \cdot)$ is in $C_c(\mathbb{R}^m)$ and

$$|f_j(x, z)| \leq \chi_{F_j}(x) \max_{i \leq j} \|\varphi_i\|_\infty \in L^1(E)$$

for every $(x, z) \in E \times \mathbb{R}^m$, where

$$F_j = \bigcup_{i=1}^{j} E_i.$$

By (8.3) for each fixed $j \in \mathbb{N}$ we have

$$\liminf_{k \to \infty} \int_E f(x, v_{n_k}(x)) \, dx \geq \lim_{k \to \infty} \int_E f_j(x, v_{n_k}(x)) \, dx$$

$$= \int_E \int_{\mathbb{R}^m} f_j(x, z) \, d\nu_x(z) \, dx.$$

By the Lebesgue monotone convergence theorem, and using the fact that $f \geq 0$, letting $j \to \infty$ in the previous inequality yields

$$\liminf_{k \to \infty} \int_E f(x, v_{n_k}(x)) \, dx \geq \int_E \int_{\mathbb{R}^m} f(x, z) \, d\nu_x(z) \, dx.$$

To prove the last part of the statement of the theorem, assume that (8.2) holds. Then for any fixed $\varepsilon > 0$ there exists $t > 0$ such that

$$\limsup_{k \to \infty} |\{x \in E : |v_{n_k}(x)| \geq t\}| < \varepsilon. \tag{8.5}$$

Let $\varphi_t : [0, \infty) \to [0, 1]$ be defined by

$$\varphi_t(s) := \begin{cases} 1 & \text{if } s \leq t, \\ 1 - s + t & \text{if } t < s < t + 1, \\ 0 & \text{if } s \leq t + 1. \end{cases}$$

Then the function

$$\psi(x, z) := \chi_E(x) \varphi_t(|z|), \quad (x, z) \in E \times \mathbb{R}^m,$$

belongs to $L^1(E; C_0(\mathbb{R}^m))$, and

$$|E| = |\{x \in E : |v_{n_k}(x)| \leq t\}| + |\{x \in E : |v_{n_k}(x)| > t\}|$$

$$\leq \int_E \varphi_t(|v_{n_k}(x)|) \, dx + |\{x \in E : |v_{n_k}(x)| > t\}|.$$

Taking the limit superior in the previous inequality and using (8.3) and (8.5) we obtain

$$|E| \leq \int_E \int_{\mathbb{R}^m} \varphi_t(|z|) \, d\nu_x(z) \, dx + \varepsilon \leq \int_E \nu_x(\mathbb{R}^m) \, dx + \varepsilon \leq |E| + \varepsilon,$$

where we have used the facts that $\varphi_t \leq 1$, $\nu_x \geq 0$, and $\nu_x(\mathbb{R}^m) \leq 1$ for $\mathcal{L}^N$ a.e. $x \in E$. Letting $\varepsilon \to 0^+$ we obtain that

$$\int_E \nu_x(\mathbb{R}^m) \, dx = |E|,$$

which, recalling again that $\nu_x(\mathbb{R}^m) \leq 1$ for $\mathcal{L}^N$ a.e. $x \in E$, implies that $\nu_x(\mathbb{R}^m) = 1$ for $\mathcal{L}^N$ a.e. $x \in E$, and we conclude that ν is a Young measure. Conversely, assume that ν is a Young measure. Then

$$|E| - |\{x \in E : |v_{n_k}| > t + 1\}| = |\{x \in E : |v_{n_k}| \leq t + 1\}|$$

$$\geq \int_E \varphi_t(|v_{n_k}(x)|) \, dx.$$

Taking the limit inferior in the previous inequality and using (8.3) we obtain

$$|E| - \limsup_{k \to \infty} |\{x \in E : |v_{n_k}| > t + 1\}| \geq \int_E \int_{\mathbb{R}^m} \varphi_t(|z|) \, d\nu_x(z) \, dx.$$

Since $0 \leq \varphi_t \leq 1$ and $\varphi_t \nearrow 1$, letting $t \to \infty$ in the previous inequality and using the Lebesgue monotone convergence theorem, we obtain

$$|E| - \limsup_{t \to \infty} \limsup_{k \to \infty} |\{x \in E : |v_{n_k}| > t + 1\}| \geq \int_E \int_{\mathbb{R}^m} d\nu_x(z) \, dx = |E|,$$

where we have used the fact that $\nu_x(\mathbb{R}^m) = 1$ for $\mathcal{L}^N$ a.e. $x \in E$. Hence

$$\lim_{t \to \infty} \limsup_{k \to \infty} |\{x \in E : |v_{n_k}| > t\}| = 0.$$

This concludes the proof.

In view of (8.2) and (8.4) we now introduce the concept of Young measure ν generated by a sequence $\{v_n\}$.

Definition 8.3. *Let $E \subset \mathbb{R}^N$ be a Lebesgue measurable set with finite measure. A sequence $\{v_n\}$ of measurable functions $v_n : E \to \mathbb{R}^m$ satisfying*

$$\lim_{t \to \infty} \limsup_{n \to \infty} |\{x \in E : |v_n| > t\}| = 0 \tag{8.6}$$

is said to generate *a Young measure $\nu \in L_w^\infty(E; \mathcal{M}(\mathbb{R}^m; \mathbb{R}))$ if for every $h \in L^1(E)$ and $\varphi \in C_0(\mathbb{R}^m)$ we have*

$$\lim_{n \to \infty} \int_E h(x) \, \varphi(v_n(x)) \, dx = \int_E h(x) \int_{\mathbb{R}^m} \varphi(z) \, d\nu_x(z) \, dx.$$

The next proposition is similar to de la Vallée Poussin's theorem.

Proposition 8.4. *Let $E \subset \mathbb{R}^N$ be a Lebesgue measurable set with finite measure and let $\{v_n\}$ be a sequence of measurable functions $v_n : E \to \mathbb{R}^m$. Then condition (8.6) is satisfied if and only if there exists a continuous, nondecreasing function $g : [0, \infty) \to [0, \infty]$ such that $g(t) \to \infty$ as $t \to \infty$ and*

$$\limsup_{n \to \infty} \int_E g(|v_n(x)|)\, dx < \infty. \tag{8.7}$$

Proof. Suppose that (8.7) holds. Since g is nondecreasing, for any $t > 0$ we have

$$g(t)\, |\{x \in E : |v_n(x)| > t\}| \leq \int_E g(|v_n(x)|)\, dx,$$

and so

$$\limsup_{n \to \infty} g(t)\, |\{x \in E : |v_n(x)| > t\}| \leq \limsup_{n \to \infty} \int_E g(|v_n(x)|)\, dx < \infty.$$

Since $g(t) \to \infty$ as $t \to \infty$ we obtain (8.6).

Conversely, assume that (8.6) holds. Then we may choose $0 < t_j < t_{j+1}$, $j \in \mathbb{N}$, with $t_j \to \infty$ as $j \to \infty$, such that

$$\limsup_{n \to \infty} |\{x \in E : |v_n(x)| > t_j\}| \leq \frac{1}{j^3}.$$

Define

$$h(t) := \begin{cases} 0 \text{ if } t \in [0, t_1), \\ j \text{ if } t \in [t_j, t_{j+1}), \ j \in \mathbb{N}. \end{cases}$$

Then

$$\int_E h(|v_n(x)|)\, dx = \sum_{j=1}^{\infty} j\, |\{x \in E : t_j \leq |v_n(x)| < t_{j+1}\}|,$$

and so

$$\limsup_{n \to \infty} \int_E h(|v_n(x)|)\, dx \leq \sum_{j=1}^{\infty} j \limsup_{k \to \infty} |\{x \in E : t_j \leq |v_n(x)| < t_{j+1}\}|$$

$$\leq \sum_{j=1}^{\infty} \frac{1}{j^2} < \infty.$$

It now suffices to replace the function h with a continuous, nondecreasing function, $0 \leq g \leq h$, such that $g(t) \to \infty$ as $t \to \infty$.

Remark 8.5. In applications we will be particularly interested in the cases in which $g(t) := t^p$, $1 \leq p < \infty$, and for $p = \infty$,

$$g\left(t\right) := \begin{cases} \dfrac{t}{\alpha - t} & 0 \le t < \alpha, \\ \infty & t \ge \alpha, \end{cases}$$

corresponding to Young measures generated by sequences bounded in $L^p\left(E; \mathbb{R}^m\right)$, $1 \le p \le \infty$, with

$$\sup_n \|v_n\|_{L^\infty} < \alpha$$

if $p = \infty$. Precisely, in view of Theorem 8.2 and Proposition 8.4, any bounded sequence in $L^p\left(E; \mathbb{R}^m\right)$, $1 \le p \le \infty$, admits a subsequence that generates a Young measure $\nu \in L^\infty_w\left(E; \mathcal{M}\left(\mathbb{R}^m; \mathbb{R}\right)\right)$ such that

$$\int_E \int_{\mathbb{R}^m} g(|z|)\, d\nu_x(z)\, dx < \infty.$$

We are now ready to prove the first main result of this chapter.

Theorem 8.6 (Fundamental theorem for Young measures). *Let $E \subset \mathbb{R}^N$ be a Lebesgue measurable set with finite measure and consider a sequence $\{v_n\}$ of measurable functions $v_n : E \to \mathbb{R}^m$ that generates a Young measure $\nu \in L^\infty_w\left(E; \mathcal{M}\left(\mathbb{R}^m; \mathbb{R}\right)\right)$. Then*

(i) for any normal integrand $f : E \times \mathbb{R}^m \to [-\infty, \infty]$ such that $\{f^-(\cdot, v_n(\cdot))\}$ is equi-integrable,

$$\liminf_{n \to \infty} \int_E f(x, v_n(x))\, dx \ge \int_E \overline{f}(x)\, dx;$$

(ii) for any Carathéodory function $f : E \times \mathbb{R}^m \to [-\infty, \infty]$ such that $\{f(\cdot, v_n(\cdot))\} \subset L^1\left(E\right)$ one has

$$f(\cdot, v_n(\cdot)) \rightharpoonup \overline{f} \text{ in } L^1\left(E\right)$$

if and only if $\{f(\cdot, v_n(\cdot))\}$ is equi-integrable;
(iii) if $K \subset \mathbb{R}^m$ is a compact set, then

$$\operatorname{supp} \nu_x \subset K \text{ for } \mathcal{L}^N \text{ a.e. } x \in E$$

if and only if $\operatorname{dist}(v_n, K) \to 0$ in measure.

Proof. (i) Since $\{f^-(\cdot, v_n(\cdot))\}$ is equi-integrable, by Theorem 2.29, for any fixed $\varepsilon > 0$ we may find a constant $M > 0$ such that

$$\int_{\{f^-(x, v_n) \ge M\}} f^-(x, v_n)\, dx \le \varepsilon$$

for all $n \in \mathbb{N}$. Let $f_M := \max\{f, -M\}$, which is a normal integrand bounded from below. By Theorem 8.2(iii),

$$\liminf_{n \to \infty} \int_E f(x, v_n) \, dx$$

$$= \liminf_{n \to \infty} \left[\int_{\{f(x,v_n) > -M\}} f_M(x, v_n) \, dx - \int_{\{f^-(x,v_n) \geq M\}} f^-(x, v_n) \, dx \right]$$

$$\geq \liminf_{n \to \infty} \int_E f_M(x, v_n(x)) \, dx - \varepsilon$$

$$\geq \int_E \int_{\mathbb{R}^m} f_M(x, z) \, d\nu_x(z) \, dx - \varepsilon$$

$$\geq \int_E \int_{\mathbb{R}^m} f(x, z) \, d\nu_x(z) \, dx - \varepsilon,$$

where in the last inequality we used the fact that $f_M \geq f$, and we may now let $\varepsilon \to 0^+$.

(ii) Assume that $\{f(\cdot, v_n(\cdot))\}$ is equi-integrable and fix $h \in L^\infty(E)$. Then the function $\pm h(x) f(x, z)$ is a normal integrand, $\{\pm h(\cdot) f(\cdot, v_n(\cdot))\}$ is equi-integrable, and so by (i) we get

$$\liminf_{n \to \infty} \left(\pm \int_E h(x) f(x, v_n(x)) \, dx \right) \geq \pm \int_E h(x) \overline{f}(x) \, dx,$$

that is,

$$\lim_{n \to \infty} \int_E h(x) f(x, v_n(x)) \, dx = \int_E h(x) \overline{f}(x) \, dx.$$

Conversely, if $f(\cdot, v_n(\cdot)) \rightharpoonup \overline{f}$ in $L^1(E)$, then $\{f(\cdot, v_n(\cdot))\}$ is equi-integrable by the Dunford–Pettis theorem.

(iii) Assume that dist $(v_n, K) \to 0$ in measure and let $\varphi \in C_0(\mathbb{R}^m \setminus K)$. Then for every $\varepsilon > 0$ there exists a compact set $K_1 \subset \mathbb{R}^m \setminus K$ such that

$$|\varphi(z)| < \varepsilon \text{ for all } z \notin K_1. \tag{8.8}$$

Thus we may find a constant $C_\varepsilon > 0$ such that

$$|\varphi(z)| \leq \varepsilon + C_\varepsilon \operatorname{dist}(z, K) \text{ for all } z \in \mathbb{R}^m \setminus K.$$

Indeed, if not, there exists a sequence $\{z_j\} \subset \mathbb{R}^m \setminus K$ such that

$$|\varphi(z_j)| \geq \varepsilon + j \operatorname{dist}(z_j, K). \tag{8.9}$$

By (8.8) it follows that $z_j \in K_1$ and up to the extraction of a subsequence if necessary, $z_j \to z \in K_1$. In particular, dist $(z, K) > 0$, and so letting $j \to \infty$ in (8.9) we reach a contradiction.

Hence $(|\varphi(v_n)| - \varepsilon)^+ \to 0$ in measure, and since φ is bounded we apply (ii) to conclude that

$$\langle \nu_x, (|\varphi(\cdot)| - \varepsilon)^+ \rangle_{\mathcal{M}, C_0} = 0 \text{ for } \mathcal{L}^N \text{ a.e. } x \in E.$$

Letting $\varepsilon \to 0^+$ we have that

$$\langle \nu_x^*, \varphi \rangle_{\mathcal{M}, C_0} = 0 \text{ for all } \varphi \in C_0 \left(\mathbb{R}^m \setminus K \right).$$

Hence $\operatorname{supp} \nu_x \subset K$ for $\mathcal{L}^N$ a.e. $x \in E$.

Conversely, assume that $\operatorname{supp} \nu_x \subset K$ for $\mathcal{L}^N$ a.e. $x \in E$. Let

$$f(x, z) := \chi_E (x) \min \left\{ \operatorname{dist} (z, K), 1 \right\}.$$

Then f is a bounded Carathéodory function and thus $\{ f(\cdot, v_n(\cdot)) \}$ is equi-integrable. By part (ii) we obtain that

$$\lim_{n \to \infty} \int_E \min \left\{ \operatorname{dist} (v_n(x), K), 1 \right\} dx$$

$$= \int_E \int_{\mathbb{R}^m} \min \left\{ \operatorname{dist} (z, K), 1 \right\} d\nu_x(z) \, dx = 0,$$

since $\operatorname{supp} \nu_x \subset K$ for $\mathcal{L}^N$ a.e. $x \in E$. By Vitali's convergence theorem we conclude that $\operatorname{dist} (v_n, K) \to 0$ in measure. $\qquad \blacksquare$

We now present several important applications of the fundamental theorem for Young measures.

Corollary 8.7. *Let $E \subset \mathbb{R}^N$ be a Lebesgue measurable set with finite measure and let $\{v_n\}$, $\{w_n\}$ be two sequences of measurable functions, $v_n, w_n : E \to \mathbb{R}^m$. If $\{v_n\}$ generates a Young measure ν and $\{w_n\}$ converges in measure to a measurable function $w : E \to \mathbb{R}^m$, then $\{v_n + w_n\}$ generates the translated Young measure $\Gamma_w (\nu)$, where $\langle \Gamma_a \nu, \varphi \rangle_{\mathcal{M}, C_0} := \langle \nu, \varphi (\cdot + a) \rangle_{\mathcal{M}, C_0}$ for $a \in \mathbb{R}^m$, $\varphi \in C_0 (\mathbb{R}^m)$. In particular, if $w_n \to 0$ in measure, then $\{v_n + w_n\}$ generates the Young measure ν.*

Proof. In view of Definition 8.3, we first prove that $\{v_n + w_n\}$ satisfies (8.6). Fix $\varepsilon > 0$ and, in view of (8.6) for $\{v_n\}$, choose $t > 0$ large enough such that

$$\limsup_{n \to \infty} \left| \left\{ x \in E : |v_n (x)| > \frac{t}{3} \right\} \right| < \frac{\varepsilon}{2}, \quad \left| \left\{ x \in E : |w (x)| > \frac{t}{3} \right\} \right| < \frac{\varepsilon}{2},$$

where for the latter inequality we used the fact that

$$\bigcap_{k=1}^{\infty} \{ x \in E : |w (x)| > k \} = \emptyset. \tag{8.10}$$

Since

$$\lim_{n \to \infty} \left| \left\{ x \in E : |w_n (x) - w (x)| > \frac{t}{3} \right\} \right| = 0,$$

we have

$$\limsup_{n \to \infty} |\{x \in E : |v_n(x) + w_n(x)| > t\}|$$

$$\leq \limsup_{n \to \infty} \left| \left\{ x \in E : |v_n(x)| > \frac{t}{3} \right\} \right| + \left| \left\{ x \in E : |w(x)| > \frac{t}{3} \right\} \right|$$

$$+ \lim_{n \to \infty} \left| \left\{ x \in E : |w_n(x) - w(x)| > \frac{t}{3} \right\} \right|$$

$$< \varepsilon.$$

Thus $\{v_n + w_n\}$ satisfies (8.6).

Now let $h \in L^1(E)$ and $\varphi \in C_0(\mathbb{R}^m)$. Fix $\varepsilon > 0$. Since φ is uniformly continuous there exists $\delta > 0$ such that

$$|\varphi(z) - \varphi(z')| \leq \varepsilon$$

for all z, $z' \in \mathbb{R}^m$ with $|z - z'| < \delta$. We have

$$\left| \int_E h(x) \varphi(v_n(x) + w_n(x))\, dx - \int_E h(x) \varphi(v_n(x) + w(x))\, dx \right|$$

$$\leq 2 \|\varphi\|_\infty \int_{E \cap \{|w_n - w| \geq \delta\}} |h(x)|\, dx + \varepsilon \int_{E \cap \{|w_n - w| < \delta\}} |h(x)|\, dx,$$

and thus

$$\limsup_{n \to \infty} \left| \int_E h(x) \varphi(v_n(x) + w_n(x))\, dx - \int_E h(x) \varphi(v_n(x) + w(x))\, dx \right|$$

$$\leq \varepsilon \int_E |h(x)|\, dx.$$

Given the arbitrariness of ε, we deduce that

$$\lim_{n \to \infty} \int_E h(x) \varphi(v_n(x) + w_n(x))\, dx = \lim_{n \to \infty} \int_E h(x) \varphi(v_n(x) + w(x))\, dx$$

$$= \int_E h(x) \langle \nu_x, \varphi(\cdot + w(x)) \rangle\, dx = \int_E h(x) \langle \Gamma_{w(x)} \nu_x, \varphi \rangle\, dx,$$

where we applied Theorem 8.6 to the Carathéodory function

$$(x, y) \mapsto h(x) \varphi(y + w(x)).$$

This concludes the proof.

The next result shows that if a sequence $\{v_n\}$ generates a Young measure ν, then there exists a sequence $\{w_n\}$ that still generates ν but has better integrability properties.

Corollary 8.8. *Let $E \subset \mathbb{R}^N$ be a Lebesgue measurable set with finite measure and let $\{v_n\}$ be a sequence of measurable functions $v_n : E \to \mathbb{R}^m$ that generates a Young measure ν and such that*

$$\limsup_{n\to\infty} \int_E g(|v_n(x)|)\, dx < \infty,$$

for some continuous nondecreasing function $g : [0,\infty) \to [0,\infty]$ with $g(t) \to \infty$ as $t \to \infty$. Then there exists a sequence $\{w_n\}$ of measurable functions $w_n : E \to \mathbb{R}^m$ that generates the same Young measure ν and such that $\{g(|w_n(\cdot)|)\}$ is equi-integrable. Moreover,

$$\int_E \int_{\mathbb{R}^m} g(|z|)\, d\nu_x(z)\, dx < \infty.$$

Proof. By Theorem 8.6(i) we have

$$\int_E \int_{\mathbb{R}^m} g(|z|)\, d\nu_x(z)\, dx < \infty.$$

By extracting a subsequence if necessary, we may assume that

$$\limsup_{n\to\infty} \int_E g(|v_n(x)|)\, dx = \lim_{n\to\infty} \int_E g(|v_n(x)|)\, dx.$$

By the decomposition lemma there exists a subsequence of $\{v_n\}$ (not relabeled) and an increasing sequence of numbers $r_n \to \infty$ such that $\{\tau_{r_n} \circ g \, (|v_n(\cdot)|)\}$ is equi-integrable and

$$|\{x \in E : g(|v_n(x)|) \neq \tau_{r_n} \circ g\,(|v_n(x)|)\}| \to 0, \tag{8.11}$$

where for $r > 0$ the truncation $\tau_r : \mathbb{R} \to \mathbb{R}$ is defined by

$$\tau_r(s) := \begin{cases} s & \text{if} \quad |s| \leq r, \\ \dfrac{s}{|s|}r & \text{if} \quad |s| > r. \end{cases}$$

Define

$$w_n(x) := \begin{cases} v_n(x) & \text{if } g\,(|v_n(x)|) \leq r_n, \\ z_0 & \text{if } g\,(|v_n(x)|) > r_n, \end{cases}$$

where $z_0 \in \mathbb{R}^m$ is such that $g(|z_0|) \in \mathbb{R}$. By (8.11),

$$|\{x \in E : v_n(x) \neq w_n(x)\}| \to 0,$$

and so by Corollary 8.7, $\{w_n\}$ generates the same Young measure ν. Observe that

$$g\,(|w_n(x)|) \leq \tau_{r_n} \circ g\,(|v_n(x)|) + g(|z_0|),$$

and equi-integrability follows.

The following corollary is particularly useful in applications (see, e.g., [FoTa89]).

Corollary 8.9. *Let $E \subset \mathbb{R}^N$ be a Lebesgue measurable set with finite measure and let $\{v_n\}$, $v_n : E \to \mathbb{R}^m$, be a sequence of measurable functions that generates a Young measure ν. Then $\{v_n\}$ converges in measure to a function $v : E \to \mathbb{R}^m$ if and only if $\nu_x = \delta_{v(x)}$ for $\mathcal{L}^N$ a.e. $x \in E$.*

Proof. Assume first that $\{v_n\}$ converges to v in measure. Let $w_n := v$ for all n. Then by Corollary 8.7, $\{v_n\}$ and $\{w_n\}$ generate the same Young measure. Since $\{w_n\}$ generates the Young measure δ_v, it follows that $\nu_x = \delta_{v(x)}$ for $\mathcal{L}^N$ a.e. $x \in E$.

Conversely, assume that $\nu_x = \delta_{v(x)}$ for $\mathcal{L}^N$ a.e. $x \in E$. Since $\nu \in L^\infty_w (E; \mathcal{M}(\mathbb{R}^m))$, the function v is measurable. Fix $\varepsilon, \eta > 0$. By Lusin's theorem choose $K \subset E$ compact such that $|E \setminus K| \leq \eta$ and $v : K \to \mathbb{R}^m$ is uniformly continuous. Find $\rho > 0$ such that

$$|v(x) - v(x')| \leq \frac{\varepsilon}{2} \tag{8.12}$$

for all $x, x' \in K$, with $|x - x'| \leq \rho$. Since K is compact there exist $x_1, \ldots, x_\ell \in K$, $\ell \in \mathbb{N}$, such that

$$K \subset \bigcup_{i=1}^{\ell} B(x_i, \rho).$$

Let $E_i := B^*(x_i, \rho) \cap K$, where

$$B^*(x_i, \rho) := B(x_i, \rho) \setminus \bigcup_{j=1}^{i-i} B(x_j, \rho),$$

and define

$$w(x) := \sum_{i=1}^{\ell} v(x_i) \chi_{E_i}(x).$$

Note that by (8.12),

$$\left| \left\{ x \in E : |v(x) - w(x)| > \frac{\varepsilon}{2} \right\} \right| \leq \eta. \tag{8.13}$$

Let

$$f(x, z) := \sum_{i=1}^{\ell} \chi_{E_i}(x) \varphi(|z - v(x_i)|),$$

where $\varphi : [0, \infty) \to [0, 1]$ is a continuous function such that $\varphi = 0$ on $\left[0, \frac{\varepsilon}{2}\right]$ and $\varphi = 1$ on $[\varepsilon, \infty)$. Then f is a bounded Carathéodory function, and thus $\{f(\cdot, v_n(\cdot))\}$ is equi-integrable. Then

$$|\{x \in E : |v_n - w| > \varepsilon\}| \le \sum_{i=1}^{\ell} |\{x \in E_i : |v_n - v(x_i)| > \varepsilon\}| + \eta$$

$$\le \sum_{i=1}^{\ell} \int_{E_i} \varphi(|v_n(x) - v(x_i)|)\, dx + \eta$$

$$= \int_E f(x, v_n(x))\, dx + \eta,$$

and so by Theorem 8.6(ii) we obtain that

$$\limsup_{n \to \infty} |\{x \in E : |v_n - w| > \varepsilon\}| \le \lim_{n \to \infty} \int_E f(x, v_n(x))\, dx + \eta$$

$$= \int_E \int_{\mathbb{R}^m} f(x, z)\, d\delta_{v(x)}(z)\, dx + \eta = \int_E f(x, v(x))\, dx + \eta$$

$$= \sum_{i=1}^{\ell} \int_{E_i} \varphi(|v(x) - v(x_i)|)\, dx + \eta$$

$$\le \left|\left\{x \in K : |v - w| > \frac{\varepsilon}{2}\right\}\right| + \eta \le 2\eta,$$

where we have used (8.13) and the fact that $\varphi \le 1$. Hence

$$\limsup_{n \to \infty} |\{x \in E : |v_n - v| > 2\varepsilon\}|$$

$$\le \limsup_{n \to \infty} |\{x \in E : |v_n - w| > \varepsilon\}| + |\{x \in E : |v - w| > \varepsilon\}|$$

$$\le 2\eta.$$

It now suffices to let $\eta \to 0^+$.

The next result deals with pairs of sequences.

Corollary 8.10. *Let $E \subset \mathbb{R}^N$ be a Lebesgue measurable set with finite measure and let $\{u_n\}$, $\{v_n\}$ be two sequences of measurable functions, with $u_n : E \to \mathbb{R}^d$ and $v_n : E \to \mathbb{R}^m$. If $\{v_n\}$ generates the Young measure $\nu \in L_w^\infty(E; \mathcal{M}(\mathbb{R}^m; \mathbb{R}))$ and u_n converges pointwise $\mathcal{L}^N$ a.e. in E to a function $u : E \to \mathbb{R}^d$, then the sequence $\{(u_n, v_n)\}$ generates the Young measure*

$$x \mapsto \delta_{u(x)} \otimes \nu_x.$$

Proof. In view of Definition 8.3, we first prove that $\{(u_n, v_n)\}$ satisfies (8.6). For $t > 0$,

$$\left|\{x \in E : |(u_n, v_n)(x)| > t\}\right|$$

$$\leq \left|\left\{x \in E : |u_n(x)| > \frac{t}{\sqrt{2}}\right\}\right| + \left|\left\{x \in E : |v_n(x)| > \frac{t}{\sqrt{2}}\right\}\right|$$

$$\leq \left|\left\{x \in E : |u_n(x) - u(x)| > \frac{t}{2\sqrt{2}}\right\}\right| + \left|\left\{x \in E : |u(x)| > \frac{t}{2\sqrt{2}}\right\}\right|$$

$$+ \left|\left\{x \in E : |v_n(x)| > \frac{t}{\sqrt{2}}\right\}\right|.$$

Due to property (8.6) for $\{v_n\}$, using an argument analogous to that of (8.10), and since u_n converges pointwise, and hence in measure, to u we deduce that (8.6) is satisfied by $\{(u_n, v_n)\}$.

Let $h \in L^1(E)$, $\psi \in C_0(\mathbb{R}^d)$, and $\varphi \in C_0(\mathbb{R}^m)$. Then

$$|h(x)\psi(u_n)| \leq \|\psi\|_\infty |h(x)|,$$

and so by the Lebesgue dominated convergence theorem, $h\psi(u_n) \to h\psi(u)$ strongly in $L^1(E)$. Since $\{v_n\}$ generates the Young measure ν, we have that $\varphi(v_n) \overset{*}{\rightharpoonup} \overline{\varphi}$ in $L^\infty(E)$, and so

$$\lim_{n \to \infty} \int_E h(x)\psi(u_n)\varphi(v_n)\, dx = \int_E h(x)\psi(u)\left(\int_{\mathbb{R}^m} \varphi(z)\, d\nu_x(z)\right) dx,$$

that is,

$$(\psi \otimes \varphi)(u_n, v_n) \overset{*}{\rightharpoonup} \langle \delta_{u(\cdot)} \otimes \nu_\cdot, \psi \otimes \varphi \rangle_{\mathcal{M}, C_0} \text{ in } L^\infty(E).$$

Since tensors of the form $\psi \otimes \varphi$ are dense in $C_0(\mathbb{R}^d \times \mathbb{R}^m)$, the proof is complete.

Next we study the relation between biting convergence and Young measures.

Theorem 8.11. *Let $E \subset \mathbb{R}^N$ be a Lebesgue measurable set with finite measure and let $\{v_n\}$ be a sequence of measurable functions $v_n : E \to \mathbb{R}^m$ that generates a Young measure ν. Then for any Carathéodory function $f : E \times \mathbb{R}^m \to [-\infty, \infty]$ such that $\{f(\cdot, v_n(\cdot))\}$ is bounded in $L^1(E)$, there exists a subsequence $\{v_{n_k}\}$ of $\{v_n\}$ such that*

$$f(\cdot, v_{n_k}(\cdot)) \overset{b}{\rightharpoonup} \overline{f} \text{ in } L^1(E).$$

Proof. By the biting lemma there exist a function $g \in L^1(E)$, a subsequence $\{v_{n_k}\}$ of $\{v_n\}$, and a decreasing sequence of Lebesgue measurable sets $\{E_j\} \subset E$, with $|E_j| \to 0$, such that

$$f(\cdot, v_{n_k}(\cdot)) \rightharpoonup g \text{ in } L^1(E \setminus E_j) \text{ for every } j \in \mathbb{N}.$$

By Theorem 8.6(ii) we have that $g = \overline{f}$ for $\mathcal{L}^N$ a.e. $x \in E \setminus E_j$ for every $j \in \mathbb{N}$. Since $|E_j| \to 0$ it follows that $g = \overline{f}$ for $\mathcal{L}^N$ a.e. $x \in E$.

Remark 8.12. As a consequence of the previous theorem, if $\{v_n\}$ is bounded in $L^1(E;\mathbb{R}^m)$, then by taking as f a projection, that is, $f(z) := z_i$, $i = 1, \ldots, m$, where $z = (z_1, \ldots, z_m) \in \mathbb{R}^m$, we obtain that there exists a subsequence $\{v_{n_k}\}$ of $\{v_n\}$ such that

$$v_{n_k}(\cdot)) \overset{b}{\rightharpoonup} v \text{ in } L^1(E),$$

where for $\mathcal{L}^N$ a.e. $x \in E$,

$$v(x) = \int_{\mathbb{R}^m} z \, d\nu_x(z).$$

Using the techniques developed in this chapter we next give a proof of the decomposition lemma stated and proved in Chapter 2, where here the domain of integration is a Lebesgue measurable set of $\mathbb{R}^N$ with finite measure.

Lemma 8.13 (Decomposition lemma, II). *Let $E \subset \mathbb{R}^N$ be a Lebesgue measurable set with finite measure and let $\{u_n\}$ be a sequence of functions uniformly bounded in $L^p(E)$, $1 \le p < \infty$. For $r > 0$ consider the truncation $\tau_r : \mathbb{R} \to \mathbb{R}$ defined by*

$$\tau_r(z) := \begin{cases} z & \text{if } |z| \le r, \\ \dfrac{z}{|z|}r & \text{if } |z| > r. \end{cases}$$

Then there exists a subsequence of $\{u_n\}$ (not relabeled) and an increasing sequence of numbers $r_n \to \infty$ such that the truncated sequence $\{\tau_{r_n} \circ u_n\}$ is p-equi-integrable, and

$$|\{x \in E : u_n(x) \ne (\tau_{r_n} \circ u_n)(x)\}| \to 0. \tag{8.14}$$

Proof. Without loss of generality, by Theorem 8.2 we may assume that the sequence $\{u_n\}$ generates a Young measure $\nu \in L_w^\infty(E; \mathcal{M}(\mathbb{R};\mathbb{R}))$ and by Theorem 8.6 (i),

$$\int_E \int_{\mathbb{R}} |z|^p \, d\nu_x(z) \, dx < \infty. \tag{8.15}$$

Using Theorem 8.6(ii) we obtain

$$\lim_{r \to \infty} \lim_{n \to \infty} \int_E |\tau_r \circ u_n|^p \, dx = \lim_{r \to \infty} \int_E \int_{\mathbb{R}} |\tau_r(z)|^p \, d\nu_x(z) \, dx$$

$$= \int_E \int_{\mathbb{R}} |z|^p \, d\nu_x(z) \, dx,$$

where the last equality has been obtained via the Lebesgue monotone convergence theorem. Therefore we may find an increasing sequence of numbers $r_n \to \infty$ such that

$$\lim_{n \to \infty} \int_E |\tau_{r_n} \circ u_n|^p \, dx = \int_E \int_{\mathbb{R}} |z|^p \, d\nu_x(z) \, dx. \tag{8.16}$$

Property (8.14) follows directly from the boundedness of the sequence $\{u_n\}$ in L^p. By Corollary 8.7 the sequence $\{\tau_{r_n} \circ u_n\}$ generates the same Young measure $\nu \in L^\infty_w (E; \mathcal{M}(\mathbb{R}; \mathbb{R}))$, and in view of (8.15), (8.16), and Theorem 8.6(ii) the sequence $\{\tau_{r_n} \circ u_n\}$ is p-equi-integrable.

8.2 Characterization of Young Measures

In this section we prove that every Young measure is generated by a sequence of measurable functions.

Theorem 8.14. *Let $E \subset \mathbb{R}^N$ be a Lebesgue measurable set with finite measure and let $\nu \in L^\infty_w (E; \mathcal{M}(\mathbb{R}^m; \mathbb{R}))$ be a Young measure. Then there exists a continuous, nondecreasing function $g : [0, \infty) \to [0, \infty]$ such that $g(t) \to \infty$ as $t \to \infty$ and*

$$\int_E \int_{\mathbb{R}^m} g(|z|) \, d\nu_x(z) \, dx < \infty. \tag{8.17}$$

Proof. We claim that

$$\lim_{t \to \infty} \int_E \int_{\{z \in \mathbb{R}^m : |z| > t\}} d\nu_x(z) \, dx = 0. \tag{8.18}$$

Indeed, since $\nu_x \in \Pr(\mathbb{R}^m)$ for $\mathcal{L}^N$ a.e. $x \in E$, by Proposition 1.7(ii) we have that

$$\lim_{t \to \infty} \int_{\{z \in \mathbb{R}^m : |z| > t\}} d\nu_x(z) = 0,$$

and since

$$0 \leq \int_{\{z \in \mathbb{R}^m : |z| > t\}} d\nu_x(z) \, dx \leq \int_{\mathbb{R}^m} d\nu_x(z) = 1,$$

(8.18) follows by the Lebesgue dominated convergence theorem. We can now proceed as in the proof of Proposition 8.4 to show that (8.18) implies (8.17). Precisely, choose $0 < t_j < t_{j+1}$, $j \in \mathbb{N}$, with $t_j \to \infty$ as $j \to \infty$, such that

$$\int_E \int_{\{z \in \mathbb{R}^m : |z| > t_j\}} d\nu_x(z) \, dx \leq \frac{1}{j^3},$$

and define

$$h(t) := \begin{cases} 0 \text{ if } t \in [0, t_1), \\ j \text{ if } t \in [t_j, t_{j+1}), \, j \in \mathbb{N}. \end{cases}$$

Then

$$\int_E \int_{\{z\in\mathbb{R}^m:\, |z|\leq t_k\}} h(|z|)\, d\nu_x(z)\, dx$$

$$= \int_E \sum_{j=1}^{k-1} \int_{\{z\in\mathbb{R}^m:\, t_j\leq |z|<t_{j+1}\}} h(|z|)\, d\nu_x(z)\, dx$$

$$= \sum_{j=1}^{k-1} j \int_E \int_{\{z\in\mathbb{R}^m:\, t_j\leq |z|<t_{j+1}\}} d\nu_x(z)\, dx$$

$$\leq \sum_{j=1}^{k} \frac{1}{j^2} \leq \sum_{j=1}^{\infty} \frac{1}{j^2} < \infty.$$

By the Lebesgue monotone convergence theorem, letting $k \to \infty$ we obtain

$$\int_E \int_{\mathbb{R}^m} h(|z|)\, d\nu_x(z)\, dx \leq \sum_{j=1}^{\infty} \frac{1}{j^2} < \infty.$$

It now suffices to replace the function h with a continuous, nondecreasing function $0 \leq g \leq h$.

Next we show that the condition (8.17) is equivalent to the existence of a sequence $\{v_n\}$ of measurable functions $v_n : E \to \mathbb{R}^m$ that generates ν and such that $\{g(|v_n(\cdot)|)\}$ is equi-integrable. We begin with the case of a homogeneous Young measure.

Definition 8.15. *Let $E \subset \mathbb{R}^N$ be a Lebesgue measurable set with finite measure. A Young measure $\nu \in L_w^\infty (E; \mathcal{M}(\mathbb{R}^m; \mathbb{R}))$ is said to be* homogeneous *if there exists a probability measure $\mu \in \mathrm{Pr}(\mathbb{R}^m)$ such that $\nu_x = \mu$ for $\mathcal{L}^N$ a.e. $x \in E$.*

Exercise 8.16. *Let $a,\ b \in \mathbb{R}^m$ and $\theta \in (0,1)$, and consider the function*

$$v(x) := \begin{cases} a \text{ if } x \in (0,\theta), \\ b \text{ if } x \in (\theta,1). \end{cases}$$

Extend v periodically to all of $\mathbb{R}$ with period 1 and let $v_n(x) := v(nx)$. Prove that the sequence $\{v_n\}$ generates the homogeneous Young measure

$$\nu_x = \theta \delta_a + (1-\theta)\, \delta_b.$$

8.2.1 The Homogeneous Case

Theorem 8.17. *Let $E \subset \mathbb{R}^N$ be a Lebesgue measurable set with finite measure, let $\nu \in \mathcal{P}(\mathbb{R}^m)$, and let $g : [0,\infty) \to [0,\infty]$ be a continuous, nondecreasing function such that $g(t) \to \infty$ as $t \to \infty$. Then the following two conditions are equivalent:*

(i) $\displaystyle \int_{\mathbb{R}^m} g(|z|)\, d\nu(z) < \infty;$

(ii) there exists a sequence $\{v_n\}$ of measurable functions $v_n : E \to \mathbb{R}^m$ that generates ν and such that $\{g(|v_n(\cdot)|)\}$ is equi-integrable.

Proof. That (ii) implies (i) follows from Theorem 8.6(ii). In the remainder of the proof we show that (i) implies (ii).

Step 1: Fix $R > 0$ and define $D_R := \overline{B_m(0, R)}$ and

$$\mathbb{H}_R := \Big\{\mu \in \mathrm{Pr}\,(\mathbb{R}^m) : \operatorname{supp}\mu \subset D_R, \ \mu \text{ is generated by a sequence } \{v_n\}$$
$$\text{with } \|v_n\|_{L^\infty(E;\mathbb{R}^m)} \le R\Big\}.$$

We claim that $\mathbb{H}_R$ is convex. Indeed, fix μ_1, $\mu_2 \in \mathbb{H}_R$ and $\theta \in (0,1)$. Let $\{v_n\}$, $\{w_n\}$ be two sequences of measurable functions v_n, $w_n : E \to D_R$ that generate μ_1 and μ_2, respectively. By Proposition 2.87 there exists a sequence $\{E_k\}$ of Lebesgue measurable sets, with $E_k \subset E$, such that $\chi_{E_k} \overset{*}{\rightharpoonup} \theta$ in $L^\infty(E)$. For n, $k \in \mathbb{N}$ define

$$w_{n,k}(x) := \begin{cases} v_n(x) & \text{if } x \in E_k, \\ w_n(x) & \text{if } x \in E \setminus E_k. \end{cases}$$

Let $h \in L^1(E)$ and $\varphi \in C_0(\mathbb{R}^m)$. Then

$$\int_E h(x)\, \varphi(w_{n,k}(x))\, dx$$
$$= \int_E \chi_{E_k}(x)\, h(x)\, \varphi(v_n(x))\, dx + \int_E (1 - \chi_{E_k}(x))\, h(x)\, \varphi(w_n(x))\, dx$$
$$\to \theta \int_E h(x)\, \varphi(v_n(x))\, dx + (1 - \theta) \int_E h(x)\, \varphi(w_n(x))\, dx$$

as $k \to \infty$, and so

$$\lim_{n\to\infty} \lim_{k\to\infty} \int_E h(x)\, \varphi(w_{n,k}(x))\, dx$$
$$= \int_E h(x)\, dx \, (\theta \langle \mu_1, \varphi \rangle_{\mathcal{M},C_0} + (1 - \theta) \langle \mu_2, \varphi \rangle_{\mathcal{M},C_0}),$$

where we have used Theorem 8.6(ii). Since $L^1(E)$ and $C_0(\mathbb{R}^m)$ are separable, we may extract a diagonal sequence $\{w_{n,k_n}\}$ for which the previous limit exists for each h and φ in a dense set of $L^1(E)$ and $C_0(\mathbb{R}^m)$, respectively (and hence by density for all $h \in L^1(E)$ and $\varphi \in C_0(\mathbb{R}^m)$). Since the sequence $\{w_{n,k_n}\}$ generates the Young measure $\theta\mu_1 + (1 - \theta)\mu_2$, we have that $\theta\mu_1 + (1 - \theta)\mu_2 \in \mathbb{H}_R$ and the claim is proved.

Step 2: We claim that $\mathbb{H}_R$ is relatively closed in $\mathrm{Pr}\,(\mathbb{R}^m)$ with respect to the weak star topology $\sigma(\mathcal{M}(\mathbb{R}^m; \mathbb{R}), C_0(\mathbb{R}^m))$ ($\sigma(\mathcal{M}, C_0)$ for simplicity), i.e.,

$$\overline{\mathbb{H}_R}^{\sigma(\mathcal{M}, C_0)} \cap \Pr(\mathbb{R}^m) = \mathbb{H}_R.$$

Indeed, let $\mu \in \overline{\mathbb{H}_R}^{\sigma(\mathcal{M}, C_0)} \cap \Pr(\mathbb{R}^m)$ and let $\{h_j\}$ and $\{\varphi_j\}$ be dense in $L^1(E)$ and $C_0(\mathbb{R}^m)$, respectively. By the definition of weak star topology in $\mathcal{M}(\mathbb{R}^m; \mathbb{R})$, for every $n \in \mathbb{N}$ there exists $\mu_n \in \mathbb{H}_R$ such that

$$\left| \langle \mu - \mu_n, \varphi_j \rangle_{\mathcal{M}, C_0} \right| < \frac{1}{2n \max\limits_{0 \le i \le n} \left(1 + \|h_i\|_{L^1(E)} \right)}, \quad j = 1, \dots, n. \quad (8.19)$$

By the definition of $\mathbb{H}_R$, the homogeneous Young measure μ_n is generated by a sequence $\left\{ v_n^{(k)} \right\}_{k \in \mathbb{N}}$ with

$$\sup_{k \in \mathbb{N}} \left\| v_n^{(k)} \right\|_{L^\infty(E; \mathbb{R}^m)} \le R. \quad (8.20)$$

Choose $k = k(n) \in \mathbb{N}$ sufficiently large that with $v_n := v_n^{(k(n))}$ we have

$$\left| \langle \mu_n, \varphi_j \rangle_{\mathcal{M}, C_0} \int_E h_i \, dx - \int_E h_i \varphi_j(v_n) \, dx \right| < \frac{1}{2n}, \quad i, j = 1, \dots, n,$$

so that by (8.19),

$$\left| \langle \mu, \varphi_j \rangle_{\mathcal{M}, C_0} \int_E h_i \, dx - \int_E h_i \varphi_j(v_n) \, dx \right| < \frac{1}{n}, \quad i, j = 1, \dots, n. \quad (8.21)$$

By (8.20), Proposition 8.4, and Theorem 8.6(iii), a subsequence $\{v_{n_k}\}$ of $\{v_n\}$ generates a Young measure ν with $\operatorname{supp} \mu \subset D_R$. From (8.21) and the density properties of $\{h_j\}$ and $\{\varphi_j\}$ it follows that $\mu = \nu$. Hence $\mu \in \mathbb{H}_R$.

Step 3: Let
$$\mathbb{M}_R := \{\mu \in \Pr(\mathbb{R}^m) : \operatorname{supp} \mu \subset D_R\}.$$

Note that $\mathbb{H}_R \subset \mathbb{M}_R$. We claim that $\mathbb{H}_R \supset \mathbb{M}_R$. Assume by contradiction that there exists $\mu \in \mathbb{M}_R$ such that $\mu \notin \mathbb{H}_R$. By Steps 1 and 2,

$$\mu \notin \overline{\mathbb{H}_R}^{\sigma(\mathcal{M}, C_0)} = \overline{\operatorname{co} \mathbb{H}_R}^{\sigma(\mathcal{M}, C_0)},$$

and therefore by the Hahn–Banach theorem there exist a linear functional $L : \mathcal{M}(\mathbb{R}^m; \mathbb{R}) \to \mathbb{R}$, continuous with respect to the weak star topology $\sigma(\mathcal{M}, C_0)$, and $\alpha \in \mathbb{R}$ such that

$$L(\nu) \ge \alpha \text{ for all } \nu \in \mathbb{H}_R, \quad \alpha > L(\mu).$$

By Theorem A.72 there exists $\varphi \in C_0(\mathbb{R}^m)$ such that

$$L(\nu) = \langle \nu, \varphi \rangle_{\mathcal{M}, C_0} = \int_{\mathbb{R}^m} \varphi \, d\nu$$

for all $\nu \in \mathcal{M}(\mathbb{R}^m; \mathbb{R})$, and so

$$\langle \nu, \varphi \rangle_{\mathcal{M}, C_0} \geq \alpha \text{ for all } \nu \in \mathbb{H}_R, \quad \alpha > \langle \mu, \varphi \rangle_{\mathcal{M}, C_0}. \qquad (8.22)$$

Define

$$\varphi_R(z) := \begin{cases} \varphi(z) & \text{if } |z| \leq R, \\ \infty & \text{otherwise.} \end{cases}$$

Since $\operatorname{supp} \mu \subset D_R$ we have that

$$\alpha > \langle \mu, \varphi \rangle_{\mathcal{M}, C_0} = \int_{\mathbb{R}^m} \varphi_R(z)\, d\mu(z) \geq \int_{\mathbb{R}^m} \varphi_R^{**}(z)\, d\mu(z) \geq \varphi_R^{**}(a), \qquad (8.23)$$

by Jensen's inequality and where $a := \int_{\mathbb{R}^m} z\, d\mu(z)$. Note that $|a| \leq R$. Since φ_R is lower semicontinuous, it follows by Theorem 4.98 that

$$\varphi_R^{**}(a) = \inf \left\{ \sum_{i=1}^{m+1} \theta_i \varphi_R(z_i) : \sum_{i=1}^{m+1} \theta_i = 1,\ \sum_{i=1}^{m+1} \theta_i z_i = a,\ \theta_i \geq 0,\ z_i \in \mathbb{R}^m \right\}$$

$$= \inf \left\{ \sum_{i=1}^{m+1} \theta_i \varphi(z_i) : \sum_{i=1}^{m+1} \theta_i = 1,\ \sum_{i=1}^{m+1} \theta_i z_i = a,\ \theta_i \geq 0,\ z_i \in D_R \right\}.$$

Fix $\varepsilon > 0$ and write

$$\varphi_R^{**}(a) + \varepsilon \geq \sum_{i=1}^{m+1} \theta_i \varphi(z_i),$$

where $\theta_i \geq 0, z_i \in D_R, i = 1, \ldots, m+1$,

$$\sum_{i=1}^{m+1} \theta_i = 1, \quad \sum_{i=1}^{m+1} \theta_i z_i = a.$$

Observe that since $\delta_{z_i} \in \mathbb{H}_R$ for all $i = 1, \ldots, m+1$, and $\mathbb{H}_R$ is convex (see Step 1), we have

$$\nu := \sum_{i=1}^{m+1} \theta_i \delta_{z_i} \in \mathbb{H}_R.$$

Therefore by (8.22) and (8.23),

$$\varphi_R^{**}(a) + \varepsilon \geq \sum_{i=1}^{m+1} \theta_i \varphi(z_i) = \langle \nu, \varphi \rangle_{\mathcal{M}, C_0} \geq \alpha > \varphi_R^{**}(a),$$

and letting $\varepsilon \to 0^+$ we obtain a contradiction.

Step 4: We are now ready to prove the general case. Since the set E has finite measure, by rescaling, we may assume that $|E| = 1$. Let $\nu \in \operatorname{Pr}(\mathbb{R}^m)$ be such that (i) holds. Since $\nu \in \operatorname{Pr}(\mathbb{R}^m)$, by Proposition 1.7(ii) we have that

$$\lim_{t \to \infty} \nu(\{z \in \mathbb{R}^m : |z| > t\}) = 0.$$

Hence for all $k \in \mathbb{N}$ sufficiently large, say $k \geq k_0$, we have that

$$\nu\left(\{z \in \mathbb{R}^m : |z| \leq k\}\right) > 0.$$

For $k \geq k_0$ consider the probability measure ν_k, defined by

$$\nu_k(F) := c_k \nu(F \cap D_k) \quad F \in \mathcal{B}(\mathbb{R}^m),$$

where

$$c_k := \frac{1}{\nu(D_k)}, \quad D_k := \overline{B_m(0, k)}.$$

Note that $\nu_k \in \mathbb{M}_k$, where

$$\mathbb{M}_k := \{\mu \in \text{Pr}(\mathbb{R}^m) : \text{supp}\,\mu \subset D_k\}.$$

By Step 3, for each fixed $k \geq k_0$ there exists a sequence $\{v_{n,k}\}$ of measurable functions $v_{n,k} : E \to D_k$ that generates ν_k. Let $\{h_j\}$ and $\{\varphi_j\}$ be dense in $L^1(E)$ and $C_0(\mathbb{R}^m)$, respectively. Since $\{v_{n,k}\}$ generates ν_k and $v_{n,k} : E \to D_k$, by Theorem 8.6(ii) for all $i, j \in \mathbb{N}$, we have

$$\lim_{n \to \infty} \int_E h_i(x)\, \varphi_j(v_{n,k}(x))\, dx = \int_E h_i(x)\, dx \int_{\mathbb{R}^m} \varphi_j(z)\, d\nu_k(z)$$

$$= c_k \int_E h_i(x)\, dx \int_{D_k} \varphi_j(z)\, d\nu(z),$$

and

$$\lim_{n \to \infty} \int_E g\left(|v_{n,k}(x)|\right)\, dx = |E| \int_{\mathbb{R}^m} g(|z|)\, d\nu_k(z)$$

$$= c_k \int_{D_k} g(|z|)\, d\nu(z),$$

where we have used the facts that $|E| = 1$ and that for each fixed $k \in \mathbb{N}$ the sequences $\{\varphi_j(v_{n,k}(\cdot))\}$ and $\{g(|v_{n,k}(\cdot)|)\}$ are equi-integrable, since $\|v_{n,k}\|_{L^\infty(E;\mathbb{R}^m)} \leq k$. By virtue of the Lebesgue dominated convergence theorem, we obtain that

$$\lim_{k \to \infty} \lim_{n \to \infty} \int_E h_i(x)\, \varphi_j(v_{n,k}(x))\, dx = \int_E h_i(x)\, dx \int_{\mathbb{R}^m} \varphi_j(z)\, d\nu(z)$$

and

$$\lim_{k \to \infty} \lim_{n \to \infty} \int_E g\left(|v_{n,k}(x)|\right)\, dx = \int_{\mathbb{R}^m} g(|z|)\, d\nu(z).$$

Extract a diagonal sequence $\{v_\ell\}$ of $\{v_{n,k}\}$ such that

$$\lim_{\ell \to \infty} \int_E h_i(x)\, \varphi_j(v_\ell(x))\, dx = \int_E h_i(x)\, dx \int_{\mathbb{R}^m} \varphi_j(z)\, d\nu(z),$$

$$\lim_{\ell \to \infty} \int_E g\left(|v_\ell(x)|\right)\, dx = \int_{\mathbb{R}^m} g(|z|)\, d\nu(z),$$

for all i, $j \in \mathbb{N}$. By the density properties of $\{h_j\}$ and $\{\varphi_j\}$ and Theorem 8.6(ii), it follows that $\{v_\ell\}$ generates ν and $\{g(|v_\ell(\cdot)|)\}$ is equi-integrable in $L^1(E)$.

Exercise 8.18. Prove that if the measurable set $E \subset \mathbb{R}^N$ is bounded, then in Step 1 it is possible to use the Riemann–Lebesgue lemma and thus avoid Proposition 2.87.

8.2.2 The Inhomogeneous Case

We are now ready to consider the inhomogeneous case.

Theorem 8.19. *Let $E \subset \mathbb{R}^N$ be a Lebesgue measurable set with finite measure, let $\nu \in L_w^\infty(E; \mathcal{M}(\mathbb{R}^m; \mathbb{R}))$ be a Young measure, and let $g : [0, \infty) \to [0, \infty]$ be a continuous, nondecreasing function such that $g(t) \to \infty$ as $t \to \infty$. Then the following two conditions are equivalent:*

(i) $\int_E \int_{\mathbb{R}^m} g(|z|)\, d\nu_x(z)\, dx < \infty$;
(ii) there exists a sequence $\{v_n\}$ of measurable functions $v_n : E \to \mathbb{R}^m$ that generates ν and such that $\{g(|v_n(\cdot)|)\}$ is equi-integrable in $L^1(E)$.

Proof. That (ii) implies (i) follows from Theorem 8.6(ii).

Step 1: To prove that (i) implies (ii) we first claim that, without loss of generality, we may consider the case that the set E is an open set with finite measure. Indeed, condition (i) implies that

$$\int_{\mathbb{R}^m} g(|z|)\, d\nu_x(z) < \infty$$

for $\mathcal{L}^N$ a.e. $x \in E$. Select any $x_0 \in E$ for which the previous integral is finite and $\nu_{x_0} \in \Pr(\mathbb{R}^m)$. Fix an open set $\Omega \subset \mathbb{R}^N$ of finite measure that contains E and define

$$\mu_x := \begin{cases} \nu_x & \text{if } x \in E, \\ \nu_{x_0} & \text{if } x \in \Omega \setminus E. \end{cases}$$

Then $\mu \in L_w^\infty(\Omega; \mathcal{M}(\mathbb{R}^m; \mathbb{R}))$ is a Young measure and (i) is satisfied. Thus if we show that there exists a sequence $\{v_n\}$ of measurable functions $v_n : \Omega \to \mathbb{R}^m$ that generates μ and such that $\{g(|v_n(\cdot)|)\}$ is equi-integrable in $L^1(\Omega)$, then the sequence $\{v_n\}$ restricted to E generates ν. Also, without loss of generality we may assume that

$$\int_{\mathbb{R}^m} g(|z|)\, d\nu_x(z) < \infty \tag{8.24}$$

for *all* $x \in E$.

Step 2: In view of the previous step we may assume that E is an open set with finite measure. Let $\{h_j\}_{j \in \mathbb{N}} \subset C_c(E)$ and $\{\varphi_j\}_{j \in \mathbb{N}}$ be dense in $L^1(E)$

and $C_0(\mathbb{R}^m)$, respectively. Set $h_0 = 1$ and $\varphi_0 = g$. For $j \in \mathbb{N}_0$ and $x \in E$ define

$$\psi_j(x) := \int_{\mathbb{R}^m} \varphi_j(z)\, d\nu_x(z) \qquad (8.25)$$

and let

$$F := \bigcap_{j \in \mathbb{N}_0} \{x \in E : x \text{ Lebesgue point for } \psi_j\},$$

where we used the fact that in view of (8.24), $\psi_j \in L^1(E)$ for all $j \in \mathbb{N}_0$. Then $|E \setminus F| = 0$. For each $k \in \mathbb{N}$ set

$$\mathcal{G}_k := \left\{ Q(x,\varepsilon) \subset E : x \in F, \, 0 < \varepsilon \le \frac{1}{k}, \right.$$

$$\left. \frac{1}{\varepsilon^N} \int_{Q(x,\varepsilon)} |\psi_j(y) - \psi_j(x)|\, dy < \frac{1}{k}, \, j = 0, \ldots, k \right\}.$$

By Vitali's covering theorem and Lebesgue's differentiation theorem we may write

$$E = \bigcup_\ell Q(x_{k\ell}, \varepsilon_{k\ell}) \cup N_k, \quad |N_k| = 0,$$

for some mutually disjoint cubes $Q(x_{k\ell}, \varepsilon_{k\ell}) \in \mathcal{G}_k$. For each $i, j \in \mathbb{N}_0$ let $k > j$. Then

$$\left| \int_E h_i(x)\psi_j(x)\, dx - \sum_\ell \psi_j(x_{k\ell}) \int_{Q(x_{k\ell}, \varepsilon_{k\ell})} h_i(x)\, dx \right|$$

$$\le \sum_\ell \int_{Q(x_{k\ell}, \varepsilon_{k\ell})} |h_i(x)(\psi_j(x) - \psi_j(x_{k\ell}))|\, dx$$

$$\le \|h_i\|_\infty \sum_\ell \int_{Q(x_{k\ell}, \varepsilon_{k\ell})} |\psi_j(x) - \psi_j(x_{k\ell})|\, dx$$

$$\le \|h_i\|_\infty \frac{1}{k} \sum_\ell |Q(x_{k\ell}, \varepsilon_{k\ell})| = \|h_i\|_\infty \frac{1}{k} |E|.$$

Hence

$$\int_E h_i(x)\, \psi_j(x)\, dx = \lim_{k \to \infty} \sum_\ell \psi_j(x_{k\ell}) \int_{Q(x_{k\ell}, \varepsilon_{k\ell})} h_i(x)\, dx \qquad (8.26)$$

for every $i, j \in \mathbb{N}_0$.

By the previous theorem and (8.24), for each fixed $k, \ell \in \mathbb{N}$ there exists a sequence $\{v_{nk\ell}\}$ of measurable functions $v_{nk\ell} : Q(x_{k\ell}, \varepsilon_{k\ell}) \to \mathbb{R}^m$ that generates $\nu_{x_{k\ell}}$ and such that $\{g(|v_{nk\ell}(\cdot)|)\}$ is equi-integrable in $Q(x_{k\ell}, \varepsilon_{k\ell})$.

Fix $k, \ell \in \mathbb{N}$. Since $\{v_{nk\ell}\}_n$ generates $\nu_{x_{k\ell}}$, we may choose $n_{k\ell}$ such that

$$\left| \int_{Q(x_{k\ell}, \varepsilon_{k\ell})} h_i(x)\, \varphi_j(v_{n_{k\ell}k\ell}(x))\, dx - \psi_j(x_{k\ell}) \int_{Q(x_{k\ell}, \varepsilon_{k\ell})} h_i(x)\, dx \right| \le \frac{1}{2^\ell k}$$

$$(8.27)$$

for all i, $j = 0, \dots, k$. Set

$$v_k(x) := \begin{cases} v_{n_{k\ell}k\ell}(x) & \text{if } x \in Q(x_{k\ell}, \varepsilon_{k\ell}) \text{ for some } \ell, \\ 0 & \text{if } x \in N_k. \end{cases}$$

We claim that $\{v_k\}$ generates ν. Indeed, fix i, $j \in \mathbb{N}_0$ and let $k > \max\{i, j\}$. By (8.26), (8.27), and the fact that $|N_k| = 0$ we have

$$\lim_{k \to \infty} \int_E h_i(x)\,\varphi_j(v_k(x))\,dx = \lim_{k \to \infty} \sum_\ell \int_{Q(x_{k\ell}, \varepsilon_{k\ell})} h_i(x)\,\varphi_j(v_{n_{k\ell}k\ell}(x))\,dx$$

$$= \lim_{k \to \infty} \sum_\ell \psi_j(x_{k\ell}) \int_{Q(x_{k\ell}, \varepsilon_{k\ell})} h_i(x)\,dx$$

$$= \int_E h_i(x)\,\psi_j(x)\,dx$$

$$= \int_E h_i(x) \int_{\mathbb{R}^m} \varphi_j(z)\,d\nu_x(z)\,dx,$$

where in the last identity we have used (8.25). In particular, taking $i = j = 0$, we get

$$\lim_{k \to \infty} \int_E g(v_k)\,dx = \int_E \int_{\mathbb{R}^m} g(z)\,d\nu_x(z)\,dx,$$

and so by Theorem 8.6(ii) we conclude that $\{g(|v_k(\cdot)|)\}$ is equi-integrable. This concludes the proof.

8.3 Relaxation

In this section we consider functionals of the form

$$v \in L^p(E; \mathbb{R}^m) \mapsto I(v) := \int_E f(x, v)\,dx,$$

where $E \subset \mathbb{R}^N$ is a Lebesgue measurable set with finite measure, $1 \le p \le \infty$, and $f : E \times \mathbb{R}^m \to (-\infty, \infty]$ is a normal integrand. We recall that $\mathcal{E}_p$ is the greatest functional below I that is sequentially lower semicontinuous with respect to weak (respectively weak star if $p = \infty$) convergence in $L^p(E; \mathbb{R}^m)$, and it coincides with $\mathcal{I}_p$, where

$$\mathcal{I}_p(v) := \inf\Big\{ \liminf_{n \to \infty} I(v_n) : \{v_n\} \subset L^p(E; \mathbb{R}^m),$$

$$v_n \rightharpoonup v \ (\overset{*}{\rightharpoonup} \text{ if } p = \infty) \text{ in } L^p(E; \mathbb{R}^m) \Big\},$$

provided f satisfies suitable coercivity conditions (see Remark 6.69(i)).

Theorem 8.20. *Let $E \subset \mathbb{R}^N$ be a Lebesgue measurable set with finite measure, $1 < p < \infty$, and let $f : E \times \mathbb{R}^m \to (-\infty, \infty]$ be a normal integrand satisfying*

$$f(x, z) \geq \frac{1}{C} |z|^p - C \qquad (8.28)$$

for $\mathcal{L}^N$ a.e. $x \in E$ and for all $z \in \mathbb{R}^m$, and for some $C > 0$. Then for every $v \in L^p(E; \mathbb{R}^m)$,

$$\mathcal{E}_p(v) = \mathcal{I}_p(v) = \int_E f^{**}(x, v(x)) \, dx \qquad (8.29)$$

$$= \inf_{\nu \in \mathcal{Y}_v} \int_E \int_{\mathbb{R}^m} f(x, z) \, d\nu_x(z) \, dx,$$

where

$$\mathcal{Y}_v := \Big\{ \nu \in L_w^\infty(E; \mathcal{M}(\mathbb{R}^m; \mathbb{R})), \ \nu \text{ is a Young measure,}$$

$$\int_E \int_{\mathbb{R}^m} |z|^p \, d\nu_x(z) \, dx < \infty,$$

$$v(x) = \int_{\mathbb{R}^m} z \, d\nu_x(z) \text{ for } \mathcal{L}^N \text{ a.e. } x \in E \Big\}.$$

Moreover, if

$$\int_E f^{**}(x, v(x)) \, dx < \infty \qquad (8.30)$$

then the infimum is realized by some Young measure $\nu^0 \in \mathcal{Y}_v$ such that

$$f^{**}(x, v(x)) = \int_{\mathbb{R}^m} f(x, z) \, d\nu_x^0(z)$$

for $\mathcal{L}^N$ a.e. $x \in E$.

Proof. We observe that the first two equalities in (8.29) follow by Remark 6.69(i).

Let $\{v_n\} \subset L^p(E; \mathbb{R}^m)$ be such that $v_n \rightharpoonup v$ in $L^p(E; \mathbb{R}^m)$ and, without loss of generality, and in view of Remark 8.5, assume that

$$\liminf_{n \to \infty} I(v_n) = \lim_{n \to \infty} I(v_n)$$

and $\{v_n\}$ generates a Young measure ν. By Remarks 8.12 and 8.5 it follows that $\nu \in \mathcal{Y}_v$. By the fundamental theorem for Young measures (i) we have that

$$\liminf_{n \to \infty} \int_E f(x, v_n(x)) \, dx \geq \int_E \int_{\mathbb{R}^m} f(x, z) \, d\nu_x(z) \, dx,$$

and given the arbitrariness of $\{v_n\}$ we deduce that

$$\mathcal{I}_p(v) \geq \inf_{\nu \in \mathcal{Y}_v} \int_E \int_{\mathbb{R}^m} f(x, z) \, d\nu_x(z) \, dx.$$

To prove the reverse inequality fix any $\nu \in \mathcal{Y}_v$. By Jensen's inequality,

$$\int_{\mathbb{R}^m} f(x,z) \, d\nu_x(z) \geq \int_{\mathbb{R}^m} f^{**}(x,z) \, d\nu_x(z) \geq f^{**}(x, v(x)) \qquad (8.31)$$

for $\mathcal{L}^N$ a.e. $x \in E$, where we have used the fact that $v(x) = \int_{\mathbb{R}^m} z \, d\nu_x(z)$ for $\mathcal{L}^N$ a.e. $x \in E$. Integrating over E and then taking the infimum over all $\nu \in \mathcal{Y}_v$, we conclude that

$$\inf_{\nu \in \mathcal{Y}_v} \int_E \int_{\mathbb{R}^m} f(x,z) \, d\nu_x(z) \, dx \geq \int_E f^{**}(x, v(x)) \, dx. \qquad (8.32)$$

By Theorem 6.68, (8.29) follows.

To prove the second part of the theorem, in view of (8.28), by replacing f with $f + C$ we can assume, without loss of generality, that $f \geq 0$. Due to the coercivity condition (8.28) it is possible to find a sequence $\{v_n\} \subset L^p(E; \mathbb{R}^m)$ converging weakly to v in $L^p(E; \mathbb{R}^m)$ and such that

$$\int_E f^{**}(x, v(x)) \, dx = \mathcal{I}_p(v) = \lim_{n \to \infty} \int_E f(x, v_n(x)) \, dx < \infty.$$

By Remark 8.12, up to a subsequence (not relabeled), $\{v_n\}$ generates a Young measure ν^0 and by the fundamental theorem for Young measures and the first part of the theorem we have

$$\inf_{\nu \in \mathcal{Y}_v} \int_E \int_{\mathbb{R}^m} f(x,z) \, d\nu_x(z) \, dx = \mathcal{I}_p(v) = \lim_{n \to \infty} \int_E f(x, v_n(x)) \, dx$$

$$\geq \int_E \int_{\mathbb{R}^m} f(x,z) \, d\nu_x^0(z) \, dx,$$

which shows that the infimum is attained. Hence by (8.30) and (8.29),

$$\int_E \left[\int_{\mathbb{R}^m} f(x,z) \, d\nu_x^0(z) - f^{**}(x, v(x)) \right] dx = 0.$$

Since by (8.31) the expression in square brackets is nonnegative, it follows that

$$\int_{\mathbb{R}^m} f(x,z) \, d\nu_x^0(z) = f^{**}(x, v(x))$$

for $\mathcal{L}^N$ a.e. $x \in E$. This concludes the proof.

Exercise 8.21. Prove that the previous theorem continues to hold in the cases $p = 1$ and $p = \infty$ provided

(a) the coercivity condition (8.28) is replaced by

$$f(x,z) \geq \frac{1}{C} g(|z|) - C$$

for $\mathcal{L}^N$ a.e. $x \in E$ and for all $z \in \mathbb{R}^m$, for some $C > 0$, and where $g : [0, \infty) \to [0, \infty)$ is a continuous, nondecreasing function such that

$$\lim_{t \to \infty} \frac{g(t)}{t} = \infty$$

for $p = 1$, and

$$g(t) := \begin{cases} \dfrac{t}{\alpha - t} & 0 \leq t < \alpha, \\ \infty & t \geq \alpha, \end{cases}$$

for some $\alpha > 0$, for $p = \infty$;

(b) $\mathcal{Y}_v$ is given by

$$\mathcal{Y}_v := \Big\{ \nu \in L_w^\infty (E; \mathcal{M}(\mathbb{R}^m; \mathbb{R})), \nu \text{ is a Young measure,}$$

$$\int_E \int_{\mathbb{R}^m} g(|z|) \, d\nu_x(z) \, dx < \infty,$$

$$v(x) = \int_{\mathbb{R}^m} z \, d\nu_x(z) \text{ for } \mathcal{L}^N \text{ a.e. } x \in E \Big\}$$

for $p = 1$, and

$$\mathcal{Y}_v := \Big\{ \nu \in L_w^\infty (E; \mathcal{M}(\mathbb{R}^m; \mathbb{R})), \nu \text{ is a Young measure,}$$

$$\operatorname{supp} \nu_x \subset K \text{ for } \mathcal{L}^N \text{ a.e. } x \in E \text{ and for some compact}$$

$$\text{set } K \subset \mathbb{R}^m, v(x) = \int_{\mathbb{R}^m} z \, d\nu_x(z) \text{ for } \mathcal{L}^N \text{ a.e. } x \in E \Big\}$$

for $p = \infty$.

Recalling the observation made at the beginning of this chapter, the next result illustrates a situation in which the minimum of a certain constraint energy functional does not exist as an L^p function but may be attained at a "generalized" function, namely the Young measure generated by a minimizing sequence.

Theorem 8.22. *Let $E \subset \mathbb{R}^N$ be a Lebesgue measurable set with finite measure, let $1 \leq p \leq \infty$, and let $f : \mathbb{R}^m \to \mathbb{R}$ be a continuous function bounded from below by an affine function. If $\alpha \in \mathbb{R}^m$, then*

$$\inf \Big\{ \int_E f(v(x)) \, dx : v \in L^p(E; \mathbb{R}^m), \frac{1}{|E|} \int_E v(x) \, dx = \alpha \Big\} \qquad (8.33)$$

$$= \inf \Big\{ \int_E \int_{\mathbb{R}^m} f(z) \, d\nu_x(z) \, dx : \nu \in \mathcal{Y}^\alpha \Big\} = f^{**}(\alpha) \, |E| ,$$

where

$$\mathcal{Y}^{\alpha} := \Big\{ \nu \in L_w^{\infty} \left(E; \mathcal{M} \left(\mathbb{R}^m; \mathbb{R} \right) \right), \ \nu \ is \ a \ Young \ measure,$$

$$\int_E \int_{\mathbb{R}^m} |z|^p \ d\nu_x(z) \, dx < \infty, \ \frac{1}{|E|} \int_E \int_{\mathbb{R}^m} z \, d\nu_x(z) \, dx = \alpha \Big\},$$

and the infimum is attained in any of the first two expressions if and only if $\alpha \in \mathrm{co} M_{\alpha}$, where

$$M_{\alpha} := \{ z \in \mathbb{R}^m : f(z) = f^{**}(\alpha) + \beta \cdot (z - \alpha) \ for \ all \ \beta \in \partial f^{**}(\alpha) \}.$$

Moreover, if f satisfies the growth condition

$$\lim_{|z| \to \infty} \frac{f(z)}{|z|} = \infty, \tag{8.34}$$

then any minimizing sequence $\{v_n\} \subset L^p(E; \mathbb{R}^m)$ admits a subsequence generating a Young measure $\nu \in \mathcal{Y}^{\alpha}$ such that $\mathrm{supp}\, \nu_x \subset M_{\alpha}$ for $\mathcal{L}^N$ a.e. $x \in E$ and

$$\inf \Big\{ \int_E f(v(x)) \, dx : v \in L^p(E; \mathbb{R}^m), \ \frac{1}{|E|} \int_E v(x) \, dx = \alpha \Big\}$$

$$= \int_E \int_{\mathbb{R}^m} f(z) \, d\nu_x(z) \, dx = f^{**}(\alpha) |E|.$$

Proof. In view of Theorem 5.38, to prove (8.33) it suffices to show that

$$\inf \Big\{ \int_E f(v(x)) \, dx : v \in L^p(E; \mathbb{R}^m), \ \frac{1}{|E|} \int_E v(x) \, dx = \alpha \Big\}$$

$$\geq \inf \Big\{ \int_E \int_{\mathbb{R}^m} f(z) \, d\nu_x(z) \, dx : \nu \in \mathcal{Y}^{\alpha} \Big\} \geq f^{**}(\alpha) |E|.$$

The latter inequality follows from Jensen's inequality as in (8.31) and (8.32), while to prove the first it suffices to observe that to each function $v \in L^p(E; \mathbb{R}^m)$ we can associate the Young measure $\nu \in \mathcal{Y}^{\alpha}$ defined by

$$\nu_x := \delta_{v(x)} \ for \ x \in E.$$

Then

$$\int_E f(v(x)) \, dx = \int_E \int_{\mathbb{R}^m} f(z) \, d\nu_x(z) \, dx$$

$$\geq \inf \Big\{ \int_E \int_{\mathbb{R}^m} f(z) \, d\tilde{\nu}_x(z) \, dx : \tilde{\nu} \in \mathcal{Y}^{\alpha} \Big\}.$$

Hence (8.33) holds.

Next we show that the infimum is attained in (8.33) if and only if $\alpha \in \mathrm{co} M_{\alpha}$. Again by Theorem 5.38 it suffices to show that the infimum

$$\inf\left\{\int_E\int_{\mathbb{R}^m}f(z)\,d\nu_x(z)\,dx : \nu\in\mathcal{Y}^\alpha\right\}=f^{**}(\alpha)\,|E| \qquad (8.35)$$

is attained if and only if $\alpha\in\mathrm{co}M_\alpha$. Suppose now that the infimum is attained at some Young measure $\nu^0\in\mathcal{Y}^\alpha$. Given any $\beta\in\partial f^{**}(\alpha)$ (recall Theorem 4.53) it follows that

$$f(z)\geq f^{**}(z)\geq f^{**}(\alpha)+\beta\cdot(z-\alpha), \qquad (8.36)$$

for all $z\in\mathbb{R}^m$, and so

$$\int_E\int_{\mathbb{R}^m}f(z)\,d\nu_x^0(z)\,dx \geq \int_E\int_{\mathbb{R}^m}f^{**}(z)\,d\nu_x^0(z)\,dx$$
$$\geq \int_E\int_{\mathbb{R}^m}[f^{**}(\alpha)+\beta\cdot(z-\alpha)]\,d\nu_x^0(z)\,dx$$
$$= f^{**}(\alpha)\,|E|\,,$$

where in the last identity we have used the fact that

$$\frac{1}{|E|}\int_E\int_{\mathbb{R}^m}z\,d\nu_x^0(z)\,dx=\alpha. \qquad (8.37)$$

It now follows from (8.35) that all the inequalities must be identities. From (8.36) we deduce that

$$\int_{\mathbb{R}^m}f(z)\,d\nu_x^0(z)=\int_{\mathbb{R}^m}f^{**}(z)\,d\nu_x^0(z)=\int_{\mathbb{R}^m}[f^{**}(\alpha)+\beta\cdot(z-\alpha)]\,d\nu_x^0(z)$$

for $\mathcal{L}^N$ a.e. $x\in E$ and, in turn,

$$\mathrm{supp}\,\nu_x^0\subset\{z\in\mathbb{R}^m : f(z)=f^{**}(z)=f^{**}(\alpha)+\beta\cdot(z-\alpha)\} \qquad (8.38)$$

for $\mathcal{L}^N$ a.e. $x\in E$. Define

$$v(x):=\int_{\mathbb{R}^m}z\,d\nu_x^0(z),\quad x\in E.$$

Since for $\mathcal{L}^N$ a.e. $x\in E$ the measure ν_x^0 is a probability measure, it follows from Lemma 5.39 and (8.38) that

$$v(x)\in\mathrm{co}\{z\in\mathbb{R}^m : f(z)=f^{**}(z)=f^{**}(\alpha)+\beta\cdot(z-\alpha)\}$$

for $\mathcal{L}^N$ a.e. $x\in E$, which, together with (8.37), yields

$$\alpha\in\mathrm{co}\{z\in\mathbb{R}^m : f(z)=f^{**}(z)=f^{**}(\alpha)+\beta\cdot(z-\alpha)\}.$$

The final part of the theorem follows exactly as in the proof of the previous theorem using Remark 8.21.

Remark 8.23. As a consequence of the previous theorem, it follows in particular that condition (8.34) implies $\alpha\in\mathrm{co}M_\alpha$. We are not aware of a direct proof of this fact.

Part IV

Appendix

A

Functional Analysis and Set Theory

In order to keep this book as self-contained as possible, in this appendix we collect without proofs several results from functional analysis that have been used throughout the book. This part is intended mostly for graduate students. Most proofs will be omitted. Basic references are [AliBo99], [Bre83], [DuSc88], [Ru91], [Yo95]. The reader should be warned that in [DuSc88] and [Ru91] the definitions of normal spaces and topological vector spaces are different from the standard ones.

A.1 Some Results from Functional Analysis

A.1.1 Topological Spaces

Definition A.1. *Let X be a nonempty set. A collection $\tau \subseteq \mathcal{P}(X)$ is a* topology *if*

(i) $\emptyset,\ X \in \tau$;
(ii) if $U_i \in \tau$ *for* $i = 1, \ldots, M$, *then* $U_1 \cap \ldots \cap U_M \in \tau$;
(iii) if $\{U_\alpha\}_{\alpha \in I}$ *is an arbitrary collection of elements of* τ *then* $\bigcup_{\alpha \in I} U_\alpha \in \tau$.

Example A.2. Given a nonempty set X, the smallest topology consists of $\{\emptyset, X\}$, while the largest topology contains all subsets as open sets, and is called the *discrete topology*.

The pair (X, τ) is called *topological space* and the elements of τ *open sets*. For simplicity, we often apply the term topological space only to X. A set $C \subset X$ is *closed* if its complement $X \setminus C$ is open. The *closure $\overline{E}$ of a set*

$E \subset X$ is the smallest closed set that contains E. The *interior* $\overset{\circ}{E}$ *of a set*
$E \subset X$ is the union of all its open subsets.

A subset E of a topological space X is said to be *dense* if its closure is
the entire space, i.e., $\overline{E} = X$. We say that a topological space is *separable* if
it contains a countable dense subset.

Given a point $x \in X$, a *neighborhood*[1] of x is any open set $U \in \tau$ that
contains x. Given a set $E \subset X$, a *neighborhood* of E is any open set $U \in \tau$
that contains E. A topological space is a *Hausdorff space* if for any $x, y \in X$
with $x \neq y$ we may find two disjoint neighborhoods of x and y.

Given a topological space X and a sequence $\{x_n\}$, we say that $\{x_n\}$ *converges* to a point $x \in X$ if for every neighborhood U of x we have that $x_n \in U$
for all n sufficiently large. Note that unless the space is Hausdorff, the limit
may not be unique.

A topological space is a *normal space* if for every pair of disjoint closed
sets $C_1, C_2 \subset X$ we may find two disjoint neighborhoods of C_1 and C_2.

Definition A.3. *Let X, Y be two topological spaces and let $f : X \to Y$ be a
function from X into Y. We say that f is* continuous *if $f^{-1}(U)$ is open for
every open set $U \subset Y$.*

The space of all continuous functions $f : X \to Y$ is denoted by $C(X;Y)$.
The next two theorems give important characterizations of normal spaces.

Theorem A.4 (Urysohn). *A topological space X is normal if and only if
for any two disjoint closed sets $C_1, C_2 \subset X$ there exists a continuous function
$f : X \to [0,1]$ such that $f \equiv 1$ in C_1 and $f \equiv 0$ in C_2.*

Theorem A.5 (Tietze extension). *A topological space X is normal if and
only for any closed set $C \subset X$ and any continuous function $f : C \to \mathbb{R}$ there
exists a continuous function $F : X \to \mathbb{R}$ such that $F(x) = f(x)$ for all $x \in C$.
Moreover, if $f(C) \subset [a,b]$ then F may be constructed so that*

$$F(C) \subset [a, b].$$

We now introduce the notion of a base for a topology. Let (X, τ) be a
topological space. A family β of open sets of X is a *base* for the topology τ
if every open set $U \in \tau$ may be written as union of elements of β. Given a
point $x \in X$, a family β_x of neighborhoods of x is a *local base at x* if every
neighborhood of x contains an element of β_x.

Proposition A.6. *Let X be a nonempty set and let $\beta \subset \mathcal{P}(X)$ be a family
of sets such that*

(i) for every $x \in X$ there exists $B \in \beta$ such that $x \in B$;

---

[1] The reader should be warned that in some texts (e.g., [DuSc88] and [Ru91]) the
definition of neighborhood is different.

(ii) for every B_1, $B_2 \in \beta$, *with* $B_1 \cap B_2 \neq \emptyset$, *and for every* $x \in B_1 \cap B_2$ *there exists* $B_3 \in \beta$ *such that* $x \in B_3$ *and* $B_3 \subset B_1 \cap B_2$.

Then the collection $\tau \subset \mathcal{P}(X)$ *of arbitrary unions of members of* β *is a topology for which* β *is a base.*

As we will see later on, the metrizability and the normability of a given topology depend on the properties of a base (see Theorems A.17 and A.40 below).

Definition A.7. *Let* (X, τ) *be a topological space.*

(i) The space X *satisfies the* first axiom of countability *if every* $x \in X$ *admits a countable base of open sets.*

(ii) The space X *satisfies the* second axiom of countability *if it has a countable base.*

The following result is used in Chapter 4.

Theorem A.8 (Lindelöf). *Let* (X, τ) *be a topological space satisfying the second axiom of countability. Then every family* $\mathcal{F} \subset \tau$ *contains a countable subfamily* $\{U_n\} \subset \mathcal{F}$ *such that*

$$\bigcup_{U \in \mathcal{F}} U = \bigcup_n U_n.$$

In the text we use several notions of compactness.

Definition A.9. *Let* X *be a topological space.*

(i) A set $K \subset X$ *is* compact *if given any open cover of* K, *i.e., any collection* $\{U_\alpha\}$ *of elements of* τ *such that* $\bigcup_\alpha U_\alpha \supset K$, *then we may find a finite subcover (i.e., a finite subcollection of* $\{U_\alpha\}$ *whose union still contains* K);

(ii) a set $E \subset X$ *is* relatively compact *(or* precompact*) if its closure* $\overline{E}$ *is compact;*

(iii) a set $E \subset X$ *is* σ-compact *if it can be written as a countable union of compact sets;*

(iv) the topological space X *is* locally compact *if every point* $x \in X$ *has a neighborhood whose closure is compact.*

Remark A.10. A closed subset of a compact topological space is compact. On the other hand, a compact set of a Hausdorff space is closed.

Note that if the topology is not Hausdorff then compact sets may not be closed.

Example A.11. Given a nonempty set X endowed with the smallest topology, any nonempty set strictly contained in X is compact but not closed.

The next theorem is used to construct cutoff functions and partitions of unity. These two notions will play a central role in [FoLe10].

Theorem A.12. *If X is a locally compact Hausdorff space and $K \subset U \subset X$, with K compact and U open, then there exists W open such that $\overline{W}$ is compact and $K \subset W \subset \overline{W} \subset U$.*

Corollary A.13. *If X is a locally compact Hausdorff space and $K \subset U \subset X$, with K compact and U open, then there exists a function $\varphi \in C_c(X)$ such that $0 \leq \varphi \leq 1$, $\varphi \equiv 1$ on K, and $\varphi \equiv 0$ on $X \setminus U$.*

The function φ is usually referred to as a *cutoff function*. To introduce partitions of unity we need the notion of a locally finite family.

Definition A.14. *Let X be a topological space and let $\mathcal{F}$ be a collection of subsets of X. Then*

(i) $\mathcal{F}$ is locally finite *if every $x \in X$ has a neighborhood meeting only finitely many $U \in \mathcal{F}$.*
(ii) $\mathcal{F}$ is σ-locally finite if

$$\mathcal{F} = \bigcup_{n=1}^{\infty} \mathcal{F}_n,$$

where each $\mathcal{F}_n$ is a locally finite collection in X.

We are now ready to introduce partitions of unity. Their existence is proved in Theorems 2.77 and 6.18.

Definition A.15. *If X is a topological space, a* partition of unity *on X is a family $\{\varphi_\alpha\}_{\alpha \in I}$ of continuous functions $\varphi_\alpha : X \to [0,1]$ such that*

$$\sum_{\alpha \in I} \varphi_\alpha(x) = 1$$

for all $x \in X$. A partition of unity is locally finite *if for every $x \in X$ there exists a neighborhood U of x such that the set $\{\alpha \in I : U \cap \operatorname{supp} \varphi_\alpha \neq \emptyset\}$ is finite. If $\{U_\alpha\}_{\alpha \in I}$ is an open cover of X, a partition of unity* subordinated to *the cover $\{U_\alpha\}_{\alpha \in I}$ is a partition of unity $\{\varphi_\alpha\}_{\alpha \in I}$ such that $\operatorname{supp} \varphi_\alpha \subset U_\alpha$ for each $\alpha \in I$.*

A.1.2 Metric Spaces

Definition A.16. *A* metric *on a set X is a map $d : X \times X \to [0, \infty)$ such that*

(i) $d(x,y) \leq d(x,z) + d(z,y)$ for all x, y, $z \in X$;
(ii) $d(x,y) = d(y,x)$ for all x, $y \in X$;
(iii) $d(x,y) = 0$ if and only if $x = y$.

A *metric space* (X, d) is a set X endowed with a metric d. When there is no possibility of confusion we abbreviate by saying that X is a metric space. If $r > 0$, the *(open) ball* of center $x_0 \in X$ and radius r is the set

$$B(x_0, r) := \{x \in X : d(x_0, x) < r\}.$$

If $x \in X$ and $E \subset X$, the *distance* of x from the set E is defined by

$$\operatorname{dist}(x, E) := \inf\{d(x, y) : y \in E\},$$

while the distance between two sets E_1, $E_2 \subset X$ is defined by

$$\operatorname{dist}(E_1, E_2) := \inf\{d(x, y) : x \in E_1, y \in E_2\}.$$

A metric space (X, d) can always be rendered into a topological space (X, τ) by taking as a base for the topology τ the family of all open balls. We then say that τ is *determined* by d. Note that (X, τ) is a Hausdorff normal space. Indeed, if $x \neq y$ then $B\left(x, \frac{d(x,y)}{2}\right)$ and $B\left(y, \frac{d(x,y)}{2}\right)$ are disjoint neighborhoods of x and y, respectively, while if $C_1, C_2 \subset X$ are disjoint and closed, then the open sets

$$U_1 := \{x \in X : \operatorname{dist}(x, C_1) < \operatorname{dist}(x, C_2)\},$$
$$U_2 := \{x \in X : \operatorname{dist}(x, C_1) > \operatorname{dist}(x, C_2)\},$$

are two disjoint neighborhoods of C_1 and C_2, respectively.

A topological space X is *metrizable* if its topology can be determined by a metric.

Theorem A.17. *A topological space is metrizable if and only if*

(i) singletons are closed;
(ii) for any closed set $C \subset X$ and for any $x \notin C$ there exist disjoint open neighborhoods of C and x;
(iii) it has a σ-locally finite base.

A sequence $\{x_n\} \subset X$ *converges (strongly)* to $x \in X$ if

$$\lim_{n \to \infty} d(x_n, x_0) = 0.$$

A *Cauchy sequence* in a metric space is a sequence $\{x_n\} \subset X$ such that

$$\lim_{n, m \to \infty} d(x_n, x_m) = 0.$$

A metric space X is said to be *complete* if every Cauchy sequence is convergent.

Theorem A.18 (Baire category theorem). *Let (X, d) be a complete metric space. Then the intersection of a countable family of open dense sets in X is still dense in X.*

A.1.3 Topological Vector Spaces

Let X be a vector space over $\mathbb{R}$ and let $E \subset X$. The set E is said to be *balanced* if $tx \in E$ for all $x \in E$ and $t \in [-1,1]$. We say that $E \subset X$ is *absorbing* if for every $x \in X$ there exists $t > 0$ such that $sx \in E$ for all $0 \le s \le t$.

Definition A.19. *Given a vector space X over $\mathbb{R}$ endowed with a topology τ, the pair (X, τ) is called a* topological vector space *if the functions*

$$X \times X \to X, \qquad \mathbb{R} \times X \to X,$$
$$(x,y) \mapsto x + y, \quad and \quad (t,x) \mapsto tx,$$

are continuous with respect to τ.

Remark A.20. (i) In a topological vector space a set U is open if and only if $x + U$ is open for all $x \in X$. Hence to give a base it is enough to give a local base at the origin.

(ii) Using the continuity of addition and scalar multiplication it is possible to show that each neighborhood U of the origin is absorbing and it contains a neighborhood of zero W such that $W + W \subset U$ and $\overline{W} \subset U$, as well as a balanced neighborhood of zero.

As a corollary of Theorem A.17 we have the following:

Corollary A.21. *A topological vector space is metrizable if and only if*

(i) singletons are closed;
(ii) it has a countable base.

Definition A.22. *Let X be a topological vector space. A set $E \subset X$ is said to be* topologically bounded *if for each neighborhood U of 0 there exists $t > 0$ such that $E \subset tU$.*

Note that when the topology τ is generated by a metric d, sets bounded in the topological sense and in the metric sense may be different. To see this, it suffices to observe that the metric $d_1 := \frac{d}{d+1}$ generates the same topology as d, but since $d_1 \le 1$, every set in X is bounded with respect to d_1.

We now define Cauchy sequences in a topological vector space.

Definition A.23. *Let X be a topological vector space. A sequence $\{x_n\} \subset X$ is called a* Cauchy sequence *if for every neighborhood U of the origin there exists an integer $\overline{n} \in \mathbb{N}$ such that*

$$x_n - x_k \in U$$

for all $k, n \ge \overline{n}$. The space X is complete *if every Cauchy sequence is convergent.*

Note that Cauchy (and hence convergent) sequences are bounded in the topological sense.

Proposition A.24. *Let X be a topological vector space and let $\{x_n\} \subset X$ be a Cauchy sequence. Then the set $\{x_n : n \in \mathbb{N}\}$ is topologically bounded.*

Proof. Let U be a neighborhood of the origin. Using the continuity of addition, construct a balanced neighborhood of the origin W such that $W + W \subset U$. Since $\{x_n\} \subset X$ is a Cauchy sequence there exists $\overline{n} \in \mathbb{N}$ such that $x_n - x_k \in W$ for all $k, n \geq \overline{n}$. In particular, $x_n \in x_{\overline{n}} + W$ for all $n \geq \overline{n}$. Since W is absorbing we may find $t > 1$ so large that $x_k \in tW \subset tU$ for all $k = 1, \dots, \overline{n}$. Moreover, for $n \geq \overline{n}$,

$$x_n \in x_{\overline{n}} + W \subset tW + W \subset tW + tW \subset tU,$$

which shows that $x_n \in tU$ for all $n \in \mathbb{N}$.

Topologically bounded sets play an important role in the normability of locally convex topological vector spaces (see Theorem A.40 below):

Definition A.25. *A topological vector space X is* locally convex *if every point $x \in X$ has a neighborhood that is convex.*

Proposition A.26. *A locally convex topological vector space admits a local base at the origin consisting of balanced convex neighborhoods of zero.*

Let X be a vector space over $\mathbb{R}$ and let $E \subset X$. The function $p_E : X \to \mathbb{R}$ defined by

$$p_E(x) := \inf\{t > 0 : x \in tE\}, \quad x \in X,$$

is called the *gauge* or *Minkowski functional* of E.

Definition A.27. *Let X be a vector space over $\mathbb{R}$. A function $p : X \to \mathbb{R}$ is called a* seminorm *if*

$$p(x + y) \leq p(x) + p(y)$$

for all $x, y \in X$ and $p(tx) = |t|\, p(x)$ for all $x \in X$ and $t \in \mathbb{R}$.

Remark A.28. Let X be a vector space over $\mathbb{R}$ and let $E \subset X$. The gauge p_E of set E is a seminorm if and only if E is balanced, absorbing, and convex.

Theorem A.29. *If $\mathcal{F}$ is a balanced, convex local base of 0 for a locally convex topological vector space X, then the family $\{p_U : U \in \mathcal{F}\}$ is a family of continuous seminorms. Conversely, given a family $\mathcal{P}$ of seminorms on a vector space X, the collection of all finite intersections of sets of the form*

$$V(p, n) := \left\{ x \in X : p(x) < \frac{1}{n} \right\}, \quad p \in \mathcal{P}, \, n \in \mathbb{N},$$

is a balanced, convex local base of 0 for a topology τ that turns X into a locally convex topological vector space such that each p is continuous with respect to τ.

We now give some necessary and sufficient conditions for the topology τ given in the previous theorem to be Hausdorff and for a set to be topologically bounded.

Corollary A.30. *Let* $\mathcal{P}$ *be a family of seminorms on a vector space* X *and let* τ *be the locally convex topology generated by* $\mathcal{P}$. *Then*

(i) τ *is Hausdorff if and only if* $p(x) = 0$ *for all* $p \in \mathcal{P}$ *implies that* $x = 0$;
(ii) a set $E \subset X$ *is topologically bounded if and only if the set* $p(E)$ *is bounded in* $\mathbb{R}$ *for all* $p \in \mathcal{P}$.

We now introduce the notion of dual space.

Definition A.31. *Let* X *and* Y *be two vector spaces. A map* $L : X \to Y$ *is called a* linear operator *if*

(i) $L(x + y) = L(x) + L(y)$ *for all* $x, y \in X$;
(ii) $L(tx) = t L(x)$ *for all* $x \in X$ *and* $t \in \mathbb{R}$.

If X and Y are topological vector spaces, then the vector space of all continuous linear operators from X to Y is denoted by $\mathcal{L}(X; Y)$. In the special case $Y = \mathbb{R}$, the space $\mathcal{L}(X; \mathbb{R})$ is called the *dual space* of X and it is denoted by X'. The elements of X' are also called continuous linear *functionals*.

The bilinear (i.e., linear in each variable) mapping

$$\langle \cdot, \cdot \rangle_{X', X} : X' \times X \to \mathbb{R}, \tag{A.1}$$

$$(L, x) \mapsto L(x),$$

is called the *duality pairing*.

The dual space $\mathcal{L}(X'; \mathbb{R})$ of X' is called *bidual space* of X and it is denoted by X''.

Definition A.32. *Let* X, Y *be topological vector spaces. An operator* $L : X \to Y$ *is* bounded *if it sends bounded sets of* X *into bounded sets of* Y.

Theorem A.33. *Let* X, Y *be topological vector spaces and let* $L : X \to Y$ *be linear and continuous. Then* L *is bounded.*

Proof. Let $E \subset X$ be a bounded set and let $W \subset Y$ be a neighborhood of zero. Since L is continuous and $L(0) = 0$, there exists a neighborhood $U \subset X$ of zero such that $L(U) \subset W$. By the boundedness of E there exists $t > 0$ such that $E \subset tU$. Hence

$$L(E) \subset L(tU) = tL(U) \subset tW.$$

This shows that $L(E)$ is bounded in Y.

Exercise A.34. On the vector space $C\left([0,1]\right)$ consider two topologies, the first, τ, given by the metric

$$d\left(f,g\right) = \int_0^1 \frac{|f\left(x\right) - g\left(x\right)|}{1 + |f\left(x\right) - g\left(x\right)|}\, dx, \quad f,\, g \in C\left([0,1]\right),$$

and the second, τ_p, given by the family of seminorms $\{p_x\}_{x \in [0,1]}$, where

$$p_x\left(f\right) := |f\left(x\right)|, \quad f \in C\left([0,1]\right).$$

Prove that

1. the identity $I\,:\,\left(C\left([0,1]\right),\tau_p\right) \to \left(C\left[0,1\right],\tau\right)$ maps bounded sets into bounded sets;
2. the identity $I\,:\,\left(C\left([0,1]\right),\tau_p\right) \to \left(C\left[0,1\right],\tau\right)$ is sequentially continuous but not continuous.
3. Deduce that the topology τ_p is not compatible with any metric.

We now present several versions of the Hahn–Banach theorem. These are heavily used in the second part of the text, starting from Chapter 4.

Theorem A.35 (Hahn–Banach, analytic form). *Let X be a vector space, let Y be a subspace of X, and let $p : X \to \mathbb{R}$ be a convex function. Then for any linear functional $L : Y \to \mathbb{R}$ such that*

$$L\left(x\right) \le p\left(x\right) \quad \text{for all } x \in Y$$

there exists a linear functional $L_1 : X \to \mathbb{R}$ such that

$$L_1\left(x\right) = L\left(x\right) \quad \text{for all } x \in Y$$

and

$$L\left(x\right) \le p\left(x\right) \quad \text{for all } x \in X.$$

The finite-dimensional version of the next theorem may be found in Chapter 4.

Theorem A.36 (Hahn–Banach, first geometric form). *Let X be a topological vector space, and let E, $F \subset X$ be nonempty disjoint convex sets. Assume that E has an interior point. Then there exist a continuous linear functional $L : X \to \mathbb{R}$, $L \ne 0$, and a number $\alpha \in \mathbb{R}$ such that*

$$L\left(x\right) \ge \alpha \quad \text{for all } x \in E \text{ and } L\left(x\right) \le \alpha \quad \text{for all } x \in F.$$

As a corollary of Hahn–Banach theorem one can prove the following result (see also Exercise 2.43).

Corollary A.37. *Let X be a topological vector space. Then the dual X' of X is not the null space if and only if X has a convex neighborhood of the origin strictly contained in X.*

In view of the previous corollary it is natural to restrict attention to locally convex topological vector spaces.

Theorem A.38 (Hahn–Banach, second geometric form). *Let X be a locally convex topological vector space, and let C, $K \subset X$ be nonempty disjoint convex sets, with C closed and K compact. Then there exist a continuous linear functional $L : X \to \mathbb{R}$ and two numbers $\alpha \in \mathbb{R}$ and $\varepsilon > 0$ such that*

$$L(x) \leq \alpha - \varepsilon \quad \text{for all } x \in C \text{ and } L(x) \geq \alpha + \varepsilon \quad \text{for all } x \in K.$$

A.1.4 Normed Spaces

Definition A.39. *A norm on a vector space X is a map*

$$\|\cdot\| : X \to [0, \infty)$$

such that

(i) $\|x + y\| \leq \|x\| + \|y\|$ for all $x, y \in X$;
(ii) $\|tx\| = |t| \, \|x\|$ for all $x \in X$ and $t \in \mathbb{R}$;
(iii) $\|x\| = 0$ implies $x = 0$.

A *normed space* $(X, \|\cdot\|)$ is a vector space X endowed with a norm $\|\cdot\|$. For simplicity, we often say that X is a normed space.

If for every $x, y \in X$ we define

$$d(x, y) := \|x - y\|,$$

then (X, d) is a metric space. We say that a normed space X is a *Banach space* if it is complete as a metric space.

A topological vector space is *normable* if its topology can be determined by a norm.

Theorem A.40. *A topological vector space X is normable if and only if it is locally convex and it has a topologically bounded neighborhood of 0.*

Two norms $\|\cdot\|_1$ and $\|\cdot\|_2$ are *equivalent* if there exists a positive constant $C > 0$ such that

$$\frac{1}{C} \|x\|_1 \leq \|x\|_2 \leq C \|x\|_1 \quad \text{for all } x \in X.$$

Equivalent norms induce the same topology on X.

Proposition A.41. *Let X and Y be normed spaces with norms $\|\cdot\|_X$ and $\|\cdot\|_Y$, respectively.*

(i) A linear operator $L : X \to Y$ is continuous if and only if

$$\|L\|_{\mathcal{L}(X;Y)} := \sup_{x \in X \setminus \{0\}} \frac{\|L(x)\|_Y}{\|x\|_X} < \infty;$$

(ii) the mapping $L \in \mathcal{L}(X;Y) \mapsto \|L\|_{\mathcal{L}(X;Y)}$ is a norm;
(iii) if Y is a Banach space then so is $\mathcal{L}(X;Y)$; conversely, if $X \neq \{0\}$ and $\mathcal{L}(X;Y)$ is a Banach space then so is Y.

As a corollary of the Hahn–Banach theorem one has the following result.

Corollary A.42. *Let X be a normed space. Then for all $x \in X$,*

$$\|x\|_X = \max_{L \in X', \|L\|_{X'} \leq 1} \frac{|L(x)|}{\|x\|_X}.$$

Remark A.43. This corollary is especially useful for L^p spaces.

Theorem A.44 (Banach–Steinhaus). *Let X, Y be Banach spaces and let $\{L_\alpha\}_{\alpha \in I}$ be a family of linear continuous operators $L_\alpha : X \to Y$ such that*

$$\sup_{\alpha \in I} \|L_\alpha(x)\|_Y < \infty$$

for every $x \in X$. Then

$$\sup_{\alpha \in I} \|L_\alpha\|_{\mathcal{L}(X;Y)} < \infty.$$

Definition A.45. *Let X and Y be normed spaces with norms $\|\cdot\|_X$ and $\|\cdot\|_Y$. A continuous linear operator $L \in \mathcal{L}(X;Y)$ is said to be* compact *if it maps every bounded subset of X onto a relatively compact subset of Y.*

In particular, if L is compact then from every bounded sequence $\{x_n\} \subset X$ we may extract a subsequence $\{x_{n_k}\}$ such that $\{L(x_{n_k})\}$ converges in Y.

Definition A.46. *We say that the normed space X is* embedded *in the normed space Y and we write*

$$X \hookrightarrow Y$$

if X is a vector subspace of Y and the immersion

$$i : X \to Y,$$
$$x \mapsto x,$$

is continuous.

Note that since the immersion is linear, in view of Proposition A.41 the continuity of i is equivalent to requiring the existence of a constant $M > 0$ such that

$$\|x\|_Y \leq M \|x\|_X \quad \text{for all } x \in X.$$

We say that X is *compactly embedded* in Y if the immersion i is a compact operator. Compact embeddings will play an important role in the study of Sobolev spaces in [FoLe10].

A.1.5 Weak Topologies

Given a locally convex topological vector space X, for each $L \in X'$ the function $p_L : X \to [0, \infty)$ defined by

$$p_L(x) := |L(x)|, \quad x \in X, \tag{A.2}$$

is a seminorm. In view of Theorem A.29, the family of seminorms $\{p_L\}_{L \in X'}$ generates a locally convex topology $\sigma(X, X')$ on the space X, called the *weak topology*, such that each p_L is continuous with respect to $\sigma(X, X')$. In turn, this implies that every $L \in X'$ is $\sigma(X, X')$ continuous.

Theorem A.47. *Let X be a locally convex topological vector space and let $E \subset X$. Then*

(i) E is bounded with respect to the (strong) topology if and only if it weakly bounded;

(ii) if E is convex then E is closed if and only if it weakly closed.

Definition A.48. *Given a locally convex topological vector space X, a sequence $\{x_n\} \subset X$ converges weakly to $x \in X$ if it converges to x with respect to the weak topology $\sigma(X, X')$.*

We write $x_n \rightharpoonup x$. In view of Theorem A.29 and (A.2), we have the following result.

Proposition A.49. *Let X be a locally convex topological vector space. A sequence $\{x_n\} \subset X$ converges weakly to $x \in X$ if and only if*

$$\lim_{n \to \infty} L(x_n) = L(x)$$

for every $L \in X'$.

Similarly, given a locally convex topological vector space X, for each $x \in X$ the function $p_x : X' \to [0, \infty)$ defined by

$$p_x(L) := |L(x)|, \quad L \in X', \tag{A.3}$$

is a seminorm. In view of Theorem A.29, the family of seminorms $\{p_x\}_{x \in X}$ generates a locally convex topology $\sigma(X', X)$ on the space X', called the *weak star topology*, such that each p_x is continuous with respect to $\sigma(X', X)$.

Definition A.50. *Let X be a locally convex topological vector space. A sequence $\{L_n\} \subset X'$ is weakly star convergent to L in X' if it converges to L with respect to the weak star topology $\sigma(X', X)$.*

We write $L_n \overset{*}{\rightharpoonup} L$. In view of Theorem A.29 and (A.3), we have the following result.

Proposition A.51. *Let X be a locally convex topological vector space. A sequence $\{L_n\} \subset X'$ converges weakly star to $L \in X'$ if and only if*

$$\lim_{n \to \infty} L_n(x) = L(x)$$

for every $x \in X$.

Theorem A.52 (Banach–Alaoglu). *If U is a neighborhood of 0 in a locally convex topological vector space X, then*

$$K := \{L \in X' : |L(x)| \le 1 \text{ for every } x \in U\}$$

is weak star compact.

Corollary A.53. *If X is a normed space then the closed unit ball of X',*

$$\{L \in X' : \|L\|_{X'} \le 1\},$$

is weak star compact.

If X is separable, it actually turns out that weak star compact sets are metrizable, and thus one can work with the friendlier notion of sequential compactness.

Theorem A.54. *Let X be a separable, locally convex topological vector space and let $K \subset X'$ be weak star compact. Then $(K, \sigma(X', X))$ is metrizable.*

Hence, also in view of the Banach–Alaoglu theorem, we have the following:

Corollary A.55. *Let U be a neighborhood of 0 in a separable locally convex topological vector space X and let $\{L_n\} \subset X'$ be such that*

$$|L_n(x)| \le 1 \text{ for every } x \in U \text{ and for all } n \in \mathbb{N}.$$

Then there exists a subsequence $\{L_{n_k}\}$ that is weakly star convergent. In particular, if X is a separable Banach space and $\{L_n\} \subset X'$ is any bounded sequence in X', then there exists a subsequence that is weakly star convergent.

For Banach spaces the converse of Theorem A.54 holds:

Theorem A.56. *Let X be a Banach space. Then the unit ball $B(0; 1)$ in X' endowed with the weak star topology is metrizable if and only if X is separable.*

Proposition A.57. *Let X be a Banach space. If a sequence $\{L_n\} \subset X'$ converges weakly star to $L \in X'$, then it is bounded and*

$$\|L\|_{X'} \le \liminf_{n \to \infty} \|L_n\|_{X'}.$$

Proposition A.58. *Let X be a Banach space. If X' is separable then so is X.*

The converse is false in general (take, for example, the separable space $L^1 (\mathbb{R}^N)$ and its dual $L^\infty (\mathbb{R}^N)$).

We now study analogous results for the weak topology. An infinite-dimensional Banach space when endowed with the weak topology is never metrizable. However, we have the following:

Theorem A.59. Let X be a Banach space whose dual X' is separable. Then the unit ball $B(0;1)$ endowed with the weak topology is metrizable.

Definition A.60. Let X be a locally convex topological vector space. A set $K \subset X$ is called sequentially weakly compact if every sequence $\{x_n\} \subset K$ has a subsequence converging weakly to a point in K.

Theorem A.61. Let X be a Banach space. If $K \subset X$ is weakly compact then it is weakly sequentially compact.

Using Banach–Alaoglu's theorem one can prove the following theorem:

Theorem A.62 (Eberlein–Šmulian). Let E be a subset of a Banach space X. Then the weak closure of E is weakly compact if and only if for any sequence $\{x_n\} \subset E$ there exists a subsequence weakly convergent to some element of X.

As an immediate application of the Hahn–Banach theorem we have the following:

Proposition A.63. Let X be a normed space and consider the linear operator mapping $J : X \to X''$ defined by

$$J(x)(L) := L(x), \quad L \in X'.$$

Then $\|J(x)\|_{X''} = \|x\|_X$ for all $x \in X$. In particular, J is injective and continuous.

Definition A.64. A normed space X is reflexive if $J(X) = X''$.

In this case it is possible to identify X with its bidual X''.

Theorem A.65 (Kakutani). A Banach space is reflexive if and only if the closed unit ball $\{x \in X : \|x\| \leq 1\}$ is weakly compact.

In view of the previous theorem and the Eberlein–Šmulian theorem we have the following corollary:

Corollary A.66. Let X be a reflexive Banach space and let $\{x_n\} \subset X$ be a bounded sequence. Then there exists a subsequence that is weakly convergent.

Proposition A.67. A normed space X is reflexive if and only if X' is reflexive.

The following proposition is used throughout the text, sometimes without mention.

Proposition A.68. *Let X be a Banach space. If a sequence $\{x_n\} \subset X$ converges weakly to $x \in X$, then it is bounded and*

$$\|x\| \leq \liminf_{n \to \infty} \|x_n\| .$$

An important family of reflexive Banach spaces that includes $L^p\left(\mathbb{R}^N\right)$, $1 < p < \infty$, is given by uniformly convex Banach spaces.

Definition A.69. *A normed space X is* uniformly convex *if for every $\varepsilon > 0$ there exists $\delta > 0$ such that for every x, $y \in X$, with $\|x\| \leq 1$, $\|y\| \leq 1$, and $\|x - y\| > \varepsilon$,*

$$\left\| \frac{x+y}{2} \right\| < 1 - \delta.$$

Theorem A.70. *Let X be a uniformly convex Banach space. Then*

(i) (Milman) X is reflexive;
(ii) if $\{x_n\} \subset X$ converges weakly to $x \in X$ and

$$\limsup_{n \to \infty} \|x_n\| \leq \|x\| ,$$

then $\{x_n\}$ converges strongly to x.

A.1.6 Dual Pairs

In Chapter 6 we will use the notion of dual pairs.

Definition A.71. *Given two vector spaces X, Y, a* duality pairing *between X and Y is a bilinear map*

$$\langle \cdot, \cdot \rangle_{X,Y} : X \times Y \to \mathbb{R}$$

with the following properties:

(i) for all $x \in X \setminus \{0\}$ there exists $y \in Y \setminus \{0\}$ such that $\langle x, y \rangle_{X,Y} \neq 0$;
(ii) for all $y \in Y \setminus \{0\}$ there exists $x \in X \setminus \{0\}$ such that $\langle x, y \rangle_{X,Y} \neq 0$.

The triple $\left(X, Y, \langle \cdot, \cdot \rangle_{X,Y} \right)$ is called a dual pair. *For simplicity we will refer to (X, Y) as a dual pair.*

Given a dual pair it is always possible to endow X and Y with locally convex topologies as follows: for every $y \in Y$ consider the seminorm

$$p_y(x) := \left| \langle x, y \rangle_{X,Y} \right|, \quad x \in X.$$

By Theorem A.29, the family of seminorms $\{p_y\}_{y \in Y}$ generates a locally convex topology on X, denoted by $\sigma(X, Y)$. Note that a sequence $\{x_n\} \subset X$ converges to some element $x \in X$ with respect to the topology $\sigma(X, Y)$ if and only if

$$\lim_{n \to \infty} \langle x_n, y \rangle_{X,Y} = \langle x, y \rangle_{X,Y}$$

for every $y \in Y$. Similarly, for every $x \in X$ consider the seminorm

$$p_x(y) := \left| \langle x, y \rangle_{X,Y} \right|, \quad y \in Y.$$

Then again by Theorem A.29 the family of seminorms $\{p_x\}_{x \in X}$ generates a locally convex topology on Y, denoted by $\sigma(Y, X)$, and a sequence $\{y_n\} \subset Y$ converges to some element $y \in Y$ with respect to the topology $\sigma(Y, X)$ if and only if

$$\lim_{n \to \infty} \langle x, y_n \rangle_{X,Y} = \langle x, y \rangle_{X,Y}$$

for all $x \in X$. The importance of duality pairs is given by the following theorem.

Theorem A.72. Let $\left(X, Y, \langle \cdot, \cdot \rangle_{X,Y} \right)$ be a dual pair. Then the topological dual of $(X, \sigma(X, Y))$ is Y, that is, for every functional $L : X \to \mathbb{R}$ linear and continuous with respect to $\sigma(X, Y)$ there exists a unique $y \in Y$ such that

$$L(x) = \langle x, y \rangle_{X,Y}$$

for all $x \in X$. Similarly, the topological dual of $(Y, \sigma(Y, X))$ is X.

The proof of the previous theorem is hinged on the following auxiliary result.

Lemma A.73. Let $L, L_1, \ldots, L_n : X \to \mathbb{R}$ be linear functionals on a vector space X. Then there exist scalars $t_1, \ldots, t_n \in \mathbb{R}$ such that

$$L = \sum_{i=1}^{n} t_i L_i$$

if and only if

$$\bigcap_{i=1}^{n} \ker L_i \subset \ker L.$$

Proof. If $L = \sum_{i=1}^{n} t_i L_i$ for some $t_1, \ldots, t_n \in \mathbb{R}$ and if $x \in \bigcap_{i=1}^{n} \ker L_i$, then

$$L(x) = \sum_{i=1}^{n} t_i L_i(x) = 0,$$

so that $x \in \ker L$. Thus $\bigcap_{i=1}^{n} \ker L_i \subset \ker L$.

Conversely, assume that $\bigcap_{i=1}^{n} \ker L_i \subset \ker L$ and define the linear operator $T : X \to \mathbb{R}^n$ by

$$T(x) := (L_1(x), \ldots, L_n(x)), \quad x \in X.$$

On the range of T define the linear functional $\Phi : T(X) \to \mathbb{R}$ by

$$\Phi(L_1(x), \ldots, L_n(x)) := L(x).$$

Note that Φ is well-defined since $\bigcap_{i=1}^{n} \ker L_i \subset \ker L$. Extend Φ to all of $\mathbb{R}^n$. Then there exist scalars $t_1, \ldots, t_n \in \mathbb{R}$ such that

$$\Phi(y_1, \ldots, y_n) = t_1 y_1 + \ldots + t_n y_n$$

for all $y \in \mathbb{R}^n$. In particular, if $y \in T(X)$ then we may find $x \in X$ such that $y = T(x)$, so that

$$L(x) = \Phi(L_1(x), \ldots, L_n(x)) = t_1 L_1(x) + \ldots + t_n L_n(x).$$

This shows that $L = \sum_{i=1}^{n} t_i L_i$.

We now turn to the proof of Theorem A.72.

Proof (Theorem A.72). Since L is continuous at zero, given $\varepsilon = 1$, by Theorem A.29 there exist $k \in \mathbb{N}$ and $y_1, \ldots, y_n \in Y$ such that for all $x \in X$ with $p_{y_i}(x) := \left| \langle x, y_i \rangle_{X,Y} \right| \leq \frac{1}{k}$ for all $i = 1, \ldots, n$ we have $|L(x)| \leq 1$. Define the linear functionals $L_1, \ldots, L_n : X \to \mathbb{R}$ by

$$L_i(x) := \langle x, y_i \rangle_{X,Y}, \quad x \in X,$$

$i = 1, \ldots, n$. If $x \in \bigcap_{i=1}^{n} \ker L_i$ and $t > 0$, then $p_{y_i}(tx) = 0 \leq \frac{1}{k}$ for all $i = 1, \ldots, n$, and so

$$t |L(x)| \leq 1$$

for all $t > 0$, which implies that $L(x) = 0$. Thus $\bigcap_{i=1}^{n} \ker L_i \subset \ker L$, and so by the previous lemma there exist scalars $t_1, \ldots, t_n \in \mathbb{R}$ such that $L = \sum_{i=1}^{n} t_i L_i$. Define

$$y := \sum_{i=1}^{n} t_i y_i.$$

Then

$$L(x) = \sum_{i=1}^{n} t_i L_i(y) = \sum_{i=1}^{n} t_i \langle x, y_i \rangle_{X,Y} = \left\langle x, \sum_{i=1}^{n} t_i y_i \right\rangle_{X,Y} = \langle x, y \rangle_{X,Y}$$

for all $x \in X$.

To prove uniqueness assume that there exists $y_1 \neq y$ such that

$$L(x) = \langle x, y_1 \rangle_{X,Y}$$

for all $x \in X$. Then $\langle x, y - y_1 \rangle_{X,Y} = 0$ for all $x \in X$, which contradicts the definition of duality.

The previous theorem helps to explain why the topologies $\sigma(X, Y)$ and $\sigma(Y, X)$ are still referred to as the weak and weak star topologies.

Example A.74. (i) The most important example of a dual pair is given of course by taking a normed space X and by taking as Y its dual, with (A.1) as duality pairing.

(ii) Given an open set $\Omega \subset \mathbb{R}^N$ with the Lebesgue measure as underlying measure, in Section 6.4.5 we will also use the duality pair

$$\left(L^1\left(\Omega; \mathbb{R}^m\right), C_0\left(\Omega; \mathbb{R}^m\right)\right)$$

under the duality

$$\langle u, \phi \rangle_{L^1(\Omega;\mathbb{R}^m), C_0(\Omega;\mathbb{R}^m)} := \int_\Omega u \cdot \phi \, dx$$

for $u \in L^1\left(\Omega; \mathbb{R}^m\right)$ and $C_0\left(\Omega; \mathbb{R}^m\right)$, as well as

$$\left(\mathcal{M}\left(\Omega; \mathbb{R}^m\right), C_0\left(\Omega; \mathbb{R}^m\right)\right)$$

under the duality

$$\langle \lambda, \phi \rangle_{\mathcal{M}(\Omega;\mathbb{R}^m), C_0(\Omega;\mathbb{R}^m)} := \int_\Omega \phi \, d\lambda$$

for $\lambda \in \mathcal{M}\left(\Omega; \mathbb{R}^m\right)$ and $C_0\left(\Omega; \mathbb{R}^m\right)$.

A.1.7 Hilbert Spaces

Definition A.75. *An* inner product *on a vector space X is a map*

$$(\cdot, \cdot) : X \times X \to \mathbb{R}$$

such that

(i) $(x, y) = (y, x)$ *for all x, $y \in X$;*

(ii) $(sx + ty, z) = s(x, z) + t(y, z)$ *for all x, y, $z \in X$ and s, $t \in \mathbb{R}$;*

(iii) $(x, x) \geq 0$ *for every $x \in X$, $(x, x) = 0$ if and only if $x = 0$.*

If for every $x \in X$ we define

$$\|x\| := \sqrt{(x, x)},$$

then X becomes a normed space. We say that a normed space X is a *Hilbert space* if it is a Banach space.

Theorem A.76. *Let $(X, \|\cdot\|)$ be a normed space. Then there exists an inner product $(\cdot, \cdot) : X \times X \to \mathbb{R}$ such that $\|x\| = \sqrt{(x, x)}$ for all $x \in X$ if and only if $\|\cdot\|$ satisfies the* parallelogram law

$$\|x + y\|^2 + \|x - y\|^2 = 2\|x\|^2 + 2\|y\|^2$$

for all x, $y \in X$.

Remark A.77. If $\|\cdot\|$ satisfies the parallelogram law, then the inner product in the previous theorem is defined as

$$(x,y) := \frac{1}{4}\left[\|x+y\|^2 - \|x-y\|^2\right]$$

for all $x, y \in X$.

A.2 Wellorderings, Ordinals, and Cardinals

(by Ernest Schimmerling)

We give an overview of some basic results from set theory and we state specific theorems that are used in the text, namely Propositions A.82 and A.84.

Consider an arbitrary set X and a binary relation $\prec$ on X. We call $(X, \prec)$ a *linear ordering* if

(i) $\prec$ is *transitive* on X:

for all $x, y, z \in X$, if $x \prec y$ and $y \prec z$ then $x \prec z$;

(ii) *trichotomy* holds:

for all $x, y \in X$ we have that $x \prec y$ or $y \prec x$ or $x = y$;

(iii) $\prec$ is *irreflexive*:

for all $x \in X$ the property $x \prec x$ does not hold.

Some authors call this a *strict linear ordering* to distinguish it from the associated relation $\preccurlyeq$, whereby

for all $x, y \in X$ we have that $x \preccurlyeq y$ if and only if $x \prec y$ or $x = y$.

We say that $(X, \prec)$ is a *wellordering* if it is a linear ordering and $\prec$ is *wellfounded* on X:

for every $E \subset X$ with $E \neq \emptyset$ there exists $x \in E$
such that $x \preccurlyeq y$ for all $y \in E$.

The element x is called the $\prec$-*least* element of S. The property that there is no infinite descending sequence

$$\ldots \prec x_n \prec \ldots \prec x_1 \prec x_0$$

is equivalent to wellfoundedness. The usual proofs by induction and recursive definitions may be extended to wellorderings in the following way.

Proposition A.78 (Proofs by induction). *Let* $(X, \prec)$ *be a wellordering. Let* $P(x)$ *be a statement about a variable* x. *Suppose that for all* $y \in X$,

if $P(x)$ *holds for all* $x \prec y$ *then* $P(y)$ *holds.*

Then $P(y)$ *holds for all* $y \in X$.

Proposition A.79 (Recursive definitions). *Let* $(X, \prec)$ *be a wellordering. Let* $F : X \times \mathcal{G} \to Y$ *be a function such that* Y *is a set and* $\mathcal{G}$ *is the family of all partial functions from* X *to* Y^2. *Then there is a unique function* $G : X \to Y$ *such that*

$$G(y) = F(y, G \lfloor \{x \in X : x \prec y\}))$$

for all $y \in X$.

The function $F : X \times \mathcal{G} \to Y$ in Proposition A.79 tells us how to define $G(y)$ based only on the knowledge of y and of the function $x \mapsto G(x)$ for $x \prec y$.

Define X to be a *transitive set* if every element of X is also a subset of X, that is,

for all $y \in X$ and for all $x \in y$ we have $x \in X$.

We say that α is an *ordinal* if α is a transitive set and $(\alpha, \in)$ is a wellordering. It is a fundamental assumption of mathematics that $\in$ is wellfounded; this is called the *foundation axiom* in set theory. Therefore, α is an ordinal if and only if α is a transitive set and $(\alpha, \in)$ is a linear ordering. Starting from $\emptyset$, we use the operation $\alpha + 1 := \alpha \cup \{\alpha\}$ at successor stages and take unions at limit stages to generate all the ordinals beginning with the natural numbers $0 := \emptyset$, $1 := \{0\}$, $2 := \{0, 1\}$, $3 := \{0, 1, 2\}$, etc. The next ordinal after all the natural numbers is the set of nonnegative natural numbers[3],

$$\omega := \{0, 1, 2, \dots\}.$$

After ω come

$$\omega + 1 := \{0, 1, 2, \dots, \omega\},$$
$$\omega + 2 := \{0, 1, 2, \dots, \omega, \omega + 1\},$$
$$\omega + 3 := \{0, 1, 2, \dots, \omega, \omega + 1, \omega + 2\},$$

etc., followed by

$$\omega + \omega := \{0, 1, 2, \dots, \omega, \omega + 1, \omega + 2, \dots\},$$
$$\omega + \omega + 1 := \{0, 1, 2, \dots, \omega, \omega + 1, \omega + 2, \dots, \omega + \omega\},$$

---

[2] We say that g is a *partial function on* X if g is a function with domain contained in X.

[3] In the text ω is denoted by $\mathbb{N}_0$.

etc. Notice that $\in$ coincides with the usual ordering $<$ on the natural numbers. For this reason, when it comes to ordinals, we often write $\alpha < \beta$ instead of $\alpha \in \beta$. It is also worth mentioning that addition, multiplication, and exponentiation on ω lift to operations on the class of ordinals. We do not give the details, but we remark that care is needed in ordinal arithmetic, since, for example, $1 + \omega = \omega \neq \omega + 1$. The relationship between ordinals and arbitrary wellorderings is summarized by the following result.

Proposition A.80. *Let* $(X, \prec)$ *be a wellordering. Then there is a unique ordinal α and a unique order isomorphism* $(X, \prec) \simeq (\alpha, \in)$.

Another fundamental assumption of mathematics is that for every set X, there exists a binary relation $\prec$ on X such that $(X, \prec)$ is a wellordering. This is the *axiom of choice* (AC) in set theory. By Proposition A.80, AC is equivalent to the statement that for every set X, there exists an ordinal α and a bijection $f : \alpha \to X$. In plain language, this says that we can list the elements of X as

$$f(0), f(1), \ldots, f(\omega), f(\omega + 1), \ldots,$$

where α is the length of the list.

The *power set of* X, written $\mathcal{P}(X)$, is defined to be the set of all subsets of X. In other words, $\mathcal{P}(X) := \{Y : Y \subset X\}$.

Proposition A.81 (Cantor). *For all X, there is no surjection from X to $\mathcal{P}(X)$.*

We call a set X *countable* if either X is finite or X is in one-to-one correspondence with ω. It is a corollary to Proposition A.81 that $\mathcal{P}(\omega)$ is uncountable and that $\mathbb{R}$ is uncountable since $\mathbb{R}$ and $\mathcal{P}(\omega)$ are in one-to-one correspondence. From this and what we said about Proposition A.80 and AC, it follows that there are uncountable ordinals, that is, infinite ordinals that are not in one-to-one correspondence with ω. The least uncountable ordinal is called ω_1. Note that the ordinals $\omega + 1$, $\omega + \omega$, $\omega \cdot \omega$, ω^ω, etc., are all countable and hence strictly less than ω_1.

The following versions of Propositions A.78 and A.79 for ω_1 are used in the text. They are easy consequences of the general theory we have outlined.

Proposition A.82 (Proofs by induction). *Let $P(\alpha)$ be a statement about a variable α. Suppose that for all $\beta < \omega_1$,*

if $P(\alpha)$ holds for all $\alpha < \beta$ then $P(\beta)$ holds.

Then $P(\beta)$ holds for all $\beta < \omega_1$.

Remark A.83. If α and β are ordinals and $\beta = \alpha + 1$, then β is called a *successor* ordinal. Nonzero ordinals that are not successor ordinals are called *limit ordinals*. Often, in applications, the verification of the hypothesis of Proposition A.82 is broken up into three cases: $\beta = 0$, $\beta = \alpha + 1$, and β a limit ordinal.

Proposition A.84 (Recursive definitions). *Let* $F : \omega_1 \times \mathcal{G} \to Y$ *be a function, where* Y *is a set and* $\mathcal{G}$ *is the family of all partial functions from* ω_1 *to* Y. *Then there is a unique function* $G : \omega_1 \to Y$ *such that*

$$G(\beta) = F(\beta, G\lfloor \beta)$$

for all $\beta < \omega_1$.

Remark A.85. Often, in applications of Proposition A.84, the definition of $F(\beta, \cdot)$ is broken up into three cases: $\beta = 0$, $\beta = \alpha + 1$, and β a limit ordinal.

Given a set X, the least ordinal in one-to-one correspondence with X is called the *cardinality of* X, which is written card X. An ordinal α is a *cardinal* if and only if card $\alpha = \alpha$. Clearly, ω is a cardinal because it is the least infinite ordinal. When we think of ω as a cardinal, we may call it $\aleph_0$. It is also clear that ω_1 is a cardinal because it is the least uncountable ordinal. When we think of ω_1 as a cardinal, we may call it $\aleph_1$. In general, if α is an ordinal, then $\aleph_\alpha := \omega_\alpha$ is the αth infinite cardinal. The distinction between ω_α and $\aleph_\alpha$ is nonmathematical (they are equal), but it is useful nonetheless because cardinal arithmetic is different from ordinal arithmetic. A few examples suffice to give the flavor of cardinal arithmetic, which does not play an explicit role in the text:

$$\aleph_0 = \text{card}\,\omega = \text{card}\,(\omega + 1) = \text{card}\,(\omega + \omega) = \text{card}\,(\omega \cdot \omega) = \text{card}\,\omega^\omega$$

and

$$\aleph_1 = \text{card}\,\omega_1 = \text{card}\,(\omega_1 + \omega_1) = \text{card}\,(\omega_1 \cdot \omega_1) = \text{card}\,\omega_1^{\omega_1}.$$

In the examples above, we mean ordinal arithmetic inside card $(\cdot)$. Cardinal exponentiation is defined so that

$$2^{\text{card}\,X} = \text{card}\{f : f : X \to \{0,1\}\} = \text{card}\,\mathcal{P}(X).$$

To see that this makes sense consider the case in which X is a natural number: there are 2^n subsets of $n = \{0, \ldots, n-1\}$. Cardinalities of sets that are familiar to analysts include:

$$\aleph_0 = \text{card}\,\mathbb{N} = \text{card}\,\mathbb{Z} = \text{card}\,\mathbb{Q},$$

$$
\begin{aligned}
2^{\aleph_0} &= \text{card}\,\mathbb{R} \\
&= \text{card}\,\mathbb{C} \\
&= \text{card}\{B \subset \mathbb{R} : B \text{ is a Borel set}\} \\
&= \text{card}\{f : f : \mathbb{N} \longrightarrow \mathbb{R}\} \\
&= \text{card}\{f : \mathbb{R} \to \mathbb{R} : f \text{ continuous}\},
\end{aligned}
$$

and

$$2^{2^{\aleph_0}} = \operatorname{card}\mathcal{P}(\mathbb{R}) = \operatorname{card}\{f : f : \mathbb{R} \to \mathbb{R}\}.$$

From Proposition A.81, we know that $2^{\aleph_0} \geq \aleph_1$. However, by deep results of Kurt Gödel and Paul Cohen, the value of α such that $2^{\aleph_0} = \aleph_\alpha$ cannot be determined from the underlying assumptions about mathematics used in this textbook except in the unlikely event that these underlying assumptions are inconsistent. Collectively, these underlying assumptions are called Zermelo–Fraenkel set theory with the axiom of choice or ZFC. For more on this, see [Ku83] or [DW87].

B

Notes and Open Problems

> Unfortunately what is little
> recognized is that the most
> worthwhile scientific books are
> those in which the author clearly
> indicates what he does not know;
> for an author most hurts his
> readers by concealing difficulties.

> Evariste Galois

Chapter 1

For more information on abstract measure theory and for the proofs omitted in this chapter we refer to [AliBo99], [AmFuP00], [DB02], [DuSc88], [Ed95], [EvGa92], [Fe69], [Fol99], [GiMoSo98], [Rao04] [Ru87]; [Z67]. The reader should be warned that in several of these books (e.g., [DuSc88], [EvGa92], [Fe69]) outer measures are called measures.

Section 1.1

Section 1.1.1: For more information on measures with the finite subset property we refer to [Rao04] and [Z67]. Theorem 1.12 is due to Hewitt and Yoshida [He-Yo52]. The present proof is due to Heider [Hei58]. Exercise 1.18 is taken from [Fol99]. In the literature there are different definitions of atoms (see Remark 1.19). The proof of Proposition 1.20 has been taken from a paper of Farkas [Fa03]. Proposition 1.22 is due to Johnson [J70] (although with a different definition of atoms).

Section 1.1.2: The standard proof of Proposition 1.52 is based on the Riesz representation theorem in C_0 (see [Fol99], [Ru87]). The approach followed here is from [Coh93]. Exercises 1.58 and 1.61 are taken from [Fol99] and [AliBo99], respectively. The De Giorgi–Letta theorem in this version may be found in

[AmFuP00], [BrDF98], and [Bu89]. Theorem 1.64 is due to Fonseca and Malý [FoMy97].

Section 1.1.3: The proof of Proposition 1.87 is adapted from a paper of Buttazzo and Dal Maso [BuDM83]. Corollary 1.90 is due to Alberti [Al93], while Proposition 1.91 is taken from a paper of Halmos [Hal48].

Section 1.1.4: The proof of Step 3 in the Radon–Nikodym theorem is taken from [Rao04], to which we refer for further extensions. Lemmas 1.102, 1.103, 1.113 and Theorem 1.114 are due to De Giorgi (see [Mo96]). The proofs presented here have been obtained in collaboration with Massimiliano Morini.

Proposition 1.110 and Theorem 1.111 are taken from [Rao04], to which we refer for more information on localizable measures. Theorem 1.118 is due to Maynard [May79].

Section 1.1.5: Proposition 1.128 is due to Luther [Lu67], while Theorem 1.130 is due to Mukherjea [Muk73].

♠ Find necessary and sufficient conditions for the validity of Fubini's and Tonelli's theorems. See [Muk72], [Muk73] for some partial results.

Section 1.1.6: Theorem 1.133 is taken from [Rog98], while the remaining proofs are taken from a paper of Leese [Lee78].

Section 1.2

Section 1.2.1: The Morse covering theorem was originally proved by Morse in [Mor47] (see also the Errata in the paper of Bledsoe and Morse [BleMor52]). The current proof is a modified version of the one presented in a paper of Bliedtner and Loeb [Bli-Lo92]; in particular, it avoids transfinite induction. We thank Danut Arama for useful discussions on this part.

Section 1.2.2: The proof of the Besicovitch derivation theorem is an adaptation of the one due to Ambrosio and Dal Maso [AmDM92] (see also [AmFuP00]).

Section 1.3

Sections 1.3.1–1.3.3: Some of the material in these sections is taken from [AliBo99], [DB02], and [DuSc88].

Section 1.3.4: Some of the material in this section is taken from [AmFuP00] and [Bi99]. We thank Gordan Zitkovic for useful conversations on this part and for suggesting Exercise 1.211.

Chapter 2

For more information on L^p spaces and for the proofs omitted in this chapter we refer to [AmFuP00], [Bar95], [DB02], [DuSc88], [Ed95], [EvGa92], [Fol99], [Ru87], [Z67].

Section 2.1

Section 2.1.1: Theorem 2.5 is due to Subramanian [S78] (see also the papers of Romero [R83] and of Villani [Vi85]). We thank Peter Lumsdaine for a clean proof of Step 3 of Theorem 2.16.

Section 2.1.2: The decomposition lemma was originally proved by Kadec' and Pelčin'ski [KP65]. The present proof may be found in a paper of Delbaen and Schachermayer [DS99] (see also the paper of Fonseca, Müller, and Pedregal [FoMuPe98]). We thank Gordan Zitkovic for bringing the reference [DS99] to our attention.

Section 2.1.3: The proof of Theorem 2.34 is due to W. Rudin (see [Le56]). Theorem 2.35 is due to Leach [Le56]. The Riesz representation theorem in L^1 is due to Schwartz [Sc51], while the proof of Corollary 2.41 may be found in [Rao04] and [Z67]. Exercise 2.43 is taken from a paper of Farkas [Fa03]. The Riesz representation theorem in L^∞ is due to Hewitt and Yoshida [He-Yo52] (see also [H94] and [Rao04]).

Section 2.1.4: The proof of the Dunford–Pettis theorem in this generality is due to Ambrosio, Fusco, and Pallara [AmFuP00] except for Step 4.

Theorem 2.59 is due to Dal Maso (see [Am89]).

Section 2.1.5: For alternative proofs of the biting lemma and for historical background we refer to [BaMu89], [GiMoSo98], [Pe97]. Exercises 2.64 and 2.69 are taken from a paper of Ball and Murat [BaMu89]. Proposition 2.71 may be found in a paper of Müller [Mu90].

♠ Find an appropriate version of the biting lemma in the case in which the measure μ is infinite.

Section 2.2

Section 2.2.2: The proof of Theorem 2.88 is taken from a paper of Serrin [Ser62].
Section 2.2.3: The material in this subsection is taken from [Ste70], [Ste93].

Section 2.3

With the exception of the proof of the Riesz representation theorem, all the material in this section is taken from [DuSc88], [DieU77], [Ed95], and [SY05].

Chapter 3

For more information on the material of this chapter we refer to [BrDF98], [Br02], [Bu89], [CaDA02], [DM93], [GiMoSo98]. The reader should be warned that we use here a different definition of coercivity.

Chapter 4

For more information on the material of this chapter we refer to [AliBo99], [CaDA02], [CasVa77], [Clar90], [Dac89], [EkTe99], [Roc97], [RocWe98].

Section 4.1

The proof of Proposition 4.7 is taken from [Bo].

Section 4.2

The proofs in this section are taken from [Bo].

Section 4.4

Remark 4.31 may be found in a paper of Artola and Tartar [ArtTa95]. Theorem 4.32 in this generality is due to C. Pucci (see [DB02]); see also the papers of McShane [Mc34] and Kirszbraum [Kir34] for the Lipschitz case.

Section 4.5

Theorem 4.36 is well known. The present proof, which provides an explicit constant, may be found in a paper of Ball, Kirchheim, and Kristensen [BaKiKr00]. Theorem 4.56 is taken from a paper of Bauschke, Borwein, and Combettes [BBC01]. Proposition 4.64 was first proved by Marcellini [Mar85].

Section 4.6

The material of this section is taken from [Roc97], [RocWe98].

Section 4.7

Theorem 4.79 is due to De Giorgi [DG68](see also the paper of Gori and Marcellini [GorMar02]). We thank Virginia De Cicco for bringing the reference [DG68] to our attention

♠ Theorem 4.79 provides an explicit formula for the affine functions approximating a given convex function f. Does an analogous formula hold for convex functions $f : \mathbb{R}^m \to (-\infty, \infty]$?

Section 4.8

Exercise 4.94, as well as Theorem 4.98 and Proposition 4.100 in Section 4.8, is due to Carbone and De Arcangelis [CaDA99] (see also [CaDA02]). Proposition 4.102 is due to Ambrosio [Am87]. The proof of Theorem 4.103 presented here is due to Kirchheim and Kristensen [KiKr01].

Section 4.9

Theorem 4.107 may be found in a paper of Toranzos [T67].

Chapter 5
Section 5.1

Theorems 5.1 and 5.6 have been adapted from a paper of Marcus and Mizel [MarMi79]. We thank an anonymous referee for significant simplifications of the original proof and for Lemma 5.2.

Section 5.2

Section 5.2.2: The blowup method was introduced in a paper of Fonseca and Müller [FoMu92] (see also the paper of Fonseca and Malý [FoMy97]). The first part of the statement of Proposition 5.16 may be found in two papers due to Ioffe [Io77], [Io77a], while inequality (5.20) is due to Giuseppe Savaré. The necessity part of Theorem 5.17 seems to be new.

Section 5.2.3: Theorem 5.27 and the necessity part of Theorems 5.19, 5.25 seem to be new.

♠ Prove a suitable version of Theorem 5.27 when E is unbounded. Note that in this case the dual of $C_b\left(\overline{E}; \mathbb{R}^m\right)$ is the space of regular finitely additive measures rba $\left(\overline{E}; \mathbb{R}^m\right)$, where it is known that Radon–Nikodym theorem fails.

Section 5.3

The proof of Theorem 5.29 is taken from a paper of Buttazzo and Dal Maso [BuDM83] (see also [Bu89]).

Section 5.4

Section 5.4.1: The proof of the second part of Theorem 5.32 has been adapted from the work of Carbone and Dearcangelis [CaDA96], [CaDA99] (see also [CaDA02]). Exercise 5.35 may be found in [CaDA96] and [CaDA02].

Section 5.4.2: The proof of Theorem 5.36 draws upon a paper of Goffman and Serrin [GoSer64] (see also the paper of Serrin [Ser61]).

♠ Is the function h in Theorem 5.36 the recession function of the function g?

Section 5.5

Theorem 5.38 is due to Friesecke [Fri94], while Theorem 5.40 seems to be new.

Chapter 6
Section 6.1

For more information on multifunctions we refer to [AliBo99], [CasVa77].

Section 6.1.1: The proof of Theorem 6.9 is taken from a paper of Leese [Lee78], while Theorem 6.13 may be found in a paper of Valadier [Va71].

Section 6.1.2: The material of this subsection is due to Michael [Mi56], except for Theorem 6.18, which is taken from a paper of Fathi [F97].

Section 6.2

For more information on integrands we refer to [Bu89], [EkTe99], and [Roc76].

Section 6.2.1: Proposition 6.24 is due to Buttazzo and Dal Maso [BuDM83].

Section 6.2.2: The material of this subsection is taken from [Bu89], [EkTe99], and [Roc76].

Section 6.2.3: Theorem 6.36 is due to De Giorgi (see [GorMar02]). Theorem 6.40 was proved in a paper of Dal Maso and Sbordone [DMSb95]. Proposition 6.42 is due to Ambrosio [Am87] (see also the paper of Fonseca and Leoni [FoLe00]). Proposition 6.43 may be found in [Roc76].

♠ Under the hypotheses of Proposition 6.42 one can prove the existence of two sequences of continuous functions

$$a_i : E \to \mathbb{R}, \qquad b_i : E \to \mathbb{R}^m,$$

such that

$$f(x, z) = \sup_{i \in \mathbb{N}} \{a_i(x) + b_i(x) \cdot z\}$$

for all $x \in E$ and $z \in \mathbb{R}^m$. On the other hand, if we assume that f is also real-valued, then De Giorgi's theorem allows one to give an alternative approximation in which the functions a_i and b_i are now given by the explicit formula (6.26). Under the hypotheses of Proposition 6.42 and when f is real-valued, are the functions (6.26) continuous?

Section 6.3

Subsection 6.3.1: Theorem 6.45 is due to Alberti [Al93].

♠ Prove the analogue of Theorem 6.45 for sets E with infinite measure. What happens if the Lebesgue measure is replaced by an arbitrary measure?
♠ Prove the analogue of Exercises 5.5 and 5.8 for integrands $f = f(x, z)$.

Section 6.4

♠ Throughout this section the underlying measure is the Lebesgue measure. Most of the proofs can be carried out for non atomic Radon measures. But what happens when the underlying measure has atoms? See the papers of Bouchitté and Buttazzo [BouBu90] and [BouBu92] for some results in this direction.

Section 6.4.1: Theorem 6.49 is due to Alberti [Al93].
Section 6.4.3: Theorem 6.54 is due to Ioffe [Io77], [Io77a]. The sufficiency part of the proof is adapted from a paper of Fonseca and Müller [FoMu92] (see also the paper the paper of Fonseca and Malý [FoMy97]). Step 1 of the necessity part of the proof is taken from [Bu89]. We thank Giuseppe Savaré for simplifying the proof of Step 2.
Section 6.4.5: The material of this subsection is taken from a paper of Bouchitté and Valadier [BouVa88] (see also the paper of Ambrosio and Buttazzo [AmBu88] and [Bu89]).

♠ Condition (6.95) in Theorem 6.57 is not necessary. Find a necessary and sufficient growth condition from below. That is, find the analogue of the growth conditions in Theorems 5.19 and 5.25 in Section 5.2.3.

♠ Without the growth condition (6.95) is (6.96) still valid? If not, what is the right formula?

♠ Characterize the class of integrands f for which condition (6.96) holds.

♠ Prove the analogue of Theorem 5.27 for integrands $f = f(x, z)$.

Section 6.5

Theorem 6.65 and Exercise 6.66 are taken from a paper of Buttazzo and Dal Maso [BuDM83].

Section 6.6

Section 6.6.1: The second part of Theorem 6.68 has been adapted from the work of Carbone and Dearcangelis [CaDA96], [CaDA99] (see also [CaDA02]).

♠ What happens in Theorem 6.68 when the growth condition (6.145) is replaced by the growth condition (6.49)?

Section 6.6.2: Theorem 6.70 is due to Bouchitté and Valadier [BouVa88] (see also the paper of Ambrosio and Buttazzo [AmBu88] and [Bu89]).

♠ If the coercivity condition (6.150) is replaced by the hypothesis that $f \geq 0$ what is the relation between the integrands h and f in Remark 6.71(i)? Does h still coincide with the function $\tilde{f}$ defined in (6.151)?

♠ More generally, what happens in Theorem 6.70 when the growth condition (6.150) is replaced by the growth condition (6.49) for $p = 1$?

♠ What is the analogue of Theorem 5.38 for integrands $f = f(x, z)$?

Chapter 7
Section 7.3

Section 7.3.2: Theorem 7.5 and Lemma 7.7 (except for Step 4) are due to Ioffe [Io77].

♠ Find a simpler proof of Lemma 7.7.

♠ What is the analogue of Theorem 6.57 for integrands $f = f(x, u, z)$?

Section 7.4

Proposition 6.44 is taken from [Bu89], while Exercise 7.14 is from a paper of Marcellini and Sbordone [MarSb80].

♠ What is the analogue of Theorem 6.70 for integrands $f = f(x, u, z)$?

♠ What is the analogue of Theorem 5.38 for integrands $f = f(x, u, z)$?

Chapter 8

For more information on Young measures we refer to the monographs [Mu99], [Pe97], [Ta79], [Y69], as well as to the papers of Ball [Ba89], Berliocchi and Lasry [BerLa73], and Tartar [Ta95].

Appendix

Basic references for the appendix are [AliBo99], [Bre83], [DuSc88], [Ru91], [Yo95].

Notation and List of Symbols

- $\mathbb{N}$ = the set of positive integers, $\mathbb{N}_0 = \mathbb{N} \cup \{0\}$; $\mathbb{R}$ = real line; $\overline{\mathbb{R}} :=$ $[-\infty, \infty]$ = the extended real line; if $x \in \mathbb{R}$, then $x^+ := \max\{x, 0\}$, $x^- := \max\{-x, 0\}$, $|x| := x^+ + x^-$, $\lceil x \rceil$ = the integer part of x.
- $\mathbb{R}^N$ = the N-dimensional Euclidean space, $N \geq 1$; for $x = (x_1, \ldots, x_N) \in \mathbb{R}^N$,

$$|x| := \sqrt{(x_1)^2 + \ldots + (x_N)^2};$$

Ω = open set of $\mathbb{R}^N$ (not necessarily bounded); $B_N(x, r)$ (or simply $B(x_0, r)$ whenever the underlying space is clear) = open ball in $\mathbb{R}^N$ of center x_0 and radius r;

$$Q_N(x_0, r) := x_0 + \left(-\frac{r}{2}, \frac{r}{2}\right)^N$$

(or simply $Q(x_0, r)$ whenever the underlying space is clear); S^{N-1} = unit sphere in $\mathbb{R}^N$; for a multi-index $\alpha = (\alpha_1, \ldots, \alpha_N) \in (\mathbb{N}_0)^N$,

$$\frac{\partial^\alpha}{\partial x^\alpha} := \frac{\partial^{|\alpha|}}{\partial x_1^{\alpha_1} \ldots \partial x_N^{\alpha_N}}, \quad |\alpha| := \alpha_1 + \ldots + \alpha_N;$$

$C^m(\Omega)$ = the space of all functions that are continuous together with their partial derivatives up to order $m \in \mathbb{N}_0$;

$$C^\infty(\Omega) := \bigcap_{m=0}^{\infty} C^m(\Omega);$$

$C_c^m(\Omega)$ and $C_c^\infty(\Omega)$ = the subspaces of $C^m(\Omega)$ and $C^\infty(\Omega)$, respectively, consisting of all functions with compact support.
- X, Y usually denote sets or spaces, card X = the cardinality of a set X, $\mathcal{P}(X)$ = the family of all subsets of a set X; $\mathcal{F}$, $\mathcal{G}$ family of sets or of functions;
- A, U usually denote open sets, K a compact set; B a Borel set, C a closed or convex set or an arbitrary constant; τ = topology;

- $\mathfrak{M}$, $\mathfrak{N}$ = algebras or σ-algebras; $\mathfrak{M} \otimes \mathfrak{N}$ = product σ-algebra of $\mathfrak{M}$ and $\mathfrak{N}$ (not to be confused with $\mathfrak{M} \times \mathfrak{N}$); $\mathcal{B}(X)$ = Borel σ-algebra;
- μ, ν, υ = (positive) finitely additive measures or positive measures; $\nu \perp \mu$ means that μ, ν are mutually singular measures; $\nu \ll \mu$ means that the measure ν is absolutely continuous with respect to the measure μ; $\frac{d\nu}{d\mu}$ is the Radon–Nikodym derivative of ν with respect to μ; $\mathrm{supp}\,\mu$ support of a Borel measure; μ^* outer measure;
- $\lfloor$ = restriction;
- dist = distance; diam = diameter;
- $\|\cdot\|$ norm or total variation; $\|\cdot\|_{L^p}$ norm in L^p spaces;
- E, F, G usually denote sets; ∂E = boundary of E; $\mathrm{co}\,(E)$ = convex hull of E; aff (E) = affine hull of E; $\mathrm{ri}_{\mathrm{aff}}(E)$ = relative interior of E with respect to aff (E); $\mathrm{rb}_{\mathrm{aff}}(E)$ = the relative boundary of E with respect to aff (E), E^∞ = the recession cone of E, $\ker E$ = kernel of a set;
- χ_E = characteristic function of the set E; I_E = indicator function of the set E;
- u, v, and w usually denote functions or variables, z usually denotes an element of $\mathbb{R}^m$;
- f, g, φ, ψ, ϕ usually denote functions; $\mathrm{supp}\,f$ = support of the function f; $\mathrm{dom}_e\,f$ = effective domain of the function f; $\mathrm{Lip}\,f$ = Lipschitz constant of the function f; $\partial f(z)$ = subdifferential of f at z; $\mathrm{osc}\,(f; E)$ = oscillation of f on E; f^∞ = the recession function of f; $\mathcal{C}\,f$ = the convex envelope of f; f^* = the polar or conjugate function of f; f^{**} = the bipolar or biconjugate function of f^1; $f * g$ = convolution of the functions f and g;
- $\mathcal{L}_o^N$ = Lebesgue outer measure, $\mathcal{L}^N$ = Lebesgue measure;

$$\alpha_N := \mathcal{L}^N\left(B\left(0,1\right)\right);$$

$|E| := \mathcal{L}^N(E)$ for $E \subset \mathbb{R}^N$ Lebesgue measurable;
- $\mathrm{M}(u)$, $\mathrm{M}_R(u)$ denote various types of maximal functions,
- det = determinant of a matrix or a linear mapping;
- $\langle \cdot, \cdot \rangle_{X,Y}$ = duality pairing;
- λ, ς usually denote signed measures or finitely additive signed measures;
- $\mathcal{M}(X;\mathbb{R})$ = space of all (signed) finite Radon measures;
- $\mathrm{ba}(X,\mathfrak{M})$ = space of all bounded finitely additive signed measures; $\mathrm{rba}(X,\mathfrak{M})$ = space of all regular bounded finitely additive signed measures; $\mathrm{ba}(X,\mathfrak{M},\mu)$ = space of all bounded finitely additive signed measures absolutely continuous with respect to μ;
- $B(X,\mathfrak{M})$ = space of all bounded measurable functions;
- $C(X)$ = space of all continuous functions; $C_b(X)$ = space of all continuous bounded functions; $C_0(X)$ = space of all continuous functions that vanish

---

[1] Unless otherwise specified, $\mathcal{C}f(x,\cdot)$, $f^*(x,\cdot)$, $f^{**}(x,\cdot)$, and lsc $f(x,\cdot)$ stand for the convex envelope, polar, bipolar, and lower semicontinuous envelope, respectively, of $f(x,\cdot)$.

at infinity; $C_c(X)$ = space of all continuous functions whose support is compact;

- $L^p(X, \mathfrak{M}, \mu)$, $L^p(X, \mu)$, $L^p(X)$ are various notations for L^p spaces; p' = Hölder conjugate exponent of p;
- spaces $\ell^p := L^p(\mathbb{N}, \mathcal{P}(\mathbb{N}), \text{counting measure})$;
- $L^p(X; Y) = L^p$ space on Banach spaces; $L^p_w(X; Y') = L^p$ space of weakly star measurable functions;
- $\rightharpoonup$ denotes weak convergence; $\overset{*}{\rightharpoonup}$ denotes weak star convergence; $\overset{b}{\rightharpoonup}$ denotes biting convergence;
- I usually denotes an interval, a function, or a functional, epi I = epigraph of the function I; lsc I = lower semicontinuous envelope of the function I; slsc I = sequentially lower semicontinuous envelope of the function I; $\mathcal{E}$, $\mathcal{E}_p$ = types of relaxed energy of the function I;
- Γ = multifunction, Gr Γ = graph of the multifunction.

Acknowledgments

This book grew out of a series of lectures, Ph.D. research topics courses and courses taught at several summer schools, and for this the authors thank the hospitality and support of Carnegie Mellon University and the University of Pavia, Italy. Exceptional working conditions during the preparation of this book were offered by the Mathematisches Forschungsinstitut Oberwolfach through the Research in Pairs program in the Spring of 2002 and by the Institute for Advanced Study, Princeton, in Spring 2003.

Several iterations of the manuscript benefited from the input of many colleagues and students in particular: Tom Bohman, Giuseppe Buttazzo, Gianni Dal Maso, Virginia De Cicco, Jan Kristensen, Giuseppe Savare, Enrico Vitali, Gordan Zitkovic and the students Danut Arama, Peter Lumsdaine, and Alex Rand. The authors are profoundly indebted to Nicola Fusco and Massimiliano Morini for their careful reading and extremely insightful comments and suggestions that greatly improved the book, to Ernest Schimmerling for writing Section A.2, and to two anonymous referees for pointing out some errors and typos in an earlier draft.

The authors thank the Center for Nonlinear Analysis (NSF Grants No. DMS-9803791, DMS-0405343) for its support during the preparation of this book. The research of I. Fonseca was partially supported by the National Science Foundation under Grants No. DMS-0103798, DMS-040171 and that of G. Leoni under Grant No. DMS-0405423.

The epigraphs were taken from the Mathematical Quotations Server of Furman University:

http://math.furman.edu/~mwoodard/mquot.html

References

[Al93] Alberti, G.: Integral representation of local functionals. Ann. Mat. Pura Appl. (4) **165**, 49–86 (1993).

[AlBu93] Alberti, G., Buttazzo, G.: Local mappings on spaces of differentiable functions. Manuscripta Math. **79**, no. 1, 81–97 (1993). Errata Manuscripta Math. **81**, no. 3-4, 445–446 (1993).

[AliBur99] Aliprantis, C.D., Burkinshaw, O.: *Problems in Real Analysis. A Workbook with Solutions*. Second edition. Academic Press, Inc., San Diego, CA (1999).

[AliBo99] Aliprantis, C.D., Border, K.C.: *Infinite-Dimensional Analysis. A Hitchhiker's Guide*. Second edition. Springer-Verlag, Berlin (1999).

[Am87] Ambrosio, L.: New lower semicontinuity results for integral functionals. Rend. Accad. Naz. Sci. XL Mem. Mat. (5) **11**, no. 1, 1–42 (1987).

[Am89] Ambrosio, L.: A compactness theorem for a new class of functions of bounded variation. Boll. Un. Mat. Ital. B (7) **3**, no. 4, 857–881 (1989).

[AmBu88] Ambrosio, L., Buttazzo, G.: Weak lower semicontinuous envelope of functionals defined on a space of measures. Ann. Mat. Pura Appl. (4) **150**, 311–339 (1988).

[AmDM92] Ambrosio, L., Dal Maso, G.: On the relaxation in $BV(\Omega; R^m)$ of quasiconvex integrals. J. Funct. Anal. **109**, no. 1, 76–97 (1992).

[AmFuP00] Ambrosio, L., Fusco, N., Pallara, D.: *Functions of Bounded Variation and Free Discontinuity Problems*. Oxford Mathematical Monographs. The Clarendon Press, Oxford University Press, New York (2000).

[ArtTa95] Artola, M., Tartar, L.: Un résultat d'unicité pour une classe de problèmes paraboliques quasi-linéaires. Ricerche Mat. **44**, no. 2, 409–420 (1995)(1996).

[Ba89] Ball, J.M.: A version of the fundamental theorem for Young measures. In: PDEs and continuum models of phase transitions (Nice, 1988), 207–215, Lecture Notes in Phys., 344, Springer, Berlin (1989).

[BaKiKr00] Ball, J.M., Kirchheim, B., Kristensen, J.: Regularity of quasiconvex envelopes. Calc. Var. Partial Differential Equations **11**, no. 4, 333–359 (2000).

[BaMu89] Ball, J.M., Murat, F.: Remarks on Chacon's biting lemma. Proc. Amer. Math. Soc. **107**, no. 3, 655–663 (1989).

[Bar95]   Bartle, R.G.: *The Elements of Integration and Lebesgue Measure*. Containing a corrected reprint of the 1966 original. Wiley Classics Library. A Wiley-Interscience Publication. John Wiley & Sons, Inc., New York, (1995).

[BBC01]   Bauschke, H.H., Borwein, J.M., Combettes, P.L.: Essential smoothness, essential strict convexity, and Legendre functions in Banach spaces. Commun. Contemp. Math. **3**, no. 4, 615–647 (2001).

[BerLa73] Berliocchi, H., Lasry, J.M.: Intégrandes normales et mesures paramétrées en calcul des variations. (French) Bull. Soc. Math. France **101**, 129–184 (1973).

[Bi99]   Billingsley, P.: *Convergence of probability measures*. Second edition. Wiley Series in Probability and Statistics: Probability and Statistics. A Wiley-Interscience Publication. John Wiley & Sons, Inc., New York (1999).

[BleMor52] Bledsoe, W.W., Morse, A.P.: Some aspects of covering theory. Proc. Amer. Math. Soc. **3**, 804–812 (1952).

[Bli-Lo92] Bliedtner, J., Loeb, P.: A reduction technique for limit theorems in analysis and probability theory. Ark. Mat. **30**, no. 1, 25–43 (1992).

[Bo]   Border, K.C.: *Convex Analysis and Economic Theory*, Lecture Notes.

[BouBu90] Bouchitté, G., Buttazzo, G.: New lower semicontinuity results for nonconvex functionals defined on measures. Nonlinear Anal. **15**, no. 7, 679–692 (1990).

[BouBu92] Bouchitté, G., Buttazzo, G.: Integral representation of nonconvex functionals defined on measures. Ann. Inst. H. Poincaré Anal. Non Linéaire **9**, no. 1, 101–117 (1992).

[BouDM93] Bouchitté, G., Dal Maso, G.: Integral representation and relaxation of convex local functionals on $BV(\Omega)$. Ann. Scuola Norm. Sup. Pisa Cl. Sci. (4) **20**, no. 4, 483–533 (1993).

[BouVa88] Bouchitté, G., Valadier, M.: Integral representation of convex functionals on a space of measures. J. Funct. Anal. **80**, no. 2, 398–420 (1988).

[BrDF98] Braides, A., Defranceschi, A.: *Homogenization of Multiple Integrals*. Oxford Lecture Series in Mathematics and Its Applications, 12. The Clarendon Press, Oxford University Press, New York (1998).

[Br02]   Braides, A.: *Γ-Convergence for Beginners*. Oxford Lecture Series in Mathematics and Its Applications, 22. Oxford University Press, Oxford (2002).

[Bre83]   Brezis, H.: *Analyse Fonctionnelle. Théorie et Applications*. Collection Mathématiques Appliquées pour la Maîtrise. Masson, Paris (1983).

[Bu89]   Buttazzo, G.: *Semicontinuity, Relaxation and Integral Representation in the Calculus of Variations*. Pitman Research Notes in Mathematics Series, 207. Longman Scientific & Technical, Harlow; copublished in the United States with John Wiley & Sons, Inc., New York (1989).

[BuDM83] Buttazzo, G., Dal Maso, G.: On Nemyckii operators and integral representation of local functionals. Rend. Mat. (7) **3**, no. 3, 491–509 (1983).

[CaDA96] Carbone, L., De Arcangelis, R.: On the relaxation of some classes of unbounded integral functionals. Matematiche (Catania) **51**, no. 2, 221–252 (1996) (1997).

[CaDA99] Carbone, L., De Arcangelis, R.: On a non-standard convex regularization and the relaxation of unbounded integral functionals of the calculus of variations. J. Convex Anal. **6**, no. 1, 141–162 (1999).

[CaDA02] Carbone, L., De Arcangelis, R.: *Unbounded Functionals in the Calculus of Variations. Representation, Relaxation, and Homogenization.* Chapman & Hall/CRC Monographs and Surveys in Pure and Applied Mathematics, 125. Chapman & Hall/CRC, Boca Raton, FL (2002).

[CasVa77] Castaing, C., Valadier, M.: *Convex Analysis and Measurable Multifunctions.* Lecture Notes in Mathematics, Vol. 580. Springer-Verlag, Berlin-New York, (1977).

[Clar90] Clarke, F.H.: *Optimization and Nonsmooth Analysis.* Second edition. Classics in Applied Mathematics, 5. Society for Industrial and Applied Mathematics (SIAM), Philadelphia, PA (1990).

[Coh93] Cohn, D.L.: *Measure Theory.* Reprint of the 1980 original. Birkhäuser Boston, Inc., Boston, MA (1993).

[Dac89] Dacorogna, B.: *Direct Methods in the Calculus of Variations.* Applied Mathematical Sciences, 78. Springer-Verlag, Berlin (1989).

[DW87] Dales, H.G., Woodin, W.H.: *An Introduction to Independence for Analysts.* London Mathematical Society Lecture Note Series, 115. Cambridge University Press, Cambridge (1987).

[DM93] Dal Maso, G.: *An Introduction to Γ-convergence.* Progress in Nonlinear Differential Equations and Their Applications, 8. Birkhäuser Boston, Inc., Boston, MA (1993).

[DMSb95] Dal Maso, G., Sbordone, C.: Weak lower semicontinuity of polyconvex integrals: a borderline case. Math. Z. **218**, no. 4, 603–609 (1995).

[DG68] De Giorgi E.: Teoremi di semicontinuità del Calcolo delle Variazioni, Istituto Nazionale di Alta Matematica, 1968-1969.

[DVP15] De La Vallée Poussin, C.: Sur l'intégrale de Lebesgue. (French) [On the Lebesgue integral] Trans. Amer. Math. Soc. **16**, no. 4, 435–501 (1915).

[DS99] Delbaen, F., Schachermayer, W.: A compactness principle for bounded sequences of martingales with applications. In: Seminar on Stochastic Analysis, Random Fields and Applications (Ascona, 1996), 137–173, Progr. Probab., 45, Birkhäuser, Basel (1999).

[DB02] DiBenedetto, E.: *Real Analysis.* Birkhäuser Advanced Texts: Basler Lehrbücher. Birkhäuser Boston, Inc., Boston, MA (2002).

[DieU77] Diestel, J., Uhl, J.J., Jr.: *Vector Measures.* With a foreword by B. J. Pettis. Mathematical Surveys, No. 15. American Mathematical Society, Providence, R.I. (1977).

[DuSc88] Dunford, N., Schwartz, J.T.: *Linear Operators. Part I. General Theory.* With the assistance of William G. Bade and Robert G. Bartle. Reprint of the 1958 original. Wiley Classics Library. A Wiley-Interscience Publication. John Wiley & Sons, Inc., New York (1988).

[Ed95] Edwards, R.E.: *Functional Analysis. Theory and Applications.* Corrected reprint of the 1965 original. Dover Publications, Inc., New York (1995).

[EkTe99] Ekeland, I., Témam, R.: *Convex Analysis and Variational Problems.* Translated from the French. Corrected reprint of the 1976 English edition. Classics in Applied Mathematics, 28. Society for Industrial and Applied Mathematics (SIAM), Philadelphia, PA (1999).

[EvGa92] Evans, L.C., Gariepy, R.F.: *Measure Theory and Fine Properties of Functions.* Studies in Advanced Mathematics. CRC Press, Boca Raton, FL (1992).

[Fa03] Farkas, B.: On the duals of L^p spaces with $0 < p < 1$. Acta Math. Hungar. **98**, no. 1-2, 71–77 (2003).

[F97]   Fathi, A.: Partitions of unity for countable covers. Amer. Math. Monthly **104**, no. 8, 720–723 (1997).

[Fe69]   Federer, H.: *Geometric Measure Theory*. Die Grundlehren der mathematischen Wissenschaften, Band 153 Springer-Verlag New York Inc., New York (1969).

[Fol99]   Folland, G.B.: *Real analysis. Modern Techniques and Their Applications.* Second edition. Pure and Applied Mathematics (New York). A Wiley-Interscience Publication. John Wiley & Sons, Inc., New York, (1999).

[FoLe00]   Fonseca, I., Leoni, G.: Some remarks on lower semicontinuity. Indiana Univ. Math. J. **49**, no. 2, 617–635 (2000).

[FoLe10]   Fonseca, I., Leoni, G.: *Modern Methods in the Calculus of Variations: Sobolev Spaces.* In preparation.

[FoMy97]   Fonseca, I., Malý, J.: Relaxation of multiple integrals below the growth exponent. Ann. Inst. H. Poincaré Anal. Non Linéaire **14**, no. 3, 309–338 (1997).

[FoMu92]   Fonseca, I., Müller, S.: Quasi-convex integrands and lower semicontinuity in L^1. SIAM J. Math. Anal. **23**, no. 5, 1081–1098 (1992).

[FoMuPe98] Fonseca, I., Müller, S., Pedregal, P.: Analysis of concentration and oscillation effects generated by gradients. SIAM J. Math. Anal. **29**, no. 3, 736–756 (1998).

[FoTa89]   Fonseca, I., Tartar, L.: The gradient theory of phase transitions for systems with two potential wells. Proc. Roy. Soc. Edinburgh Sect. A **111**, no. 1-2, 89–102 (1989).

[Fri94]   Friesecke, G.: A necessary and sufficient condition for nonattainment and formation of microstructure almost everywhere in scalar variational problems. Proc. Roy. Soc. Edinburgh Sect. A **124**, no. 3, 437–471 (1994).

[GiMoSo98] Giaquinta, M., Modica, G., Souček, J.: *Cartesian Currents in the Calculus of Variations. I. Cartesian Currents.* Ergebnisse der Mathematik und ihrer Grenzgebiete. 3. Folge. A Series of Modern Surveys in Mathematics, 37. Springer-Verlag, Berlin (1998).

[GoSer64] Goffman, C., Serrin, J.: Sublinear functions of measures and variational integrals. Duke Math. J. **31**, 159–178 (1964).

[GorMar02] Gori, M., Marcellini, P.: An extension of the Serrin's lower semicontinuity theorem. Special issue on optimization (Montpellier, 2000). J. Convex Anal. **9**, no. 2, 475–502 (2002).

[Hal48]   Halmos, P.R.: The range of a vector measure. Bull. Amer. Math. Soc. **54**, 416–421 (1948).

[Ha16]   Hardy, G. H.: Weierstrass's non-differentiable function. Trans. Amer. Math. Soc. **17**, no. 3, 301–325 (1916).

[Hei58]   Heider, L. J.: A representation theory for measures on Boolean algebras. Michigan Math. J. **5**, 213–221 (1958).

[H94]   Hensgen, W.: An example concerning the Yosida-Hewitt decomposition of finitely additive measures. Proc. Amer. Math. Soc. **121**, no. 2, 641–642 (1994).

[He-Yo52] Yosida, K., Hewitt, E.: Finitely additive measures. Trans. Amer. Math. Soc. **72**, 46–66 (1952).

[Io77]   Ioffe, A.D.: On lower semicontinuity of integral functionals. I. SIAM J. Control Optimization **15**, no. 4, 521–538 (1977).

[Io77a]   Ioffe, A.D.: On lower semicontinuity of integral functionals. II. SIAM J. Control Optimization **15**, no. 6, 991–1000 (1977).

[J70]   Johnson, R.: A. Atomic and nonatomic measures. Proc. Amer. Math. Soc. 25, 650–655 (1970).

[KP65]   Kadec', M.Ĭ., Pelčin'ski, A.: Basic sequences, bi-orthogonal systems and norming sets in Banach and Fréchet spaces. (Russian) Studia Math. 25, 297–323 (1965).

[KiKr01]   Kirchheim, B., Kristensen, J.: Differentiability of convex envelopes. C. R. Acad. Sci. Paris Sér. I Math. 333, no. 8, 725–728 (2001).

[Kir34]   Kirszbraun, M.D.: Uber die zusammenziehenden und Lipschitzchen Transformationen. Fund. Math., 22, 77–108 (1934).

[Kr76]   Krasnosel'skiĭ, M. A., Zabreĭko, P. P., Pustyl'nik, E. I., Sobolevskiĭ, P. E.: Integral operators in spaces of summable functions. Translated from the Russian by T. Ando. Monographs and Textbooks on Mechanics of Solids and Fluids, Mechanics: Analysis. Noordhoff International Publishing, Leiden (1976).

[Ku83]   Kunen, K.: Set Theory. An Introduction to Independence Proofs. Reprint of the 1980 original. Studies in Logic and the Foundations of Mathematics, 102. North-Holland Publishing Co., Amsterdam (1983).

[Le56]   Leach, E.B.: On a converse of the Hölder inequality. Proc. Amer. Math. Soc. 7, 607–608 (1956).

[Lee78]   Leese, S.J.: Measurable selections and the uniformization of Souslin sets. Amer. J. Math. 100, no. 1, 19–41 (1978).

[Lu67]   Luther, N.Y.: Unique extension and product measures. Canad. J. Math. 19, 757–763 (1967).

[Mar85]   Marcellini, P.: Approximation of quasiconvex functions, and lower semicontinuity of multiple integrals. Manuscripta Math. 51, no. 1-3, 1–28 (1985).

[MarSb80] Marcellini, P., Sbordone, C.: Semicontinuity problems in the calculus of variations. Nonlinear Anal. 4, no. 2, 241–257 (1980).

[May79]   Maynard, H.B.: A Radon-Nikodym theorem for finitely additive bounded measures. Pacific J. Math. 83, no. 2, 401–413 (1979).

[MarMi79] Marcus, M., Mizel, V.J.: Complete characterization of functions which act, via superposition, on Sobolev spaces. Trans. Amer. Math. Soc. 251, 187–218 (1979).

[Mc34]   McShane, E. J.: Extension of range of functions, Bull. Amer. Math. Soc. 40, 837–842 (1934).

[Mi56]   Michael, E.: Continuous selections. I. Ann. of Math. (2) 63, 361–382 (1956).

[Mo96]   Mora, M.G.: Approssimazione non locale in problemi con discontinuità libera, Thesis, University of Parma, a.a. 1996-97.

[Mor47]   Morse, A.P.: Perfect blankets. Trans. Amer. Math. Soc. 61, 418–442 (1947).

[Muk72]   Mukherjea, A.: A remark on Tonelli's theorem on integration in product spaces. Pacific J. Math. 42, 177–185 (1972).

[Muk73]   Mukherjea, A.: Remark on Tonelli's theorem on integration in product spaces. II. Indiana Univ. Math. J. 23, 679–684 (1973/74).

[Mu90]   Müller, S.: Higher integrability of determinants and weak convergence in L^1. J. Reine Angew. Math. 412, 20–34 (1990).

[Mu99]   Müller, S.: Variational Models for Microstructure and Phase Transitions. Calculus of Variations and Geometric Evolution Problems (Cetraro, 1996), 85–210, Lecture Notes in Math., 1713, Springer, Berlin (1999).

[Pe97]   Pedregal, P.: *Parametrized Measures and Variational Principles*. Progress in Nonlinear Differential Equations and Their Applications, 30. Birkhäuser Verlag, Basel (1997).

[Rao04]   Rao, M.M.: *Measure Theory and Integration*. Second edition. Monographs and Textbooks in Pure and Applied Mathematics, 265. Marcel Dekker, Inc., New York (2004).

[Roc76]   Rockafellar, R.T.: Integral functionals, normal integrands and measurable selections. In: Nonlinear operators and the calculus of variations (Summer School, Univ. Libre Bruxelles, Brussels, 1975), 157–207. Lecture Notes in Math., Vol. 543, Springer, Berlin (1976).

[Roc97]   Rockafellar, R.T.: *Convex Analysis*. Reprint of the 1970 original. Princeton Landmarks in Mathematics. Princeton Paperbacks. Princeton University Press, Princeton, NJ (1997).

[RocWe98] Rockafellar, R.T., Wets, R.J.-B.: *Variational Analysis*. Grundlehren der Mathematischen Wissenschaften [Fundamental Principles of Mathematical Sciences], 317. Springer-Verlag, Berlin (1998).

[Rog98]   Rogers, C. A.: *Hausdorff Measures*. Reprint of the 1970 original. With a foreword by K. J. Falconer. Cambridge Mathematical Library. Cambridge University Press, Cambridge (1998).

[R83]   Romero, J.L.: When is $L^p(\mu)$ contained in $L^q(\mu)$? Amer. Math. Monthly **90**, no. 3, 203–206 (1983).

[Ru87]   Rudin, W.: *Real and Complex Analysis*. Third edition. McGraw-Hill Book Co., New York (1987).

[Ru91]   Rudin, W.: *Functional Analysis*. Second edition. International Series in Pure and Applied Mathematics. McGraw-Hill, Inc., New York (1991).

[SY05]   Schwabik, S., Ye, G.: *Topics in Banach Space Integration*. Series in Real Analysis, 10. World Scientific Publishing Co. Pte. Ltd., Hackensack, NJ (2005).

[Sc51]   Schwartz, J.: A note on the space L^p. Proc. Amer. Math. Soc. **2**, 270–275 (1951).

[Ser61]   Serrin, J.: On the definition and properties of certain variational integrals. Trans. Amer. Math. Soc. **101**, 139–167 (1961).

[Ser62]   Serrin, J.: Strong convergence in a product space. Proc. Amer. Math. Soc. **13**, 651–655 (1962).

[Ste70]   Stein, E.M.: *Singular Integrals and Differentiability Properties of Functions*. Princeton Mathematical Series, No. 30 Princeton University Press, Princeton, N.J. (1970).

[Ste93]   Stein, E.M.: *Harmonic Analysis: Real-Variable Methods, Orthogonality, and Oscillatory Integrals*. With the assistance of Timothy S. Murphy. Princeton Mathematical Series, 43. Monographs in Harmonic Analysis, III. Princeton University Press, Princeton, NJ (1993).

[S78]   Subramanian, B.: On the inclusion $L^p(\mu) \subset L^q(\mu)$. Amer. Math. Monthly **85**, no. 6, 479–481 (1978).

[Ta79]   Tartar, L.: Compensated compactness and applications to partial differential equations. In: Nonlinear analysis and mechanics: Heriot-Watt Symposium, Vol. IV, pp. 136–212, Res. Notes in Math., 39, Pitman, Boston, Mass.-London (1979).

[Ta95]   Tartar, L.: Beyond Young measures. Microstructure and phase transitions in solids (Udine, 1994). Meccanica **30**, no. 5, 505–526 (1995).

[T67]   Toranzos, F.A.: Radial functions of convex and star-shaped bodies. Amer. Math. Monthly **74**, 278–280 (1967).

[Va71]   Valadier, M.: Multi-applications mesurables à valeurs convexes compactes. (French) J. Math. Pures Appl. (9) **50**, 265–297 (1971).

[Vi85]   Villani, A.: Another note on the inclusion $L^p(\mu) \subset L^q(\mu)$ Amer. Math. Monthly **92**, no. 7, 485–487 (1985).

[Yo95]   Yosida, K.: *Functional Analysis.* Reprint of the sixth (1980) edition. Classics in Mathematics. Springer-Verlag, Berlin (1995).

[Y69]   Young, L.C.: *Lectures on the Calculus of Variations and Optimal Control Theory.* Foreword by Wendell H. Fleming W. B. Saunders Co., Philadelphia-London-Toronto, Ont. (1969).

[W04]   Willard, S.: *General Topology.* Reprint of the 1970 original. Dover Publications, Inc., Mineola, NY (2004).

[Z67]   Zaanen, A.C.: *Integration.* Completely revised edition of An introduction to the theory of integration North-Holland Publishing Co., Amsterdam; Interscience Publishers John Wiley & Sons, Inc., New York (1967).

Index

Springer Monographs in Mathematics

This series publishes advanced monographs giving well-written presentations of the "state-of-the-art" in fields of mathematical research that have acquired the maturity needed for such a treatment. They are sufficiently self-contained to be accessible to more than just the intimate specialists of the subject, and sufficiently comprehensive to remain valuable references for many years. Besides the current state of knowledge in its field, an SMM volume should also describe its relevance to and interaction with neighbouring fields of mathematics, and give pointers to future directions of research.

Abhyankar, S.S. **Resolution of Singularities of Embedded Algebraic Surfaces** 2nd enlarged ed. 1998
Alexandrov, A.D. **Convex Polyhedra** 2005
Andrievskii, V.V.; Blatt, H.-P. **Discrepancy of Signed Measures and Polynomial Approximation** 2002
Angell, T.S.; Kirsch, A. **Optimization Methods in Electromagnetic Radiation** 2004
Ara, P.; Mathieu, M. **Local Multipliers of C*-Algebras** 2003
Armitage, D.H.; Gardiner, S.J. **Classical Potential Theory** 2001
Arnold, L. **Random Dynamical Systems** corr. 2nd printing 2003 (1st ed. 1998)
Arveson, W. **Noncommutative Dynamics and E-Semigroups** 2003
Aubin, T. **Some Nonlinear Problems in Riemannian Geometry** 1998
Auslender, A.; Teboulle, M. **Asymptotic Cones and Functions in Optimization and Variational Inequalities** 2003
Banagl, M. **Topological Invariants of Stratified Spaces** 2006
Banasiak, J.; Arlotti, L. **Perturbations of Positive Semigroups with Applications** 2006
Bang-Jensen, J.; Gutin, G. **Digraphs** 2001
Baues, H.-J. **Combinatorial Foundation of Homology and Homotopy** 1999
Böttcher, A.; Silbermann, B. **Analysis of Toeplitz Operators** 2006
Brown, K.S. **Buildings** 3rd printing 2000 (1st ed. 1998)
Chang, K.-C. **Methods in Nonlinear Analysis** 2005
Cherry, W.; Ye, Z. **Nevanlinna's Theory of Value Distribution** 2001
Ching, W.K. **Iterative Methods for Queuing and Manufacturing Systems** 2001
Chudinovich, I. **Variational and Potential Methods for a Class of Linear Hyperbolic Evolutionary Processes** 2005
Coates, J.; Sujatha, R. **Cyclotomic Fields and Zeta Values** 2006
Crabb, M.C.; James, I.M. **Fibrewise Homotopy Theory** 1998
Dineen, S. **Complex Analysis on Infinite Dimensional Spaces** 1999
Dugundji, J.; Granas, A. **Fixed Point Theory** 2003
Ebbinghaus, H.-D.; Flum, J. **Finite Model Theory** 2006
Edmunds, D.E.; Evans, W.D. **Hardy Operators, Function Spaces and Embeddings** 2004
Elstrodt, J.; Grunewald, F.; Mennicke, J. **Groups Acting on Hyperbolic Space** 1998
Engler, A.J.; Prestel, A. **Valued Fields** 2005
Fadell, E.R.; Husseini, S.Y. **Geometry and Topology of Configuration Spaces** 2001
Fedorov, Yu. N.; Kozlov, V.V. **A Memoir on Integrable Systems** 2001
Flenner, H.; O'Carroll, L.; Vogel, W. **Joins and Intersections** 1999
Gelfand, S.I.; Manin, Y.I. **Methods of Homological Algebra** 2nd ed. 2003 (1st ed. 1996)
Griess, R.L. Jr. **Twelve Sporadic Groups** 1998
Gras, G. **Class Field Theory** corr. 2nd printing 2005
Greuel, G.-M.; Lossen, C.; Shustin, E. **Introduction to Singularities and Deformations** 2007
Groah, J; Smoller, J; Temple B. **Shock Wave Interactions in General Relativity** 2007
Hida, H. **p-Adic Automorphic Forms on Shimura Varieties** 2004
Ischebeck, F; Rao, R.A. **Ideals and Reality** 2005
Ivrii, V. **Microlocal Analysis and Precise Spectral Asymptotics** 1998
Jakimovski, A.; Sharma, A.; Szabados, J. **Walsh Equiconvergence of Complex Interpolating Polynomials** 2006
Jech, T. **Set Theory** (3rd revised edition 2002)
Jorgenson, J.; Lang, S. **Spherical Inversion on SLn (R)** 2001
Kanamori, A. **The Higher Infinite** corr. 2nd printing 2005 (2nd ed. 2003)

Kanovei, V. **Nonstandard Analysis, Axiomatically** 2005
Khoshnevisan, D. **Multiparameter Processes** 2002
Koch, H. **Galois Theory of p-Extensions** 2002
Komornik, V. **Fourier Series in Control Theory** 2005
Kozlov, V.; Maz'ya, V. **Differential Equations with Operator Coefficients** 1999
Lam, T.Y. **Serre's Problem on Projective Modules** 2006
Landsman, N.P. **Mathematical Topics between Classical & Quantum Mechanics** 1998
Leach, J.A.; Needham, D.J. **Matched Asymptotic Expansions in Reaction-Diffusion Theory** 2004
Lebedev, L.P.; Vorovich, I.I. **Functional Analysis in Mechanics** 2002
Lemmermeyer, F. **Reciprocity Laws: From Euler to Eisenstein** 2000
Malle, G.; Matzat, B.H. **Inverse Galois Theory** 1999
Mardesic, S. **Strong Shape and Homology** 2000
Margulis, G.A. **On Some Aspects of the Theory of Anosov Systems** 2004
Miyake, T. **Modular Forms** 2006
Murdock, J. **Normal Forms and Unfoldings for Local Dynamical Systems** 2002
Narkiewicz, W. **Elementary and Analytic Theory of Algebraic Numbers** 3rd ed. 2004
Narkiewicz, W. **The Development of Prime Number Theory** 2000
Neittaanmaki, P.; Sprekels, J.; Tiba, D. **Optimization of Elliptic Systems. Theory and Applications** 2006
Onishchik, A.L. **Projective and Cayley–Klein Geometries** 2006
Parker, C.; Rowley, P. **Symplectic Amalgams** 2002
Peller, V. (Ed.) **Hankel Operators and Their Applications** 2003
Prestel, A.; Delzell, C.N. **Positive Polynomials** 2001
Puig, L. **Blocks of Finite Groups** 2002
Ranicki, A. **High-dimensional Knot Theory** 1998
Ribenboim, P. **The Theory of Classical Valuations** 1999
Rowe, E.G.P. **Geometrical Physics in Minkowski Spacetime** 2001
Rudyak, Y.B. **On Thorn Spectra, Orientability and Cobordism** 1998
Ryan, R.A. **Introduction to Tensor Products of Banach Spaces** 2002
Saranen, J.; Vainikko, G. **Periodic Integral and Pseudodifferential Equations with Numerical Approximation** 2002
Schneider, P. **Nonarchimedean Functional Analysis** 2002
Serre, J-P. **Complex Semisimple Lie Algebras** 2001 (reprint of first ed. 1987)
Serre, J-P. **Galois Cohomology** corr. 2nd printing 2002 (1st ed. 1997)
Serre, J-P. **Local Algebra** 2000
Serre, J-P. **Trees** corr. 2nd printing 2003 (1st ed. 1980)
Smirnov, E. **Hausdorff Spectra in Functional Analysis** 2002
Springer, T.A.; Veldkamp, F.D. **Octonions, Jordan Algebras, and Exceptional Groups** 2000
Székelyhidi, L. **Discrete Spectral Synthesis and Its Applications** 2006
Sznitman, A.-S. **Brownian Motion, Obstacles and Random Media** 1998
Taira, K. **Semigroups, Boundary Value Problems and Markov Processes** 2003
Talagrand, M. **The Generic Chaining** 2005
Tauvel, P.; Yu, R.W.T. **Lie Algebras and Algebraic Groups** 2005
Tits, J.; Weiss, R.M. **Moufang Polygons** 2002
Uchiyama, A. **Hardy Spaces on the Euclidean Space** 2001
Üstünel, A.-S.; Zakai, M. **Transformation of Measure on Wiener Space** 2000
Vasconcelos, W. **Integral Closure. Rees Algebras, Multiplicities, Algorithms** 2005
Yang, Y. **Solitons in Field Theory and Nonlinear Analysis** 2001
Zieschang, P.-H. **Theory of Association Schemes** 2005